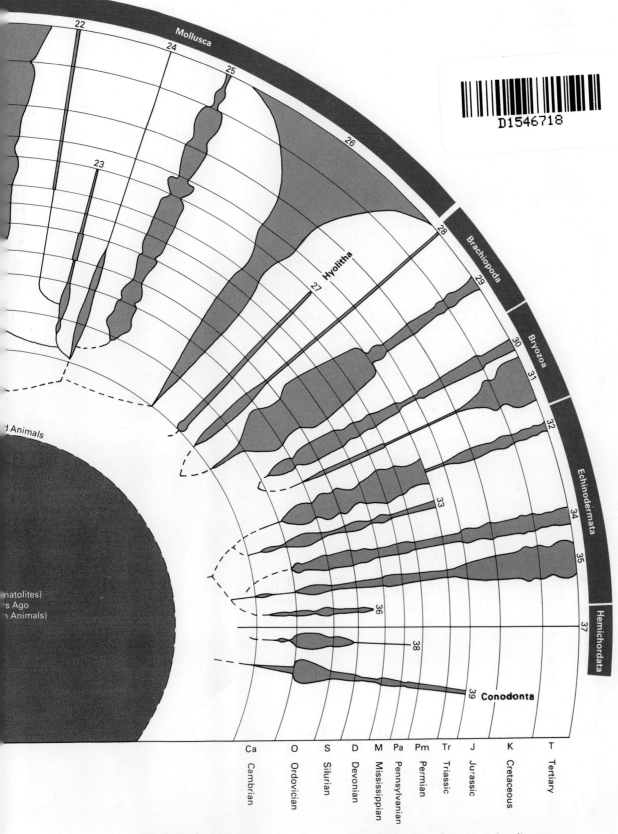

Triploblastic coelomate grade, pseudometamerous

Phylum Mollusca
20 Polyplacophora
21 Pelecypoda
22 Scaphopoda
23 Rostroconchia
24 Monoplacophora
25 Cephalopoda
26 Gastropoda

27 *Phylum Hyolitha*

Triploblastic coelomate grade, oligomerous

Phylum Brachiopoda
28 Inarticulata
29 Articulata

Phylum Bryozoa
30 Stenolaemata
31 Gymnolaemata

Phylum Echinodermata
32 Crinozoa
33 Blastozoa
34 Asterozoa
35 Echinozoa
36 Homalozoa

Phylum Hemichordata
37 Pterobranchia
38 Graptolithina

39 *Phylum Conodonta*

Fossil Invertebrates

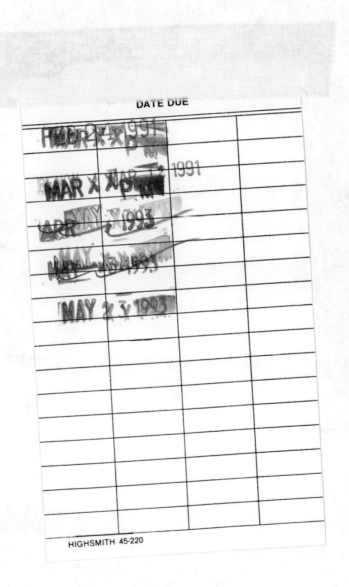

Frontispiece. A reconstruction of the famous Burgess Shale fauna and flora from the Middle Cambrian of Yoho Park, British Columbia is on exhibit at the Smithsonian's National Museum of Natural History. The fossils depict a teeming shallow-marine community of organisms—many of them fragile—that are not found elsewhere. The community contains many species of unknown biological affinities as well as sponges, trilobites, and brachiopods. The diorama recreates the muddy bottom at the base of an algal reef where delicate species were buried by the slumping reef face. (Photo by Chip Clark, the Smithsonian Institution.)

Fossil Invertebrates

SENIOR EDITOR

Richard S. Boardman
Smithsonian Institution, Washington

EDITORS

Alan H. Cheetham
Smithsonian Institution, Washington

Albert J. Rowell
University of Kansas

BLACKWELL SCIENTIFIC PUBLICATIONS

PALO ALTO OXFORD LONDON
EDINBURGH BOSTON MELBOURNE

© 1987 by Blackwell Scientific Publications
Editorial offices:
667 Lytton Avenue, Palo Alto, California 94301, USA
Osney Mead, Oxford, OX2 0EL
8 John Street, London, WC1N 2ES
23 Ainslie Place, Edinburgh, EH3 6AJ
52 Beacon Street, Boston, Massachusetts 02108, USA
107 Barry Street, Carlton, Victoria 3053, Australia

First published 1987

Set by Enset Photosetting,
Midsomer Norton, Bath, Avon

Sponsoring Editor: John Staples
Manuscript Editor: Deborah Gale
Interior and Cover Design: Gary Head

DISTRIBUTORS

USA and Canada
   Blackwell Scientific Publications Inc
   P O Box 50009, Palo Alto
   California 94303
   Tel: (415) 965−4081

Australia
   Blackwell Scientific Publications
   (Australia) Pty Ltd
   107 Barry Street,
   Carlton, Victoria 3053
   Tel: (03) 347 0300

United Kingdom
   Blackwell Scientific Publications
   Osney Mead, Oxford OX2 0EL
   Tel: 0865 240201

Library of Congress
Cataloging in Publication Data

Fossil invertebrates.

   Includes index.
   1. Invertebrates, Fossil.   I. Boardman, Richard S.
   II.   Cheetham, Alan H.   III. Rowell, A.J.
   QE770.F67 1985 562 84−28403

ISBN 0−86542−302−4

British Library
Cataloguing in Publication Data

Fossil invertebrates.
   1. Invertebrates, Fossil
   I. Boardman, Richard S.   II. Cheetham, Alan H.
   III. Rowell, Albert J.
   562 QE770

ISBN 0−86542−302−4

Contents

List of Contributors

Peter D. Ashlock *Snow Entomological Museum, University of Kansas, Lawrence, Kansas*

Anna K. Behrensmeyer *Department of Paleobiology, National Musuem of Natural History, Washington D.C.*

William B.N. Berry *Department of Paleontology, University of California, Berkeley, California*

Richard S. Boardman *Department of Paleobiology, National Museum of Natural History, Washington, D.C.*

Martin A. Buzas *Department of Paleobiology, National Museum of Natural History, Washington, D.C.*

Alan H. Cheetham *Department of Paleobiology, National Museum of Natural History, Washington, D.C.*

David L. Clark *Department of Geology & Geophysics, University of Wisconsin, Madison, Wisconsin*

Anthony G. Coates *Department of Geology, George Washington University, Washington, D.C.*

Raymond C. Douglass *Paleontology & Stratigraphy Branch, U.S. Geological Survey, Washington, D.C.*

Roland A. Gangloff *Merritt College, Oakland, California*

Mackenzie Gordon Jr. *Paleontology & Stratigraphy Branch, U.S. Geological Survey, Washington, D.C.*

Richard E. Grant *Department of Paleobiology, National Museum of Natural History, Washington, D.C.*

Roger L. Kaesler *Department of Geology, University of Kansas, Lawrence, Kansas*

Porter M. Kier *Department of Paleobiology, National Museum .of Natural History, Washington, D.C.*

William A. Oliver Jr. *Paleontology & Stratigraphy Branch, U.S. Geological Survey, Washington, D.C.*

Richard G. Osgood Jr.* *Department of Geology and Geography, The College of Wooster, Wooster, Ohio*

John S. Peel *Geological Survey of Greenland, Copenhagen, Denmark*

John Pojeta Jr. *Paleontology & Stratigraphy Branch, U.S. Geological Survey, Washington, D.C.*

J. Keith Rigby *Department of Geology, Brigham Young University, Provo, Utah*

Richard A. Robison *Department of Geology, University of Kansas, Lawrence, Kansas*

Albert J. Rowell *Department of Geology, University of Kansas, Lawrence, Kansas*

Bruce Runnegar *Department of Geology, University of New England, Armidale, N.S.W., Australia*

Charles C. Smith *Tenneco Oil Company, Houston, Texas*

James Sprinkle *Department of Geological Sciences, University of Texas, Austin, Texas*

Michael E. Taylor *Paleontology & Stratigraphy Branch, U.S. Geological Survey, Denver, Colorado*

Kenneth M. Towe *Department of Paleobiology, National Museum of Natural History, Washington, D.C.*

James W. Valentine *Department of Geology, University of California, Santa Barbara, California*

* Deceased

Preface

Fossil Invertebrates is designed to be a primary textbook for college courses in invertebrate paleontology. It combines a balanced treatment by research specialists of the current state of knowledge with flexible organization that permits its use in courses ranging from one quarter to two semesters. Students are expected to have a knowledge of high school biology and to have completed a course in historical geology.

The immense, diffuse, and specialized literature of invertebrate paleontology makes understanding the current state of the science a formidable task. The combined research experience of twenty-seven authors gives the book a unique richness in information, interpretation, and evaluation of controversies, doubts, and unanswered questions that are necessary to present the current state of invertebrate paleontology.

The specific goals of *Fossil Invertebrates* are:
● To impart an understanding of the basic nature of the science, what is known, controversial, or unknown. The several sides of controversial questions are discussed.
● To provide an understanding of the physical nature of the invertebrate fossil record, from its excellent preservations to its deficiencies.
● To present enough biology for students to appreciate fossil invertebrates as once living organisms, to understand the levels of invertebrate organization, and to visualize the fossil record as a series of living communities through time.
● To include enough information about morphology, terminology, and classification for each phylum to enable students to enter the professional literature, especially through summary works such as the *Treatise on Invertebrate Paleontology*.

● To provide adequate principles for a foundation for graduate training in paleontology.
● To provide a sufficient base for graduates entering commercial work to apply invertebrate paleontology to the solution of geologic problems.
● To give enough information on stratigraphic distribution throughout the fossil record of invertebrates to allow determination of approximate geologic age.

Existing textbooks emphasize either principles or morphology and classification. *Fossil Invertebrates* attempts to combine the two approaches. This book begins with chapters written by specialists on the principles of invertebrate organization, ecology and paleoecology, evolution, preservation, classification, and biostratigraphy and biogeography. The phylum chapters that follow contain examples of many of these principles and introduce others. This emphasis on principles is consistent with the current state of the science and should increase student interest while reducing the traditional burden of memorization.

The thirteen chapters on invertebrate phyla (including one on the kingdom Protista) are arranged in the sequence determined by the levels of organization outlined in Chapter 2 'Invertebrate Organization: A Review.' These chapters are followed by a chapter on trace fossils. Each phylum chapter is organized with major topics as nearly as possible in the same sequence. These topics include morphology, classification, biology (modes of life, growth, reproduction), origin and evolution, functional morphology and ecology, and stratigraphic and geographic distribution.

To accommodate courses of different lengths, major phylum chapters are divided into two parts. Part I,

'Phylum Overview,' introduces the phylum and its second-level taxa and covers some of their features. This section can serve either as the entire assignment for a phylum in a short course or as an introduction to Part II of the chapter, called 'Additional Concepts.' The 'Additional Concepts' sections cover lower-level taxa in greater detail. The two parts together are designed to provide enough material for a two-semester course.

Some phyla demonstrate certain aspects of the science better than others. A section of 'Phylum Overview' of most major phylum chapters is 'Features of the Phylum,' which emphasizes concepts, principles, or applications that are especially well demonstrated by the particular phylum.

Relative lengths of phylum chapters are the results of several approaches to the problem. Guidance for first estimates of relative chapter lengths came from exisiting textbooks and the *Treatise on Invertebrate Paleontology*. These initial estimates were modified first by allowing additional space for recent increases in knowledge about a few major taxa that were inadequately covered in existing textbooks. An attempt was made to achieve approximately the same amount of detail in each phylum chapter (with the exception of the Protista), but this is highly subjective and difficult to determine. The chapter on Protista is less detailed because we assume that they will be covered in more detail in micropaleontology courses. Finally, the lengths of a few of the chapters were determined by a compromise between the three editors and individual authors. The two-part division of the major phylum chapters should provide flexibility in modifying the relative emphasis on phyla to suit individual courses.

Each author, of course, has a different writing style. These differences have been softened by editing. Editorial efforts have attempted to simplify the language and shorten sentences as much as possible. Much remains of individual writing styles, but we hope that each chapter will be equally understandable.

Several editorial policies were established at the beginning of the project. The most important terms are in bold face type: these have been reduced in number as much as possible—again a compromise between authors and editors. Generally, they are terms thought to be of first importance by authors and used more than once in a chapter. Every effort has been made to boldface, define, and illustrate terms where first used. Terms not boldfaced and those in parentheses are thought to be of lesser importance .

Taxonomic diagnoses are in a brief telegraphic style for ready reference and cross comparison. The index is detailed enough to permit use of *Fossil Invertebrates* both as a textbook and as a reference book. In place of a glossary, boldfaced page numbers in the index indicate where boldfaced terms are defined. Some terms have different meanings in different phyla, requiring that they be boldfaced more than once.

Authors of specific research referred to in the text are not indicated or referenced. Supplementary reading lists are short and generally include only classic or summary papers that will introduce students to more advanced reading. It is assumed that all students will have access to the *Treatise on Invertebrate Paleontology* and standard bibliographic tools.

Photographs of specimens are emphasized throughout the phylum chapters so that illustrations are as realistic as possible. Also, photographed specimens are from all over the world, to emphasize the international flavor of invertebrate paleontology as a science.

The manuscript for *Fossil Invertebrates* was thoroughly reviewed by both specialists and generalists. A spirited reading group was organized within the Department of Paleobiology of the National Museum of Natural History, made up largely of geology majors who had recently finished undergraduate training. They did a thorough job of critiquing each chapter, and established some of the editorial policies that were adopted. At different times the group included: Michael Brett-Surman, Timothy Collins, Kathleen Bryant, Virginia Gonzalez, Linda Jacobsen, Cathy McNair, Natalie Marovelli, Laurel Smith, Mark Symborski, and Karen Wetmore.

The following colleagues contributed to the book either through technical reviews or by supplying information or illustrations: T.W. Amsden, W.C. Banta, R.L. Batten, F.M. Bayer, Stefan Bengston, J.M. Berdan, J. Bergström, S.M. Bergström, A.G. Beu, Patricia Blackwelder, D.B. Blake, P.S. Boyer, M.A. Bradshaw, S.D. Cairns, F.M. Carpenter, Franco Cati, Micheline Clocchiatti, T.M. Collins, S.M. Conway Morris, P.L. Cook, G.A. Cooper, Paul Copper, Richard Cowen, R.A. Davis, J.W. Durham, J.T. Dutro, Jr., A.A. Ekdale, R.N. Eldredge, C.H. Ellis, R.L. Ethington, J.A. Fagerstrom, R.F. Fleming, Stefan Gartner, Jr., M.F. Glaessner, A.G. Harris, W.D. Hartman, J.E. Hazel, L.F. Hintze, J.B.C. Jackson, O.L. Karklins, S.A. Kling, Judith Lang, N.G. Lane, R.M. Linsley, Jere Lipps, H.A. Lowenstam, D.B. Macurda, Jr., R.F. Maddocks, C.E. Mason, P.A. Maxwell, F.K. McKinney, D.I. McKinnon, J.F. Mello, D.L. Meyer, R.B. Neuman, D.J. Nichols, Denise Noel, A.R. Palmer, D.L. Pawson, R.K. Pickerill, R.A. Pohowsky, J.E. Repetski, E.S. Richardson, Jr., B.M.

Rickards, Joseph Rosewater, P.A. Sandberg, K.B. Sandved, N.J. Silberling, N.F. Sohl, J.E. Sorauf, S.M. Stanley, C.W. Stearn, P.D. Taylor, Erik Thomsen, John Utgaard, Erhard Voigt, T.R. Waller, L.W. Ward, R.R. West, H.B. Whittington, Alwyn Williams, T.S. Wood, E.L. Yochelson, Renjie Zhang.

A textbook written by twenty-seven authors is an experiment. The editors request criticism from users that will contribute to an improved second edition. Royalties from the sale of this book will go to the Paleontological Society to support the science of paleontology. It should be noted the Society had nothing to do with the development of the book nor has it ever been a sponsor.

R.S.B.
A.H.C.
A.J.R.

Nature and Scope of Invertebrate Paleontology

Richard S. Boardman
Alan H. Cheetham
Albert J. Rowell

What is invertebrate paleontology?

Paleontology is the study of life through geologic time and, therefore, is based on the principles and methods of both biology and geology. **Fossils,** the remains, traces, or imprints of once-living organisms preserved in the earth's crust, provide the direct evidence of prehistoric life. By studying fossils, paleontologists are able to trace the course of evolution of life and habitats on the earth from their remote beginnings to the seemingly endless diversity of the present (see front endpaper).

The earth is approximately 4.6 billion years old. The oldest-known fossils are sedimentary structures (stromatolites) from Western Australia presumed to have been made by microorganisms that lived about 3.5 billion years ago. For the next billion years or more, evolution proceeded relatively slowly as primitive algal life began to contribute to the accumulation of oxygen in the earth's atmosphere. Rates of evolution may have increased markedly with the invention of sexual reproduction, possibly as early as 1.5 billion years ago. Clear fossil evidence for the earliest animals has been found in rocks between 650 and 750 million years old. Such fossils, representing several animal groups, have now been found around the world, in the Ediacara Hills of South Australia, Africa, the Soviet Union, England, Canada, and the United States. These early animals were marine invertebrates, including jellyfish and soft corals of the Cnidaria (Chapter 11), annelids (Chapter 12), arthropods (Chapter 13), and specimens of unknown affinities. Mineralized skeletons of invertebrates first appeared about 600 million years ago, marking the beginning of

the **Phanerozoic,** that part of geologic time represented by rocks in which the evidence of life is abundant: the Cambrian period and later (see back endpapers).

The Phanerozoic fossil record is dominated in both abundance and diversity by organisms collectively called **invertebrates.** Traditionally defined as animals without backbones, invertebrates were once regarded as a formal group of high rank in classifications of past and present life. In modern usage, invertebrate is a term of convenience in both paleontology and biology, referring to a heterogeneous assemblage ranging from mostly animal-like unicellular organisms to numerous major groups of multicellular animals, excluding the vertebrates. At one end this grouping includes forms that share some characteristics with simple plants (Chapter 8). At the other end are forms that share important properties with the vertebrates (Chapter 19). Connecting these extremes is a series of major levels of biologic organization (Chapter 2), each characterized by distinctive forms and structures, adaptations, and evolutionary history (Chapters 8 to 20).

It is convenient to study invertebrates collectively for a couple of reasons. First, similarities and differences in invertebrates' form and structure can be understood to a certain extent as being the result of progressive modifications in levels of biologic organization, from unicellular organisms to animals with organized tissues, organs, or systems of increasing complexity (Chapter 2). Second, an invertebrate group's survival and evolution is significantly tied to that of other invertebrate life forms around them. Most of the fossil record is contained in sedimentary rocks of marine and marginal marine origins. These fossils represent the incomplete and commonly physically altered evidence of complexly interacting marine or marginal marine assemblages, such as those presently forming coral reef communities. Numerous invertebrate groups are typically represented in living reef communities including attached corals, sponges, and bryozoans; mobile crabs, snails, starfish, and sea urchins; and floating and bottom-dwelling unicellular organisms with both plantlike and animal-like properties. The form, adaptations, and even evolutionary histories of these diverse groups can in part be better understood if their interactions, such as competition for food and space or predator-prey relationships, are considered. The organisms and their interactions also must be considered in reconstructing the physical environments of the geologic past.

Fish and especially sea plants also form important parts of many marine and marginal marine communities. In the future, the study of invertebrate paleontology separated from vertebrate paleontology and paleobotany could become a serious limitation in many kinds of paleontologic investigations.

Relations of paleontology to biology and geology

To understand fossils as the remains, traces, or imprints of once-living organisms, the paleontologist must employ a knowledge of biology. To understand fossils as a record of evolutionary history, extending over 3.5 billion years, geologic principles must also be applied. The ranges in time of the various fossil groups are established by painstaking documentation of relative positions of fossils in sequences of rocks. Chronological tie points within rock sequences largely based on dating by radioactive decay provide the absolute time scale on which evolution has occurred. Equally important to consider are the effects on fossil sequences in different areas of long-term climatic changes and shifting positions of continental and oceanic plates. Once the evolutionary history of a fossil group becomes sufficiently well known, it can serve in detailed studies of relative ages of fossiliferous rocks and in reconstruction of past sedimentary environments and their changing geographic distribution through geologic time.

The fossil record offers both advantages and disadvantages in the study of the evolutionary history of life. It is the only source of direct evidence of the nature of past life and, therefore, of how fossils and modern organisms fit into evolutionary history. Also, paleontologists can investigate biologic changes occurring at time scales of many thousands to millions of years, obviously out of the reach of investigations on modern organisms. However, at this scale, many biologic changes may be condensed into short intervals of rock sequences and appear to be abrupt events within which more detailed changes cannot be distinguished.

A disadvantage of the fossil record is that fossils are unevenly distributed, both chronologically and geographically, and they represent an estimated tiny percentage of the kinds of organisms that have lived. Some kinds of organisms are more readily fossilized than others, and thus the relative proportions of different organisms in the fossil record do not necessarily reflect the composition of once-living assemblages. Furthermore, parts of some organisms are more readily preserved than others (Chapter 5). As a result, fossilized remains of different groups of invertebrates reflect different amounts of the whole organism. For example, echinoderm fossils (Chapter 18) reflect most of the animal, so that many extinct forms can be understood in great detail. Conodont elements (Chapter 20), however, are such a small part of that animal that their position in the invertebrates is controversial. These biases must be considered in any paleontologic investigation. New fossil evidence in groups only partly preserved, therefore, often results in major revisions of interpretations. The fossil record of many groups is so rich, however, that the broad features of their evolutionary histories can be reconstructed with reasonable confidence.

Many aspects of the biology of fossil groups can be inferred only by comparison with living organisms. The starting point in paleontologic investigation is **morphology**, the form and structure of fossil and living organisms. Morphologic study of fossils usually involves detailed comparisons of preserved parts of fossils with corresponding parts of living organisms.

Knowledge of how preserved parts, such as shells, are related to soft tissues that secreted them or were attached to them is also important. An understanding of the relations between preserved and nonpreserved parts generally is based on recognition of the organism's level of organization, beginning with cells and their parts in simpler organisms and continuing through tissues, organs and organ systems in more complex organisms (Chapter 2). For example, fossil brachiopod shells (Chapter 16) usually have internal markings that, by comparison to those of living brachiopods, are interpreted as the attachments for muscles that open and close the two-part shell. These muscles form a complex system passing through the body cavity of a brachiopod and distinguish it as a complex, higher invertebrate.

In comparing the morphology of fossils with one another and with living organisms, the paleontologist recognizes major changes in form and structure. Living organisms change with individual growth and with differentiation by sex or other functions (Chapter 2). These changes can be profound, as in the metamorphosis of insects (Chapter 13). Records of growth or functional differentiation can be preserved in fossils, for example, by growth lines in mollusk shells (Chapter 14), but in many groups must be inferred by less obvious comparisons with living organisms. Differences in morphology within and between fossil groups are also the result of the evolutionary process whereby changes in form, structure, and associated functions and behavior are passed from ancestral to descendant populations. The recognition of evolutionary changes in morphology requires an understanding of how the genetic material controlling form, structure, function, and behavior deviates from the ancestral condition, both generation by generation and over geologic time (Chapter 4).

One effect of evolution is to make the form and structure of living organisms an inexact guide to the whole-organism morphology of fossils. Preserved parts may document evolutionary change in great detail but tell little about changes that might have occurred in missing parts, especially soft tissues that determine the organism's level of biologic organization. Preserved parts may be so different from those in any living organisms that little can be inferred about the nature of the whole organism, such as the archaeocyathans (Chapter 9), graptolites (Chapter 19), and conodonts (Chapter 20). Such limitations can obscure the relationship of some fossil groups to an overall evolutionary history, but do not in themselves prevent the inference of detailed evolutionary trends within a group and its use in relative dating of fossiliferous rocks.

Even with great evolutionary differences, parts of organisms may be similar in form or structure because of similar functions, as in feeding or locomotion. By studying the functional or behavioral significance of form and structure, called **functional morphology**, the paleontologist can make inferences not only about functions in fossils but also about whole-organism morphology in some groups without close living relatives. For example, certain structural similarities suggest that the extinct archaeocyathans (Chapter 9) may have fed by currents passing through porous walls, as do living sponges (Chapter 10). Some organisms' activities result in sedimentary structures, such as burrows or feeding tracks, that can be preserved in rocks as **trace fossils** (Chapter 21) and provide direct evidence of function or behavior.

The general purpose of morphologic study in paleontology is to infer the form and structure of whole organisms from preserved remains, traces, or imprints. This information is then used by the paleontologist, in conjunction with the distribution of the fossils in space and time and the physical nature of their enclosing rock, as the basis for other paleontologic pursuits. These are the study of **paleoecology**, the relations between fossil organisms and their environments (Chapter 3); the inference of evolutionary patterns (Chapter 4); the development of **classifications**, hierarchical systems of groups of organisms, both living and fossil, generally reflecting their evolutionary history (Chapter 6); and the development of **biostratigraphy**, the separation and correlation of rock units on the basis of contained fossils (Chapter 7).

Supplementary reading

Dodd, J. R.; Stanton, R. J. Jr. *Paleoecology, concepts and applications.* New York: Wiley-Interscience; 1981. Introduction to the principal features of paleoecology and its literature. Although intended as an advanced text, even the less experienced can gain by reading it.

Gray, J.; Boucot, A. J.; Berry, W. B. N., editors. *Communities of the past.* Stroudsburg, PA: Hutchinson Ross Publishing Co.; 1981. Collection of 18 essays on various aspects of community distribution and structure.

Raup, D. M.; Stanley, S. M., *Principles of paleontology*, 2nd ed. San Francisco: W. H. Freeman & Co.; 1978. More detailed treatment of many paleontologic principles presented in this book.

Schopf, J. W., editor. *Origin and evolution of earth's earliest biosphere: An interdisciplinary study.* Princeton, NJ: Princeton University Press, 1983.

2

Invertebrate Organization: A Review

James W. Valentine

The invertebrate cell

Invertebrates are constructed of a hierarchy of components of which cells are the fundamental building blocks. **Cells** are discrete bodies of protoplasm surrounded by a membrane. Some organisms such as the Protozoa are formed of only a single cell, but animals contain numerous cells organized into **tissues**, such as nervous or muscular tissues. Tissues, in turn, may be associated to form **organs** (for example, a muscle or the liver). Organs may in turn be organized into **organ systems**, such as the digestive system, which perform major physiologic activities. Complex organisms are composed of a harmoniously integrated set of organ systems.

Cells range in size from smaller than a micron in diameter to as bulky as the yolk of a large bird's egg and vary widely in form and function (Figure 2.1). Each cell contains an array of biochemical machinery; it is a unit of metabolism. The **protoplasm,** the living material of an invertebrate cell, consists of one or more **nuclei** surrounded by **cytoplasm**. Cytoplasm contains structures such as mitochondria, which perform energy-releasing processes and tend to cluster in regions of high metabolic activity. Other cytoplasmic structures are **ribosomes**, bodies on which proteins are synthesized. Ribosomes are minute, visible only with electron microscopy, but many thousands are present in each cell.

The nucleus contrasts with the cytoplasm in being a denser body. It is also surrounded by its own membrane. As invertebrate cells divide, their nuclei contain **chromosomes** composed of the nucleic acid **DNA**

(deoxyribonucleic acid) and proteins. Nuclear DNA is arranged in very long molecules made up of two strands coiled into a spiral or helix. Each strand is composed of a chain of chemical units called **nucleotides**; a nucleotide consists of a nitrogenous base attached to a sugar and a phosphate (Figure 2.2). The nitrogenous bases in DNA are adenine (A), guanine (G), cytosine (C) and thymine (T). Thus there are four types of nucleotides. So far as their chemical properties are concerned, all four types may occur in any order within a DNA strand. When a cell reproduces, the order of the nitrogenous bases in the DNA strands is usually preserved, although mistakes do happen. The mistakes are termed **mutations**.

It is important that the order of the bases normally be preserved because it forms a code, the genetic code, which contains the information required for the development of a mature organism. When an invertebrate reproduces, its biologic legacy is encoded in its DNA and is passed on to its offspring. The nuclear DNA also functions as an executive suite for the cell, directing metabolic activities that largely govern the form and function of the cell itself and its role in the tissue of which it is a part.

The individual units of DNA that perform these functions are called **genes**; they are segments of DNA strands containing specific nucleotide sequences. There are several types of genes, including structural and regulatory genes.

Structural genes specify the structure of **proteins**, the large organic molecules that perform or catalyze most of the biochemical reactions in an organism. Proteins are composed of series of units called **amino acids**; the

4

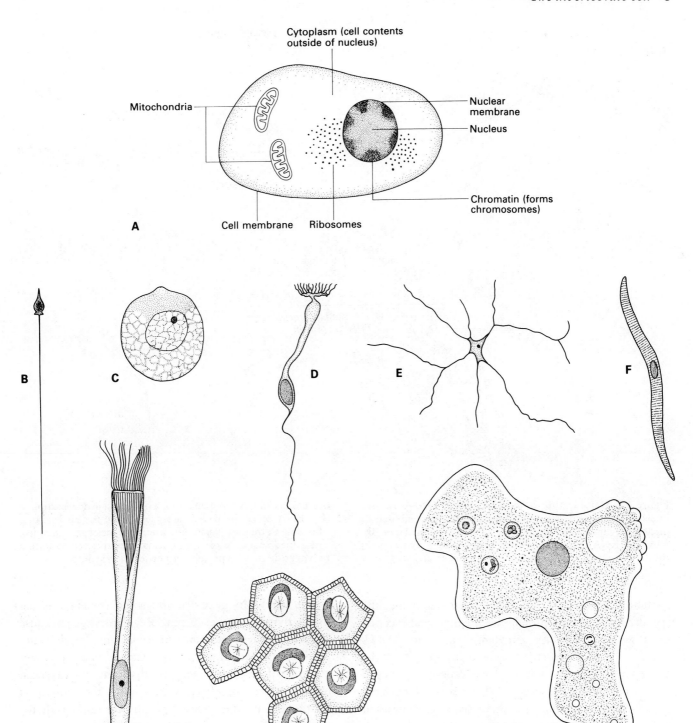

Figure 2.1. Diversity in invertebrate cell form and function. **A,** Generalized cell with major structures. Mitochondria carry enzymes associated with energy-releasing processes. **B,** Sperm (gamete) of the clam *Mytilus*. **C,** Egg (gamete) of the annelid *Myzostoma*. **D,** Sensory cell from the snail *Helix*, with sensory filaments. **E,** Connective cell from a snail. **F,** Striated muscle cell from a cnidarian. **G,** Ciliated cell from clam intestine. **H,** Cells from the pharynx of a tunicate. **I,** Amoeba *Chaos,* a single-celled organism. The nucleus is heavily stippled, the light-colored areas are vacuoles (regions containing water or food).

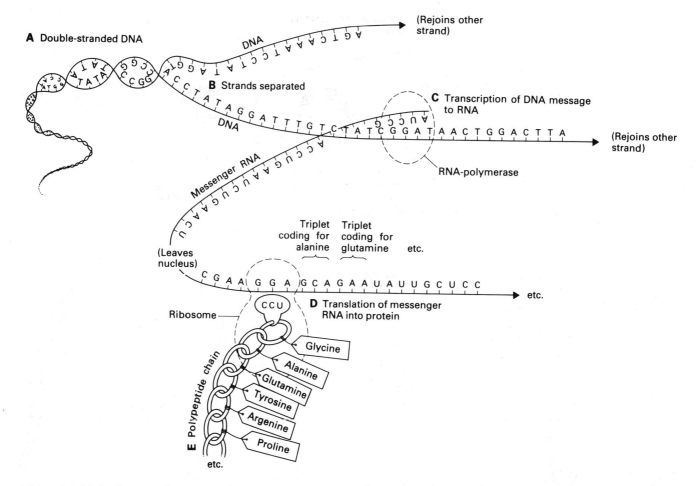

Figure 2.2. Major features of protein synthesis. DNA exists in a nucleus as a double-stranded, helical molecule composed of nucleotide chains (**A**). Strands are separated by enzymes (**B**). Next, an RNA nucleotide sequence complementary to one of the DNA sequences is formed through enzyme activity (partly the activity of RNA-polymerase), a process called transcription (**C**). RNA takes the coded message outside the nucleus to a ribosome in the cytoplasm (**D**), where the message is decoded. The coded sequence of amino acids is synthesized to form a polypeptide chain, a protein, or portion of protein (**E**).

particular arrangement of amino acids determines the specific function of a protein. Each amino acid is coded by a sequence of three nitrogenous bases in a DNA strand. Thus, a structural gene is composed of the sequence of nitrogenous bases that encodes a sequence of amino acids to form a protein. When a particular protein is needed by a cell, the appropriate sequence of bases in one DNA strand is transcribed to another nucleic acid, **messenger RNA** (ribonucleic acid), which carries the coded message to the cytoplasm. RNA is also composed of nucleotides, but the base uracil (U) substitutes for thymine. The transcription is possible because A in the DNA always is replaced by U in the messenger RNA, G by C, C by G, and T by A, respectively (Figure 2.2). The message is translated into a protein product on the RNA-rich ribosomes. Many proteins function as **enzymes**, which catalyze the main biochemical reactions of development and metabolism.

The **regulatory gene** determines the agents governing the activity of other genes. Regulatory genes cause structural genes to become active when their products are required or sometimes to be inactive when a proper amount of structural gene product has already been manufactured. In effect, they control the pattern of expression of other genes and thus are responsible for important differences in form between different kinds of organisms. A gene may be both structural and regulatory if, for example, it indicates a protein that affects the production of one or more other genes.

Still another kind of gene specifies molecules such as RNA, which is used in the manufcture of gene products.

Thus the nucleus controls cellular function chiefly by exercising control of the protein-producing machinery. Both the control and the proteins are specified by genes. The sum of all the hereditary factors (genes) possessed by an organism is termed the **genotype**. The appearance of the organism, including its morphology, physiology, and biochemistry, is termed the **phenotype** and results from interaction between the genotype and the environment. A range of phenotypes may develop from a given genotype, depending upon the environmental circumstances.

Reproduction

Sexual reproduction

Sexual reproduction occurs at some time among virtually all invertebrate groups; indeed, it is evidently the primitive reproductive mode in the plant and animal kingdoms, having arisen in some common plant or animal ancestor. In sexual reproduction, each of two parents contributes to an offspring about half of its genetic complement. This parental half is called a **genome**. In most invertebrates the maternal genome is contained in an egg and the paternal genome, in a sperm; reproductive cells (eggs or sperm) are called **gametes** and a fertilized egg, a **zygote**.

A sexually reproducing, multicellular animal may contain billions of cells, which develop from a zygote by cell division. The chromosomes in a cell duplicate to produce two identical sets. The cell then divides to produce two daughter cells, each with a chromosome set like the original parent cell. This process is termed **mitosis**. In any individual, the cells produced by mitosis normally have genotypes identical to the zygote. The chromosomes occur in **homologous** pairs; the members of a pair resemble each other in shape and size. An exception sometimes occurs in the chromosomes that determine sex; in some species the pair is dissimilar in appearance.

The only cells not produced by mitosis are gametes, which undergo another system of cell division called **meiosis** or **reduction division**. During meiosis, two cell divisions produce four daughter cells, but the chromosomes duplicate only once. Thus each of these four cells has only half the chromosome complement of the zygote. One chromosome from each pair is included in every gamete. Thus for the offspring, one chromosome in each pair has come from each parent. When these offspring reproduce, their gametes also contain one each of their chromosome pairs (from one or the other of the original parents, now the grandparents-to-be). However, the different pairs assort independently, so that different gametes contain different mixtures of the grandparental chromosomes.

There is another source of genetic variability. During meiosis the paired chromosomes actually join for a short time and exchange some DNA. When this occurs, some genes in one chromosome may be transferred to the other, a process called **crossing over**. Therefore the first offspring already have, in their chromosomes, combinations of genes that were not present in their parents. New gene combinations occur in the grandchildren because of genetic contributions from other grandparents.

Ordinarily the genes in each chromosome of a homologous pair have identical functions, so that each gene is present in a double dose. The genes commonly occur in the same order on each homologous chromosome; the position of a gene within a chromosome (and therefore within the DNA helix) is called the **gene locus**. However, the genes at the same locus in homologous chromosomes may not be precisely alike, but may produce somewhat different amino acid sequences, owing to the accumulation of past mutations in either or both. The protein produced by one gene will be slightly different from that produced by the other and may have a different functional range and physiologic and morphologic effect. Such variants of a gene (occurring at the same locus) are termed **alleles**. If the same allele occurs at a given locus in homologous chromosomes, the genotype is said to be **homozygous** for that gene; if the alleles are different, the genotype is said to be **heterozygous**.

Because of the many gene loci in an invertebrate, anywhere from a few thousand to tens of thousands, and because the grandparents of an individual may have different alleles at some loci, the possible gene combinations are immense. In fact no two zygotes would ordinarily be genetically identical. For example, an invertebrate might contain 10,000 structural gene loci and be heterozygous at 10 per cent of them. The number of different allele combinations that may appear in its gametes is then 2^{1000} or 10^{301}. All the invertebrates in the world have not produced that many gametes since the beginning of time.

Asexual reproduction

Asexual reproduction has evolved in many different animal and plant groups, although of the phyla discussed in this text only a few Protozoa (Table 2.1) seem to have dispensed with sexual reproduction entirely. In some single-celled organisms a parental cell simply divides by mitosis to produce two identical daughter cells. A more complicated procedure occurs when many nuclei, each with an identical set of chromosomes, are produced within a single parental cell. Such a cell may sometimes divide into as many offspring cells as there are nuclei.

Asexual reproduction in multicellular animals

Table 2.1. Grades of body plans of living animal phyla in order of increasing complexity

Phylum	Remarks
Unicellular grade	
Protozoa	Amoeba, foraminifera
Primitive multicellular grade	
Porifera	Sponges
Diploblastic grade	
Cnidaria	Jellyfish, corals
Ctenophora	Comb jellies
Triploblastic acoelomate grade	
Platyhelminthes	Flatworms
Nemertina	Proboscis worms
Gnathostomulida	Microscopic
Triploblastic pseudocoelomate grade	
Entoprocta	Resemble bryozoans
Acanthocephala	Parasitic worms
Rotifera	Microscopic rotifers
Gastrotricha	Microscopic
Kinorhyncha	Microscopic
Nematoda	Roundworms
Priapulida	Possibly amerous coelomates
Triploblastic coelomate grade	
Metamerous body plans	
Annelida	Earthworms, fan-worms
Arthropoda	Crabs, shrimp, insects
Amerous body plans	
Sipunculida	Peanut worms
Echiuroida	Sausage shaped
Pseudometamerous body plans	
Mollusca	Clams, snails
Oligomerous body plans	
Phoronida	Horseshoe worms
Brachiopoda	Lamp shells
Bryozoa	All colonial
Chaetognatha	Arrow worms
Pogonophora	Deep sea only today
Echinodermata	Starfish, sea urchins
Hemichordata	Acorn worms
Urochordata	Sea squirts (tunicates)
Chordata	*Amphioxus*, vertebrates

frequently occurs by **regeneration** or **budding**; these processes are most common in the more simply organized animals. Sponges, if fragmented, can usually regenerate from each piece. Even earthworms can regenerate certain fragments to produce new individuals; in one species it is theoretically possible for one individual to produce 15,000 new individuals in this way in about two months.

In the reproductive activities of sponges and some bryozoans, asexual multicellular buds are grown by the adults and detach from the parent body to develop into young. In some Cnidaria, young arise as outgrowths of parents, often in a special zone. Some of these young separate from the parents by simply pinching off to start

life on their own. Others remain united to their parents and may bud off new individuals themselves, which also remain united. Such a population of physically united, genetically identical individuals is termed a **colony**. There are other patterns of budding, but all of them are accommodated on the cellular level by divisions that provide daughter cells with identical copies of their parental cell chromosomes. Colonies produced by budding occur in several animal groups including the tunicates (phylum Urochordata), which are closely allied with the chordates (Table 2.1).

One type of asexual reproduction, **parthenogenesis**, occurs even among complex animals that ordinarily reproduce sexually. Parthenogenesis occurs when eggs develop into adults without benefit of fertilization, owing to some stimulus. Parthenogenetic offspring thus bear the chromosomes of their mother only. **Clones** are genetically identical individuals that are descended from a single founding parent by any form of asexual reproduction.

Alternation of generations

A major advantage of sexual reproduction is that each generation is usually genetically and therefore functionally variable and somewhat different from the preceding one. This means that in a world where the environment is varied and changeable, at least some individuals of a new generation are likely to be well enough adapted to the environment to persist. The best gene combinations among sexually reproducing organisms, however, may be broken up by the genetic recombination that accompanies their reproduction. Populations of asexually produced organisms, by contrast, may all be like their parents genetically, and when conditions are just right for them, they can produce many identical copies of themselves, expanding twice as rapidly as similar sexual organisms to exploit the favorable environment. On the other hand, a whole population of clones may be swept away by an inclement environmental change. Each system has its advantages and disadvantages. In general, asexual reproduction seems most advantageous in forms with very short generation spans, which allow them to respond most readily to short-term favorable conditions.

Many organisms can reproduce either sexually or asexually, and some display a more or less regular **alternation of generations**: a sexually produced generation reproduces asexually to form a generation that reproduces sexually. Alternating generations of the same species may differ morphologically as well as reproductively. The alternate generations of some species even pursue different modes of life; some cnidarians (jellyfish and their allies) have an asexually produced generation of solitary, free-floating individuals and a sexually produced generation of bottom-dwelling colonies. In some protozoans both generation types seem to live similarly, with asexual reproduction used to exploit favorable conditions and sexual reproduction to ensure a supply of new gene combinations when environments change—a mixed strategy for survival.

Development

Differentiation

The entire life history of an organism from fertilization (or asexual production) to death is called its **ontogeny**. The changes that lead from a zygote to an adult individual are termed **development**. In unicellular organisms most development is accomplished by the prereproductive processes that lead to the partitioning of parental cellular contents among daughter cells. After cell division, only growth may be required to prepare the daughter cells for their own reproduction.

In multicellular animals, however, a more complicated process of development follows reproduction. The cells must be multiplied from the zygote; the bodies of complex organisms contain billions of cells. The cells must also be **differentiated**; all multicellular animals have at least a few different types of cells, and the more complex have about 200 cell types. Finally, cell multiplication and differentiation must produce an ordered pattern of growth that leads through viable stages to an adult.

Cell divisions that result in the development of multicellular animals are similar to those that divide a protozoan into identical daughter cells. Mitosis produces cells that are genetically identical. The differentiation of animal cells arises in more than one way. During cell division, the material outside the nucleus may divide unequally, and the daughter cells may thus begin with different contents. Also, regulatory genes of different cells may specify the operation of different structural genes, in different associations, or at different times. Even though two cells have identical genotypes, a different subset of genes may function in each cell, resulting in different cell phenotypes. The regulatory genes are themselves regulated by the products of previously active genes, by secretions (such as hormones) from tissues formed by previous gene activity, or by other special stimuli. Still other differentiations arise from processes outside the nucleus that are not under direct nuclear control; presumably these are attributable to gene activities. The whole system is set in motion by events associated with fertilization in sexually reproducing organisms or in asexually reproducing organisms, by some other cues, presumably derived from parental cells.

Developmental types

Invertebrates employ several distinctive developmental patterns (Table 2.2). Most deposit or release eggs in the sea, but some retain fertilized eggs in protected spaces within or upon their bodies until the young hatch.

In **direct development**, the young animal emerges from the egg as a juvenile that is more or less a miniature adult in appearance: development is greatly advanced at hatching. The egg contains nutriment in the form of yolk to feed the young during this relatively protracted development. Yolky eggs are larger and require more energy to produce than do nonyolky eggs, other things being equal, so that the number of large yolky eggs that a female can produce during a given reproductive period is usually small (although in invertebrates, small numbers may mean hundreds or thousands).

In many marine invertebrates, the young have at least one life stage, different from that of the adult, interposed between fertilization and attainment of the adult body form. This is the **larval** stage.

More than one developmental mode is represented among larvae. Many larvae, called **lecithotrophic**, do not feed and therefore have eggs that are provided with yolk. These forms may spend their larval lives floating and swimming weakly in the surface layers of the oceans (**pelagic** lecithotrophic) or may live on or near the bottom (**demersal** lecithotrophic larvae) before settling to the bottom to metamorphose into adult forms. Still other lecithotrophic larvae are brooded, protected in some special structure or body space of their parent.

The largest number of invertebrate species today have larvae that do feed, and most of these are pelagic forms living in the surface layers of the oceans (**planktotrophic** larvae). Some larvae that lead demersal lives near or upon the sea floor evidently feed there; some deep-sea species may follow this pattern, but little is known about the extent of larval feeding in the demersal mode.

These different developmental modes confer different properties upon their species and seem to be adapted to different circumstances. Since planktotrophic larvae feed, they do not require yolky eggs, and therefore much less energy needs to be expended per egg and many more larvae may be produced per female. Furthermore,

Table 2.2. Common developmental patterns in benthic (bottom-dwelling) marine invertebrates

Developmental class	Embryonic or larval habitat	Energy supply	Chief occurrence
Direct	Eggs (masses on bottom or within parental body or structure)	Yolk in eggs	Large proportion of fauna on high-latitude shelves, common in the deep sea
Indirect	Demersal (larvae on or just above sea floor)	Yolk in eggs (some brooded)	Common on high-latitude shelves and in the deep sea
		Larvae feed	Probably in deep sea
	Pelagic (floating larvae)	Yolk (lecithotrophic)	Small proportion everywhere, probably relatively more important on high-latitude shelves
		Larvae feed (planktotrophic)	Most common in tropics and in shallow water

as the larvae feed, they may prolong their larval stage for weeks, in some cases up to several months. Some species produce such vast numbers of long-lived planktotrophic larvae that some of them survive transport across the entire Atlantic Ocean in the equatorial current system. Larvae that do not feed, on the other hand, must metamorphose soon or starve; they commonly live for only hours to a week or two before metamorphosing.

Planktotrophic larvae are most common in the tropics, where they are found in the majority of invertebrate species and where food supplies for these larvae are the most reliable. Toward the poles, where the food supply is highly seasonal, and in the deep sea, where food items are scarce in the water column, nonfeeding larvae predominate.

Growth and morphogenesis

Growth in multicellular organisms is due chiefly to cell multiplication. **Morphogenesis**, or the development of body form, occurs in several ways. Differences in the direction of cell division can lead to differential growth; for example, if all cells in a tissue divide in one plane, the tissue will obviously grow at right angles to that plane since all new cells are added in that dimension. Nearly any shape can be achieved by controlling the direction and amount of cell division. **Morphogenetic movements** also occur when cells migrate from their site of formation to a growing site elsewhere. Localized **growth zones** can occur wherein new cells are proliferated. These zones may retain their positions in the body, as with plants or with some animal colonies that are grown at the tips of branches. As a result of these kinds of cellular growth, different portions of bodies may grow at different rates to produce shape changes.

Comparative growth rates can be expressed as follows:

$$y = bx^k$$

where y is the size of a structure; x, the size of the body or some other structure; and k, the ratio of the **specific growth** rates of y and x. Specific growth is a measure of the percentage increase in living substance per unit of living substance present. Thus if the specific growth rates of a small and a large organ are equal, the larger will actually be adding mass faster, but not faster per unit of mass of the organs. The term b represents the size difference between y and x to balance the equation. When k = 1, the specific growth of x and y is equal and is termed **isometric growth**; when k ≠ 1, y will grow disproportionately larger (k > 1) or smaller (k < 1) than x as growth continues. This is **allometric growth**, and as long as k remains constant, the change in proportions remains constant with size change.

Colonies

Colonial growth has features that solitary animals do not display, involving the patterns of populations of united individuals. Since colonies are clones, all individuals normally have identical genotypes, barring mutations. Each individual in a colony has its own ontogeny, and therefore there will be a pattern of difference among individuals in that the earlier formed will represent older ontogenetic stages than the later formed. The entire colony may also display a sequence of morphologic stages, so that the earlier individuals may be different from later ones, even when equivalent ontogenetic stages are compared; this sort of change is **astogeny**. Finally, frequently form and function differ among individuals in a colony—a **polymorphism**.

Individuals in colonial animals display a wide range of interdependence. In some colonies there are no polymorphs and few functions that benefit the colony as opposed to the individual. In other colonies (the Portuguese man-of-war, for example), polymorphs are extremely specialized; for example, some feed, others digest food, and still others reproduce. The individuals in such a colony function as organs in the body of a solitary animal, and are entirely submerged or integrated into a colonial organization. Such a colony is virtually a superorganism.

Adult body plans

Lower grades and body plans

The organisms considered in this book range from the relatively simple protists to forms as complex as the octopus and insects. There are two aspects to such differences in complexity. One is the level of organization of body construction, or **grade**; the other, the variations in anatomic **body plans** within a grade. Different grades are indicated by such criteria as whether the organisms are unicellular or multicellular, whether they have two or three well-differentiated tissue layers, and whether a body cavity is present. At any level of organization or grade, several distinctive body plans are common. Animals that share a major type of body plan are classed into a **phylum** (Chapter 6). Between 25 and 30 living phyla are recognized by most authorities, and extinct groups are considered to represent additional phyla. In this chapter the major grades are reviewed briefly (Figures 2.3 and 2.4) and the living phyla listed (Table 2.1). Those phyla with significant fossil records are discussed in detail in following chapters.

Unicellular grade Body plans among unicellular organisms are defined by the presence or absence of photosynthetic organelles, cilia, and similar criteria. Only one group is treated here—the protozoans (Figure 2.3, *A* and *B*), which are the most animal-like of unicellular organisms and are probably ancestral to the animal kingdom. Some protozoans possess mineralized skeletons.

Primitive multicellular grade The early, simpler multicellular organisms that were ancestral to the more complex living animals are all extinct. The more primitive body plan possessed by living invertebrates is found among the sponges (phylum Porifera, Figure 2.3, *C*). Sponges are not ancestral to the more complex animals, however; their developmental pathway is unlike that of all other animals. Instead they probably represent a group that arose from unicellular protozoans independently of the ancestors of the rest of the kingdom Animalia (animals); they are thus often classed in a separate subkingdom, the **Parazoa**. Their body walls resemble a tissue—an aggregate of cells that forms a structural part of a plant or animal. Two layers of cells are separated by a gelatinous substance that contains wandering cells (amoebocytes). Sponges hve no well-defined organs.

Diploblastic grade **Diploblastic** means having two germ layers, that is, having two differentiated tissue layers each proliferated by mitosis from separate embryonic regions. These layers are termed the **ectoderm** (outer layer) and **endoderm** (inner layer). Jellyfish (phylum Cnidaria) are an example (Figure 2.3, *D* and *E*). All the more complex animal phyla also have ectoderm and endoderm (sponges do not), and thus the diploblastic animals, together with all the more complex ones, are believed to be related and are placed in the subkingdom **Metazoa**. Since cnidarians are the simplest living metazoans in body plan, most authorities regard them as the most primitive. The tissue layers are separated by a gelatinous, locally fibrous layer termed **mesogloea**, the "jelly" of jellyfish. Primitive organs (such as gonads) are present.

Triploblastic acoelomate grade The phyla at this grade (for example, Platyhelminthes and Nemertina) are wormlike animals that have three cellular germ layers: ectoderm and endoderm, as in diploblastic forms, plus **mesoderm**, which lies between them (Figure 2.3, *F* and *G*). They are thus termed **triploblastic**. The animals at this grade possess no body cavities, but have series of duplicated organs bilaterally disposed down their elongated bodies. Some have primitive circulatory systems. Some advanced groups have cores of mesodermal tissues that are sufficiently incompressible to act as **hydrostatic skeletons**, bulging out in spaces between muscular contractions. The bulges can act as anchors and, if made to flow backward over the body surface by coordinated muscular activity, permit the worm to flow forward. This locomotory method, **peristalsis**, permits

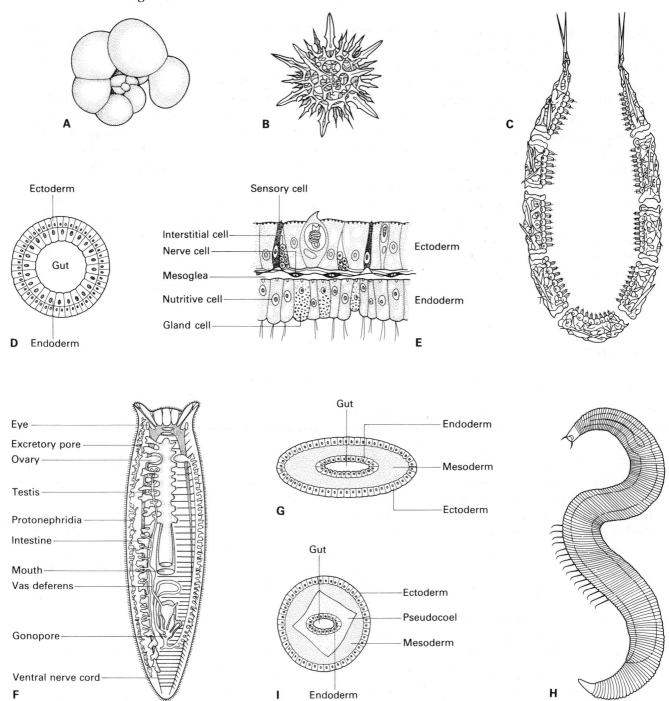

Figure 2.3. Lower invertebrate grades. Unicellular grade: **A**, skeleton of a foraminiferan; **B**, skeleton of a radiolarian (both protozoans belonging to the same major subgroup as the amoebae in Figure 2.1, *I*). Primitive multicellular grade: **C**, cross section of a simple sponge showing cell layers. Diploblastic grade: **D**, diagrammatic cross section through a simple cnidarian to show two-layered body wall; **E**, close-up of part of a cnidarian body wall to show details of tissues and cell types.

Triploblastic acoelomate grade: **F**, longitudinal section of a complex flatworm to show serial duplication of internal organs; **G**, diagrammatic transverse section of a flatworm to show three-layered body wall. Triploblastic pseudocoelomate grade: **H**, free-living marine nematode; **I**, diagrammatic transverse section of a nematode to show the body cavity (pseudocoel) between the gut and mesodermal tissues.

actual burrowing by sediment displacement. None of these animals is known to have possessed a mineralized skeleton.

Triploblastic pseudocoelomate grade During early development in many multicellular animals, there is a stage when the dividing cells form a hollow, fluid-filled sphere, the **blastula**; its interior space is the **blastocoel** (Figure 2.5). Usually the blastocoel is eventually filled with tissues. However, some phyla of triploblastic animals (for example, Nematoda) have a body space appearing in the adult as a fluid-filled body cavity between the digestive tract and body wall. This space is sometimes interpreted as a blastocoel. It may function as a hydrostatic skeleton and is termed a **pseudocoel** (Figure 2.3, *I*). Many pseudocoelomates are microscopic and lack respiratory and circulatory systems. None is known to have ever possessed a mineralized skeleton. (For a slightly different usage of pseudocoel, see Chapter 17.)

Triploblastic coelomate grade and body plans

The **coelom** or "true" body cavity does not arise from the blastocoel but is a different space, entirely surrounded by mesodermal tissue. In some animals the coelom arises as an outpocketing from the gut that becomes lined with mesodermal tissue; in others it arises as an entirely new space by splitting apart of mesoderm. Filled with fluid, the coelom communicates to the exterior only through special ducts that convey wastes or reproductive products. Organs are situated in the coelom (heart, kidneys, gonads, and others). A variety of coelomic architectures is found (Figure 2.4); most of the phyla that share each type seem closely related, for they share other features as well.

Metamerous coelomates In some wormlike groups (Annelida) the coelom is divided along its length into a number of **segments** or **meres**, primitively separated by transverse partitions called **septa**. Each segment regularly contains a number of the same paired organs (such as kidneys, nerve ganglia, lateral blood vessels, and gonads) and often bears paired appendages (Figure 2.4, *A* and *B*). This is a **metamerous** coelom, used to antagonize body-wall muscles in peristaltic locomotion. Earthworms are a familiar metamerous animal. The effects of peristaltic contractions are decreased by the septa and are not transmitted freely throughout the length of the coelom. Therefore body shape is more easily maintained, and more than one peristaltic contraction can be present at the same time. One metameric group (Arthropoda) has developed a stiffened outer skeleton so that locomotion is by appendages. The septa are obsolete, the coelom itself is greatly reduced, and body shape is maintained by blood and outer skeleton. Nevertheless, the modular repetition of paired organs indicates that these forms are metamerous.

Amerous coelomates Simple coelomic cavities that are neither segmented nor otherwise divided into regions may be called **amerous**. An example is the Sipunculida (Figure 2.4, *C*). In these forms the anterior body is commonly provided with a retractable organ for burrowing or feeding. No amerous form is known to have possessed a mineralized skeleton.

Pseudometamerous coelomates **Pseudometamerous** animals (which include the Mollusca: clams, snails, squids, and so on) have irregularly duplicated organ systems (Figure 2.4, *D*), but their coelomic spaces are not divided into regions. The coelom does not envelop the digestive tract; it chiefly permits movement of contractile organs or forms ducts. Although the coelomic spaces are mesodermal, they may have evolved independently of those in the other coelomates.

Oligomerous coelomates The **oligomerous** plan consists of a coelom that is divided into two or three longitudinal regions that have separate functions. Three somewhat distinctive groups of oligomerous phyla exist. One group (Phoronida, Brachiopoda, Bryozoa) has elongate to saclike bodies surmounted by a hollow, ciliated tentacular crown that is supported on a ribbon-like structure (Figure 2.4, *E* and *F*); this organ, known as the **lophophore**, is employed in feeding and respiration. A second group (Hemichordata, Urochordata), which may possess a tentacular crown also, has in some members paired **gill slits** that permit water, ingested through the mouth, to be expelled without regurgitation. Such slits, developed in primitive chordate groups and early fishes, eventually evolved into vertebrate jaws. A third group (Echinodermata) has a tentacular system that is elaborated into an extensive series of canals, the **water-vascular system**, from which arise branches that can function for locomotion, respiration, and feeding.

Interrelations of major invertebrate types

Paleontologists are much interested in the family tree of life represented by fossil organisms; the ancestral history of a group of organisms is termed its **phylogeny**. In principle the fossil record permits the evolution of groups to be traced from one to another. In practice this has not yet proven possible for phyla; no phylum can be traced

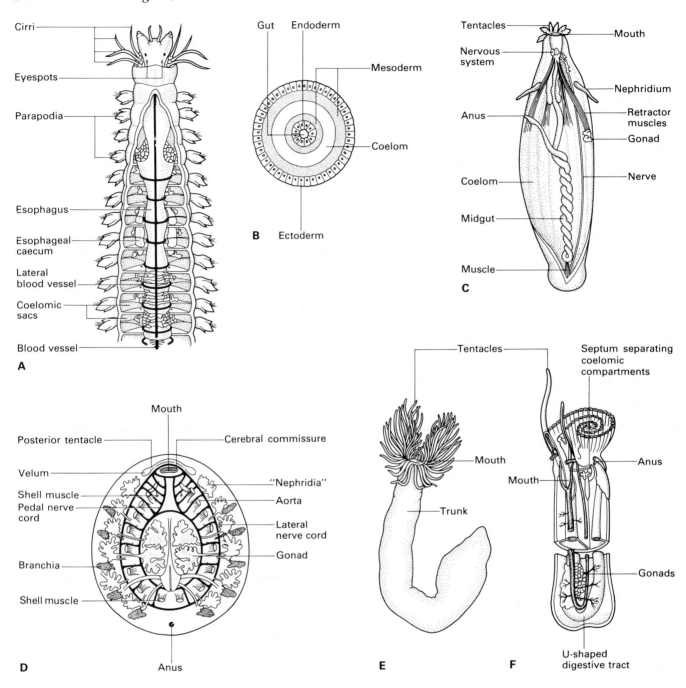

Figure 2.4. Coelomate body plans. Metamerous plan: **A,** section of anterior end of a marine annelid (*Nereis*) to show segmented coelom and regularized repetition of organs; **B,** diagrammatic transverse cross section of an annelid to show coelomic architecture. Amerous plan: **C,** anatomy of a typical sipunculid; the coelom lacks segmentation. Pseudometamerous plan: **D,** a primitive mollusk, *Neopilina,* with ventral anatomy showing seriated organs. (From Wingstrand, K. G. *Neopilina galatheae* Lemche, 1957. *Galathea Report 3*; 1959.) Oligomerous plan: **E,** phoronid, a primitive oligomerous form; **F,** schematic view of the interior of a phoronid showing division of coelom into tentacular and trunk regions.

through fossils from an ancestral phylum. Thus their interrelations—their phylogenies—must be inferred from their similarities of development and form. This is no easy task, since patterns of similarity vary from character to character. Some similarities indicate close common ancestry, while others are due to independent evolution of similar solutions to the same problems or opportunities. A few examples will clarify the problem, although the solution remains difficult.

Larval resemblances

In general, each phylum has larvae that are characteristic in form. However, some phyla that are distinctive in adult form have larvae that are rather similar to each other. It then becomes necessary to decide whether the larval or adult resemblances more nearly reflect relationships between phyla. Recall that invertebrates have a variety of developmental histories (Table 2.2). These differences occur within phyla—arthropods, mollusks, and echinoderms all have chiefly pelagic larvae in the tropics but display direct development in high latitudes. Thus larval forms are obviously sensitive to environmental influences and may be modified or even dispensed with independently of the adult structure.

Developmental patterns

Similarities in developmental pathways have been given much weight in estimating the closeness of ancestry among phyla. This is partly because of a once widely held hypothesis, now shown to be false, that the stages in the development of an organism indicate the succession of adult stages in its ancestry ("ontogeny recapitulates phylogeny"). If this were true, the evolutionary history of an organism could be read directly from its developmental stages. What is in fact inherited is not a succession of ancestral adult stages, but an entire developmental sequence; evolutionary history is a succession of developmental sequences, not just a succession of adult stages. Developmental stages may evolve (obviously the larval stages mentioned previously do so) as adaptations to conditions of development, just as adult stages evolve as adaptations to conditions of adult life.

Developmental stages nevertheless provide evidence of ancestry, just as adult stages do; groups with similar developmental stages are related unless the stages have arisen independently. Three developmental features have played particularly important roles in interpretation of invertebrate phylogeny. One is the pattern of very early cell division. When the zygote first divides it simply splits into two smaller daughter cells, and those into even smaller cells, and so on. This early stage is termed **cleavage**. In some animals the resulting cells are packed in vertical tiers, one above the other (Figure 2.5, *A*); this is **radial cleavage**. In other animals the cells are packed in alternating positions (Figure 2.5, *B*); this is **spiral cleavage**. When cleavage ends, the cells form a sphere termed the blastula (Figure 2.5, *C*); if the interior is hollow, it is the blastocoel.

A second important developmental feature is the way

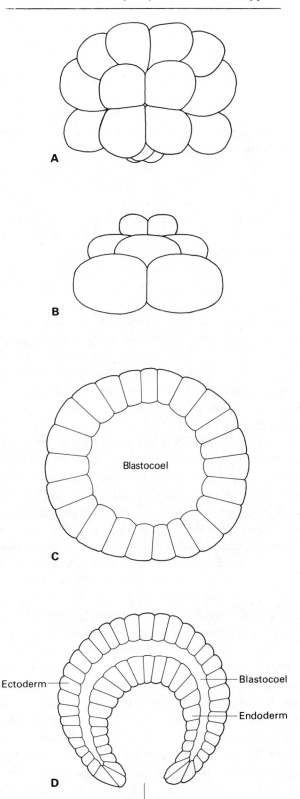

Figure 2.5. Early developmental patterns in invertebrates. **A,** Cleavage of the radial type. The zygote subdivides with cells in straight rows vertically. **B,** Cleavage of the spiral type. The zygote subdivides with cells in alternate rows vertically. **C,** Blastula stage showing blastocoel. **D,** Gastrula stage showing blastocoel, blastopore, and ectodermal and endodermal layers.

in which the mouth forms. As development proceeds from the blastula, one side of the sphere may collapse inward (invaginate) to produce a double-walled sphere called the **gastrula** (Figure 2.5, *D*). The inner wall of the gastrula becomes endoderm and the outer wall, ectoderm. The space between them is the blastocoel, while the opening into the endoderm-lined inner cavity is termed the **blastopore** (Figure 2.5, *D*). The endoderm-lined cavity becomes the gut as development proceeds. If the blastopore becomes the site of the mouth, the animals are termed **protostomous**. If, on the other hand, the mouth forms in another location, the animals are termed **deuterostomous** (in some of these the anus forms at the site of the blastopore).

A third important developmental feature, found in coelomates, is the method by which the coelomic cavity appears. Mesoderm is proliferated from endoderm into the blastocoel, which becomes filled in coelomates. In some animals, outpocketings from the gut form spaces that are then surrounded by mesoderm to become the coelomic cavity; this is the **enterocoelous** method. In other animals, a split simply appears within the mesodermal tissue and expands to form the coelomic cavity—the **schizocoelous** method.

These and other developmental features have been used to form a model of animal phylogeny. There are two assemblages of coelomates that differ in cleavage pattern, the formation of the mouth, and the formation of the coelomic cavity. One assemblage is protostomous and schizocoelous with spiral cleavage. Included are amerous, metamerous, and pseudometamerous coelomates (Table 2.1). The triploblastic acoelomates (Table 2.1) are also protostomous with spiral cleavage and are usually placed with this assemblage (many authorities believe they are ancestral to the protostomous coelomates). This group of phyla is called the Protostomia. Another assemblage is deuterostomous and enterocoelous with radial cleavage. This group of phyla includes many of the oligomerous coelomates (Table 2.1) and is termed the Deuterostomia.

In fact there are difficulties with such a subdivision of the Metazoa. Many phyla do not display all the characters of either assemblage clearly; sometimes the characters are mixed, varying independently. This is particularly well illustrated by some of the oligomerous phyla. Hemichordates are enterocoels, and have radial cleavage and are deuterostomous. Phoronids are enterocoels, have radial cleavage but are protostomous. Urochordates probably have no coelom at all, but have radial cleavage and are deuterostomous. All three phyla are Deuterostomia. The questions of how the phyla are related are not simply solved by study of their developmental features.

Adult body plans and grades

Resemblances among adult body plans may but need not suggest close relationships. A pseudocoelomate phylum, the Entoprocta, so closely resembles a coelomate phylum, the Bryozoa, that they continue to be considered closely allied by some authorities. To other authorities their different grades of construction indicate that there is no close relationship between them. Similarity in grade of construction, on the other hand, does not necessarily certify close relationship; some grades (such as the primitive multicellular grade) have been evolved more than once.

Skeletal plans

Skeletons are the supporting and protecting structures of animals. As such they are coadapted with the nonskeletal anatomic features to form a harmoniously integrated body plan. In their supporting role they may control body shape and transmit the action of muscles so that certain body parts may work.

Some invertebrates employ tissue or fluid skeletons. Some nemertines, for example, use fairly incompressible tissues to complement their muscular systems in locomotion. Fluid skeletons (as in pseudocoelomates and some coelomates) may also function in this manner, when they are termed hydrostatic skeletons. In some fluid skeletons (as in some mollusks), fluid is pumped from one body region to another as required, to form a **hydraulic skeleton**. Other skeletons, which are thought of as the classic sort of fossils, are of rigid, durable substances. They may be organic (arthropods and some hemichordates) or mineralized (sponges, corals and many coelomate phyla). Some animals (as bryozoans) possess both rigid and fluid skeletons.

The rigid skeleton has some advantages over the others, the most important of which is that it holds its shape without the assistance of muscles. Thus, although a rigid skeleton takes some energy for an animal to secrete, it saves energy thereafter and is an efficient, evolutionary device. Rigid skeletons also have other functions, such as protection. They have been invented independently by each phylum that has one, and some phyla have invented rigid skeletons more than once. Therefore, skeletal plans displayed among different animal groups differ because skeletons are really organs evolved to accommodate and enhance the functioning of the particular body plan of each group. It is possible to speak of **skeletal plans**, each of which is coadapted with a body plan; they can be no more helpful than can body plans in establishing the interrelationships of phyla.

Most rigid invertebrate skeletons are secreted from tissues on body surfaces so that the skeleton forms outside the ectodermal layer as an **exoskeleton**. A few invertebrates have **endoskeletons**, which are secreted within tissues.

Since skeletons must accommodate growth, their range of form is constrained. Many groups secrete basically conical exoskeletons (Figure 2.6, *A* and *B*), which grow by accretion at the large open end, either in rings or chambers of new growth. Often such skeletons are coiled, usually in a spiral, since that permits an increase in size without a change in shape. Arthropod exoskeletons (Figure 2.6, *C*) are not accretionary but are secreted by nearly the whole body surface to fit "like a glove". They must be discarded—**molted**—and larger skeletons secreted periodically as growth continues.

Echinoderm skeletons (Figure 2.6, *D* and *E*) are secreted within tissues and are technically endoskeletons, but surround the main body elements to function as external skeletons. They are composed of numerous platelike or spicular elements, often articulated so as to cover the body and appendages like armor plates. Growth may occur both by enlargement of the plates and by the addition of new ones. Some sponges (and members of other phyla) secrete individual spicules. They are scattered through tissues for support or protection; they may be fixed into a skeletal framework. Some other rarer skeletal plans are described in succeeding chapters.

Colonies have skeletal plans also—not only the plan of each individual, but the plan of their arrangement is involved. The following are the most basic types of colonial arrangement: one dimensional, as in stolonal tubelike colonies wherein individuals are arranged in a line; two dimensional, as when colonies are sheetlike, usually encrusting; and three dimensional, as when colonies have bushy or other branching growth forms.

Fossil record of grades and body plans

The fossil record of living invertebrate phyla is summarized in Figure 2.7. The earliest fossils that we can be sure represent invertebrates occur nearly 700 million years ago; they are burrows. Burrowing implies that animals with hydrostatic skeletons had appeared. Although it is possible to imagine a burrower at any grade from diploblastic up, these fossil burrows most likely represent

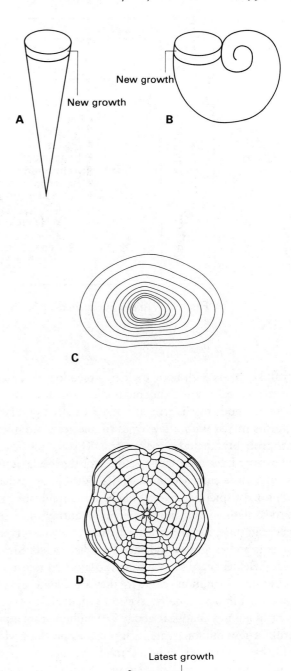

Figure 2.6. Some skeletal plans to accommodate growth. **A,** Expanding conical exoskeleton that grows by accretion at the aperture. **B,** Coiled cone that accommodates growth in the same manner but that is more compact. **C,** Molt stages of an arthropod (Crustacea: Ostracoda); mature left valve and series of nine successively larger shells that were molted during growth; modified from Moore, R. C., editor. *Treatise on invertebrate paleontology*, part Q, vol. 3; 1961. After Kesling, R. V., Terminology of ostracod carapaces, *Univ. Michigan, Mus. Paleont. Contr.*, 9: 45; 1951; and Sars, G. O., *An account of the Crustacea of Norway*, vol. 9; 1925. **D,** Echinoderm endoskeleton of discrete plates (the echinoid *Clypeaster*). **E,** Single plate from *Clypeaster* showing growth lines.

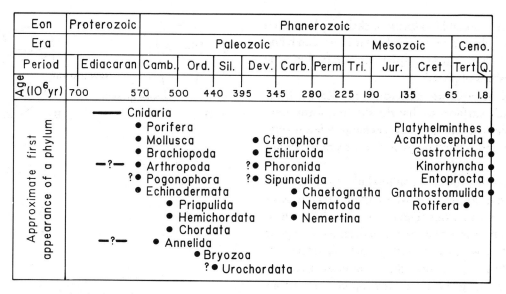

Figure 2.7. Approximate times of the first appearances of living animal phyla.

wormlike forms with body cavities—pseudocoelomates or (since the burrows are not minute) coelomates.

The first body fossils are soft-bodied, chiefly diploblastic forms mixed with a few types of uncertain affinities that probably include coelomates. This is the Late Proterozoic **Ediacaran fauna**, named for the locale of its best-known occurrence in Australia. With the possible exception of spicules, it seems likely that durably mineralized skeletons were not yet evolved during most of Ediacaran time, for they are much more easily preserved than soft bodies and should have been discovered. Rigid skeletal elements are known from the latest Proterozoic, just before 570 million years ago; they are minute and of unknown affinity. Finally, skeletons that clearly represent living phyla appear about 570 million years ago. Within a few million years such phyla as brachiopods, mollusks, arthropods, and echinoderms appear, representing the main coelomate grades. Soft-bodied phyla are found only under exceptional conditions of preservation. A number are known from the Cambrian, chiefly because of one unusual fossil locality in the Burgess Shale of Canada, and a few others appear at younger horizons, particularly in late Carboniferous rock in Illinois. Several soft-bodied phyla have no known fossil records (Figure 2.7).

Thus, the order of appearance of major fossil groups is related only partly to the order of their origin. The presence or absence of durable skeletons and other preservable features and the chance discovery of exceptionally well-preserved fossils can be important in determining first appearances in the fossil record.

Supplementary reading

Brusca, G. J. *General patterns of invertebrate development.* Eureka, CA.: Mad River Press; 1975. Well-illustrated survey of developmental patterns in invertebrate phyla, with short accounts of their possible significance in animal classification.

Clark, R. B. *Dynamics in metazoan evolution: the origin of the coelom and segments.* Oxford: Clarendon Press; 1964. Account of the functional significance of coelomic architectures, with special attention to the annelids.

Clark, R. B. Radiation of the metazoa, pp. 55–102. In: House, M. R., editor. *The origins of major invertebrate groups.* London: Academic Press, Inc.; 1979. Up-to-date review of evidence and opinion on the relationships between the major groups of animals.

Fretter, V.; Graham, A. *A functional anatomy of invertebrates.* London: Academic Press, Inc.; 1976. The anatomy of representatives of living invertebrate phyla.

Strickberger, M. *Genetics.* 2nd ed. New York: Macmillan Publishing Co., Inc.; 1976. Perhaps the most comprehensive account of genetics in a single volume.

3

Ecology–Paleoecology

Martin A. Buzas
Anna K. Behrensmeyer

Even the most casual observer notices that animals and plants are not uniformly distributed over the earth. Polar bears live in the arctic and coral reefs in warm, shallow waters. Deserts are barren and rain forests lush. Without thinking about it, we associate groups of organisms with each other and with particular physical environments. We notice that the distribution and abundance of organisms are not haphazard but form patterns. Through experience we learn to recognize these patterns in nature. In a sense, we are all ecologists because ecology seeks to explain these patterns in the distribution and abundance of organisms.

Ecology is defined as the study of the mutual relations between organisms and their environment. The word ecology is derived from a Greek word meaning house, and therefore ecology literally means the study of organisms where they live. The place where they live is called their **habitat**.

We expect to find seashells at the seashore and not in the mountains. Thus, when seashells were first discovered in the mountains of Italy, they presented a dilemma. Were they the works of the devil left to confuse their discoverers or had the mountains once been a sea floor? Notice acceptance of the latter explanation depends on our observation of patterns of ancient organisms to ancient environments is given the prefix *paleo* (ancient) and is called **paleoecology**. Paleoecology can be more complicated than ecology because the original relations of the organisms and environments often are changed by the processes of preservation. To reconstruct the original paleoecology, it is important to understand the taphonomy of a particular fossil deposit. **Taphonomy** is the study of the processes of burial and preservation (Chapter 5) and how they create samples of the original populations of living, interacting organisms. Many organisms are not preserved as fossils, and others may be preserved with biases for or against certain species or morphological types. Sometimes organisms are not buried in the same environment where they originally lived. These taphonomic alterations or "overprints" may be corrected for if they are understood, but in other cases preservational biases severely limit paleoecologic interpretations.

Because organisms are adapted to particular environments and have been throughout the geologic past, paleoecology provides a method for inferring past environments and habitats from the remains of organisms. Indeed, this ability to reconstruct ancient environments has contributed greatly to our understanding of earth history. For example, given the known physiologic constraints on the growth of modern coral reefs, a great deal can be inferred about water temperature and climate from coral reefs in the fossil record. At the same time, we wish to document the distribution and abundance of organisms in such habitats over long periods of time to gain understanding of the patterns observed today. A few million years ago the species inhabiting the earth were much like the ones alive now. Their associations and habitat preferences lie within the range of our experience. The task becomes more difficult the farther we go back in time. Creatures without modern counterparts inhabited the earth hundreds of millions of years ago, and although their remains still exhibit patterns, these are difficult to decipher.

Unispecies ecology

Study of the ecology of a single species is called **autecology**. Generally, the autecology of an organism refers to the way it makes a living and is studied using various methods such as functional morphology, comparisons with habitats of living relatives, and sedimentary context. These help to define its original ecological role. Another aspect of autecology is the study of populations of single species. Populations have parameters or characteristics not applicable to individuals: one of these is density.

Density

Density is defined as the average number of individuals in a given area or volume. While density is a straightforward measure of population abundance, it is difficult to measure accurately because species have different spatial patterns. Densities as measured in the fossil record may not correspond to original densities because of taphonomic overprints.

Spatial Patterns

In modern ecosystems, some species are randomly distributed. **Random** means that if an area is divided into contiguous cells, the probability of an individual being in any one of the cells is the same and equal to 1/N, where *N* is the number of cells.

The most common spatial pattern is termed patchy, clumped, or **aggregated**. In a random distribution the arithmetic mean of the number of individuals in a cell, $\bar{x}$, equals the variance (a measure of spread about the mean), s^2. In an aggregated distribution the variance is greater than the mean, $s^2 > \bar{x}$. Consequently, the ratio $s^2/\bar{x}$ will be one for a random distribution, but greater than one for an aggregated distribution.

For some organisms, the terms patchy, aggregated, or clumped are appropriate. For example, cows under a tree in a pasture are clumped, and oysters occur in reefs that, even in a small bay, are aggregated. For other organisms, however, patches or clumps are not so easy to define. The variance is still greater than the mean, and these organisms are not distributed at random, but neither are they highly aggregated so that patches are easily identified. Figure 3.1 gives examples of random and slightly aggregated spatial patterns. The figure illustrates that spatial patterns cannot be confidently determined by eye but must be statistically determined. In aggregated distributions more observations will be required to measure the density with any given degree of confidence than in a

random distribution. The number of observations required will depend on the degree of aggregation.

While it is relatively easy to determine if organisms are aggregated, it is not so easy to determine why. Different species are aggregated for different reasons. Oysters must find a hard substrate on which to settle and so their distribution depends upon the distribution of adequate substrates. Some animals reproduce asexually and the

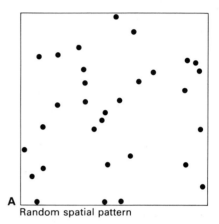

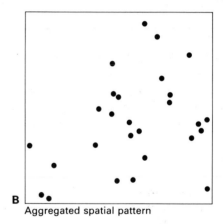

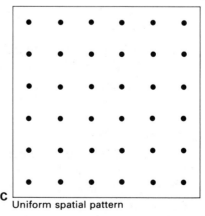

Figure 3.1. Three types of spatial patterns exhibited by organisms.

young are grouped about a parent. Many fish form "schools" and are aggregated because of their behavior.

Over a relatively small area some organisms have a uniform or even distribution (Figure 3.1). In a **uniform** distribution the variance is less than the mean. Trees in an orchard provide an easily visualized although unnatural example. Some clams who live in the sediment feed through a long siphon or tube by ingesting the sediment in a circular pattern around themselves. This mode of feeding encourages adjacent individuals to remain outside of the feeding range of their neighbors, and results in an even distribution like a planted orchard.

Understanding of spatial patterns in modern organisms helps provide models for the past. We can expect that fossil organisms had random, aggregated, or uniform distributions when they were alive, and such patterns can be sought in fossiliferous deposits. But we must always keep in mind that taphonomic processes may have altered the original patterns, and look for evidence that eliminates or corrects for this possibility. If taphonomic biases can be accounted for, a great deal can be learned about the ecologic distributions of organisms that may have no living relatives.

Intrinsic rate of natural increase

Long ago Malthus predicted that if human populations were not checked, a great famine would result. The problem is easily understood. Two individuals give rise to several offspring, each of whom gives rise to several more, and so on. Within a few generations the resulting population becomes enormous. Mathematically the relationship can be expressed by the equation

$$N_t = N_0 e^{rt}$$

where N_0 is the number of individuals existing at some arbitrary time; t, the number of time units; N_t the number of individuals t time units later; e, the base of the natural logarithms or 2.72; and r, the instantaneous rate of increase. All species have the ability to reproduce at an exponential rate and the population parameter r is called the **intrinsic rate of natural increase.** Obviously, such growth cannot go unchecked for long or a single species would fill the earth. No species can maintain an exponential growth rate, and the many reasons why they cannot explain the distribution and abundance of organisms.

Niche

Anyone who has studied nature marvels at how many different ways species "make their living." Among marine animals, large carnivores are familiar to everyone, but species called **suspension feeders** filter minute organisms from the water. Still others called **deposit-feeders** ingest sediment and extract nourishment from the organic matter contained therein. **Sessile species** live attached to rocks and other organisms; **motile species** move about. Bottom dwelling or **benthic species** are either **epifaunal** (living at the surface) or **infaunal** (living within the sediment). Species floating in the water are **planktic**, and those who swim, **nektic.** Almost every imaginable mode of life is occupied by some species.

Each species has certain requirements necessary to maintain itself. The collective requirements each species needs to remain alive and reproduce itself are called the species **niche.**

Limiting variables

Each species density is limited by **abiotic** (nonliving) and **biotic** (living) variables. For marine organisms, common abiotic variables include temperature, salinity, and oxygen. These variables measure the "weather" in a marine environment. Tolerances to these variables vary from species to species depending on the dimension or breadth of its niche. Some marine species have wide tolerances, others very narrow. Those with wide tolerances for salinity can occupy most estuaries where salinities are less than those in the open ocean. Even these species are limited by their tolerance, however, and a heavy rainstorm with flooding may so reduce salinities that entire populations can be decimated.

Common biotic variables include food, disease, predation, and competition. These variables are complex and unlike most abiotic variables cannot be measured by a simple instrument. For example, if there is no food, clearly the organism will die. Food may, however, be ample but for a variety of reasons the organism is unable to reach it. Some predation is probably beneficial, but excessive predation may reduce population density to zero.

If two species compete for the same resource and one species is a superior competitor, and if competition proceeds uninterrupted, only one species will survive (competitive exclusion principle). Such competition is analogous to a card game in which the dealer always has a slight advantage, like blackjack or 21 where the dealer wins ties. The player may win for a while, but providing the dealer has enough money (resources), and the game goes on uninterrupted, the dealer must win. Consequently, to coexist, species must avoid prolonged competition. To do so, they partition resources, and this is why all the different modes of life cited earlier exist.

Sometimes two potential competitors unwittingly avoid competition through a third party. In this situation a predator may so reduce the population of a superior competitor that it can never effectively compete with an inferior competitor and so they coexist. Many different interwoven mechanisms have evolved to ensure the coexistence of species.

We assume that although species were different in the past, they were affected by biotic and abiotic variables similar to those that can be observed in modern ecosystems. This is a uniformitarian or **actualistic assumption**. Although we must depend upon the present to provide models for the ecology of the past, it is also important to realize that this has limitations. It is all too easy to practice "me too ecology," in which we set out to find ways that show the past was just like the present. Modern analogues are more useful if they are simply compared with the past, with equal emphasis on both the evidence that conforms to the analogue and the evidence that does not. In reconstructing the biotic and abiotic variables that were important to an extinct organism, one must also consider the effects of taphonomic processes that may have affected the fossil sample. The taphonomy of a fossil assemblage not only indicates potential biases affecting paleoecologic interpretations, but it also provides additional information concerning paleoecology. Thus the patterns of damage on the shells of bivalves can show that their major predator was a carnivorous snail, even though the snail itself may not be represented in the fossil assemblage.

Ecology of the hard-shelled clam

To illustrate some of the principles we have been discussing, let us examine the ecology of the delectable hard shelled clam or quahog called *Mercenaria mercenaria*.

To understand the ecology of any species, we must know something about its life history. The quahog reproduces by ejecting sperm and eggs into the water where fertilization takes place. Sperm are ejected first, and somehow the females sensing their presence emit their eggs. Typically, spawning takes place in early summer and larvae are formed by the millions. The larvae become part of the plankton (floaters) for about a week after which they settle to the bottom to take up a benthic life style. Initially, they attach themselves to the sediment with small threadlike strands (byssus) and are epifaunal, but soon they develop a foot and burrow into the sediment and remain infaunal for the rest of their lives, filter

feeding on microscopic algae by pumping sea water through their siphons. If left alone, they have a life span of eight to ten years.

Now, let us look at the quahog's ecology. Geographically the quahog is distributed in shallow waters from Nova Scotia to the Gulf of Mexico. Its exclusion from higher latitudes may be regulated by temperature because a temperature of about 25°C is required to induce spawning and eggs fail to develop below 15°C. Within any particular bay or estuary, quahogs are not randomly distributed because they require salinities of at least 17 parts per thousand (normal sea water is 35 parts per thousand) and larvae actually select areas where currents and eddies will ensure sufficient food supply.

In those areas where the quahog has successfully settled, densities are usually two or three clams per square meter. Of the millions of larvae formed, only enough survive to maintain a stable population. During their planktic existence, the larvae are voraciously eaten by a myriad of organisms. Once they settle to the bottom, the juveniles, having soft shells, are eaten by crabs and snails. As the clams become larger, larger snails, crabs, fish, starfish, birds, and humans continue to harvest them. Predation, therefore, accounts for much of their density regulation.

Competition probably also limits quahog densities. Several other species of clams and snails also filter the water for the same microscopic algae, and other species of clams make the same substrate their home. Little is known about how effective these competitors are in regulating quahog density.

This brief look at the ecology of the quahog illustrates the complexities involved in trying to explain the distribution and abundance of a single species.

In interglacial sediments of the Pleistocene epoch in southern Maryland, the quahog is found abundantly along with other fossils. The fossils are well preserved and their shells articulated and in natural living positions. This is taphonomic evidence that the quahog fossils are preserved in their original life context; bivalve shells normally come apart soon after death and cannot be transported and reburied in any configuration resembling their life positions in the substrate. The specimens found in the deposit are morphologically similar to those found today in the shallow marine waters. The assumption is made that morphologic similarity in the hard parts implies physiological similarity in terms of adaptation to water chemistry and climatic conditions. The assumption seems warranted because the shape of the shell is such an integral part of the quahog's life-style.

It can be concluded, therefore, that the climate in

Maryland before the last glaciation was similar to the present climate, and had there been natives they would have enjoyed the same "shellfish" that we do. If we use similar lines of reasoning to interpret environments from fossils of the more distant past, however, we have to be increasingly careful about assumptions concerning similarities of the adaptations of the fossil organisms and their nearest living relatives. It is often necessary to find independent evidence for environmental conditions such as sedimentary structures, geochemical indicators, or patterns of paleogeographic distribution.

Multispecies ecology

From the review of the ecology of the quahog it is evident that studying the ecology of a single species is impossible without considering the other organisms sharing the habitat. Most ecologic studies concentrate on more than one species, and the study of multispecies ecology is called **synecology**.

Ecosystem

To study the structure and function of an area like an estuary, it is classified into abiotic and biotic components or functional units. The entire ensemble of interacting abiotic and biotic units is called an **ecosystem**. The energy to drive the ecosystem comes from the sun. Plants photosynthesize the sun's energy to produce their cells. These self-nourishing or **autotrophic** organisms are the primary producers and supply the energy for the rest of the organisms. On shallow sea bottoms these plants often form dense masses of sea grass. The vast majority of marine plant species are, however, microscopic and are never seen by the casual observer. Most of the invertebrates in this book are **heterotrophs** or "other nourishing." Heterotrophs consuming plants are called **herbivores**. The herbivores are in turn eaten by **carnivores** (flesh-eaters), other carnivores eat these carnivores, and they themselves are eaten by the top carnivores.

This **food chain** cannot go on indefinitely because at each succeeding level some energy is lost. Consequently, a higher level must have a lower biomass (amount of organic matter) or productivity than the next level. For this reason the food chain is often referred to as a **food pyramid**.

The autotrophs require not only sunlight but also nutrients for their growth and reproduction. A final group is required by the ecosystem to break down and decay dead organisms into the elements required for growth. The bacteria and fungi are the **decomposers** who perform this indispensable task. Some plants and animals, usually those containing hard parts made of silica, carbonate, or tough materials such as cellulose (Chapter 5), avoid being recycled by the decomposers and instead become fossils. Quick burial and removal from the area of biologic activity is key to the formation of fossils. The selective activities of the predators, scavengers, and decomposers impose taphonomic biases on what gets preserved. The depositional regime also can control preservation to some extent, depending on whether sedimentation is episodic or continuous. Usually all the soft, organic parts of organisms are recycled, but sometimes organic matter accumulates on the bottom, is buried, and becomes fossil fuel.

A simple diagram of an ecosystem is shown in Figure 3.2. In reality the ecosystem is much more complex because different roles are played by the same species during its life cycle (remember the quahog). Many of the pathways are not in the form of a simple pyramid, but instead have many feedbacks and interactions. Nevertheless, in a qualitative way, the ecosystem is composed of different interdependent functional units. The relationships are not simple, nor is the system in a simple state of equilibrium. Rather the ecosystem is in a **dynamic equilibrium**, composed of ever changing participants, and is (unfortunately for the scientist) unpredictable in many ways.

There is little doubt that ecosystems of the geologic past operated much as they do today, with autotrophs and heterotrophs forming complex food chains. Yet the types of organisms in these different roles and the overall structures of the food chains were very different. The study of these past ecologic relationships is one of the more fascinating areas of paleoecology because it provides glimpses of biologic realms impossible to imagine without the fossil record.

Community

The living part of an ecosystem is called a **community**. Such a definition, however, has little utilitarian value. The system is so complicated that no one has successfully isolated all the species involved or worked out the complex interdependent relations existing between them. Instead, we commonly study parts of an ecosystem and refer to them as a community anyway. For instance, we may study the larger invertebrates inhabiting the bottom and refer to them as the benthic community even though we know the benthic community is also composed of a multitude of species undetected without the aid of a microscope. For practicality, then, a community is often

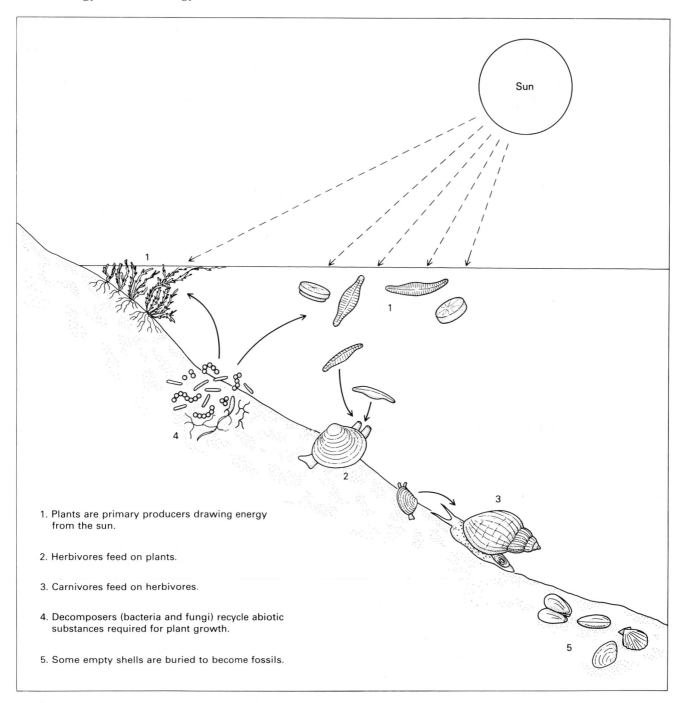

Figure 3.2. Diagram of a simplified ecosystem.

1. Plants are primary producers drawing energy from the sun.

2. Herbivores feed on plants.

3. Carnivores feed on herbivores.

4. Decomposers (bacteria and fungi) recycle abiotic substances required for plant growth.

5. Some empty shells are buried to become fossils.

referred to as a group of organisms living together at the same time and place. What the group consists of is largely up to the discretion of the observer. Clearly, an ecologic community is not like an organism consisting of functional parts like arms and legs. A community typically includes a wide spectrum of interdependencies among organisms but is much more loosely integrated than a single organism. Many species can come and go without destroying the community as an entity.

Just as populations have characteristics or parameters inapplicable to individuals, so also multispecies populations or communities have parameters inapplicable to a unispecies population. When the density of a single species is measured in a series of samples, the result is a column of numbers, one for each sample. If instead of recording the density of only one species, the density of many species is recorded in each sample, the result is a table of numbers called a matrix. The number of rows in the matrix equals the number of samples taken and the number of columns, the number of species observed. Various statistical techniques are available to analyze densities within a matrix or between matrices. Most of them, however, are beyond the scope of this book.

Sometimes ecologists find it convenient to express their data as percentages rather than as densities. The relative abundance of a species in a community is obtained as a percentage by totaling the number of individuals in a sample, dividing the number observed for that species by the total, and multiplying by 100. These species percentages are often referred to as **relative abundances** or **species frequencies**. Percentages are often used instead of densities because they are easier to visualize than densities, which vary greatly from sample to sample, and experience has shown that they reliably characterize a community.

A major problem in paleocommunity studies is that the parameters used by ecologists to characterize their modern communities are difficult to measure in fossil assemblages. Taphonomic processes alter the numbers of organic remains of different species, affecting relative abundances and spatial density counts. Paleontologists cannot always sample standardized areas or volumes because of poor preservation and/or exposures. Consequently, species may be arbitrarily recorded as abundant, common, and rare. Sometimes even such categorization is impossible and species are recorded as present or absent. In this case the matrix of observations would simply consist of 1's and 0's or X's and O's. While the presence or absence data do not contain as much information as actual counts from carefully measured areas or volumes, they can be successfully used to characterize and discriminate between communities. Often it helps to recast information available from modern communities to make it simpler and more readily comparable with the fossil record.

Theoretically, if more was known about taphonomic processes that affect species abundances during preservation, we might be able to reconstruct original abundances for fossil communities. One strategy has been to study modern ecosystems and their potential fossil record—to compare the abundances of species represented by skeletal debris on the bottom with abundances in the living community. This has shown that even for modern communities, abundances vary tremendously over the course of a year, so that it is difficult to characterize what is normal. The modern skeletal debris represents a time-averaged sample over years to decades. In fossil deposits, time-averaging may occur over much longer periods, hundreds to thousands of years. In spite

of this, similar abundance patterns show up in samples of fossil communities from different areas. Paleocommunities are often characterized by these consistent patterns, even though we are not sure how they represent the workings of the living community at any particular point in time.

Within habitat studies

Some ecological studies concentrate on species living within a single kind of habitat such as shallow muddy bottoms or sea-grass beds. Often these habitats are studied over a period of months or years to elucidate the dynamics of the community. In other studies, similar habitats are studied in different geographic areas or fossil and modern examples are compared.

An example is provided by a study of deposit-feeding clams from modern sediments of Cape Cod and from Silurian sediments in Nova Scotia. Deposit-feeding clams can be divided into siphonate and nonsiphonate forms. The siphonate forms burrow into the sediment and extrude a siphon or tube from the burrow to ingest sediment. Nonsiphonate forms burrow through the sediment ingesting it as they go. At Cape Cod soupy sediments are dominated by nonsiphonate forms because the respiratory siphon of siphonate forms becomes clogged. In areas where the sediment becomes more compact, both forms are present. In even more compacted sediments, siphonate forms dominate. The clams are also stratified with depth, some species feeding near the surface, some slightly deeper, and others as deep as 14 cm. There is a consistent relationship of morphology, feeding adaptation, and depth below the sediment-water interface. In Silurian deposit-feeding communities of Nova Scotia the same relationships of morphology and relative depth are found, indicating the same kinds of feeding adaptations for pelecypods of 400 million years ago and the present. Over this long period the basic structure of these deposit-feeding communities has not changed, even though many different species have come and gone in the various feeding niches. This is an unusually clear case of ecologic uniformity that can be supported by the functional morphology of the animals and their life/death context within the sediments. Typically it is more difficult to establish comparability between modern analogues and Paleozoic communities.

Between habitat studies

Traditionally the ocean depths are classified into **neritic** (0 to 200 m), **bathyal** (200 to 2000 m), and **abyssal**

(> 2000 m) **zones**. The communities occupying these zones are, as might be imagined, vastly different. Most groups of organisms exhibit a much finer zonation with depth, some having five or more zones within neritic depths alone.

Long ago paleontologists recognized that over a sizable area a formation (a geologically mappable unit) may change with respect to the fossils it contains. The name **biofacies** (meaning biologic aspect) was given to these recognizable changes. By comparing biofacies in Tertiary sediments with modern samples, paleontologists demonstrated that many biofacies consist of the remains of organisms from various depth zones. The biofacies are, in a sense, fossil communities. It is important to remember, however, that taphonomic processes may have altered relationships among the preserved organisms and between them and their original habitats. Organisms preserved together cannot always be assumed to have lived together. Biofacies is a useful term applied to distinctive fossil assemblages whose paleocommunity relationships must then be interpreted using taphonomic,

sedimentary, and other contextual evidence. Some argument exists over how biofacies and paleocommunities should be defined. Today, however, recognitions of groups of organisms distributed in a manner that can be mapped are effectively accomplished with the aid of a computer. The particular method employed mathematically constitutes the definition of the resulting groups no matter what we call them.

Biofacies can be grouped into larger areas covering substantial portions of the earth. For example, the species found in the Arctic are much different from ones found in the tropics. These areas of large geographical extent whose organisms differ greatly are called **faunal provinces**.

Along the eastern seaboard of North America boundaries for modern macrofaunal provinces are approximately at southern Newfoundland, Cape Hatteras, and Southern Florida. Small crustaceans called ostracods collected off Virginia and North Carolina illustrate how organisms group into mappable units. Figure 3.3 shows the results of a cluster analysis on 159 species from 85

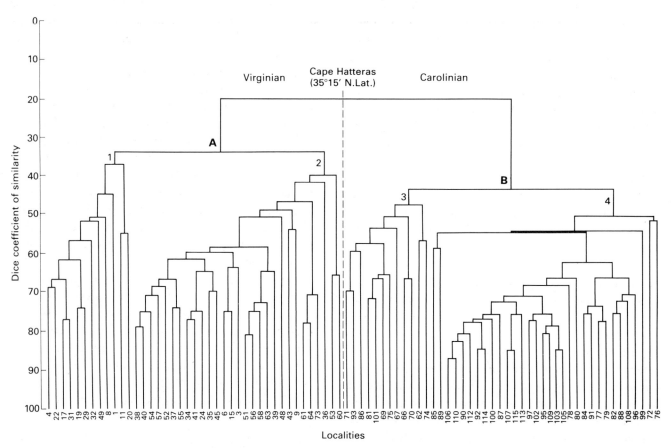

Figure 3.3. Dendrogram showing faunal provinces (**A** and **B**), and biofacies (*1, 2, 3,* and *4*). (Modified from Hazel, J. E. Paleoclimatology of the Yorktown formation (Upper Miocene and Lower Pliocene) of Virginia and North Carolina. In Oertli, H. J., editor. *Bull. Centre Recherche Pau SNPA*, p. 367; 1970).

localities. The dendrogram is broken into two large clusters coinciding with the latitude of Cape Hatteras. These large areas are termed the Virginian and Carolinian faunal provinces. Within each of these provinces are two smaller clusters or biofacies. Identification of these modern patterns and their relationship to temperature provides a powerful analogue for the past, allowing reconstruction of paleoclimatic trends in the Pliocene when ostracod species were much the same as today. Applying modern distributional patterns of biofacies to the more distant past is more difficult when species were different, although sometimes there are helpful morphologic clues for environmental parameters (e.g., test thickness and architecture) that transcend taxonomic groupings.

Another example of how a modern fauna can be used to recreate an ancient depositional basin is given in Chapter 8.

Species diversity

The number of species observed in a given area or volume is referred to as species richness, species density, or most commonly as **species diversity**. At first glance a measure such as the number of species encountered in an area seems simple and straightforward. It is not. This is so because the number of species, S, is a function of the number of individuals, N, counted. If S is plotted on an arithmetic scale and N, on a log scale, the resulting relationship is nearly linear. Such a plot for marine protozoans (foraminifera) from Jamaica is shown in Figure 3.4. Notice for an N of 10,000, S is about 110; for an N of 1000, S is only 65. Consequently, when comparing species diversities from different habitats, or within the same habitat with time, care should be taken to maintain an approximately constant N.

Taphonomic processes can also affect species diversities by eliminating the smaller or more delicate ones and biasing the fossil assemblage toward the robust species with more readily preserved hard parts. Theoretically, the most common (and ecologically important) organism in a community could be entirely absent in biofacies if taphonomic processes worked against its preservation. It is difficult if not impossible to correct for such biases, but we can hold unknown variables such as this constant by comparing diversity in biofacies that have similar taphonomic histories. This strategy is typically taken for granted when making comparisons of diversities between modern and fossil biofacies from the same general environment.

Modern marginal marine environments such as

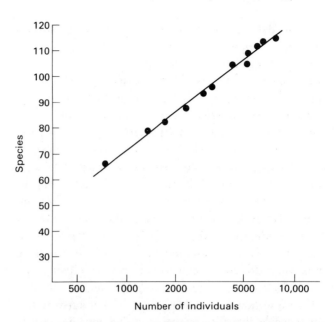

Figure 3.4. Plot of number of species versus number of individuals. Note linear relationship when individuals are plotted on a log scale.

marshes, estuaries, and bays usually have fewer species than open-ocean environments. In the open ocean the species diversity for many groups increases with depth. A few years ago (except for coral reefs) maximum diversities were assumed to be at the edge of the continental shelf and decreased thereafter so that abyssal depths were thought to contain very few species. Careful sampling in recent years, however, indicates that, for several groups, species diversity at abyssal depths is as high as or higher than that in shallow waters of the same faunal province.

Long ago naturalists realized that within any habitat, species diversity increases for virtually all groups as latitude decreases. In other words, maximum diversities occur in the tropics. Many ideas have been evoked to explain this phenomenon. Among the most popular are stability of environmental variables over long periods. This hypothesis reasons that in the tropics and also in the deep sea, abiotic variables are nearly constant so that organisms are freed from destruction by vagaries of weather. Given sufficient time, they partition available resources through specialization. Theoretically, without disturbance these habitats should evolve toward higher and higher species diversities. Predation and competition, however, have also been shown to be important in regulating species diversity. Probably no single hypothesis can account for all the observations. To assess the importance of the stability-time hypothesis, however, the species diversity of ancient communities must be documented.

Some data are available for comparison of species diversities in modern and ancient habitats. Species diversities of benthic foraminifera (bottom-dwelling protozoans having a small shell) have been examined from all marine environments. In Pliocene deposits of the east coast of North America, habitats interpreted from faunal composition as ranging in depth from about 30 to 1000 m have been identified. The species diversities in these habitats are the same as those found at similar depths off the east coast today. In the subsurface Oligocene of Texas, species diversities of benthic foraminifera are strikingly similar to those found in the Gulf of Mexico today. Evidently, for these animals each habitat is able to support only a certain number of species; the same maximum diversity seems to hold for 30 million years ago as well as for Pliocene and recent times.

The question as to when particular habitats reached the species diversity observed today remains largely unanswered. The farther back we go in time the more difficult it becomes to identify the habitat the species lived in. We find our experience with associations in modern habitats has less and less relevancy as we examine the remains of organisms with progressively fewer modern counterparts.

Studies comparing species diversities of Paleozoic and Mesozoic sediments with those of the Recent have produced equivocal results. Nevertheless, we do see patterns in the distribution and abundance of organisms even in the earliest sediments. We need only to discover how to decipher them.

Supplementary reading

Ager, D. K. *Principles of paleoecology.* New York: McGraw-Hill Co.; Inc., 1963. The first English language text book written on paleoecology. Although now dated, it still contains much useful information.

Carriker, M. R. Interrelation of functional morphology, behavior, and autecology in early stages of the bivalve *Mercenaria mercenaria. Journal of the Elisha Mitchell Scientific Society* 77: 168–241, 1961. Excellent account of the autecology of the hard shelled clam.

Gray, J.; Boucot, A. J.; Berry, W. B. N., editors. *Communities of the past.* Stroudsburg; PA: Hutchinson Ross Publishing Co.; 1981. Diverse collection of papers on case studies of marine and nonmarine paleocommunities.

Imbrie, J.; Newell, N. D., editor. *Approaches to paleoecology.* New York: John Wiley & Sons, Inc.; 1964. Collection of papers on biologic, sedimentologic, diagenetic, and statistical approaches to paleoecology. A milestone in the evolution of paleoecologic thought.

Laporte, L. F. *Ancient environments.* 2nd ed. Prentice-Hall Foundations of Earth Science Series. Englewood Cliffs, NJ: Prentice-Hall Inc.; 1979. Introduction to interpretations of ancient environments through both sedimentary and paleontologic methods.

Levinton, J. S.; Bambach, R. K. A comparative study of Silurian and Recent deposit-feeding bivalve communities. *Paleobiology* 1: 97–124; 1975. Outstanding comparative within-habitat study of communities separated by 400 million years.

MacArthur, R. H. *Geographical ecology.* New York: Harper & Row; 1972. Classic book seeking to explain patterns in nature. Many ideas are presented that should provide fruitful thought for the paleoecologist.

Odum, E. P. *Fundamentals of ecology.* Philadelphia: W. B. Saunders Co.; 1971. Basic introduction to ecology that is well written and easy to understand.

Reyment, R. A. *Introduction to quantitative paleoecology.* New York: Elsevier Pub. Co.; 1971. Concepts and examples of statistical approaches to an inherently quantitative subject are well illustrated.

Scott, R. W.; West, R. R., editors. *Structure and classification of paleocommunities.* Stroudsburg, PA: Dowden, Hutchinson, and Ross, Inc.; 1976. Collection of papers on community paleoecology. Some papers classify parts of the biosphere, others compare living-dead-fossil assemblages, and still others study various aspects of paleocommunities.

Valentine, J. W. *Evolutionary paleoecology of the marine biosphere.* Englewood Cliffs, NJ: Prentice-Hall, Inc.; 1973. Advanced text bringing together theory and observation to explain the marine fossil record.

4

Evolution

James W. Valentine

General processes

Evolution is the complex of processes through which living organisms originated from nonliving matter and have become diversified and modified through sustained changes in form and function. To sustain such modification, the changes must be heritable. Thus on one level, evolution is a change in the heredity of organisms, in their genes. Recall that in most sexually reproducing populations each individual is genetically unique. The number of different alleles and gene arrangements present in such populations is far larger than in any one individual. All the genotypes of all the individuals in a population constitute the pooled genetic variability in a population; the next generation of genotypes must be assembled from this **gene pool**. Each gene or gene arrangement in the gene pool has a given frequency at any time; evolution can be regarded as a sustained change in these frequencies.

Natural selection is the chief process that gives direction to gene frequency changes. Animal breeders have altered the properties of domestic animals and plants by selecting for reproduction those individuals with the most desirable qualities (large milk production in cows, for example); this is artificial selection. In nature, individuals that best survive environmental rigors or that have greater reproductive potential tend to leave a higher proportion of offspring than others. Such individuals are said to have higher Darwin **fitness**. The heritable qualities that contribute least to fitness will become progressively less common; this is natural selection at work.

Another way in which gene frequencies may be changed in a gene pool is by **mutation**, a change in the nucleotide sequence of a gene (point mutation) or in the arrangement of genes along a DNA strand or in the number of strands (chromosomal mutations). These changes are commonly evoked by radiation or by chemicals that cause changes in the bases that form parts of the nucleotides. In point mutations, one or a few nitrogen bases are added, deleted, or substituted so that the protein product of the gene has a different amino acid sequence than formerly. A change in amino acid sequence often changes the function of a gene, and a different allele is created. Since genes and gene arrangements are replicated during reproduction, these changes are hereditary.

The chance failure of a generation to maintain the alleles in its gene pool in exactly the same frequencies as the parental generation can also cause an evolutionary change. A rather rare allele may happen not to be present in any offspring, just by chance. For some rare alleles, the chances might be 10 to 1 against this happening; in this case 10% of such alleles, on the average, will be lost from the gene pool in each generation. Once these alleles have left a species' gene pool, they are gone forever unless created anew by mutation. Rare alleles may also become more common by chance than selection would ordinarily permit. These types of random gene frequency changes are termed **genetic drift**. Drift is most effective in small populations, where significant chance fluctuations from parental frequencies are more likely.

One last way in which populations may have their gene frequencies altered is through the transfer of genes from one population to another, termed **gene migration** (or **gene flow**). Immigrant individuals may bring different alleles or gene arrangements into a population if they arrive in large numbers. Passing individuals may breed and leave genes behind in fertilized eggs. Emigrants may also change a gene pool by leaving to breed elsewhere. In the sea, gene migration is commonly accomplished by floating eggs or larvae.

Adaptation is the acquisition of characteristics (adaptations) that benefit a population or species. **Adaptedness** is a measure of the adjustment of an organism, populations, or species to an environment. All species are better adapted to some environments than to others. Well-adapted individuals tend to be more fit than poorly adapted ones and thus leave more offspring, so that natural selection usually enhances adaptation. Sometimes, however, selection can lower adaptedness. For example, some alleles raise the reproductive potential of their carriers because sperm that bear them are more likely to be successful in fertilization, but they have deleterious effects on the adults that bear them. Thus they enhance fitness and increase in frequency relative to competing alleles, but lower the adaptedness of the population in which they spread. Unless the effects of such alleles are modified or masked in some way, they can cause extinction.

Of the ways in which significant changes in gene pools may occur, it is selection that acts most consistently to maintain or improve adaptations. Mutation and drift occur at random with respect to the adaptation of organisms. Because of the natural genetic variation that is maintained in most populations and the segregation and recombination that occurs during meiosis, the number of potential gene combinations in an average invertebrate population may be about 10^{300}. Mutation rates are commonly around 10^{-5} per gene per generation, or fewer than one per average individual invertebrate and perhaps 10^5 or fewer per average invertebrate population. Clearly, most change in gene frequencies will result from the selection of fitter genotypes from among the many possible allele combinations rather than from the incorporation of new mutations. The indispensable role of mutation, insofar as gene frequency changes are concerned, is to provide variability to the gene pool, from which selection can mold succeeding generations.

In a relatively unchanging environment, selection (called **stabilizing selection**) acts chiefly to winnow out those gene combinations and mutations that are less fit, and thus largely to promote the status quo. In a changing environment, some individuals with new gene combinations and arrangements may be fitter than those with the old ones, and thus environmental change tends to evoke evolutionary changes that promote adaptation to new conditions (**directional selection**). So far as selection is concerned, the environment is composed of everything to which an organism must be adapted—patterns of physical-chemical conditions, food, predators and other kinds of organisms, and even other members of the same population. Environmental requirements and opportunities direct and shape the pathways of gene frequency change.

Speciation

Reproductive aspects

A **species** is composed of all the individuals that draw their genes from the same gene pool and, if they reproduce, return genes there. Usually a species is represented by many **populations**. A population is a group of individuals of the same species, usually living together and somewhat removed from other populations, within which interbreeding is common. Gene flow between populations has less effect on gene frequencies than interbreeding within each population, but some interpopulation migration usually occurs to maintain gene flow and continuity of the species' gene pool. Individuals belonging to different species cannot ordinarily interbreed (or do not have fertile offspring) so that the gene pools of different species are isolated. Species can come into being when reproductive isolation is established between populations and thus their evolutionary futures become separated, each determined by the changes within a different gene pool.

In the majority of animal species, reproductive isolation is believed to arise in association with the geographic isolation of populations. Each species occupies a particular geographic region, ordinarily circumscribed by conditions that act as barriers to the species' further spread—the termination of the species' habitat (as at the shoreline for shallow marine species) or a climatic boundary may form a barrier. Particularly near the borders of a species' geographic range, suitable habitats are likely to be patchy and separated by areas of unsuitable conditions.

Migration between populations separated by some form of barrier is obviously restricted, and a slight environmental change may prevent migration and therefore sever gene flow altogether. Once they are geographically isolated, populations may evolve along different lines, especially if subjected to somewhat different conditions. Gene combinations that are relatively rare or absent in one population often become common in another. At some point the isolated populations become sufficiently different that the members of one can no longer interbreed successfully with members of the other, even if given the opportunity. Two species are now present.

Populations (or species) that have overlapping geographic ranges are called **sympatric**; those with distinct, nonoverlapping ranges, **allopatric**. All degrees of reproductive isolation are found to exist between members of closely related allopatric populations when they are brought into contact. Some interbreed freely with abundant vigorous hybrid offspring; in others the hybrids have reduced fertilities; and still others are infertile and must be classed as separate species. Thus, various stages of **allopatric speciation** can be studied in living populations.

Another mode of speciation (that need not be allopatric) is termed **quantum speciation**. As the name implies, this describes a sudden rather than a gradual rise of new species. Several different mechanisms are included under quantum speciation, all of which act to nearly or completely eliminate gene flow between a newly arisen genetic variant and the population from which it originated. One is by key mutations at a few loci (or even at only one) that permit a significant change in function. Highly specialized invertebrates such as parasites have been regarded as particularly susceptible to such a speciation mode; if a few genetic changes permitted a parasite to occupy a new host, gene flow with the parental population could be severed.

Others sorts of quantum speciation mechanisms involve chromosomal mutations—duplications or rearrangements that reduce or eliminate fertility between the mutants and normal members of their parental population. In plants that can self-fertilize, a single individual could be infertile with the species from which it arose and yet father an entire new species.

Populations (or species) that share a common range boundary are called **parapatric**. It is likely that populations which have low rates of gene flow and are subjected to strong selection gradients can be severed into two genetically distinctive gene pools. Genetic differences tend to accumulate across such gene pool boundaries; once begun they can lead to **parapatric speciation.**

The amount of change in structural genes that accompanies speciation can vary. Theoretically, species that arise by chromosomal mutations may be identical to their parental or sister species in structural genes, even though they are infertile when crossed with them. Some species are known that differ by as few as 5 per cent of their structural genes; other populations that differ by as much as 25 per cent can still interbreed.

Morphologic aspects

Species are normally distinguished from each other on the basis of their morphologic distinctiveness. Sometimes allopatric populations are morphologically distinctive even though their members can interbreed successfully if brought together. If an isolated population is founded by only a few migrant individuals that happen to be somewhat unusual genetically and morphologically, succeeding generations of the isolate will inherit this peculiarity and be morphologically distinctive themselves; this is termed the **founder effect**.

It is not uncommon for species evolved allopatrically from a common ancestor to become sympatric later. Such species would usually be closely similar in habits and habitat requirements, so that competition between them might be especially strong. In this event, selection should favor individuals in either species that are least like the other species, for they would be subject to the least interspecific competition. Thus, evolution would tend to cause these species to diverge even further. When this divergence is recorded in morphologic change, it is termed **character displacement**.

Not all species are morphologically distinctive. Some populations that cannot be told apart morphologically cannot crossbreed even though they may be sympatric. They are termed **sibling species**. Despite their morphologic similarity, some differ in over half of their structural genes. Thus they are distinctive biochemically if not morphologically.

It is not possible to determine reproductive isolation patterns directly among fossil populations. Fossil populations that are morphologically rather distinctive are deemed to be separate species. Fossil species, and living species that are recognized morphologically only, are termed **morphospecies**. In using the morphospecies concept, a species that happens to possess some freely interbreeding morphologic variants may be divided erroneously into two or more morphospecies. Furthermore, sibling species may be erroneously lumped into the same morphospecies. There are nonmorphologic criteria that are helpful in distinguishing fossil species. For example, populations that occur in such different places or that lived at such different times that they could not represent a continuous lineage can be considered as distinct fossil species, even if they are morphologically indistinguishable. On the other hand, contemporary fossil populations that are somewhat distinctive morphologically but that are found to form a geographic trend are open to interpretation as geographic variants of the same species.

Morphologic patterns in time

By tracing morphospecies in the fossil record, paleontologists study patterns of species origin and evolution over vast lengths of time. Such studies cannot be duplicated in the laboratory, but are essential in developing inferences concerning the evolutionary processes that have been most important during the history of life and the circumstances under which different processes occur. The chief difficulty in these studies is the incompleteness of the fossil record. Perhaps of all the invertebrate species that have ever lived, only 1 in 30 to 50 are preserved as fossils. Therefore, tracing morphologic lineages is very difficult because of gaps in their fossil record. In interpreting fossil occurrences, it is difficult to be sure that the observed patterns are not merely patterns of preservation. However, by studying particularly well-preserved morphospecies patterns, paleontologists have reached conclusions as to their most probable significance.

The allopatric and sometimes the rapid (quantum) origin of many species also makes it difficult to trace actual speciation events in the fossil record; ancestral species have usually lived, not where descendant species are collected, but elsewhere. Nevertheless, several allied morphospecies lineages may sometimes be traceable through some interval of geologic time, and when this is done, two contrasting patterns stand out. The morphology of some lineages changes gradually through time (Figure 4.1, *A*). This **gradualistic pattern** can be interpreted as a response by natural selection to gradual changes or at least to stresses in the physical or biotic environment. Time series of fossil populations, each grading into the succeeding population morphologically, are termed **chronoclines**.

Another pattern is for species to appear abruptly without known intermediate forms to link them to their ancestors, to endure without displaying significant morphologic change even over millions of years, and then to disappear from the fossil record. This is called a **punctuational pattern** (Figure 4.1, *B*). The morphologic **stasis** or lack of change implies that the morphology is in some way buffered against change, perhaps through stabilizing selection to a fairly constant range of habitat conditions. Adaptive responses that do occur must be chiefly physiologic rather than morphologic. The fossil

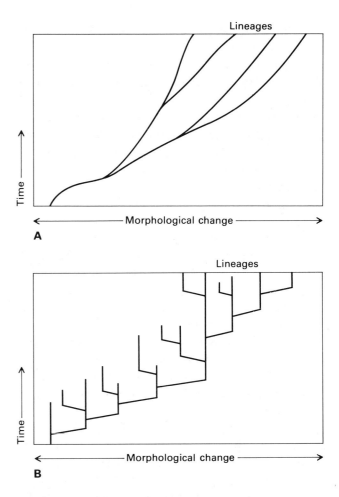

Figure 4.1. Two modes of morphologic change. **A**, Gradualistic mode: lineages change gradually although not necessarily evenly through time. **B**, punctuational mode: lineages usually exhibit little or no change but then display an evolutionary spurt to evolve rapidly to a new morphologic type. Branching is represented in each mode.

record indicates that prolonged maintenance of species morphologies is surprisingly common.

The abrupt appearance of morphospecies in the punctuational pattern is often attributed to quantum speciation; thus the morphologic change that characterizes a new morphospecies is assumed to orginate in close association with the speciation event itself. As mentioned earlier, the term "quantum speciation" covers a variety of processes, and we do not yet know which of these are the more common. Most explanations of abrupt appearances appeal to the rapid evolutionary change possible in very small isolated populations, less than 100 or so individuals. A gene may spread quickly to all members of a very small population in just a few generations, whereas in populations with millions of members, a gene or gene combination would obviously require many more generations to spread thoroughly.

Chromosomal or other mutations or unusual gene combinations might therefore become quickly fixed in the gene pool of a small population, perhaps aided by drift. Since to create the punctuational pattern, a significant morphologic change must be associated with the new genes or gene arrangements, some researchers have suggested that regulatory gene changes are often involved (Chapter 2). This is because many morphologic factors (size, proportions, and numbers of many characters) are functions of the regulatory process, and because a single change in a regulatory mechanism could alter the pattern of expression of large numbers of structural genes.

Morphologic change through selective changes in single genes with small effects can also be rapid under appropriate circumstances, regardless of population size. Small populations would most often be undetected, however. Given a reasonable amount of genetic variability, supplemented by mutations at rates that are known to occur commonly, populations can achieve morphologic change sufficient to create a morphospecies within only hundreds to thousands of generations. For most invertebrate fossil groups, this means that new morphospecies may be evolved within a thousand to a few tens of thousands of years. This is so rapid compared with the ordinary resolution of sampling in the fossil record that it would certainly appear as abrupt under most conditions.

If any of these processes of rapid morphospeciation is successful in creating a species well adapted to widespread conditions, the population may grow to large size and become broadly distributed. Then the species might be likely to enter the fossil record, and it would not be surprising if the intermediate forms connecting the new morphospecies with its parental species, being rare and present for only a short time in any event, were not to be found.

Transspecific evolution

Origin of animal groups

Living animals can be classed into groups of related species that are discrete. Living invertebrates, for example, can be placed into such groups as snails, crabs, or earthworms, each with a distinctive body plan, between which there are no living series of intermediate forms. The origin of such groups of species is **transspecific evolution**. The question then arises, what is the origin and significance of such groups and of the gaps between them?

The answer, foreseen by Charles Darwin, relates to the structure of the earth's environment and to the survival of the better-adapted lineages. The environment is not a continuum with gradual changes between all environmental conditions, but rather a mosaic of distinctive conditions. Between the mosaic patches there are changes in the conditions of life. Marine animals that float in the water and those that live on the sea bed face distinctive environmental challenges, as do animals that burrow in the ocean floor. Visualize each major adaptational mode, such as swimming or burrowing, as an **adaptive zone**. Each zone can be further subdivided: some parts of the sea floor are sandy, some muddy, some rocky; some parts are always dark, others lit by sunlight daily; and so on, with each set of conditions entailing a different adaptive response. Thus the environmental mosaic creates numerous **adaptive subzones**.

Distinctive species groups, then, may arise in the following manner. An animal group that is evolving in a given adaptive zone (as zone *A*, Figure 4.2), gives rise to a lineage that is adapted to conditions in the subzone nearest the boundary between adaptive zones (lineage *f*, Figure 4.2). Since this subzone is marginal to the main zone *A*, the adaptation of lineages there may be atypical for its group, for they have evolved in response to atypical conditions. Occasionally some of these unusual adaptations prove to be useful in coping with conditions in the neighbouring zone as well (zone *B*, Figure 4.2); in this case they are called **preadaptations**. Invasion of the new zone may then be possible to a pioneering branch of this lineage, although it will probably not be very well adapted there. However, if competitors are few or absent, the lineage may be able to survive until further evolutionary changes give rise to more extensive adaptations to the newly encountered conditions called **postadaptations**. The ill-adapted pioneering population is likely to be small, and genetic drift may aid in the appearance of novel gene combinations; selection can then favor any that are useful. Even so, early extinction must be the usual fate of pioneering lineages. Those few that do achieve a reasonable degree of adaptation in the newly occupied zone will have established a novel sort of species with distinctive adaptations in a zone with a few competitors. This provides an opportunity for diversification into the subzones of the new zone, creating an array of allied species, they will share through inheritance a unique complex of characters, preadaptive and postadaptive, that were evolved by the pioneering lineage. A distinctive group of species has appeared.

In the fossil record, distinctive animal groups commonly appear suddenly and without a clear immediate ancestor; recall the way that the animal phyla appear (Chapter 2). This has led some authorities to postulate

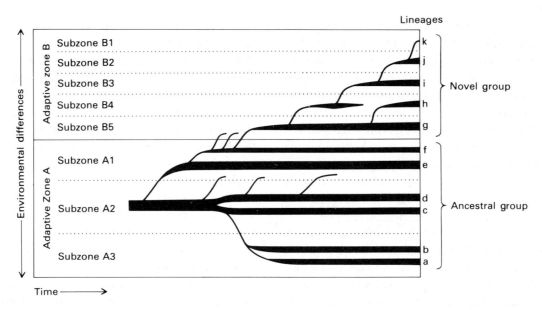

Figure 4.2. Adaptive zone model. *A* and *B* are adaptive zones, between which lies a profound environmental discontinuity. Each is subdivided into subzones, separated by less marked discontinuities; even within the subzones, several distinctive environments may be found. The adaptedness of lineages is indicated by the width of their traces. A well-adapted lineage in subzone *A2* diversifies, and one marginal branch (lineage *f*) gives rise to a series of lineages that are well enough preadapted to conditions in zone *B* to cross the bordering discontinuity and invade that zone. Most become extinct but one lineage (*g*) achieves an adequate postadaptation to prevail and eventually it diversifies to produce a novel group of organisms, all of which will share with *g* a suite of characters that are preadaptive and postadaptive to zone *B*.

that unknown evolutionary mechanisms were responsible. On the other hand, many workers believe that the several types of evolutionary processes that occur during speciation (recall that there are different speciation patterns) are sufficient to account for the rise and diversification of all the various groups of organisms. Certainly, different aspects of genetic change must predominate in different circumstances. A significant increase in complexity must entail the enlargement of the genome. Single-celled organisms need only enough genes to carry on cell metabolism and cope with the environment; a multicellular organism requires all such genes and, in addition, needs genes to control the number, position, and morphology of the many cells. Simple organisms with 15 cell types need fewer genes than complex ones with 200. Therefore, the evolutionary trends that lead to new, higher levels of complexity involve adding new genes (and thus more DNA) to the genome.

In evolving new body plans at about the same level of complexity, important changes should occur in the patterns of gene expression to give rise to different anatomic patterns. Changes in the structural genes, which have been the most-studied genetic aspect of evolution, must eventually be involved in virtually all evolutionary change, for new structural alleles would be required to maximize fitness in the face of myriad new physiologic and behavioral demands.

Trends through time within groups

Selection acts to adapt species to their prevailing environments. Therefore, the nature of the environments largely determines evolutionary trends, given the nature of the species themselves. Changes within species are chiefly in response to environmental pressures and opportunities, which cause higher fitness to shift from one gene combination and arrangement to another. Species are not adapted by selection to the environments they will encounter in the future. Whether a species will prosper in a different environment is largely a matter of chance. Environmental changes, and new environments encountered when species disperse, are largely random with respect to the previous adaptations of the species. Therefore, evolutionary trends have a large random element. This is true for extinctions as well as for diversifications.

Since speciation is largely at random with respect to future environmental opportunities, trends within invertebrate groups are created partly by differential survival of those species that happen to prove preadapted to new environments they encounter, compared with those that are not so preadapted. A succession of species may produce a morphologic trend as they follow a trend in environmental change. In Figure 4.1, *B*, for example, there are just as many speciation events to the left as to the right, yet a conspicuous morphologic trend to the right is created by the differential success of new morphologies in that direction. Morphologic trends may also be created by differential rates of speciation among lineages.

For morphologic trends that are created through differential speciation or extinction rates, selection is acting on a different level than for trends created by processes acting within gene pools; this level is termed **species selection**. Distinctive morphologic innovations can arise through a succession of speciation events. In the fossil record of such innovations we can expect to learn much about transspecific evolution.

Interspecies or intergroup patterns

Often the evolutionary trends of a given lineage occur independently of change in other lineages, so far as we can tell. However, lineages sometimes display interesting patterns relative to each other (Figure 4.3). When lineages become morphologically more similar through time, they are **convergent**; when less similar, **divergent**. Usually the interpretation is that the lineages are evolving adaptations to more similar or less similar modes of life, respectively. Sometimes three or more lineages will diverge in an **adaptive radiation**. This is the pattern that commonly occurs when a lineage successfully invades a different adaptive zone; the pioneer lineage undergoes a radiation as it gives rise to branches within different subzones. Finally, lineages may change in concert so as to neither increase nor decrease their morphologic distance. This is **parallelism** and is usually interpreted as indication that an environmental change has evoked similar responses in those lineages.

These patterns are easiest to recognize among closely related, simultaneously existing lineages, between which morphologic distance is easiest to judge. Sometimes they are found among lineages that are **heterochronous**, or separated in time. Heterochronous convergence and parallelism, for example, suggest a temporal repetition of environmental conditions. In heterochronous convergence, lineages at each time evolve toward a similar morphologic solution to the repeated challenges. In heterochronous parallelism, lineages respond with nearly identical morphologic changes.

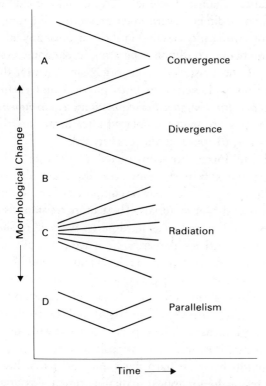

Figure 4.3. Intergroup patterns of morphologic change. **A,** Convergence: groups become more similar. **B,** Divergence: groups become less similar. **C,** Radiation: several groups diverge from a common point. **D,** Parallelism: groups evolve but maintain a certain morphologic distance, sometimes despite a change in evolutionary direction.

Supplementary reading

Bush, G. L. Modes of animal speciation. *Ann. Rev. Ecol. Syst.* 6: 339–364; 1975. Balanced discussion of the different ways in which animal species can originate.

Dobzhansky, T. H.; Ayala, F. J.; Stebbins, G. L.; Valentine, J. W. *Evolution.* San Francisco: W. H. Freeman & Co.; 1977. Relatively comprehensive account of evolutionary processes, with some examples from the fossil record.

Eldredge, N.; Gould, S. J. Punctuated equilibria: an alternative to phyletic gradualism, pp. 82–115. In: Schopf, T. J. M., editor. *Models in paleobiology.* San Francisco: W. H. Freeman & Co.; 1972. The essay that brought attention to the punctuated pattern.

Mayr, E., *Animal species and evolution.* Cambridge: Harvard University Press; 1963. Beautifully organized account of processes of evolution from a zoologic viewpoint; a classical work.

Simpson, G. G. *The major features of evolution.* New York: Columbia University Press; 1953. Another classic, dealing particularly well with transspecific evolution with many examples from the fossil record.

Stanley, S. M. *Macroevolution: pattern and process.* San Francisco: W. H. Freeman & Co.; 1979. Exploration of the processes and consequences of evolution at the level of species selection.

5

Fossil Preservation

Kenneth M. Towe

The paleontologist's task is to trace and interpret the history of life back through time. Fossils, the record of past life and its activities, are thus the currency of paleontology. The factors involved in preserving this record of life are basic to much paleontologic interpretation. It can be said that fossil preservation is to the paleontologist as tissue preservation is to the biologist: without it there is no record, and the better it is, the better is the information obtainable.

The doctrine of uniformitarianism asserts that knowledge of the present can provide insight into the past. All over the earth, from mountain lakes to the deep ocean floor, one finds evidence of living organisms and their activities. But what percentage of these organisms in today's varied environments will survive to become tomorrow's fossils? It has been estimated that of the more than one million living animal species less than 10 per cent are likely to be preserved as fossils. The fact is that fossil preservation is a rare event, and taken as a whole, the fossil record is exceedingly incomplete and biased, even under the best of circumstances.

What factors influence preservation? What are the types of preservation? How does preservation affect the science of paleontology? Although some of the answers to these and related questions seem obvious, others are complex, making the whole subject a potentially treacherous area for the paleontologist.

The natural preservation of any organism or assemblage of organisms is dependent on the interaction of several factors. Among the more important are (1) organism composition and structure, (2) numerical abundance, (3) sedimentary environment, and (4) post-depositional change.

Organism composition and structure

If there is a key word in fossil preservation, it is resistance: physical and chemical resistance. There is the resistance against bacterial action, acid conditions, abrasion, sediment overburden pressure, oxidation, reduction, water movement—in short, against any and all environmental conditions operating to eliminate even a trace of the existence of an organism. Clearly, those organisms endowed with both physical and chemical resistance *for the specific sedimentary environment* in which they are living, or placed after death, are those most likely to appear in the fossil record.

All life forms are constructed of delicate cellular tissues, the **soft parts**. However, there are also noncellular tissues, which may be composed of or reinforced by more resistant materials. These more resistant tissues are loosely referred to as **hard parts**. We read of shells, cysts, tests, cuticles, perisarcs, carapaces, and, of course, skeletons—exoskeletons, endoskeletons, and hydrostatic skeletons. The list seems endless, and both usage and definition can vary from one taxonomic group to the next. Ideally, the invertebrate world could be divided into those organisms with hard parts and those without. Out of context, however, it is difficult to make a clear-cut distinction between hard parts and soft parts. Indeed, these terms do not appear in dictionaries or glossaries of geology in spite of their wide if not uniform usage among paleontologists. Unfortunately, for too many people, hard parts are simply any biomineralized structures—those parts of any organism that are constructed of minerals such as the carbonates, phosphates, or silicates, among others. In addition to the calcified tissues and

Table 5.1. Distribution of the more important compositional materials among various taxonomic groups. X indicates a major occurrence; x indicates a lesser occurrence.*

Taxa	Inorganic				Organic			
	CARBONATES	PHOSPHATES	SILICA	IRON OXIDES	CHITIN	CELLULOSE	COLLAGEN	KERATIN
Procaryotes	X	x		x		x		
Algae	X		x		x	X		
Higher plants	x		x	x		X		
Protozoa	X		X	X	x	x		
Fungi	x	x		x	X	X		
Porifera	X		X	x			X	
Cnidaria	X				x		x	
Bryozoa	X	x			X		x	
Brachiopoda	X	X			X		x	
Mollusca	X	x	x	x	x		x	
Annelida	X	X		x	x		X	
Arthropoda	X	X	x	x	X		x	
Echinodermata	X	x	x				X	
Chordata	x	X		x		x	X	X

*Modified from Lowenstam, H. A. In: Goldberg, E. D., editor. *The sea, Marine Chemistry* (vol. 5). New York: John Wiley & Sons; 1974, p. 716.

other mineralized structures, however, there are hard parts that are completely unmineralized, being made instead of resistant organic materials such as chitin, cellulose, keratin, or collagen. Table 5.1 shows the distribution of some of these primary constituents among various taxa. The significant point is that while the presence of primary mineral matter is certainly important, it is not a prerequisite for a hard part or for the preservation of a hard part. Some of the most spectacular fossil finds are of organisms that were unmineralized. The graptolites (Chapter 19) are a notable example of an entire fossil group that is completely lacking in mineralized hard parts.

If the composition of hard parts is important to their chances of fossilization, so too is their structure and microstructure. Clearly, a delicate thin-walled shell is less likely to make it into the fossil record than a more massive thick-walled shell. Apart from these obvious relationships is the question of what may be described as the organic-inorganic ratio. This ratio between the mineralized and unmineralized portions of hard parts is often crucial to preservation. In general, a low organic-inorganic ratio will favor more complete fossilization. For example, the microscopic calcareous tests of many foraminifera (Chapter 8) contain very small quantities of

organic material, the entire skeleton being a densely calcified structure. On the other hand, the coccosphere (Chapter 8) and the echinoderm test (Chapter 18) are constructed of numerous densely calcified individual plates, but each plate is sutured to the next with organic material that is subject to rapid postmortem degradation. Thus individual coccoliths and echinoderm plates are frequently encountered as fossils while the complete coccosphere or echinoderm skeleton is more rare.

The trilobite exoskeleton has a low organic-inorganic ratio. Trilobite fossil remains are common. In crabs, lobsters, and shrimps the ratio is high, and their remains are seldom found. Furthermore, most of their mineralized portions are so morphologically indistinctive that even if the pieces were preserved few would recognize them as such. In other words, the density of mineralization and the micromorphology of the hard parts are both important when it comes to preservation.

Numerical abundance

All other things being equal, it would seem obvious that those species with large numbers of individuals are more likely to be preserved as fossils than those that are rare. But all things are seldom equal, and the abundance factor must be used with caution. Even the most abundant, long-lived, or rapidly reproducing species, if it lacks an appropriate composition and structure, may only rarely appear as a fossil. Therefore, in interpreting any fossil assemblage, it may not be correct to assume that the relative abundance of those organisms found is a true representation of the relative abundance of the living population. For example, scientists studying the planktic foraminifera of the modern oceans note that the relative abundance of those species caught alive in plankton tows sometimes differs from their dead tests dredged from sediments below. This can be related to the differential preservation of the delicate, thin-walled forms versus the robust thicker-walled species.

Sedimentary environment

Included within the sedimentary environment as a factor in preservation are other influences such as sediment type, rate of sedimentation, and organism mode of life, all of which are interrelated. Each sedimentary environment has its own chances of preservation, which influences the chances of preservation of those organisms living in that environment.

The majority of the sedimentary rocks, and hence the majority of the fossils, found in the geologic record were deposited in the marine environment, and of these, most represent the shallow water neritic zone. The sediments of deeper water environments, the bathyal and abyssal zones, are much less commonly exposed on the continents. As a result, fossils of organisms that lived in these environments are seldom found. Among the nonmarine terrestrial and transitional environments are deposits associated with streams, lakes, swamps, deltas, lagoons, and beaches—each with its own faunal and floral elements and each with its own chances of preservation.

Within any of these environments are the additional factors of sediment type and rate of sedimentation. Among the clastic deposits, it is the finer-grained rocks: the sandstones, siltstones, and shales that one looks to for fossil material. The coarse clastic deposits, the conglomerates and coarse sandstones, are not conducive to fossil preservation. The rate of clastic sedimentation is also important. The slow day-to-day accumulation of sediment and shelly materials is one mechanism. How-

ever, it is commonly observed that better preservation, especially of benthic assemblages, takes place during one-time catastrophic conditions of rapid sediment movement and organism burial. The coarse clastics, while representing the rapid depositional situation, are also indicative of a higher energy environment requiring a mode of life in which fewer organisms are successfully adapted to live and in which few hard parts are physically resistant enough to survive.

The carbonate sediments, the limestones, have still other characteristics and chances of preservation. Because the majority of these calcareous rocks result from life activities, these rock types are themselves fossils, and detailed study of such rocks is often a paleontologic bonanza where recrystallization and other postdepositional modifications have not been active. The reef-building activities of the corals, coralline algae, bryozoans, and other members of this community are well-known. The planktic realm consisting of such

Figure 5.1. *Tentaculites carteri* from the Devonian Gaspé sandstone in Canada (USNM 99562). This sample illustrates original shell material, both the internal surface (*sis*) and the external surface (*ses*), as well as internal molds, or steinkerns (*sti*). External molds (*sxm*) include one of a chonetid brachiopod at left center. (Photo by K. Sandved.) (×4).

calcified groups as the coccoliths, discoasters, and foraminifera has also made a substantial contribution to vast quantities of fine-grained carbonate deposits. Within these deposits are often found fossils of larger benthic organisms that were either living in the waters below them or were washed in from adjacent areas and preserved encased within these accumulations of microscopic hard parts.

Sediment size and sorting also exert an influence on the type of preservation. Organisms with resistant mineralized skeletons can have their hard parts directly preserved, the most resistant even in coarse clastic sediments. The less physically resistant organisms and those that are unmineralized are directly preservable only in the very fine-grained deposits, usually as carbonaceous films on the rock surfaces. In addition to these direct fossils, there are other indirect fossils that are body impressions of organisms made in the fine-grained sediments (Figure 5.1). These are termed **molds**. External molds show the outer surfaces of hard parts. Internal molds reveal the inner surfaces and sometimes are referred to as **steinkerns**. Other indirect fossils are the tracks, trails, and burrows made by various organisms and generally called **trace fossils**. A full discussion of trace fossils can be found in Chapter 21.

Equally important to fossil preservation is the postmortem chemical resistance of organisms and the chemistry of the sedimentary environment of deposition. The word postmortem is important here because, in life, many organisms create their own internal microenvironment to form and maintain compounds that are otherwise not stable in the external environment in which the organism itself is living. Figure 5.2 illustrates some of the more important chemical influences in a general way. Two geochemical thresholds or "fences" are emphasized: the limestone fence and the organic matter fence. As is evident from the diagram, any calcareous hard parts subjected after death to the environments to the left of the limestone fence (lower pH) are subject to dissolution and are less likely to be preserved. At higher pHs, siliceous hard parts such as those of the radiolarians, sponges, and diatoms are subject to dissolution. These reactions may be retarded by the presence of external organic linings and organic matrices within skeletal tissues. The oxidizing environment above the organic matter fence (positive Eh) is unfavorable for preservation of organic structure. Thus the presence of carbonized remains and sulfides is usually an indication of reducing conditions (negative Eh) in the depositional environment. In addition to these chemical factors, biochemical influences exerted by bacterial decay can occur over a broad range

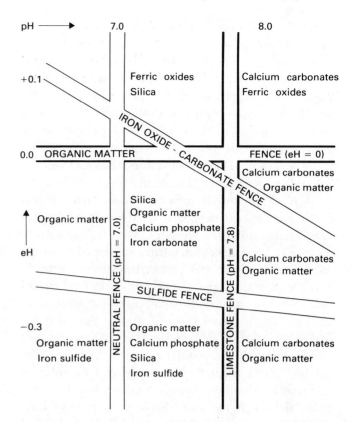

Figure 5.2. Eh-pH boundaries defining the presence or absence of some common sedimentary materials with fossil-forming potential. (Modified from Krumbein, W. C.; Garrels, R. M. *J. Geol.* 60: 26; 1952.)

of environmental conditions. This is a major factor controlling the preservation of fossils lacking mineralized hard parts.

Postdepositional change

The fact that an organism becomes incorporated as part of a sedimentary deposit does not ensure its immortality as a fossil. A variety of postdepositional changes can operate to alter it, sometimes to enhance its preservability but most often to reduce it.

As we have seen, chemical and biochemical influences can act to eliminate the remains of various organisms even before sedimentary burial. These same factors are also of importance in the postdepositional milieu. If the sediment is unconsolidated, the action of ground waters may remove all traces of incipient fossils. On the other hand, if the sediment is well-indurated, the dissolution of shelly material can still leave indirect evidence in the form of complete molds of the shells (Figure 5.1), which can later be filled in with other substances to form **casts**. Sometimes the original hard parts are replaced with mineral matter of another type, this **replacement** generally taking on the name of the secondary mineral material (for example, pyritization, phosphatization, silicification, or dolomitization). This process of almost atom-for-atom replacement is poorly understood from a geochemical standpoint but has produced some of the world's most exquisite fossils. In other situations mineral matter is added to the existing hard parts to create **petrifactions**. This type of preservation usually acts on more porous hard parts such as wood tissues, bones, and echinoderm fragments, to mention a few.

More destructive is **recrystallization**. Here the less stable hard part mineralogies are transformed through time, temperature, and pressure to more stable minerals. Aragonite and magnesium calcites are altered to calcite, amorphous silica and opal-cristobalite eventually become quartz, and the hydrous iron oxides tend toward hematite. Nonequilibrium biomineralogic shapes such as laths or needles, even those of calcite, may be transformed to a more equidimensional mosaic of interlocking grains. Normally, recrystallization destroys most of the original microstructure, but the gross appearance of recrystallized hard parts can be retained. In general, the older the fossil the more likely it is to have been recrystallized to some degree, especially if the original mineralogy was not the stable form.

Aragonite hard parts, common in Tertiary deposits, are usually recrystallized or dissolved from Paleozoic rocks. Paleozoic gastropods with original aragonite shell material are rare as are pelecypods with pearly nacre (Chapter 14). Indeed the very existence of this aragonite nacre in some of the very early pelecypods is questionable because of preservation problems. On the other hand, the fossil record of Paleozoic corals (Chapter 11) is much better. Most of them appear to have precipitated calcite (or magnesium calcite) in contrast to their post-Paleozoic descendants who preferred the metastable aragonite and whose ultimate fossil record is therefore subject to the certainty of recrystallization.

Another potentially destructive force acting on organisms is that of **distortion**. Here the biogenic remains are compressed or flattened by sediment overburden. Older fossils may be deformed by metamorphism and other structural processes. The distortion can vary from slight changes in shape to complete obliteration.

Finally, the inevitable process of erosion destroys many otherwise well-preserved fossils. Unless the hard parts are resistant enough to survive a second cycle of burial to become reworked into younger deposits, they will be doomed to oblivion.

One commonly reads of "good" or "poor" preservation, either with respect to individual fossils or entire assemblages. Good preservation, like beauty, is in the eye of the beholder. Thus one person's ideal preservation may be another person's worthless fossil. Some stratigraphic and systematic paleontologists working in very old or highly deformed rocks may be delighted with the discovery of casts and molds that others might ignore. In an area of recrystallization, even metamorphically distorted fossils can be significant. On the other hand, the paleobiologist looking for microstructural or paleobiochemical information will be dismayed by evidence of replacement or recrystallization regardless of how exquisite the gross morphology may be. For them an ancient shell fragment still composed of original aragonite may be worth more than an entire assemblage of silicified material.

Considering all the difficulties associated with an organism becoming and remaining a fossil, it should be no surprise that the fossil record is so poor. At this point it is traditional to mention the rare and spectacular fossil finds such as insects in amber, mammoths in ice, the Burgess Shale fauna of Canada, or the Solnhofen Limestone of Germany. These outstanding examples of fossil preservation are indeed instructive, for they tend to humble us by pointing out the diversity of organisms and the mass of information that is surely missing from the work-a-day paleontologic collection. It is only those more resistant hard parts that are normally found in the other "average" sediments. A high percentage of the organisms recorded from the exceptional or rare deposits is unique to those deposits, being found nowhere else. These deposits provide us with glimpses of what might have been for the rest of the fossil record.

With collections and published works, the zoologist records for the future a glimpse of the Holocene. Thus, in addition to the 10 percent or so of living species that stand a chance of being naturally preserved in tomorrow's sediments, some zoology collection of today may be the Burgess Shale of the future. Years from now an invertebrate paleontologist from another world, discovering one of these collections, would be able to plot a marked increase in diversity from the Tertiary to the Holocene. If our visitor chose to interpret this plot as the result of rapid evolutionary diversification, he would be wrong. It would be the poor Tertiary preservation that created this apparent increase in diversity, not the zoologic collection. Most of the organisms were there in the Tertiary—they just were not preserved.

Taxonomy, phylogeny, biostratigraphy, functional morphology, diversity, paleogeography, paleoecology, origins, and extinctions—they are all at the mercy of and subject to revision by preservation. Given better preservation, we would know more about the systematic position of those odd groups whose true affinities remain a mystery. Phylogenies would become clearer and stratigraphy more precise. Such perennial problems as the origins of the eucaryotic cell, the Metazoa, or even life itself would become better known. In short, with more and better preservation, many of the problems of paleontology would disappear. One might exclaim "That's obvious!" But sometimes the obvious is overlooked—the preservation factor is all too often taken for granted.

Supplementary reading

Conway Morris, S. The Burgess Shale (Middle Cambrian) fauna. *Annual Review of Ecology and Systematics* 10:327–349; 1979. Succinct review of the composition, preservation, and significance of this famous "soft-bodied" fauna.

Durham, J. W. The incompleteness of our knowledge of the fossil record. *Journal of Paleontology* 41:559–565; 1967. Text of a Presidential Address to the Paleontological Society of America concerning preservation and the fossil record.

Kier, P. M. The poor fossil record of the regular echinoid. *Paleobiology* 3:168–174; 1977. Discussion of the impact of differential preservation on the diversity of regular and irregular echinoids.

Krumbein, W. C.; Garrels, R. M. Origin and classification of chemical sediments in terms of pH and oxidation-reduction potential. *Journal of Geology* 60:1–33; 1952. Classic paper developing the "geochemical fence" concept and its predictive value for interpreting common sedimentary environments and their associated mineralogies.

Lowenstam, H. A. Impact of life on chemical and physical processes, pp. 715–796. In: Goldberg, E. D., editor *The sea. Marine Chemistry*, vol. 5. New York: John Wiley & Sons; 1974.

Nicol, D. The number of living animal species likely to be fossilized. *Florida Scientist* 40:135–139, 1977. Estimate, by phyla, of the numbers of animal species both suitable and unsuitable for fossilization.

Rhoads, D. C.; Lutz, R. A. editors. *Skeletal growth of aquatic organisms: biological records of environmental change.* New York: Plenum Press; 1980. Carefully edited selection of papers relating the influence of environmental factors to the structural, morphological, and chemical features of skeletal parts of organisms, especially the Mollusca.

Wainwright, S. A.; Biggs, W. D.; Currey, J. D.; Gosline, J. M. *Mechanical design in organisms.* New York: John Wiley & Sons; 1975. Extended treatment of biomechanics with a useful bibliography.

6

Classification: Philosophies and Methods

Peter D. Ashlock

Although rudimentary biologic classification may predate civilization and our first formal classifications are those of Aristotle, the questions of how classifications are to be constructed and even to what use they should be put are by no means settled. Indeed, the past 35 years have seen more discussions and arguments about the methodology and philosophy of classification than any other time. A substantial literature is evidence of this intellectual ferment. Agreement, however, has not yet been reached, and the methods and philosophies of many systematists differ little from those of the great naturalist and taxonomist, Linnaeus, whose *Systema Naturae*, written in 1758, marks the beginning of modern classification of animals.

We honor Linnaeus by beginning the discussion of classification with a review of his work because, in a period of chaos and complexity, he devised practical techniques for the naming of groups of organisms and their ranking and ordering.

When Linnaeus was a student, the scientific name of a species consisted of several Latin terms and was, in effect, a short description. Linnaeus simplified matters by consistently using **binominal nomenclature**. In Linnaeus's system, the scientific name of an organism consists of a collective or **generic name** and a specific or **species name**. He standardized the synonymy, a shorthand method of recording the various names applied to a species by different workers or in different parts of the world. Linnaeus developed elaborate morphologic Latin vocabularies for describing plants and animals so that, for example, a worker in France could know clearly what a worker in Sweden had observed. He also developed and

rigorously applied a telegraphic style in his descriptions so that any student could rapidly and accurately compare the characteristics of different but related creatures.

Linneaus was enormously popular not only because his methods were useful to his contemporaries and students—they are still in use today—but because his ideas reflected the general philosophy of western Europe at that time.

Philosophically, Linnaeus was an **essentialist**. He strove to find one feature or very few features that expressed the "essence" of each **taxon** (any named group of organisms; plural **taxa**). In doing so, he prided himself that he was discovering the "plan of the Creator." A classification can be recognized as essentialist when taxa are characterized by one or two discrete features that are both sufficient and necessary for recognition. As a result of Linnaeus's immense popularity, naming and classifying almost completely dominated as principal activities in natural history for many years.

In the 19th century naturalists took expeditions abroad, and enormous quantities of incredibly diverse specimens, both neontologic (**neontology** is the study of living organisms) and paleontologic, flooded into the great museums of Europe. As a direct result of this fuller exploration of the earth, Linnaean influence waned. Not only were new groups of organisms discovered, but gaps between what were thought to be distinct groups began to fill in. It became apparent that establishing the essence of each species and each genus was impossible because no such essence exists.

The great task of ordering and ranking continued, but in guiding it the essentialist philosophy was mostly replaced by a **nominalist** philosophy, which holds that the only real and concrete things in nature are individuals. That is, the various categories into which individual specimens are grouped—species, genus, family, order, class, and phylum—are human constructs whose only purpose is to make it possible to speak of a great many individuals collectively.

The nominalists were also **empiricists**. An empirical view of science relies principally on observation and experiment, finding such theories and explanations as "the Creator's plan" and evolution irrelevant. Instead of pinpointing one or a few essential characters for each taxon (**monothetic classification**), the empiricists attempted to correlate many characters (**polythetic classification**).

The empirical method proved useful in the case of Radiata, which was proposed as a phylum early in the 19th century. The phylum grouping was based monothetically on a single necessary and sufficient character or attribute—that each animal included was radially symmetrical. Before the century was half over, careful observation of specimens by empiricists showed the Radiata to be an unnatural group containing two distinct phyla, the Cnidaria and the Echinodermata, each of which possessed a distinct set of highly correlated characters.

Publication of Darwin's *Origin of Species* in 1859 finally provided an acceptable explanation for the existence of so-called natural groups of organisms: Each group is the product of descent from a common ancestor. His writings are rich in theory and advice that apply to the process of classification.

Darwin's ideas, of course, created a great stir among both biologists and the general public, principally because of their inescapable implications about human origins. Taxonomists of the time were as divided as everyone else, but even those who agreed with Darwin did not change their classification methods to the degree that, in hindsight, the new theories allowed. Evolution actually provided the reasons for what systematists were already doing: the *why* for their *how*. For example, the empiricist principle that classifications should be based on many correlated characters was reinterpreted in an evolutionary context: this observed correlation is the result of common descent. Now, natural groups of organisms should be **monophyletic**; that is, they are descendants of a single ancestral species. Unnatural groups, such as the Radiata, were seen to be **polyphyletic**, or descendants of different ancestors mistakenly grouped by use of noticeably convergent characters. **Convergent characters** are similar, often adaptive features handed down by different ancestors who perhaps, in the process of surviving, had to cope with similar environmental problems and did so in a similar way (Chapter 4).

For the two decades that followed the publication of Darwin's work, it was the fashion to draw phylogenetic trees, treelike diagrams that depict the postulated evolutionary history of organisms. Some systematists had the goal of connecting all groups of plants and animals, both recent and fossil, in one great tree. They thought that when the **phylogeny** (the evolutionary history) of all taxa was understood, their relationship would be understood and from such understanding the most useful classifications could be produced.

Not only comparative anatomy but comparative embryology showed great promise in solving difficult questions of relationship. The theory of **biogenetic law** is often stated as "ontogeny recapitulates phylogeny." This means that any individual organism during development from egg to adult passes through stages similar to adult stages of ancestors in the entire evolutionary history of its group. The study of ontogeny was highly productive, and some major problems of phylogeny were solved. For example, the notochord discovered in the larva of the tunicates (sea squirts) showed them not to be mollusks but instead related to the Chordata. We now know that the biogenetic law is not true as it was originally understood, but that it can be accepted in a highly modified form. After all, related forms usually do go through similar embryologic stages, but these may have little to do with the adult stages of their ancestors.

In the last decade or so of the 19th century, scientists gradually lost interest in problems of phylogeny and classification above the species level. Most of the major groups of organisms had been found and described, and the phylogenetic methods of the Darwinian period had failed to answer too many questions about relationship, many of which remain unanswered today. In the early part of the 20th century, some of the facts about mutation were discovered, and Mendel's laws of inheritance were rediscovered. Darwin's theory of natural selection, once thought to provide the entire answer about evolution, fell into temporary disrepute, for it was now believed that mutation was the operating principle behind evolution. Much useful research was done on the nature and role of mutation in speciation, and gradually it became clear that evolution (change) takes place when natural selection acts on a population in which genes and chromosomes are recombining, mutating, or both. The process is more complex than anyone anticipated. A modified school of evolutionary thought known as neo-Darwinism emerged, which concentrated on population dynamics and speciation. These problems interested natural historians most, and major concern about taxa above the species level did not return until the early 1950s.

Even before attention returned to the theoretical basis for higher classification, there was a constant conflict between lumpers, who wanted a few large taxa, and splitters, who wanted many small ones. In general, the splitters prevailed because of the need to keep track of the enormous number of described animals (approaching one million species). Some splitters, in their enthusiasm, made a mockery of the purpose of classification by recognizing too many monotypic (single-species) genera and far too many families, orders, and other higher taxa. A great many classifications were, and are, one-character, monothetic identification schemes, which the next specialist too often changed with little or no justification. Different workers studying the same organisms produced radically different classifications, depending on which character they emphasized. Those who had to use these classifications—other biologists and occasionally the public—had no way to judge one against the other and, with good reason, many wondered if classification had anything at all to do with science. The problem still exists. Many taxonomists recognized, some with discomfort, that there was a large element of art in classification as practiced.

In the last thirty years, there have been several strong reactions to the traditional methods of classification characteristic of the 20th century's first fifty years. Two new approaches have developed; numerical phenetics and cladistics. Both approaches have transformed and improved the methods of those who continue to pursue the traditional evolutionary goal of recognizing homogeneous taxa descended from a common ancestor.

Phenetics

The object in a **phenetic** study is to establish degrees of overall similarity among the organisms under study. A simplified model can demonstrate the general principle and processes of such a study. Figure 6.1 shows five imaginary trilobites from the same time period. For the purposes of this demonstration, assume that these five are all the known species of their group; that is, they are more closely related to one another than any one is to any other trilobite. A glance at the figure should permit division of the five species into two groups, but more study is needed to establish the relationships within the two groups and to determine just how the groups are related to each other.

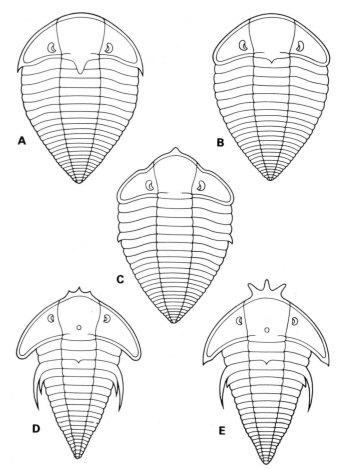

Figure 6.1. Five taxa of imaginary trilobites whose distinctive characters are used to demonstrate three methods of classification.

The first step in a phenetic study is to find and tabulate all characteristics of each species. Each feature that differs in at least one of the five species is given a code (Table 6.1). Typically, numerical taxonomists simply number or letter these features, but a mnemonic code (EyP = eye position) is easier to remember and thus easier to use. These taxonomists usually call a feature such as eye position a character and the variations character states (eye posterior; eye anterior). In this they depart from standard practice, in which **character** is used to mean not the feature but its variation, or that which distinguishes. Each state of each feature is then given a distinguishing code, here numerical, so that every variant has a distinctive code designation. For example, CkW is cheek width, and CkW_0 is the character in which the cheek (lateral lobe of the head) is about as wide as the glabella (central lobe). Most features listed in Table 6.1 have only two variations or characters, but some have more. HdSp (head spines) has four. Characters are numbered in order of similarity; although they could have been numbered in reverse order, "no spines" can be

Table 6.1. List of characters of imaginary trilobites

Code	Character
CkEm	0 = Cheek (side lobe of head) rounded anteriorly
	1 = Cheek emarginate (indented) anteriorly
CkLa	0 = Cheek with rounded lateral angle
	1 = Cheek with spine on lateral angle
CkW	0 = Cheek width about equal to width of glabella (middle lobe of head)
	1 = Cheek wider than glabella
EyP	0 = Eye nearer posterior margin of cheek
	1 = Eye nearer anterior margin of cheek
HdNd	0 = Head without median node
	1 = Head with median node
HdSp	0 = Head without spines anteriorly
	1 = Head with one spine anteriorly
	2 = Head with three short spines anteriorly
	3 = Head with three long spines anteriorly
OcSg	0 = Occipital (last head) segment without spines
	1 = Occipital segment with short spine
	2 = Occipital segment with long spine
Th2Sp	0 = Second thoracic (body) segment without median spines
	1 = Second thoracic segment with median spine
Th3Sp	0 = Third thoracic segment without lateral spines
	1 = Third thoracic segment with long lateral spine
Th4Sp	0 = Fourth thoracic segment without lateral spines
	1 = Fourth thoracic segment with short lateral spine
	2 = Fourth thoracic segment with long lateral spine
ThW	0 = Thorax with side lobe wider than middle lobe
	1 = Thorax with side and middle lobes about equal in width

Table 6.2. Data matrix

Code	A	B	C	D	E
CkEm	0	0	1	0	0
CkLa	1	0	0	0	1
CkW	0	0	0	1	1
EyP	0	0	0	1	1
HdNd	0	0	0	1	1
HdSp	0	0	1	2	3
OcSg	2	1	0	0	0
Th2Sp	0	0	0	1	1
Th3Sp	0	0	0	1	1
Th4Sp	0	0	1	2	1
ThW	0	0	0	1	1

Table 6.3. Similarity matrix

	A	B	C	D	E
A	X	2	6	13	12
B	2	X	4	11	12
C	6	4	X	9	10
D	13	11	9	X	3
E	12	12	10	3	X

The similarity matrix lists the species both across the top and down the left side of the table. The intersections of the same species, of course, play no part in the analysis. The column under each species is identical to the row for that species, and the upper right of the table matches the lower left. Filling in the numbers on both sides of the diagonal simplifies the process of cluster analysis discussed next.

The similarity matrix contains all the information any phenetic study can provide, but it is in a form that is difficult to grasp, especially when many taxa are analyzed. **Cluster analysis** enables one to display the information in the similarity matrix in a more easily understood form. Several methods of cluster analysis have been devised, and two will be demonstrated here. It should be noted that all methods of cluster analysis involve some information loss.

Each number in the similarity matrix represents the taxonomic distance between a pair of species. The value for pair AB, for example, is calculated from the data matrix (Table 6.2) by comparing each character in columns A and B. Differences are found only in CkLa (1 and 0) and OcSg (2 and 1). The total of two differences, each equal to 1, is entered on the similarity matrix for the distance between species A and B. Similarly, the sum of

only 0 or 3, first or last, or the measure of degree of similarity is destroyed.

After feature and character code for each variable in the study group has been assigned, the codes are arranged in a **data matrix** (Table 6.2). Species are commonly listed at the top, and feature codes run down the left side. The body of the matrix contains the numerical character code of each feature of each species.

Any of several coefficients of similarity may be applied to the completed data matrix to obtain a **similarity matrix**. These coefficients are divided into two distinct groups. One group is based on correlation, in which higher values indicate closer relationship. The other group includes various distance measures, in which lower values indicate closer relationship. Distance measures are somewhat more reliable and easier to interpret than correlation coefficients, and they are much easier to calculate. A simple distance measure, which will serve the purposes of this demonstration, involves counting the differences (always positive numbers) between each possible pair of species. These are then assembled into a similarity matrix (Table 6.3).

the differences between species C and E is 10. (Note that the difference between character 1 of a feature and character 3 is 2, which explains why characters must be carefully numbered in order of similarity.)

Once the similarity matrix is completed, species may be grouped into higher taxa. In one method, grouping is accomplished with a cross-averaging process called UPGMA cluster analysis (Unweighted Pair Group Method with Averaging). The similarity matrix is searched for the pair of species that differs in the smallest number of characters. In the example, A and B differ by only two characters and form the first primary cluster. Next the matrix is searched for another species that differs from members of the primary cluster AB by the smallest average amount. Species C differs from A and B by 6 and 4, respectively, for an average value of 5; D and E differ from A and B by much greater average values. C then joins the cluster AB, unless it has a closer relationship to another as yet unplaced species (that is, D or E). Since C differs from D by 9 and from E by 10, C should join AB to form a larger cluster AB-C. D and E, which differ from one another by only 3, form a separate cluster of their own. All species having been clustered, it remains to join the two clusters. This is done by averaging the six values that represent all the possible cross relationships between the two clusters (A-D, A-E; B-D, B-E; C-D, C-E). The average is 11.2. These clusters can then be drawn in a treelike diagram called a **phenogram** (Figure 6.2), which shows the phenetic relationships among the five members of the study group.

The phenogram, however, has several faults, one of which is that it closely resembles a phylogenetic tree, which, as we shall see, it is not. Furthermore, the UPGMA cluster analysis becomes less and less accurate as clusters become more inclusive, and it can be a misleading or incorrect representation of the information in the similarity matrix. The phenogram, then, in spite of its popularity, is the least desirable format for a phenetic study.

Probably the simplest cluster method is the **Prim network**, or as it is more properly called, the **shortest spanning tree**. Prim developed his clustering method in 1957 to find the most efficient way to connect cities by telephone wires. This nearest-neighbor technique does not have the faults of the UPGMA analysis. Working from the similarity matrix (Table 6.3), the object is to connect all the species in a single network with no closed loops. First one finds the two closest species. A and B, you will remember, are separated by a value of 2. Then one searches for the next largest value in the similarity matrix, which is the distance between D and E. D and E may be connected at a distance of 3 to form a separate piece of the network. B and C differ by 4, and so C is added to the AB network at this distance. The next highest number is 6, the distance between A and C, but connecting A and C would create a closed loop, and this value must be ignored. The next highest number is 9, the distance between C and D. The network is completed, then, by connecting the two pieces of the network at a distance of 9. The completed network is shown in Figure 6.3.

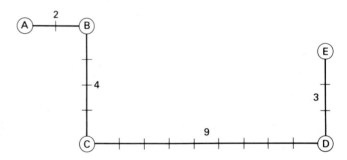

Figure 6.3. Prim network, or shortest spanning tree, showing the relative similarity of five taxa of imaginary trilobites.

A great many other systems for phenetic study are available. In contrast to cluster analysis, various methods of ordination, which portray taxa in two- or three-dimensional scattergrams, involve various mathematically complex methods known as **principal component analysis**. Instead of clustering taxa by characters as in the preceding examples (called a Q-mode analysis), a principal component analysis begins by clustering characters (R-mode analysis). Groups or suites of correlated characters form the components. The component that accounts for the greatest amount of variation in the study group becomes the first prinipical component, and those accounting for progressively less

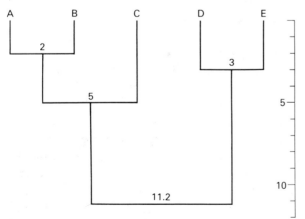

Figure 6.2. Phenogram, showing the relative similarity of five taxa of imaginary trilobites.

of the variation become the second and third components and so on. The first two or three components are each used to arrange the taxa into linear arrays, with the least similar forms placed at opposite ends of each array. These arrays are placed on different axes to form the scattergram. The most sophisticated approach produces a three-dimensional scattergram, which requires a model, a perspective drawing, or even a stereoscopic picture for the results to be appreciated. The required graphics can be produced by the computer.

Any of these phenetic methods will group taxa into clusters. The question one must ask at this point is what these clusters represent. Each combination of similarity coefficient and cluster analysis can and does group taxa, but too often the groups produced by one combination are substantially different from those produced by another. If the organisms are actually related, it is difficult to account for these differences in clusters. The problem may be that the only way one can define "phenetically similar" is in terms of the phenetic method used. There seems no single definition of what constitutes phenetic similarity. In its heyday numerical taxonomy was supported as an empirical method featuring absolute objectivity. Pure empiricism, however, is as impossible as complete objectivity. In spite of these problems, phenetic techniques can be useful, especially in work on such closely related forms as distinctive populations of the same species.

Cladistics

Unlike phenetic taxonomy, cladistic studies employ evolutionary theory in their methods. This approach was termed phylogenetic systematics in 1950 by its founder, Willi Hennig. The object of a **cladistic analysis** is to produce a credible hypothesis of the sequence of development of new lineages—that is, the phyletic branching—in the evolutionary history of a higher group of organisms. This branching component of phylogeny is termed **cladogenesis,** the order in which new lineages appear, and the progressive evolution between branches is termed **anagenesis,** the relative amount of change between speciation events. Cladists generally place little emphasis on anagenesis.

Hennig and his followers introduced new terms and changed the meaning of old ones; these must be defined before the process of cladistic analysis can be discussed. The term monophyletic was coined by Haeckel in the 1860s in discussion of the single (monophyletic) or multiple (polyphyletic) origin of life on earth. It has since been used to describe taxa whose members are descend-

ants of a single ancestor species (monophyletic groups) and contrasted with polyphyletic, used for hodgepodge assemblages of species that are superficially similar but derived from more than one ancestor.

Hennig restricted the meaning of the word monophyletic to describe groups that include *all* descendants of that ancestor. He coined another word, **paraphyletic,** for groups with a single common ancestor species that do *not* include all descendants of that ancestor. Paraphyletic groups are judged by cladists to be invalid, that is, to be neither classified nor named. Because Hennig's use of monophyletic represents a change in the meaning of a term of long standing and wide general use, the term **holophyletic** was later coined to replace Hennig's monophyletic. The result is that monophyletic can retain its traditional application to groups descendant from single ancestors. Monophyletic groups, however, can be of two kinds, those that contain all the descendants of the common ancestor (holophyletic groups) and those that do not (paraphyletic groups). Mammals, birds, and the Echinodermata are all examples of monophyletic groups that are also holophyletic; reptiles and primitively wingless insects (Apterygota) are examples of monophyletic groups that are also paraphyletic. The Reptilia are a paraphyletic group because the birds, who are much changed and not a member of the Reptilia as it is traditionally conceived, share a common ancestor with the crocodilians that is not an ancestor of other living reptiles. Although the Apterygota had a common ancestor, the group does not contain the winged insects, which evolved from within the group.

Another feature of Hennig's cladistic theory is that each group must have a **sister group.** Thus a completely resolved cladogram is completely dichotomous. At any branching point, there are a single ancestral lineage and two descendant lineages or sister groups. A little thought will show that any such tree may easily be resolved into nested series of holophyletic groups. Cladistic analysis, then, is the process of recognizing holophyletic groups of species. A group is recognized as holophyletic if all the members of the group possess in common at least one unique evolutionary innovation, that is, a character found in no other organism. Examples are the feathers of birds and the water vascular system and tube feet of recent Echinodermata. Hennig termed unique evolutionary innovations found in two or more species **synapomorphous** characters. An **apomorphous** character is an advanced (derived) character; a **plesiomorphous** character is a more primitive expression of the same feature.

The recognition of synapomorphous characters is

essential to cladistic analysis. To identify a unique evolutionary innovation, the investigator must have a knowledge of the characteristics of related organisms outside the group under study. True synapomorphous characters, which evolve only once, are often less common than parallel or convergent characters, which have evolved more than once and are useless for cladistic analysis.

Another simple demonstration can illustrate cladistic analysis and distinguish it from phenetic analysis. The five imaginary trilobites and the characters listed for them in Table 6.1 may serve again. However, it must be understood that in cladistic analysis, the characters are numbered 0 through 3 not simply to place most similar expressions of each feature together, but also to place them in a **primitive to derived sequence**, the assumed order in which they evolved. That is, $HdSp_0$ (head without spines) has been determined to be the most primitive condition through comparison of the study group A through E with its relatives, none of whom have head spines of any sort. In the study group, species A and B

have the most primitive state (no spines or $HdSp_0$), species C has one short spine ($HdSp_1$), D has three short spines ($HdSp_2$), and E has three long spines ($HdSp_3$). Each feature is compared with the same feature in relatives of the study group, and the characters of each feature have been similarly sequenced to produce the number codes in Table 6.1. Paleontologists usually assume that organisms from lower strata are more primitive than those from higher strata. Although this is generally true, it is also true that a primitive form may survive relatively unchanged for long periods. In lineages whose fossil records are incomplete, the derived character may appear to be earlier than the primitive character.

As should be clear from the previous discussion, all that is actually known, cladistically speaking, about any group that has been shown to be holophyletic is that its members have a common ancestor (Figure 6.4, *A*). The cladistic configurations of the group still require analysis. The derived character in one of the trilobite features listed in Table 6.1, $CkEm_1$, is **autapomorphous**; it is unique for the taxon that possesses it and does not serve to group any taxa. $CkEm_1$ has been entered on the tree in Figure 6.4, *A*, whose configuration shows that $CkEm_1$ does not contribute to the cladistic analysis.

All the remaining features have derived characters that group two or more of the five imaginary trilobites. They produce a series of five feature trees (Figure 6.4, *B* to *F*). Each of six features produces an identical tree (Figure 6.4, *C*). The characters $HdSp_3$ (Figure 6.4, *D*), $OcSg_2$ (Figure 6.4, *E*), and $Th4Sp_2$ (Figure 6.4, *F*), each of which is found in one species only, are other autapomorphous characters that add nothing to the cladistic analysis.

Four feature trees (nine features) (Figure 6.4, *C* to *F*) are compatible with one another; that is, no tree is impossible in view of any of the others. Only one, Figure 6.4, *B* for CkLa, the cheek lateral angle, groups A with E, in distinct conflict with the other trees. Perhaps the cheek lateral spine is a simple feature, easily evolved. For that reason, $CkLa_1$ is judged to be a parallel character, independently evolved because of a genetically similar heritage.

Parallel and convergent characters give incorrect results in cladistic analysis precisely because they were independently evolved. Hence the recognition of the all-too-common parallel character is an important part of cladistic analysis. Usually it represents the loss of a derived character or is of minimal complexity. Employing the relatively more complex characters available in the analysis first (those that would seem least likely to evolve more than once), and using simple, easily

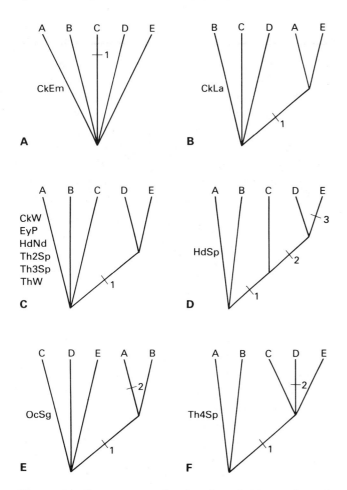

Figure 6.4. Feature trees, showing the primitive to derived sequences of characters in five taxa of imaginary trilobites.

evolved ones as a last resort, and then only when they are not in conflict with complex ones, will serve to avoid muddying the analysis with parallel characters.

In Figure 6.4, *C*, six features show that D and E form a holophyletic group. Figure 6.4, *D* also shows that species C, D, and E form another, larger holophyletic group. Finally, Figure 6.4, *F* for feature Th4Sp indicates again that species C, D, and E form a holophyletic group. (Character $Th4Sp_2$, however, is an autapomorphous character.) These trees can be combined into a single cladogram (Figure 6.5). In a more complex study, it is helpful to break the assembly process into several steps.

The final cladogram (Figure 6.5) diagrams a series of inferred speciation events (nodes 1, 2, 3, and 4) that led to the five imaginary trilobites, species A through E. Trilobite B has no character on its clade (branch) from node 2, and it is entirely possible that B and node 2 are one and the same creature, but it is also possible that characters unique to B have been missed (for example, those on appendages, which are unknown).

Hennig and the cladists have shown how to use character evidence to establish a rigorous sequence of branching in a phylogeny. They derive the classification from the cladogram, however, and this method results in classifications that differ greatly from traditional ones. Since only holophyletic groups are valid, one could recognize DE as a taxon, but not ABC. Rather, one would classify and name four groups: (1) ABCDE; (2) AB; (3) CDE; and (4) DE. The advantage of this procedure is that one can derive the sequence of cladogenesis from the formal classification. The disadvantage is that such classifications tend to group rather dissimilar organisms (trilobite C with D and E) and to separate similar ones (trilobite C from A and B), making it difficult to identify organisms and remember how they are grouped. Such classifications may be unstable as well, since correction of a minor error in cladistic analysis may cause great changes in classification.

Evolutionary systematics

The traditional approach to classification, as distinct from the recently developed phenetic and cladistic schools, has been termed evolutionary systematics because it follows closely Darwin's concepts of classification. Evolutionary systematists begin with the fact that species of plants and animals may be sorted into collective groups whose members are similar to one another and differ more or less distinctly from their relatives. Members of a group—birds, mammals, or trilobites—share the same general way of making a living and show distinctive levels of organization. Such groups are called **grades** (see Chapter 2).

Darwin recognized the evolutionary origin of grades and said that named groups of organisms should be based on genealogy but need not include all descendants of any given ancestor. Simpson, the vertebrate paleontologist, put it another way: classifications should be consistent with phylogeny but need not rigidly reflect it. Both Darwin and Simpson recognized that organisms do not evolve at an even rate and that a rapidly evolving lineage may change enough to constitute a new grade. In more modern terms, evolutionary systematists recognize as valid taxa those grades that are sufficiently dissimilar, that is, separated from their near relatives by distinct anagenetic gaps in phylogeny.

The methodologic approach has been to start with groups of organisms that seem similar, often groups recognized as taxa by earlier systematists. Most such groups have been named, or if newly discovered, may be named with the understanding that all named taxa are provisional. Untested groups of similar organisms have been termed **phena** (singular, **phenon**). The problem of the evolutionary systematist is to demonstrate that a phenon under study is indeed a valid evolutionary taxon.

Procedures in the past have suffered from the lack of a clear idea of how to establish the needed inferences. The pre-Darwinian nominalists and their descendants the pheneticists have a part of the answer: to look at as many features as possible rather than to try finding an essential few. Having tabulated as many features as possible, the modern evolutionary systematist establishes the evolutionary history (phylogeny) of the study group by determining the branching pattern (cladogenesis) and the relative progressive evolution (anagenesis) of the group. Major gaps in the resulting phylogram are evolutionary gaps and are used to delimit taxa. Thus the new procedure begins with the cladistic analysis already outlined and continues by adding as many of the observed characters as possible to the analysis to achieve a measure of anagenesis.

In the imaginary group of trilobites, a cladistic analysis has already been accomplished. Because this study used all the available characters, the anagenetic analysis has also been done, but it is not yet expressed. The cladogram (Figure 6.5) must be redrawn as a phylogram (Figure 6.6), changing both the length of branches and the angles between them to show the amount of change that has been inferred for each branch. The resulting phylogram shows that the original groups ABC (paraphyletic) and DE (holophyletic) are distinct as

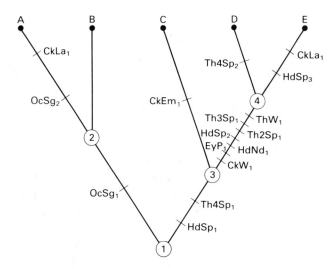

Figure 6.5. Cladogram, showing the evolutionary branching sequence of five taxa of imaginary trilobites.

demonstrated by the major gap between nodes 3 and 4.

A comparison of Figure 6.2 (the phenogram), Figure 6.5 (the cladogram), and Figure 6.6 (the phylogram) is instructive. Both the phenogram and phylogram demonstrate that similarities group ABC and DE, but their branching sequences are different. The cladogram and phylogram have identical branching sequences, but the similarities particularly within group ABC are not obvious on the cladogram. The phylogram is especially suited to portraying both cladogensis and anagenesis.

The characters may be weighted, or given greater importance in deriving a classification, if some rationale for their relative taxonomic importance can be found. A character may be given greater weight because it is relatively complex or because many species possess it. A character found in only a few species may not have contributed much to its lineage, whereas a character established in a large group of organisms deserves greater weight. Parallel characters deserve lesser weight.

Since the beginning of systematics, a major goal has been to produce classifications that are "natural." However, the phrases "natural groups" and "natural classifications" are, in an important sense, meaningless. To an evolutionary systematist, a natural group is a monophyletic group, be it paraphyletic or holophyletic, that is consistent with an inferred phylogeny, and contains relatively similar organisms. To a cladist, a natural group is one that is strictly monophyletic (that is, holophyletic), whether or not its members resemble one another. A pheneticist recognizes natural groups as those based solely on similarity assessed through a maximum number of features without regard to phylogeny. Linnaeus and his predecessors called a group natural if its members possessed characters thought to be "placed there by the Creator" so that the group might be recognized by humankind. In other words, all schools of systematics have called the products of their own methods natural.

Classifications in and of themselves are neither theories nor hypotheses. They should aid the memory, facilitate identification, and be as stable as possible. They must be used by geologists and biologists who are not systematists. A good classification permits the largest number of generalizations to be drawn about the included groups. Classifications are perhaps best regarded as indexes to the research and analysis that produced them. They are conveniences. The analysis itself, however, can be called a theory, built from many individual hypotheses.

The choice of method of analysis too often depends on current fashion; a method might better be chosen to suit the problem. While Linnaean essentialism has no place in modern systematics, nominalism in its modern form, numerical phenetics, has particular application to the study of very similar forms. Phenetic studies have not worked with highly dissimilar organisms, where slightly different analytic tools have produced grossly different results, but these tools admirably serve the purpose of fine discrimination.

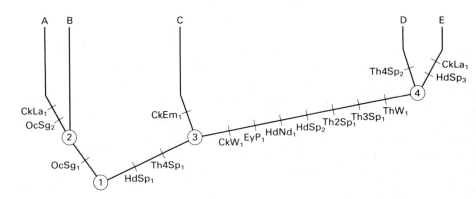

Figure 6.6 Phylogram, showing how a cladogram is modified by the number of characters on each internode. Note that the phylogram groups taxa *ABC* and taxa *DE*, as does the phenogram and the Prim network, while retaining the branching sequence of the cladogram.

Cladistic analysis is essential in any evolutionary study that, working on larger differences, seeks to group species into higher taxa. Cladistic classification, however, uses only one of the two major processes of phylogeny, cladistic branching, which recognizes change but ignores rate of change. By recognizing holophyletic groups only, cladists are often forced to ignore major anagenetic gaps, and their classifications may differ greatly from traditional ones. Groups dependent on a few or weak characters are inherently unstable; thus corrections of minor errors in cladistic analysis sometimes result in major reclassifications.

A modern evolutionary classification based on major anagenetic gaps results in taxa that are relatively homogeneous. They are easier to characterize and remember, and more can be said about them. Because these gaps are partly the basis for division of the geologic record into time periods, a paleontologist cannot afford to ignore them.

One inevitable result of the application of new tools and old-fashioned rigor to the problem of how to build and use a classification properly has been a great increase in general interest in the problem. The near future, one hopes, will see consensus.

Supplementary reading

Hennig, W. *Phylogenetic systematics.* Urbana, Ill.: University of Illinois Press; 1966. The classic work on cladistic analysis and classification.

Hull, D. Contemporary systematic philosophies. *Annual Review of Ecology and Systematics* 1: 19–54; 1970. Review of the three modern schools of thought from a philosopher's viewpoint.

Mayr, E. *Principles of systematic zoology.* New York: McGraw-Hill Book Co.; 1969. The most recent text by an evolutionary systematist. (A new edition is in preparation.)

Mayr, E. *The growth of biological thought: diversity, evolution, and inheritance.* Cambridge, MA: Belknap Press, Harvard University Press; 1982. The section on diversity is a history of systematic theory.

Sneath, P. H. A.; Sokal, R. R. *Numerical taxonomy.* San Francisco: W. H. Freeman & Co.; 1973. Latest edition of the basic text on phenetic systematics.

Wiley, E. O. *Phylogenetics: the theory and practice of phylogenetic systematics.* New York: John Wiley & Sons, Inc.; 1981. Recent review of cladistic analysis and classification.

7

Biostratigraphy and Paleobiogeography

Michael E. Taylor

Biostratigraphy and paleobiogeography are disciplines of paleontologic research that deal with the stratigraphic and geographic distribution of fossil taxa. Both kinds of studies yield information of importance to the understanding of the biologic history of the earth and provide a basis for the application of paleontologic information to the solution of geologic problems.

Biostratigraphy affords the critical tools necessary for recognition of relative ages of strata in geographically remote areas, such as the next mountain range or across oceans. However, to some degree every organism is restricted in geographic distribution. For example, local restrictions of marine organisms may result from the substrate being too soft or too hard, the water turbulence being too weak or too strong, or differences in oxygen and light. Regional restrictions in biotic distribution may result from barriers to dispersal, such as climatic differences, positions of continents, width of ocean basins, or differences in physiochemical properties of water. These factors produce limitations to the time correlation of strata by fossils, limitations that can be partially overcome by considering the paleoecological and paleobiologic principles that apply to the particular faunas used to correlate strata. Attaining the goals of biostratigraphy requires analysis of controls on local paleoenvironmental distribution (discussed in the section "Facies"), geographic distribution (discussed in the section "Paleobiogeography"), and organic evolution (discussed in the next section).

Biostratigraphy

Biostratigraphy is the study of the stratigraphic distribution of fossils. Principal objectives include (1) vertical zoning of sediments or sedimentary rocks according to their fossil content, (2) description and analysis of lateral variations in coeval taxonomic assemblages, and (3) time correlations of strata based on the recognition of nonrepetitive isochronous events within biostratigraphic sequences. The study of fossils for time correlation is sometimes called **biochronology** to distinguish it from studies that emphasize other aspects of biostratigraphy.

Achieving the objectives of biostratigraphy requires (1) careful collection of samples from measured stratigraphic sections, or cores, (2) systematic identification of species (3) plotting of vertical ranges and co-occurrences of identified taxa, and (4) division of the stratigraphic sequence into zones based on differences in the contained biotas. Numerous methods of defining zones are employed in different situations depending on the quality of data and the purposes for which the zonation is to be used. Zonation schemes are most commonly used for the time correlation of strata, but not all methods of biostratigraphic zonation are equally valid for time correlation. Some methods result in zonal schemes that are predominantly paleoecologic in significance. Paleoecologically oriented biostratigraphic classifications may contribute information on paleocommunities and sedimentary environments, information that is in turn important to comprehensive analyses of paleobiogeography and paleogeography.

The development of biostratigraphy was closely tied to the need of 19th-century geologists to understand the geologic history of the earth's crust and to classify sedimentary strata into schemes that reflect relative age.

William Smith, an English civil engineer living in the late 18th and early 19th centuries, was among the first to successfully use the unique character of fossil assemblages to determine superpositional succession, equate (or correlate) strata of equivalent age, and order strata into a relative time scale. Smith discovered that fossil assemblages are not randomly distributed in sedimentary strata, but rather, as a general rule, the same strata are found always in the same order or superposition and contain the same characteristic fossils. This discovery—sometimes referred to as the law of biotic succession—allowed Smith to determine the correct succession of strata and, in turn, to apply that knowledge and, in 1815 to produce the first regional geologic map, a map of England and Wales.

Organic evolution

Empirical biostratigraphy was practiced by Smith and others in the early 19th century without a clear understanding of the underlying processes that made it work. Darwin's theory of organic evolution through natural selection of inheritable characteristics, published in 1859, 44 years after publication of Smith's geologic map of England and Wales, provided a process for explaining the changes in biotic succession observed by Smith in the stratigraphic record. Maturation of the understanding of biologic evolution, attendant with knowledge of the ecologic parameters that affect species distribution, has put biostratigraphy on a modern scientific foundation.

The observational data with which biostratigraphers work are morphospecies, species defined and recognized on the basis of morphologic characters shared among individual specimens (Chapter 4). Because of the central role that species play in zonal biostratigraphy, an understanding of the processes of speciation and the manner in which speciation patterns are reflected by the fossil record is foremost in importance to our understanding of the strengths and limitations of biostratigraphic correlation.

Modern theory of species evolution is dominated by two models that reflect emphases on different aspects of the evolutionary process (Figure 7.1 and Chapter 4). The first model, **phyletic gradualism**, originated with Darwin's concept of organic evolution. The model holds that changes in frequency of morphologic characters occur gradually in response to selection pressure on species populations. Abrupt changes in morphology in

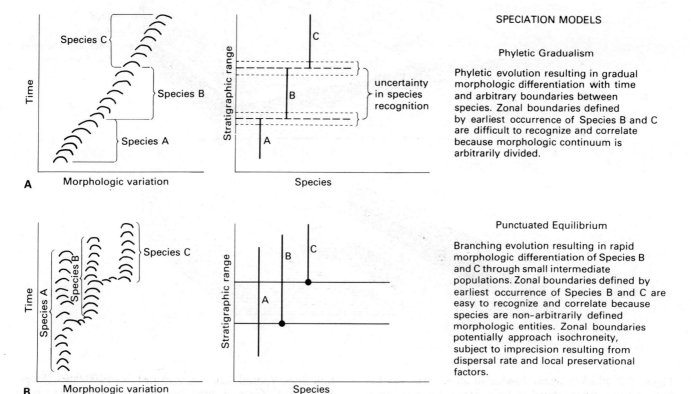

Figure 7.1. Implications of species evolution models to the definition of biochronologic zones.

the fossil record are viewed as something other than speciation events, such as sampling, preservational failures, or migration. If a complete record was preserved, it would be expected to show continuous, gradual morphologic change from one species to the next. According to this model, species are viewed as arbitrary segments of a morphologic continuum (Figure 7.1, *A*).

The second model, called **punctuated equilibrium**, characterizes species as real natural entities dominated for most of their history by morphologic stability, or stasis, and characterized by abrupt origins and terminations in space and time. Speciation is accomplished by the relatively rapid (perhaps 5000 to 50,000 years as an order of magnitude) development of reproductive isolation in some geographically isolated population of the parent species, followed by a longer period of stasis in the daughter populations. Continuing research is needed to test and refine the two evolutionary models. If phyletic gradualism is the predominant pattern in the evolutionary process, the gradational changes in morphology will

lead to zonal boundaries defined on stratigraphically lowest and highest observed occurrences of arbitrarily defined species. Zonal boundaries will be inherently imprecise and errors will be introduced in attempted time correlations because of the subjective basis for the definition of species limits. On the other hand, if punctuated equilibrium is predominant, many of the observed discontinuities in morphologic continua may be the result of a natural expression of the evolutionary process that abruptly produces new species. The punctuated equilibrium alternative potentially allows stratigraphic paleontologists an opportunity to objectively recognize species near their point of evolutionary origin—a unique point in time that can be used effectively for time correlation.

Facies

Since the mid-19th century, geologists have recognized that the petrographic and biologic characteristics of contemporaneously formed strata may change from place to

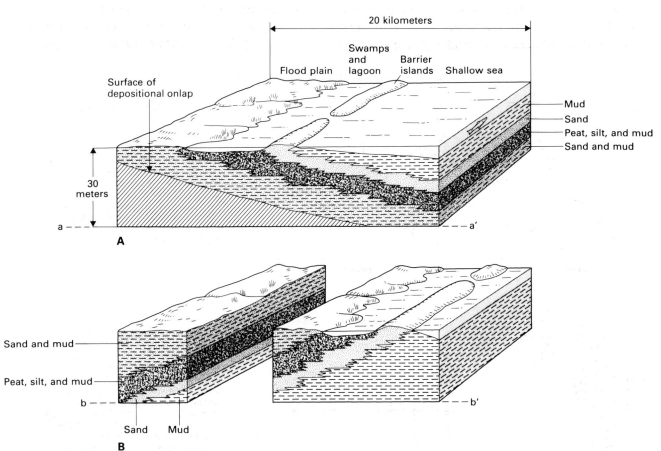

Figure 7.2. Block diagram showing idealized facies development along a marine shoreline during transgressive conditions (**A**) and regressive conditions (**B**). Any line parallel to *a–a'* or *b–b'* is isochronous and shows that sediment type cuts across time lines. Vertical scale is exaggerated about 500 times. (From Eicher, D.L. *Geologic time,* 2nd ed., p. 43. Reprinted by permission of Prentice-Hall, Inc., Englewood Cliffs, NJ.)

place, just as the products of contemporaneous sedimentary environments commonly change laterally along a shoreline or from shallower to deeper parts of a depositional basin. **Facies** is the aspect and characteristics of a rock or sediment that differentiate it from other contemporaneously formed rocks or sediments. Rocks or sediments differentiated by petrographic characteristics are called **lithofacies,** whereas rocks of sediments differentiated by their organic content are called **biofacies.** The depositional environment within which the lithofacies formed is called the **lithotope,** whereas the **biotope** is the habitat or habitats of the living assemblage that now characterizes a biofacies. Figure 7.2 shows the relationship between depositional environments along a marine shoreline and development of stratigraphic facies.

The occurrence of fossils in sediments or rocks is determined by the ecologic requirements of the once-living organisms, the transportation of the organismal remains after death, burial of the remains, and preservation of the fossils and strata in which the remains are buried (Chapter 5). Biofacies composition results from an interplay between the organisms' original ecologic requirements and the degree to which transportation and preservation have altered the relative abundance of individuals and taxonomic composition of the once living community.

Biofacies have some characteristics in common with biologic communities and are sometimes confused with them. However, the concept of biologic community usually implies some degree of ecologic interdependency of the elements that make up the community assemblage. Alternatively, the term community is sometimes used for assemblages of species that occupied a contiguous geographic space in mutual response to physical conditions rather than species interdependence. Some biofacies approach the living community in degree of fidelity by which they preserve the original community composition. However, most biofacies contain fossils of only part of the total assemblage of the contributing living community or communities and usually with distorted representation of relative abundances resulting from hydraulic sorting, scavenging, and differential preservation. For example, different assemblages of planktic foraminifers may live in separate depth-stratified natural communities in oceanic water masses. After death the separate assemblages may become mixed and accumulate in bottom sediments, thereby partially or completely losing their community identity. Similarly, some benthic animals that lived in shelf-edge or upper basin-slope biotopes may be redeposited in deeper water and mixed with members from indigenous deeper water communities.

Biostratigraphic zones

The zone is the basic unit of biostratigraphic classification. The essential data on which zones are based are the occurrences of taxa in a stratigraphic sequence and the superpositional arrangement of occurrences relative to one another. In practice, several different concepts of zonation are employed by stratigraphic paleontologists. Broadly speaking, the different zonal concepts can be grouped into three kinds. One group emphasizes the objective definition of zonal boundaries and uses units called **interval zones.** A second group emphasizes the taxonomic content of the zonal units and uses units called **assemblage zones.** A third kind of zonal unit, called **abundance zones,** is based on the maximum and/or minimum abundance of taxa and therefore emphasizes paleoecologic or sedimentologic influences on the organisms or their remains. Abundance zones may emphasize the relative abundance of certain taxa among taxonomic assemblages or the stratigraphic interval of maximum or minimum abundance of a single taxon. Many names have been proposed for the various types of zones based on fossils. Some of the more common types are shown in Figure 7.3.

Interval zones may be objectively defined and characterized, and relatively easy to identify away from the geographic areas of the type section, or primary reference section, where they were originally defined. Boundaries of interval zones may be strongly influenced, however, by local preservational conditions, minor interruptions in sedimentary record (diastems), and local fluctuations in environmental conditions that may have controlled the lowest or highest observed stratigraphic occurrence of a taxon. These factors limit the usefulness of interval zones in biochronology, unless the evolutionary history of the defining taxon is sufficiently understood. If a species' lowest observed occurrence coincides approximately with the horizon of evolutionary origin, such a temporally unique event may become the basis for converting some interval zone boundaries into horizons useful for time correlation.

In some instances, zonal boundaries are defined at the highest occurrences of taxa (for example, consecutive range zones type II in Figure 7.3). This practice is common in subsurface studies when microfossils are recovered from well cuttings and lowest occurrences are difficult to determine because of the possibility of sample contamination by slumping from higher horizons in the drill hole. Zonal boundaries defined by highest occurrences of taxa should not be considered isochronous without corroborative evidence because of numerous factors that can result in apparent local extinction (see

INTERVAL ZONES

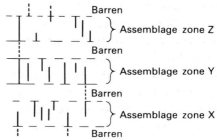

Local range zone

Concurrent range zone

Consecutive range zone
(type I)

Consecutive range zone
(type II)

ABUNDANCE ZONE

Acme zone

ASSEMBLAGE ZONES

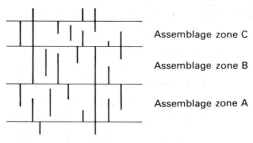

Barren
} Assemblage zone Z
Barren
} Assemblage zone Y
Barren
} Assemblage zone X
Barren

Non-contiguous assemblage zones

Assemblage zone C

Assemblage zone B

Assemblage zone A

Contiguous assemblage zones

Figure 7.3. Kinds of biostratigraphic zones. Interval zone boundaries are the most easy to objectively define and recognize, but they are also the most susceptible to diachronism because of ecological and preservational factors. In general, contiguous assemblage zones are most effective for time correlation, especially in instances where evolutionary history is understood for a large proportion of the taxa occurring in the zones. Chronozones (not indicated) are any zones with isochronous boundaries. (Adapted from the *North American stratigraphic code* by the North American Commission on Stratigraphic Nomenclature, 1983. American Association of Petroleum Geologists. Bulletin Vol. 67, No. 5. Used by permission.)

the section "Implications of species distribution to biostratigraphy").

Assemblage zones are useful in applied biostratigraphy because isolated collections of fossils can usually be assigned to an assemblage zone with a high degree of confidence. This may be important for deducing general age relations of poorly fossiliferous strata in structurally disturbed terrains; particularly in those instances when isolated assemblages can be compared with biostratigraphic reference sections developed in undisturbed areas. A shortcoming of assemblage zones is that their boundaries may be only vaguely defined by numerous overlapping ranges in transitional intervals (contiguous assemblage zones), or boundaries may be defined by nonfossiliferous intervals between fossil-bearing ones (noncontiguous assemblage zones). Boundaries so defined should not be expected to be the same age wherever they occur.

Noncontiguous assemblage zones are often employed in reconnaissance work, but through more detailed sampling, they may become contiguous assemblage zones. Zonation for time correlation is most effective when contiguous assemblage zones are developed by integration of data from numerous closely spaced measured sections so as to ameliorate local facies differences, and when a high number of observed lowest occurrences of species approximate the stratigraphic level of their evolutionary origin.

Resolution of time

Species duration is important to zonal biostratigraphy because it ultimately affects the theoretical limits of resolution in biochronological correlation. For example, duration of invertebrate species in the Cretaceous of the North American Western Interior averages approximately 2.5 Ma (mega-annum or 10^6 years), whereas relatively rapidly evolving species are estimated to have durations of roughly 1 Ma. One million years may be the correct order of magnitude for average duration of some other rapidly evolving species, but data are few and poorly documented because of the imprecision of isotopic ages upon which averages must be calculated. However, if 1 Ma average duration for rapidly evolving species is assumed to be a valid estimate for the whole of the Phanerozoic, an assumption that by no means has been proved, biochronologic zones of 1 Ma or less duration should be generally possible. Indeed durations of some zones in the North American Cretaceous, based on detailed studies of ammonite evolutionary lineages, range from 0.2 to 0.9 Ma.

Currently isotopic dating has limited precision because of analytical laboratory methods, varying quality of samples, and other factors. For example, state-of-the-art potassium-argon ages of Cretaceous bentonites and glauconites typically contain uncertainty limits of 5 to 6 per cent or more. The uncertainty limits translate to 3.2 to 6.8 Ma or more for isotopic ages measured for the Cretaceous period. This represents five to 25 times less precision than is potentially possible for correlations by fossils. Figure 7.4 shows the potential relative precision in time correlation by biochronological and isotopic methods based on an assumed species duration of 1 Ma and uncertainty limits of 1 per cent and 10 per cent in

isotopic dating. The graph shows that biochronological precision exceeds that of isotopic dating methods for most of the Phanerozoic eon. Despite these limitations, isotopic dating can calibrate the age of biostratigraphic zones and convert zonal schemes to an equal-interval time scale, an objective that cannot be achieved by fossils alone.

Relation between biostratigraphy and stratigraphy

The application of biostratigraphy to stratigraphic problems is enhanced by considering the relation between biostratigraphy and other kinds of stratigraphic studies. Concepts and procedures in stratigraphic studies are closely tied to the technical terms, or nomenclature, used by stratigraphic geologists.

Beginning with the first International Geological Congress in Paris in 1878, international committees have attempted to bring uniformity in usage of stratigraphic terminology to the geologic community. This has been attempted through a series of stratigraphic codes, or guides, that represent some degree of consensus among stratigraphic geologists on stratigraphic nomenclature and procedures. The most comprehensive document of this kind produced to date is the *International Stratigraphic Guide, A Guide to Stratigraphic Classification, Terminology, and Procedure,* edited by Hollis D. Hedberg and published in 1976. Another important guide is

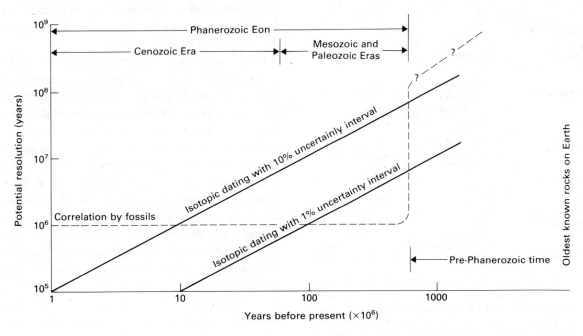

Figure 7.4. Potential resolution in time correlation by biochronologic and isotopic dating methods. Dashed line is minimum potential resolution by biochronologic methods assuming a one million year average duration for rapidly evolving species. (Adapted from diagram by Norman J. Silberling, used by permission.)

the *North American Stratigraphic Code* (1983). The code expands on some aspects of the *International Stratigraphic Guide* and is more reflective of current practices in North America. None of the stratigraphic guides is without flaw, however. Each has strengths and weaknesses resulting not only from a pragmatic need to compromise the differing philosophies of the authors, but also from the complexity of natural systems—it simply is not possible to write a brief series of rules consistently applicable to all natural situations. However, the present guides, and their predecessors provide milestones in the development of stratigraphic concepts. The guides should be recognized as such—applied when appropriate and quickly modified or abandoned when the complexities of new research problems cannot be adequately resolved by their recommendations.

Four major categories of units are used for stratigraphic classification: lithostratigraphy, biostratigraphy, chronostratigraphy, and geochronology (Figure 7.5). **Lithostratigraphy** involves the stratigraphic division of the earth's crust into mappable units that are defined and characterized by their consistency of lithologic character or by combination of characteristics and their position in a stratigraphic succession. The formation is the basic unit of lithostratigraphy. Biostratigraphy, as discussed earlier, involves the stratigraphic division of the earth's crust into units that are defined and characterized by

differences in observed fossil content. The zone is the basic unit of biostratigraphy and, as defined, is independent of lithostratigraphic units. Lithostratigraphy and biostratigraphy are parallel and analogous disciplines. Both rely on objective, observable characteristics contained within the rocks.

Chronostratigraphy is the classification of strata into bodies of sediments or sedimentary rocks that were deposited during a specific interval of geologic time. Chronostratigraphic units are interpretive and require discovery of criteria from lithostratigraphic and biostratigraphic studies that have a unique time of origin and can be recognized from place to place. Lithostratigraphic criteria for establishing chronostratigraphic boundaries include, for example, volcanic ash beds and other isochronous marker beds that can be traced with confidence over an area of significant size. Similarly, paleomagnetic properties of strata may show alternations of reversed and normal polarities of the earth's dipolar magnetic field at the time that strata were deposited, criteria that may be applied to chronostratigraphic correlation.

Biostratigraphic evidence for defining and correlating chronostratigraphic units is usually the most effective criterion in fossiliferous strata. Biostratigraphic evidence relies on the temporal uniqueness of species and the irreversibility of the process of organic evolution. As indicated previously, the use of paleontologic data to define and correlate chronostratigraphic units is called biochronology to distinguish such studies from other objectives of biostratigraphy, such as biofacies analysis. The **chronozone**, a zone with isochronous boundaries, is the basic unit of biochronology.

Geochronology is the arrangement and correlation of events in earth history in a purely time framework. The time scales by which geochronological units are calibrated may be an **ordinal scale** (a time scale measured in units that describe relative position in sequence but not duration), as in biochronology and magnetostratigraphy; or an **interval scale** (a time scale in which units of measurement are of equal duration), as in the various types of geochronometric dating. **Geochronometry** is the direct division of geologic time based on quantitative dating methods, for example, radiometric decay of certain parent isotopes such as uranium-lead, rubidium-strontium, and potassium-argon. Fission track dating is another geochronometric method of major importance to stratigraphy because it often can be applied to minerals, such as apatite or zircon in some volcanic ash falls, that formed more or less at the same time as the sediments. Geochronometry measures geologic time in years and is the most effective method of transforming

Major Stratigraphic Categories	Units of Stratigraphic Classification	
Lithostratigraphic	Group Formation Member Bed(s)	
Biostratigraphic	Kinds of Zones: Abundance Zones Assemblage Zones Interval Zones	Equivalent geochronologic (time) Units
Chronostratigraphic (time-stratigraphic)	Eonothem Erathem System Series Stage Chronozone	Eon Era Period Epoch Age Chron

Figure 7.5. Categories and units of stratigraphic classification. Biochronologic units (not indicated) are chronostratigraphic units that are based on the interpretation of fossils for their significance in time correlation. (Modified from the Hedberg, H. D., editor. *International stratigraphic guide: a guide to stratigraphic classification, terminology, and procedure*, p. 10. © 1976. Reprinted by permission of John Wiley & Sons, Inc., New York.)

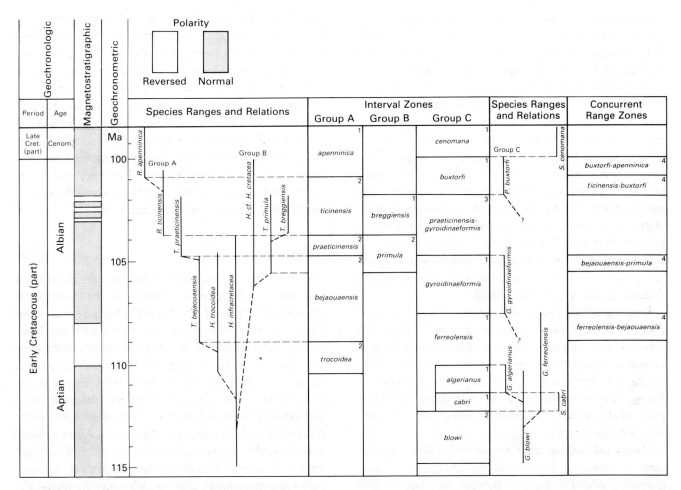

Figure 7.6. Integrated biostratigraphic, geochronologic, geochronometric, and magnetostratigraphic scale for the early Cretaceous Aptian, Albian, and early Cenomanian ages. Faunal zones are constructed from ranges of planktic foraminifers. Several ways are shown for integrating species ranges to form zones. Kind of zone is indicated: *1*, local range zone; *2*, consecutive range zone (type I); *3*, consecutive range zone (type II), and *4*, concurrent range zone (see Figure 7.3 for definitions).

Diagonal dashed lines indicate evolutionary origin of species where known. Horizontal dashed lines are eye guides. Generic names shown are abbreviations of *Rotalipora* (R.), *Ticinella* (T.), *Hedbergella* (H.), *Planomalina* (P.), *Globigerinelloides* (G.), and *Schackonia* (S.). (Modified from Van Hinte, J. E. A Cretaceous time scale. © 1976 by the American Association of Petroleum Geologists. Bulletin Vol. 60, No. 4. All rights reserved.)

the geologic time scale from an ordinal to an interval scale. Integration of lithostratigraphic, biostratigraphic, and isotopic ages has resulted in development of the modern geologic time scale (see back endpages)—a geochronologic scale within which geologic events can be classified and correlated on a worldwide basis. An example of some biostratigraphic zones, how they are constructed, and their relation to geochronologic, geochronometric and magnetostratigraphic scales is given in Figure 7.6.

Paleobiogeography

Species and higher taxa form geographic distribution patterns in relation to the area of the habitat to which they are adapted, in relation to their ability to disperse

within that geographic area, and in relation to their time and place of evolutionary origin. **Biogeography** is the study of the geographic distribution of species, higher taxa, and associations of taxa. It involves the descriptive classification of biotic distribution patterns, analysis of the evolutionary origin of taxa forming the distributional patterns, and the interpretation of environmental conditions that control or maintain boundaries between biogeographic units. The paleontologic counterpart of biogeography—paleobiogeography—adds the dimension of time to the study of biotic distribution. Paleobiogeography gives paleontologists an opportunity to study the development and changes in distribution patterns in space and time that have occurred during earth history and the relationship between biotic patterns and the changing geography of the earth's crust.

Biogeographic studies employ a loose hierarchy of units in the classification of biotic distribution patterns. The **province** is the basic unit of classification and is normally defined by the degree of endemism, that is, the proportion of taxa restricted to a given area. The particular proportion of geographically restricted species or genera used to characterize provinces may vary markedly. A **region** is a group of several taxonomically related provinces, and a **realm** is the largest subdivision of the biosphere that reflects major biotic differences among the higher taxonomic categories (Figure 7.7). Biotic regions and realms may be recognized by the degree of endemism of genera, families, or, in some cases, orders of plants or animals that may distinguish one area from others. The definition and recognition of biogeographic units, especially provinces, differ markedly among biogeographers, depending on their purpose of study.

Biogeographic studies may focus on the developmental history of provinces, regions, or realms; or they may focus on the changes in environmental conditions and ecologic responses to those conditions reflected by changes in distribution patterns of biotic assemblages. These different emphases may be generally grouped into two subdisciplines: historical biogeography and ecological biogeography.

Historical biogeography stresses the evolutionary history of provincial biotas and the changes in biogeographic distribution patterns through time. Such studies normally incorporate information on the history of evolving lineages occupying geographic areas and on the origin, persistence, or extinction of provinces through time.

Historical biogeographic studies alone are inadequate to elucidate the causal factors that produce, maintain, and eventually obliterate biotic provinces. **Ecologic biogeography** deals with the causal relations in biogeography, environmental conditions, and ecologic responses. These are normally discovered only by combining information on biotic evolution with analysis of the ecologic interaction between biotic species and communities and their habitats and with environmental fluctuations in the habitats through geologic time.

Ecologic biogeography studies may classify species associations into provinces based on several different criteria. For example, the degree of taxonomic difference between two or more geographic areas is commonly used in quantitative studies of species distribution. Alternatively, provinces may be distinguished from other provinces by the degree of change in taxonomic composition of assemblages along a bathymetric or latitudinal gradient. Changes along a gradient may be especially useful in discovering the nonarbitrary boundaries between provinces, which in turn can be studied to help determine the environmental factors that allow provincial biotas to maintain their distinctive characteristics.

Knowledge of the occurrence and distribution of marine organisms is greater for the near-shore habitats and markedly decreases with water depth. Most modern concepts of marine biogeography, therefore, have developed from studies of faunas occupying shallow water habitats, areas representing only 4 to 5 per cent of the earth's surface. We will now examine some of the general factors that influence species distribution.

Habitats

The biosphere is divisible into several natural divisions that may represent major barriers to the dispersal of organisms. The most abrupt, and most obvious, is the barrier between nonmarine and marine habitats. Nonmarine habitats are divided between aquatic and terrestrial, each of which has a plethora of subdivisions that reflects the great diversity of nonmarine environments. The marine realm is divided into **pelagic** (the water mass) and **benthic** (bottom) **habitats**. These habitats are further divided on the basis of water depth, availability of light, and other physiographic criteria (Figure 7.8). The major benthic habitats of shelf seas are the **littoral**, the area exposed between high and low tides, and the **neritic**, the subtidal area occupied by the continental shelves. Benthic oceanic habitats include the **bathyal**, corresponding approximately with the continental slope; the **abyssal** for the broad flat plains of the deep-sea ocean basins; and the **hadal habitat** for the deep-sea trenches.

The earth's present climatic regime, ranging from cold polar regions to warm equatorial regions, has an average temperature gradient of about 30°C in marine surface waters. The strong thermal gradient is paralleled by the north-south arrangement of marine faunal provinces on shelf seas (Figure 7.7), although locally the latitudinal position of province boundaries may be modified by the particular patterns of cold and warm marine currents. Boundaries between Holocene provinces (Figure 7.7) typically are located at topographic features such as bays or peninsulas, which tend to localize boundaries between different oceanic currents. Water masses associated with different current systems usually have different physical and chemical properties. If a boundary is present in one location for a sufficient length of time, the local fauna may be differentiated by adapting to the environmental conditions of each water mass.

The Holocene coincidence of narrow shelf seas, pre-

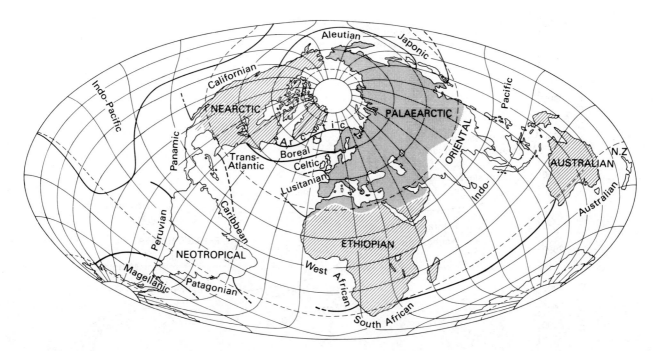

Figure 7.7. World map showing major biogeographic land regions and marine faunal provinces. The marine provinces are defined mainly by littoral and neritic mollusks. (From Davies, A. M. *Tertiary faunas,* vol. 2. © 1975 [revised edition] by George Allen & Unwin Ltd, Herts, England. All rights reserved.)

dominantly north-south orientation of continents, and high pole-to-equator thermal gradient, tend to localize modern major environmental discontinuities in a predominantly north-south direction. However, through most of the Phanerozoic eon, east-west barriers to ocean circulation were less well developed than at present. Shelf seas occupied much broader areas, and the latitudinal thermal gradient was probably much lower, in part due to the ameliorating effects of unrestricted oceanic circulation. Such conditions resulted in patterns of distribution of faunal provinces somewhat different than at present. Latitudinal environmental differentiation was more subdued and faunal provincial differentiation produced, in some instances, bandlike patterns arranged concentrically around continental land masses.

The area of the earth's surface occupied by the major habitats can be estimated by plotting the habitats on a **hypsographic curve** (Figure 7.8). The area under the curve represents the approximate cumulative percentage of the total surface of the earth occupied by a particular habitat at the present time. Significantly, the marine realm occupies nearly 71 per cent of the earth's surface.

Examination of the hypsographic curve shows that a relatively minor rise in sea level would result in a major increase in surface area occupied by the shelf seas. This in turn would greatly increase the habitat area available for development and colonization by marine communities.

The latter condition is characteristic of most of Phanerozoic time, whereas the present relatively low stand of sea level and narrow continental shelves are atypical of geologic history as a whole.

The most effective barrier to biotic dispersal within the marine realm is temperature. Temperature may affect species dispersal in accordance with the adaptive range of the adult population. Perhaps more importantly in many invertebrate phyla, temperature may affect the timing of seasonal reproduction or larval development, since larvae have a relatively narrower range of temperature tolerances. Thus, province boundaries are often drawn where the limits of a number of co-occurring species coincide with abrupt temperature changes between different oceanic water masses as the associated currents impinge on the neritic or littoral zones of a marine shelf.

A major physical barrier to faunal dispersal within the modern marine realm is between the thermosphere and psychrosphere. The **thermosphere** is that part of the marine realm occupied by warm or seasonally variable hydroclimates that are concentrated in low and middle latitudes. The **psychrosphere** consists of the cold, temperature-stable marine hydroclimates below the thermosphere in low and intermediate latitudes and at all depths in polar regions. The boundary between the thermosphere and psychosphere forms a **thermocline**, a transitional area within a water column where the water

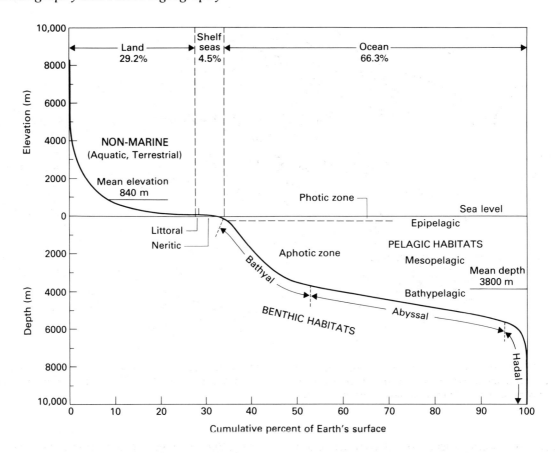

Figure 7.8. Major earth habitats plotted on a hypsographic curve. The curve shows the percentage of the earth's solid surface occupied by each major habitat. (Modified from Sverdrup, H. U.; Johnson, M. W.; Fleming, R. H. *The oceans: their physics, chemistry, and general biology,* p. 19. © 1942, renewed 1970. Reprinted by permission of Prentice-Hall, Inc., Englewood Cliffs, NJ.)

temperature changes markedly over a relatively short distance. Position of the thermocline may vary in depth according to latitude, bottom topography, and location of marine currents. In low and intermediate latitudes, the thermocline is often associated with the change in marine circulation regimes at the transition from shelf seas to oceanic regions. At depth, the boundary commonly occurs in the bathyal zone near or below the shelf-slope break. In low and intermediate latitudes, the thermocline is a major barrier to faunal dispersal, resulting in highly distinct shallow and deep water faunas. In high latitudes, shallow water benthic faunas show relatively higher levels of taxonomic similarity with those of the deep sea, and depth-related faunal differences are more subdued.

Species dispersal

Species vary markedly in their distributional characteristics. Species that are geographically widespread are said to be **cosmopolitan**, a general term that lacks genetic connotations. In biogeographic analyses of cosmopolitan species, it is important to distinguish, when possible, between species that are widespread because the environmental conditions to which they are adapted are widespread (for example, many bathyal and abyssal species), and those species that are widespread because they have wide environmental tolerances and efficient dispersal mechanisms, such as long-lived planktic larvae. To clarify these concepts, some additional terms are introduced:

Species adapted to a narrow range of environmental conditions are said to be **stenotopic**, whereas species with wide environmental tolerances are **eurytopic**. Species also may be classified by their geographic distributions: the term **stenogeographic** is used for species with limited geographic ranges and **eurygeographic**, for species with wide geographic ranges. Some species are geographically widespread but occur in isolated patches of favorable habitat rather than throughout the whole geographic range of the species. Such species are called **disjunct eurygeographs**. Many estuarine species belong to this category. **Endemic species**, or higher taxa, are those taxa restricted to a particular geographic area, such as a single province or part of a province. Endemism may result when species are "new" in an evolutionary sense and still undergoing initial dispersal, or when the habitat to which they are adapted occupies a relatively small area. Alternatively it may result from environmental restriction of an originally larger geographic range to a smaller range, called a **refugium**.

The ability of a species to disperse throughout suitable habitats some distance from its area of evolutionary origin affects its usefulness in biogeographic and biostratigraphic studies. Common barriers to migration of species dispersal include distance, unfavorable environment to be crossed, larvae of low mobility or short duration of viability, narrow environmental tolerances, and biological interactions such as competition and predation. Tectonic barriers such as mountains, oceanic ridges, and distances between margins of continents may represent formidable barriers to species dispersal. As discussed earlier, differences in climatic and other physiochemical characteristics of adjacent oceanic water masses—both between surface currents and at depth—may also limit species dispersal.

Implications of species distribution to biostratigraphy

Stratigraphic implications of species dispersal patterns are important to interpretations of the biostratigraphic record and to biochronological correlations. Figure 7.9, *A* shows a hypothetical distribution pattern for hypothetical species A and changes in distribution patterns in space and time that may result from fluctuations of ecological barriers. The distributional history of a species can be described in three phases:

1 Initial dispersal from the area of evolutionary origin until ecologic barriers to adaptive limits are encountered.
2 Restriction of range or breakdown of barriers in response to climatic or other environmental changes.
3 Extinction or migration as a result of environmental changes.

As environmental changes exceed the adaptive limits of the species, local extinction or migration may occur at different times in different parts of the range thereby eventually reducing the species range to a small refugium. Alternatively, extinction may occur more or less isochronously throughout the species range if environmental changes are widespread, sudden, and of sufficient magnitude to permanently disrupt the reproductive cycle simultaneously throughout the species range.

The potential stratigraphic expression in measured sections A through H of the species distribution pattern outlined in Figure 7.9, *A* is shown in Figure 7.9, *B*. The probability of sampling from the area of evolutionary origin (stratigraphic section E) or terminal extinction of a species (stratigraphic section D) is slight because of the small areas involved. The rate of initial dispersal may be slow for benthic invertebrates that brood their young or

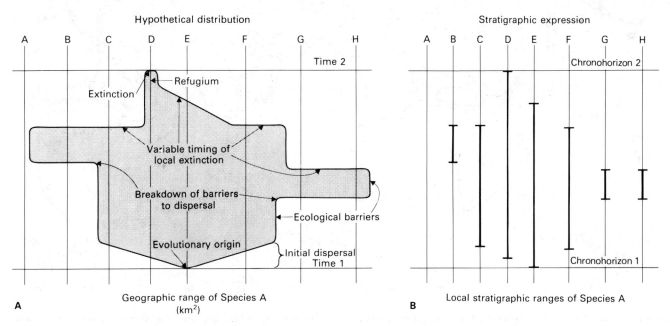

Figure 7.9. Diagrams showing geographic range (**A**) and potential stratigraphic expression (**B**) of hypothetical species *A* sampled from stratigraphic sections *A* through *H*. Sources of potential error in biochronologic correlation based on first occurrences of species are evident.

have benthic larvae, whereas initial dispersal may be rapid or geologically instantaneous for pelagic species or benthic species with long-lived planktic larvae. Biostratigraphic correlations based on few stratigraphic sections (for example, between sections B and G or H in Figure 7.9) may be strongly diachronous or fortuitously precise as between sections C and F or G and H. However, the probability of accurately estimating the total chronostratigraphic duration of species A can be increased by sampling numerous stratigraphic sections and integrating local range data into a composite species range chart (for example, Figure 7.6).

Plate tectonics and paleobiogeography

Development of the plate tectonic model in the last two decades has resulted in a major revolution in our understanding of the processes that work in the earth's crust. The model states that the earth's crust consists of a series of rigid lithospheric plates that are in motion relative to one another (Figure 7.10). A crustal plate may consist entirely of either continental or oceanic crust, or it may consist of both continental and oceanic crust functioning

as a single plate. Crustal plates may move relative to one another by lateral shear, they may move away from one another concomitant with generation of new crust along oceanic ridges, or plates may converge and override one another at subduction zones.

The plate tectonics model has far-reaching implications for paleobiogeographic studies. For example, convergence or divergence of the shallow marine shelves surrounding the continents may result in the breakdown or establishment of new barriers to migration of species; or collision of continents may result in new dispersal routes for terrestrial biotas. The breakup of geographic regions after organisms are buried may result in fragmentation of biotic provinces on separate plates which in turn may become widely removed from their original position. Alternatively, crustal fragments may be tectonically juxtaposed in accretionary terrains that record several different paleobiogeographic histories.

Figure 7.11, *A* shows the inferred ancient land mass of Pangea as it may have looked during the late Paleozoic. During the early Mesozoic, Pangea is thought to have begun to break up, producing a northern continental mass, Laurasia, and a southern land mass, Gondwana

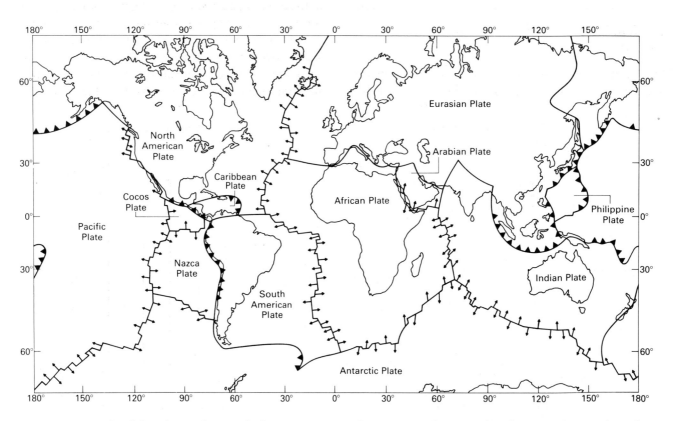

Figure 7.10. Twelve lithospheric plates and their present motions. Triangles indicate position of overriding plates at converging plate boundaries. Small arrows indicate approximate direction of motion along diverging oceanic ridge systems. The oceanic ridges are also the sites of generation of new oceanic crust. (From Uyeda, S. *The new view of the earth.* © 1978 by W. A. Freeman & Co., San Francisco, CA. All rights reserved.)

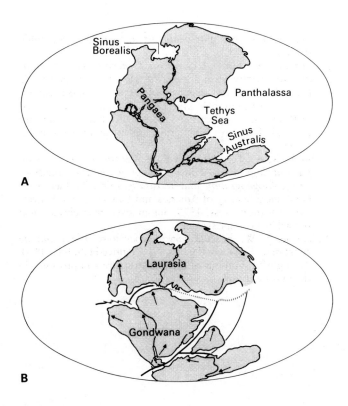

Figure 7.11. A, The ancient land mass of Pangea as it may have looked near the end of the Paleozoic era. **B,** World geography in the early Mesozoic era, after about 20 million years of drift. Between **A** and **B** the Tethys Sea expanded so as to circulate freely through the entire equatorial region, thereby providing an avenue for dispersal of tropical marine species different from dispersal routes that exist in modern oceans. (From Uyeda, S. *The new view of the earth.* © 1978 by W. H. Freeman & Co., San Francisco, CA. After Dietz, R. S.; Holden, J. C. The breakup of Pangea. © 1970 by Scientific American, Inc. All rights reserved.)

(Figure 7.11, *B*). The two were separated by an inferred spreading center that allowed the Tethys Ocean to circulate uninhibited through equatorial regions. This freely circulating equatorial sea resulted in worldwide dispersal of tropical marine faunas, a condition quite the opposite from the north-south obstructions to longitudinal dispersal that exist today.

Strategy in paleobiogeography

Paleobiogeographic research uses, with varying degrees of success, information on biotic provinces to reconstruct paleogeography. For such studies to be meaningful, the habitats occupied by ancient biotas must also be known, since the potential for geographic dispersal among marine biotas varies greatly in different habitats along both climatic and bathymetric gradients. Failure to take these relationships into account has led to numerous contradictory reconstructions of ancient paleogeography.

For the biogeographer working with Holocene distribution patterns, stability of position of the modern continents can be more or less taken for granted. Problems encountered in interpreting distributional patterns are usually related to climatic fluctuations, migrations, minor sea level fluctuations, and resulting changes in competition between species. On the other hand, paleobiogeographers working on biotic distribution problems, particularly for times during the early two-thirds of the Phanerozoic eon, must contend with changes in position of continents, major changes in oceanic circulation patterns, growth and demise of whole ocean basins, and potential fragmentation of biotic provinces after the organisms have been preserved as fossils.

How are these latter problems to be overcome? A strategy that seems clear is that paleobiogeographic studies must incorporate the best geologic information available on the past positions and movements of lithospheric plates. This should include paleomagnetic studies to determine the paleolatitudinal position of plates, structural and petrographic information that may bear on location of ancient suture zones, and regional stratigraphic and sedimentologic information that can resolve the spatial continuity of depositional basins and the environments occupied by their contained biotas. On the paleobiologic side, paleobiogeographic studies require thorough understanding of the evolutionary relationships of biotas. Detailed systematic studies are an essential underpinning for modern paleobiogeographic investigations. Paleoecologic information is grossly incomplete or lacking for most Phanerozoic biotas but must be developed before accurate causal explanations can be formulated and tested. In addition, better distributional models for living groups are needed to provide the basis for comparison with the fossil record. For example, living biotas of the shallow marine habitats are relatively well known, but those of the deep sea, which occupy about 66 per cent of the earth's surface, are known only in the sketchiest of terms. In all, much more remains to be learned about the geography of adaptive radiation and dispersal, more than has been accomplished during the last 100 years. Some of the most challenging intellectual puzzles that lie ahead for paleontologists are to be found in paleobiogeographic research.

Supplementary reading

Berry, W. B. N. *Growth of a prehistoric time scale.* San

Francisco: W. H. Freeman & Co.; 1968. Excellent summary of the historical development of the geologic time scale and the principles on which it is based.

Brown, J. H.; Gibson, A. C. *Biogeography*, St. Louis, MO: C. V. Mosby Co.: 1983. An excellent review of principles of biogeography.

Cohee, G. V.; Glaessner, M. F.; Hedberg, H. D., editors. *Contributions to the geologic time scale*. Tulsa, OK: American Association of Petroleum Geologists, Studies in Geology No. 6; 1978. Collection of technical reports on the present status of the chronostratigraphic and geochronometric bases for the geologic time scale.

Ekman, S. *Zoogeography of the sea*. London: Sidgwick & Jackson; 1967. The Bible of marine zoogeography—old but inspiring!

Gray, J.; Boucot, A. J., editors. *Historical biogeography, plate tectonics, and the changing environment*. Corvallis, OR: Oregon State University Press; 1976. Collection of reports that reflect the state-of-the-art of paleobiogeographic studies.

Kauffman, E. G.; Hazel, J. E., editors. *Concepts and methods of biostratigraphy*. Stroudsburg, PA: Dowden, Hutchinson & Ross, Inc.; 1977. Collection of reports by specialists on zonal biostratigraphy and biofacies analysis. Concepts, methods, and applications in particular biotic groups are covered. Extensive bibliogrphy up to about 1974 is included.

Robison, R. A.; Teichert, C., editors. *Treatise on invertebrate paleontology. Part A: Introduction, fossilization (taphonomy), biogeography and biostratigraphy*. Boulder, CO: Geological Society of America and Lawrence, KS: University of Kansas Press; 1979. Summarizes invertebrate fossil record by geologic system.

Valentine, J. W. *Evolutionary paleoecology of the marine biosphere*. Englewood Cliffs, NJ: Prentice-Hall, Inc.; 1973. Thoughtful synthesis of organic structure and processes in the marine realm.

8

Kingdom Protista

Martin A. Buzas
Raymond C. Douglass
Charles C. Smith

Part I Kingdom overview

Martin A. Buzas

Most of this book is about multicellular invertebrate animals. For them there is no need to consider a category of classification higher than phylum. This chapter surveys microorganisms, however, which requires use of the highest level of classification: the Kingdom.

As children, we learn to divide the world of living organisms into plants and animals. Similarly, the earliest naturalists divided the living world into two great kingdoms—Kingdom Planta and Kingdom Animalia. Plants are placed in one kingdom because they have the ability to photosynthesize, are autotrophic, and most are sedentary. Animals are heterotrophic, and most are mobile. In the world of larger organisms the scheme works quite well; a horse is an animal, and a tree is a plant.

After the invention of the microscope, a previously unseen world of microorganisms unfolded. Most of the microorganisms studied were single celled. Those containing **chloroplasts** (organelles within the cytoplasm containing chlorophyll), and, therefore, capable of photosynthesis, were classified as plants. Those without chloroplasts but with cilia or **flagella** (whiplike structures) for locomotion were classified as animals. Problematic organisms such as bacteria, blue-green algae, and fungi were arbitrarily placed in the plant kingdom.

With the invention of more sophisticated instruments and advances in microbiology and biochemistry, the two-kingdom system began to collapse. Studies of cells re-

vealed that although amoebas, flowers, clams, tigers, and lions are morphologically distinct, the biochemistry and structure of their cells are remarkably similar. All these organisms have cells with one or more nuclei surrounded by a well-defined nuclear membrane with DNA arranged in long strands along chromosomes (Chapter 2). Such cells are called **eucaryotic**. The bacteria and blue-green algae have cells without a nuclear membrane, and DNA is scattered throughout the cytoplasm. These cells are called **procaryotic**. The basic division between procaryotic and eucaryotic cells represents the most fundamental difference between organisms. Consequently, the bacteria and blue-green algae are placed together in the Kingdom Monera.

Among organisms with eucaryotic cells, the fungi are unique. They are heterotrophic, obtaining nutrition through absorption of organic matter. Most scientists now separate them from the plant kingdom and place them in a fourth kingdom, the Fungi.

As studies of single-celled "plants and animals" progressed, a number of organisms were discovered to contain chloroplasts, but in the dark when photosynthesis is not possible, these adaptable organisms become heterotrophs. Thus, within a single cell lies the ability to ingest organic matter and also photosynthesize energy from the sun. A few species belonging to groups originally classified as plants have gone even farther and lost their chloroplasts altogether to become obligate heterotrophs. These species exist solely on organic nutrients associated with decaying seaweed.

While larger plants are sedentary, many single-celled autotrophic species are motile. They have developed a variety of structures to move about in the water or to glide over surfaces such as sand grains and other organisms.

Some single-celled species originally classified as animals have found a way to use the sun's energy through **symbiosis** (two or more organisms living in close association). Many species harbor algal cells within their cells as symbionts. The photosynthesis of the algae provides the host with nutrients while the algae are provided a "greenhouse" in which to live. Each organism retains its single-celled identity, and in concert they achieve the best of both the world of autotrophs and heterotrophs.

Studies of life cycles also show a mixture of plant and animal attributes. Some single-celled organisms originally classified as animals (heterotrophic and mobile) have an alteration of generations with a chromosome number similar to plants rather than animals.

The distinction between plant and animal, so readily recognizable in larger organisms, becomes blurred in the world of single-celled creatures. Plant and animal attributes can occur in a single cell. Consequently, the Kingdom Protista was established for single-celled eucaryotic organisms. At present, then, the Kingdoms Monera, Fungi, Protista, Animalia, and Planta are recognized.

Although many protist groups exist, only a few are of interest to paleontologists. These groups are mainly marine, and their combined fossil record extends from the Precambrian to the Holocene.

Of the marine protists studied by paleontologists, Part I surveys the radiolaria, foraminifera, diatoms, and coccolithophores. Of these the most thoroughly studied fossils are the foraminifera.

Radiolaria

Radiolaria are heterotrophs with cytoplasm divided into inner and outer portions by a membrane. They possess **pseudopodia** (nonpermanent extensions of cytoplasm for purposes of locomotion and feeding) and a shell or **test** made of silica. The radiolarian test is composed of an intricate lattice-like network of bars and spines. Their tests are either spherical (Figure 8.1) or helmet or bell shaped (Figure 8.2), and are about 50 to 200 μm in size. Asexual reproduction is common, and sexual reproduction has been reported but not confirmed. Their life cycle is from one to several months.

All modern radiolaria are planktic and occur only in the open ocean as part of the marine zooplankton. Their pseudopodia trap bacteria and other protists for nourishment. Abundances are commonly from hundreds to thousands of cells per liter of water. Some species are

stratified in the water column, and different assemblages occur at different depths. Living cells have been reported in water from depths in excess of 1000 m, but most species live in the uppermost few hundred meters. Like most plankton their abundances vary seasonally. Radiolaria are most abundant in equatorial water masses, but some species are also abundant in subpolar seas. As with most groups of organisms the greatest number of species is recorded from the tropics with fewer species at the higher latitudes.

After radiolarian cells reproduce asexually (leaving the parent test empty) or die, the tests fall to the ocean floor and accumulate in the sediments. In deep ocean basins such as that of the North Pacific, where calcareous tests are dissolved, the silica tests of the radiolaria are the principal components of the sediment. Some fossil deposits of chert are composed almost entirely of radiolarian tests. Radiolarians are used extensively for biostratigraphic correlation of oceanic sediments, particularly in those situations where calcareous tests have been removed through dissolution. Radiolaria are reported from the Precambrian, but their undisputed stratigraphic range is Cambrian to Holocene.

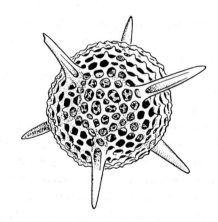

Figure 8.1. Spherical radiolarian, *Actinomma*.

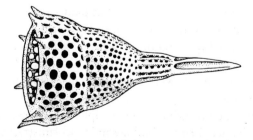

Figure 8.2. Helmet-shaped radiolarian, *Lamprocyclase*.

In older classifications the radiolaria were placed in the Kingdom Animalia, Phylum Protozoa. In modern classifications they are placed in the Kingdom Protista. Some authors place the radiolaria in the Phylum Sarcodina (a former class of the Protozoa), whereas others elevate the radiolarians to phylum status. Because they were originally classified in the animal kingdom, systematists follow the rules of nomenclature for animals.

Foraminifera

Foraminifera are heterotrophs with pseudopodia. They secrete a test composed of calcium carbonate or build a

Figure 8.3. Calcareous perforated foraminiferan, *Cibicides*.

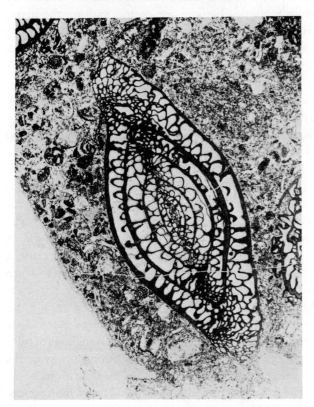

Figure 8.4. Thin section of a fusulinid, *Schwagerina*.

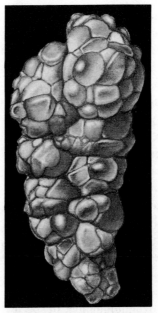

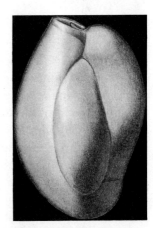

Figure 8.6. Calcareous imperforate foraminiferan, *Triloculina*.

Figure 8.5. Agglutinated foraminiferan, *Gaudryina*.

test composed of detrital sand grains, other foraminifera, sponge spicules, or whatever they can find. For study purposes they are divided into smaller foraminifera (about 200 to 600 μm) and larger foraminifera (about 0.5 mm to 10 cm). The smaller foraminifera are usually identified by their external characters (Figure 8.3), and the larger foraminifera by characters shown only in thin section (Figure 8.4).

The foraminifera of interest to paleontologists have tests with four basic kinds of wall structure. **Calcareous perforate** foraminifera have walls of secreted calcium carbonate with perforations (Figure 8.3). **Agglutinated** foraminifera have walls built of sand-sized material (Figure 8.5). **Calcareous imperforate** foraminifera have walls of secreted calcium carbonate without perforations (Figure 8.6). A fourth type of wall structure is called **microgranular** (Figure 8.4). These walls are composed of one or more layers of calcium carbonate particles cemented with a calcareous cement.

Reproduction is both asexual and sexual. Usually several asexual generations precede a sexual one. The reproduction time for a generation varies from a few weeks to more than a year, depending on the species and the suitability of environmental conditions.

Foraminifera are marine and either benthic or planktic. Modern benthic foraminifera are ubiquitous in all marine environments. Abundances of living benthic foraminifera are commonly about a million cells per square meter of sea floor. Abundances are not uniform throughout the year and exhibit maxima every few months. Different species assemblages characterize marshes, shelf, slope, and abyssal environments. Agglutinated foraminifera are the most common constituents of marsh and some abyssal environments. The calcareous perforate forms dominate shelf and slope environments in the higher latitudes. They are joined by numerous species of calcareous imperforate foraminifera in the lower latitudes. Consequently, the highest species diversity occurs in the tropics. The distribution of benthic foraminifera is probably better documented than that of any other kind of marine organism.

Planktic foraminifera are found only in the open ocean. Abundances are usually in the thousands of cells per liter of water and vary seasonally and geographically. The distribution of species assemblages has been correlated with the major water masses in the oceans. Far fewer species of planktic foraminifera exist than benthic species. Nevertheless, they exhibit the same trend in species diversity, with the maximum number of species living in tropical waters.

After death or reproduction the tests of benthic foraminifera become incorporated into the sediments and serve as useful paleoecologic and biostratigraphic indicators. In the search for oil, determination of sea level changes plays an important role. Because the benthic foraminifera are such good indicators of water depth, they are used extensively by petroleum geologists to determine the depth of water in which ancient sediments were deposited. The benthic foraminifera have a fossil record extending back to the Cambrian. They are more useful as biostratigraphic indicators in the upper Paleozoic to the Holocene.

Tests of planktic foraminifera like those of radiolaria sink to the bottom of the ocean after death or reproduction. Vast areas of the ocean floor are composed principally of the remains of planktic foraminifera. These deposits are referred to as *Globigerina* ooze. Based on their modern distribution and oxygen isotope analysis, the planktic forminifera are used to analyze ancient ocean currents and climates. The planktic foraminifera first appear in the Mesozoic but are most useful in the Cenozoic where their abundance and worldwide distribution have permitted an extensive biostratigraphic zonation unparalleled by any other group of organism.

Like the radiolaria, the foraminifera were originally classified in the Kingdom Animalia, Phylum Protozoa. Today most researchers place them in the Kingdom Protista and either give them phylum rank or place them in the Phylum Sarcodina.

Diatoms

Diatoms are generally autotrophic and synthesize energy from the sun through photosynthesis. They have bivalved shells of opalline silica called **frustules.** The frustule is made up of two unequal valves so that one fits over the other like a pill box or a Petri dish (Figure 8.7). Diatoms are either **centric** (circular) in shape (Figure 8.7) or **pennate** (elongated) with featherlike markings (Figure 8.8). Frustules are about 1.5 μm to 2 mm.

Asexual reproduction is accomplished by simple division of the frustule. The valves separate, and a new valve grows inside each of the separated valves. After a number of asexual generations, a sexual generation follows through formation of spores.

Diatoms are both planktic and benthic and occur in freshwater as well as in marine environments. They are a major source of food for heterotrophs, and are probably the most thoroughly studied of the living marine protists. Most planktic diatoms are centric and are the major primary producers in the photic zone (0 to 200 m) of the oceans. In offshore waters usually hundreds or thousands of cells occur per liter of water; nearshore and in estuaries the concentration of cells often reaches a million per liter. Their abundance varies greatly during the year, with large peaks in abundance referred to as blooms. Often a succession of species is observed, each bloom dominated by a different species.

Benthic diatoms are mostly pennate and live on sediments or on other organisms. They are among the first to colonize any available surface in the photic zone. They occur on top and bottom surfaces of polar sea ice where enough sun penetrates to permit photosynthesis and on rocks, shells and blades of sea grass. They cause a yellowish infestation on the skin of whales which sometimes penetrates the epidermis. Many species have the ability to glide over substrates, and they move horizontally as well as vertically. A square meter of sea floor in the photic zone usually contains millions of cells. Although their abundance varies during the year, the variation is not as marked as in the planktic forms.

The fossil record of diatoms extends from the Cretaceous to the Holocene. When environmental conditions are

Figure 8.7. Centric diatom, *Coscinodiscus.*

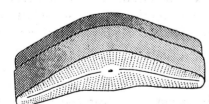

Figure 8.8. Pennate diatom, *Cymbella.*

favorable, diatoms can be extremely abundant, and their remains form deposits called diatomites. These deposits contain millions of frustules per gram of sediment. Such deposits are being formed today in the subarctic and subantarctic, and in some marginal marine basins overlain by nutrient-rich waters. In the north and equatorial Pacific, diatoms have been especially useful for biostratigraphic correlation of sediments from the Miocene to the Holocene. Terrestrial sequences have also been studied, but paleontologists have not given the diatoms the attention they deserve, and much research remains to be done.

Coccolithophores

Coccolithophores are generally autotrophic with a pair of equal-sized flagella and tests made up of many minute shields called **coccoliths** (Figure 8.9). The variation in structure of the coccoliths is complex and forms the basis for identification of species. Coccolithophores are about 3 to 25 μm, and individual coccoliths range from less than 1 to about 15 μm. A variety of calcareous planktic groups considered to be of plant origin that are between 5 to 60 μm are referred to as **nannoplankton**, and the coccolithophores are among these.

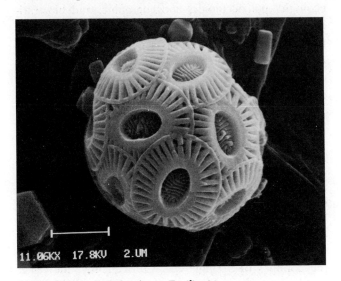

Figure 8.9. Coccolithophore, *Emiliania.*

The life cycle of coccolithophores is poorly known. The mode of reproduction most often observed is asexual, but sexual phases may occur also. In some species, alternate generations have morphologically dissimilar shields, making the designation of species extremely difficult.

Coccolithiphores are mostly planktic and nearly exclusively marine. Along with the diatoms they are the main constituents of the oceanic phytoplankton. Because they require the sun's energy for photosynthesis, they are restricted to the photic zone (0 to 200 m). Like most groups of organisms, cocccolithophores are most diverse in tropical waters, and only a few species occur in colder waters. In tropical waters, abundances are usually in the thousands of cells per liter of water while in colder waters a liter may contain a million cells. Studies on the distribution of coccolithophores are still in their infancy, and much remains to be learned about their distribution, abundance, and seasonal variation.

After life the coccolithophores fall apart into individual tiny shields, which, after reaching the bottom, become abundant constituents of marine sediments. Large areas of the ocean floor contain concentrations of coccoliths with millions per gram of sediment. A simply prepared laboratory slide contains thousands of shields and so these fossils are ideally suited for quantitative studies. Since the scanning electron microscope has become a standard laboratory instrument, the coccolithophores have received great attention. Their stratigraphic range is from Jurassic to Holocene, and zonations based on these fossils already exist for the Mesozoic and Cenozoic. Future studies will greatly refine and extend their usefulness.

Like the diatoms the coccolithophores were originally placed in the Kingdom Planta, Division Chrysophyta. Although now generally regarded as protists, the rules of nomenclature for plants are used because of tradition.

Origins

Procaryotic cells of the Kingdom Monera are the oldest known fossils. Evidence of blue-green, algal-like cells and possibly bacteria have been identified in Precambrian cherts from Australia that are 3500 million years old. Many disoveries of procaryotic cells from the fossil record have been reported, but documentation is most difficult because of the nature of preservation and possible contamination.

Eucaryotic cells probably evolved from procaryotic cells in the Late Precambrian, but exactly how is a matter of speculation. To fully understand the hypotheses of how such evolution might have taken place requires a knowledge of bacteriology and biochemistry beyond the scope of this book. A plausible hypothesis suggests that a symbiotic relationship between aerobic and anaerobic bacteria produced the first amoeba-like eucaryotic cells. Similar symbiotic relationships involving photosynthetic bacteria may have produced the first cells with true chloroplasts. From these early eucaryotic cells evolved the shelled protists and all multicellular organisms.

The earliest radiolaria were spherical. During the long history from the Cambrian until the present, the spherical radiolaria have remained rather conservative in their evolution. Helmet-shaped radiolaria do not appear until the Mesozoic, and their great diversification took place during the Cenozoic.

The foraminifera appear in the Cambrian as single-chambered, agglutinated species. In the Devonian, multi-chambered species and calcareous species evolved. By the Carboniferous, calcareous species were in abundance. Larger foraminifera called fusulinids underwent a rapid radiation in the Carboniferous and Permian, evolving into thousands of species. Like so many organisms, they became extinct at the end of the Permian. All the Paleozoic foraminifera were benthic. The planktic foraminifera evolved in the Mesozoic and underwent several radiations during the Cenozoic.

Centric diatoms first appeared in the Late Mesozoic and pennate forms first appeared in the Paleocene. During the Cenozoic, pennate diatoms evolved into thousands of species and today far surpass the centric forms in number of species.

Coccolithophores also first appear in the Early Jurassic. Their rapid evolution during the Cenozoic surpasses that of any other protist group. Consequently, they are ideal for biostratigraphy and are being intensively studied.

In modern oceans the diatoms, coccolithophores, helmet-shaped radiolaria, and foraminifera are abundant constituents of the plankton. They made their first appearance as plankton in the Late Mesozoic. What physical-chemical or biologic events prompted their emergence in the Mesozoic is an unsolved paleontologic problem. Certainly, the Paleozoic and Early Mesozoic seas must have been strikingly different from today's seas.

Supplementary reading

Brasier, M. D. *Microfossils*. London: George Allen & Unwin; 1980. Succinct summary of all the microfossil groups.

Haq, B. U.; Boersma, A., editors. *Introduction to marine micropaleontology*. New York: Elsevier North-Holland, Inc.; 1978. Provides a more detailed coverage of microfossils and their living counterparts.

Margulus, L. The origin of plant and animal cells. American Scientist **59**: 230–235; 1971. Easily understood review suggesting a mechanism for the origin of multicellular organisms and the need for more than two kingdoms.

Part II   Additional concepts

Smaller foraminifera

Martin A. Buzas

Living smaller foraminifera are overwhelming in diversity and density. In most places, a few milliliters (a tablespoonful) of marine sediment yields hundreds of foraminiferal tests. The majority of species are benthic, but some are planktic. Benthic species (infaunal and epifaunal) live in all marine environments from marshes and estuaries to the deepest abyssal basins. Planktic species live only in the open ocean but are worldwide in their distribution. When foraminifera die, their tests accumulate in the sediments. Indeed, large areas of the ocean floor are covered mainly by tests of planktic species (*Globigerina* ooze). Many tropical beaches are composed almost entirely of tests from shallow-water benthic species.

The geologic record of foraminifera begins in the Cambrian, but they did not become abundant until the Pennsylvanian. Their great abundance, worldwide distribution, and long fossil record make foraminifera ideal for ecologic, paleoecologic, zoogeographic, and biostratigraphic studies. Because of their small size, they are readily recoverable by subsurface drilling and have been the major group of fossils used by the oil industry for dating rocks. For these reasons, more paleontologists work on foraminifera than any other group.

The living cell

Single-celled protists were once thought to be simple and primitive. We now realize these organisms are not simple at all. Multicellular animals carry on the complex functions of locomotion, feeding, digestion, and reproduction through a division of labor among many different kinds of cells. Protists must carry on all these functions within the confines of a single cell. Consequently, the cell is highly complex, and great changes take place within it as the organism's activites change.

The living material of the cell is the **protoplasm.** Within the protoplasm are single or multiple **nuclei.** The nucleus is important in cell development and reproduction. Protoplasm exclusive of the nucleus, the **cytoplasm,** occurs both inside and outside of foraminiferal tests (Figure 8.10).

Feeding and locomotion are accomplished by pseudopodia made up of cytoplasm outside the test. Sometimes pseudopodia form single threadlike strands and at other times intricate nets in the same organism (Figure 8.10). They are often many times longer than the diameter of the test. Pseudopodial activity varies from species to species and at different times for the same individual. An individual may go for weeks without extruding any pseudopodia at all and then, for some unknown reason, become highly active. Viewed under a microscope the pseudopodia appear to be streaming because granules within the cytoplasm can be seen moving along the strands. In this way the cytoplasm can extend or retract and take on various shapes for locomotion and feeding.

Food consists mainly of diatoms and bacteria, but some species have been observed capturing small crustaceans in their pseudopodial net. Some species move food along the pseudopodia and into the test for digestion, whereas others digest their food outside of the test.

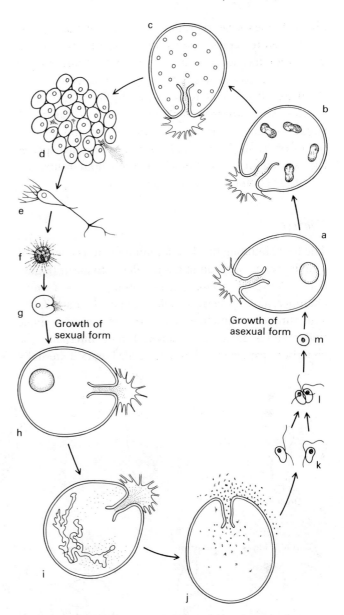

Figure 8.11. Life cycle of a species of *Iridia*. (Modified from LeCalvez, J. Andre des foraminiferes. In P. Grasse, editor. *Traite de zoologie*, Tome I. Paris: Libraries de L'Academie de Medecine; 1953, pp. 149–265.)

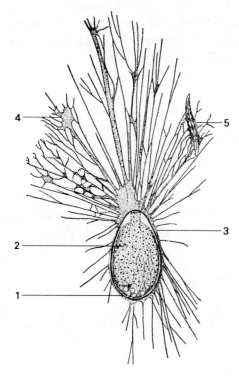

Figure 8.10. *Allogromia.* *1,* Test; *2* and *3,* cytoplasm; *4,* pseudopodia; *5,* entrapped diatom. (Modified from Schultze, M. S. Ueber den organismus der polythalamisn (foraminiferin) vesbt bemerkungen über die rhizopodin in allgemeinem, Leipzig. 68 p.)

Foraminifera reproduce sexually and asexually. Ideally alternation of generations is orderly (Figure 8.11). The adult asexual form (Figure 8.11, *a*) divides its nucleus into several nuclei just prior to asexual reproduction (Figure 8.11, *b* and *c*). The cytoplasm is portioned out among the nuclei and expelled from the test (Figure 8.11, *d*). For several hours to a day, the young have a planktic

existence after which they settle to the bottom to become adult sexual forms (Figure 8.11, *h*). When ready, they reproduce sexually. The nucleus disintegrates (Figure 8.11, *i*) to form millions of **gametes** (reproductive cells), which are ejected into the water (Figure 8.11, *j* and *k*). The gametes from different parents unite to form a zygote (fertilized cell) (Figure 8.11, *m*), which grows into an adult, and the process repeats itself. The complete life cycle requires about one year.

Although life cycles are known for only about 20 or 30 species, each has a variation on the basic theme presented above. Some species reproduce asexually for many generations before reproducing sexually. In others a sexual generation has never been observed. Individuals of some species unite to exchange gametes; others eject them into the water. Planktic phases occur in the life cycle of some species, while others remain totally benthic or planktic. A single nucleus occurs in both the sexual and asexual forms of *Iridia* (Figure 8.11), but in many other species the asexual form is multinucleate throughout its life, and only the sexual form has a single nucleus.

The test

The basic building block of foraminifera tests consists of a cavity with a surrounding wall called a **chamber** (Figure 8.12). Although a few species consist of only a single chamber, most species are multichambered (Figure 8.12). The simplest multichambered arrangement is a single linear series forming a **uniserial** test. Internally the chambers are separated by walls called **septa**. Externally

a line or junction forms where the septa meet the chamber walls. This external line formed between two chambers is called a **suture**. In addition to a uniserial arrangement, **biserial** and **triserial** chamber arrangements are common. Instead of forming a straight series of chambers, some foraminifera coil. Each volution in a coil (through 360 degrees) is called a **whorl**. If the test coils in a single plane (that is, the chambers are centered on the plane), the coil is called **planispiral**. Because of the bilateral symmetry, both sides of the test will appear identical. If the test coils in a spire, like a snail, the coil is called **trochospiral**. A raised area in the center of a coil is called an **umbo** and a depression, an **umbilicus**. A test in which earlier chambers become enveloped by later ones is called **involute**. One in which chambers from a previous whorl are visible is called **evolute**. In many trochospiral forms the spiral side is evolute and the umbilical side is involute. Some foraminifera add their chambers in several planes. A common arrangement in which five chambers are visible is called **quinqueloculine**.

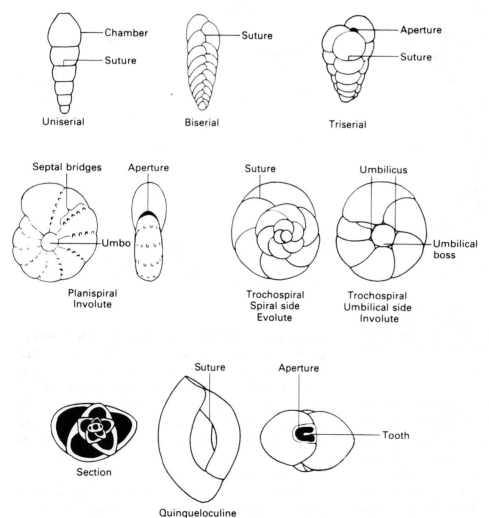

Figure 8.12. Terminology used in describing the test.

Characteristically foraminifera have an opening or **aperture** on the terminal chamber through which cytoplasm can be extruded. As new chambers are added, the aperture of the previous chamber remains open providing a continuous pathway throughout the test. The aperture can have a variety of shapes and sometimes is multiple. In some cases a platelike structure called a **tooth** occurs within the aperture.

The composition and structure of foraminiferal walls are important in the classification of foraminifera. Foraminifera have either organic, agglutinated, or calcareous walls. Organic walls occur in relatively few species (Figure 8.10). They are usually brown in color, semitransparent, and fragile. Organic walls are usually thin, and in some species they are destroyed simply by drying. Because researchers commonly dry samples as part of their preparation, species with organic walls are often not recorded from recent assemblages even though they are present. Because they contain no hard parts, the fossil record of foraminifera with organic walls is poor.

Many species have agglutinated walls. These organisms usually cement mineral grains together to form their chambers. A few species select more exotic substances such as sponge spicules or other foraminifera to make their tests.

Calcareous walls occur in the greatest number of species. These walls are formed through the secretion of calcium carbonate by the organisms. Calcareous walls may be either perforate or imperforate. Perforate walls have holes or pores through the chamber walls. Earlier workers believed these pores served as passageways for pseudopodia. Recent observations using electron microscopy reveal, however, that the pores are plugged by an organic membrane, which sometimes is itself perforated. The function of pores is poorly understood. They may serve as an exchange for gaseous or other material outside the test or may be important in the calcification process. Imperforate walls do not contain pores and are often called **porcelaneous** because when viewed under reflected light the walls look like porcelain china.

Examination of perforate walls by polarizing optical microscopes and electron microscopes indicates the crystals making up the walls have a preferred orientation in some species and in others a haphazard one. When viewed under the electron microscope, porcelaneous walls appear like a brick pavement with a haphazard arrangement of crystals. The kinds of wall structures and their significance are an active area of research.

Classification

Numerous classifications have been proposed since d'Orbigny first classified the foraminifera in 1826. The most recent and most widely used is in the *Treatise on Invertebrate Paleontology* (1964). A recent revision (1974) of that classification contains five suborders, 19 superfamilies, and 118 families. The number of names for genera is estimated to exceed 3000 and for species, 40,000. A recent survey indicates about 40 genera are being added per year.

Table 8.1. Classification of Foraminifera.

Allogromids	Test organic. Upper Cambrian to Holocene. Example: *Allogromia* (Figure 8.10).
Textularids	Test composed of agglutinated material cemented together by animal. Cambrian to Holocene. Examples: *Trochammina* (Figure 8.16), *Ammobaculites* (Figure 8.17), *Reophax* (Figure 8.27).
Fusulinids	Larger foraminifera with test composed of microgranular calcium carbonate, some with two or more wall layers. Ordovician to Permian.
Rotalids	Test calcareous, perforate. Permian to Holocene. Examples: *Globigerina* (Figure 8.13), *Globorotalia* (Figure 8.14), *Orbulina* (Figure 8.15), *Ammonia* (Figure 8.18), *Elphidium* (Figure 8.19), *Rosalina* (Figure 8.20), *Cassidulina* (Figure 8.22), *Nonion* (Figure 8.23), *Nonionella* (Figure 8.24), *Bolivina* (Figure 8.25), *Lagena* (Figure 8.28).
Miliolids	Test calcareous, imperforate, porcelaneous. Pennsylvanian to Holocene. Examples: *Quinqueloculina* (Figure 8.21), *Pyrgo* (Figure 8.26).

At the suborder (Table 8.1) and superfamily levels the current classification is based on the structure of the foraminiferal wall. Both microstructural and optical properties are taken into account. Research since the publication of the *Treatise* indicates wall structures are much more complex than previously thought. Indeed, species within the same genus may possess very different wall structures. Under the current classification these species would be placed in separate superfamilies.

This is so because, although few foraminiferal researchers agree on a classification, little research is being conducted on systematics above the genus level. Most foraminiferal research is directed toward solving geologic, ecologic, paleoecologic, and biostratigraphic problems. The taxon most frequently used in this research is the species. Table 8.1 lists, defines, and gives examples of genera for each of the five suborders proposed in the *Treatise on Invertebrate Paleontology*. At this level most foraminiferal workers would not object to the classification, but many would prefer to regard the suborders as informal groups.

Distribution of modern foraminifera

The distribution of modern species is well documented, and comparison of ancient faunas, benthic and planktic, with their modern counterparts facilitates the reconstruction of ancient environments.

Distribution of modern planktic foraminifera Planktic foraminifera are distributed throughout the world oceans. They occur only in the open ocean, however, and are not found living near shore or in estuaries or lagoons. When they die, their tests accumulate on the bottom along with those of the benthic foraminifera. The ratio of planktic to benthic tests increases with distance from shore, and in deep sea sediments, planktic tests outnumber benthic ones by one or two orders of magnitude.

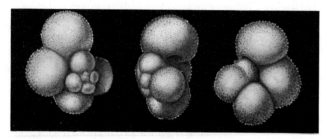

Figure 8.13. *Globigerina.* Test trochospiral, evolute on spiral side; chambers spherical; wall calcareous, perforate; surface smooth, pitted or spinose; aperture large, umbilical. (Paleocene to Holocene.)

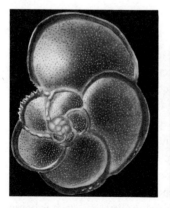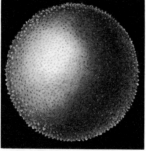

Figure 8.14. *Globorotalia.* Test trochospiral, evolute on spiral side, evolute on umbilical; chambers angular, periphery keeled; wall calcareous, perforate; aperture a marginal umbilical arch. (Paleocene to Holocene.)

Figure 8.15. *Orbulina.* Test trochospiral, globigerine in juvenile, adult spherical composed of a single chamber enveloping previous ones; wall calcareous, perforate; aperture not visible in adult. (Lower Miocene to Holocene.)

Only three or four species belonging to the planktic genus *Globigerina* (Figure 8.13) are abundant in Arctic waters. In tropical waters, species of *Globigerina* are

joined by species belonging to the planktic genera *Globorotalia* (Figure 8.14), *Orbulina*, (Figure 8.15), and others. Planktic genera have spherical or flattened chambers to facilitate their floating lifestyle. Because planktic organisms are floaters, they necessarily must go where oceanic currents carry them. A huge clockwise circulatory pattern exists in the North Atlantic, extending from tropical waters northward along the coast of North America and then eastward at about the latitude of Labrador toward the British Isles. From the British Isles, currents flow southward toward Africa. Because of this gyre, tropical forms are consistently found at higher latitudes on the western side of the North Atlantic than on the eastern side. Today, tropical forms such as *Globorotalia* are found in the North Atlantic gyre from about 25 degrees north off Africa to about 40 degrees north off Newfoundland.

Distribution of modern benthic foraminifera In the open ocean the number of benthic species is much greater than in estuaries and lagoons. As with most groups of organisms, species diversity increases with a decrease in latitudes, and highest species diversity occurs in the tropics. The higher latitudes are dominated by species belonging to the rotalids and textularids (Table 8.1) in the lower latitudes species belonging to the miliolids are also abundant. Species diversity also increases with depth. A typical pattern exhibits an increase in species diversity from shallow waters to the edge of the continental shelf, a decrease on the continental slope, and another increase at abyssal depths.

Marshes, estuaries, and lagoons are usually referred to as marginal marine environments. The typical marine marsh is covered by vegetation and is barely under water at high tide. Changes in abiotic variables such as temperature and salinity are extreme. Marsh environments are inhabited by only a few species, and most of them belong to the agglutinated genera *Trochammina* and *Ammobaculites* (Figures 8.16 and 8.17).

Abiotic variables also vary greatly in estuaries and lagoons. Species diversity in these environments is greater than in marshes, and calcareous species are usually more abundant than agglutinated ones. Species belonging to the perforate genera *Ammonia*, *Elphidium*, and *Rosalina* are common (Figures 8.18 to 8.20).

In the open ocean benthic foraminifera exhibit a striking depth zonation. The near-shore part of the neritic zone extends from low tide to about 30 to 50 m in depth. Here foraminifera are subjected to turbulence and high variability of abiotic variables. Not surprisingly, the fauna closely resembles those found in estuaries and

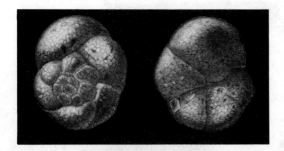

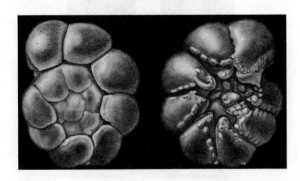

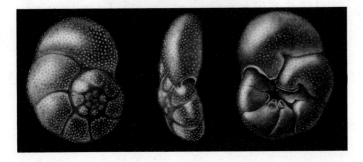

lagoons. Species belonging to the perforate genera *Elphidium, Rosalina, Ammonia* and the imperforate genus *Quinqueloculina* usually dominate. (Figures 8.18 to 8.21).

Farther offshore in the neritic zone (50 to 200 m) seasonal effects become minimal. Perforate genera such as *Cassidulina, Nonion,* and *Nonionella* are often dominant (Figures 8.22 to 8.24).

The bathyal zone extends from about 200 to 2000 or 3000 m in depth, and variations in abiotic variables are small. Species belonging to the perforate genera *Bolivina* (Figure 8.25) and *Cassidulina* (Figure 8.22) and the imperforate genus *Pyrgo* (Figure 8.26) are abundant.

The abyssal zone extends from about 2000 or 3000 m to the great oceanic depths of over 7000 m. At these great depths abiotic variables are constant over long periods of time. In the deepest parts of the ocean, calcium carbonate is unstable, and the foraminiferal fauna occupying these depths is often dominated by agglutinated species. Of the agglutinated genera illustrated here, species belonging to *Reophax* are present (Figure 8.27). At those abyssal depths where calcium carbonate is not dissolved, calcareous species belonging to the perforate genera *Cassidulina* (Figure 8.22) and *Lagena* (Figure 8.28), among others, are important constituents.

Evolution

The fossil record of foraminifera extends throughout the Phanerozoic. Like most organisms, their history is not one of continuous radiation from simple to more complex and diverse forms. Rather, they exhibit explosive periods of evolution followed by massive extinctions.

Figure 8.16. *Trochammina.* Test trochospiral, evolute on spiral side, involute on umbilical side; wall agglutinated; aperture a low arch. (Carboniferous to Holocene.)

Figure 8.17. *Ammobaculites.* Test close coiled, planispiral in early portion, later uncoiled; wall agglutinated; aperture terminal, rounded. (Carboniferous to Holocene.)

Figure 8.18. *Ammonia.* Test biconvex, trochospiral; wall calcareous, perforate; umbilicus with boss or bosses, irregular granules around umbilicus and sutures; aperture on margin. (Oligocene to Holocene.)

Figure 8.19. *Elphidium.* Test planispiral, involute; chambers with distinct sutures having septal bridges; wall calcareous, perforate; aperture single or multiple. (Eocene to Holocene.)

Figure 8.20. *Rosalina.* Test trochospiral, evolute on spiral side, flattened and involute on umbilical side; wall calcareous, perforate; aperture an arch at base of final chamber with flaps covering portions of umbilicus. (Holocene.)

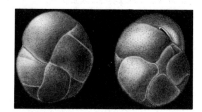

Figure 8.21. *Quinqueloculina.* Test coiled, chambers alternating in five planes so that three chambers are externally visible on one side of test and four on other; wall calcareous, porcelaneous, imperforate; aperture terminal, often with tooth. (Jurassic to Holocene.)

Figure 8.22. *Cassidulina.* Test biumbonate, biserially coiled; wall calcareous, perforate; aperture an elongate slit. (Eocene to Holocene.)

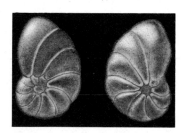

Figure 8.23. *Nonion.* Test planispiral, involute; wall calcareous, perforate; aperture an arched slit. (Paleocene to Holocene.)

Figure 8.24. *Nonionella.* Test slightly trochospiral, evolute on spiral side, involute on umbilical side; wall calcareous, perforate; aperture an arched slit. (Upper Cretaceous to Holocene.)

Benthic evolution The first known foraminifera appeared in the Cambrian and were agglutinated with a single chamber. Over 100 million years passed before species with multichambers and a secreted calcareous wall evolved in the Devonian. During the Carboniferous and Permian an explosive radiation took place, especially among larger foraminifera (see "Larger Foraminifera"). At the end of the Paleozoic, massive extinctions occurred, and the evolution of foraminifera was stalled until the Late Triassic-Jurassic. The calcareous groups so prevalent today made their appearance during the remainder of the Mesozoic. Unlike many other groups, the benthic foraminifera did not undergo drastic changes at the Cretaceous–Paleozoic boundary. The early Cenozoic was, however, a time of massive extinctions, and by the Miocene the modern fauna had evolved.

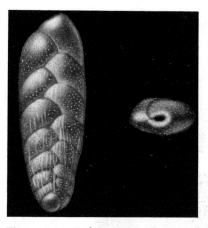

Figure 8.25. *Bolivina.* Test elongate, biserial; wall calcareous, perforate; aperture terminal, forming an elongate loop. (Cretaceous to Holocene.)

Figure 8.26. *Pyrgo.* Test quiqueloculine in juvenile, two chambered in adult; wall calcareous, porcelaneous, imperforate; aperture terminal with tooth. (Jurassic to Holocene.)

Figure 8.27. *Reophax.* Test elongate, straight to slightly curved, uniserial, chambers few; wall agglutinated; aperture terminal, at end of a neck. (Pennsylvanian to Holocene.)

Figure 8.28. *Lagena.* Test with single flask-shaped chamber, rarely two; wall calcareous, perforate; aperture terminal, often on an elongate neck. (Jurassic to Holocene.)

Planktic evolution Although the evolution of planktic foraminifera is highly complex, the salient features can be understood by considering the evolution of three basic **morphotypes** (morphologic forms) represented by three genera studied in this chapter. The first is called the globigerine morphotype and consists of a trochospiral coil with an umbilical aperture like *Globigerina* (Figure 8.13). The second morphotype is called orbuline and consists of an early globigerine stage followed by envelopment of the entire test by the last chamber of spherical shape like *Orbulina* (Figure 8.15). The third morphotype is called globorotalid and consists of angular chambers with a keel like *Globorotalia* (Figure 8.14).

During the Paleocene only globigerine morphotypes existed. In the Late Paleocene and Early Eocene a great radiation took place, and many species of the orbuline and globorotalid morphotypes flourished (Figure 8.29). Extinctions began in the Late Eocene and persisted throughout the Oligocene. The globigerine morphotypes were unaffected by the mass extinctions, and only the orbuline and globorotalid morphotypes disappeared. In the Early Miocene another great radiation began, and orbuline and globorotalid morphotypes evolved again. They continue to flourish in modern seas. The precise documentation of the species involved in these great radiations and extinctions provides a biostratigraphic zonation of the Cenozoic.

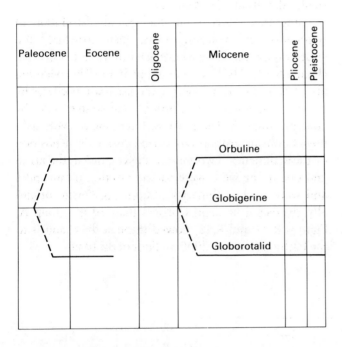

Figure 8.29. Distribution of planktic morphologies in the Cenozoic. (Modified from Cifelli, R. *Systematic Zoology,* vol. 18. Washington, D.C.: Smithsonian Institution; 1969, p. 162.)

This kind of evolution is called **iterative** and is an example of heterochronous convergence because the same ancestral stock (globigerines) gave rise to morphologically similar descendants at different times. Iterative evolution is common among other phyla as well. The reasons why the same morphotypes reappear is obscure because the adaptive significance of the test morphology of the foraminifera is unknown. The distribution of these morphotypes in modern seas suggests, however, that these extinctions and radiations are related to global climatic patterns and oceanic circulation.

Examples of paleontologic studies

Foraminifera have been applied widely to problems in biostratigraphy, in the reconstruction of ancient environments, or in combinations of the two. Biostratigraphic zonation of the Cenozoic using foraminifera, for example, is based largely on knowledge of the evolution of planktic foraminifera just described. The correlation of these foraminiferal zones with those of other organisms and radiometric dates provides an accurate chronology for the Cenozoic. Reconstruction of ancient sedimentary environments is based on our detailed knowledge of the distribution of modern planktic and benthic foraminifera and comparison with ancient faunas.

During the past few decades an increasing number of investigations have concentrated on the history of the world oceans. Cores have been taken from numerous sites at oceanic depths all over the world. Analyses of foraminiferal faunas have contributed greatly to correlating between cores, reconstructing ancient environments, and unraveling the history of the world oceans.

Thousands of studies using foraminifera have been published. Two examples of their usefulness follow.

Oceanic circulation Miocene sediments from cores drilled at two sides of the Atlantic, one near Greenland

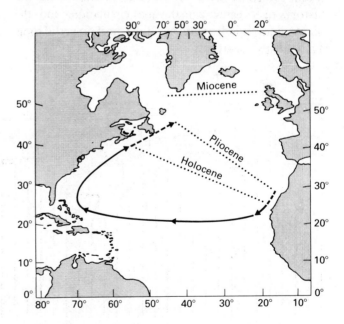

Figure 8.30. Clockwise displacement of tropical species of planktic foraminifera occurs in Pliocene and Holocene. In Miocene, tropical species extended much farther northward. The dotted lines indicate northern extent of tropical planktic species for each series. (After Cifelli, R. Reprinted by permission from *Nature,* Vol. 264, No. 5585, p. 431. Copyright © 1976 Macmillan Journals Limited.)

and the other near the British Isles, contain many tropical planktic species (Figure 8.30). Tropical species are also found abundantly in the Mediterranean during the Miocene.

No tropical species are found in Pliocene and younger sediments from the Greenland and British Isles sites or in the Mediterranean. Similarly, a core drilled in the Bay of Biscay off France contains no tropical Pliocene species. Tropical species are abundant south of these localities, however, in Pliocene sediments from cores drilled near Labrador (about 55 degrees north) and just south of the Mediterranean (about 30 degrees north). Tropical species are found from about 25 degrees north off Africa to about 40 degrees north off Newfoundland (Figure 8.30). As discussed earlier, the asymmetrical distribution of planktic species is due to the huge clockwise circulatory pattern of the North Atlantic gyre.

These observations indicate that during the Miocene the North Atlantic had a much wider dispersal of tropical species, and possibly their distribution was symmetrical on both sides of the Atlantic. A great reduction in the distribution of tropical species had taken place by the Pliocene and the east-west distribution was asymmetrical although farther north than today. The reasons for this major change in the circulation pattern of the North Atlantic are unclear, but it does represent a long-term trend. Much remains to be done in unraveling the history of oceanic climate and circulation, and the planktic foraminifera are one of the most powerful tools available to researchers.

Reconstruction of an ancient marine basin   Miocene deposits from New Zealand provide an example of how an ancient ocean basin can be reconstructed using benthic foraminifera. Numerous samples containing over 300 species were analyzed by a cluster analysis procedure similar to that in Chapter 6, using species as characteristics and samples as taxa. Five sample groups are evident in the dendrogram shown in Figure 8.31. Many species in these groups have living counterparts, and in modern oceans these assemblages live within particular depth ranges. The species diversity of modern assemblages within these depth ranges is also similar to the species diversity of the groups in the dendrogram. The inescapable conclusion is that the groups in the dendrogram represent species assemblages that lived at particular depths in an ancient sea.

Using this information, the spatial distribution of sediments and volcanic rocks, their structure, and stratigraphy, the Early Miocene basin was reconstructed (Figure 8.32). The basin was about 50 to 80 km wide and about 100 to 150 km long. On the east it was bordered by a land mass and on the northern and western sides by volcanic piles. Volcanic rocks intermingled with sediments indicate eruptions occurred during the deposition of the sediments. Two marine shelves, one in the north and one in the west, were adjacent to the volcanic piles and were cut by submarine channels. Sediments originally deposited at depths of less than 50 m (group A, Figures 8.31 and 8.32) flowed through the channels to the bathyal (1000 to 2000 m) floor of the basin.

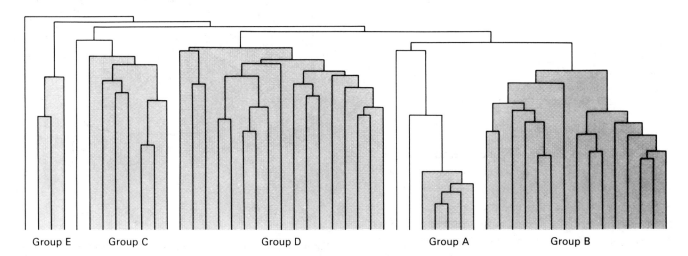

Group E   Group C   Group D   Group A   Group B

Figure 8.31. Dendrogram of benthic foraminiferal samples from Early Miocene of New Zealand. Paleobathymetric analyses infer water depths at the time of original deposition as follows: group A, 0 to 50 m; group B 50 to 200 m; group C 200 to 1000 m; group D, 150 to 2000 m; group E, 20 to 100 m.

(Modified from Hayward, B. W.; Buzas, M. A. Taxonomy and paleoecology of Early Miocene benthic foraminifera of northern New Zealand and the North Tasman Sea. *Smithsonian Contributions to Paleobiology*, No. 36, pp. 1–154; 1979.)

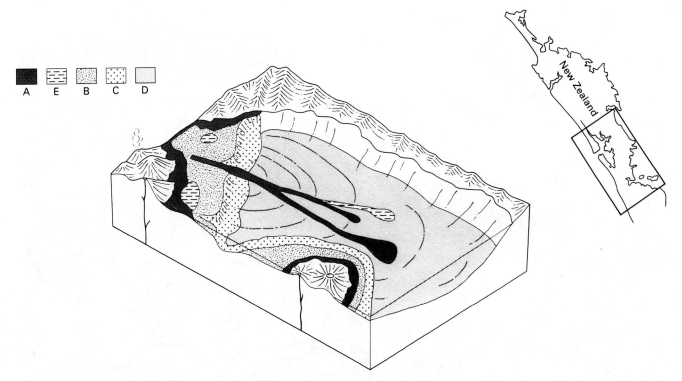

Figure 8.32. Interpretation of paleogeography in Early Miocene of New Zealand. Groups are the same as those shown in Figure 8.31. Note the downward displacement of shallow water sediments (group A and E) to the bathyal sea floor. (Modified from Hayward B. W.; Buzas, M. A. Taxonomy and paleoecology of Early Miocene benthic foraminifera of northern New Zealand and the North Tasman Sea. *Smithsonian Contributions to Paleobiology*, No. 36, pp. 1–154; 1979.)

Before the foraminifera were studied, geologists believed all the sediments in this Miocene basin were deposited in relatively shallow water (neritic). Analysis of the benthic foraminifera provided the necessary information to reconstruct the basin with fidelity.

Supplementary reading

Boltovskoy, E.; Wright, R. *Recent foraminifera*. Dr. Junk b.v., The Hague; 1976. The most up-to-date text on modern foraminifera. All aspects are treated and the bibliography is excellent.

Broadhead, T. W., editor. *Foraminifera: notes for a short course*. University of Tennessee Department of Geological Sciences Studies in Geology 6; 1982. Experts from the various fields of foraminiferal studies summarize current progress for the nonexpert.

Cushman, J. A. *Foraminifera, their classification and economic use*. Cambridge, MA: Harvard University Press; 1955. The classic text on foraminifera written by the "father" of North American foraminiferal studies.

Haynes, J. R. *Foraminifera*. London: Macmillan Publishers Ltd; 1981. The most up-to-date text book on all aspects of foraminifera.

Hedley, R. H.; Adams C. G., editors. *Foraminifera*, vols. 1–3. New York: Academic Press, Inc.; 1974, 1976, 1978. Ongoing series dealing with all aspects of foraminiferal research. Specific subjects are summarized by experts in the field.

Loeblich, A. R., Jr.; Tappan, H. *Treatise on invertebrate paleontology, c: Protista 2, Sarcodina chiefly "Thecamoebians" and Foraminiferida*, Vols 1 and 2. Lawrence, KS: University of Kansas Press, 1964. The most accepted classification of foraminifera and an indispensable reference to foraminiferal genera.

Murray, J. W. *Distribution and ecology of living benthic foraminiferids*. New York: Crane, Russak & Co., Inc.; 1973. Summary of all the literature on ecology of living foraminifera.

Phleger, F. B. *Ecology and distribution of recent foraminifera*. Baltimore: Johns Hopkins Press; 1960. An excellent summary of modern foraminiferal distributions by the pioneer of foraminiferal ecology.

Larger foraminifera

Raymond C. Douglass

The larger foraminifera include a wide variety of forms that are not closely related taxonomically. The distinction between the larger and smaller foraminifera is made primarily on traditional methods of study. The larger foraminifera include those that are of such internal complexity that they are studied in thin section, whereas

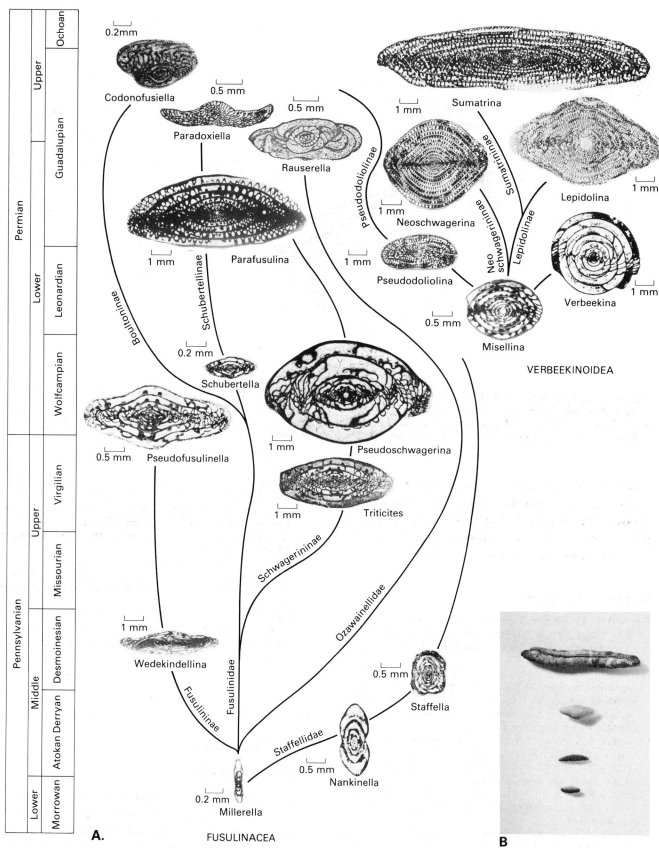

Figure 8.33. Fusulinid phylogeny. **A,** Simplified phylogeny of two superfamilies illustrated with few representative genera, all shown in axial section. Smaller forms shown at greater magnification. **B,** Exterior views of typical fusulinids shown at natural size.

the smaller foraminifera are studied primarily as whole specimens. Some Paleozoic larger foraminifera attain lengths of at least 60 mm; and some Tertiary larger foraminifera have diameters of about 100 mm.

Larger foraminifera are widely distributed geographically. The late Paleozoic forms show a distribution in modern geography from above the Arctic Circle to 52 degrees south, but the actual latitudinal range may have more closely approximated the distribution of Holocene larger foraminifera that are found only in tropical to temperate seas. Most larger foraminifera are of post-Paleozoic age. The fusulinids include one of the largest groups of larger foraminifera and are exclusively of Late Paleozoic age (Figure 8.33). They first appeared in Early Pennsylvanian time and died out before the end of the Permian. Their widespread distribution and rapid evolution during that time provide a basis for detailed correlations. A zonal scheme based on the fusulinid genera has proven especially useful for rocks of middle Pennsylvanian through middle Permian age. The other large foraminifera started during the Mesozoic and developed abundantly through the middle Tertiary with some continuing to the present. They are most useful for correlation of rocks of Late Cretaceous to Miocene age.

The living cell

The biology of the larger foraminifera is similar to that of foraminifera in general as described in the preceding section. The larger size of the individuals does not appear to hamper locomotion, and specimens up to 3 cm in diameter are known to move around on the substrate and to climb vegetation, like turtle grass, for feeding on diatoms and algae. Some larger foraminifera ingest photosynthetic algae (zooxanthellae), which then live within the cytoplasm of the foraminifera in a symbiotic relationship.

The test

Most larger foraminifera are planispiral and are either evolute (*Millerella* of Figure 8.33) or involute (the remainder of Figure 8.33). Some exceptions are uniserial, as in some textularids (Figure 8.34, *C*).

The composition and structure of the foraminiferal walls are especially important in the larger foraminifera. Agglutinated, microgranular, perforate, and imperforate walls all occur in larger foraminifera and are studied in great detail. The agglutinated matter in textularids (Figure 8.34) is held together by organic cements. The microgranular walls of fusulinids are simple or multilayered (Figure 8.35).

Fusulinids are all multichambered. Growth starts with an initial chamber called the **proloculus** (Figure 8.36, *B* and *C*) that tends to be spherical. Specimens of sexual generations start with a small **microspheric** proloculus (Figure 8.36, *B*) and specimens of asexual generations start with a large **megalospheric** proloculus (Figure 8.36, *C*). Chambers are added to form the coiled shell by extension of the spiral wall, the **spirotheca** (Figure 8.36, *C*), outward then down to enclose a chamber. As each chamber is added, the leading wall becomes the **antetheca** and the newly enclosed antetheca becomes a **septum** (Figure 8.36, *C*). The antetheca has no aperture, but there are many **septal pores** through which the pseudopodia extend for feeding and locomotion (Figure 8.35, *A*). After several chambers are added, a central **tunnel** is resorbed through the septa (Figure 8.36), allowing freer communication between chambers. The resorbed shell material is commonly redeposited as ridges called **chomata**, on each side of the tunnel (Figure 8.36, *A* and *B*). The septa are essentially straight in many of the primitive fusulinids. They develop **fluting** in more advanced forms (Figure 8.37), and **chamberlets** develop where opposing folds touch and partly divide the chambers. Resorption of the septa where opposing folds touch opens **cuniculi** as passages between chambers (Figure 8.37).

Classification

The larger foraminifera include representatives of every suborder of the foraminifera except the allogromids (Table 8.1). Classification of the larger foraminifera has proceeded at an unequal pace among the various groups because some forms are more widely used in stratigraphy than others. For example the fusulinids (Figure 8.33) are important in Late Paleozoic biostratigraphy and have, therefore, been studied intensely. They are assigned to five families, eight subfamilies, and about 80 genera. The classification in this group is relatively stable with minimal changes and few additions at the generic level.

The textularids (Figure 8.34) include larger foraminifera in two families, five subfamilies, and 32 genera of Mesozoic and Cenozoic age. Few forms are of significant stratigraphic interest, and the group is relatively neglected in the published record.

Larger foraminifera among the miliolids (Figure 8.38) are assigned to three families, four subfamilies, and 32 genera of Mesozoic and Cenozoic ages. Some of this group (the alveolinids, Figure 8.38, *B*) bear a striking

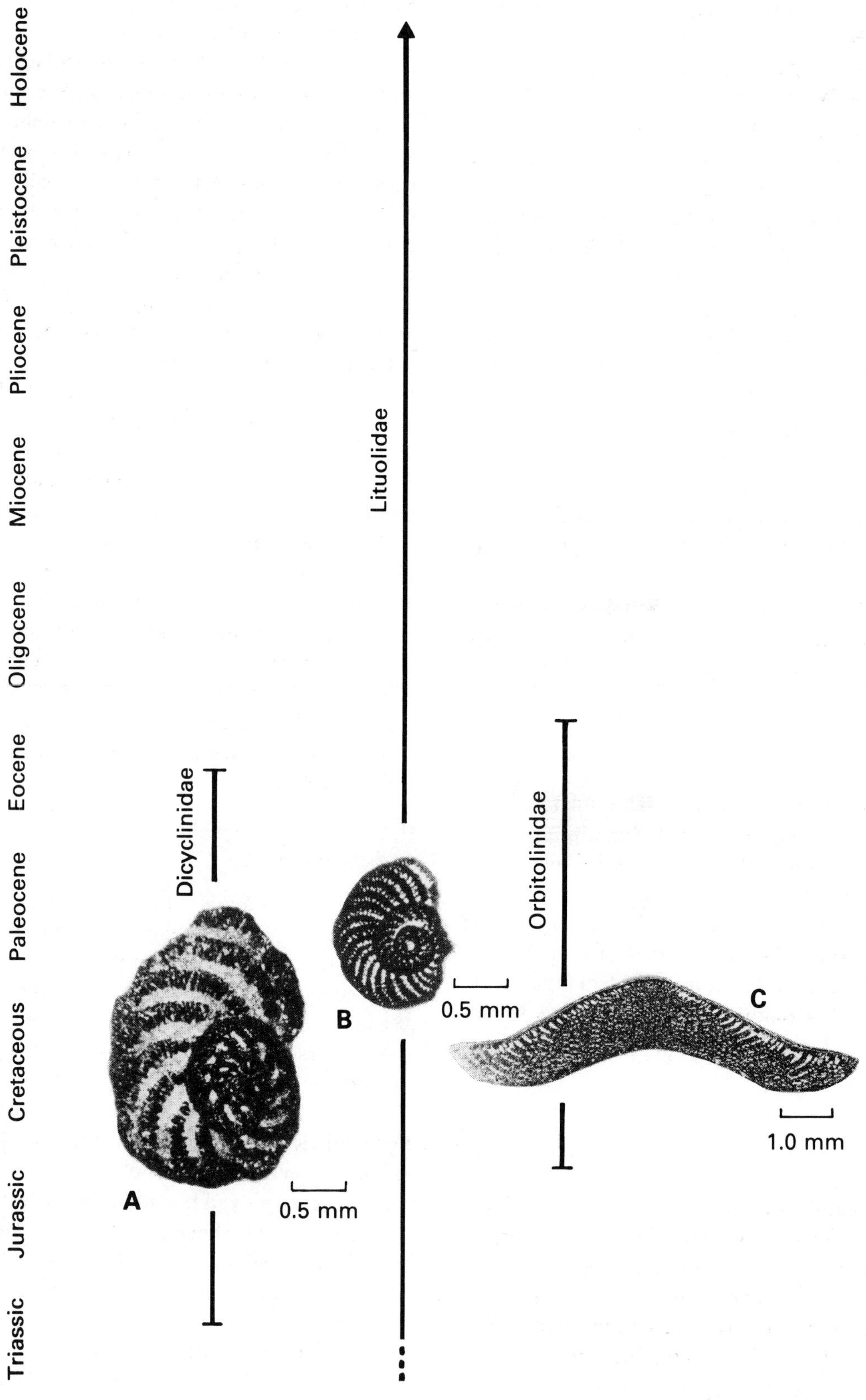

Figure 8.34. Textularids. **A,** Sagittal section of *Pseudocyclammina* (Jurassic). **B,** Sagittal section of *Choffatella* (Cretaceous). **C,** Axial section of *Orbitolina* (Early Cretaceous).

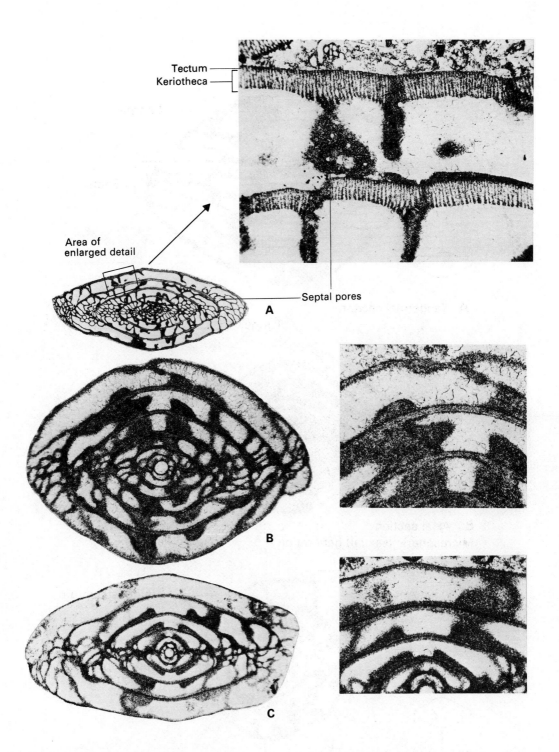

Tectum
Keriotheca

Area of
enlarged detail

Septal pores

A

B

C

Figure 8.35. Wall structures. Detailed structure of wall changes progressively through time. Three stages in development illustrated. **A,** Axial section of *Triticites* and enlarged detail of one area showing complex wall with thin upper layer and thick layer of honeycomb structures. Septal pores common in this specimen. (Upper Pennsylvanian.) **B,** Axial section of *Fusulinella* showing wall composed of middle lighter layer with darker layers above and below. (Middle Pennsylvanian.) **C,** Axial section of *Profusulinella* showing relatively simple wall with thin dark layer and lighter inner layer. A thin deposit on the outer surface in some volutions is a secondary deposit. (Lower part of Middle Pennsylvanian.)

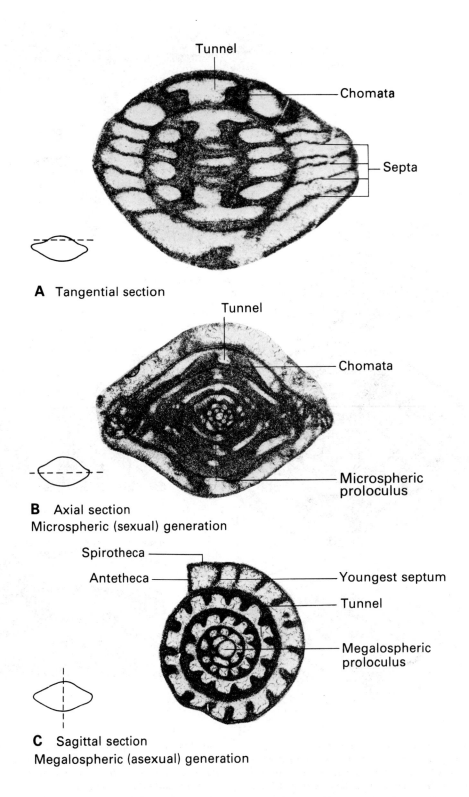

A Tangential section

B Axial section
Microspheric (sexual) generation

C Sagittal section
Megalospheric (asexual) generation

Figure 8.36. Microspheric and megalospheric generations of *Fusulinella*. Tangential (**A**), axial (**B**) and sagittal sections (**C**). Tangential section is similar in both microspheric (sexual) and megalospheric (asexual) forms because diagnostic proloculus is not in section. Microspheric form starts with small proloculus. Coiling commonly occurs in one plane; then the axis rotates about 90 degrees and coiling continues in new plane. Megalospheric form starts with a relatively large proloculus and coiling generally is all in one plane.

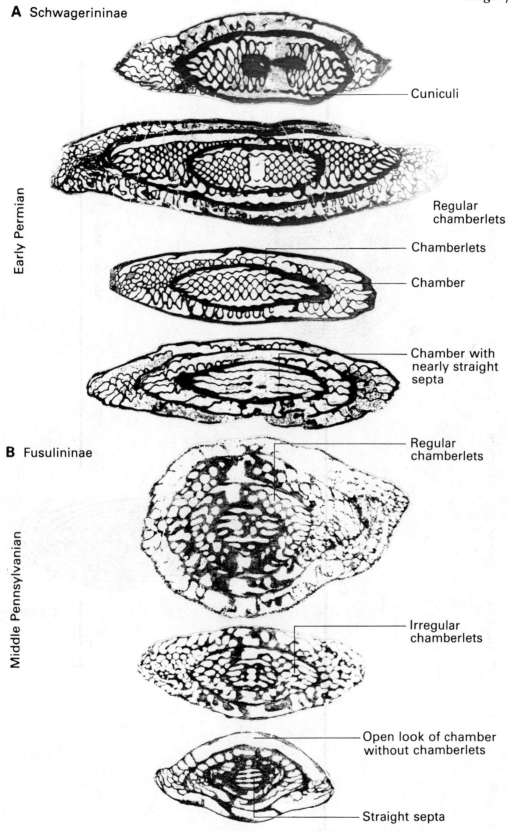

A Schwagerininae

Early Permian

— Cuniculi

Regular
chamberlets

— Chamberlets

— Chamber

Chamber with
nearly straight
septa

B Fusulininae

Middle Pennsylvanian

Regular
chamberlets

Irregular
chamberlets

Open look of chamber
without chamberlets

Straight septa

Figure 8.37. Septal fluting. Evolutionary trend toward increased intensity in fluting of septa repeats in many lineages. Here fluting is illustrated for two lineages, one in Middle Pennsylvanian and other in Early Permian. **A**, Schwagerininae. Tangential sections showing increasing intensity in septal fluting from nearly straight septa through increasingly regular chamberlets to development of cuniculi in Permian species. **B**, Fusulininae. Tangential sections showing increasing intensity in septal fluting from nearly straight septa through increasingly regular chamberlets in Pennsylvanian species.

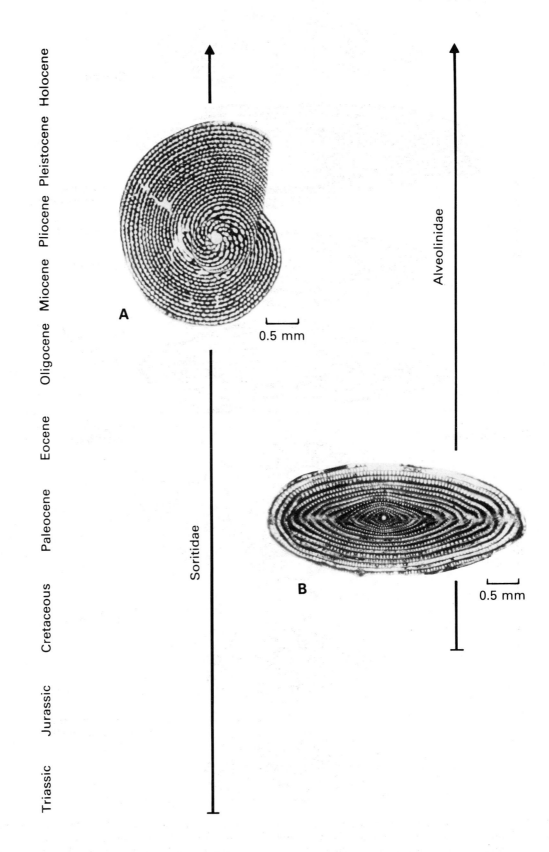

Figure 8.38. Miliolids. A, Sagittal section of *Archaias* (Holocene). **B,** Axial section of *Alveolinella* (Eocene).

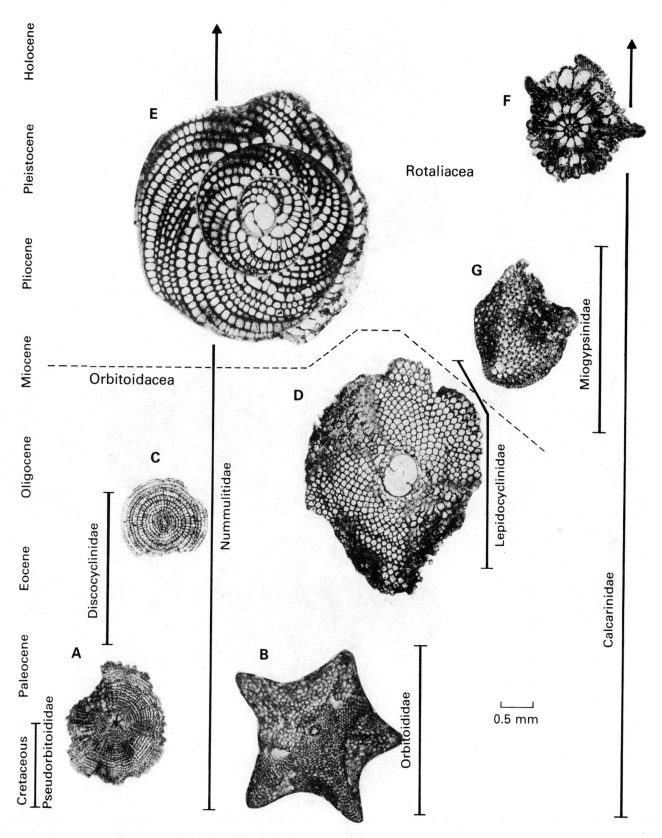

Figure 8.39. Rotalids. All specimens are shown in sagittal sections. **A**, *Vaughanina* (Late Cretaceous). **B**, *Asterorbis* (Late Cretaceous). **C**, *Pseudophragmina* (Middle Eocene). **D**, *Lepidocyclina* (Late Eocene). **E**, *Spiroclypeus* (Early Miocene). **F**, *Calcarina* (Holocene). **G**, *Miogypsina* (Oligocene).

A Schwagerininae

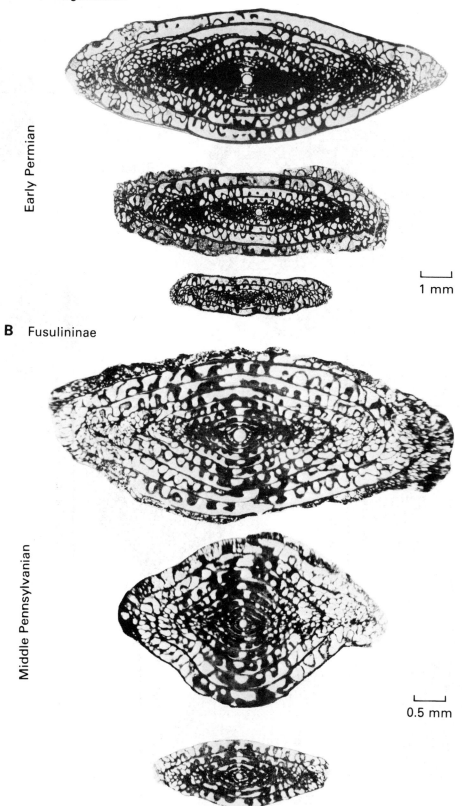

B Fusulininae

Figure 8.40. Size and shape through time. Tendency in many lineages to increase in size and to change shape through time illustrated by two series, one of Fusulininae from the Middle Pennsylvanian and other of Schwagerininae from the Early Permian. **A,** Developmental sequence of schwagerinids showing increases in size of proloculus, number of volutions, overall size, and complexity of septal fluting. **B,** Developmental sequence of fusulinids showing increases in size of proloculus, number of volutions, and overall size.

external resemblance to fusulinids, but the nature of the wall material and the internal structures are totally dissimilar.

The rotalids (Figure 8.39) include an important group of larger foraminifera assigned to six families, four subfamilies, and 20 genera. This group includes most of the forms commonly referred to as larger foraminifera or orbitoids and includes many forms that are stratigraphically useful in rocks of early to middle Tertiary age. Their apparent restriction to warm water environments limits their geographic distribution and, therefore, their usefulness.

Ecology and paleoecology

Modern larger foraminfera live in shallow marine waters on the continental shelf and on all reefs. They are abundant in water as shallow as 1 m and live as deep as 100 m where water is clear enough for light penetration. Light is important not only to species with symbiotic algae but also to those that feed on microscopic plant material. Living habits for the fossil larger foraminifera can only be inferred from the living forms and from the lithologic setting and associated faunas.

The Cenozoic and Mesozoic larger foraminifera seem to have lived in tropical to temperate waters and are limited in occurrence to the Indo–Pacific areas, the Mediterranean borders, the Caribbean, and the Gulf Coast with only minor exceptions. The Paleozoic forms may have had similar restricted habitats, but the rocks containing them are now distributed from the Arctic Circle to southern South America and throughout most of the present temperate zone. Many of the fusulinids were cosmopolitan, having representatives throughout the world distribution of the group. Other fusulinids are more restricted and are only found in Tethyan deposits presumed to have been deposited in a tropical or subtropical belt from the Mediterranean through the Middle East to China, Japan and the Western Pacific border areas.

Evolutionary trends in fusulinids

The fusulinids evolved from small lenticular forms in Early Pennsylvanian time, developed a variety of shapes, and attained large dimensions by Permian time (Figure 8.33). The rapid evolution and widespread geographic distribution make them particularly good stratigraphic indicators for divisions of Pennsylvanian and Permian time. Several evolutionary trends are repeated in different lineages, at different times, and at different rates,

including:

1. Increase in the size of the initial chamber (proloculus)
2. Increase in overall size
3. Increase in complexity of the wall
4. Increase in intensity of septal fluting.

The small (0.5 mm in diameter) and lenticular fusulinids found in rocks of Early Pennsylvanian age (Figure 8.33) evolved by Late Pennsylvanian time to fusiform fusulinids (Figure 8.40), and the length of specimens increased from about 2 mm to more than 10 mm. The earliest prolocular diameters were less than 10 μm and they increased to almost 200 μm during Middle Pennsylvanian time. The wall structure of the Early and Middle Pennsylvanian forms is of the simpler kinds (Figure 8.35, *B* and *C*).

Beginning in Late Pennsylvanian time the more complex keriothecal wall (Figure 8.35, *A*) was developed by many forms. Forms with this wall structure (Schwagerininae of Figure 8.33) diversified rapidly with increasing prolocular size and test size (Figure 8.40), intensification of septal fluting (Figure 8.37) and a variety of shapes. By Late Permian time some elongate subcylindrical forms attained lengths up to 140 mm. Prolocular diameters in the Schwagerininae increased from about 80 μm in the early Late Pennsylvanian to near 250 μm by Early Permian time, and some forms developed prolocular diameters near 1 cm in Late Permian time.

The Verbeekinoidea, a superfamily of the fusulinids, developed in the Permian. The earliest forms are relatively small and nearly round (Figure 8.33). Diversification in shape and increases in size and complexity were rapid in middle to Late Permian time, and they provide a detailed zonation for that part of geologic time.

Supplementary reading

Brasier, M. D. *Microfossils*, Boston: George Allen & Unwin; 1980. Includes illustrations of 20 groups of microorganisms, mostly marine, that occur as fossils. Short sections on biology, occurrence, and methods of collection and preparation for study are presented.

Cole, W. S. *Larger foraminifera from Eniwetok Atoll drill holes, 1957.* U.S. Geological Survey Professional Paper 260-V; 1958, pp. 743–784, pl. 231–249. Describes and illustrates typical assemblages of Tertiary larger foraminifera.

Douglass, R. C. *The development of fusulinid biostratigraphy.* pp. 463–481. In: Kauffman, E. G., Hazel, J. E., editors. *Concepts and methods of biostratigraphy.* Stroudsburg, PA: Dowden, Hutchinson & Ross, Inc.; Summary of use of fusulinids in biostratigraphy.

Ozawa, T. Notes on the phylogeny and classification of the superfamily Verbeekinoidea, 1970. *Memoirs Faculty Science Kyushu University Series D, Geology* 20:58; 1970. Details the morphologic changes on which the phylogeny of a fusulinid superfamily is based.

Ross, C. A. Development of fusulinid (Foraminiferida) faunal realms. 1967. *Journal of Paleontology* 41: 1341–1354; 1967. Paleogeographic study of distribution and evolution of the fusulinids.

Ross, C. A. Biology and ecology of *Marginopora vertebralis* (Foraminiferida), Great Barrier Reef. *Journal of Protozoology* 19:181–192; 1972. Study of living larger foraminifera and the influence of environment on their growth and development.

Thompson, M. L. Studies of American fusulinids. *Protozoa* (Article 1), University of Kansas Paleontological Contributions. University of Kansas Publications, 1948, pp. 1–184. Introduction to the fusulinids and illustrations of all genera known at that time.

Vaughan, T. W.; Cole, W. S. *Preliminary report on the Cretaceous and Tertiary larger foraminifera of Trinidad British West Indies.* Geological Society of America Special Paper 30; 1941. Descriptions and illustrations of a large variety of post-Paleozoic larger forams.

Coccolithophores

Charles C. Smith

Coccolithophores are unicellular, photosynthetic, flagella-bearing organisms that produce coccoliths, calcareous skeletal plates, during one or more phases in their life cycle. The entire external skeletal test of a coccolithophore is constructed of interlocking coccoliths. The test is usually subspherical or ellipsoidal in shape and is called a **coccosphere** (Figure 8.41). Living coccolithophores are generally less than 25 μm in maximum diameter, and their calcareous skeletal plates range from less than 1 μm to perhaps 15 μm in maximum dimension. The coccolithophores are generally assigned to the Class Coccolithophyceae, established to include all members of the Division Chrysophyta, or golden-brown algae, that secrete coccoliths (Figure 8.41).

Despite their small size, coccolithophores have been among the most abundant members of the oceanic phytoplankton from Early Jurassic time to the present. Like other photosynthetic organisms, they are dependent on the energy of sunlight, and most species flourish only in the upper layers of the water column. Fossil remains of coccolithophores are generally included in the broader categories of calcareous nannoplankton or calcareous nannofossils, both terms referring to any of the small, variously shaped, calcified fossil skeletal elements thought to be of plantlike origin.

Although fossil coccoliths were first described in 1836, serious interest in nannofossil research was initiated during the 1950s as a result of technologic improvements in transmitted light microscopes and the development of sophisticated electron microscopes. Relatively recent studies established for the first time the complex structure of the coccolith and the vast diversity of nannoplankton assemblages in marine sediments. Today, the calcareous nannoplankton rank with the planktic foraminifera, radiolaria, ammonites, and other pelagic groups of fossils as excellent biostratigraphic indices. Their planktic habit, cosmopolitan distribution, rapid change during Mesozoic and Cenozoic time, great diversity, and extreme abundance in the smallest of samples have proven the calcareous nannoplankton to be superb biostratigraphic indicators. These combined qualities have been used to construct detailed systems of zonation applicable to the worldwide correlation of marine strata. Although knowledge of both coccolithophores and other calcareous nannoplankton has grown appreciably during the past few years, intensive biologic, morphologic, taxonomic, phylogenetic, and biostratigraphic investigations are just beginning.

Morphology

Cell morphology Little is known about the morphology of coccolithophores. With the establishment of the value of fossil coccoliths in the biostratigraphic correlation and dating of marine sediments, it is difficult to understand why so few scientists are studying the living organisms. However, from the few studies that have been conducted, we have recently begun to learn some of the details regarding the basic morphology of the soft parts of the coccolithophores and the mechanics of coccolith production.

Living coccolithophores are unicellular organisms ranging in size from about 3 to 25 μm in maximum dimensions. They are generally subspherical to subcylindrical in shape, taper toward one or both ends, and normally possess a primary radial symmetry. Coccolithophores are usually free-swimming during one stage in their life cycle and may become benthic or attached during a nonmotile stage. The living cells have two elongate, whiplike flagella (singular, flagellum), which are filaments of uncertain composition whose rapid movement results in cell rotation and propulsion. Most coccolithophores also have a third appendage called a **haptonema**. The haptonema is a coiled, threadlike structure located between the two flagella and when fully extended may be 10 to 20 times the length of the cell body. Although usually tightly coiled during locomotion, the haptonema may be partly or fully extended and used in attaching the cell to the substrate.

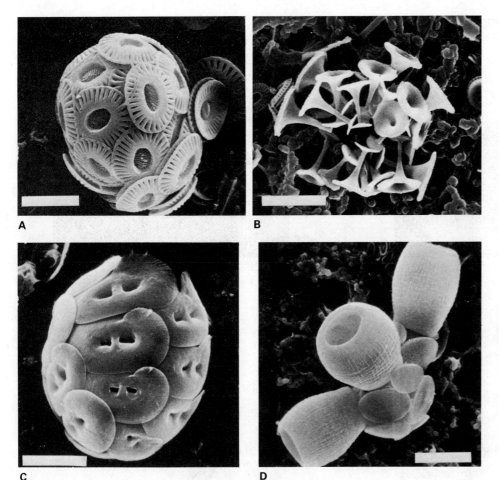

Figure 8.41. Scanning electron micrographs of some Holocene coccospheres illustrating variation in size, shape, and morphology of constituent coccoliths. **A,** Spherical coccosphere of *Emiliania* constructed of two layers of interlocking coccoliths. Bar = 3 μm. **B,** Partially disintegrated coccosphere of *Discosphaera* with tubelike coccoliths. Bar = 6 μm. **C,** Broadly elliptical coccosphere of *Helicosphaera* showing the helical shape of individual coccoliths. Bar = 6 μm. **D,** Curiously shaped coccosphere of *Scyphosphaera* constructed in part of basket-shaped coccoliths. Bar = 10 μm. (**B** courtesy Dr. Stanley Kling, Scripps Institution of Oceanography, University of California, San Diego, California; **C** and **D** courtesy Dr. Franco Cati, Instituto di Geologi e Paleontologia, Universita di Bologna, Bologna, Italy.)

Although the ultrastructures of few species of coccolithophores have been studied, most biologists agree on the basic structure and organization of the principal parts of the cell (Figures 8.42, 8.43). The cell, which may be naked or encased by an external coccosphere covering, is normally bounded by two double membranes that enclose the cytoplasm and single, relatively large nucleus. The cells have two chloroplasts that are conspicuous and easily recognizable by their large size, laminate structure, and location along the periphery at either side of the cell. The chloroplasts contain the chlorophyll-like pigments that transform radiant energy of sunlight into carbohydrates, the substance from which the coccolithophore synthesizes all other compounds within its cell. Mitochondria, which are closely associated with each chloroplast, are the principal bodies in which respiration, or oxidation of the carbohydrates, takes place. A large pulsating **vacuole**, occupying a fourth to a half of the cell body, consists of a fluid solution believed to function in the removal of photosynthetic waste products. Toward the anterior part of the cell is a single, irregularly shaped **Golgi apparatus** consisting of parallel aligned vesicles or tubules whose function is believed to be related to the formation of the cell wall membranes.

Even though the mechanisms of coccolith formation are poorly understood, most biologists agree that the Golgi apparatus is the principal organelle involved in the process. Recent studies indicate that the precursors of coccoliths, called **organic matrix scales**, are formed within the Golgi apparatus. The scales, which act as a template for coccolith formation, are extruded into open vacuoles within the cytoplasm where they are mineralized to form the mature coccolith. The mature coccoliths are then extruded through the cytoplasm to the cell membrane to form the external coccosphere covering.

Types of coccoliths The mature coccoliths, although generally in the form of calcareous plates or disks, exhibit a great variety of structural forms (Figure 8.44 C to F, H, and I). Relating the disaggregated coccoliths to coccosphere genera is often impossible as more than one type of coccolith may be found on a single coccosphere. Furthermore, many types of coccoliths are completely unlike those of living species.

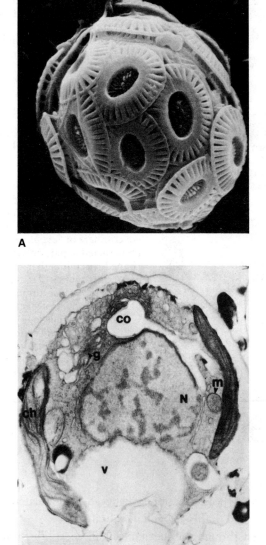

B

Figure 8.42. A, Scanning electron micrograph of Holocene coccosphere of *Emiliania*. Bar = 3 μm. **B,** Transmission electron micrograph of thin section of cultured cell. *co,* coccolith; *ch,* chloroplast; *g,* Golgi apparatus; *m,* mitochondria; *n,* nucleus; *v,* vacuole. Bar = 2 μm. (Courtesy Patricia Blackwelder, Ocean Sciences Center, Nova University, Dania, Fl.)

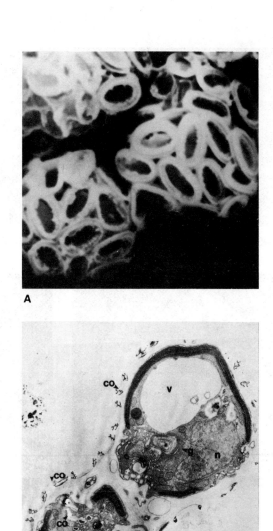

B

Figure 8.43. A, Scanning electron micrograph of coccosphere of *Cricosphaera*. Bar = 4 μm. **B,** Transmission electron micrograph of a thin section of a cultured cell during cell division. *co,* coccolith; *ch,* chloroplast; *g,* Golgi apparatus; *n,* nucleus; *m,* mitochondria; *v,* vacuole. (Courtesy Patricia Blackwelder, Ocean Sciences Center, Nova University, Dania, Fl.)

Many names are used in describing the general shapes of fossil and recent nannoplankton. To a limited extent, the names are descriptive and serve to assemble the great variety of shapes into broad categories that show externally similar morphologic forms, although they have no application in the description of true ultra-structure. Some of the most commonly recognized forms, their names, and brief descriptions are found in Figure 8.44.

Morphologic terminology of coccoliths The terminology used in the study of coccoliths is in a state of flux. Detailed investigations of the ultrastructural details of coccoliths have only recently begun, and many different terms have been introduced to designate their component parts. The few general terms illustrated in Figure 8.45 include only those that are normally encountered in the literature.

Figure 8.44. Scanning electron micrographs of some calcareous nannoplankton showing diverse morphology. Bar = 2 μm. **A,** Ceratolith, horseshoe-shaped nannofossils usually constructed of two slightly curved arms. **B,** Sphenolith, a type of nannofossil having a prism-shaped base constructed of radial elements surmounted by a conical spine. **C,** Placolith (proximal surface), a coccolith constructed of two shields which are connected by a central tube. **D,** Discolith, a coccolith constructed of a single perforate or imperforate circular or elliptical shield having a thickened outer peripheral margin or a raised rim. **E,** Discolith having a helical flange. **F,** Distal surface of a coccolith called a placolith. **G,** Asterolith, a type of nannofossil having rossette, snowflake, or star-shaped external form. **H,** Rhabdolith, a type of coccolith having a relatively long stem attached to its distal surface. **I,** Lopdolith, basket-shaped coccolith having a distal opening. (Courtesy C. Howard Ellis, Marathon Oil Company, Denver, CO)

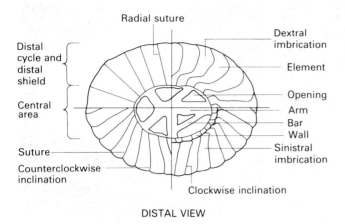

DISTAL VIEW

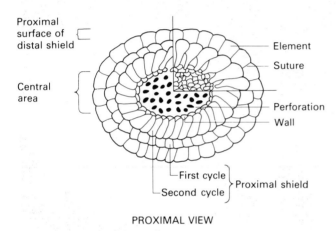

PROXIMAL VIEW

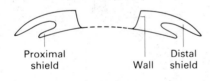

EDGE VIEW

Figure 8.45. Diagrammatic sketches of a theoretical coccolith showing some commonly used descriptive morphological terms (From Farinacci, A. *Proceedings of the Second Planktonic Conference, Roma 1970*, Edizioni Tecnoscienza, 1971.)

Reproduction

The life histories of coccolithophores are not well known. Of the approximately 250 species of extant coccolithophores, no more than ten have been studied, and of these, the life histories of only three or four are known in any detail. Of the few studies that have been conducted, the results have proven to be complex and unexpected.

Reproduction normally takes place asexually by two principal methods:

1   Longitudinal fission of the cell with regeneration of two identical cells

2   Multiple fission of the cell within the coccosphere and subsequent release of numerous daughter cells that may be motile and may regenerate coccoliths

Although reproduction is normally asexual, sexual fusion within one or more stages in the complicated life cycle has been suggested but not adequately confirmed.

A remarkable life cycle has been described for a common species inhabiting colder North Atlantic waters. The motile, flagellated, and actively swimming form has its cell in a flexible case consisting of **holococcoliths**, referring to the uniform size and shape of the rhombohedral calcite crystallites that constitute the coccosphere covering. When maintained in culture, the species reproduces by longitudinal binary fission, usually producing two daughter cells that are initially retained within the holococcolith-encrusted case. The daughter cells later emerge from their coccosphere covering, leaving it behind on the bottom of the culture flask. The daughter cells are nonflagellate and benthic and also accumulate on the bottom of the flask where they undergo cell enlargement. Although lacking flagella, the naked daughter cells appear to retain their haptonema as organelles of attachment. After a short period of growth, the benthic daughter cells undergo double fission, resulting in four, nonflagellate, benthic cells that produce coccoliths of a significantly different type from those produced in the motile phase. The coccoliths produced during this benthic stage consist of imbricating, often interlocking calcite crystals of varying sizes and shapes termed **heterococcoliths**. The heterococcoliths produced in the benthic nonmotile stage are so different from the holococcoliths produced in the motile flagellate stage that they were previously assigned to an entirely different genus and species. The liberation of motile flagella-bearing cells from the heterococcolith-encrusted benthic cell completes this particular type of two-stage life cycle. Although workers have consistently searched for fusion stages during culture experiments involving these forms, all attempts to observe sexual phases have proved unsuccessful.

Other laboratory-cultured species are known in which the coccolith-bearing motile stage produces benthic sessile stages that are multicellular or filamentous and devoid of coccoliths. All these attached phases are so different from typical coccolith-bearing phases that they once were thought to belong to entirely different families. Thus, many familiar benthic filamentous algae may prove to be stages in life cycles of coccolithophores once their life histories have been documented. Future culture studies of coccolith-bearing algae may confirm types of life-history relationships already known. More likely, they will reveal others that are totally unexpected.

Classification

One of the earliest problems to face nannofossil workers was simply to which kingdom the coccolithophores should be assigned. These organisms were considered to be animals (Protozoa) by many early workers, whereas others regarded them as plants, and, because many coccolithophores possess characteristics of both animals and plants, they were regarded as Protista by still other workers. A general characteristic of these organisms is that most possess chloroplasts and, like other plants, synthesize carbohydrates from carbon dioxide and water through the energy of sunlight. Other forms are known that lack chloroplasts, are capable of locomotion, are free-swimming during their motile phase, and have been observed to ingest bacteria and other small solid food particles, all characteristics of the animal kingdom. The majority of workers, therefore, consider the coccolithophores as sharing many affinities with both plants and

Figure 8.47. Scanning electron micrograph of a coccosphere consisting of two distinctly different types of coccoliths. *a*, Species of *Gephyrocapsa*. *b*, species of *Emiliania*. Bar = 2 μm. (Courtesy Dr. Micheline Clocchiatti, Laboratoire de Geologie du Museum National d'Histoire Naturelle; Paris, France. © Gauthier-Villars, Paris.)

animals and thus assign them to Kingdom Protista. Furthermore, many workers assign the coccolithophores to the Division Chrysophyta, or golden-brown algae, on the basis of the characteristics of their cell structure, pigmentation, and flagellation, and to the Class Coccolithophyceae, or coccolith secreting chrysophytes.

A phylogenetic scheme of classification is the most desirable, since it relies heavily on total morphology as well as an understanding of the life history of the organism that produced the fossil. The nannofossil micropaleontologist, however, must rely on isolated fossil coccoliths because the coccosphere normally separates into its individual constituent platelets soon after incorporation into the sediment. This problem is complicated by platelet dimorphism within the same coccosphere. For some time we have known that some coccolithophores possess one type of coccolith around the flagellar pole of the coccosphere and another type on the remainder of the coccosphere (Figure 8.46). This phenomenon is now well known in many living species, having been repeatedly documented in studies of living coccolithophores from the Mediterranean and the Pacific and northern Atlantic Oceans. Furthermore, recent studies have shown the existence of two morphologically distinct types of coccoliths uniformly distributed over the same coccosphere (Figure 8.47). Less serious, perhaps,

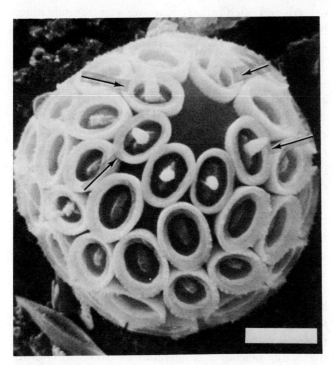

Figure 8.46. Scanning electron micrograph illustrating platelet dimorphism in *Syracosphaera*. Bar = 4 μm. Note the difference in morphology of rhabdoliths surrounding the oral opening (arrows) as compared with the majority of discoliths on the remainder of the coccosphere. (Courtesy Dr. Franco Cati, Instituto di Geologia e Paleontologia, Universita di Bologna, Bologna, Italy.)

are the subtle changes in coccolith morphology resulting from variations in nutrient supply and differences in water temperature in which the coccolithophores are living.

Another of the several problems encountered in assigning fossil coccoliths to suprageneric hierarchies is that the few living coccolithophores that have been studied in detail exhibit dimorphic two-stage life cycles as discussed above. Each stage produces coccoliths so different that they once were thought to belong to entirely different families.

The tremendous variation in individual coccolith morphology resulting from dimorphic, two-stage cycles, as well as documented variability of coccolith morphology on a single coccosphere, suggests that the assignment of isolated fossil coccoliths to generic (and certainly higher) taxonomic categories is even more artificial than was previously supposed. Although many workers apply and even encourage the use of suprageneric classifications, there appears to be little or no phylogenetic significance, either real or implied, in any of the existing schemes.

Most workers use the *International Code of Botanical Nomenclature* in the description and classification of calcareous nannoplankton. Inherent in the botanical code is the concept of form genera. This concept expressly provides for the naming and organization of taxa consisting of isolated, fragmentary fossil parts lacking known relationships to other fossil parts or to the specific parent organism. At present, it does not seem necessary to convert potentially valuable biostratigraphic fossil forms into a phylogenetically meaningless and seemingly endless array of reshuffled categories. Thus, the concept of form genera is followed herein for the classification of coccoliths. Until the knowledge of living coccolithophores as well as fossil coccolith floras is significantly advanced beyond present understanding, it seems best to use a purely artificial classification based on overall morphologic similarities to arrange species into categories no higher than the generic level.

Finally, it should be stressed that the problems mentioned with regard to classification of isolated fossil coccoliths should not detract from the importance of coccoliths as extremely valuable biostratigraphic indexes. Although the micropaleontologist routinely assigns morphologically different forms to separate species when in fact they may belong to the same biologic organism, the consequences are negligible. The worst that can happen is the belief that two or more species are thought to exist rather than one, although all will have precisely the same geologic range.

Ecology and biogeography

As with all living organisms, coccolithophores have certain environmental limits within which they can survive and reproduce. Being photosynthetic, the great majority of these organisms live within the upper 100 m of oceanic waters within the photic zone of light penetration. The geographic distribution of open-ocean coccolithophores appears to be most strongly controlled by light intensity, water temperature, and surface and near-surface water circulation patterns. Other types of coccolithophores are known to inhabit only shallow water coastal margins such as estuaries, bays, and lagoons having restricted water circulations. In these environments, the coccolithophores seem to be most strongly controlled by variable water salinities.

Our earlier knowledge of the relationships between environmental factors and coccolithophore distribution has come from laboratory experiments on selected species. Generally, the laboratory studies relate environmental variables either to coccolithophore growth rate (expressed in numbers of fissions per day) or to the ability of the coccolithophore to produce coccoliths. For example, several studies that have been conducted on a ubiquitous species (Figures 8.41, A, and 8.42) show that growth and calcification occur over a temperature range between 7°C and 27°C, but that the maximum growth rate is two to three times greater between 18°C and 24°C. This species is also found to have a wide tolerance for salinity, having its greatest growth rate between about 20‰ and 40‰ (parts per thousand). Another experiment showed that at a temperature of 15°C and salinity of 35‰, the growth rate of the same species was uniform as long as the population did not exceed 300,000 cells per milliliter. Of particular interest to the paleoecologist is the documentation of abnormally shaped coccoliths at the extremes of temperature and salinity tolerances and the fact that laboratory variables resulting in both normally and abnormally shaped coccoliths are in excellent agreement with observations on cells of this species collected from North Atlantic waters.

Laboratory experiments with other coccolithophores have shown that the benthic phase is able to survive in water temperatures as great as 37.5°C and, after being frozen for several days, releases normal motile cells after the water reaches a temperature of 16°C. Survival of benthic cells of other species has been documented in salinities as low as 3‰, with liberation of motile cells in salinities between 3‰ and 90‰. Further experimentation has shown that benthic cells are able to survive encasement in salt residues from evaporated sea water and to liberate motile cells when returned to normally saline waters (35‰). Still other observations have shown that some benthic stages are able to survive and reproduce in water having a pH ranging from 3.4 to 10.15, although at a pH of less than 5.7 the liberated motile cells are devoid of a coccolith cover.

Although early laboratory experiments such as these are of theoretical interest to biologists, ecologists, and taxonomists, as well as others, and indicate survival under wide tolerances, more recent studies in natural environments indicate that some coccolithophores live within narrow ranges of environmental variables. Thus, their potential as paleoenvironmental indicators is great, although their application in paleoenvironmental studies has only recently begun.

Much of our knowledge regarding the ecology of coccolithophore species has also come from studies of their horizontal and vertical distribution in the oceans. These studies are generally conducted in combination with conventional temperature, salinity, and chemistry analyses of the water from which the coccolithophores were collected. The studies indicate that coccolithophores may be the most abundant phytoplankton in the oceans and are probably the major contributor to the food chain in oceanic environments.

Living coccolith-producing organisms are known to be present in all ocean waters, and although they are predominantly marine, a few brackish- and fresh-water forms are known. Like most other marine organisms, coccolithophores are most diverse in tropical and temperate waters, although the few species present in colder waters often reach astounding abundances. In surface tropical waters, their concentrations generally range from a few thousand to a few tens of thousands of cells per liter, whereas waters at high latitudes contain a few hundred thousand to several million coccolithophores per liter.

Detailed studies of coccolithophore distribution have been conducted most thoroughly in the central and northern parts of the Atlantic and Pacific Oceans. These data are normally presented in map form, showing geographic distribution of individual species or floral assemblages, or presented in graphic form, showing variation in numbers of cells per liter with latitude (Figure 8.48, *A*) and physical and chemical properties of the water column (Figure 8.48, *B* and *C*).

Although there are more than 250 species of living coccolithophores, the geographic distribution of only a few species has been studied in any detail, the majority of studies being limited to single ocean basins (Figure 8.49). A significant discovery resulting from these studies is that the latitudinal boundaries of geographic provinces defined by the distribution of living coccolithophore species in the water column generally are farther apart than the boundaries of provinces as defined by the same assemblage of coccolithophores recovered from the bottom sediments. Because sediment accumulation rates in the deep ocean basins amount to only a few centimeters per thousand years, the sediment biogeographic province boundaries are probably offset by bottom current and turbidity transport, as well as by burrowing and mixing of the sediment by benthic organisms. However, most workers attribute this discrepancy in province boundaries to the rapid warming of the ocean basins since the last glacial stage some 12,000 years ago, with subsequent expansion of living coccolithophore species to their present geographic province within the last few thousand years. A warming trend in the water column would result in warm water species having a greater latitudinal distribution in the water column than in the bottom sediments (Figure 8.49), whereas cold water species would have more narrow province boundaries delineating the living population.

Several distinct coccolithophore assemblage provinces have been defined for the surface water as well as for several subsurface water layers from studies conducted in the Pacific Ocean. One study resulted in the definition of six distinct surface floral provinces, or coccolithophore zones, believed to be controlled by water temperature. Of particular interest is the discovery that species used in defining surface water coccolithophore provinces are not necessarily maintained in the underlying photic water column (Figure 8.50). For example, within the central and northern Pacific Ocean, three distinct coccolithophore zones have been delineated for the upper 200 m water column (Figure 8.51). Latitudinal boundaries of surface floral provinces are believed to be related to temperature as well as to major oceanic current patterns, whereas the dissimilar coccolithophore provinces of the middle and lower photic water layers may be more strongly influenced by salinity fluctuations and reduced temperature and light intensity. Horizontal and vertical distribution of coccolithophores is probably much more complex than this rather simplistic view, and much more detailed sampling will be necessary to attain a better understanding of environmental influences that affect recent coccolithophore distribution.

Biostratigraphy

Many characteristics of the calcareous nannofossils make them superb fossils to use in developing detailed

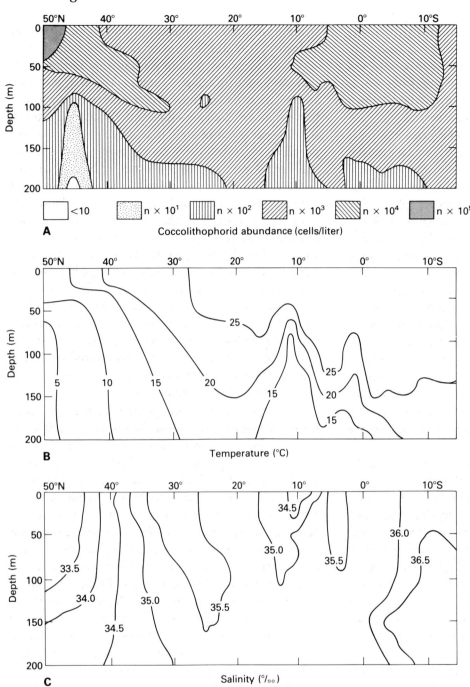

Figure 8.48. A, Cross section showing the abundance of coccolithophore species throughout the 200 m water column along a north-south traverse (155 degrees west longitude) extending from 50 degrees north to 16 degrees south latitude, northern and central Pacific Ocean. **B,** Temperature distribution and, **C,** salinity of same cross section. Note the high concentration of coccolithophores in shallow waters between 41 and 50 degrees north, extending southward to 30 degrees north as a distinct middle photic layer province. This high concentration area corresponds well with water temperatures between 10°C and 19°C. A similar relationship between coccolithophore abundance and water temperature exists between about 10 degrees north and 10 degrees south, where two cold subsurface currents affect cell abundances. Variations in water salinity apparently have no significant influence on coccolithophore concentrations along this traverse (Reprinted with permission from Okada, H.; Honjo, S. *Deep-sea research;* © 1973, Pergamon Press Ltd.)

systems of biostratigraphic zonation applicable in the dating and correlation of marine sediments. Characteristics that give these fossils distinct advantages over other types of marine organisms include: (1) their small size and extraordinary abundance in marine sediments; (2) their planktic habit and resulting worldwide geographic distribution; and (3) their rapid morphologic change, which allows subdivision of the geologic column into narrow biostratigraphic intervals.

The constituent coccoliths from many coccolithophores are incredibly minute. For example, many coccoliths range from less than 1 to about 3 or 4 μm in maximum length or diameter. The small size and extraordinary abundance of coccolithophores in the upper photic layers of the water column result in their contribution to marine sediments in astronomically large numbers. Estimates of a trillion (10^{12}) nannofossils per cubic centimeter of oceanic ooze have been recorded. Many shallower marine rocks contain a few hundred million to perhaps a billion (10^9) nannofossils per cubic centimeter of sediment (Figure 8.52) while a few seem to have none. Thus, only a few milligrams of samples are

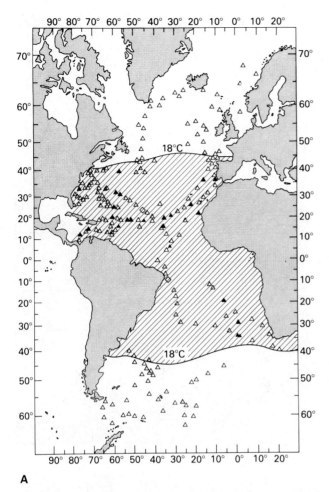

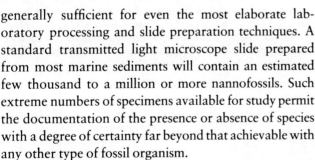

A

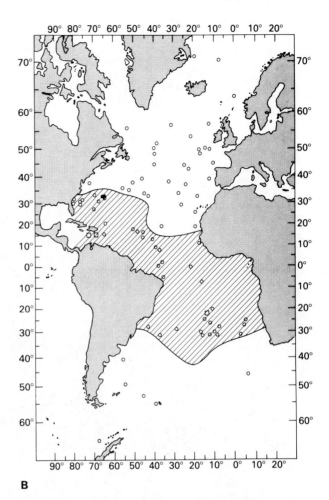

B

Figure 8.49. Biogeography of the coccolithophore *Discosphaera* in the Atlantic Ocean. **A,** Distribution of living plankton in the photic water column. **B,** Distribution in bottom sediment samples (Reprinted with permission from McIntyre, A.; Be, A. *Deep-sea research;* © 1967, Pergamon Press Ltd.)

generally sufficient for even the most elaborate laboratory processing and slide preparation techniques. A standard transmitted light microscope slide prepared from most marine sediments will contain an estimated few thousand to a million or more nannofossils. Such extreme numbers of specimens available for study permit the documentation of the presence or absence of species with a degree of certainty far beyond that achievable with any other type of fossil organism.

Because living coccolithophores are passively floating organisms, they are widely distributed geographically. Surely their fossil counterparts had a similar life habit. Documented reports of Mesozoic and Cenozoic nannofossil floras from widely separated localities have established that most species in an assemblage are surprisingly similar to those in geographically remote assemblages of equivalent age. Evidently, many fossil species were more tolerant to wide ranges in the physical and chemical properties of the water column than their living counterparts. Furthermore, the general worldwide similarity of nannofossils in sediments of equivalent age may also be due to their small size, extremely slow settling rate through the water column, and widespread distribution by prevailing oceanic current patterns.

Although gradual evolutionary change has been documented in some nannofossils, detailed paleontologic investigations have shown that many morphologic changes occurred abruptly. That such sudden change can occur is understandable because most coccolithophores reproduce asexually. Exchange of nuclear material between genetically modified and unmodified relatives, typical in sexual reproduction, need not occur in the coccolithophores. Thus, any successful genetic change can be reproduced asexually in vast numbers in a short period of time. The abrupt appearance (as well as disappearance) of many fossil coccoliths lacking recognizable ancestors is well documented in the paleontologic literature and makes nannofossils superb indices for

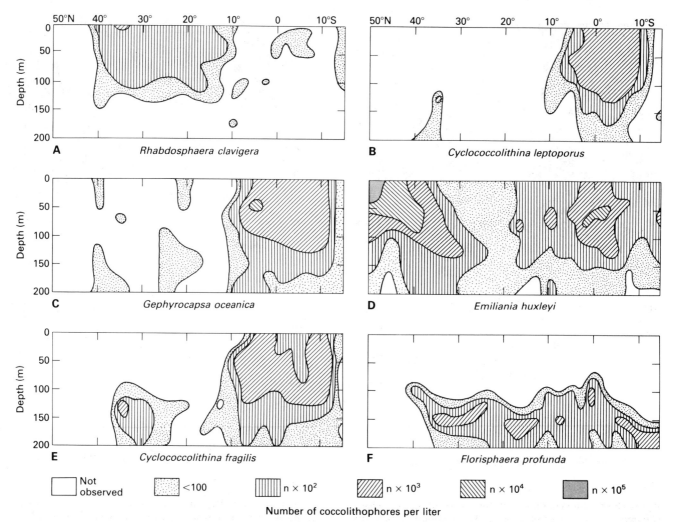

Figure 8.50. Vertical distribution of coccolithophore species in the 200 m water column along a north-south traverse (155 degrees west longitude) extending from 50 degrees north to 16 degrees south latitude, northern and central Pacific Ocean. **A,** *Rhabdosphaera clavigera* is an upper photic layer species that is rare below 100 m. **B,** *Cyclococcolithina leptoporus* dominates the upper and middle photic layers in equatorial latitudes. *Gephyrocapsa oceanica* (**C**) and *Emiliania huxleyi* (**D**) are widespread upper and middle photic layer species, although *G.*

oceanica is most abundant at equatorial latitudes and *E. huxleyi* dominates high latitude coccolithophore floras. **E,** *Cyclococcolithina fragilis* is an upper and middle photic layer species in equatorial latitudes, whereas in transitional latitudes it is a middle to lower photic zone species. **F,** *Florisphaera profunda* is a typical lower photic layer species most abundant in water layers below 100 m (Reprinted with permission from Okada, H.; Honjo, S. *Deep-sea research;* © 1973, Pergamon Press Ltd.)

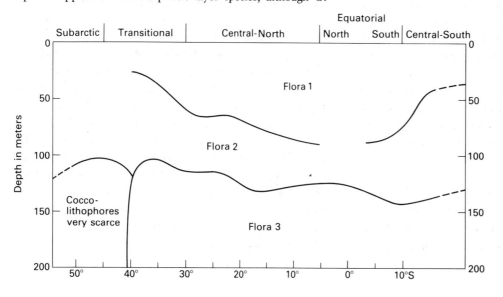

Figure 8.51. Vertical distribution of coccolithophore floras in the 200 m water column along a north-south traverse (155 degrees west longitude) extending from 50 degrees north to 16 degrees south latitude, northern and central Pacific Ocean (Reprinted with permission from Okada, H.; Honjo, S. *Deep-sea research;* © 1973, Pergamon Press Ltd.)

Figure 8.52. Scanning electron micrograph of a fracture surface from a sample of Kimmeridge Clay (Late Jurassic). Dorsetshire, England. Bar = 10 μm. (Courtesy Dr. Denise Noel, Laboratoire de Géologie du Muséum National d'Histoire Naturelle, Paris, France. © Gauthier-Villars, Paris.)

developing finely divided biostratigraphic subdivisions of the geologic column.

Additional characteristics of calcareous nannofossils give them advantages over many other types of fossils. The ease and rapidity of sample preparation prior to examination by the micropaleontologist is especially important, particularly when rapid decisions are required regarding the biostratigraphic assignments and age of a sample. Furthermore, the coccoliths have a relatively greater resistance to solution than do many other types of calcareous fossils. Their presence in sediments that accumulated well below the carbonate compensation level (the oceanic depth below which the rate of solution of calcium carbonate exceeds the rate of supply) is an established and well documented fact. Although the phenomenon is poorly understood, their presence below the oceanic carbonate solution depth may be due to one of the following:

1. The crystal form of individual elements that make up the nannofossils.

2. A protective organic coating acquired during extrusion of the coccolith through the cytoplasm.

3. Incorporation of a complex, cellulose-like polysaccharide material within the lattice of skeletal calcite elements.

4. Their incorporation into fecal pellets.

5. A combination of these or other unknown processes.

Fossil calcareous nannoplankton occur most commonly in fine-grained rocks such as dark mudstone, calcareous shale, marl, chalk, and other sediments that normally accumulate in neritic and open-ocean environments of deposition. These skeletal remains are first abundant in rocks of Early Jurassic age, where their diversity and rather complex morphology suggest that they must have had their origin at some earlier time. Although nannofossils have been reported from rocks of Silurian and Devonian age in North America, Devonian strata in Morocco, and Carboniferous rocks in Algeria, many of the reported forms have such striking similarities to Jurassic and later species that contamination seems a distinct possibility. The discovery of Pennsylvanian nannofossils in Missouri seems to represent the most reliable report of early nannofossils.

Although the discovery of coccoliths and associated calcareous nannofossils dates from the middle 1830s, the first biostratigraphic zonation of Mesozoic strata using calcareous nannofossils was published in 1963. Since the middle 1960s, considerable progress has been made in establishing the stratigraphic ranges of many Mesozoic nannofossil taxa. Figure 8.53 represents a greatly condensed version of one of the more recent calcareous nannofossil zonations of Mesozoic strata.

The end of the Mesozoic is marked by a dramatic worldwide reduction in calcareous nannofossil floras. Of the approximately 240 species of Maastrichtian nannofossils, only four or five survived into the earliest Tertiary. Although causes for this mass extinction have been debated, the explanations have not proved to be entirely satisfactory. However, from the few survivors, diversification progressed at a rapid rate. A flourishing nannoplankton community was reestablished by middle Paleocene time, and an early Eocene flora of almost 300 species marks the greatest diversity of calcareous nannoplankton during either Mesozoic or Cenozoic time.

The stratigraphic ranges and geographic distribution of Cenozoic nannofossils are much better known than those of the Mesozoic, resulting in well-defined, established, high-resolution biostratigraphic zonal schemes for this era. Their great biostratigraphic value in the petroleum industry has resulted in the development of several Cenozoic nannofossil zonations, which are principally applicable to near-shore and continental shelf areas. Because of the great number of Cenozoic sections recovered during the several phases of the Deep Sea Drilling Project, many other zonations or partial zonations have been proposed for Cenozoic deep ocean basin sediments. Figure 8.54 is a greatly condensed version of

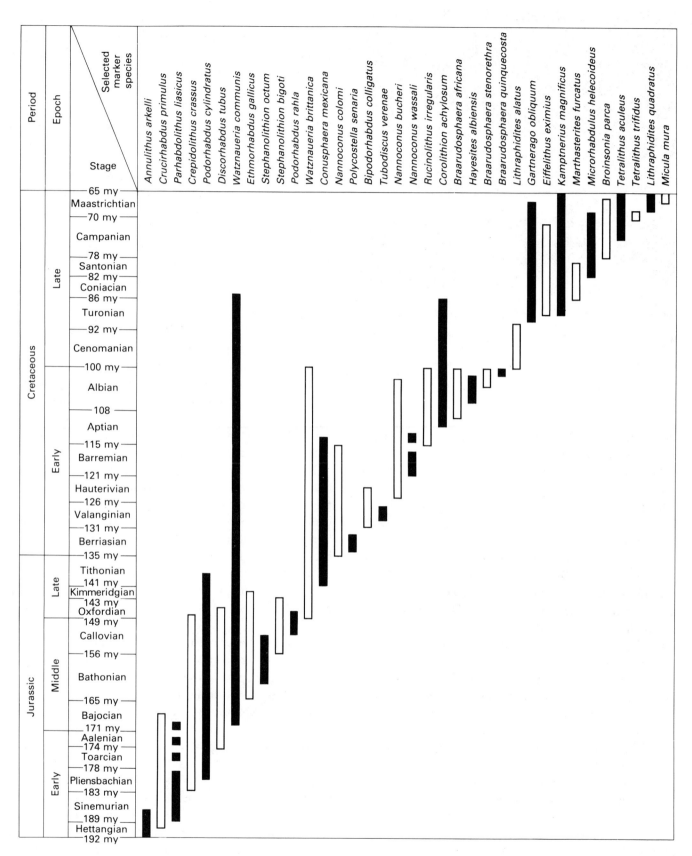

Figure 8.53. Chart showing biostratigraphic ranges of a few important Jurassic and Cretaceous nannofossil taxa (From Thierstein, H.R. *Marine Micropaleontology*, Amsterdam: Elsevier Scientific Publishing Co.; 1976, pp. 338–339.)

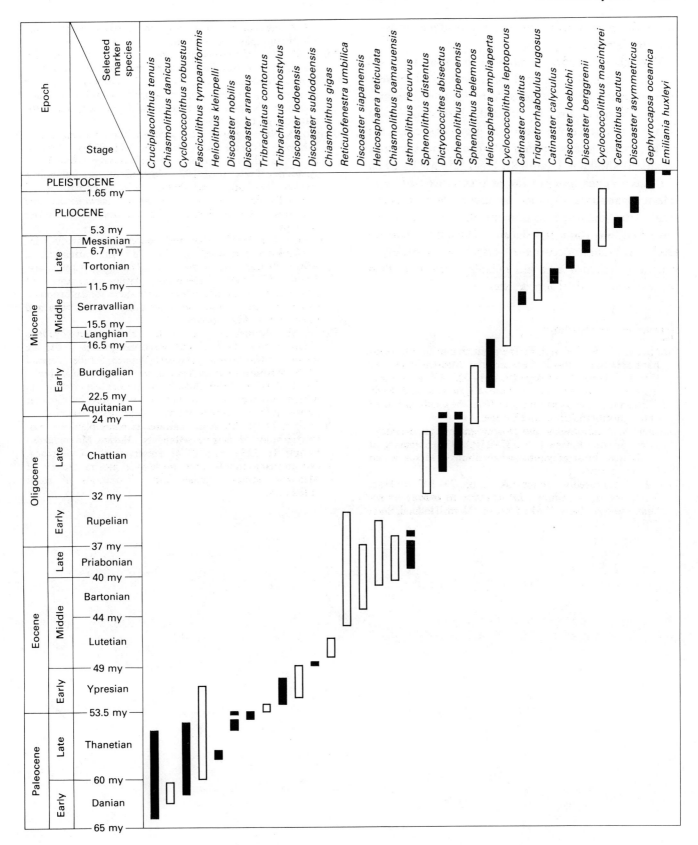

Figure 8.54. Chart showing biostratigraphic ranges of a few important Cenozoic nannofossil taxa. Note the relatively short ranges of Cenozoic species as compared with those of many Mesozoic forms (From Gartner, S. Nannofossils and biostratigraphy: an overview. *Earth-science reviews* 13:238–239; 1977.)

one of the more comprehensive nannofossil zonations for Cenozoic strata.

Most contemporary nannofossil zonations are based on the concept of an interval zone, the interval between two biostratigraphic horizons. Although these biostratigraphic horizons may be either the highest or lowest occurrence surface of a species (or group of species), the range charts and simplified Mesozoic and Cenozoic zonations shown in Figures 8.53 and 8.54 are based on the initial appearance, or lowest occurrence surface of selected nannofossil species. Because of their extremely small size, the calcareous nannoplankton are easily reworked into younger sediments. Thus, the extinction level, or highest occurrence surface of a species, is generally regarded as a less reliable datum level than those based on initial appearances.

Supplementary reading

Bramlette, M. N.; Martini, E. The great change in calcareous nannoplankton fossils between the Maastrichtian and Danian. *Micropaleontology* 10: 291–322; 1964. Comprehensive documentation of Late Cretaceous and Early Tertiary calcareous nannofossils from the southern United States, northern Africa, and Europe.

Gartner, S. Nannofossils and biostratigraphy: an overview. *Earth-Science Review* 13: 227–250; 1977. Summary of nannofossil biostratigraphy and zonations of Mesozoic and Cenozoic strata.

Haq, B. U. Calcareous nannoplankton, pp. 79–107. In: Haq, B.; Boersma, A., editors. *Introduction to marine micropaleontology*. New York: Elsevier, North-Holland, Inc.; 1978. Introduction to biology, morphology, ecology, and biostratigraphic distribution.

Hay, W. W. Calcareous nannofossils, pp. 1055–1200. In: Ramsay, A. T. S., editor. *Oceanic micropaleontology*. New York: Academic Press, Inc.; 1977. Through review of procedures, techniques, geographic distribution, and stratigraphic zonations, together with profusely illustrated nannofossil taxonomy.

Honjo, S. Biogeography and provincialism of living coccolithophorids in the Pacific Ocean, pp. 951–972. In: Ramsay, A. T. S., editor. *Oceanic micropaleontology*. New York: Academic Press, Inc.; 1977. Patterns of coccolithophore species distribution and abundance in the equatorial and north Pacific Ocean, and relationships between diversity and physical and chemical characteristics of the water column.

Martini, E. Standard Tertiary and Quaternary calcareous nannoplankton zonation, pp. 739–785. In: Farinacci, A., editor. *Proceedings of the Second Planktonic Conference, Roma 1970*, 1971. Definition, datum indicators and associated species, and reference localities for 26 proposed Cenozoic nannofossil zones, including range charts and illustrations of key species.

Parke, M.; Adams, A. The motile (*Crystallolithus hyalinus* Gaarder and Markali) and nonmotile phases in the life history of *Coccolithus pelagicus* (Wallich) Schiller. *Journal of the Marine Biological Association of the United Kingdom* 39: 263–274; 1960. Results of culture experiments in which two types of coccoliths are produced by different phases in the life history of one coccolithophore species.

Thierstein, H. R. Mesozoic calcareous nannoplankton biostratigraphy of marine sediments. *Marine Micropaleontology* 1: 325–362; 1976. Preservation, paleoecology, evolutionary trends, and worldwide biostratigraphy of Mesozoic strata together with illustrations of zonal indicators.

9

Phylum Archaeocyatha

J. Keith Rigby
Roland A. Gangloff

Archaeocyatha are an extinct phylum of calcareous, conical marine organisms (Figure 9.1) that first appear in the geologic record near the base of the Cambrian on the Siberian Platform. They spread rapidly and by the middle part of the Early Cambrian had essentially worldwide distribution (Figure 9.2). The phylum was markedly restricted at the end of Early Cambrian time, however, so that Middle Cambrian archaeocyathans are known only from northern Australia and the fold belts of the southern Urals and Siberia in the Soviet Union. The phylum probably became extinct early in Middle Cambrian time. Reports of later archaeocyathans are questionable.

Archaeocyathans were among the first phyla to use calcium carbonate as a skeletal material and formed the first reeflike deposits of the Paleozoic. Archaeocyathans are used as stratigraphic index fossils for the Lower Cambrian in the Soviet Union, particularly in carbonate facies.

Taxonomic position of the archaeocyathans is unsettled, although the prevailing current opinion is to treat them as a separate phylum. They were first classed with corals and sponges, then with foraminifera by the mid-1800s, then variously as stony corals, peculiar sponges, or calcareous algae. Extensive study of the group during the 1930s and 1940s and later investigations have encouraged worldwide acceptance of archaeocyathans as a separate phylum. They have been considered intermediate in position between sponges and corals or between colonial protozoans and sponges, as treated here. As such, archaeocyathans belong to the primitive multicellular grade (Chapter 2).

Internal structure is critical in studies of archaeocyathans. As a consequence, thin sections are necessary, even though fossils may weather or be etched free of matrix. **Transverse sections**, cut horizontally across the skeleton (Figure 9.3), and **longitudinal sections**, cut vertically along the skeletal axis, are the minimum sections necessary to adequately describe or identify specimens. **Tangential sections**, cut parallel to the outer or inner wall or to the radial structures, may be required. However, most archaeocyathan skeletons curve and twist, making all these sections difficult to obtain. Unoriented slices of archaeocyathan-bearing rocks usually produce oblique, nearly transverse and longitudinal sections. These oblique sections often allow determination of critical details of both inner and outer wall pore patterns from single sections. Because unoriented sections through specimens are usually numerous and easy to produce, they are heavily depended upon in identification of genera and families. Gross relationships may be observed where the fossils are silicified and can be etched free of matrix. Most archaeocyathans that occur in limestone are calcareous, however, and are difficult to remove from matrix.

Morphology

The typical archaeocyathan skeleton, or cup, is a solitary, highly porous, double-walled, inverted cone (Figure 9.4). Characteristically, an **inner wall** surrounds and defines the axial **central cavity**. Between the inner wall and the

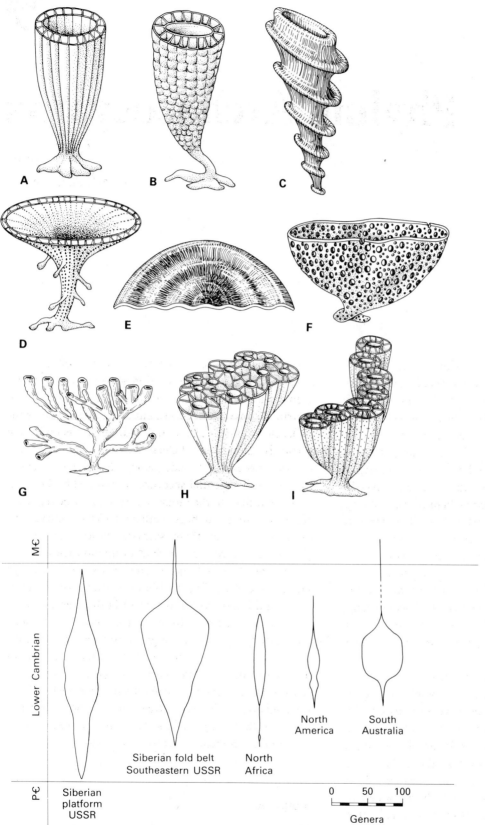

Figure 9.1 External form of solitary and colonial archaeocyathans. **A,** Simple conical and erect form of *Ajacicyathus*, which becomes cylindrical above. **B,** Somewhat sculptured, curved, and steeply conical form of *Kotuyicyathus*. **C,** Transversely annulate, elliptical, and conical form of *Orbicyathus*. **D,** Broadly flaring conical form of *Paranacyathus*, with rapid expansion in the upper part of the skeleton. **E,** Wavy, discoidal, and flabellate form of *Okulitchicyathus*. **F,** Porous, open, bowl-shaped *Cryptoporocyathus*, with relatively thin walls. **G,** Branched colonial form of *Archaeolynthus*. **H,** Massive colonial form of *Ajacicyathus*. **I,** Chainlike colonial form of *Ajacicyathus*.

Figure 9.2 Geological range and diversity of Archaeocyatha. The phylum first appears in the geological record almost at the base of the Cambrian and is recorded from Middle Cambrian rocks only from the southern Urals and Siberia in the Soviet Union and from Northern Territory in Australia. Generic variability of the faunas is shown by width of the bars and is greatest in southern and southeastern Russian faunas.

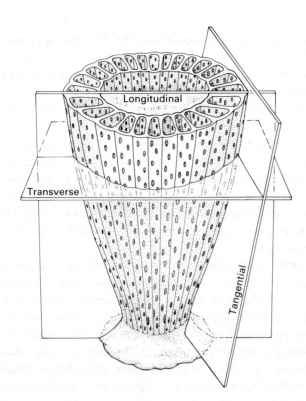

Figure 9.3 Generalized archaeocyathan showing orientation of longitudinal and transverse sections through the skeleton. One possible tangential section is shown as tangent to the exterior, but such sections may be cut at other orientations and tangential to other structures. Oblique sections are those cut diagonally through the skeleton.

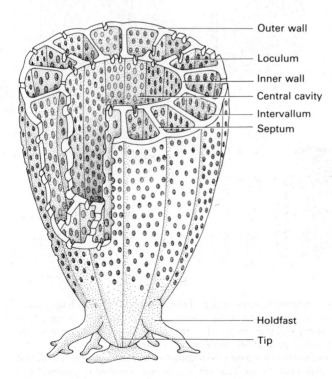

Figure 9.4 Generalized archaeocyathan showing relationships and nomenclature of the porous skeletal elements of a two-walled form.

outer wall is a region called the **intervallum**. It is subdivided into **loculi** or segments by porous or aporous radial partitions or **septa**. The larger or upper end of the cup is usually open, and the lower end or **tip** is closed and often buried in the substrate or held by rootlike **holdfasts**.

Average cups range from 10 to 25 mm in diameter and up to 150 mm tall, although some species have mature cups only 2 or 3 mm across. Very large cups may attain diameters up to 600 mm and heights up to 300 mm. Smaller and more highly porous cups generally occur in genera and species of older rocks, while larger less porous cups are found in taxa of younger rocks. Archaeocyathans show their greatest diversity and abundance in the upper part of the Lower Cambrian sequence, where they are often associated with bioherms and inter-biohermal sediments.

Three major skeletal plans are recognized among archaeocyathans: single-walled forms (Figure 9.5), double-walled forms with regular bladelike radial septa (Figures 9.4 and 9.10, *C* to *F*), and double-walled forms with very irregular curved and wavy plates or subradial curved partitions, called **taenae** or **pseudosepta**, between the inner and outer walls (Figure 9.10, *A*). This later group is the most difficult and challenging to reconstruct or identify.

Most archaeocyathan cups are steeply conical in earliest growth stages (Figure 9.1, *B*) but commonly become more gently conical or subcylindrical in later stages (Figure 9.1, *A*), although a great diversity of skeletal forms are possessed by archaeocyathans as a

Figure 9.5 *Archaeolynthus,* a small single-walled fossil and one of the simplest archaeocyathan genera. It shows the porous nature and common conical form of archaeocyathan skeletons. This specimen is silicified, and the surrounding matrix has been partially removed by etching in acid. It is from the Lower Cambrian of southern Australia. (×7.) (From Hill, D. In Teichert, C., editor. *Treatise on invertebrate paleontology, Part E,* 2nd ed. New York and Lawrence, KA: Geological Society of America and University of Kansas Press; 1972.)

whole. Some taxa, for example, are bowl shaped (Figure 9.1, *F*), and others are fan shaped or discoidal (Figure 9.1, *E*). Transverse rings or annuli may be present in some (Figure 9.1, *C*), and other cups may exhibit rapid (Figure 9.1, *D*) or irregular expansions in width along their length.

Branched, massive, and chainlike colonies (Figures 9.1, *G* to *I*) have been described and are generally considered to be uncommon. However, dendroid colonies, when developed in a number of different taxa, sometimes became locally dominant and characteristic. For example, dendroid colonies are common in some beds in Siberia and in the Great Basin in the western United States.

Outer walls (Figure 9.6) are porous, except at the tip, and generally are more complexly porous in later growth stages. Where walls are thick, the openings are elongated and are termed **pore-tubes** or **pore-canals** (Figure 9.6, *D*). These canals or tubes can be simple or branched. Some outer walls are constructed of intersecting vertical and horizontal elements (Figure 9.6, *C* and *G*) rather than

porous in the usual sense (Figure 9.6, *B*). Outer walls may also develop louvres (Figure 9.6, *F* and *G*), **V**-shaped annuli (Figure 9.6, *E*), or long spines (Figure 9.6, *H*) that project beyond the main wall surface.

In addition to these primary outer walls, some taxa developed secondary walls outside the outer walls that were imperforate or finely porous. These modifications often coincide with abundant imperforate bubblelike structures and secondary thickenings of walls and other skeletal components. Recent observations indicate that these secondary walls may reflect reactions to changes in ambient environment and are of questionable taxonomic importance.

Inner walls (Figure 9.7) are generally more coarsely perforate than outer ones and show almost endless variety. The wall may be only a simple thin or thick porous sheet (Figure 9.7, *A*) or slightly more complex and composed of horizontal lintels or laths, which are elongate boardlike tangential skeletal elements that connect the inner edges of septa and tabulae. More complex inner walls are composed of elongate louvres

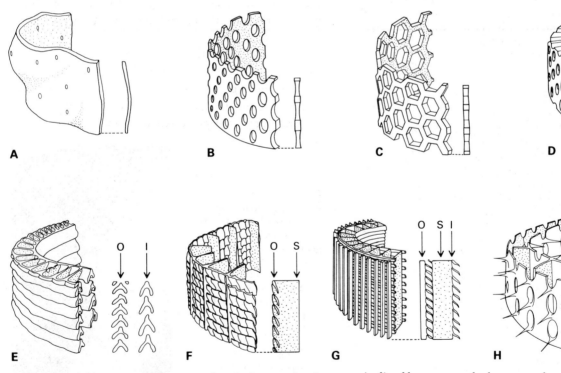

Figure 9.6 Structures of the outer wall in Archaeocyatha. **A,** Thin, simply perforated wall with scattered pores in *Bicyathus*. **B,** Moderately coarsely perforate, moderately thick wall with circular pores in *Ajacicyathus*. **C,** Retiform outer wall (pores closely spaced) with large hexagonal pores in *Erismacoscinus*. **D,** Simple thick wall with alternating elongate pore-canals in *Archaeolynthus*. **E,** Outer wall (*O*) composed of inverted **V**'s separated by ringlike pores at the outer edges of septa in *Annulocyathus*. (*I* is inner wall). **F,** Outer wall of flat **S**-shaped or inclined louvres attached to outer edges of the radiating septa (*S*) in *Kijacyathus*. **G,** Outer clathrate wall of longitudinal (vertical) laths attached to the outer edge of inclined annulate louvres in *Botomocyathus*. **H,** Coarsely perforate, thin outer walls with elongate radial spines in *Robustocyathus*. (Modified from Hill, D. In Teichert, C., editor. *Treatise on invertebrate paleontology, Part E,* 2nd ed. New York and Lawrence, KA: Geological Society of America and University of Kansas Press; 1972.)

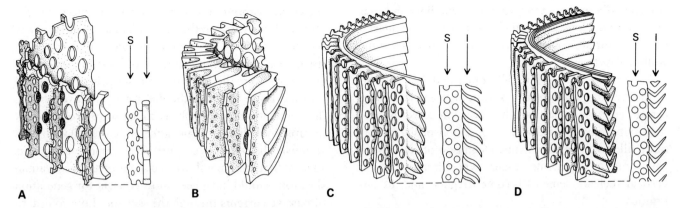

Figure 9.7 Structure of the inner wall in Archaeocyatha. **A,** Simple, coarsely perforate inner wall (*I*) attached to the edges of radiate septa (*S*). **B,** Thick inner wall with gently inclined and coarse pore-canals. Radiating fins are attached segments of septa. **C,** Annulate inner wall with louvrelike annuli (flat **S**-shaped cross section) attached to segments of porous radiating septa. **D,** Annulate inner wall (V-shaped cross section) attached to inner edges of porous septa. (Modified from Hill, D. In Teichert, C., editor. *Treatise on invertebrate paleontology, Part E,* 2nd ed. New York and Lawrence, KA: Geological Society of America and University of Kansas Press; 1972.)

(Figure 9.7, *C*) or more platelike scales that may be inclined, horizontal, or resupinate, like a flattened horizontal **S**. In addition, shelves or annuli of **U, S,** or **V** shapes (Figure 9.7, *D*) may be attached to the inner end of the radial septa to make up the inner wall. Some thick walls are perforated by long pore-canals (Figure 9.7, *B*) that are terminated by various kinds of spines or deflecting devices.

The open intervallum in double-walled forms (Figure 9.4) can be subdivided by different structures. Septa are vertical, more or less regularly spaced, platelike, porous dividers in early genera but may be more irregular and complex in advanced forms. The intervallum is also divided horizontally in some genera by porous **tabulae** (Figure 9.8, *B*) that interconnect the inner and outer walls and septa, or by imperforate vesicular bubblelike **dissepiments** (Figure 9.8, *A*). Dissepiments may also extend into the central cavity, but other intervallar structures normally do not.

Synapticulae (Figure 9.8, *C*) are transverse rods that

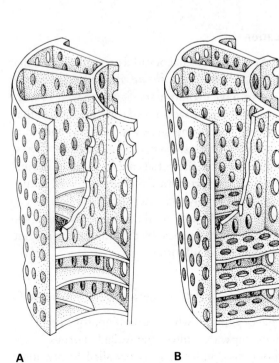

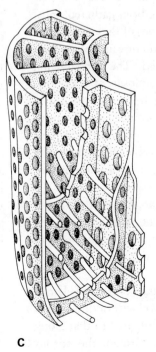

Figure 9.8 Segments of archaeocyathan skeletal cups showing intervallar structures. **A,** Inner nonporous, bubbly dissepiments bridging the outer and inner walls and occurring between septa. **B,** Horizontal porous tabulae extending between the outer and inner wall and between septa. **C,** Horizontal and concentric spikelike synapticulae connecting porous septa in the intervallum.

connect adjacent septa in the intervallum. **Bars** and **rods** are also thin, elongated, cylindrical-to-prismatic skeletal elements but are commonly radial and may connect the walls across the intervallum.

The central cavity is usually open and well defined inside the inner wall. In some advanced genera and under certain ecologic conditions, however, the lower part of the cavity may be lined with secondary deposits or partially filled with dissepiments. Definition of the central cavity may also become obscure where horizontal and vertical skeletal elements become irregular and discontinuous.

Soft-part anatomy is unknown. No modern analogues are known for the phylum, and impressions of soft parts have not been recognized even in enclosing matrix. Methods of respiration, feeding, digestion, waste removal, and reproduction are all unknown. If the hypothesis of archaeocyathans intermediate between protozoans and sponges is correct, life processes were probably on a cellular rather than tissue or diploblastic level.

The intervallum rather than the central cavity is considered to have been the center of biologic activity. The porous skeleton suggests free water or nutrient circulation throughout the organism. There is no general agreement about the direction of flow, but currents probably moved in through the smaller openings in the outer wall and out through the larger openings in the inner wall, with the central cavity functioning as an excurrent opening essentially like the spongocoel in sponges.

Because few specific clues regarding the soft-part anatomy of archaeocyathans have been gathered, most workers concerned with the affinities of these organisms have studied their skeletal microstructure in the hope that relationships might be established. The fundamental pattern of archaeocyathan cups is a mosaic of thick polyhedral plates, commonly ranging from 5 to 10 μm in diameter. This pattern has not provided any positive evidence of affinities with other organisms. It has, however, provided a basis for possible elimination of a relationship to sphinctozoan sponges, those with spicular skeletons, or corals with fibrous microstructure. Although no known calcareous algae possess the variety and complexity of skeletal elements found among the Archaeocyatha, the microstructure of algae may prove to be closest to the Archaeocyatha and, therefore, suggests some affinities.

Experiments with models of archaeocyathan skeletons indicate that differences in current velocity or direction at the top of the central cavity, as compared to that along the walls, produces passive flow through the porous skeleton, without a pumping force produced by the animal. Flow through the intervallum and central cavity requires that they be open and suggests that the living tissue was in the intervallum, or possibly as a thin covering over the skeleton as suggested by healed injuries. Lower parts of the skeletons that have tabulae, dissepiments, or other obstructions were probably not occupied by living tissue after water circulation was seriously limited or cut off. The same experiments suggest that the general size and shape of archaeocyathan skeletons would have been most efficient for generation of passive currents through the skeleton. Low conical to saucer-shaped forms needed higher velocity currents around them than did the deep conical to cylindrical forms, if passive current generation through the skeleton was indeed the way of life of the archaeocyathans.

Colonial forms in Archaeocyatha developed by budding from the ancestral cup. Catenulate colonies (Figure 9.1, *I* and 9.11) have individual cups arranged in chainlike rows. The outer walls did not develop between cups so that the entire colony is included within a single outside wall. Massive colonies (Figure 9.1, *H*) developed where cups grew irregularly, again without outer walls between cups. In branching dendroid colonies, however, (Figure 9.1, *G*) each branch is a cup that is isolated by a complete outer wall, except the first-formed, basal, or proximal cups of the colony.

Colonies developed asexually by buds growing from the outer part of the outer wall, or from the intervallum in some Irregulares (see the next section).

Classification

The classification of Archaeocyatha presented here is that of the *Treatise on Invertebrate Paleontology* and is considered to have evolutionary significance. It is based upon studies by Soviet workers of early archaeocyathan developmental stages and shows conclusions and ontogenetic concepts of these paleontologists. Class characteristics appear first in the initial cup, followed by ordinal and subordinal characteristics, that are followed in turn by appearance of family characteristics, all before the cup is approximately 1 mm in diameter. Subfamily and generic characteristics appear by the time the cup is 2 mm in diameter in almost all forms.

CLASS REGULARES (Figures 9.9, *A*, 9.10, *C* to *F*)

Archaeocyatha in which earliest aporous tip of cup widens upward into one-walled porous stage without dissepiments. In two-walled forms, inner

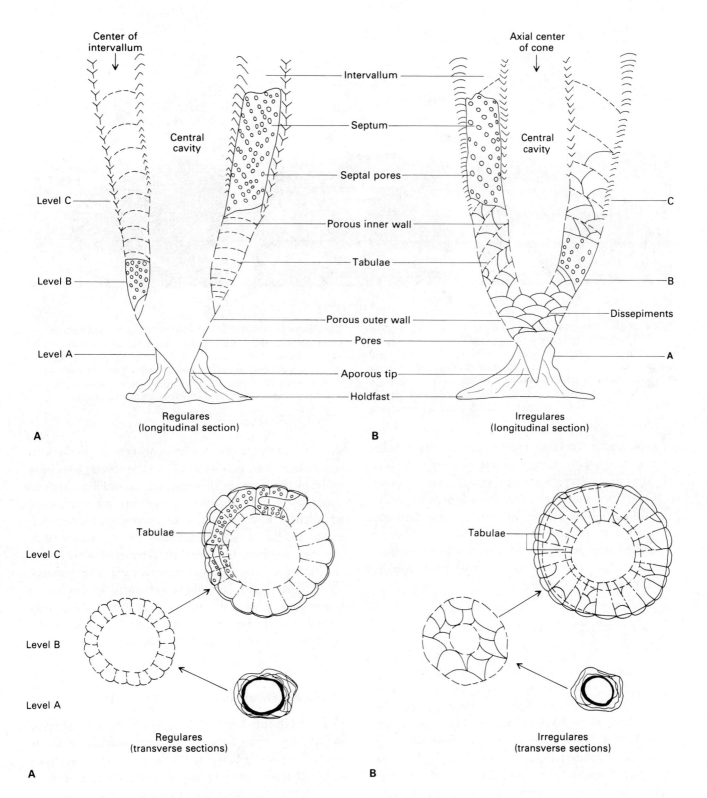

Figure 9.9 Longitudinal and transverse cross sections through cones of a regular (**A**) and an irregular (**B**) archaeocyathan, showing some of the characteristic stages and the typical onto-genetic appearance of various skeletal structures of the two classes. **A,** Perforate inner wall appears in the Regulares before other structures. Septa and tabulae appear at level *B,* still early in the ontogeny. **B,** Dissepiments and other intervallar structures, such as irregularly oriented rods, appear before the inner wall in the Irregulares. Both classes show increasing complexity of wall structure in upper parts of the cups.

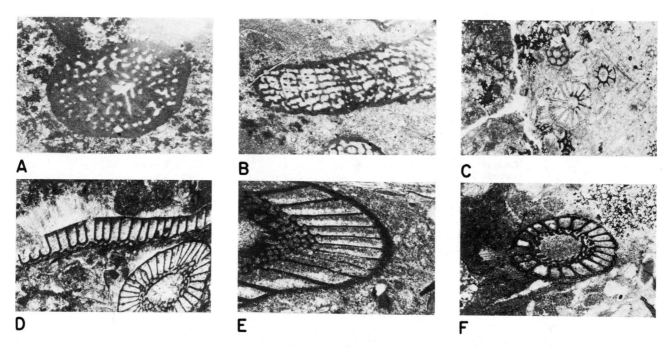

Figure 9.10 Photomicrographs of irregular and regular Archaeocyatha. **A**, Transverse section, and **B**, oblique section of *Protopharetra*, irregular archaeocyathan with ill-defined central cavity. (×5.) **C**, *Robustocyathus*, a regular archaeocyathan. Transverse sections at various levels through four separate individuals, all partially replaced with pyritic silica. (×5.) **D**, *Palmericyathus* (the upper platelike form) and *Cordilleracyathus* (the lower specimen) cut in oval near transverse section. Both are regular genera. Complex inner walls show well on both genera. **E**, Oblique section of *Cordilleracyathus*, a regular genus, shows complex porous inner wall, simple smooth septa, and a simple outer wall. (×5.) **F**, Approximately transverse section of *Cordilleracyathus*. (×5.)

wall and intervallum follow with septa or tabulae alone, or with septa and flat or convex tabulae together; dissepiments and synapticulae in some. Septal pores in longitudinal rows diverge upward in fanlike fashion toward both inner and outer walls. Curvature of convex tabulae symmetrical about center of intervallum. Other intervallum structures may also occur. (Lower Cambrian to Middle Cambrian, 173 genera.)

CLASS IRREGULARES (Figures 9.9, B; 9.10, A and B)

Archaeocytha in which earliest aporous tip of cup widens upward into one-walled porous stage with dissepiments and, in some, irregularly oriented rods and bars. In two-walled forms, inner wall and intervallum follow, always with dissepiments and commonly with tabulae and either rods and bars or septa. Septal pores in rows that arch upward and outward toward outer wall from inner wall. Curvature of tabulae symmetrical about axial center of cone. (Lower Cambrian to Middle Cambrian; 60 genera with one questionably placed genus from Upper Cambrian rocks of Antarctica.)

Recent attempts by Soviet and French workers at archaeocyathan taxonomy above the species level have utilized the concept of homologic variability. Skeletal patterns in even distantly related lineages often show parallelism in development of homologous structures. Development in one lineage may allow predictions in another. In some instances skeletal patterns of previously undiscovered genera have been predicted. The Archaeocyatha seem to be particularly well suited for this type of study and, among the relatively primitive phyla, may prove to be the best group for testing the validity of this interpretation of evolutionary patterns.

Origin and Distribution

Origin of the phylum is obscure, but it appears as the first major step from complex colonial unicellular forms toward higher organisms. Archaeocyathans were probably not in the line of descent of other multicellular phyla but represent an early evolutionary sideline or offshoot that flourished in the Early Cambrian and became extinct in Middle Cambrian. The phylum became extinct at about the same time that the filter-feeding sponges proliferated in the marine environment. Archaeocyathans show provincialism, with distinctive faunas or subprovinces in Australia and Antarctica, in the Appalachian and Cordilleran troughs of North America, in Europe and Africa, and in the Soviet Union (Siberia) (Figure 9.2). Such provincialism is to be expected in rapidly evolving, geographically isolated benthic forms. Distribution must have been accomplished by early larval stages and determined by ecologically suitable space and vagaries of current motion within favorable environments.

Paleoecology

Archaeocyathans were an exclusively marine group of organisms. Their remains are most variable and most abundant in carbonate and argillaceous sediments that were apparently deposited in depths of 20 to 30 m. The depth estimates are based upon the occurrences with blue-green algae, dimensions of bioherms that the algae and archaeocyathans were able to build, primary sedimentary structures produced by currents, and the fragmentation of many archaeocyathan skeletons. In the Siberian Platform sequence archaeocyathans are not found in sediments thought to have been deposited below 100 m, but their largest numbers are found in sediments 20 to 30 m deep where they formed reefs. Archaeocyathans are most abundant where carbonate rocks are nearly pure, with only 10 to 20 percent terrigenous debris. They are also limited to rocks with relatively low levels of magnesium carbonate and are lacking in lagoonal deposits where magnesium carbonate is an important part of the rocks. They do occur locally in rocks suggestive of hypersalinity on the Siberian Platform. Archaeocyathans apparently were somewhat tolerant of turbidity, for some species flourished where carbonate rocks contain up to one-third insoluble debris. They do occur in interbioherm shaly rocks but are restricted there in size, growth form, abundance, and diversity. Some have been found in fetid limestones in Siberia.

The consistent and close association of archaeocyathans with the algae *Renalcis* and *Epiphyton* suggest a possible symbiotic relationship.

Much remains to be learned about the evolution, stratigraphic zonation, origin, and paleoecology of archaeocyathans, particularly in areas outside the Soviet Union. Faunas in North America, in particular, need to be examined more thoroughly in light of the great expansion of knowledge of the phylum elsewhere. Many new localities from Mexico to Alaska need to be studied.

Supplementary reading

Debrenne, F. A revision of Australian genera of Archaeocyatha. *Royal Society of South Australia, Transactions,* 94: 21–49: 1970. Landmark revision of Australian genera with emphasis on the class Irregulares. Useful taxonomic chart included.

Hill, D. Archaeocyatha from Antarctica and a review of the phylum. Trans-Antarctic Expedition 1955–1958, *Scientific Report,* (Geology No. 3), 10: 1–151, 1965. Landmark review of the taxonomy of Archaeocyatha that became a standard reference prior to the *Treatise* revision.

Hill, D. Archaeocyatha, pp. 2–158. In: Teichert, C., editor. *Treatise on invertebrate paleontology, Part E,* 2nd ed. New York and Lawrence, KS: Geological Society of America and University of Kansas Press; 1972. Single-most comprehensive and scholarly reference on Archaeocyatha available.

Okulitch, V. J. *North American Pleospongia,* Special Paper 48, New York: Geological Society of America; 1943. Contains important pioneer work on the taxonomy of North American Archeocyatha.

Öpik, A. A. Cymbic Vale fauna of New South Wales and Early Cambrian biostratigraphy. Australia Department of Mines and Energy, Bureau of Mineral Resources, Geology and Geophysics Bulletin 159; 1975. Provocative discussion of the possible algal affinities for the Archaeocyatha.

Rozanov, A. Y. Homological variability of archaeocyathans. *Geological Magazine* 111: 107–120; 1974. Significant discussion of the theory ("law") of homologic series as demonstrated in regular Archaeocyatha. Usful taxonomic chart included.

Zhuravleva, I. T. Porifera, Sphinctozoa, Archaeocyathi—their connections, pp. 25 and 41–59. In: Fry, W. G., editor. *The biology of the Porifera,* Symposium Zoological Society of London; 1970. Provocative discussion of the possible relationships of Archaeocyatha that also proposes a new division of the Metazoa, the Archaeozoa.

10

Phylum Porifera

J. Keith Rigby

Part I   Phylum overview

The Porifera, or sponges, are among the simplest of living multicellular organisms (Figure 10.1). They are sedentary **filter-feeders** utilizing flagellated cells to pump water through their canal systems. Sponges are aggregates of somewhat differentiated cells that show some rudimentary interdependence and are weakly to poorly arranged into layers. The organisms are thus essentially at a cellular grade of organization. Many cells in sponges are totipotent; that is they are mobile and may change functions during their lifetime.

The Porifera are the only phylum in which the largest body opening in the animal is entirely excurrent. Sponges lack a mouth and organs associated with higher organisms. Multiple incurrent openings provide access for currents bearing nutrients and oxygen. Sponge bodies are perforated with canals connected to the incurrent openings, and the canal system is so arranged and growth so oriented that nearly each cell is in contact with the circu-

Figure 10.1. Large tubular chimney sponges and thick leathery cnidarians in shallow water at the east end of Grand Cayman Island in the western Caribbean. The chimney sponges are approximately 0.6 m high. A deep spongocoel penetrates each of the sponge branches.

116

lating, food-bearing, and waste-removing water. Skeletons are made of needlelike **spicules**, organic fibers, or calcareous laminae and are the only parts preserved in fossils. The skeleton is internal, in the sense of being surrounded by cells, and basal, in the sense of supporting the lower parts of the animal. Fossil sponges are separated into major taxa largely on the basis of their skeletons.

Modern sponges inhabit aqueous environments ranging from polar to tropical, fresh to marine, and shallow to abyssal. Some modern sponges have even been gathered from trees in warm humid areas where there is great seasonal fluctuation in water levels. **Gemmules**, small cystlike reproductive bodies, develop during low-water stages. They produce sponges at the next suitable high-water stage. Some sponges also bore into calcareous substrates, such as shells of bivalves and gastropods, or colonial skeletons of corals. Generally, however, the phylum is characterized by sessile, benthic, aquatic organisms.

Sponges tend to be gregarious and may be abundant locally, yet rare elsewhere. They are relatively rare fossils throughout much of the geological column from the Cambrian to the Holocene and commonly occur only as isolated spicules. Where abundant, skeletons of some sponges may form reefs, but elsewhere may have broken apart to produce **spiculites** or **spiculitic chert**, rocks made in large part from skeletal elements. In some formations they may have contributed considerable easily dissolved silica to the sedimentary sequence and so be preserved only as molds or replacements.

Introductory morphology

Soft parts

The cellular part of the sponge body (Figure 10.2) consists of a radially symmetrical to irregularly massive or sheetlike group of cells arranged about a canal system. Water enters the sponge through the pores, called **ostia**, and flows through tubular or chamberlike canals and out through large excurrent pores or openings, each called an **osculum**. In tubular forms, the water may pass through a large cavity, the **spongocoel** or **atrium**, and out through an osculum. The spongocoel ranges from wide and

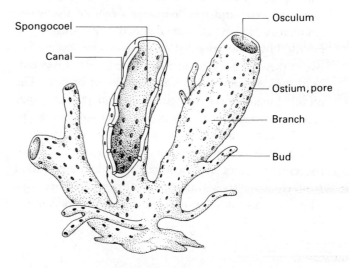

Figure 10.2. Generalized simple sponge showing gross morphology of a branching thin-walled form.

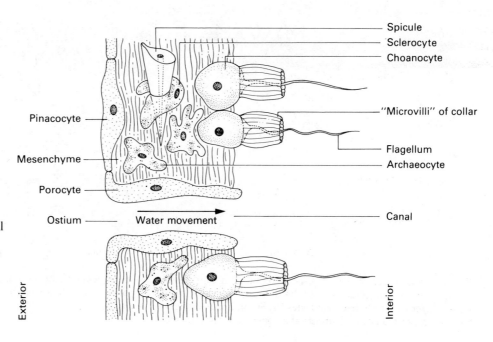

Figure 10.3. Diagram of a microscopic section of a sponge wall showing characteristic cells and their distribution. Choanocytes are the principal food gatherers and pumping cells in the sponge canal system. Sclerocytes are the cells responsible for secreting the skeleton.

saucer shaped to deep and tubular. In massive hemispherical or sheeted forms, the circulation may be through numerous incurrent and excurrent canals without development of a spongocoel.

Three general layers of cells are differentiated in most modern sponges (Figure 10.3). The outer layer in most living sponges consists of closely spaced leathery cells termed **pinacocytes**. These cells also line the spongocoel and canals in all but the simplest sponges. In some forms, however, these cells may be widely separated or lacking. The inner layer in simple sponges is characterized by variously spaced flagellate-collared cells, termed **choanocytes**. In complex sponges the choanocytes are clustered into separate chambers. Choanocytes are both the food-gathering cells and the "pumping" cells of the water circulatory system. They are distinctive of sponges. The middle layer, the **mesohyl** or **mesenchyme**, occurs between the choanocyte and pinacocyte layers. It is gelatinous and contains several different kinds of cells. The mesohyl includes **sclerocytes** and **spongocytes**, cells that secrete the skeleton; it also includes **archaeocytes**, totipotent cells that appear to be undifferentiated amoeboid cells capable of changing into all other kinds of cells of the same sponges. Archaeocytes ingest and digest food from choanocytes and transport nutrients to other cells. They may also develop into sex cells. Other cells in the

mesohyl include **collenocytes**, cells that form connective masses, and myocytes and porocytes, contractile cells that surround pores or canal openings and with collenocytes function as sphincters in the sponge canal system. Myocyte cells combine to produce openings, but porocytes are single cells through which ostia are developed. Collenocytes also move completed spicules into place in the skeleton.

Movement of water through the sponge body is produced by uncoordinated back-and-forth beating or rotary motion of the flagella of the choanocyte layer, either in isolated flagellate chambers or in the spongocoel lining.

Three structural grades of canal systems are recognized in fossil and modern sponges (Figure 10.4). All three may occur in each of several distantly related lineages and seem to show parallel evolution. The simplest sponges are of **ascon** grade (Figure 10.4, *A*). Asconoid sponges are those in which water enters the sponge through ostia in the outer wall, passes through short small canals between cells in the thin wall, and exits through the large choanocyte-lined spongocoel and the osculum. Many thin-walled fossil sponges may be of ascon grade, although proof is generally wanting. This structural grade is somewhat dependent upon recognition of the choanocyte lining of the spongocoel, and impressions of soft parts are

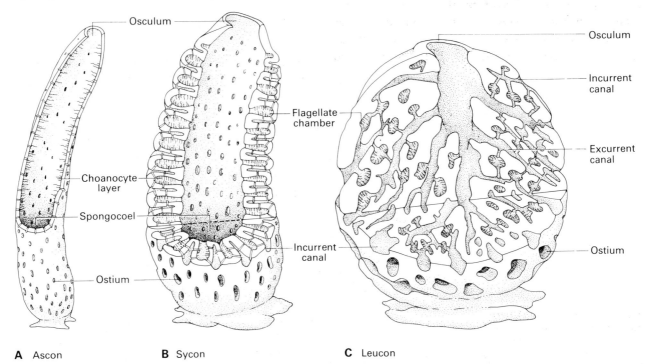

A Ascon   **B** Sycon   **C** Leucon

Figure 10.4. Three structural grades of modern and fossil sponges. **A**, Ascon grade sponges have the spongocoel lined with a layer of flagellate choanocytes. **B**, Sycon grade sponges have separate flagellate chambers that open directly into the spongocoel. C, Leucon grade sponges have flagellate chambers connected to incurrent and excurrent canals by numerous tiny canals.

rarely preserved in fossil sponges. The ascon grade is known only in calcareous living sponges, but the grade is thought to have been more widespread in fossil sponges.

Intermediate grade sponges are of **sycon** type (Figure 10.4, *B*). Water enters the sponge through porocyte-controlled ostia into radially arranged, small, incurrent canals (Figure 10.5). Water next enters separate small flagellate chambers through pores and is stirred out by choanocytes through excurrent canals into the spongocoel. The spongocoel is not lined with choanocytes. The water finally exits at the osculum. A sycon-grade sponge appears like a cluster of ascon-grade sponges. All living sycon sponges are calcareous, although some siliceous fossil sponges could have been of sycon grade.

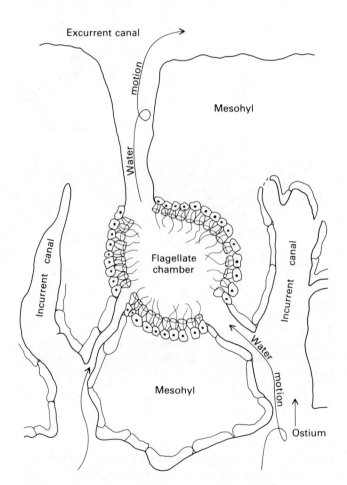

Figure 10.5. Cross section through a leucon sponge wall showing relationships of a flagellate chamber to the incurrent and excurrent canals.

Sponges of greatest complexity are of **leucon** grade (Figure 10.4, *C*) and have thick walls composed mostly of mesohyl penetrated by intricate canal systems. Flagellate chambers are commonly subdivided into tiny globular openings that are connected by numerous fine intricate canals that empty into larger excurrent canals (Figure 10.5). Leuconoid sponges commonly lack a spongocoel and may be massive, branching, or even laminar in general growth form.

Food particles brought to the sponge by water currents may be ingested by almost any cell. Much food is filtered out of the water by the sticky microvilli surrounding the flagellum of the choanocyte (Figure 10.3). These food particles are passed by streaming protoplasm to the base of the collar where some are ingested into the cell and digested there. Most particles, however, are passed to archaeocytes or collenocytes where digestion takes place. Waste products are egested and washed away by the same circulating water, generally on a cell-by-cell basis. Some limited transfer of nutrients, however, is apparently accomplished by amoeboid archaeocyte cells. These same archaeocytes also transport solid waste products from other cells of the mesohyl and discharge the debris into water in the canal system.

Skeleton

Skeletons of sponges are secreted by specialized cells, the spongocytes and sclerocytes. The skeleton is formed of either crystalline-to-microgranular calcium carbonate, silica in an opaline form, secreted by the sclerocytes, or **spongin** fibers secreted by spongocytes. Spongin, the material of commercial sponges, is chemically a relatively inert and insoluble sulfur-containing scleroprotein. Some sponges possess both spongin and siliceous spicules. Spongin is not associated with calcareous sponges. Some spongin fibers are made more rigid or are strengthened by inclusion of foreign particles within the fibers during their secretion. That condition has not been recognized in the fossil record to date, and recognition of totally spongin skeletons in the geological record is also suspect. Calcium carbonate-based skeletons may be spicular and calcitic or, if layered, granular-to-crystalline calcite or aragonite. Cross-supporting structures are also present in some major taxa. Only one group within the phylum is known to have intermixed calcareous and siliceous skeletal elements in the same animal. In some modern sponges (the sclerosponges), siliceous spicules occur in an otherwise layered-to-chambered, massive calcareous skeleton.

Each siliceous spicule is secreted intracellularly, in hours, initially by a single sclerocyte around an initial triangular or rectangular spongin fiber (Figure 10.6, *A* to *E*). Subsequent decay or loss of the fiber produces an opening, termed an **axial canal** (Figure 10.7, *A*) or **crepidal canal** (Figure 10.7, *B*) in different sponge groups. Calcareous spicules are secreted over fine needles of calcium carbonate initially in a single sclerocyte, but later secretion is extracellular by a "team" in the space between cells (Figure 10.6, *A'* to *E'*). All fossil and

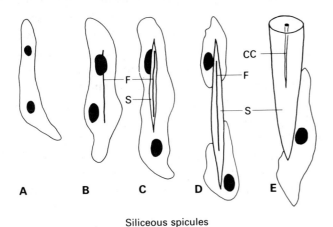

Siliceous spicules

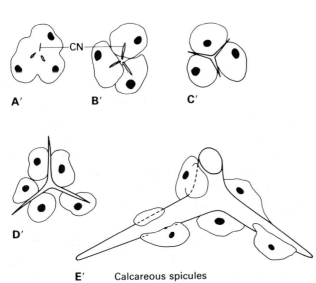

E'   Calcareous spicules

Figure 10.6. Secretion of spicules. **A** to **E** show the stages in secretion of siliceous spicules. **A,** Single nucleus of a sclerocyte divides into two. **B,** Thin fiber of spongin (*F*) is secreted along the long axis of the cell, generally between the two nuclei. **C,** Opaline silica (*S*) is secreted over the initial spongin fiber within the sclerocyte. **D,** When the spicule becomes too large to be contained within the single cell, the sclerocyte subdivides. The spicule is then enlarged by a team of cells that gradually secrete additional silica over the spongin fiber. **E,** In due time the spongin fiber will decompose to produce the crepidal canal (*CC*) after the spicule has been enlarged.

A' to *E'* show the stages in secretion of calcareous spicules. **A',** Calcareous spicules also have their origin in a single cell, which is generally trinucleate, in which three tiny needles of calcium carbonate (*CN*) are precipitated. **B',** Trinucleate cell divides, and the calcium carbonate needles are gradually enlarged. **C',** Needles are fused. **D',** Spicule is enlarged by a team of sclerocytes that work back and forth along the rays. **E',** Spicule reaches normal size for the species.

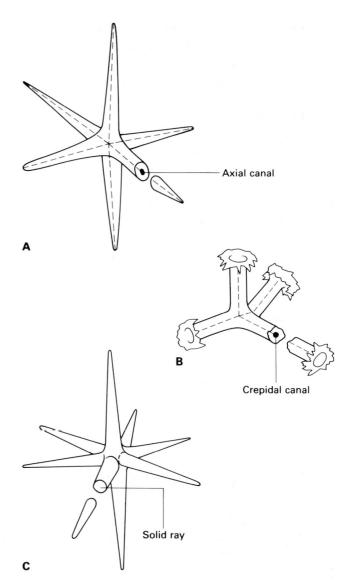

Figure 10.7. Structure of various types of sponge spicules. **A,** Siliceous hexactine. Each of the six rays was constructed around a spongin fiber that produced the axial canal. **B,** Siliceous tetractine. Each of the four rays is built around a crepidal canal formed by the initial spongin fiber. **C,** Solid-rayed octactine. Spicules such as these are characteristic of more advanced heteractinid calcareous sponges.

modern calcareous spicules lack a central canal (Figure 10.7, *C*). Irregularities in water chemistry may cause deformed spicules, particularly if secretion is blocked or modified by unusual concentrations of germanium or possibly other as yet unidentified rare elements.

Nomenclatures of spicules depend on the numbers of **axes** (**monaxon, triaxon,** or **tetraxon**) (such as *A–A'* of Figure 10.8), the numbers of **rays** or needlelike terminations corresponding to directions of growth (**monactine, diactine, triactine,** etc.) (such as *D–B* or *D–B'* of Figure 10.8), or size (**megasclere, microsclere**) (Figure 10.9).

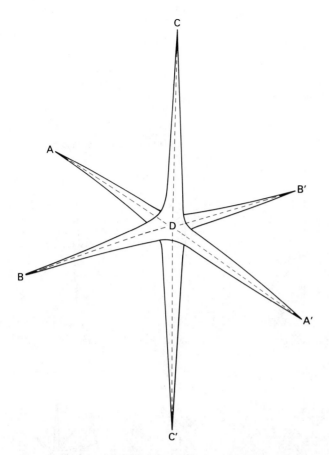

Figure 10.8. Characteristic hexactine, showing relationships of axes to rays of the spicule. Three axes (*A-A', B-B', C-C'*) are mutually arranged at right angles to each other. Individual rays extend along each axis from their common junction at *D* to their tips. For example, one ray extends along line *A-D*, and another ray along line *A'-D* so that six rays grow along the three axes.

Simplest megasclere spicules are monaxial (Figure 10.10, *A* to *H*) and are formed by growth in one (monactine) or two directions (diactine) along that axis to produce needlelike, straight, or curved spicules.

Triaxons (Figure 10.10, *I* to *N*) may be either calcareous or siliceous and may have axes arranged at 120 degrees to each other, or like prongs on a tuning fork, or at right angles to each other. Spicules with 90-degree axes (Figure 10.10, *L* and *N*) are initially siliceous. Tuning-fork types (Figure 10.10, *I* and *J*) are initially calcareous. Triaxial spicules may be triactine, with three rays (Figures 10.10, *I* to *K*); **hexactine**, with six rays (Figure 10.10, *L*); or some derivative such as a pentactine (Figure 10.10, *M*) where one ray is aborted or a **stauract** where two of the opposing rays are aborted (Figure 10.10, *N*).

Tetraxial spicules are almost consistently tetractines (Figure 10.10, *O* to *V*) and include such initially siliceous spicules as smooth calthrops, tricranoclones, rhizoclones, or dicranoclones that characterize particular genera or families. **Octactine** (Figure 10.10, *W*) spicules are special calcareous tetraxons in which three of the axes are in a plane normal to the other axis.

Some spicules have many axes or rays and are termed **polyaxons** or **polyactines** and **sphaeractines** (Figure 10.10, *X* and *Y*).

Two classes of Porifera, the Sclerospongea (Figure 10.11) and Stromatoporata (discussed further in the next section) are characterized by basal skeletons of massive or layered calcareous material. In addition, sclerosponges secrete siliceous monaxial spicules and spongin fibers as part of their compound skeleton. Their siliceous spicules may be included in the calcareous base (Figure 10.11, *B*) as additional carbonate material is secreted during growth. The basal skeleton may be trabecular, in which the calcareous crystals are arranged in an upward diverging pattern, or spherulitic, in which the calcareous crystals radiate from numerous centers.

Outer living surfaces of sclerosponges may be marked by **astrorhizae**, starlike or rootlike grooves that radiate from oscular openings (Figure 10.11, *C*). In addition,

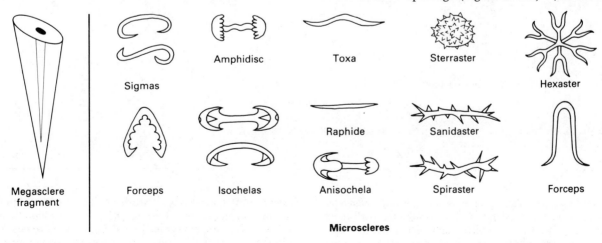

Microscleres

Figure 10.9. Nomenclature of common microscleres. Spicules are shown in comparison to a ray fragment of a megasclere.

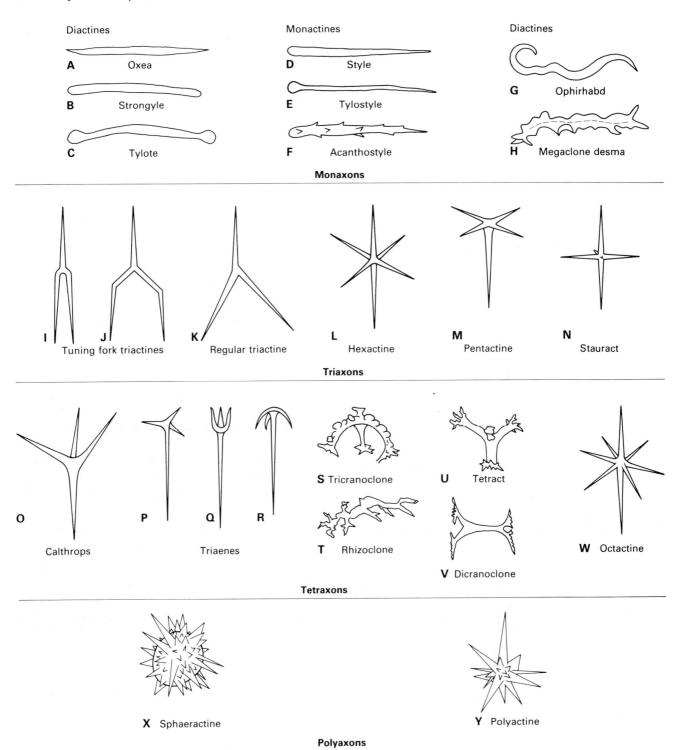

Figure 10.10. Nomenclature of common megascleres in fossil and modern sponges.

Figure 10.11. Recent sclerosponges. **A** and **B,** *Acanthochaetetes* from Saipan, Marianas Islands. **A,** Encrusting form perforated by calicles. (×1.) **B,** Photomicrograph of exterior of same specimen showing subprismatic calicles in calcareous skeleton with embedded monaxons. (Approximately ×25.) **C,** *Astro-* *sclera* from Fiji Islands. Astrorhizal-like canals are on low mounds in calcareous skeleton. (×2.) **D,** *Ceratoporella* from Jamaica. Calicles and surficial canals show in calcareous skeleton that encrusts calcareous worm tubes (×3.) (Courtesy W. D. Hartman.)

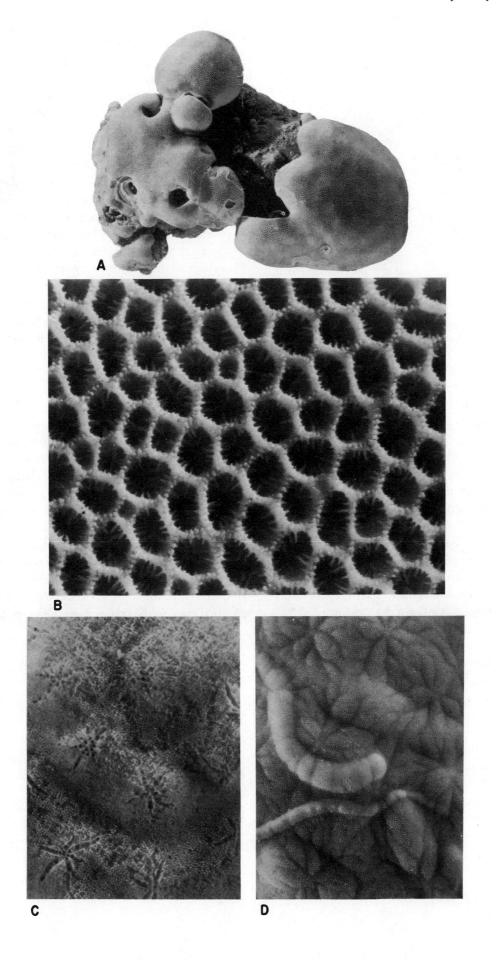

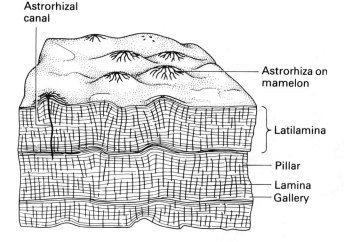

Figure 10.12. Upper surface and side of a block of the stromatoporate *Actinostroma* showing a characteristic longitudinal section. The surface is marked with low mounds, mamelons, and radiating rootlike astrorhizae. Latilaminae are regular groupings of laminae and are differentiated by spacing or thickness variations of laminae. Pillars are vertical elements and, with laminae, differentiate openings (galleries) in the skeleton. (×5.)

irregular small depressions and meandering furrows occur in the living layer of the skeleton. **Calicles** are tiny prismatic-to-cylindrical openings occupied by living layers, that extend into the calcareous skeleton (Figure 10.11, *D*). The more interior part of the skeleton is sealed off by massive calcareous deposits that fill older parts of the calicles or by **tabulae,** flat partitions or floors of the tubes. In one order, outer tabulae are perforated, and tissue extends some distance into the skeleton.

Stromatoporates (Figures 10.12 to 10.15) secreted an open calcareous skeleton, or **coenosteum,** composed of a regular network of structural elements parallel and perpendicular to the growth surface. They grew in domal, tabular, encrusting, dendroid, or digitate forms.

Surfaces of coenostea may be smooth or bear small and distinctive elevated mounds, termed **mamelons** (Figure 10.12). Skeletons of approximately one half of the species are marked also with the irregular, radially branching grooves of astrorhizae (Figures 10.12 and 10.13). These horizontal grooves may converge upon more or less vertical tubes, **astrorhizal canals,** which penetrate into the layered skeleton as irregularly branching openings.

Single, sheetlike layers parallel to the growth surface of stromatoporate coenostea are called **laminae** (Figure 10.12). Spaces between laminae are called **galleries.** Galleries are filled with calcite spar or rarely infiltrating sediment in fossils, but in life were either filled with sea water or possibly with the soft tissue of the organism. **Pillars** are rodlike elements that are oriented perpendicular to laminae and growth surfaces. They may pass through several laminae (Figure 10.14) or connect two adjacent laminae (Figure 10.15, *A*). **Dissepiments** are curved, cystlike plates that may occur in early-formed parts of the skeleton and partially fill galleries or may make up most of the skeleton in Early Paleozoic forms.

Growth may have been cyclic because laminae may be grouped into thicker-compound units, termed **latilaminae,** which look almost like seasonal growth-rings in plants (Figures 10.12 and 10.14). Latilaminae are most obvious in species with prominent laminae and limited pillars. Latilaminae are generally marked by distinctive changes in density of the coenosteum attributable either to variations in thickness of laminar layers or abrupt cyclic variations in spacing of the laminae. Causes of latilaminar development are not certain; they are not present in all stromatoporates, not even in the same locality. Laminae, however, are thought to mark changes in the environment, such as tidal, seasonal, temperature, or food cycles.

Classification

Shapes of spicules are primary data for classification in those forms with spiculate skeletons. Microscleres have been visualized as conservative elements in the skeleton and, as a consequence, show long-ranging relationships. Classifications of modern sponges, therefore, have been developed using microscleres as the key morphologic features. These classifications are of limited use in paleontology, however, for microscleres are not common in the fossil record and are unknown from nearly all exclusively fossil genera and species. Therefore, megascleres and, for some taxa, the shape of the sponge are relied upon in the taxonomy of fossil sponges.

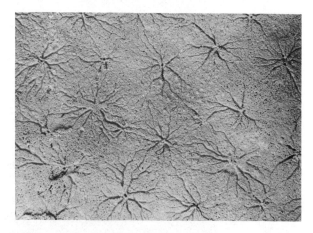

Figure 10.13. Astrorhizae, rootlike grooves, on the surface of a stromatoporate, *Syringostroma,* from the Middle Devonian of southern Ontario.

Figure 10.14. Longitudinal sections of stromatoporate skeleton (coenosteum) composed of laminae and pillars in varying developments. *Actinostroma*, shown here, has equal development of horizontal laminae and vertical pillars. Latilaminae are set off by variations in spacing or thicknss of laminae.

Figure 10.15. Thin sections of *Anostylostroma* from the Middle Devonian of southern Ontario. **A,** Longitudinal section showing prominent laminae and pillars, which are generally limited between laminae. Opening on the left is an astrorhizal canal. (×4.) **B,** Tangential section in which pillars are cut transversely and appear as small dark spots. Laminae are curved sweeping dark, but somewhat fuzzy, lines. (×4.)

A

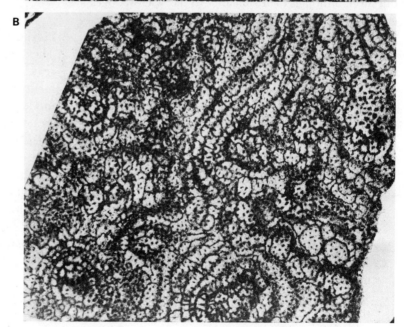

B

Some individual sponges may have numerous kinds of spicules, and some kinds of spicules occur in many genera or families. In general, taxonomy from a collection of loose spicules is difficult, if not impossible, in many groups. Complex spicules are more useful for identification than generalized ones.

Several major taxa are considered to be polyphyletic. For example, complexly shaped spicules of one order of Demospongea are thought by some workers to be sufficiently different that the order may be of multiple origins. Monaxonid demosponges have such a great variety of microscleres that they also are probably polyphyletic. At present, the accepted sponge classification is based upon morphologic similarity, however, rather than upon phylogeny. Some parts of the classification of the phylum may be more indicative of phylogeny, but the fossil sponge record is so incomplete that most taxonomic relationships must be built upon widely scattered geographic and chronological evidence. Understanding of fossil sponges is still in the early descriptive phases of investigation.

Even the classification of living sponges is difficult because of the lack of experimental and observational data. Sponges are difficult to culture in the laboratory and to sustain as whole organisms. Sponges show great environmental plasticity, and much of their morphologic variation may be unrelated to genetic differences. Such problems are compounded for fossils from which most biologic detail has long been lost.

Four classes are included within the Porifera: Demospongea, Hexactinellida, Calcarea, and Sclerospongea. The Stromatoporata are classified as Porifera with some question, although their similarity to the Sclerospongea certainly indicates close relationships. Some biologists have suggested that the Hexactinellida be excluded from the Porifera, since living organisms of this class generally lack a **pinacoderm** (outer layer of pinacocytes) and have other minor differences. Without a pinacoderm, true pores are impossible and definition of individuality is difficult. Lack of biologic and structural information on the Hexactinellida, as compared to the Demospongea or Calcarea, makes detailed evaluation of the class difficult. The conservative and historical view is to include the class within the phylum, as is done here.

CLASS DEMOSPONGEA (Figure 10.16)

Poriferans with skeletons composed of spongin, mixed spongin and siliceous spicules, or siliceous spicules built upon monaxons or tetraxons in which the rays are not at right angles to one another. (Cambrian to Holocene; 390 fossil genera.)

Demosponges are the dominant class of living sponges, with approximately 600 living genera, as compared to 60 living genera of calcisponges and 130 genera of hexactinellids. However, only 50 of these demosponge genera have solidly fused skeletons that are likely to leave a fossil record. This compares with 150 fossil genera with solidly fused skeletons known from Cretaceous rocks, when the phylum probably was at its peak of diversity. If the proportions of solid-to-loosely spiculate or aspiculate skeletons has been constant, there may have been as many as 1800 Cretaceous demosponge genera. Orders are differentiated upon the basis of skeleton composition, shapes of spicules, and whether the spicules are fused or separate.

Demosponges are dominantly marine, but the class also includes all the known freshwater sponges. Oldest representatives of the class are from Lower Cambrian rocks, but the greatest expansion in the fossil record of the Paleozoic is in Lower and Middle Ordovician rocks where sponges form reefs. From the Silurian on, however, demosponges are common in both clastic and carbonate rocks and in deep and shallow marine deposits. A few sponges with skeletons of only isolated spicules are known through miracles of preservation and discovery, like that which produced the trove of the Cambrian Burgess Shale.

CLASS HEXACTINELLIDA (Figure 10.17)

Poriferans with skeletons composed of siliceous hexactines, hexactine-derived spicules, or stauracts (rays at mutual right angles). Some with adjacent spicules fused to form rigid skeletons. (Cambrian to Holocene; 295 fossil genera.)

Hexactinellid sponges are an exclusively marine class. They are the common deep-marine sponges of modern seas, although they apparently moved into deep environments only during the middle and late Mesozoic. Many modern hexactinellids are not attached to a hard substrate but are anchored by ropy root tufts or mats of spicules. They are thus able to occupy loose or muddy bottoms. Their geologic record suggests a similar adaptation as far back as the Middle Cambrian; tufted sponges and traces of ropy rootlike tufts are moderately common from then on in black shale or fine-grained limestone assemblages.

Classification of modern hexactinellids is based in part upon soft parts and characteristic microscleres. Fossil hexactinellids, on the other hand, are classified using larger features of the skeleton. The most primitive and oldest order appeared in the Cambrian and is characterized by skeletons of loose or only weakly fused spicules. A

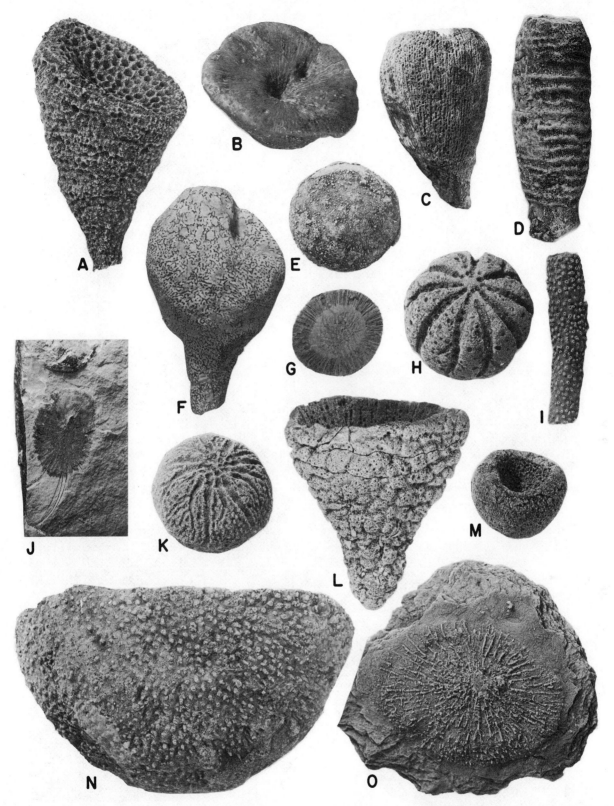

Figure 10.16. Representative demosponges, approximately natural size. **A,** *Aulocopoides* (Devonian, Western Australia). **B,** *Zittelella* (Ordovician, Illinois). **C,** *Calycocoelia* (Ordovician, Texas). **D,** *Nevadocoelia* (Ordovician, Nevada). **E,** *Scheiia* (Permian, Arctic Canada). **G,** Natural cross section of *Scheiia* showing straight radiating canals. **F,** *Phymatella* (Cretaceous, Germany). **H,** *Caryospongia* (Silurian, Tennessee). **I,** *Coelocladia* (Pennsylvanian, Texas). **J,** *Choia* (Cambrian, Utah). **K,** *Astylospongia* (Silurian, Tennessee). **L,** *Camellaspongia* (Ordovician, Minnesota). **M,** *Palaeomanon* (Silurian, Tennessee). **N,** *Haplistion* (Permian, Arctic Canada). **O,** *Haplistion* (Pennsylvanian, Utah).

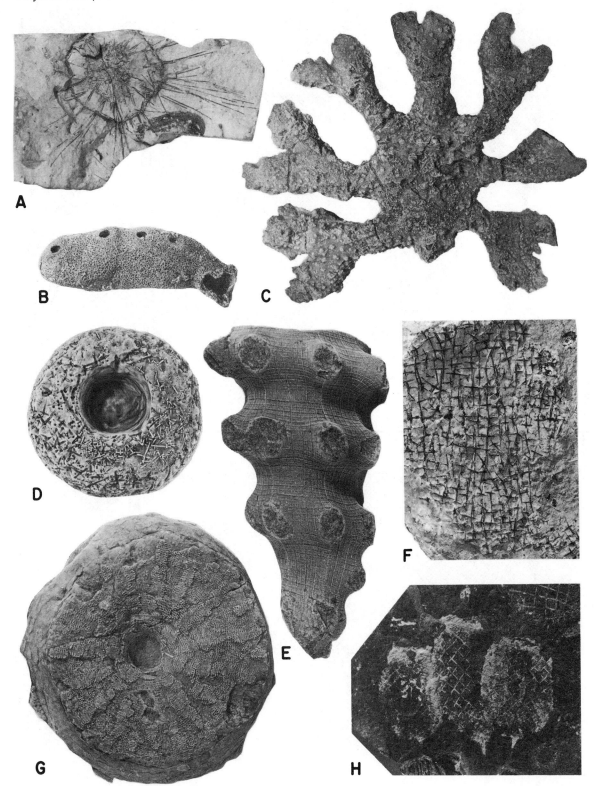

Figure 10.17. Representative hexactinellid sponges. Natural size unless noted. **A,** *Dierespongia* (Ordovician, Oklahoma). **B,** *Eureta* species (Eocene, North Carolina). **C,** *Brachiospongia* (Ordovician, Indiana). This thick-walled sponge has irregularly oriented and unevenly spaced spicules (approximately one-half size). **D,** *Twenhofelella* (Silurian, Anticosti Island, Quebec). **E,** *Hydnoceras* (Devonian, New York). The fossil is the filling of the spongocoel but shows molds of regularly oriented spicule straps in a rectangular pattern. **F,** *Mattaspongia* (Devonian, Alberta). Coarse hexactines are regularly oriented and form the thin wall. **G,** *Coeloptychium* (Cretaceous, France), showing a shallow spongocoel and dictyonine skeletal net. **H,** *Diagoniella* (Cambrian, Utah). The sponges are preserved as limonite-stained clay impressions. Root tufts show at the base of the diagonally arranged, stauract-based spicule net.

second order appeared in the Devonian and is characterized by a solidly fused, rectangularly based, three-dimensional skeleton. The third order appeared in the Triassic and has a similarly fused three-dimensional skeleton but with additional cross-bracing buttresses at the spicule centers.

Hexactinellid sponges are common elements of Cambrian faunas worldwide but are known from many localities only as dissociated spicules. They are common in clastic and calcareous deposits of quiet-water marine environments of the Paleozoic. The class in North America was probably at its apex during the Devonian and Mississippian when the famed glass sponge faunas of Indiana and New York accumulated. The hexactinellids probably reached their greatest diversity and numbers in Jurassic and Cretaceous rocks of southern Europe where they are common reef, interreef, and basin elements.

CLASS CALCAREA (Figure 10.18)

Poriferans with skeleton of calcareous spicules or porous calcareous walls lacking spicules. Aspiculate sponges beadlike or cystose linear, branched, or massive. Spongin lacking. (Cambrian to Holocene; 115 fossil genera.)

Calcareous sponges are exclusively marine organisms in modern seas, where they are most common in shallow tropical waters. Distributions in the geological record indicate that they lived in similar environments in the past.

Living calcareous sponges are classified using larval development, canal patterns, and skeletal structures. Only the skeletal structures are available to paleontologists, and thus a somewhat different classification is applied to fossils. Four major groups of calcareous sponges are recognized. The earliest known and only extinct one is the heteractinids, those sponges with octactine-based and related skeletons. They are major elements of Cambrian, Silurian, and Carboniferous faunas. Late Paleozoic heteractinids developed secondarily thickened skeletal tracts by overgrowth of spicules. This development apparently gave rise to the second group, the pharetrone sponges, of the Mesozoic and Cenozoic. These latter sponges have triactine to monactine spicules commonly welded into tracts and a solid framework by aspiculate calcareous matrix. A third and principally Cenozoic group of calcareous sponges is characterized by loose skeletons of isolated triactine and related spicules. These sponges may have had a long history but, like many demosponges with loose skeletons, have left a poor record. The fourth main group of calcareous sponges, the sphinctozoans, are characterized by skeletons that are segmented, beadlike, or bubblelike; aspiculate; and possibly external. Although their skeletons lack spicules, their canal or pore systems, growth form, and lack of specialized structures characteristic of more complex phyla suggested placement of the sphinctozoans in the Porifera. That placement is now fully supported by living representatives. They were questionably present in the Cambrian but certainly appeared by the Ordovician and reached major diversity and abundance in the Permian and Triassic, producing extensive reefs. Sphinctozoan sponges were thought to have been extinct from the Late Cretaceous, until a single living species was described in 1977.

CLASS SCLEROSPONGEA (Figure 10.11)

Poriferans with compound skeleton of siliceous spicules, spongin fibers and basal massive laminated fibrous aragonite or calcite layer, or basal layer with calicles. Lack dissepiments, laminae, and latilaminae. (Ordovician to Permian?, Triassic to Holocene; 10 genera.)

The sclerosponges are a minor class that would probably be ignored in paleontology courses were it not for their striking similarity to stromatoporates and to some fossils traditionally grouped with the tabulate corals. The sclerosponges were first described in 1911, but it was not until their rediscovery in the 1960s that the class received much attention. Living species are exclusively marine, distinctly tropical, **cryptic** (hidden under shells or rocks) forms, and are now considered possible relics of groups more common in the Paleozoic and Mesozoic. Some researchers of modern sponges would include the stromatoporates within the Sclerospongea because of the similarity of their basal calcareous skeletons. Living tissue and siliceous spicules of sclerosponges are like demosponges, although somewhat more divided into living units. If the minutely tubular fossils normally classified as chaetetids are sclerosponges, as is the contemporary interpretation, then the geological history of that class dates from the Middle Ordovician.

CLASS STROMATOPORATA (Figures 10.12 to 10.15)

Presumed poriferans with a basal, laminated-to-vertically tubular, calcareous, aspiculate skeleton. Earlier forms with laminae, latilaminae, and pillars or thickened skeletal elements. Later forms with

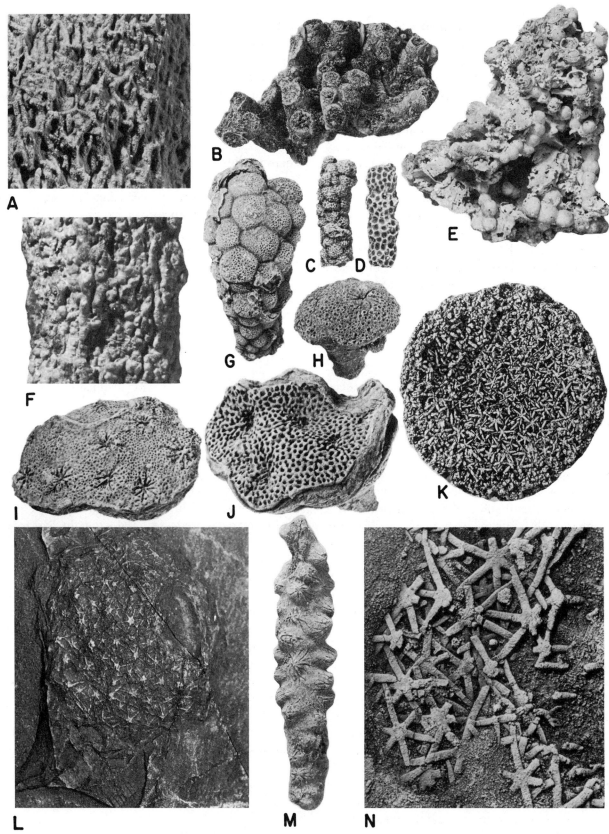

Figure 10.18. Representative calcareous sponges, approximately natural size, except where noted. **A,** *Wewokella* (Pennsylvanian, Oklahoma, ×5). **B,** *Peronodella* (Cretaceous, Great Britain). **C** and **G,** *Cystothalamia* (Permian, Tunisia). **D,** *Pseudoguadalupia* (Permian, Tunisia, ×2). **E,** *Girtyocoelia* (Permian, Texas). **F,** *Regispongia* (Pennsylvanian, Texas, ×5). **H,** *Precorynella* (Permian, Tunisia). **I** and **M,** *Stellispongia* (Permian, Tunisia). **J,** *Himatella* (Permian, Tunisia). **K,** *Astraeospongium* (Silurian, Tennessee). **L,** *Eiffelia* (Cambrian, British Columbia, ×3). **N,** *Ensiferites* (Devonian, New York, ×5).

cystose and vertically tubular skeletons, most perforated by astrorhizal canals, many with tabulae, dissepiments, and latilaminae. (Ordovician to Cretaceous; 70 genera.)

Stromatoporates are an extinct group of lower invertebrates of doubtful affinities that produced large, coral-like calcareous skeletons. They are important elements of mid-Paleozoic and mid-Mesozoic reefs. Their skeletons range from ramose or branching fingerlike masses only a few millimeters across to laminar encrusting sheets several meters wide or massive domes nearly 5 m in diameter. Growth form was apparently related mainly to environmental parameters; the organisms show plastic growth habits. Stromatoporates occur worldwide in middle Paleozoic rocks but have proven difficult to identify because of their variability in internal structure and growth form.

Origin and taxonomic relationships

The origin of the Porifera is lost in the vague and incomplete record of the Proterozoic. Sponges probably arose from flagellate protozoans, but that conclusion is still largely speculation. The sponges are considered as an evolutionary side issue or "dead end" because they did not produce other kinds of organisms. They have been and continue to be, however, an exceptionally successful phylum in marine environments.

Sponges are generally placed in a separate subkingdom, the Parazoa (as in Chapter 2), because of interpreted differences in embryologic development. It was once thought that the locomotor flagellate cells on the anterior of sponge larvae migrated inward during metamorphosis to become the choanosomal feeding layer. The initially interior cells thus become ectodermal by that invagination. Such a development is not seen in metazoan larvae and suggests that sponges may have had a separate origin.

Later workers have pointed out, however, that the theory of inversion of layers was based upon observations of very complex larvae and that more recent detailed studies on larvae of primitive demosponges do not show such development. Instead, these studies show that potential ectodermal and endodermal cells are distinguishable, even at very early stages in development. Larvae of one genus, for example, settle on the flagellate end, and the osclum opens in the free nonflagellate end. No inversion takes place. Flagellate cells disperse; some are destroyed or shed, and others may migrate inward singly or as clusters to form the choanocyte layer.

Inversion of layers thus does not appear to be a basic pattern in the Porifera, and developmentally there seems little reason to exclude the sponges from the Metazoa. Sponges, however, have remained at a cellular grade of organization, and most sponge biologists conclude that they represent a unique evolutionary branch from an unknown although probably common ancestor to the Cnidaria and other metazoans.

Part II Additional concepts

Classes traditionally included in the sponges

The following discussions generally refer to those classes traditionally included within the Porifera, that is, all those except the Stromatoporata, about which little biology is known. The Stromatoporata are treated separately in a later section.

Methods of study

Fossil sponges are generally studied free of matrix. Many well-known faunas are silicified, such as in the Cretaceous and Jurassic of northern Germany, the Permian of West Texas (Figure 10.18, *E*), and the Eocene of Australia. Silicified sponges are removed from the enclosing matrix by etching with dilute hydrochloric or acetic acid. Other faunas occur in argillaceous matrix and individual sponges weather free, such as in the Ordovician and Silurian of northern Europe or eastern and central United States.

Some sponge faunas also occur as three-dimensional casts and molds in sandstone and siltstone, such as the glass sponge assemblages of Devonian and Mississippian age in Indiana, Ohio, Pennsylvania, and New York. Many Cambrian sponges, such as those in the Burgess Shale of western Canada, are flattened impressions on bedding planes of limestone or argillaceous rocks.

Most reef-dwelling sponges, such as those in the Triassic of Italy and Germany and in much of the Permian of Texas and Sicily, must be studied in thin sections or laboriously mechanically freed of matrix. Each preservation, in each kind of matrix, requires slightly different treatment to reconstruct the three-dimensional relationships of the skeleton and canal systems.

Reproduction

Reproduction in sponges is both sexual and asexual. Egg cells develop from archaeocytes in the mesohyl, and sperm cells develop from choanocytes. Individual sponges may be dioecious and also show sex reversal from one year to the next. Fertilized eggs develop into a cell cluster, with small flagellate cells at one end and larger epithelial-like cells at the other. Larvae settle, become attached, and develop into young sponges.

Sponges reproduce asexually by budding or development of gemmules. Most noticeable buds in marine sponges are small subspherical masses of a few archaeocytes and related cells. These buds extend from the surface of the parent sponge on stalks or long filaments. Once the bud is detached, the stalk is usually resorbed. Many demosponges probably produce less obvious but perhaps more common buds that are merely small and low extensions of the parent sponge (Figure 10.2). These buds contain all the necessary cell types and are "pinched" off by the parent sponge. Other types of buds develop from basal stolons of the parent sponge. These types may remain attached to the parent.

Gemmules are subspherical, cyst-like, armored clusters of food-laden archaeocytes, and are particularly common in freshwater forms, but also occur in some marine forms. During stressful periods of unusual temperature, chemical modification, or exposure by drought, the parent sponge may degenerate into numerous gemmules. Gemmules are remarkably resistant to dehydration and other environmental stresses. Some dried gemmules have revived and produced adult sponges after being out of water for 10 years. The fossil record of gemmules probably does not reflect the actual range and abundance of these structures in the past.

All species of sponges probably are able to regenerate into fully developed animals from small fragments. The phylum generally shows great regenerative and reassociative powers. For example, after storms, repopulation of commercial sponge fishing grounds by planting sponge fragments has been scientifically although not economically successful. Among encrusting or branching forms different parts may continue to grow even though intervening sections may have died or been removed. This has led some scientists to view these sponges as colonies rather than as individuals.

Individual sponges may reassociate after having been separated into single cells or cell clusters. Cells of each type are species specific, and mixtures of cells from several species do not produce hybrids but numerous small sponges of each of the several species. Individual cells are capable of limited separate existence but do not develop into new sponges by themselves. Sponges may also heal themselves by regeneration or by lateral movement of adjacent undamaged cells.

Ecology and paleoecology

As a phylum, sponges show great ecologic variation and adaptability. Most individual species and genera are considerably more restricted environmentally, however. Temperature, salinity, currents, turbidity tolerance, and suitable substrate limit some species.

Modern sponges have varying tolerance to turbidity. Some survive in highly turbid environments and have developed back-flushing circulation to keep from being buried in smothering sediments. Modern tubular sponges generally occur where currents are moderately strong. These organisms are typically upright or oriented normal to the direction of current flow. The chimneylike pattern may also be an adaptation to remove waste-contaminated water in environments where currents are weak.

Sponges are able to produce a unidirectional flow of water though their canal systems by the beating of flagella, but that flow is enhanced where water currents move over the sponge. Experiments with sponge-shaped models show that the volume of water moving through a sponge increases when velocity of water flowing past the osculum increases, just as the amount of smoke and gas moving up a chimney increases when the wind blows. Conical or cylindrical sponges have ideal configurations for producing such induced flow for they have numerous small inlets in the lower sponge wall and a large osculum at the summit. Growth is apparently enhanced when water flow, and thus food, is increased because of the Bernoulli effect "sucking" water through the organisms. Additional experiments related to water motion in sponges are critical, since few data exist.

Sponges are able to pass great quantities of water through their systems. For example, a black loggerhead sponge, 50 cm in diameter and 30 cm tall, may draw approximately 1000 l of water through its canal systems in a single day. Other sponges may pass 10,000 to 20,000 times as much water as their volume through their canals in a single day.

Sponges are generally sessile benthic forms, achieving their main geographic distribution during larval stages. Some spherical species, however, may have been planktic, and some may have rolled free. Some may have been floaters or attached to other floating organisms, particularly some of the smaller forms described from graptolitic shales in New York or Alaska. Stalks apparently allowed certain genera to occupy muddy or soft substrates.

Paleozoic and earliest Mesozoic fossil sponges were essentially shallow-water groups. Sponges were important reef builders during Paleozoic and Mesozoic time. Demosponges (Figure 10.16, *B* to *D*) were important frame builders in reefs during early and middle Ordovician time in Utah, Nevada, New York, Vermont, Texas, Oklahoma, and southeastern Canada. Sponges are also important elements between the Silurian reefs of the Great Lakes area, in northern Europe, and along the flanks of Devonian reefs in western Australia and western Canada. Permian reefs of Texas, New Mexico, Tunisia, Sicily, and Indonesia and Triassic reefs of northern Italy and southern Germany were formed, in significant part, by beadlike calcareous sponges (sphinctozoans) (Figure 10.18, *E*) and siliceous demosponges. Hexactinellid sponges were important frame builders in Jurassic reefs of Germany, Switzerland, and France.

In the Jurassic, hexactinellid sponges apparently became successful in deeper waters, perhaps concurrently with opening of the Atlantic Ocean basin as they occupied that new territory. Modern hexactinellids are relatively rare in depths less than approximately 200 m, reaching greatest variability at intermediate oceanic or shelf depths of 200 to 600 m, based upon collections made from Indonesian waters. They are moderately common to at least 5000 m, and observations suggest that they extend to the deepest parts of the trenches and deeps.

Most calcareous sponges, however, are most numerous in water less than 100 m deep, with a depth distribution pattern that has persisted since the Carboniferous. Some living calcareous sponges do occur, however, in water up to approximately 400 m deep, and some are known in depths down to 900 m. One calcareous species has been recorded in water 2195 m deep.

Modern sclerosponges are shadowed-recess dwellers and are most common in caves and tunnels. They occur on exposed surfaces only at depths of more than approximately 60 m. Sclerosponges can be found at depths ranging from 5 to approximately 200 m. They could range into even shallower water in darkened habitats and certainly may occur in water deeper than 200 m. They are relatively rare and have been discovered widely only recently. Currently they are known principally from the Caribbean Sea area and from the Indo-Pacific tropical region. Possible sclerosponges of the Paleozoic and sclerosponges of the Mesozoic apparently thrived in more exposed conditions than preferred by modern relicts; fossil examples are commonly associated with reefs. If some of the chaetetid tabulate corals are early

sclerosponges, as some workers have suggested, certainly these organisms occupied much more prominent and exposed positions in the past than they do today.

Predators of sponges are limited, but starfish in Antarctica and fish, nudibranch snails, starfish, chitons, and turtles eat sponges in tropical water. Snails in shallow midlatitude coastal waters, plus various caddis fly larvae in fresh water, also feed to some extent on sponges. The phylum as a whole, however, is not normally considered as a standing crop to be harvested by predators. Sponges are commonly looked upon as havens of refuge for smaller organisms, and some predation may be related to organisms feeding on commensal animals within the sponge rather than on the sponge itself. Various worms, arthropods, fish, mollusks, and protozoans apparently thrive within the canal system of sponges. A single black loggerhead sponge was reported to contain over 10,000 organisms within its canals and skeleton. Shrimp pairs trapped in the spongocoel of the siliceous sponge *Euplectella* are prized as wedding gifts in the Orient. Mollusks and other animals with hard parts, trapped inside sponges, have considerable negative effect on commercial values of individual sponges and are a problem in the sponge industry.

Evolution

Evolutionary patterns in the Porifera are obscure, partly because of the generalized nature of the organisms, but more because of the inadequacy of the fossil record. A few patterns are becoming evident, however. For example, parallel trends are observed in some calcareous and hexactinellid sponges. Sponges in both groups started in the earliest Paleozoic with thin walls of regularly arranged skeletal nets one or two spicules thick (Figure 10.17, *H* is a primitive hexactinellid and Figure 10.18, *L* is a calcisponge of comparable early development). Two diverging lineages apparently developed from each group during the Paleozoic. In the hexactinellids, one lineage started with a thin geometrically arranged wall and later developed vase-shaped sponges with thickened walls of regularly arranged spicule tracts (Figure 10.17, *E*). The other lineage of hexactinellids led to thick-walled sponges in which the net became irregular (Figure 10.17, *C* and *D*). In one order of the calcisponges, a nearly similar pattern is evident. Early forms were thin-walled open conical sponges with regularly arranged skeletal nets (Figure 10.18, *L*). One lineage maintained this thin wall and regularity, but another lineage developed massive hemispherical to low conical thick-walled sponges in which the skeletal nets became irregular (Figure 10.18, *K* and *N*).

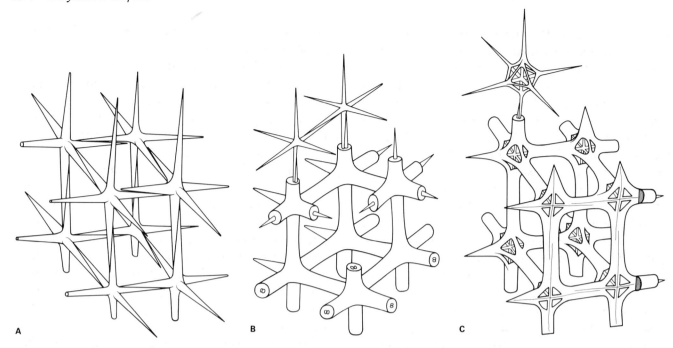

A   B   C

Figure 10.19. Skeletal structures of lyssacid, dictyid, and lychniscid hexactinellid sponges. **A**, Hexactines of lyssacid skeletal nets are discrete unfused spicules. Ray tips commonly overlap and may have been held together by an organic substance. **B**, Hexactines of dictyid skeletal nets have overlapping rays and are solidly fused into a three-dimensional net. Ray junction in the center of each spicule is simple. **C**, Hexactines of lychniscid sponges have overlapping rays fused as in dictyid sponges, but also have triangular "lanterns" produced by cross-bracing diagonal buttresses at the center of each spicule.

Within the hexactinellid sponges, skeletal complexity increases in a generalized sequence. The simplest sponges are common in Paleozoic rocks and appear to be loosely spiculed (Figure 10.19, A). These grade into forms first appearing in the Devonian, in which adjacent spicules are fused into a three-dimensional skeleton (Figure 10.19, B). The most structurally complex hexactinellid sponges develop cross-bracing buttresses at the spicule centers (Figure 10.19, C), beginning in the Triassic, adding more rigidity to the three-dimensional skeleton.

Some spicule types and patterns were remarkably stable for long periods in the geologic record. For example, a ladderlike skeleton made of dendroclones, spicules with smooth axes but complexly branched ray tips (Figure 10.20, C), evolved during the Cambrian and by the early Ordovician was the dominant skeletal pattern in the class Demospongea. As other skeletal schemes appeared, however, this skeletal type played a lesser role but did persist as a common pattern until the Late Permian, a duration of 250 million years. The tricranoclone pattern (Figure 10.20, B) also has an Ordovician to Permian range. Sphaeroclones (Figure 10.20, D) appeared first in the Ordovician and reached maximum diversity and abundance in the Middle Silurian when astylospongid faunas from northern Europe and eastern North America were dominated by these sponges. The spicule pattern lasted until Late Devonian time, a dura-

tion of over 130 million years. Other spicule patterns that characterize major groups persisted even longer. These must have met the structural needs of the sponges, but bioengineering studies have not been done on the various skeletal patterns.

Stratigraphic distribution

Fossil sponges show marked provincialism on a worldwide scale in most systems, as would be expected from sessile benthic forms. In fact, their own distribution is so discontinuous that all but regional patterns have been obscured. Cambrian sponges, however, appear to be much the same over the world, with loosely spiculated and thin-walled demosponges and hexactinellids (Figure 10.17, H), or primitively multirayed calcisponges dominating (Figure 10.18, L). Many genera are known only from the Burgess Shale of western Canada and from similar shale units in the western United States, which reduces their stratigraphic utility.

Ordovician sponge faunas have a significantly different character because thick-walled demosponges (Figure 10.16, B to D) are widespread. Areas as geographically distant as eastern Australia and western North America have the same families represented in their sponge faunas, but most genera and species are different. The same general architectural styles are present in sponges in

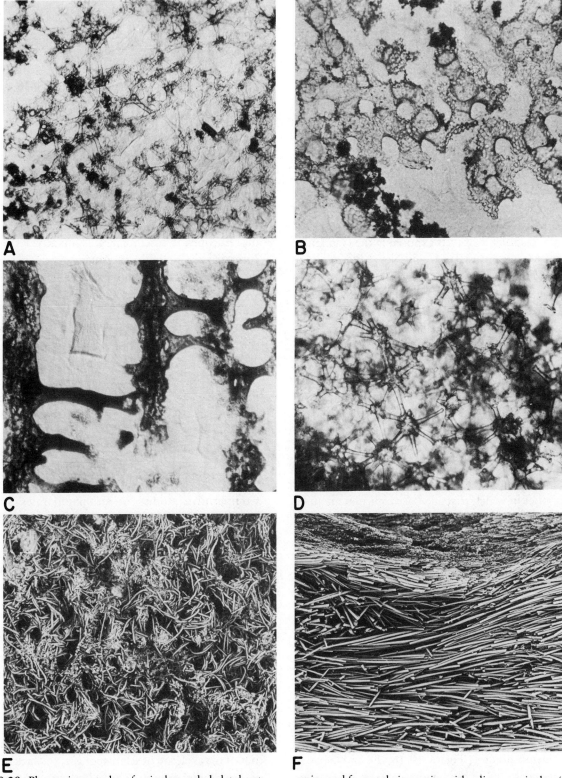

Figure 10.20. Photomicrographs of spicules and skeletal patterns in sponges. **A,** Chiastoclones of *Chiastoclonella* (Silurian, Tennessee) are X- or H-shaped spicules that are united with adjacent spicules at their ray rips and build a somewhat irregular skeletal net. (×75.) **B,** Tricranoclones of *Scheiia* (Permian, Arctic Canada) are robust spicules that have a somewhat palmate ray tip fused to the dorsal surface of the next interior spicules. Round tubercles mark the dorsal surfaces. (×75.) **C,** Dendroclones of *Aulocopium* (Silurian, Tennessee) are typical of early Paleozoic lithistid sponges. Spicules occur in ladderlike series and fuse at their ray tips with adjacent spicules. (×75.) **D,** Sphaeroclone spicules as *Astylospongia* (Silurian, Tennessee) have a spherical centrum from which radiate the rays that fuse to adjacent spicules to produce a solid skeleton. (×75.) **E,** Ophirhabds of *Ophiraphidites* (Eocene, North Carolina) are serpentine monaxons that mat together to produce strong skeletal tracts. (×20.) **F,** Root tuft of principally monaxial spicules from a hexactinellid sponge (Eocene, North Carolina). Similar root tufts occur throughout the geologic column. (×20.)

both areas but only the spherical genus *Hindia* (which externally appears like Figure 10.16, *F*) is represented in both areas.

Although the same general skeletal patterns persist, Silurian sponge faunas are dominated by demosponges containing sphaeroclones (Figure 10.16, *H, K* and *M*). In addition, calcareous sponges with star-shaped octactine spicules (Figure 10.18, *K*) are common across North America and Europe.

Calcareous sponges experienced a great burst of evolution in the Permian and developed considerable diversity (Figures 10.18, *C* to *E, G* to *J,* and *M*). They are the principal constituents of reefs exposed in the Gaudalupe and Glass Mountains in Texas and New Mexico and in contemporaneous rocks of northern Africa and Indonesia. Chaetetid sclerosponges appear in Permian reefs of Tunisia and show considerable diversity.

Triassic sponges are virtually unknown in North America outside of Alaska, the Yukon Territory, and Nevada, but they are common in exposures in southern Europe where they produced thick reefs in Italy and Germany. They are similar to Permian calcareous sponges. Sclerosponges in which impressions of spicules have been preserved occur in reef-dwelling faunas of southern Europe.

Jurassic sponge faunas are dominated in northern Europe by an abundance of slightly different calcareous sponges, ones in which triactine-based spicules are fused into ropy tracts. Jurassic sponges are particularly common in northern France and southern England. Upper Jurassic faunas are known mainly from the Tethyan Seaway, across southern Europe, southern Asia, and northern Africa where great numbers and great diversities of demosponges and various hexactinellids occur in reefs, in reef-related, or in slightly deeper shelf environments. Jurassic sponges are rare in North America.

Cretaceous sponges are rare in North America but are abundant in Europe. They include a second proliferation of calcareous sponges (Figure 10.18, *B*) in the lower part of the system. The fauna becomes dominantly hexactinellid (Figure 10.17, *G*), with lesser numbers of demosponges and calcareous sponges in the upper part of the system. Demosponges (Figure 10.16, *F*) and hexactinellid sponges, in which individual spicules were thoroughly fused together, dominate in the Cretaceous of Germany, France, and Spain where abundant and diverse assemblages have been described. Sclerosponges with spicule impressions have been described from Cretaceous rocks of Germany, Czechoslovakia, and Russia.

Tertiary fossil sponges are represented by few faunas. A moderately diverse assemblage has been described from Eocene rocks of southwestern Australia and from Miocene rocks from New Zealand. Tertiary sponge faunas are also known from eastern Caribbean islands and the Gulf and Atlantic Coastal Plains of North America but as yet are largely undescribed (Figure 10.17, *B*).

Much provincialism and discontinuous sponge distribution may be related to irregular distribution of favorable environments. For example, Cretaceous seas in the interior of North America were probably too turbid and shallow for the dominant kinds of demosponges and hexactinellid sponges to thrive, judging from depth data on European assemblages. Similarly, Triassic and Jurassic rocks in North America may have been deposited in seas too shallow and hypersaline for sponges, which may explain why sponges are virtually unknown from those strata.

Because of their strong provincialism and interrupted record, sponges have not been used extensively in biostratigraphic zonation. Currently sponges appear to have little value in regional correlation between provinces but may prove valuable on a more limited scale, particularly in those areas where they persist through a moderate stratigraphic section.

Class Stromatoporata

Comparative studies of stromatoporates and sclerosponges have recently revived earlier suggestions that the stromatoporates belong in the phylum Porifera. Many specialists still regard them as hydrozoan cnidarians, and they were described with this group in the *Treatise on Invertebrate Paleontology.* The only other group with which they have been seriously compared is the encrusting foraminifera, but this possible relationship is now rejected by paleontologists.

Major points in favor of including stromatoporates within the Porifera include: similarity in gross skeletal pattern to the sclerosponges, and similarity of the astrorhizal canal system to the canal systems in sclerosponges. Points that are merely suggestive of the phylum assignment include: similarity of fibrous microstructure with that of the skeleton of sclerosponges, and lack of evidence of the colonial growth habit in both stromatoporates and sclerosponges (that is, no repeated combinations of skeletal features that might have reflected physically connected individuals).

The major point against including stromatoporates within the Porifera is the lack of known spicules in stromatoporates. In addition, structures such as laminae, and latilaminae in stromatoporates are unknown in

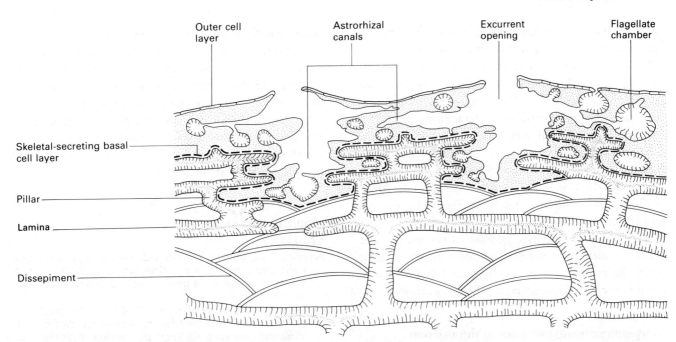

Outer cell layer   Astrorhizal canals   Excurrent opening   Flagellate chamber

Skeletal-secreting basal cell layer

Pillar

Lamina

Dissepiment

Figure 10.21. Restoration of soft parts of a stromatoporate, *Stictostroma,* following the sponge model of relationships. The cellular part of the animal blankets the coenosteum. Flagellate chambers and interconnected canals pump water in through openings in the outer cell layer and out through excurrent openings. A basal membrane or layer of specialized cells secretes the fibrous calcareous skeleton, which here is composed of laminae, pillars, and dissepiments. The restoration is modeled after modern sclerosponges. (Modified after Stearn, C. W. *Lethaia* 8: 89–100; 1975.)

sclerosponges, presumably the closest class. Currently, it seems best to include the stromatoporates within the Porifera, but as a separate class, distinct from the Sclerospongea and other sponges.

Reconstructions of the soft-part anatomy of the stromatoporate animal (Figure 10.21) have been based upon modern sclerosponges. The animal is visualized as having a thin, laterally expanded layer that encrusted the surface of the coenosteum and penetrated only a short distance into the skeleton. The skeleton was secreted by the basal layer of the animal. Astrorhizae are visualized as indentations in the skeleton beneath the excurrent canal system of the animal. Some astrorhizal canals appear as tubelike cavities in the skeleton, and some flagellate chambers may have extended down into the upper layer of the skeleton. Most of the animal, however, was exposed above the calcareous hard parts. Lower parts of the skeleton were sealed off by calcareous platelike tabulae or cystose calcareous dissepiments or were coated over with succeeding laminae.

Stromatoporates are studied in thin sections and peels, and some structure is also amenable to study with scanning electron microscopy. Most definitive thin sections are longitudinal ones (Figures 10.14 and 10.15, *A*), or those cut normal to the laminar structure, and tangential ones (Figure 10.15, *B*), or those cut parallel to the laminae.

Paleoecology

Stromatoporates apparently were exclusively marine. They are particularly common in rocks of carbonate-rich, shallow-water environments and apparently did not thrive in oxygen-starved or muddy, clastic-rich environments. They are common constituents in bioherms or reefs from Ordovician until Late Devonian and again in Jurassic and Early Cretaceous rocks. Silurian reefs on the island of Gotland in the Baltic Sea contain an abundance of stromatoporates and occur as biohermal masses 10 to 20 m thick and several hundred meters across. Stromatoporates there are intergrown with algae, colonial rugose, and tabulate corals. They show marked ecologic zonation within reefs.

Stromatoporates perhaps had their greatest diversity and reef-building capability in Devonian rocks. In western Canada, dendroid forms produce extensive biostromal debris beds in marginal lagoonal rocks behind the reefs. The reef facies is composed mostly of hemispherical and tabular forms that decrease in size from shallow water into turbulent zones that capped the reefs. Smaller hemispherical masses also occur in the frontal reef debris apron. Similar patterns have been well described for extensive Devonian reefs of the American Midwest and Far West, western Australia, and northern Europe. Much of the petroleum production from Silurian and Devonian barrier and pinnacle reefs is from porous limestone and dolomite of stromatoporate origin.

Mesozoic stromatoporates are less common throughout the world than are Paleozoic ones and have not played a major role in reef development, although they are commonly associated with scleractinian corals that are important reef builders. Mesozoic stromatoporates still appear to have been shallow, normal marine, warmwater organisms and were particularly characteristic of the Gulf of Mexico and the Tethyan region of the Mediterranean and southern Asian areas. They did not spread into the cryptic deep-water habitats like those occupied by modern sclerosponges, at least as presently recognized. Mesozoic stromatoporates do occur commonly in oolitic limestones, which also suggests a shallow-water, moderately turbulent habitat for those organisms.

Morphologic trends and diversity through time

The origin of stromatoporates is obscure. Cambrian stromatoporates have been reported from the Soviet Union, but the taxonomic position of these forms is in doubt. The oldest unquestioned stromatoporate, at present, is *Pseudostylodictyon* from Middle Ordovician rocks of Vermont and New York. The class may have descended from sponges or spongelike organisms that did not deposit calcified skeletons until after the Cambrian.

Most Middle and Late Ordovician stromatoporates are composed of cystlike laminae. Stromatoporates with well-developed simple laminae and short pillars appeared in Late Ordovician time and are abundant in Silurian strata. Forms with thick cellular elements forming a network in which neither laminae or pillars are distinct rose in mid-Silurian time and were common until late in the Devonian. Stromatoporates reached their maximum development in North America concurrently with reef development in Middle and Late Devonian time.

In the middle of Late Devonian time, stromatoporates began a marked reduction in abundance and diversity and were virtually extinct by the beginning of the Mississippian. The last stage in the evolution of Paleozoic stromatoporates in latest Devonian time was characterized by a return of forms having dissepiments like those that had dominated Ordovician faunas.

A long gap occurs between the late Devonian and early Mesozoic records, suggesting that the Mesozoic and Paleozoic forms may be only distantly related. Their skeletal structures do indicate, however, that both assemblages should be included within the class Stromatoporata. Affinities must be evaluated by reexamination of the enigmatic forms of intermediate age.

During the Jurassic the forms included in the stromatoporates developed dominantly vertical rather than horizontal skeletal structures. Species increased in abundance and variability, approaching that of the Middle Devonian. Again the class underwent marked reduction during the Cretaceous and became extinct by the end of the Cretaceous. A reported Tertiary form is questionable and is probably not a stromatoporate.

Supplementary reading

Sponges

Berguist, P. R. *Sponges*. Berkeley and Los Angeles: University of California Press; 1978. Excellent, very readable and current summary of biology of sponges with minor treatment of the fossil record.

DeLaubenfels, M. W. Porifera, pp. 21–112. In: Moore, R. C., editor. *Treatise on invertebrate paleontology, Part E*. New York and Lawrence, KS: Geological Society of America and University of Kansas Press; 1955. Only overall treatment of fossil sponges. Genera described until approximately 1953 are listed and arranged into a somewhat difficult classification.

Finks, R. M. Late Paleozoic sponge faunas of the Texas region: The siliceous sponges. *Bulletin of the American Museum of Natural History* 120; 1960. Classic paper on the most varied Permian sponge fauna known, with synthesis relating these faunas to earlier and later faunas.

Hall, J.; Clarke, J. M. A memoir on the Paleozoic reticulate sponges constituting the family Dictyospongiidae. *Memoirs of the New York State Museum* 2: 350; 1899. Monumental volume on famous and significant faunas of mainly Devonian and Mississippian ages.

Rigby, J. K. Sponges and reef and related facies through time, pp. 1374–1388. In: Rigby, J. K., editor. *Reefs through time, Part J*. North American Paleontological Convention, Chicago, 1969. Lawrence, KS: Allen Press; 1971. Summary of occurrences of sponges in reefs and their frame, binding, and accessory functions in the fabric.

Seilacher, A. Die Sphinctzoa, eine Gruppe fossiler Kalkschwamme. *Abhandlungen Akademie der Wissenshaften und der Literatur in Mainz, Mathematisch-naturwissenshaftliche Klass* 10: 721–790; 1961. Summarized relationships of all known genera of the group and their morphology and distribution. A standard reference.

Walcott, C. D. Middle Cambrian Spongiae. *Smithsonian Miscellaneous Collections* 67: 261–364; 1920. Description of the sponges from the Burgess Shale, the most varied Cambrian assemblage known; well illustrated. A classic.

Stromatoporata

Flugel, E.; Flugel-Kahler, E. Stromatoporoidea. *Fossilium Catalogus* I(116): 1–681; 1968. Useful summary of morphology, distribution, and uses on worldwide basis.

Galloway, J. J.; St. Jean, J. Ordovician Stromatoporoidea of North America. *Bulletin of American Paleontology* 43(194): 5–90; 1961. Standard reference work on Early Paleozoic genera.

Hartman, W. D.; Goreau, T. E. Jamaican coralline sponges: their morphology, ecology and fossil representatives. *Zoological Society of London* Symposium 25: 205–243; 1970.

Pivotal paper comparing modern sponges to stromatoporates.

LeCompte, M. Stromatoporoidea, pp. 107–144. In: Moore, R. C., editor. *Treatise on invertebrate paleontology, Part F.* New York and Lawrence, KS: Geological Society of America and University of Kansas Press; 1956. Excellent review of morphology and stratigraphic distribution, with a list of all then known genera.

Mori, K. Stromatoporoids from the Silurian of Gotland, 2. *Stockholm Contributions in Geology* 22: 1–150; 1970. Classic study of Silurian forms from northern Europe.

Stearn, C. W. Devonian stromatoporoids from the Canadian Rocky Mountains. *Journal of Paleontology* 25: 932–948; 1961. Extensive description of morphology, facies, and stratigraphic relationships of fossils from the reef belts of western Canada.

Stearn, C. W. The stromatoporoid animal. *Lethaia* 8: 89–100; 1975. Summarizes possible organism relationships to the calcareous skeleton based upon sclerosponges as a model.

11

Phylum Cnidaria

William A. Oliver, Jr.
Anthony G. Coates

Part I Phylum overview

The Cnidaria are a medium-sized phylum that contains a large variety of both solitary and colonial invertebrates, including hydroids, jellyfish, sea anemones, and corals. Although these animals are simply constructed, they have evolved a variety of cell types that occur in only two body layers. The two layers develop from the embryo and are separated by a middle noncellular layer (for example, the jelly of jellyfish). There is no mesoderm or coelom. Particular functions are undertaken by epithelial, muscular, nervous, and connective tissues that are differentiated within the primary body layers.

Cnidarians contain no organs composed of specialized cells grouped in definite structures like those found in higher animals. Cnidarians also differ from all other higher Metazoa in having a primary radial symmetry and specialized **cnidoblasts** (stinging cells) that contain and fire **nematocysts** (stinging structures). Cnidoblasts occur in almost all cnidarians, giving the phylum its name. The sting of some tropical forms, such as the sea wasp, may be fatal to humans; that of many others, such as the Portuguese man-of-war and the Atlantic coast sea nettle, can be extremely painful. Another general feature of Cnidaria is **polymorphism**. This means that different body forms occur within the same species either sequentially, at different stages of the life cycle (not including the larval stage), or simultaneously within a colony.

Cnidarians have no specific excretory, respiratory, or circulatory structures. The digestive system consists of a central mouth leading to a closed gastrovascular cavity, or **enteron** (Figure 11.2), within which digestion is both extracellular and intracellular. Undigested material is ejected through the mouth. **Tentacles**, which are primarily food gathering structures, surround the mouth and may contain hollow extensions of the enteron. A series of muscle cells is present in both cellular layers, allowing the animal to contract both longitudinally and transversely as well as to bend, eject matter, and activate tentacles. Nerve cells are arranged in an irregular network, generally at the base of the outer cell layer, and are adequate to coordinate movements within individuals and throughout colonies.

The cnidarian skeleton, which serves as a support when present, may be external or internal and either chitinous, horny, or calcareous. Because many groups have no skeleton or one that easily disintegrates, the fossil record of the phylum is uneven. The most important cnidarian fossils are the corals because they secrete massive calcareous skeletons, which are common in Ordovician to Holocene sedimentary rocks.

Cnidarians are abundant in most aquatic environments, although fossils are known only from marine sedimentary rocks. Many cnidarians are sessile, but a few attach themselves to floating plant material, and some of these are able to move short distances along the substrate. Other cnidarians are adapted to a planktic or nektic life in the marine realm.

The geologic record of Cnidaria extends from the late Proterozoic (Ediacaran stage) to the Holocene. Groups having calcareous skeletons are sufficiently common to furnish data for studies of their evolution and their life

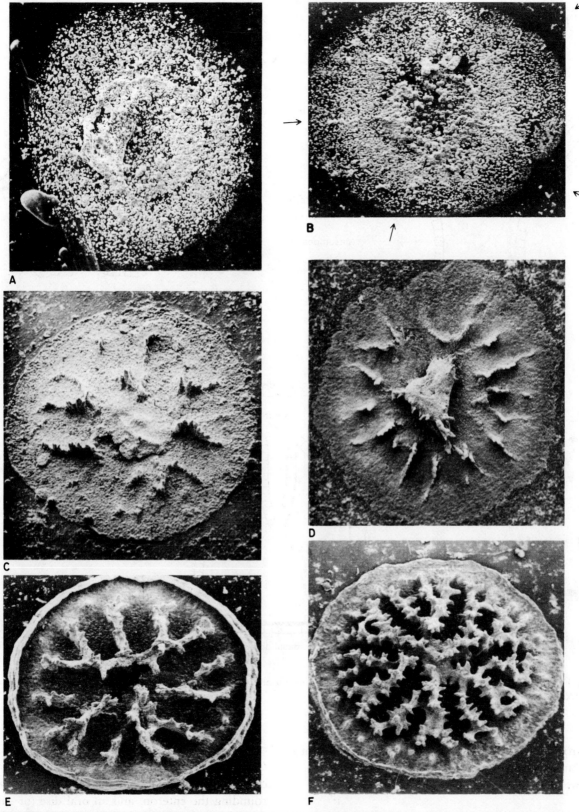

Figure 11.1. Six stages in the growth of the coral skeleton during the first 21 days after the settlement of the planula larva in the coral *Porites*. **A,** Basal plate, formed mostly of separated "seed" crystallites (5 hours, ×85). **B,** Basal plate now more compact, especially toward the center, showing the first stages of upward trabecular growth at the location of the six protosepta (arrowed; 12 hours, ×85). **C,** Basal plate with three cycles of septa showing individual trabeculae that are free at their upper ends (septal teeth) but are fused basally (5 days, ×54). **D,** Skeleton showing three cycles of septa more clearly defined than in **C** and with a central tabula (9 days, ×36). **E,** Skeleton showing basal plate turning upward at the rim (the beginning of the epitheca); the septa now have trabeculae that are beginning to diverge laterally (vepreculae; 12 days, ×54). **F,** Skeleton now has well-developed septa with numerous vepreculae, some of which have fused with those on adjacent septa to form synapticulae (21 days, ×43). (Courtesy J. S. Jell.)

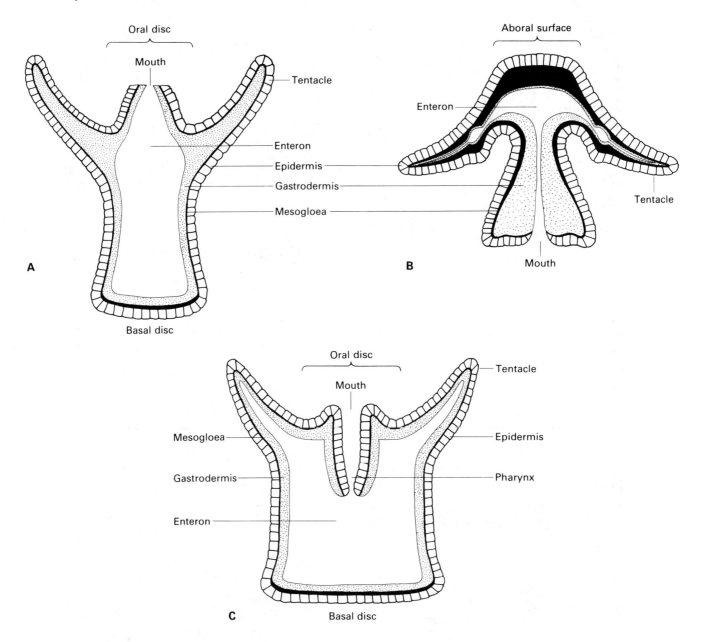

Figure 11.2. Basic body forms in Cnidaria. **A**, Simple polyp without pharynx, as in Hydrozoa. **B**, Medusa. **C**, Anthozoan polyp with pharynx.

and growth habits. They are and have been important rock builders since the early Paleozoic, especially in the reef environment.

Introductory morphology

Soft tissues

Two basic body forms predominate in cnidarians (Figure 11.2), both characterized by stinging cells and tentacles. Individual cnidarians are always modified versions of one of these body types. The **polyp** form (Figure 11.2, *A* and *C*) is somewhat cylindrical, consisting of a **basal disc** commonly attached to a substrate, a **body wall** surrounding the enteron, and an **oral disc** (or surface) in which a central mouth is surrounded by one or more circlets of tentacles.

The second body form is the **medusa** (Figure 11.2, *B*), which is shaped like a bell or inverted bowl with tentacles on the margin. The body plan of the medusa is like that of the polyp but is adapted to free living (swimming or floating) with the tentacles hanging down in the water.

The medusa enteron is generally more complex in shape than that of the polyp, and the mouth is at the end of a tubular projection extending down from the interior of the bell. Some medusae have a muscular shelf, the **velum**, projecting inward from the bell margin.

In many cnidarians, polyp and medusa stages alternate with each other (Figure 11.3). In this **alternation of generations**, the medusae commonly reproduce sexually and the polyps asexually.

The body wall of both polyp and medusa is divided into an inner **gastrodermis** (endoderm) and an outer epidermis (ectoderm), which are separated by a third layer, the gelatinous **mesogloea**. The mesogloea is not cellular but may contain horny connective tissue and fibers as well as scattered cells (Figure 11.2).

The simple polyp shown in Figure 11.2, *A*, is characteristic of hydroids and their relations. Polyps of corals and **anemones** (Figures 11.2, *C* and 11.4) are more complex in having more-or-less elaborate internal structures that divide the enteron into smaller compartments. Here, as in other sections of this chapter, more detailed information on the coral-anemone group is

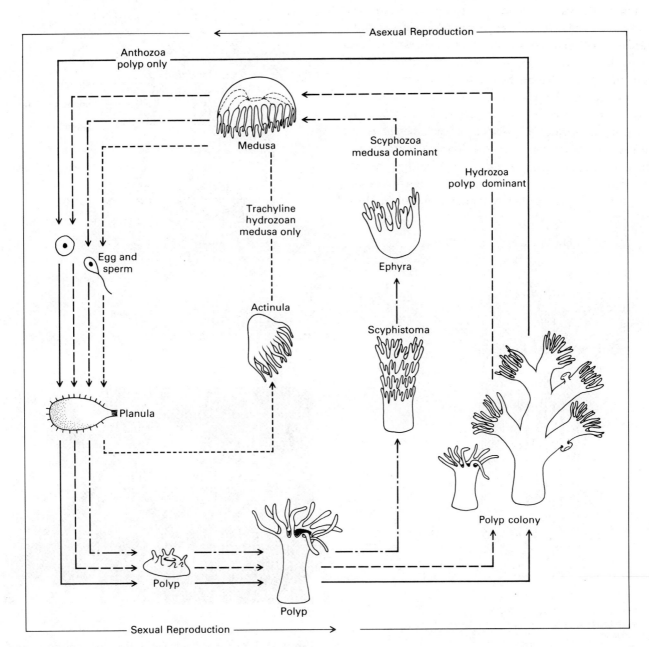

Figure 11.3. Life cycles of the three classes of cnidarians. The special actinula larva of the hydrozoan order Trachylinida may represent the type of animal from which polyps evolved. (Modified from Hill, D.; Wells, J. W. In: Moore, R. C., editor. *Treatise on invertebrate paleontology, Part F.* New York and Lawrence, KS: Geological Society of America and University of Kansas Press; 1956.)

provided because of the importance of the corals as fossils.

Coral and anemone polyps have a ciliated **pharynx**, a tube extending from the mouth down into the enteron. In most anemones, two ciliated grooves, the **siphonoglyphs**, run down opposite sides of the pharynx, and direct currents into the enteron. Radiating vertical partitions in the enteron are called **mesenteries**; these are attached to the body wall of the polyp and the oral disc (Figure 11.4). Some of the mesenteries are also attached at their inner ends to the pharynx (Figure 11.5, *A*) and are said to be complete; others are not and are incomplete. All are free at their inner edges below the pharynx. Mesenteries move food within the enteron by ciliary action, digest it, excrete particulate matter, and, in some species, bear nematocysts to kill prey that have entered the enteron; they also commonly bear the gonads.

Mesenteries consist of an infold of gastrodermis enclosing mesogloea. On one side of each mesentery are vertical muscles within the gastrodermis. Contraction of these muscles shortens the polyp and inwardly contracts and folds the tentacles and oral disc. Mesenteries are symmetrically arranged on each side of the polyp. Those

in corresponding positions on each side are called **couples** (Figure 11.7, *B*); those arranged in twos are said to be **pairs** (Figure 11.7, *B*). The mesenteries next to the ends of a compressed pharynx (Figure 11.7, *B* and *D*) have muscle bands that face away from each other; these are called **directive mesenteries**. The directive pairs are bisected by the plane of symmetry. The position and

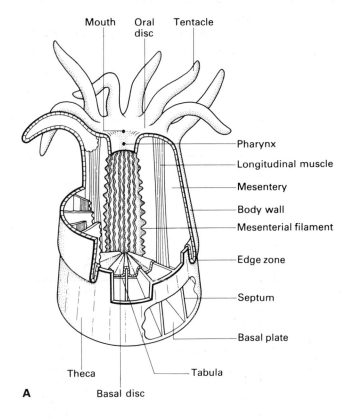

A

B

Figure 11.5. Relation of polyp to skeleton in scleractinian coral. **A,** Schematic, cutaway view of young solitary coral; mesenteries not shown in foreground. **B,** Living coral colony with polyps expanded. (**B,** courtesy K. Ruetzler.)

Figure 11.4. Cutaway view of anthozoan polyp showing the pharynx and radially arranged mesenteries that distinguish anthozoan from hydrozoan polyps.

number of mesenterial pairs can be deduced in fossil corals because skeletal partitions (septa) grow upward in positions corresponding to the space between the mesenteries of each pair.

Arrangement of early mesenteries in the larva defines a plane of symmetry by which all other adult structures are oriented. Subsequent mesenteries appear in couples symmetrically about the plane in a definite sequence, position, and number that define the different subclasses and orders.

Skeletons

Many polyps develop organic or mineral exoskeletons, and some have internal organic skeletons or mineral spicules. Organic skeletons are composed of chitin or horny material (collagen) and are rarely preserved as fossils. Some polyp colonies secrete an external sheath of chitin, which may cover only certain parts of the colony or may provide cup-shaped units into which the individual polyps can withdraw. In other groups, internal organic skeletons take the form of a flexible columnar axis, which allows growth above the substrate.

Mineral skeletons are invariably calcareous, composed of either calcite or aragonite. Internal calcareous skeletons occur in some groups as spicules within the mesogloea. These spicules may be discrete or variously united to form supporting rods or axes. Corals (and a few other cnidarians) form massive calcareous exoskeletons that are easily preserved and important as fossils.

The massive coral skeleton is external. It is secreted by the epidermis at the base of the polyp so that structural elements of the skeleton accurately reflect the outer morphology and growth of the polyp. The coral polyp initially rests upon a skeletal **basal plate** (Figures 11.1 and 11.5, *A*). With growth, the base of the polyp forms a series of radial folds, within each of which is secreted a **septum** (pl. **septa**) oriented vertically and radially. Septa are extended upward as the polyp grows, joined at their outer ends by a **theca** (skeletal wall) that is extended upward with the septa (Figures 11.5 and 11.6). The lower tapering part of each mesentery hangs down into a space between two septa (Figure 11.5, *A*). As the skeleton grows upward, the polyp lays down a series of new skeletal floors at its base. These may be single transverse plates called **tabulae** or a series of smaller domed plates called **dissepiments** (Figure 11.6). During growth, the polyp occupies only the **calice**, the upper, cup-shaped depression on top of the skeleton that is bounded by the theca, the last formed tabula or dissepiments, and the septa. The skeleton of one coral unit, whether solitary or colonial, is a **corallum**; that of any one polyp in a colony is a **corallite**. Skeletal tissue deposited between corallites in a colonial coral is called **coenosteum** and may consist of tabulae, dissepiments, or septa extended beyond the corallite wall. It is deposited by interpolypal tissue called **coenosarc**.

Solitary coralla have various shapes ranging from flat (discoid) to broadly conical to cylindrical (Figure 11.8). Coralla of colonial corals occur in a variety of forms, and a complex terminology exists to define them. Some of these forms are illustrated in Figure 11.9, and terms are introduced later where needed.

Symmetry

Cnidarians are distinguished by radial, biradial, or radiobilateral symmetry. **Radial symmetry** occurs when two or more planes, passing through the axis of symmetry that runs from mouth to base, produce identical halves (Figures 11.7, *A* and *C*). **Tetrameral radial symmetry** occurs when two planes of symmetry at right angles to each other pass through the axis of symmetry to divide a polyp or medusa into identical quadrants (Figure 11.7, *A*). Radial symmetry is characteristic of the cnidarian larva and, in varying degrees, many simple polyps and medusae. More complex polyps have elongated mouths and bilateral muscle and mesentery patterns. Some are **biradially symmetrical**; they can be divided into two equal parts by either of two planes of symmetry at right angles to each other (Figure 11.7, *B*). Others are **radiobilaterally symmetrical**, having only one plane of symmetry (Figure 11.7, *D* and *E*). All corals belong to one of the last two categories, but their skeletons commonly have so many apparently equal radiating septa that they may appear to be truly radial.

Radially symmetrical organisms seem to be well adapted to both a sessile and planktic way of life because they are equally functional in all directions.

Classification

Three living classes of Cnidaria are universally recognized: Hydrozoa, Scyphozoa and Anthozoa. The Anthozoa have a long and rich fossil record. The other classes have much poorer but still significant records. Extinct nominal classes of the phylum are minor in occurrence but of interest in broadening our knowledge of the general form and geologic history of cnidarians or cnidarian-like animals (Figure 11.10). Here only the extant classes will be discussed; the extinct ones will be discussed later in the chapter. Table 11.1 lists all the

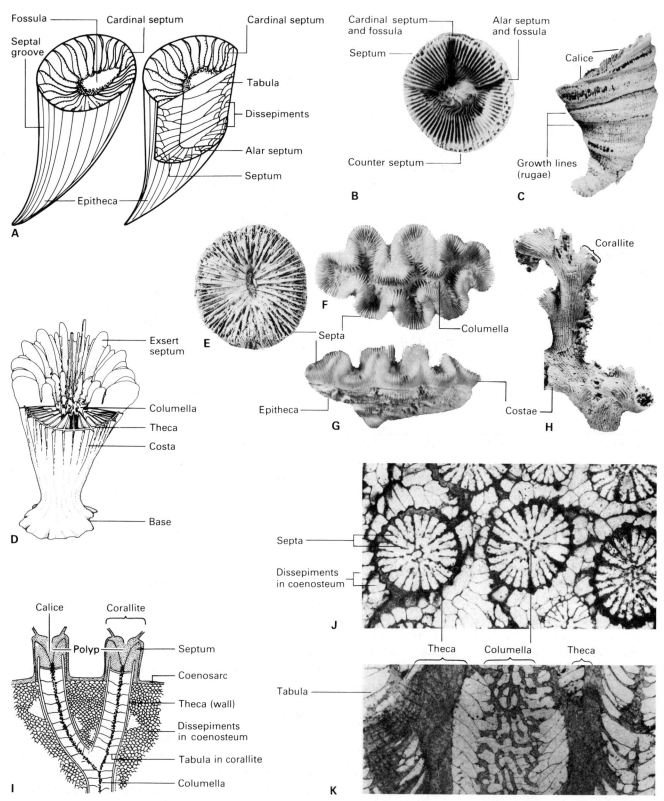

Figure 11.6. Principal morphologic features of the skeleton of rugose and scleractinian corals. **A,** Stylized solitary rugose coral skeleton (minor septa omitted for clarity); the polyp lived in the calice at the top of the corallum. A cutaway view shows the relationship of septa, tabulae, and dissepiments; note that the outer surface of the wall is grooved at the position of internal septa. **B,** Calice view of silicified corallum of *Kionelasma*

(Devonian rugosan, ×1) showing bilaterally symmetrical arrangement of septa along the cardinal-counter plane. Two cardinal sectors or quadrants are clearly marked on each side of the cardinal fossula as are the two alar fossulae; the two counter sectors (quadrants) are not as clearly separated, due to the lack of a counter fossula. **C,** Side view of the same specimen showing periodic rejuvenation or growth cyclicity (rugae) and septal classes of Cnidaria and their major subdivisions in what is generally thought to be the order of increasing complexity.

CLASS HYDROZOA (Figure 11.11)

Marine and freshwater Cnidaria in which polyp stage usually dominates in alternation of generations (Figure 11.3), although it can be absent. Both polyps and medusae radially symmetrical with simple undivided enteron and no pharynx. Mesogloea without scattered cells; eggs generally in epidermis and always shed externally. Polyps greatly variable, usually small and colonial, consisting of a stalk supporting terminal tentaculate hydranth. Hydranth with lower cup rimmed with tentacles within which central conical projection bears mouth at apex. Stalk and hydranth possibly sheathed in chitinous skeleton or by massive calcareous skeleton. Colonies arise by upright branching of polyps from horizontal, basal stolons. Polyps within colonies often polymorphic and highly specialized in structure and function. Medusae bell shaped, small and frequently with peripheral muscular shelf projecting inward from margin. (Late Proterozoic [Ediacaran] to Holocene; about 500 genera.)

The Hydrozoa (Table 11.1) comprise a particularly wide range of polypoid and medusoid animals including the following:

1   Branching, bushlike polyp colonies, most with flexible, external, tubular, chitinous (or chitinoid) skeletons (the hydroids, Figure 11.11)

2   Colonies having calcareous, coral-like skeletons (the hydrocorallines, Figure 11.42)

3   Primitive pelagic medusae with no polyp generation (the trachylines, Figure 11.3)

4   "Oceanic hydrozoans," with remarkably varied numbers of both medusoid and polypoid polymorphs clustered in one colony (the siphonophores)

5   The minute, solitary hydrozoans that live in the pore spaces within marine sands and have neither a true polyp nor a medusa (the actinulids).

The most numerous of these varied groups today are the hydroids, but they have a poorly known fossil record because the chitinous skeletons are not easily found. Most hydrozoans are marine, including most of the familiar or common forms (except *Hydra*), all those having calcareous skeletons, and most known fossils. Freshwater hydrozoans include the solitary hydra, which lacks a medusa stage, and also a few polyp colonies and medusae.

CLASS SCYPHOZOA (Figure 11.12)

Exclusively marine Cnidaria in which major part of life cycle is spent as medusa (Figure 11.3). Tetrameral radial symmetry displayed by both polyp and medusa. In most groups, four partitions of infolded gastrodermis, emphasizing tetrameral symmetry, dividing enteron and generally also marking position in epidermis of longitudinal muscle bands. Tentacles often short, frequently in multiples of four, or absent. Mesogloea often thick, with fibers and scattered cells. Polyps solitary, square in cross section. Planula metamorphoses to polyplike larva

---

grooves. D, Diagram of partially cutaway solitary scleractinian coral with septa standing high above the rim of the calice (exsert) and continued on the outside of the theca (wall) as ridges (costae); note also the anastomosing spongy (trabecular) extensions of the inner ends of the septa that form an axial columella (see also I and K). E, Top view of calice of *Vaughanoseris* (Cretaceous scleractinian, ×1), showing a deep central calical pit (fossette) and cyclically arranged septa. F and G, Top and side views of corallum of *Manicina* (Holocene colonial scleractinian, ×0.5). The valley-shaped calice is a result of linear budding of partially formed individuals joined along the valley by an axial columella. The valley is called a series, and in this specimen, secondary valleys are beginning to develop. Septa are ranged only on each side of the valley. Note the presence of costae (G) and, external to them, the remnants of a thin fragile outer layer, the epitheca, now partially worn off. H, Silicified coral *Cladocora* (Cretaceous scleractinian, ×0.5), a dendroid branching colony with prominent costae. I, Diagram of a longitudinal section through a colonial scleractinian coral in which corallites are surrounded by coenosteum. The dotted pattern shows the position of the soft tissue, with the polyp overlying the corallites and coenosarc overlying the coenosteum. In this example, tabulae occur within the corallite, and dissepiments form the coenosteum. J, Transverse thin section of *Solenastrea* (Pliocene scleractinian, ×3). Note cyclic arrangement of septa, which coalesce to form a circular wall (septotheca) and axially are loosely associated into a spongy columella (see K). Corallites are joined by an open box-work of dissepiments, which form the coenosteum (see K). K, Longitudinal thin section of *Montastrea* (Pliocene scleractinian, ×4), showing thick walls and concave thin tabulae extending to and within the spongy columella. (D from Cairns, S. D. U.S. Department of Commerce, *NDAA Technical Report NMFS* 438: 2; 1981. I after Ogilvie, 1897, as modified by Wells, J. W. In: Moore, R. C., editor. *Treatise on invertebrate paleontology, Part F.* New York and Lawrence, KS: Geological Society of America and University of Kansas Press; 1956. J and K courtesy of A. B. Foster.)

(scyphistoma) which either directly transforms into medusa or, by repeated transverse fission at one end (strobilization), gives off many medusae (Figure 11.3). Medusae variable, saucer to helmet shaped, peripheral shelf (velum) always absent. Mineralized skeleton absent. (Late Proterozoic [Ediacaran] to Holocene; about 90 genera.)

Scyphozoa include the largest known solitary cnidarian, a medusa that can reach a diameter of more than 2 m. Several fossil genera are known, including some from the Proterozoic (Ediacaran stage). The best-preserved fossils come from the Jurassic of Germany.

Scyphozoans are exclusively marine and dominantly planktic or nektic, although many have sessile polyp stages in their life cycles. Medusae swim by rhythmic pulsations of the bell, but large-scale migration takes place by drifting in response to winds or currents.

The extinct Conchopeltida and the Conulariida (considered to be scyphozoans by some specialists) are discussed later.

CLASS ANTHOZOA (Figures 11.13 to 11.16)

Solitary or colonial, exclusively marine Cnidaria, medusa absent, only free-swimming larva (planula) intervenes between repeated polyp stages (Figure 11.3). Polyp mouth central and elongated with pharynx extending deep into enteron. Ring of retractile tentacles into which the gastrovascular cavity extends surrounding the mouth. Enteron divided into vertically elongate wedge-shaped spaces by mesenteries, which alternate in position with septa when these are present. Mesogloea moderately thick, often with scattered fibrous material. Skeletons external or internal; chitinous, collagenous, calcareous or absent. (Late Proterozoic [Ediacaran] to Holocene; about 2300 genera.)

The Anthozoa include the vast majority of all fossil cnidarians. The class is also the largest one today, containing more than 6000 living species. The number and symmetry of the mesenteries in the enteron, and the type and composition of the skeleton if there is one, serve to divide this class into many groups, such as the soft corals, horny corals or sea fans, stony or true corals, and the sea anemones. The true **corals** may be defined as those members of the class Anthozoa that secrete massive external calcium carbonate skeletons. In this chapter, the corals are given special attention because of their paleontologic importance.

Coral skeletons consist basically of calcareous cones or

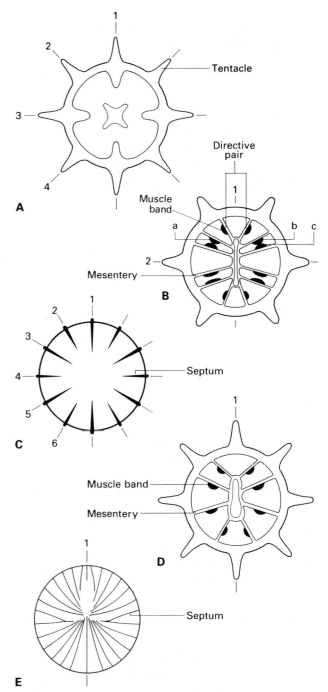

Figure 11.7. Diagrammatic transverse sections of cnidarians, illustrating symmetry and arrangements of mesenteries. Numbers mark the planes of symmetry. **A,** Fourfold radial symmetry as in scyphozoans and some hydrozoans. Planes *1* and *3* mark one system of symmetry; planes *2* and *4* define a second system. **B,** Biradial symmetry; hexamerous as in many living anthozoans. Note two planes of bilateral symmetry at right angles to each other; the two pairs of directive mesenteries are each bisected by one plane of bilateral symmetry (plane *1*), the second plane of symmetry (plane *2*) is at right angles to this. Mesenteries *a* and *b* are a couple, *b* and *c* are a pair. **C,** Apparent sixfold radial symmetry of simple scleractinian coral skeleton. **D,** Octamerous radiobilateral symmetry of octocoral; note single plane of bilateral symmetry. **E,** Radiobilateral symmetry of rugose coral skeleton.

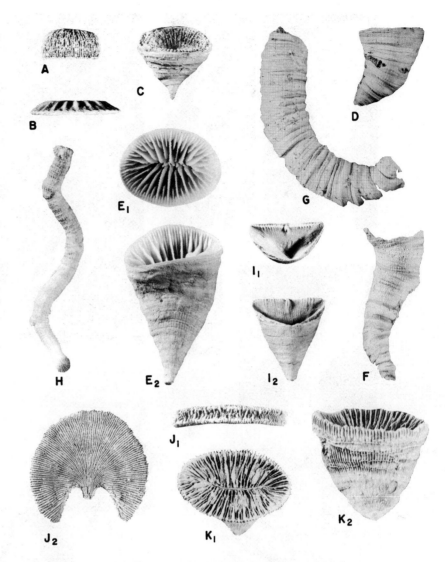

Figure 11.8. Shape in solitary corals. Most illustrations are side views of corals with circular cross sections. Where two views are shown, the calice view is *1*, and the side view is *2*. Examples include both rugosan (R) and scleractinian (S) corals, but all the shapes can be found in both groups of corals. **A,** Discoid (S, ×2). **B,** Discoid (R, ×2). **C,** Turbinate (R, ×1). **D,** Trochoid (R, ×0.66). **E,** Trochoid with oval calice (S, ×1). **F,** Ceratoid (R, ×0.66). **G,** Cylindrical (R, ×0.66). **H,** Scolecoid and cylindrical (S, ×2). **I,** Calceoloid (R, ×1). **J,** Parallel-sided flabellate (S, ×2). **K,** Slightly flabellate (S, ×0.66).

cylinders, the upper part occupied by the living polyps. Partitions that are longitudinal (septa) or transverse (tabulae) are well developed. In the earliest forms, which were an exclusively colonial group (the tabulate corals shown in Figure 11.13 and listed in Table 11.1), septa are poorly developed if at all, and the tubes have conspicuous tabulae (Figure 11.13, *J* and *L*). Another group of Paleozoic corals have those characteristics plus coenosteum (Figure 11.14). Colonies of these two groups constructed or helped construct reefs during the Ordovician, Silurian, and Devonian. A contemporaneous group, the rugose corals (Figure 11.15), were both solitary and colonial; their more varied skeletons typically have prominent septa that are grouped into four quadrants often separated by gaps or **fossulae** (Figure 11.6, *B*). They became important in the formation of reefs during the Devonian. The Paleozoic groups were replaced in the

Triassic by the modern corals, the scleractinians, shown in Figure 11.16, which have participated in the building of much larger reef structures. They have lighter, more porous, and complex skeletons in which the septa have a sixfold symmetry and no fossulae and elaborate skeletal elements (coenosteum) are built between corallites. These forms are more efficient skeleton-builders than the earlier corals, and we will discuss later why this may be important in the construction of modern atolls and barrier reefs, the largest structures ever made by non-human organisms.

Anthozoans are exclusively marine and dominantly sessile. Some anemones are capable of walking on their basal disc or tentacles, but dispersal over long distances only takes place during the planula stage, the medusa stage being absent. Anthozoans live in most marine environments from tide pools to depths of 6000 m or more.

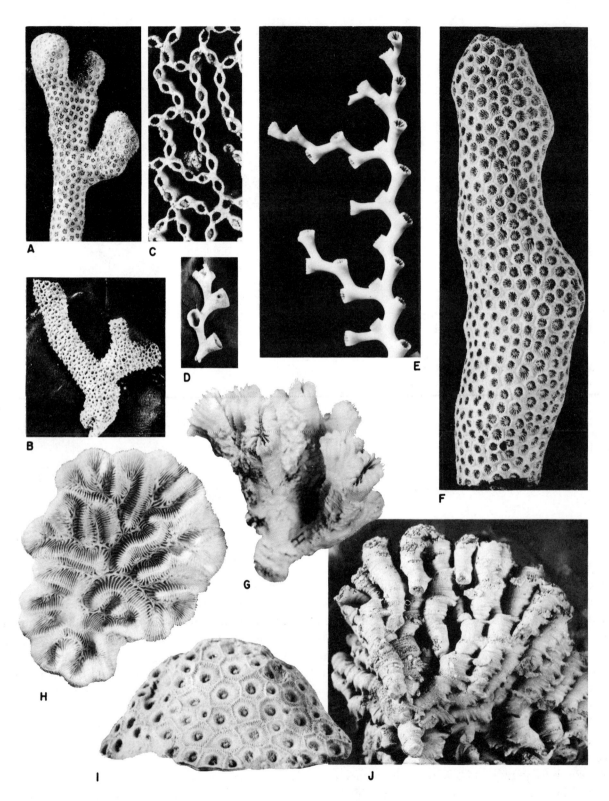

Figure 11.9. Form in colonial corals. Examples include rugosan (R), tabulate (T) and scleractinian (S) corals, but most of the forms are found in all three subclasses. Ramose—branching form of closely packed corallites: **A** (S, ×1); **B** (T, ×1), **F** (S, ×1). Dendroid—irregular branching, each branch composed of one corallite: **D** (T, ×1); **E** (S, ×1); **G** (S, ×0.66). Phaceloid—like dendroid but corallites are subparallel: **J** (R, ×1). Cateniform—chain coral: **C** (T, ×2). Massive meandroid: **H** (S, ×0.66). Massive cerioid: **I** (R, ×0.66). (See also Figures 11.13 to 11.16.)

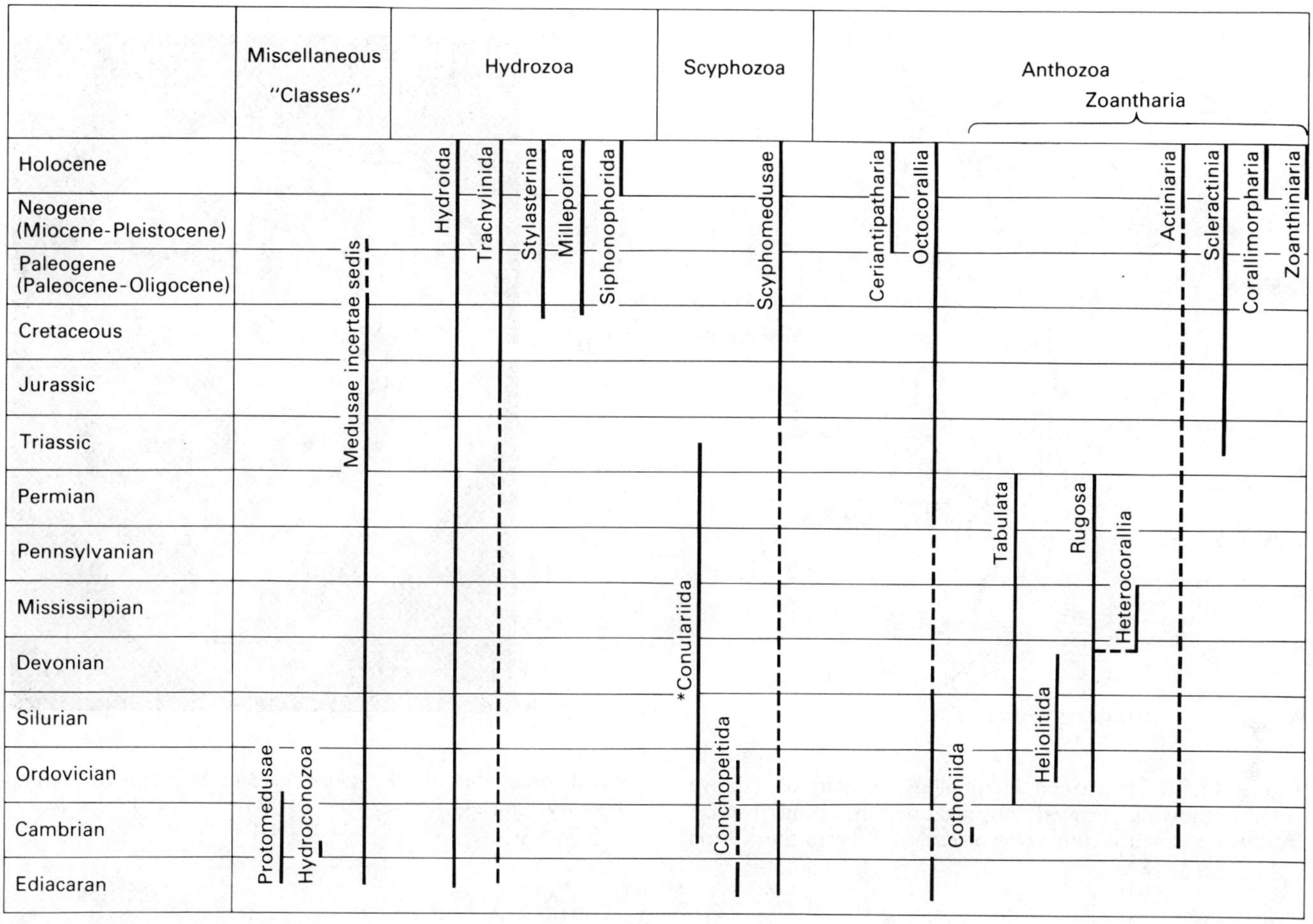

Figure 11.10. Geologic record of major groups of cnidarians discussed in chapter. An asterisk marks the Conulariida, which are probably not cnidarians.

Features of the phylum

Coloniality

Definition The word colony means different things to different people. Even among biologists, definitions depend upon the organisms being studied and, at least partly, on the predilections of the scientist. To many scientists, a colony is simply a population of organisms from a single species living in close proximity and interacting to some extent (such as colonial sea birds, bees, and others). To the specialist on marine invertebrates however, a colony implies a closer genetic relationship between individuals. Such a colony is a clone, which is defined as a genetically identical set of individuals derived by asexual reproduction from a single founding individual. In cnidarian colonies, the founding individual and the one or more generations of asexually produced off-

spring are usually not separated from each other. Normally, therefore, the individuals within colonies are genetically alike, although environmentally induced variation certainly exists.

For the paleontologist, a practical definition of a colony must include the fact that colonial individuals have preservable skeletons that hold together after the removal of all soft parts. The nature of the living connection between individuals is not necessarily indicated by fossils because such connections can be completely outside the skeletal mass. Continuity of skeletal material itself, however, and the presence of pores or openings within skeletons, connecting living chambers, strongly suggest connected soft tissue, especially if analogous connections are known in living forms. A recognizable colony in paleontology, therefore, is a preserved skeleton in which units built by individuals can be identified and reasonably interpreted as members of a single clone.

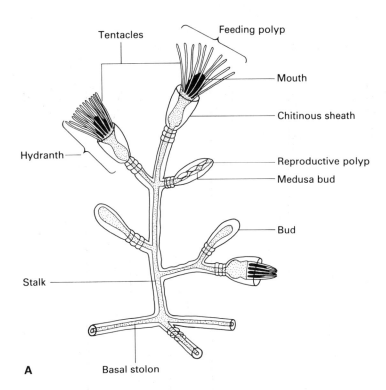

A

B

Figure 11.11. Hydrozoan morphology. **A,** Part of a polyp colony showing general organization. **B,** Pennsylvanian freshwater polyp colony (*Drevotella* ×0.9) From the Mazon Creek area, Illinois. (**B,** courtesy Field Museum of Natural History, Chicago.)

Table 11.1. Classification of the Phylum Cnidaria (adapted from various sources).

Classification	Geologic Periods
Class Protomedusae	Late Precambrian(?), Cambrian–Ordovician
Class Hydroconozoa	Early Cambrian
Class Hydrozoa	Ediacaran–Holocene
Order Trachylinida	Ediacaran(?), Cambrian(?), Jurassic–Holocene
Order Hydroida	Ediacaran, Paleozoic, Mesozoic–Holocene
Order Milleporina } Hydrocorallines	Cretaceous–Holocene
Order Stylasterina }	Cretaceous–Holocene
Order Siphonophorida	Holocene
Class Scyphozoa	Ediacaran–Holocene
Group Scyphomedusae	Ediacaran, Cambrian(?), Jurassic–Holocene
*Order Conchopeltida	Ediacaran–Ordovician
*Order Conulariida	Cambrian–Triassic
Class Anthozoa	Ediacaran–Holocene
Subclass Ceriantipatharia	Miocene–Holocene
Subclass Octocorallia	Ediacaran, Ordovician, Silurian(?), Permian(?). Triassic(?), Jurassic–Holocene
Subclass Zoantharia	Cambrian(?), Ordovician–Holocene
Orders Zoanthiniaria, Corallimorpharia, and Actiniaria	Cambrian(?), Holocene
Order Cothoniida	Middle Cambrian
Order Tabulata	Early Ordovician–Permian
Order Heliolitida	Middle Ordovician–Devonian
Order Rugosa } "True" corals	Middle Ordovician–Permian
Order Heterocorallia	Late Devonian–Mississippian
Order Scleractinia	Middle Triassic–Holocene

*Probably not cnidarian

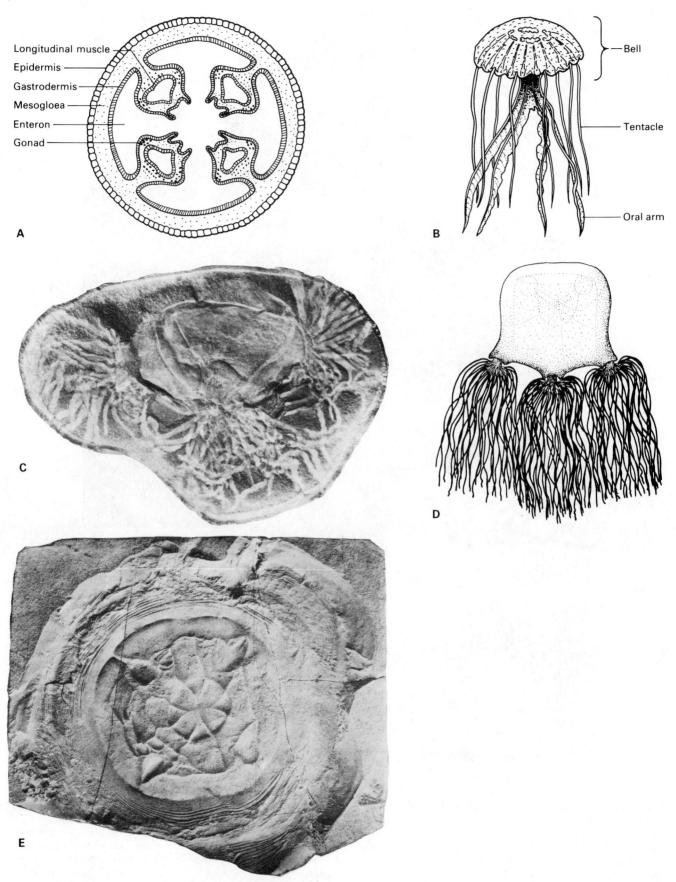

Longitudinal muscle
Epidermis
Gastrodermis
Mesogloea
Enteron
Gonad

A

Bell

Tentacle

Oral arm

B

C

D

E

Figure 11.12. Scyphozoan morphology. **A,** Diagrammatic transverse section of a medusa. **B,** Generalized view of a medusa. **C,** Fossil medusa (*Anthracomedusa,* ×0.8) from the Pennsylvanian Mazon Creek fauna, Illinois. **D,** Restoration of fossil medusa shown in **C. E,** Fossil medusa (*Rhizostomites,* ×0.25) from the Jurassic, Solnhofen Limestone, Germany. (**A** and **B** from Hyman, L. H. *The invertebrates.* New York: McGraw-Hill Book Co.; 1940; by permission. **C,** courtesy Field Museum of Natural History, Chicago. **D** from Foster, M. W., *Mazon Creek fossils.* New York: Academic Press, Inc.; 1979. **E** from Walcott, C. D. Fossil Medusae. U.S. Geological Survey, Mon. 30, pl. 41; 1898.)

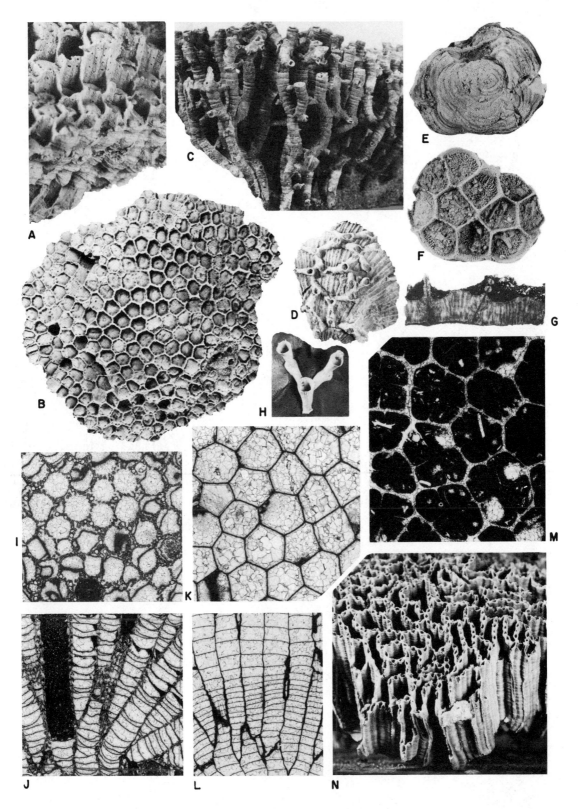

Figure 11.13. Tabulate corals. **A** and **B**, Cerioid *Favosites* (Middle Silurian, Iowa): oblique view (**A**; ×2) and top view (**B**; ×1) showing mural pores. **C**, Phaceloid *Syringopora* (Lower Devonian, New York ×1); side view showing lateral offsetting and lateral connecting tubes. **D**, Encrusting *Aulopora* colony (Middle Devonian, Ohio ×1). **E** to **G**, Cerioid *Cleistopora*: Botton view (**E**), showing holotheca, and top view (**F**) of Early Mississippian colony from Montana (×1); **G**, Vertical thin section of Devonian specimen from the Western Sahara, showing thick base, shallow calices, and mural pores (×2). **H**, Dendroid *Cladochonus* (Early Permian, Texas, ×1):
fragment of a colony showing septal ridges in lower calice. **I** and **J**, Cerioid *Calapoecia* (Late Ordovician, Kentucky, ×2.5): transverse (**I**) and vertical (**J**) thin sections showing porous walls; small black circles in **I** are cross sections of commensal "worm" tubes. **K** and **L**, Cerioid *Foerstephyllum* (Late Ordovician, Kentucky, ×2.5): transverse (**K**) and vertical (**L**) thin sections of primitive form lacking pores. **M**, Cerioid *Tetradium* (Late Ordovician, Kentucky, ×10): transverse thin section showing incipient fourfold axial increase. **N**, Oblique view of chain coral *Halysites* (Middle Silurian, Kentucky, ×1). (For other tabulates, see Figure 11.9.)

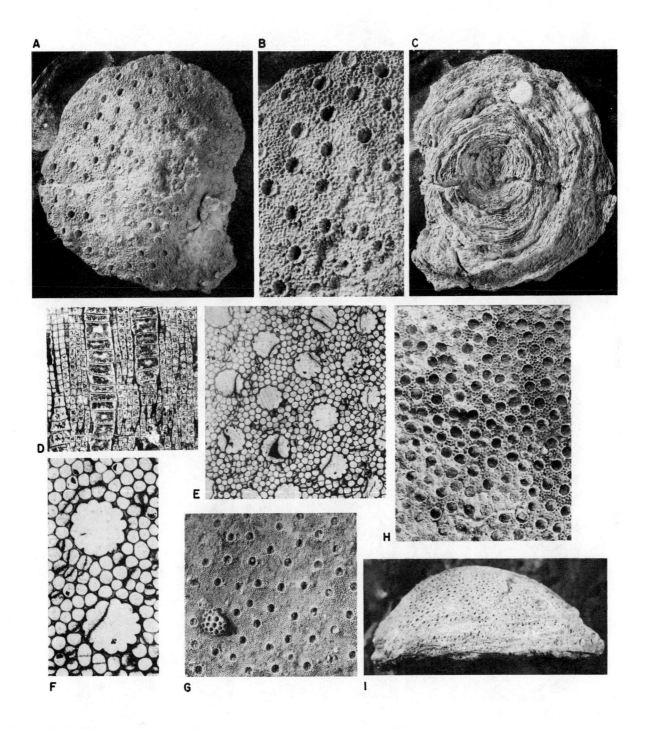

Figure 11.14. Heliolitid corals. **A** to **C**, *Plasmopora* (Early Silurian, Sweden): upper (**A**, ×1; **B**, ×2) and lower (**C**, ×1) surfaces of colony, showing septal ridges, radiating structure around calices, and basal holotheca. **D** to **I**, *Heliolites*: Vertical (**D**, ×5) and transverse (**E**, ×5; **F**, ×10) thin sections of colony from Middle Devonian of Alaska, showing tabulae and 12 equal septa in corallites. **G**, Surface view of colony from Middle Silurian, Tennessee, with widely spaced corallites (×2). Surface (**H**, ×2) and side view (**I**, ×0.66) of massive colony from the Middle Silurian, Sweden, with closely spaced corallites.

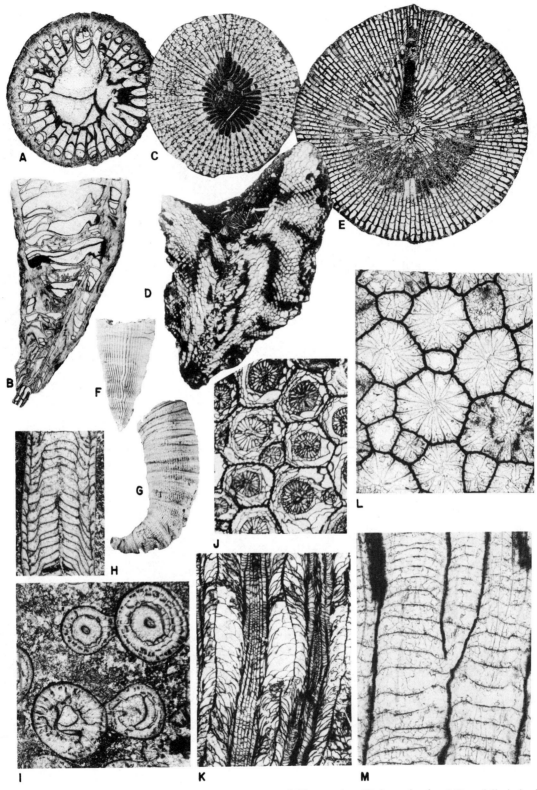

Figure 11.15. Rugose corals (**A** to **G**, solitary; **H** to **M**, colonial).
A and **B**, *Barytichisma* (Late Mississippian, Wyoming, ×2):
transverse (**A**) and longitudinal (**B**) thin sections, showing
cardinal fossula, thick wall, and complete tabulae without
dissepiments. **C** and **D**, *Heliophyllum* (Middle Devonian, New
York, ×1.5): transverse (**C**) and longitudinal (**D**) thin sections,
showing carinae, a thin wall, a wide zone of dissepiments, and
incomplete tabulae. **E**, Transverse thin section of *Fabero-
phyllum* (Late Mississippian, Utah, ×2), showing cardinal
fossula. **F**, Cardinal side of exterior of lophophyllid coral (Late
Pennsylvanian, Texas, ×1). Grooves mark position and extent
of septa. **G**, Side view of ceratoid-cylindrical form of *Brevi-
phrentis* (Middle Devonian, Kentucky, ×0.66). **H** and **I**,
Lithostrotion (Siphonodendron) (Late Mississippian, Arizona,
×4): longitudinal (**H**) and transverse (**I**) thin sections of
phaceloid colony, showing well-developed columella and single
row of peripheral dissepiments. **J** and **K**, *Lonsdaleia (Actino-
cyathus)* (Late Mississippian, Wyoming, ×2): transverse (**J**) and
longitudinal (**K**) thin sections of cerioid colony in which septa
are not developed in zone of dissepiments but well developed in
axial zone of tabulae. **L** and **M**, *Favisitina* (Late Ordovician,
Kentucky: transverse (**L**, ×5) and longitudinal (**M**, ×2.5) thin
sections of simple form lacking dissepiments, showing several
lateral (intermural) offsets within colony. (For other rugose
corals see Figures 11.6, 11.8, 11.17, and 11.18.) (**A**, **B**, **E**, and **H**
to **K**, courtesy W. J. Sando.)

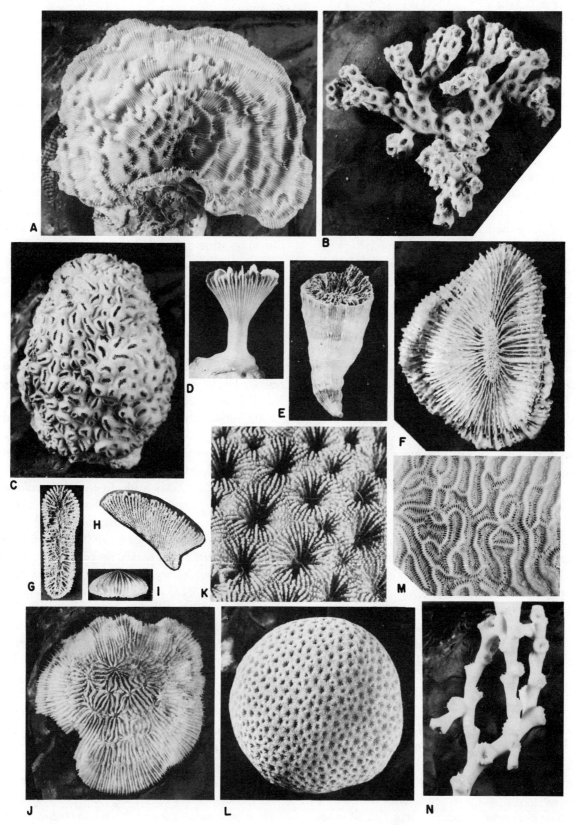

Figure 11.16. External views of scleractinian corals (**A** to **C**, **J** to **N**, colonial; **D** to **I**, solitary). **A**, Upper surface of platelike corallum of *Leptoseris* (Holocene, Caribbean, ×0.66). **B**, Ramose corallum of *Oculina* (Holocene, Atlantic, ×1). **C**, Massive corallum of *Dichocoenia* (Holocene, Jamaica, ×0.5). **D**, Side view of *Caryophyllia* (Holocene, Venezuela, ×1), showing broad base of attachment. **E**, *Asterosmilia* (Pliocene, Jamaica, ×1), trochoid. **F**, Calice view of *Scolymia* (Holocene, Jamaica, ×1), showing spongy axial structure. **G** and **H**, *Thysanus* (Pliocene, Jamaica, ×1): calice (**G**) and side (**H**) views of laterally compressed form. **I**, Side view of discoidal form of *Deltocyathus* (Holocene, Bahamas, ×1) lacking basal epitheca. **J**, Surface of colony of *Mycetophyllia* (Holocene, Belize, ×0.5) showing circumoral offsetting. **K** and **L**, Massive astreoid colony of *Siderastrea* (Holocene, Florida): surface detail (**K**, ×5) and top view (**L**, ×1). **M**, Part of surface of massive meandroid colony of *Diploria* (Holocene, Jamaica, ×1). **N**, Part of dendroid colony of *Lophelia* (Holocene, Atlantic, ×0.5).

Significance Coloniality has been an exceptionally successful strategy for many cnidarians. The reasons are complex and can be evaluated in both evolutionary and ecologic contexts. Asexual reproduction permits rapid increase in numbers (and therefore colony size) without the dangers of larval and immature life. If this is combined with the ability to grow laterally in any direction (the animal has no fixed shape), these colonies are highly adapted to conditions in which there is intense competition for limited living space (for example, on a reef). In vertical colony growth, skeletons provide mechanical strength and a platform that raises the animal off the substrate (adaptive for suspension feeders especially). Colony living also allows increased cooperation between individuals.

Asexual reproduction, however, minimizes genetic variation and thereby adaptability because new individuals are genetic replicas of their single parent. In contrast, sexual reproduction results in large numbers of genetically different larvae, most of which will not survive, but among which favorable mutants may occur. From an evolutionary standpoint, sexually reproducing species will be more variable and therefore more adaptable to environmental change through time. Colonial cnidarians get the advantages of both systems by sexually reproducing new individuals, which then bud asexually to build colonies rapidly.

Growth habits of colonies The general form of colonies may be similar in unrelated organisms because form represents a basic response to long-term environmental pressures. Common colonial coral forms are illustrated in Figure 11.9. Elaborate nomenclatures exist to describe form, but more general terms are adequate in most cases.

Colony form can be reduced to several essential geometric types (runners, vines, plates, bushes, mounds, and sheets). By virtue of these different shapes colonies have different ecologic properties with respect to ability to occupy space, define perimeters, mechanical strength, and other characteristics. For example, sheets occupy surface area efficiently, and in low-light levels, all polyps are directed toward maximum light. Bushes are mechanically efficient in sediment-laden water and allow many polyps to grow in the water column with only a small substrate perimeter to occupy and defend. Mounds are a compromise between sheets and bushes. Figure 11.9 shows that coral colonies display all these forms and may be interpreted paleoecologically using this model.

Integration The evolutionary loss of individuality in colonies is **integration**, whether referring to the polyps or the skeleton. Colonies range from aggregates of individuals that are biologically independent (polyps isolated on different parts of the skeleton after budding and initial growth) to colonies in which structure and function have been so extensively fused that individual polyps and corallites are barely perceptible.

Although both extremes occur in corals, most colonies are in some intermediate position. Thus it is useful to calibrate the degree to which integration has taken place; this is referred to as **level of integration**. Increasing level of integration involves modification of individual function and structure with a complementary increase in colony-wide cooperation and structural patterns. One view conceives this process as part of a cycle proceeding from a single individual to an aggregate of individuals and then to a sequence of higher and higher levels of integration within colonies until, with final loss of individuality of the original units, a new more complex individual is produced. Two pathways can be envisaged, one leading to increasing tissue between the polyps and the other to fusion of the polyps.

Integration within colonies can be expressed in a variety of ways:

1 *The degree of soft tissue continuity between individuals in the colony.* This can range from none, to polyps that share body walls, to organisms consisting of mouths and tentacles, all of which share a common enteron and nervous and other systems.

In many Paleozoic corals (most branching rugose and perhaps some tabulate corals) the polyps were essentially solitary animals connected only by skeletal tissue. Soon after budding, each new polyp was separated from its parent by a solid wall; as growth proceeded, corallites lengthened and separated, and polyp isolation became complete. Many branching colonies in which the corallites appear juxtaposed or fused nevertheless employed the same strategy, except with clusters of polyps instead of one. During life these branching colonies had living tissue (clusters of polyps) only at each branch tip; each living subcolony, therefore, was isolated from the others (common in many scleractinians and probably in many or most branching tabulates).

Many corals have perforate or porous walls permitting various degrees of soft part connection or communication within the colony. This is reflected in the skeleton by mural pores (in many tabulates), or by varying degrees of porosity or degeneration of the corallite wall (in a few rugosan and many scleractinian corals). A sequence of increasing connection between enterons can be inferred from Figure 11.17, *A* to *G*, in which corallite walls are first complete (Figure 11.17, *A* and *B*), become dis-

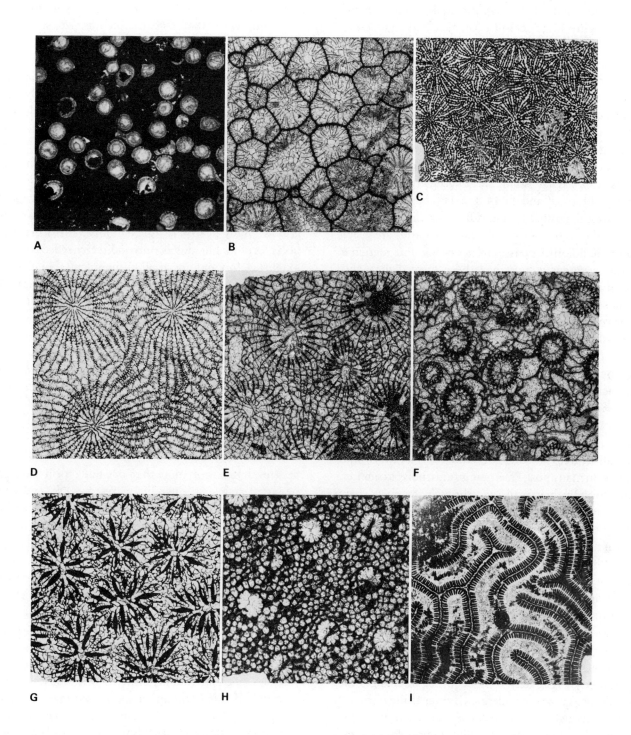

Figure 11.17. Integration in colonial corals. Transverse thin sections arranged in a morphologic series from nonintegrated (**A**) to high levels of integration (**F** to **I**). (H, heliolitid; R, rugosan; S, scleractinian; T, tabulate.) **A**. Phaceloid (R, ×1); each polyp was isolated. **B**, Cerioid (R, ×2.5); corallites separated by wall, but polyps may have joined on surface of corallum. **C**, Astreoid (S, ×2.5); no walls. **D**, Thamnasterioid (R, ×2.5); no walls, septa continuous from corallite to corallite.

E, Thamnasterioid (R, ×2); some septa discontinuous. **F**, Aphroid (R, ×2); septa limited to calice area; corallites have an inner wall and are united by dissepiments only. **G**, Aphroid (S, ×1); basically like **F** but without wall. **H**, Coenosteoid (H, ×5); corallites separated by coenosteum. **I**, meandroid (S, ×2.5); chains of polyps closely integrated, separated from other chains by coenosteum.

continuous (Figure 11.17, *C*), and finally disappear altogether (Figure 11.17, *D* to *F*). The final and highest level of integration of soft tissue is expressed by a labyrinthine body cavity having a series of mouths with no polyp boundaries, so that neither individual polyps nor corallites are distinguishable (see Figure 11.17, *I*).

2 *Development of colony-wide tissue.* Many colonies have tissues or structures that are the result of colony building and not parts of one individual. Many colonial corals in which the corallites are in lateral contact have an extensive external solid sheath called a **holotheca** (Figures 11.13, *E* and 11.14, *C*). This basal sheet must have been deposited by individuals on the lower outer edge of the colony. Generally, no boundaries are visible where the holotheca passes from one corallite margin to another, so that it must have been formed through a certain amount of cooperation between adjacent polyps. Coenosteum, common skeletal deposits within the mass of the colony, represents an advanced level of integration. Coenosteum is known in only one group of Paleozoic corals, the heliolitids (Figures 11.14 and 11.17, *H*), but it is common and striking in many post-Paleozoic species (scleractinians). In the Paleozoic species the corallite walls are solid; therefore organic connections in the colony can only have been at the surface. Post-Paleozoic species, however, combine the coenosteoid condition with all degrees of soft tissue continuity between individuals. In many post-Paleozoic groups, polyps communicate through their walls and through the coenosteum below the surface of the colony, which indicates a very high level of integration. A special kind of integration is seen in the familiar scleractinian brain coral. The valleys of the "brain" are linear series of polyp buds that share a common enteron. However, each linear series is separated from its neighbors by varying types of wall structures or by coenosteal tissue, which forms the ridges (collines) between the valleys. The sinuous interlocking pattern of the series produces the brainlike (meandroid) condition (Figure 11.17, *I*).

3 *Astogeny.* **Astogeny** is the sequence of changes that the colony as a whole undergoes during its life history. It may be thought of as analogous to the ontogeny of individuals. Polyp and corallite morphology within colonies can appear to be completely uniform, or there can be successive changes as the colony develops. Each corallite records its own ontogenetic changes, and the first (founder) corallite commonly differs from all others, reflecting its origin from a sexually produced larva. Later polyps and their corallites commonly lack the earliest stages of mesenterial and septal insertion shown by the founder. In most coral colonies, except for this founder, polyps and their corallites are remarkably uniform and such differences that do exist, such as size, seem to be random or a function of environment (local position on the colony) rather than astogenetic change.

4 *Polymorphism within colonies.* In many colonies, all individuals are morphologically and functionally alike. Other colonies have two or more kinds of individuals. In living colonies, these individuals are specialized for feeding, reproduction, defense, and other functions. In fossil colonies skeletal polymorphism is assumed to indicate similar functional differentiation, although this can seldom be demonstrated. Among the stony corals, within-colony polymorphism, evidenced in the skeleton, is rare. Many post-Paleozoic Anthozoa (octocorals) are polymorphic, but this is only indirectly reflected in the axial skeleton and not at all in the spicules. In contrast, within-colony polymorphism is well shown by the skeletons of stony hydrozoans and is greatest in other hydrozoans where dimorphism and trimorphism are common and where siphonophore colonies may include as many as three kinds of polyps and four kinds of medusae. Polymorphism is also common in scyphozoans, but both hydrozoans and scyphozoans have little or no fossil record.

Other types of polymorphism

Species that exist in more than one body form (exclusive of larval states) in their life cycles are also said to be polymorphic. Some cnidarian examples have already been noted. Both the alternation of polyp and medusa in the life cycle, and the presence of different kinds of polyps and medusae within colonies may be termed clonal polymorphism because, in each case, the different forms are produced asexually and belong to the same clones. Polyp-medusa alternation is characteristic of most hydrozoans and scyphozoans but is absent in anthozoans.

Some kinds of polymorphism reflect genetic differences. Sexual dimorphism is the most familiar example of this (although a specialized one) but is not known in the Cnidaria. In principle, any pair of alleles can produce dimorphism. However, in fossils it is difficult to prove that this is dimorphism and not differences between species, since the morphs are not physically connected. One example is illustrated from Paleozoic reef corals in which assemblages include many morphologic types, most of which are considered to be distinct species because they differ in many characters. Two forms, however, differ only in branching pattern. The two patterns occur in approximately a 3:1 ratio, suggesting

Mendelian control and that both belong to a population drawn from one species (Figure 11.18).

Polymorphism commonly represents a division of labor, different polymorphs being specialized for different functions, either within a colony or in different parts of a life cycle. Interpretation of polymorphism in fossils is largely based on knowledge of living forms.

The species problem

Studies of both living and fossil corals show them to be extremely variable, so much so that morphologic separation of closely related taxa may be difficult or, in some fossil corals, almost impossible. An extensive literature deals with this problem. As would be expected, shape and gross skeletal characters were most important in early descriptions of species of living reef corals, and species names proliferated. Recognition of the effect of environment on the corallum by taxonomists working in the field rather than in the laboratory, precipitated an argument that culminated around the turn of the century in statements to the effect that species could not be recognized at all. Many workers supported this position, emphasizing the effect of environment on morphology.

The relationship of form and other skeletal characters to environment was put on a more solid basis in the 1920s and 1930s by field scientists doing experimental work. Living colonies were divided and parts transplanted to differing situations on reefs; variation in growth rates and form were observed. In these and more recent studies, species were found to have ecotypes, distinctive varieties in particular habitats. Some species are not variable, and these tend to be restricted to one habitat; others are highly plastic and have several different morphologic expressions or ecotypes, each commonly controlled by different habitats (within-habitat variation also occurs). This implies two different ecologic strategies in the two groups of species and has important consequences in the species' potential for evolutionary change and geographic expansion.

The species problem involves defining the limits of morphologic variation in any given coral species. Given several ecotypes (variants adapted to local environmental conditions), how much of the variation is genetically and how much is environmentally controlled? This can be determined by taking samples of various species living in one habitat and transplanting them to others, carefully measuring a variety of morphologic features before and after transplanting, and observing the degree and extent of change. Many biologists and paleontologists, unaware of the capacity for environmental (ecotypic) variation in many corals have certainly created many new "species," which are in fact ecotypes and not genetically distinct units. This can occur especially when only poorly pre-

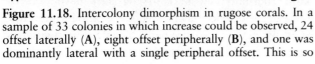

A   **B**

Figure 11.18. Intercolony dimorphism in rugose corals. In a sample of 33 colonies in which increase could be observed, 24 offset laterally (**A**), eight offset peripherally (**B**), and one was dominantly lateral with a single peripheral offset. This is so close to the 3:1 Mendelian ratio that it suggests simple genetic control with lateral offsetting dominant. (*Cyathocylindrium*, Middle Devonian, New York, ×0.5.)

served or few specimens are available for study. Coral paleontologists, and perhaps those dealing with lower Metazoa in general, should tend to be conservative in erecting new species, given the high amount of ecotypic variation at this level of organization compared with that of higher invertebrates and vertebrates. Most current taxonomic work on living corals is based on skeletal morphology combined with polyp morphology, environmental situation, behavior, and physiology.

Paleontologists have tended to follow the patterns set by students of living corals, and the history of the species problem in fossil corals is similar to that described for living ones. Some studies of variation have been made, and within-species variation should be a standard part of any good systematic study. However, while the problem of species in living corals is solvable given adequate work, that in some fossil corals is not. The skeleton in some earlier corals, especially in the Paleozoic groups, is too simple. Given plentiful variation in several species that are closely related (congeneric at least), the skeletons form a complete morphologic spectrum that defies subdivision. Some examples of Paleozoic corals that have simple morphology are shown in Figures 11.13, *K* and *L* and 11.15, *L* and *M*. More complex morphologies present lesser problems.

The problem in all corals is basically the same: great phenotypic plasticity causes overlap in morphologic boundaries of species. The student of living corals working on a reef now has other tools (such as behavior and biochemical reactions) to aid in discriminating true species. Potentially, at least, the problem can be solved. The paleontologist too has some criteria other than morphology. The fossil environment can be interpreted to see whether morphologic variation is correlated with rock type or with other preserved physical characters, and geographic and stratigraphic position can be of some use. Still, recognition of biologic species in many of the simpler Paleozoic corals is impractical, and, without question, a quantity of useful information is lost. At present, interregional studies are made at the generic level, since the identification of species beyond the local population is so uncertain.

Coral ecology and reefs

All living corals require water movement to supply nutrients and dissolved oxygen and to remove waste products; few can tolerate heavy influxes of sediment. Modern corals are divided into two ecologic groups: **hermatypic** or **reef corals** that have **zooxanthellae** (symbiotic algae) in their gastrodermal tissues, and **ahermatypic** or **nonreef corals** that lack zooxanthellae in their tissues.

Hermatypic corals are restricted to shallow tropical waters by the light requirements of their zooxanthellae. These corals and symbionts flourish within the temperature range 25 to 29°C and in depths up to 90 m and can survive water as cold as 16°C or less.

Hermatypic corals often grow to large size and may occur in great abundance with many different species. Most are colonial and are best known for their role in the construction of reefs and atolls. Their success in this milieu is partly explained by the mutually beneficial association (symbiosis) of the algae with the corals. The rate of coral metabolism is increased by the presence of the symbionts, although the process is not fully understood.

The second ecologic group, ahermatypic corals, is less restricted environmentally. About half of them are delicate, small, and solitary; colonies are slender branching forms, much less massive than hermatypes. Ahermatypes range from shallow water, some in association with reef hermatypes in the tropics, to deep cold waters in all oceans and seas. They are found to depths of 6000 m and at temperatures as low as 1°C. Branching ahermatypes have a limited capacity to form sediment-trapping banks in deeper water on the continental shelves, and some of these are large enough to be a hazard to the fishing nets of trawlers. Fossil bank structures of ahermatypes are known as far back as the Cretaceous.

No definitive morphological character identifies corals as being with or without zooxanthellae. However, there are distinctive morphological patterns that separate an assemblage of zooxanthellate corals from a fauna which is non-zooxanthellate. Based on measurement of species of living corals the following patterns seem clear: the average corallite size is smaller; the average level of integration is higher; and the dominant growth forms are colonial sheets, mounds and trees with multiple series of corallites. Such measurements have been made on several classic "reef" faunas of the Mesozoic and Tertiary, most of which exhibited a typical zooxanthellate pattern. It is interesting to note that some Cretaceous coral faunas, which have all the morphological features of a modern zooxanthellate coral fauna, occur in strata that show no independent evidence of reef building, suggesting the possibility that coral faunas that possess zooxanthellae, while today strongly associated with the reef building habit (hermatypic), may in the past have occurred without the development of reefs. It would be interesting to apply similar measurements to Paleozoic rugose and tabulate corals to see whether patterns typical of zooxanthellate corals existed in the early part of cnidarian evolutionary history.

History and relationships of cnidarians

Geologic history of classes

The principal groups of corals left an excellent record from the Middle Ordovician to Holocene (Figure 11.10). This has already been noted, and the coral group relationships are developed more fully later in the chapter. The subject of this section is the relationship of the Cnidaria to the other phyla and the relationships and geologic records of the major classes as a whole.

Impressions and compressions of polyps and medusae are scattered through the geologic record. The earliest undisputed cnidarians are in the late Proterozoic Ediacaran fauna (Figure 11.19), where two-thirds of all specimens collected are thought to belong to the phylum. Several kinds of medusae are recognized, including hydrozoan, scyphozoan, and others of uncertain affinities. In addition, several anthozoans belonging to the subclass Octocorallia (see p. 177 and Table 11.1) have been described. Specialists on living medusae and soft corals will dispute some of these assignments (and have accepted some), but undoubted cnidarians that probably represent all three of the living classes are present in the Ediacaran fauna. Objects identified as medusae have been described from even older rocks, but none are certainly cnidarian or even organic, and none are of help in understanding the early history of the phylum.

Hydrozoa Three genera of hydrozoan medusae have been reported from the Ediacaran fauna, and fragmented polypoid chitinous skeletons are known from Ordovician and younger rocks. Except for these, fossil hydrozoans are almost limited to those forms that had calcareous skeletons. Modern stony hydrozoans (Milleporina and Stylasterina) appeared in the Late Cretaceous.

Scyphozoa Two genera of Scyphozoa have been recorded from the Ediacaran fauna. The later record of scyphomedusae is spotty and inadequate for analysis of evolutionary pattern. One important group of fossils (the Conulariida), ranges from Cambrian to Triassic and may belong to the Scyphozoa (see p. 192). If so, these represent a conservative line and throw little light on scyphozoan evolution. A medusoid group, Conchopeltida (Ediacaran and Ordovician) has a distinct four-fold radial symmetry. These are logically classed as scyphozoans and might have given rise to the Conulariida, but there is no good evidence bearing on their relationship to other scyphozoans.

Anthozoa Octocorals are probably present in the Ediacaran, where they are represented by impressions of sea pen-like forms (Figure 11.19, *E* to *H*). The earliest known octocoral skeletal part is from the lower Ordovician of Sweden (Figure 11.31, *A* and *B*). This is the axis of a gorgonian, probably originally horny but preserved as calcium phosphate. The importance of the Swedish fossil is in giving the hard-to-fossilize octocorals an early record and in lending credence to the interpretation of the Ediacaran forms as octocorals.

Other possible fossil octocorals are known from the Silurian, Permian, and Triassic, but a reasonably continuous record does not begin until the Early Jurassic. Even in Cretaceous and Tertiary rocks, the fossil record is spotty because of the spicular or noncalcareous nature of skeletal structures in most of the group. Living octocorals are abundant and widespread associates of scleractinian corals in both reef and nonreef environments, and the group must have been far more common, at least in the Mesozoic and Cenozoic, than is indicated by the known fossil record.

The earliest anemone-like fossils are from the Middle Cambrian Burgess Shale (identified as Actiniaria; see subclass Zoantharia, p. 177), but these are questionable. The Middle Cambrian also includes the first skeletonized coral-like forms (the Cothoniida). These may represent an early experiment in skeleton building by anthozoan polyps. The earliest fossils generally accepted as corals are Early Ordovician (tabulate corals).

Octocorals then, are probably present in the Ediacaran fauna and Zoantharia (see subclass Zoantharia and Table 11.1) possibly in the Middle Cambrian. It seems unlikely on morphologic grounds that octocorals gave rise to the zoantharian corals, and this suggests that the Octocorallia-Zoantharia divergence took place in the Late Precambrian. Preservation of soft-bodied cnidarians in the Ediacaran, in the Middle Cambrian, and sporadically through the geologic record is important in demonstrating the great age and early diversity of the phylum, but the record is inadequate to suggest the pattern of evolution of the classes. However, it is compatible with the simple progressive theory outlined in the next section.

Origin of phylum and classes

The geologic record attests to the great age and early diversity of the cnidarians but does not yet provide any

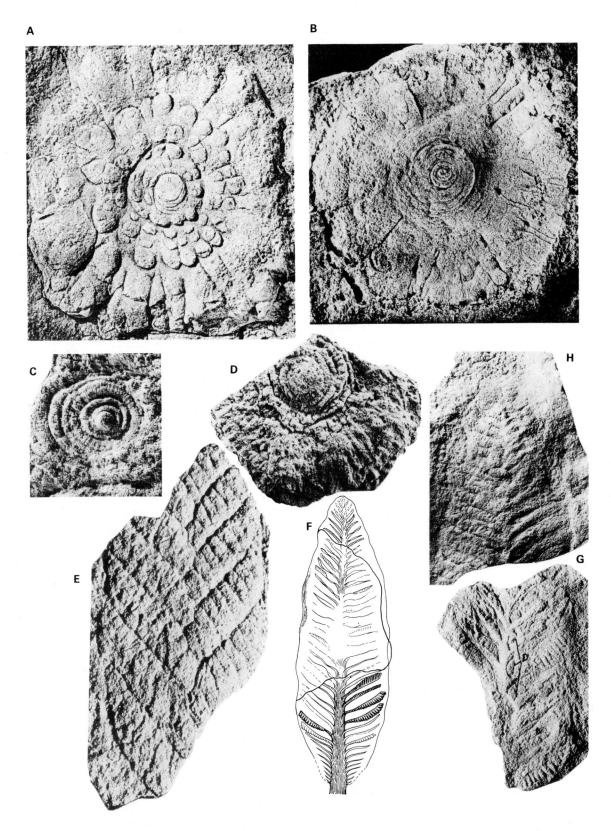

Figure 11.19. Ediacaran fauna. **A** to **C**, Medusae: **A**, *Mawsonites* (×0.66) and **B** and **C**, *Cyclomedusa* (×0.75); impressions of vertically compressed specimens. **D**, *Eoporpita*, oral view of probable hydrozoan pelagic colony (×0.5). **E** to **H**, Sessile colonies of possible octocorals. **E**, *Glaessnerina*, impression (×0.5). **F** to **H**, *Charniodiscus arborens*; restoration and two partial specimens (×0.66) on which restoration is based. (From Glaessner, M. F.; and Wade, M. J. *Palaeontology* 9; 1966; and Wade, M. J. *Palaeontology* 15; 1972.)

direct information on the origin of the phylum or the relationship of the classes. These problems must be dealt with theoretically, using information derived principally from living animals.

Cnidarians could have been derived from colonial protists through the development of specialized cells and the organization of these cells into tissues. The Porifera and other parazoans are unsuitable as cnidarian ancestors.

The radial symmetry, two-layered (diploblastic) body, and tissue level organization of digestive, nervous, and muscular systems suggest that the Cnidaria are the most

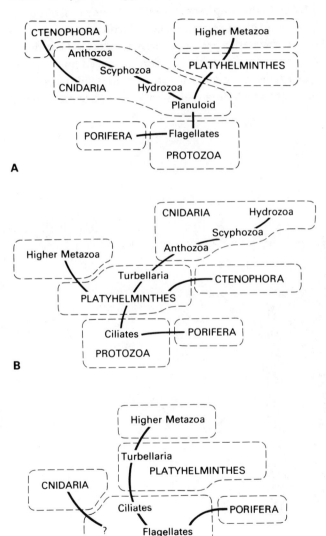

Figure 11.20. Possible evolutionary position of the Cnidaria relative to the Protozoa and the higher Metazoa. **A,** Simple progressive theory, based on Hyman (1940); both the phyla and the cnidarian classes are arranged in order of increasing complexity. **B,** Turbellarian theory, based on Hadzi (1963); proposes a bilateral ancestry for the Cnidaria and simplification within the phylum; **C,** Alternative theory, suggesting that the Cnidaria and other Metazoa arose separately from the Protozoa, based on Hanson (1977). (See text discussion.)

primitive and presumably the oldest of the metazoan phyla (Figure 11.20, *A*). All other metazoans are bilateral and three layered (triploblastic) and have organs, but they all pass through a diploblastic stage and were presumably derived from early cnidarian-like metazoans.

In trying to understand the evolution of the phylum, much depends on whether the polyp or the medusa is considered ancestral. Most biologists consider that the original sexual adult was a medusoid and that the asexual polyp represents a modified larval stage; this corresponds to the sexual roles of the two forms in most living cnidarians that have both body forms. Presumably the adult medusa originally formed by development of tentacles on a protocnidarian that was diploblastic, radially symmetrical, polarized, ciliated, and free-swimming—in other words, resembling the universal cnidarian larva called the **planula** (Figure 11.3). Such a stage exists in some hydrozoan medusae (the actinula). Hydrozoans could have arisen by the prolonged attachment of this actinula stage, which would have produced a polyp form, the asexual budding of which released medusae. Scyphozoans would then represent the evolution of the medusoid stage adapted particularly to a pelagic mode of life. This involved some division of the enteron to produce tetramerous radial symmetry and a reduction in the polyp stage. Finally, the Anthozoa might have been derived from the Scyphozoa by retaining the polypoid stage as a sexually mature adult (neoteny) producing eggs and sperm and abandoning the medusa. Anthozoa have bilateral symmetry and numerous divisions of the enteron and are well adapted to a sessile mode of life. This scheme is summarized in Figure 11.21.

If the polyp was the ancestral adult, the Anthozoa are primitive, and from them must be derived the Scyphozoa and Hydrozoa. The phylum might then have been derived from a bilaterally symmetrical ancestor (one theory

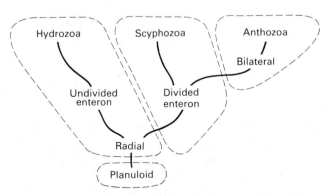

Figure 11.21. Evolution within the Cnidaria. Possible relationships of three living classes, showing arguments favoring the simple sequence of Figure 11.20, *A*.

suggests the flatworm phylum Platyhelminthes; see Figure 11.20, *B*), and radial symmetry and detachable medusae subsequently evolved.

Although the first of these hypotheses is the most widely accepted, the question is still strongly disputed and involves arguments that are beyond the scope of this book.

Part II   Additional concepts

Biology of the cnidarians

Sedentary forms, indeed the corals alone, qualify the Cnidaria as a major marine ecologic group. In addition, cnidarians form 30 percent of the species of large predators of the marine plankton. Siphonophores, the hydro-

zoan colonies consisting of polymorphs, are known to form major swarms of midwater predators across wide areas of all oceans, and medusae are the most common of marine plankton predators in inshore waters.

All cnidarians demonstrate a striking simplicity of function and structure when compared with higher Metazoa. Their body layers are thin and many functions can be accomplished simply by exchange of water between the animal and its surrounding medium. The two body layers contain various digestive, muscular, nervous, and other cells and are separated by mesogloea (Figure 11.22). In the simplest members of the phylum, the mesogloea is an almost structureless gel, consisting of some 94 per cent water, 5 per cent inorganic salts, and less than 1 per cent organic matter. In more advanced forms, mesogloea gives many cnidarians their remarkable mechanical properties including the ability to absorb sudden shocks

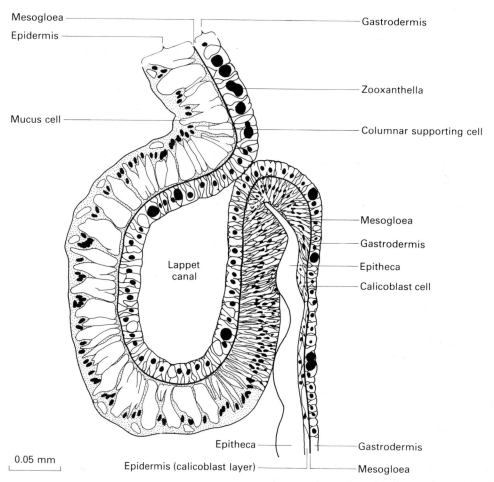

Figure 11.22. Longitudinal thin section of the skeleton and soft parts of *Manicina*, taken at the upper margin of the epitheca (including edge zone). Various cells of the epidermis and gastrodermis are indicated together with the mesogloea and the symbiotic algae (zooxanthellae) that live within the gastrodermis. Also indicated is the calicoblastic body wall, that part of the epidermis which secretes the calcareous skeleton. The

epitheca appears undulating because at different times of the day the body wall is either folded over the upper margin of the epitheca (as in the diagram) so that the epitheca grows outward or the body wall of the polyp is stretched upward so that the epitheca grows slightly inward. (From Barnes, D. J. *Proceedings of the Royal Society,* Series B 182:331–350; 1972.)

and to change size and shape. These properties stem from the elasticity of the mesogloea and from the use of the water enclosed in the gastrovascular cavity of polyps as a hydrostatic skeleton against which muscles can contract and expand. Although the mesogloea is elastic, it is also highly viscous, which means that it rebounds from shocks of short duration, such as wave surge, but deforms to many times its resting size when small but continuous pressures are applied. The mesogloea also seems to act as a lubricating substratum over which cells can move.

Musculature

Muscular activity is undertaken by epitheliomuscular cells. In these cells, elongate contractile processes called myofibrils form as extensions of a normal epithelial cell. The myofibrils combine end to end into layers, or fields, to form different muscle bands. They may lie within the epithelium or be folded to protrude into the mesogloea.

The muscle system includes longitudinal and circular fibers or bands within the outer wall. Longitudinal muscles are more or less parallel to the oral-aboral axis; circular muscles are in a plane perpendicular to these. In the Anthozoa, longitudinal muscle bands are located on the surface of the mesenteries as well as in the body wall.

Anemones and corals inflate by pumping water into the enteron. This is effected by inward-beating cilia located along the siphonoglyphs and by coordinated opening and closing of the mouth. If the volume of gastrovascular water is held constant, contraction of the circular muscles causes the polyp to become thinner but taller and/or the tentacles to thin and extend. If the longitudinal muscles contract, the polyp becomes shorter and fatter. If circular muscles are held firm, contraction of selected longitudinal muscles will bend the polyp. Thus the polyp can move and react to its environment. Peristaltic muscular movements along the polyp are also a means of bringing food from the tentacles into the gastrovascular cavity.

In medusae, the elasticity of the mesogloea combined with muscular action produces the typical rhythmic swimming movements.

Nervous system

Nerve systems in cnidarians are poorly understood but seem to be more complex than originally thought. Conduction of stimuli may be entirely epithelial with only scattered or no nerve cell involvement; this occurs frequently in the Hydrozoa. Nerve nets, when present, may be one or several independently functioning systems.

Commonly, cnidarians have a nerve net in the inner epidermis consisting of a loose two-dimensional network of nerve cells. Contacts are wherever cells happen to cross; hence, conduction tends to be in proportion to the density of the nerve net. Impulse conductors at the surface respond to light, chemicals, and vibration, as well as to stimuli that activate nematocysts (described in the next section). Conductors also include mechanoreceptors such as statocysts, which orient the organism relative to gravity, and specialized photoreceptors. Some cnidarians have multiple independent nerve systems occurring in layers within the same or different body layers or in the mesogloea. One system, for example, may act rapidly and control swimming movement or retraction of the polyp; another may act more slowly and activate tentacles.

Feeding and digestion

Most cnidarians are carnivorous, capturing a variety of prey including fish, crustaceans, and zooplankton, by paralyzing them with neurotoxins injected by the nematocysts. However, some species apparently have become nutritionally dependent upon photosynthetic algae in their tissues and have lost the usual structures and functions of their predatory relatives.

Nematocysts are unique to cnidarians. They consist of an epithelial cell containing a large space (the nematocyst capsule) that opens to the exterior. The capsule contains an introverted tube, thick and barbed at one end but continued as a thin tube coiled back around itself (Figure 11. 23, *A*). Inside the thin part of the tube are numerous small barbs. When the nematocyst is stimulated, the coiled tube is fired out at great speed, turning inside out and whipping each barb through 180 degrees (Figure 11.23, *B*). In addition, the wall of the tube has a triple set of spiral pleats so that while firing outward, the everting terminal edge of the tube also rotates. The combination of rapid eversion and rotation makes the nematocysts effective at penetrating prey, even those with hard exoskeletons such as crustaceans. Because of their nematocysts, cnidarians are able to feed efficiently despite their functional simplicity.

Explanation of the mechanism of discharge of nematocysts is still incomplete. It seems to be primarily a direct response to stimuli and not a reaction controlled by the animal. Many commensal fish certainly cause mechanical stimulation but are not stung, presumably because they have a substantial mucous coating to protect them.

Captured food is generally passed through the mouth

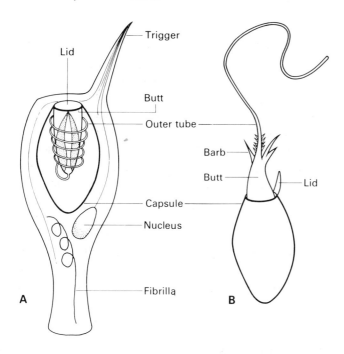

Figure 11.23. Nematocysts of *Hydra*. **A,** Undischarged cell (cnidoblast). **B,** Discharged nematocyst showing thick barbed part of tube and thin, spirally muscled, outer tube, which also has numerous fine barbs not visible at this scale.

by the tentacles, tracts of cilia, or peristaltic movement of the body. The gastrovascular cavity is capable of great distension. Within it, many glands secrete enzymes to disintegrate the food that may be too large for direct amoebic ingestion. The feeding response is activated when chemical receptors recognize certain categories of molecules released by the prey. After disintegration, ingestion and early stages of digestion take place intracellularly in vacuoles.

In anthozoans, digestion takes place within the mesenteries. In corals, mesenterial strands may be extruded from the polyp, either through the mouth, the porous wall, or oral surface, to digest food externally. In addition to mesenterial and tentacular activity in feeding, many corals also trap Protozoa and nannoplankton on surface mucous layers, which are then moved by ciliary action to the mouth. This is particularly common among hermatypic corals and makes them efficient feeders because, in effect, their total surface area is a feeding zone. Some photosynthetic products of the zooxanthellae are absorbed by the host coral cells into metabolic pathways. Whether the symbiosis primarily aids calcification or is equally important in nutrition is under discussion.

Discharge of wastes takes place at the same sites as ingestion. It may be either by diffusion of nitrogenous products, carbon dioxide, ammonia, and other material, or by the discharge of migrating amoeboid cells that carry larger particles.

Reproduction

Sexual reproduction takes place in all groups. When produced by medusae, eggs may be fertilized in the enteron or ejected through the mouth to be fertilized externally in the seawater. Sexual reproduction is mainly dioecious (eggs and sperm produced by different individuals) but if monoecious, eggs and sperm commonly ripen at different times to ensure cross fertilization. Life histories vary greatly, but in all groups some version of a planula larva is produced, which is distinctive of the phylum. When released, the planula may take from a few hours to several weeks to attach itself to a suitable substrate and develop into a polyp (Figure 11.3). Since there are no medusae in the Anthozoa, in all corals both eggs and sperm are produced in the enteron of the polyp, where the eggs are generally fertilized and then released to form planulae.

Asexual reproduction also takes place in all groups. When animals asexually split off new polyps or medusae, the process is called **budding**. In colonial corals where the budded individual remained attached to the parent, the budding process is reflected in the skeleton and is used to interpret budding patterns in fossil corals. The result of budding, the multiplication of corallites, is called **increase**.

In many cnidarians, polyps bud off free-swimming medusae. In other forms, modified medusa buds are retained as gonad-bearing individuals, which produce gametes. Medusae may also be produced by complete transverse fission of a solitary polyp (horizontal constriction of the polyp so that the upper part buds off). When this happens, the budded individual generally becomes an intermediate larval stage (ephyra) before developing into the final medusa (Figure 11.3).

Coral polyps bud asexually in a variety of ways. Budding is said to be marginal or **extratentacular** when the new bud forms outside of the central ring of tentacles, either within or outside the polyp wall. Budding is **intratentacular** when constriction of the mouth within the ring of tentacles results in longitudinal division of the parent polyp. The new polyps either separate, stay juxtaposed, or remain permanently connected in some incomplete state of separation. Buds may also form on **stolons** (tubular extensions from the parent polyp). Finally, in some species, budding of polyps is by transverse fission.

In corals, these patterns are reflected in the various types of increase. **Axial increase** (Figure 11.24, *J*) results

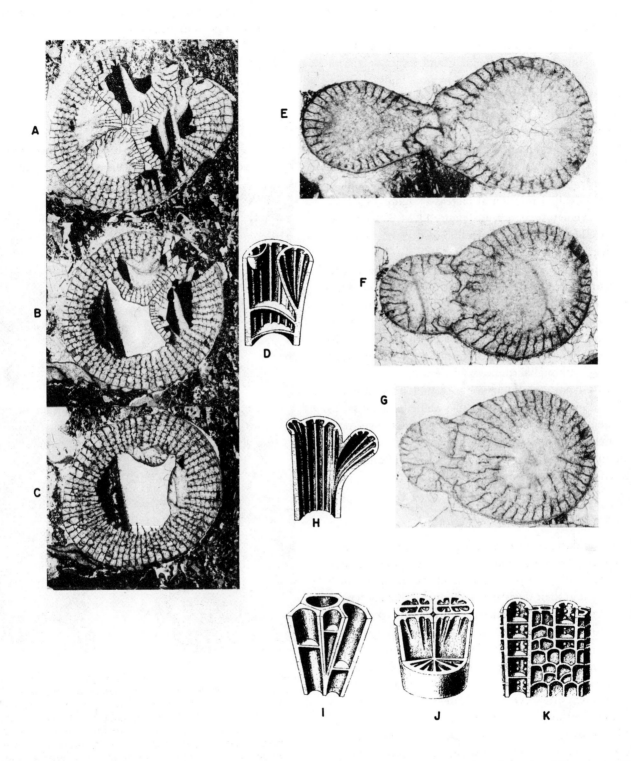

Figure 11.24. Increase. **A** to **D**, Peripheral increase: **A** to **C**, higher-to-lower serial transverse thin sections (×2.5), showing first appearance of two offsets (**C**), their initial growth (**B**, 3 mm higher than **C**), and later stage (**A**, 3 mm higher than **B**) with four offsets just beginning to separate from each other; **D**, cutaway, three-dimensional diagram. **E** to **H**, Lateral increase: **E** to **G**, higher-to-lower serial transverse thin sections (×10), showing early stage of offset development (**G**) to near separation from parent (**E**, 3 mm above **G**); intermediate stage (**F**) is 1 mm above **G**; **H**, cutaway, three-dimensional diagram. **I** to **K**, Cutaway, three-dimensional diagrams of intermural, axial, and coenosteal offsetting. Thin sections are of Middle Devonian *Cylindrophyllum* (**A** to **C**) and *Actinophyllum* (**E** to **G**) from New York. (**A** to **C** and **E** to **G** from Oliver, W. A., Jr. Noncystimorph colonial rugose corals. New York: U.S. Geological Survey, Professional Paper 869, pls 14, 31; 1976. Diagrams after Koch, G.V. Palaeontographica 29: pl. 43; 1883.)

from division of the original corallite into one or more smaller corallites called **offsets**, which occupy the original cross-sectional area of the corallite. This process involves intratentacular budding and results in the replacement of the parent by two or more buds. If offsets are formed near the periphery of the corallite but within its wall, it is called **peripheral increase** (Figure 11.24, *A* to *D*). **Lateral increase** (Figure 11.24, *E* to *H*) occurs when offsets project from the side of a corallite. If corallites are closely spaced in a colony, lateral offsets will be in a confined space surrounded by adjacent corallite walls, a condition called **intermural increase** (Figure 11.24, *I*). Finally in **coenosteal increase** (Figure 11.24, *K*), corallites arise in the coenosteum. Lateral and coenosteal increase involve extratentacular budding.

In addition, Scleractinia may show colony-wide patterns of budding. For example, intratentacular or extratentacular budding may take place in concentric rings (circumoral budding) or linear series (meandroid, Figure 11.17, *I*). These patterns are clearly indicated in the skeleton and ultimately control the shape of many colonies. Although terminology differs, many of the basic modes and patterns of increase are found in several major colonial metazoan groups, so that animals as diverse as archaeocyathans, corals, bryozoans, and sponges have geometrically similar colony forms. These forms presumably have universal adaptive significance for attached colonial animals.

Coral aggression

Many cnidarians have remarkable aggressive chemical reactions to each other and to other organisms. Modern reef-dwelling corals provide an excellent example of such a reaction in their digestive mechanism. When polyps of different coral species touch each other, one species may extrude its mesenterial filaments over the other, dissolving its tissues much as it might digest food (Figure 11.25, *A*). These corals have a pecking order, or a hierarchy of digestive dominance or aggression. Any given species either always dissolves, is not affected by, or is always dissolved by another given species.

The most dominant corals seem to be those that are smaller, more slowly growing, and usually minor elements of the reef community. Most of the rapidly expanding primary reef builders are not very chemically aggressive. Apparently smaller slowly growing corals are able to defend themselves against fast-growing corals by their digestive dominance. In addition, some corals have extralong sweeper tentacles, scattered especially around the margins of colonies, which may be able to sense and

dissolve approaching species at some distance from the colony margin and before the two colonies grow together. If the approaching colony is the same species but a different clone, these sweeper tentacles maintain a buffer zone between colonies (Figure 11.25, *B*).

Algal symbiosis

The symbiosis of hermatypic corals and zooxanthellae has already been mentioned. This is only one example of

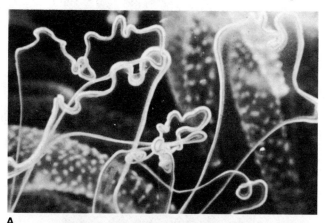

A

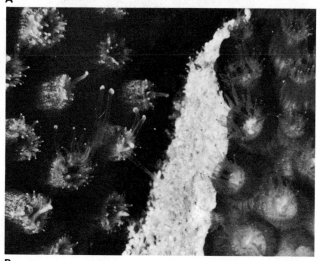

B

Figure 11.25. Coral aggression. **A,** *Meandrina,* a hermatypic coral from Jamaica, extruding mesenterial filaments. The darker line within the filaments is caused by zooxanthellae within the tissue. In the background can be seen expanded tentacles on which the white dots are the locations of batteries of nematocysts or stinging cells. The filaments may be extruded through the mouth or the body wall and can sweep the substrate in search of food or, in dominant corals, can kill adjacent polyps of other corals. **B,** Colonies of the same *Montastrea* species are here separated by a buffer zone in which longer marginal tentacles that sense the adjacent colony and are called sweeper tentacles can be seen. If the colony is not the same species, killing by digestion may occur; if it is the same species but a different clone, a buffer zone between clones is maintained as in this case. If the two colonies are the same clone they will completely merge.

a widespread endosymbiosis of cnidarians and algae. More than 100 genera have a symbiotic relationship with algae either as zoochlorellae (green-colored algae) or zooxanthellae (brown-colored algae). Zoochlorellae are well known from their worldwide occurrence in green hydra; in this relationship the role of algae in the hydra is not fully clear.

In Cnidaria, zooxanthellae are found only in marine forms, the most common hosts being anemones, corals, gorgonians, and zoanthids. These symbionts are generally restricted to the gastrodermal tissues of the host's tentacles and oral discs. They are thought to enter the cnidarian by being ingested intracellularly at the same sites as food and then to multiply rapidly. The zooxanthellae apparently interact with the host in at least three ways:

1 To translocate reduced organic carbon and nitrogen to the host.

2 To recycle certain nutrients such as phosphorous and nitrogen within the host-symbiont system.

3 To greatly accelerate the rate of calcium deposition in skeletonized forms.

The coral skeleton

Carbonate secretion

The construction of a coral skeleton is an additive process in that it preserves a record of the changes that took place in the polyp during growth (ontogeny). Since 1970, many processes of calcification (skeleton formation) have become clear because of the very high resolution of detailed morphology permitted by scanning electron microscopes. Certain processes of crystal growth in coral skeletons seem universal. The part of the epidermis that secretes the skeleton is called the **calicoblastic layer** (Figure 11.22). A space between it and the skeleton is filled with a fluid rich in organic compounds (amorphous layer in Figure 11.26). These compounds allow calcium ions (from the seawater) and bicarbonate ions to be united to form $CaCO_3$. One postulated reaction is: $Ca^{2+} + 2HCO_3^- = CaCO_3 + H_2O + CO_2$ (Figure 11.26). The fluid thus beomes saturated with $CaCO_3$, which is deposited as tiny crystals.

It has also been hypothesized that in hermatypic corals, the action of zooxanthellae in removing carbon dioxide effectively increases the rate at which the reaction can proceed, hence allowing faster skeletal growth. Some experiments indicate that skeletons may grow 20 times faster during the day when the algae are photosynthesizing than at night when they are not.

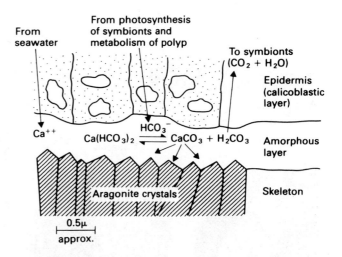

Figure 11.26. Postulated model for precipitation of the coral skeleton. Bicarbonate and calcium ions are transported to an amorphous layer lying between the epidermis (calicoblast layer) and the skeleton, where they are precipitated as skeletal aragonite needles in continuity with the existing crystals. (From Sorauf, J. *Paleontology* 15; 1972.)

In living corals, $CaCO_3$ is deposited in a series of elongate, slightly wedge-shaped (acicular) microcrystals of aragonite, approximately 0.3 to 1.5 μm wide and 8 to 15 μm high. These microcrystals are arranged into tufts or three-dimensional fans called **sclerodermites** (Figure 11.27, B). These range up to 25 μm high and 5 μm wide. Most crystals emanate from a point near the base of the sclerodermite called a center of calcification, but others are intercalated between diverging crystals. The fan-shaped pattern of the crystals in a sclerodermite is apparently the result of competition for space between individual crystals. The space competed for is the cavity below the calicoblastic layer. New crystals nucleate on the surfaces of preexisting ones. Assuming that they grow randomly into the space, those that grow laterally soon interfere with adjacent crystals, whereas those that extend into the cavity continue to grow. These sub-parallel crystals, which expand and multiply, form the sclerodermites. Adjacent fans finally interlock laterally with each other to form a stable layer of calcium carbonate (Figure 11.27, B). Because the coral deposits $CaCO_3$ much more rapidly during the day than at night, the layer of sclerodermites that are formed during each day frequently appears on the surface of the skeleton as a growth band analogous to annual rings in trees.

Skeletal elements and microstructure

Microstructure is the arrangement and physical relationship of the mineral crystals within the parts of the

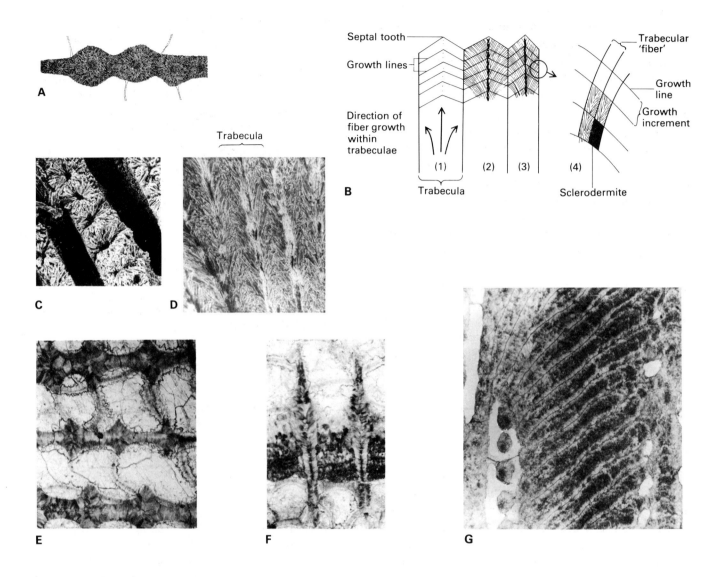

Figure 11.27. Microstructure of coral septa. **A**, Diagram of a cross-sectional view of the septum of *Myriophyllum* (Triassic scleractinian, ×50). Five trabecular cross sections are shown, varying in size, but each displaying a distinctive radiating fibrous crystal structure. **B**, Longitudinal view of idealized coral septum, showing upper edge zigzagging to form teeth whose apices mark the axis of each trabecula and the point of maximum upward growth. Because upward growth is incremental, well-preserved specimens may show a series of zigzag growth lines (*1*). More frequently, recrystallization destroys the fine detail, and a series of fibers "bouqueting" upwards is preserved (*2* and *3*), representing the upward growth track of successive tufts of crystallites (microcrystals) called sclerodermites (*4*). Sclerodermites are bounded laterally by the line of interference with the adjacent fiber and vertically by the growth lines, which in many modern corals represent a 24-hour growth period. **C**, Scanning electron micrograph of cross sectional view of trabeculae showing radiating fibers in the septa of *Rhopalophyllia* (Triassic scleractinian, ×85). **D**, Scanning electron micrograph of a longitudinal view of a septum of *Rhopalophyllia* (×100), showing laterally contiguous trabeculae with fibers "bouqueting" upward away from a central axis and laterally intergrowing at boundary of trabeculae. **E**, Cross section of a pair of septa of *Heliophyllum* (Devonian, New York, ×25); here, large trabeculae are spaced out and form lateral ridges (carinae) on the sides of the septa. Spaces between the large trabeculae are occupied by numerous small crystallites (see also Figure 11.29). **F**, Longitudinal section of a septum of *Heliophyllum* (×25), showing spaced trabeculae with "bouqueting" upward fibers and a central axis joined in the plane of the septum by poorly defined microcrystals (see also Figure 11.29). **G**, Longitudinal thin section of *Siderastrea* (Holocene scleractinian, ×64), mostly in the plane of a septum. Obliquely inclined trabeculae are fused to form a solid septum. At the left margin the theca (wall) of the corallite is visible. To the right of the theca in the lower left corner, the plane of the thin section is between two septa so that trabecular cross bars (synapticulae) are seen joined to the theca and septa by dissepiments. Farther right, the laterally fused septal trabeculae are seen in longitudinal section. At the right margin of **G** the septum is fused to the columella, leaving occasional pores. (**A**, **C**, and **D**, courtesy J.-P. Cuif. **G**, courtesy A. B. Foster.)

skeleton (Figure 11.27). In a general way, these arrangements are similar in most corals, but significant differences in detail are of some value in classification. Details of microstructure tend to be consistent within genera or families, although some families are essentially identical in microstructure. The principal contribution of microstructural studies has been an understanding of coral growth and the way skeletons are formed. Growth is periodic, and the nature and formation of both fine and coarse growth banding have become understood through detailed studies of microstructure. The interpretation of growth lines is discussed in a later section.

Coral skeletons are built of four primary elements. The first is the basal plate (Figures 11.1 and 11.5), which is the initial platform secreted after the planula has settled and to which all other skeletal parts are added. The second element consists of the vertical parts, mainly septa, which are formed within radially disposed upfolds of the calicoblastic layer above the basal plate (Figures 11.1 and 11.5). The third element includes horizontal structures such as tabulae or dissepiments (Figure 11.6); these are formed as upward growth of the polyp takes place and as the basal calicoblastic layer secretes new floors. Finally, there is the **epitheca**, an outer skeletal sheath laid down by the **edge zone**, a fold of the body wall extending over the top edge of the corallum in living solitary corals homologous to the coenosarc in colonial forms (Figure 11.5, *A*). The epitheca is a secondary thin outer sheath to the theca in most corals (Figures 11.6, *G* and 11.28).

The basal plate and epitheca have similar microstructure. Although considered as separate elements of the coral skeleton, the epitheca is simply an extension upward of the edge of the basal plate. Both elements are thin and bilayered. An outer primary layer is ultrathin (1 to 2 μm thick in studied living corals) and composed of very fine crystals. The inner (secondary) layer is thicker and composed of crystals roughly formed into fan-shaped bundles. The primary layer undulates to form the growth lines; the secondary layer fills in the undulations and presents a smoother inner surface and is thus variable in thickness. These observations are based on detailed studies of relatively few living corals but seem to be generally valid for most corals.

The septa grow upward from the basal plate. Growth is fastest around a series of axes located where the calicoblastic layer was raised to form a space above the basal plate. Sclerodermites grow into these spaces, their daily increments advancing as cone- or tent-shaped projections. The individual columns of vertically stacked cones are called **trabeculae** (Figure 11.27, *A* to *F*) and are easily seen with an optical microscope. They commonly coalesce in the median plane of the septum like stakes in a palisade fence (Figure 11.27, *B* to *D*); the tips of the trabeculae, either singly or in groups, may emerge along the upper edge of the septum as serrations (septal teeth or dentations), which mark the leading or growing edge of the septum (Figure 11.27, *B*). The microarchitecture of the septum thus accurately reflects its underlying microstructure. Each daily growth increment of the sclerodermite layers forms zigzagging, horizontally disposed growth bands on the sides of the septa. Frequently, in partly recrystallized fossils, the upward traces of sclerodermites from one increment to the next are accentuated as an upward fan-shaped series of fibers that may be the most obvious microstructure within the septa (Figure 11.27, *D* and *F*).

If the trabeculae are aligned in the plane of the septum and laterally interlocked, the septa will be **laminar** (imperforate, smooth; most rugose corals). In many corals, however, the trabeculae are not fused completely, so that gaps or pores (fenestrae) appear between them (perforate septa; many scleractinians; Figures 11.28, *C* and 11.29, *E*). Secondary trabeculae may emerge at right angles to the median plane of the septum and form spines on the septal surface (Figure 11.1, *F*). These spines frequently outline the spacing, thickness, and orientation of the major trabeculae within the septum and are an important feature in coral classification.

Horizontal elements, including dissepiments and tabulae, form from sclerodermites that grow laterally toward each other so that the sclerodermites are oriented horizontally within a thin primary layer. The dissepiment or tabula is then thickened by sclerodermites that grow at right angles to the primary layer on its upper surface.

Many Paleozoic corals show well-preserved microstructure, apparently because their original mineralogy was calcitic (orders Tabulata, Heliolitida, and Rugosa). In the Mesozoic to Cenozoic Scleractinia, the microstructure is frequently lost or obscured either because of the recrystallization of aragonite to calcite or the dissolution of aragonite in diagenesis.

In some Paleozoic corals, septal microcrystals are perpendicular to the median plane of the septum, and no trabecular structure is evident. In the walls, tabulae, and dissepiments of many Paleozoic corals, short fibrous crystals are perpendicular to the surface of the skeleton. With only slight diagenesis, this fibrous structure is lost, and simple lamellae result. Much discussion of the microstructure of fossil corals centers on which structures are original and which are diagenetic. If diagenetic patterns are closely controlled by original microstructure, they are still valuable taxonomic features.

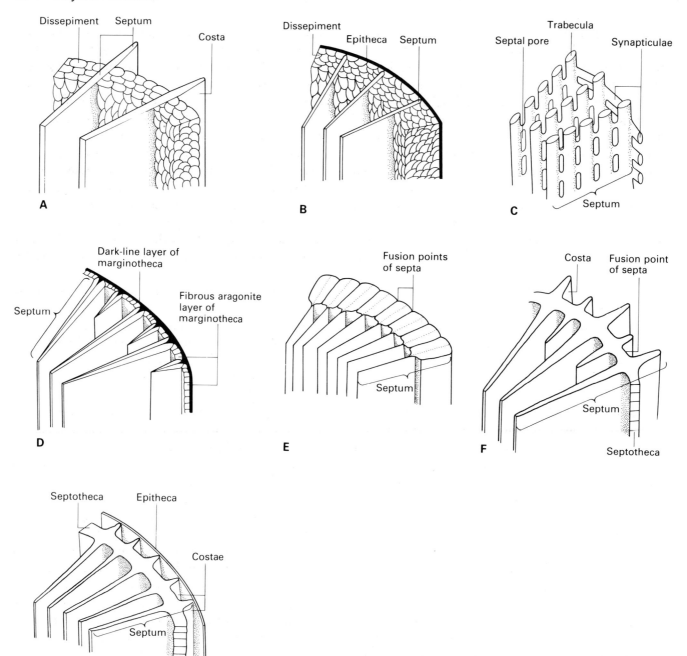

Figure 11.28. Schematic diagrams of some types of theca or wall structure in the Scleractinia, Rugosa, and other corals. **A,** Theca (costate paratheca) formed by a zone of dissepiments traversed by septa, which extend through to the exterior as ridges or costae. A thin external epitheca may also be present as in **G**; this is typical of some scleractinians. **B,** Theca (epithecate paratheca) parathecal as in **A,** but septa extend varying distances into the zone of dissepiments. The paratheca and the ends of the septa (if they extend that far) are covered by a thin epitheca. This type is typical of Rugosa. **C,** Theca (synapticulotheca) formed by a zone of synapticulae and typical of colonial scleractinians. **D,** Theca (marginotheca) apparently growing up from the basal plate independently of the septa. This theca generates the septa, and its outer dark layer appears continuous with the axial dark line of calcification of the septa. In addition to the outer dark layer, an inner fibrous aragonite layer is clearly separate from the septa. This is a primary wall structure, not an epitheca. The latter is formed after the septa and other primary wall structures have been laid down, as a thin secondary outer coating. This structure is found in some scleractinians, but apparently similar walls occur in rugosans. **E,** Theca (septotheca) formed by massive swelling of outer ends of the septa to form a broad peripheral wall. In this example, suture lines between adjacent septa are clear and the fairly smooth external surface has no costae. Epitheca may be present or absent. The septotheca is typical of scleractinians and rugosans. **F,** Theca (costate septotheca) is formed as a well-defined zone of fusion between adjacent septa; in many cases no line of suture is visible between the septa. Costae are commonly developed with this type of wall, which is typical of scleractinians. **G,** Costate septotheca, with an outer epithecal sheath (epithecate costate septotheca), occurs in scleractinians.

Septal insertion

The arrangement, symmetry, and sequence of appearance of the septa are fundamental in classifying rugose and scleractinian corals. In the Tabulata, septa are relatively unimportant, consisting of isolated trabeculae forming spines that extend only a short distance from the periphery of the corallite wall inward towards the axis. Little is known of their pattern of insertion. The Heliolitida have better and more regularly developed septa, but the insertion sequence is also unknown. In the orders Rugosa and Scleractinia however, septa are major skeletal features whose pattern and symmetry of insertion during ontogeny are the main classificatory feature.

Septa first reached a high level of development in the Rugosa. The first six septa to be laid down on the rugosan basal plate are the **protosepta**. The first two were laid down at opposite poles, delineating a plane of bilateral symmetry; they are called the **cardinal** and **counter septa** (Figure 11.30, A). Generally, two septa then appeared on either side of the cardinal septum—the **alar septa**; finally, two appeared on each side of the counter septum—the **counterlateral septa** (Figure 11.30, B and C). After the protosepta, subsequent septa (metasepta) are inserted serially at only four points—on either side of the cardinal septum and on the counter side of each alar septum (Figure 11.30, E to H). The septa thus cluster into four quadrants, which may be further demarcated by gaps, the fossulae (Figure 11.6, A and C). The septal arrangement and fossulae emphasize the bilateral symmetry of the Rugosa (Figure 11.30, H). In Rugosa, as the longer, serially inserted **major septa** appear (protosepta plus metasepta), short **minor septa** are intercalated between them (probably serially also) and thus equal them in number.

The Scleractinia, which replaced the Rugosa as the major coral order after the Paleozoic, are characterized by a different septal insertion plan. Six protosepta are laid down on the basal plate (Figure 11.1) in a manner comparable but not identical with that of the Rugosa. The subsequent metasepta are inserted in cycles of 6, 12, 24, 48, and so on, the septa of each new cycle centered in the interseptal spaces of the previous cycle (Figure 11.30, I). The result is a sixfold (hexameral) symmetry, although the third and later cycles of septa may be incomplete (Figure 11.30, J).

Supplementary septal structures

A series of modifications of the basic septal structure just outlined may take place. In both Rugosa and Scleractinia, the inner ends of the septa often extend and coalesce to form a vertical axial structure called a **columella** (Figure 11.6 D, I and J). In some groups, lateral trabecular extensions may form complex balcony-like ledges on sides of septa known as **pennulae** in scleractinians or **carinae** in rugosans (Figure 11.29). One or more trabeculae may also extend laterally from two adjacent septa until they fuse to form crossbars termed **synapticulae** (Figure 11.29, D, E and H). Other complex microarchitectural features of corals are still poorly understood, which is one reason for the lack of agreement on the classification within the orders.

Theca

The skeletal deposit that surrounds the polyp and unites the outer edges of the septa is termed the theca (or wall), an important feature of coral classification. In colonial corals, individual corallites are commonly but not necessarily separated by skeletal walls; these thecae may be wholly or partly absent in more advanced types, reflecting the high level of integration of the polyps. Various kinds of thecae are illustrated in Figure 11.28. In some corals, the theca may be a primary structure called a marginotheca, formed by upturn of the basal plate. Surrounding any of these types of theca, may be the secondary outer layer, the epitheca (Figure 11.28).

Coenosteum

Colonial skeletal deposits formed between corallites are termed coenosteum. The coenosteum is composed of a variety of structures that are essentially similar to those that are within corallites, for example, extensions of the septa beyond corallite walls (costae), synapticulae, dissepiments and tabulae, or an irregular spongy-to-solid mass of skeletal tissue. Coenosteum is deposited by coenosarc, interpolypal tissue in which individual polyp contributions are not recognizable.

Only one Paleozoic order, the Heliolitida, consistently formed coenosteum, but it became widespread and elaborate in the post-Paleozoic Scleractinia.

Coral classification

We have defined the true corals as those members of the Class Anthozoa that possess massive, external, calcareous skeletons. Other anthozoans are the sea anemones and a variety of soft corals with or without other kinds of skeletons. Three subclasses are recognized, but true corals are limited to one of these, the Zoantharia.

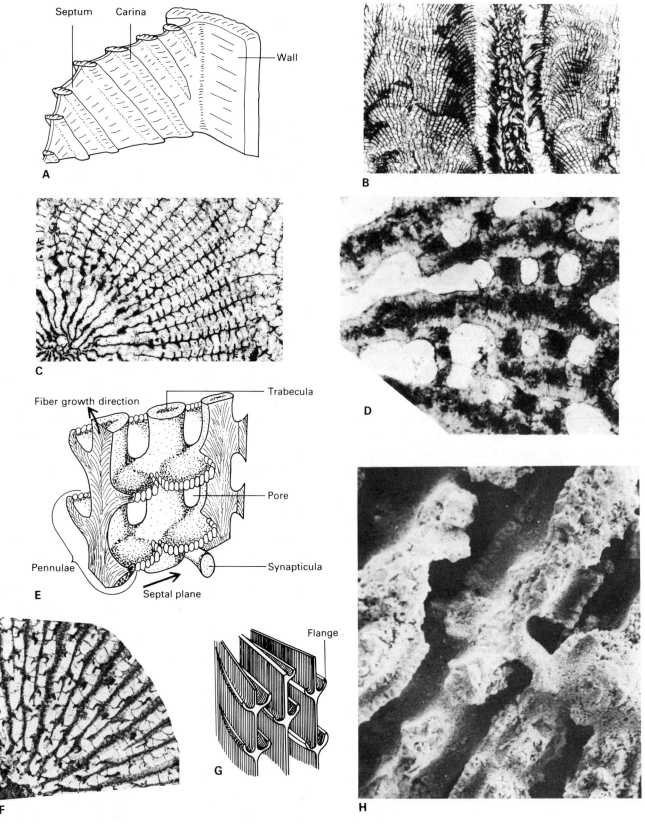

Figure 11.29. Supplementary septal structures. **A,** Geometry of septal carinae in *Heliophyllum* (Devonian rugosan; see Figure 11.27 for microstructure). **B** and **C,** Longitudinal (×2) and transverse (×5) thin sections of septal carinae (crossbars on septa) in *Heliophyllum*. In **B,** coral axis is right center; axis is lower left in **C. D,** Transverse thin section of *Siderastrea* (modern, ×64), showing septa joined by bars (synapticulae) that have the same structure as the septa (see Figure 11.27, **G,** for view of synapticulae in a longitudinal section). **E,** Reconstruction of the septum of *Chomatoseris* (Jurassic), showing

SUBCLASS *CERIANTIPATHARIA*

Solitary and colonial Cnidaria with unpaired mesenteries and weak mesenterial muscles. New mesenteries inserted only in dorsal intermesenteric spaces; after insertion, mesenterial couples shift ventrally so that insertion is serial. (Miocene to Holocene; about 50 genera.)

Ceriantipatharia include the precious black coral and a group of burrowing sea anemones. They are very rare as fossils.

SUBCLASS *OCTOCORALLIA* (Figure 11.31)

Colonial Anthozoa with eight tentacles and eight unpaired mesenteries. Symmetry bilateral, superficially radial, with elongate mouth, one siphonoglyph, and one pair of directive mesenteries. Skeleton commonly internal, spicular, horny, or calcitic; external horny or aragonitic skeletons known. (Ediacaran, Ordovician, Silurian(?), Permian(?), Triassic(?), Jurassic to Holocene; about 250 genera.)

Variation in the colonial form and composition of Octocorallia has produced a remarkable variety of living groups, which include the sea pens, soft corals, sea fans, organ pipe corals, and precious coral. Fossils of whole skeletons are exceedingly rare, but attachment structures (holdfasts) are fairly definitely known back to the Ordovician, the axis of a sea fan is known from the Ordovician, and impressions of apparent octocorals are known from the Ediacaran (Figures 11.19, *E*, to *H* and 11.31).

Calcareous spicules produced in the mesogloea of some octocorals are found in great numbers in many modern sediments, especially in reef environments. Curiously, few fossil occurrences have been recorded (Figure 11.31). Certain types of spicules are highly diagnostic of species, but when disaggregated, most are too generalized to identify even to genus.

SUBCLASS *ZOANTHARIA*

Solitary or colonial Anthozoa with coupled and paired mesenteries inserted at two, four, or six positions. Skeletons calcareous or absent. (Cambrian(?), Ordovician to Holocene; about 2000 genera.)

The subclass Zoantharia is the largest and paleontologically the most important anthozoan group because it includes most of the corals in addition to important groups of sea anemones. Polyps have paired mesenteries that are inserted in accordance with definite patterns in two, four, or six positions. In fossil orders, arrangement and insertion of septa is assumed to reflect the arrangement and insertion of mesenteries, as in living forms. Different patterns characterize the various living orders.

The subclass includes three orders of sea anemones (Zoanthiniaria, Corallimorpharia, and Actiniaria), which are important in theories regarding the evolutionary relationships of the coral orders. Fossil sea anemones are rare, but specimens preserved as impressions and carbonized films have been so identified. None can be assigned to an order but most have been classified as Actiniaria.

The principal coral orders are the Tabulata, Heliolitida, and Rugosa (all Paleozoic) and the Scleractinia (Mesozoic to Cenozoic). Two other orders, Cothoniida and Heterocorallia, are minor but add interesting areas for speculation.

ORDER *COTHONIIDA* (Figure 11.32)

Small calcitic solitary and colonial corals; cone or dish shaped, operculate, with weak septa. Insertion

---

three trabeculae, two of which are cut away to show fibrous structure (compare with Figure 11.27). Gaps between trabeculae are seen as pores. Lateral trabecular expansions of the septa form shelflike features called pennulae that may have elaborate dentate margins. Pennulae are ranged at the same level from trabecula to trabecula on the same side of a septum, but on the other side of the same septum they are commonly offset vertically. **F,** Transverse thin section of *Craspedophyllia* (Triassic, ×8), showing transverse, slightly oblique sections of pennulae. Pennulae and septa are connected by thin dissepiments. **G,** Diagram of the septa of the Paleozoic rugose coral *Metriophyllum,* showing flanges, similar to pennulae, but not related to trabeculae. **H,** Scanning electron micrograph of the septa of *Chomotoseris* in oblique view, showing pennulae with dentate rims and a synapticula (×125). (**A,** modified from Sorauf, J.; Oliver, W. A., Jr. 1976 and **G,** modified from Holwill, F. J. W. 1964 by Hill, D. In: Teichert, C., editor. *Treatise on invertebrate paleontology, Part F.* Boulder, CO, and Lawrence, KS: Geological Society of America and University of Kansas Press; 1981. **E,** from Gill, G. A. and Coates, A. G. *Lethaia,* 10: 28; 1977. **D,** courtesy A. B. Foster. **F,** courtesy J.-P. Cuif. **H,** courtesy G. A. Gill.)

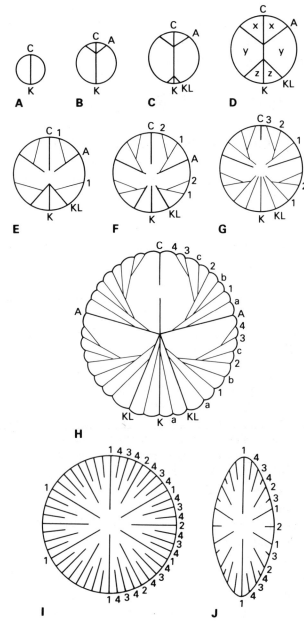

Figure 11.30. Patterns of septal insertion in rugose and scleractinian corals. **A** to **H**, Serial transverse sections of early-to-late ontogenetic stages in rugose corals. Protosepta are labelled as follows: *C*, cardinal septum; *K*, counter septum; *A*, alar septa; *KL*, counterlateral septa; sectors are labelled as follows: *x*, cardinal sectors; *y*, alar sectors; *z*, counter sectors. **A** to **C**, Early ontogenetic stages with the insertion of protosepta; **D**, Protosepta delimit six sectors (*x, y, z*), in four of which (*x, y*) subsequent metasepta will be serially inserted. **E** to **G**, Stages in the serial insertion of the metasepta in each sector; metasepta are labelled *1, 2, 3,* and *4* in order of insertion. **H**, Mature stage showing the protosepta and metasepta together with the minor septa (lower-case letters) and fossulae (gaps between sectors). **I** and **J**. Scleractinian corals. **I**, The cyclic pattern of insertion in scleractinian corals; six protosepta are labelled *1*, with three more cycles inserted in the order *2, 3,* and *4*. **J**, Compressed corallum in which septa are inserted cyclic-bilaterally due to higher cycles (*3* and *4*) developing at the poles before or without developing laterally. (From Oliver, W. A., Jr. *Paleobiology* 1980.)

pattern not known but septal arrangement compatible with that of Rugosa. Rare. (Middle Cambrian; one or two genera.)

Cothoniids are a newly discovered group of probable corals that are anomalous in morphology and time. They are most similar to certain simple rugose corals (Rugosa first appeared in the Middle Ordovician) but are known only from the Middle Cambrian and were therefore much too early and too morphologically advanced to be directly linked to that group. They may represent an early group of zoantharian polyps that secreted skeletons (and were therefore corals) but that soon became extinct, an unsuccessful experiment, not uncommon in the early history of several metazoan phyla.

The coralla of cothoniids were apparently composed of calcite and tend to be infilled with fibrous calcite; no tabulae or dissepiments have been seen. The sequence of septal insertion is not known, but visible arrangements are compatible with a rugosan pattern of development. An operculum (lid) was present (as in some rugosans), and this also is septate. The way in which the operculum was formed by the polyp and the mechanics of its articulation are not known but the relationship of the two skeletal parts is clear, since many opercula have been found in place. Colonial cothoniids are known but contain relatively few individuals. Their corallite morphology is the same as that of the solitary forms.

ORDER TABULATA (Figure 11.13)

Calcitic, exclusively colonial corals with slender corallites. Pores or tubes between corallites common. Septa absent or short; commonly expressed as rows of spines or low ridges. Tabulae common; dissepiments present or absent. (Early Ordovician to Permian; about 280 genera.)

Tabulata comprise a variety of coral groups a few of which may be unrelated and possibly not even cnidarians. Tabulates include sheet-to-massive moundlike, erect branching, and chainlike forms of various sizes and shapes (Figures 11.9, *B* to *D* and 11.13). Corallites are commonly elongate and gently tapering or cylindrical with numerous transverse tabulae. Corallites are generally small (0.5 to 5 mm); large individuals are rare. Coralla may be small and composed of few corallites, but colonies as large as 4 m in diameter containing several million corallites are known. Asexual increase in most tabulates is lateral (including intermural), but axial and peripheral increase is known.

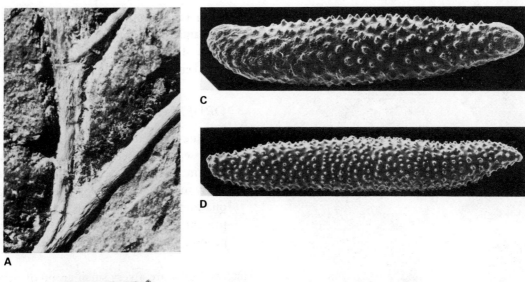

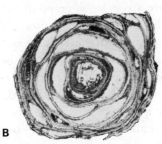

Figure 11.31. Octocorals. **A** and **B**, Unusual, well-preserved fossil sea fan (*Nonnegorgonides*, Early Ordovician, Sweden): **A**, part of branching axis (×3); **B**, transverse thin section of axis (×33). **C**, Spicule of Middle Silurian octocoral (*Atractosella*, Sweden, ×40). **D**, Spicule of Holocene octocoral at same scale for comparison (*Sinularia*, Funa Futi Atoll). (**A** and **B** from Lindström, M., *Geologica et Palaeontologica* 12: pl. 1; 1978, **C** and **D** from Bengston, S. *Journal of Paleontology* 55: 290; 1981.)

Tabulate corals have been reported from Cambrian rocks, but all specimens are of uncertain relationship. Distinct tabulates are present in the Early Ordovician and are common in rocks of Middle Ordovician to Devonian age. They are less common in the Mississippian to Permian. Reported Triassic occurrences are unconfirmed and probably erroneous.

ORDER HELIOLITIDA (Figure 11.14)

Calcitic, exclusively colonial corals with slender corallites separated by extensive coenosteum. Septa generally present, commonly 12 in number, either spinose or laminar. Tabulae common. (Middle Ordovician to Middle Devonian; about 70 genera.)

Heliolitida are similar to tabulates in being exclusively colonial, in having skeletons that were probably originally calcitic, and in having corallites of small diameter and many tabulae. They differ significantly in that the corallites are separated in the coralla by coenosteum that is apparently the deposit of colonial tissue (coenosarc) developed between individual polyps. Skeletal evidence of intercommunication between corallites (pores, connecting tubes) is lacking, although soft tissue probably covered the surface of the living colony, joining all of the

polyps in an integrated system (see the earlier section "Coloniality").

Asexual increase was commonly coenosteal, although lateral increase also took place. Coralla are massive but may be laminar to lenticular or hemispherical or even branching in overall shape. Heliotidids are limited to, but common in, rocks of Ordovician to Middle Devonian age.

ORDER RUGOSA (Figures 11.8, *B* to *D*, *F*, *G* and *I*; 11.9, *I* and *J*; 11.15)

Calcitic, solitary and colonial corals with major and minor septa. Major septa inserted serially in four positions; minor septa short and inserted between major septa, probably serially also. Tabulae almost always present; dissepiments commonly so. Axial structure present or absent, generally of septal origin. Walls formed by laterally fused septa, dissepiments, or additional stereoplasm; external walls epithecate. (Middle Ordovician to Late Permian; about 800 genera.)

Rugosa are distinct in structure but include a more varied group of corals than do the other Paleozoic orders. They are solitary or colonial and embrace most of the

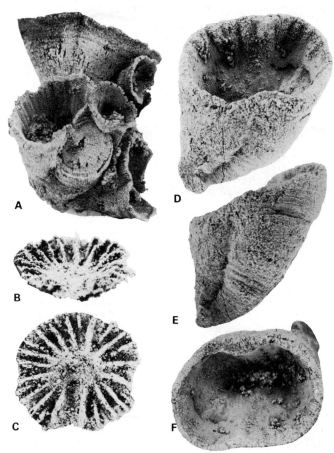

Figure 11.32. Cothoniida. **A,** Small colony showing septal ridges (left center) and operculum (lower center). **B** and **C,** Oblique and inside views of large, circular operculum. **D** to **F,** Oblique, side, and calice views of solitary specimen; note septa (**D**) and external growth lines (**E**). (*Cothonion,* Middle Cambrian, Australia, ×9.) (From Jell, P. A. and Jell, J. S.: *Alcheringa* 1: Figures 2, 3, and 9; 1976, by permission of Association of Australasian Paleontologists.)

shapes and colonial forms mentioned in the general discussion. Solitary rugose corals range in size from a few millimeters up to giants of 14 cm in diameter and nearly 1 m in length. Colonies as large as 4 m in diameter are known. The general shape and morphology of typical rugose corals are shown in Figures 11.8, *B* to *D*, *F*, *G* and *I*; 11.9, *I* and *J*; 11.15. Internal structures are best seen in either transverse (at right angles to axis) or longitudinal (cut so as to include axis) sections cut through the coral, and most Paleozoic corals are illustrated in this way.

Rugosa are characterized by a high level of septal development. Well-preserved rugose corals are composed of calcite; probably this was the original mineral composition of all rugosans. Septa are commonly trabecular in structure, although laminar septa are also known.

Rugosa range from the Middle Ordovician to Late Permian and were common through most of that time. For much of the Paleozoic they serve as good indicators of age and environment.

ORDER HETEROCORALLIA (Figure 11.33)

Calcitic elongate corals with four axially joined protosepta; subsequent septa in each quadrant remain attached to earlier septa, dividing successively to form Y-shaped patterns, although adjacent quadrants differ in degree of splitting. Tabulae present. (Late Devonian to Mississippian; five genera.)

Heterocorallia are a very small group of corals that are known from only a short period of time, although they were worldwide in distribution. They are included in the Zoantharia because they have septa and because their microstructure is similar to that of other corals. Only four of the five genera have been described, and one of these may be the immature stage of another because all known specimens are fragments. These corals are most similar to the Rugosa and may represent an aberrant offshoot of that order.

ORDER SCLERACTINIA (Figures 11.8, *A, E, H, J, K*; 11.9, *A, E* to *H*; 11.16)

Aragonitic, solitary and colonial corals with six protosepta and successive cycles of septa inserted in all interseptal spaces (so that successive cycles tend to have 6, 6, 12, 24, and so on septa). Walls commonly porous; horizontal structures varied. Coenosteum common in colonial forms. (Middle Triassic to Holocene; about 600 genera.)

Scleractinia (or Madreporaria) are often called hexacorals, after the sixfold symmetry of the septal insertion. They include all post-Paleozoic fossil and living true corals. They are both solitary and colonial and are even more varied than the Rugosa in shape and colony form. Individuals within colonies may be very small, but a solitary discoidal coral having a diameter of 1 m is also known.

Scleractinian skeletal parts have a complex microarchitecture and microstructure. The microarchitecture may be difficult to reconstruct from transverse and longitudinal sections. Extensive development of coenosarc has produced an increasingly varied range of coenosteal skeleton, which is a striking feature of the

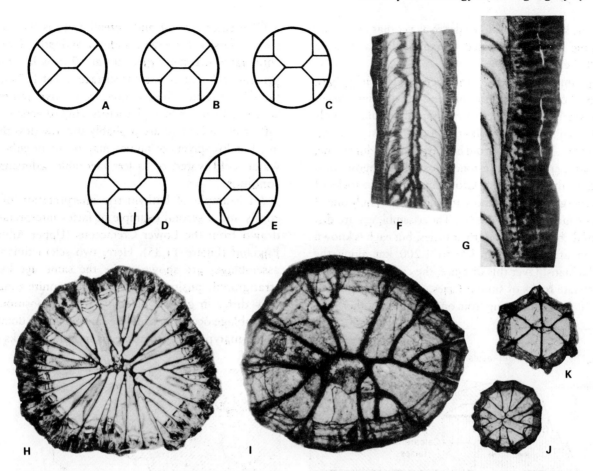

Figure 11.33. Heterocorallia. **A** to **E**, Steps in septal insertion (based on *Oligophylloides*, Late Devonian, Poland). Diagrams represent transverse sections from earliest (**A**) to later (**E**) stages (see text discussion and compare with Figure 11.30). **F** to **K**, Thin sections of heterocorals. **F**, Longitudinal section (×4) and enlarged detail (**G**, ×14) of *Heterophylloides*. **H** to **K**, Trans- verse sections through mature and immature parts of *Hetero- phyllia* (×10), which characteristically has only six septa. All specimens from Lower Carboniferous, Europe. (**A** to **E**, from Sutherland, P. K.; Forbes, C. L. *Acta Palaeontologica Polonica* 25: 499; 1981. **F** to **K** from Schindewolf, O. H. *Palaeontolog. Zeitschrift* 22: pls. 10, 11, 14; 1941.)

group's evolution that contrasts with that of most Paleozoic corals. Skeletal composition is aragonite. Many scleractinian polyps have an edge zone, which permits the addition of skeletal materials on the outside of the corallite; this is often reflected in the presence of extensive attachment structures and in other parts generally not found in the Rugosa.

The Scleractinia are known as fossils in rocks of Middle Triassic to Holocene age and are abundant in modern seas, not only in coral reefs but in varied en- vironments over large parts of oceanic space.

Coral paleoecology and biogeography

Facies analysis

Essentially all fossil corals were benthic; possible excep- tions are unimportant. Most coral genera and species tend to be associated with a particular type of rock, even though corals as a whole have been found in almost every marine sedimentary lithology.

On a regional scale, rugose corals can be separated into normal shallow marine (platform) and deeper (basinal) cephalopod-associated marine assemblages. The former are associated with carbonate rocks, the latter with shales (including black shales). The platform assemblage includes all the structurally more complex corals and most colonial corals. Basinal assemblage individuals tend to be small, morphologically simple, and long ranging in both time and space. The platform (carbonate) assemblage can often be divided into reef and nonreef assemblages.

The twofold facies division of the rugose corals is loosely analogous to the ecologic division of the sclerac- tinians where one group (hermatypes) includes the structurally more complex and most colonial forms, and the ahermatypes are structurally simple and mostly solitary. The analogy should not be carried too far, however, since there is no evidence as yet that any of the Paleozoic corals contained zooxanthellae and were actually hermatypic.

An example of facies control occurs in the Helderberg Group (Lower Devonian) of the Appalachian Basin, where small assemblages of rugose and tabulate corals are known from four different rock facies: bioherm (patch reef), stromatoporoid bed, argillaceous limestone, and coarse crinoidal limestone. Most of the rugose coral species in the Helderberg Group are limited to rocks of Helderbergian age, and most are known from only one of the four facies (Figure 11.34). The assemblages are distinct and characteristic of their facies, but each is known over an area extending 800 to 1,200 km along the Appalachians. Over this distance, the corals can be used to correlate rocks of similar facies, although they are of little value in correlating from one facies to another.

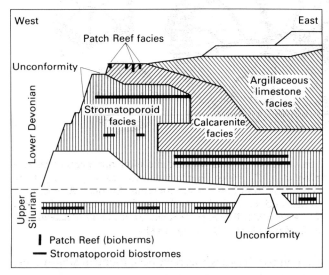

Figure 11.34. Facies control of corals. West to east section of uppermost Silurian and Lower Devonian rocks in New York, showing time relationship of four coral assemblages. Time lines are horizontal; diagonal and vertical boundaries within diagram mark facies changes. Of approximately 25 species of rugose corals in the four coral assemblages, only one species occurs in three assemblages, and three or four species occur in two assemblages; all other species are restricted to a single assemblage. (Modified from Oliver, W. A., Jr. *Alberta Soc. Petrol. Geol.,* Symposium on the Devonian System, 2:738; 1968.)

Bathymetry and facies

The depth ranges of many living genera and species are known. It can be tentatively assumed that many fossil scleractinian taxa had the same ranges and this can suggest the depth at which some fossil assemblages lived. This may be unreliable for individual taxa but becomes more meaningful with larger assemblages.

Of greater interest and potential use is the fact that coral diversity decreases rapidly with depth. With living coral faunas, generic diversity at 50 m is less than 50 percent of that at or near the surface and at 100 m, is less than 20 percent. These figures are relative and may be confused by other ecologic factors. High-diversity faunas (30 or more genera) are probably shallow (less than 50 m), but low-diversity faunas may occur in either deep water, cold water, or in less favorable sedimentologic conditions.

An example of bathymetric interpretation of fossil corals and a second example of facies interpretation is drawn from the Lower Cretaceous (Upper Albian) of England (Figure 11.35). Here, two scleractinian coral assemblages are shown to be the same age by their stratigraphic positions relative to ammonite zones, but they differ in generic diversity and composition. One assemblage occurs in sandy sediments and is dominated by hermatypes (H); the other is in muddy rocks and is

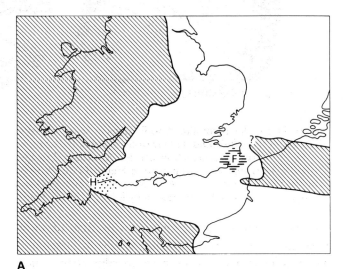

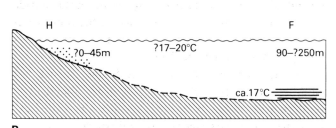

Figure 11.35. Bathymetry and facies. **A,** Paleogeography of the Upper Albian Stage (Lower Cretaceous) showing land (striped) and sea. *H* is Haldon locality of near-shore, diverse coral fauna; *F* is Folkestone locality of offshore, restricted coral fauna. **B,** Cross section from *H* to *F* (in **A**), showing calculated depths and water temperatures of the Albian Sea derived by comparing fossil coral diversity with depth curves of coral diversity in modern oceans. (See text for more detailed explanation.) (From Rosen, B. R. *Bureau de Recherches Géol. et Minières,* 89: 514; 1977.)

ahermatypic (F). By analogy with modern genera, the former represents near-shore shallow water conditions while the ahermatypic assemblage lived offshore in muddy, deeper water. By computing diversity changes with depth in a series of modern reef environments, the Albian offshore coral assemblage was estimated to have lived at a depth of 90 to 250 m.

Biogeography

Dispersal of attached corals can take place only during the part of the larval stage before settlement and attachment. The lifespan of the planula larva is from a few hours to several weeks, but most seem to settle within a few days. Theoretically, a planula might float a significant distance in hours or days, resulting in worldwide distribution of species within a few generations. Because of the temperature and other environmental requirements of the larva and the need for a proper substrate on which to settle and attach, widespread geographic ranges are rare, and dispersal is often restricted even though controlled by ocean water movements.

Corals therefore are excellent indices of biogeography. Their low dispersal rate means that a relatively high percent of the taxa in any given large area are endemic. This endemism is marked for species, very few of which achieve worldwide distribution or are even able to migrate beyond the basin in which they apparently orginated. Biogeographic studies of corals have been made of only a few of the geologic systems, but examples drawn from contrasting parts of the geologic column can illustrate the kinds of analysis that are possible.

Three biogeographic realms have been recognized in the Devonian (Figure 11.36, *A*). One of these was apparently at high Devonian latitudes (south of 45 degrees south) and was almost without corals. The other two had abundant and diverse coral assemblages in many places and of many ages. These two realms are sharply defined by the rugose corals. A stage-by-stage analysis of the Eastern American Realm (Figure 11.36, *B*) shows endemic genera generally increasing in percentage through the Early Devonian, then falling off to zero in the Late Devonian. In the Emsian stage, three-fourths of all genera known to have occurred in the Eastern American Realm are not known to have occurred anywhere else in the world in that or any earlier stage; this is an unusually high rate of endemism. The waxing and waning of endemism in the realm can be related to the late Paleozoic closing of the ocean between North America and Europe-Africa and the building of the Appalachian Mountains.

Corals can be used to divide the coral-rich Devonian realms into provinces (Figure 11.36, *C*), also on the basis of coral genera. The changing biologic relationship of these provinces through the Devonian is shown using an endemism level of 15 per cent as the criterion for recognizing each province in each stage.

In some cases coral paleobiogeography can be used to test independently the movement of continents and the evolution and development of oceans. It is generally thought that the existing North Atlantic Ocean was first opened in the Late Jurassic, was narrow during the Early Cretaceous, and expanded to approximately 4500 km by Late Cretaceous time. Figure 11.37 shows that in the tropical Tethyan Ocean Realm during the Early Cretaceous, there were no corals within the Caribbean region that were not also known in the Mediterranean region. As the Atlantic Ocean grew wider, it became a more significant barrier to east-west larval migration, so that increasing numbers of endemic coral genera evolved in the newly formed Caribbean province.

In both the Devonian and Cretaceous examples, coral biogeography reflects and therefore helps to confirm the plate tectonic model for the evolution of the Atlantic Ocean.

Coral evolution

The geologic record

Many coral-like skeletal fossils have been described from Cambrian rocks. Most of these are small in both individual and colony size, are of unknown original composition, and lack tabulae, septa, and other distinctive coral structures. Two groups, the Hydroconozoa and Cothoniida are important and are separately described in this chapter. A few other septate or tabulate tubular fossils found in Cambrian rocks may be corals or cnidarians but cannot be satisfactorily related to any of the principal coral orders.

The earliest definite Tabulata appeared in the Early Ordovician. These are massive colonial corolla with walls of adjacent polygonal corallites closely united (cerioid, Figure 11.13). They had rare tabulae but lacked septa and mural pores. A chain of morphologically intermediate forms connects these first tabulates with more advanced forms in the Middle and Late Ordovician that are distinctly coral in character. By Middle Ordovician time, tabulate corals were common and diverse elements of the normal marine fauna.

The other important Paleozoic orders, the Rugosa and Heliolitida, first appeared in Middle Ordovician rocks. All three groups participated in the building of Paleozoic

PANGAEA (LESS ASIA)

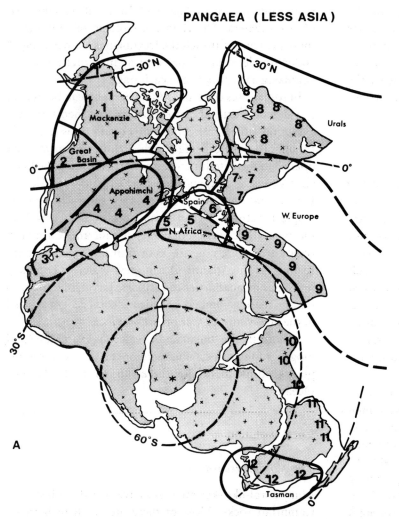

A

B

Stage	Eastern Americas realm		World
	No. of genera	Percent endemic	No. of provinces
Frasnian	11	0%	2?
Givetian	36	22%	3–4
Eifelian	28	39%	3–5
Emsian	25	76%	8–9
Siegenian	9	44%	7
Gedinnian	14	57%	5

GEOGRAPHIC AREA	EARLY DEVONIAN			MIDDLE DEVONIAN	
	GEDINNIAN	SIEGENIAN	EMSIAN	EIFELIAN	GIVETIAN
1 N.W. CANADA	Urals–W. Europe		Mackenzie		Urals–W. Europe
2 GREAT BASIN	Great Basin	Great Basin	Great Basin	Mackenzie	Urals–W. Europe
4 E.N. AMERICA	Appohimchi				Michigan Basin
5 N. AFRICA		N. Africa	N. Africa–Spain	Urals–W. Europe	
6 SPAIN			N. Africa–Spain	Urals–W. Europe	
7 W. & C. EUROPE	Urals–W. Europe				
8 URALS, ETC.	Urals–W. Europe				
9 TURKEY, ETC.				Urals–W. Europe	
12 E. AUSTRALIA	Urals–W. Europe	Tasman		Urals–W. Europe	

▨ OLD WORLD REALM   ▨ EASTERN AMERICAS REALM   ☐ INADEQUATE DATA

C

reefs, but the reefs in which they seem to have played a key role as reef constructors are small. Tabulate corals have been reported from post-Paleozoic rocks, but none of these occurrences are certain, and all these orders probably became extinct before the end of the Paleozoic.

Scleractinian corals, first known from Middle Triassic rocks, have increased in diversity almost continuously to the present time. The evolution of the great modern coral reefs has paralleled the rise of the scleractinians very closely, starting with very small Triassic reefs.

Phylogenetic relationships

The probable and possible relationships of the coral orders are diagrammed in Figure 11.38. In the Middle Ordovician, the Heliolitida probably evolved from a tabulate coral that had a similar wall structure, through the development of the coenosteoid habit. Such a tabulate genus is known, and this relationship has been suggested.

The Rugosa also may have evolved from the Tabulata and may be diphyletic, as the origins of two rugosan suborders from different groups of tabulates have been logically argued. Although Silurian and later rugose and tabulate corals are distinct, some Middle Ordovician genera and species have morphologic features of both orders. A close relationship at this time of some rugosans to tabulates has been suggested, but many details are still to be worked out or interpreted. Possibly the Rugosa are polyphyletic, having origins in skeletonless anemones (Zoanthiniaria?) as well as in the Tabulata.

Heterocorallia are few, but their general similarity to the Rugosa suggests their origin in this order during the Late Devonian.

The origin of the Scleractinia is more controversial,

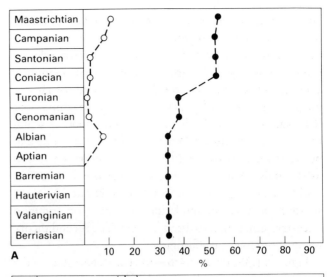

A

B

Figure 11.37. A, Graph of coral endemism for each Cretaceous stage for the Mediterranean (solid symbols) and the Caribbean regions (open symbols). Note that Caribbean coral endemism did not begin until the Albian and that Mediterranean coral endemism increased from Albian to Maastrichtian. **B,** Map showing endemic centers for the Mediterranean and Caribbean regions and the configuration of the Atlantic Ocean for the Maastrichtian (65 million years ago). (From Coates, A. G. *Special Papers in Palaeontology* 12; Palaeontological Association; 1973.)

Figure 11.36. Biogeography of Devonian rugose corals. **A,** Map of Pangaea (Asia, not shown, was in several disconnected parts) during late Early Devonian (Emsian) time. Arabic numerals mark general locations of large coral faunas. Heavy lines surround areas considered to have been faunal provinces during one or more Devonian stages; three Devonian faunal realms are recognized. The Eastern Americas Realm (EAR) included areas *3* and *4*; the Old World Realm (OWR) included all other numbered areas plus areas in Asia; the Malvinokaffric Realm (almost no corals) included central and southern South America, southern Africa, and Antarctica. **B,** Table of number of rugose coral genera in the EAR and the percent of these that are endemic for each Devonian stage. The number of recognized coral provinces throughout the world per stage is also shown (provinces are defined for this purpose as large areas in which 15% or more of the genera are endemic). Note that the waxing and waning of endemism in the EAR was paralleled by the

increase and decrease in the total number of rugose coral provinces. **C,** Diagram of changing realm and province boundaries through the Early and Middle Devonian stages. Geographic area numbers on left correspond to numbers used on map (A); province names are in body of diagram. Reading from left to right, area *2*, the Great Basin was initially a province of the OWR. Then in the Siegenian, it became an EAR province. In the Emsian the Great Basin switched affinities back to the OWR, still as a distinct province, then merged with the Mackenzie Province (OWR) in the Eifelian, and finally with the generalized Urals-Western Europe Province (OWR) in the Givetian. The Late Devonian is not shown because few coral genera were endemic at this time and provincialism was at a very low level. (Modified from Oliver, W. A., Jr.; Pedder, A. E. H. *Geological Survey of Canada*, Paper 84-1A, pp. 450-451; 1984.)

and two schools of thought are prominent. One notes the obvious similarities to the Rugosa and suggests direct descent, possibly polyphyletic, with different scleractinian suborders arising from different rugosan families. The second school notes the differences and the lack of morphologic intermediates between Scleractinia and Rugosa, especially in the sequence and mode of septal insertion and in mineral composition, and the time gap between the two groups (there are no known Early Triassic corals). This school suggests that the Scleractinia evolved from a group of sea anemones (Figure 11.38, *A*) so that the rugosan similarities are due to a common ancestry rather than to direct descent. The second theory seems more likely at this time.

Figure 11.38, *B* summarizes the basic characters of the living anthozoan subclasses and orders that bear on these relationships. Figure 11.38, *A* suggests that the whole complex of Paleozoic coral orders could have been derived from a skeletonless stock that gave rise to the living Zoanthiniaria.

Morphologic trends

Some progressive morphologic changes through time have taken place in two or more of the coral orders. These similarities suggest parallel evolution and are termed trends. Principal morphologic trends are: (1) increasing complexity of septal structure and of skeletal structures generally; (2) tendency to evolve from solitary to colonial forms; and (3) general increase in the level of integration within colonies.

A trend from simple to complex septal structure has been noted in several families of both Rugosa and Scleractinia. Some hermatypic scleractinians show a trend toward more porous or fenestrate septa and to more elaborate secondary septal structures. Porous septa are known in only one rugosan group, but many show increasing septal complexity.

Trends from solitary to colonial habits are recognized in the Rugosa and Scleratinia and from simple to more complex colonies in these groups and the Tabulata. In general, the sequence seems to have been as follows (Figure 11.17): solitary species to colonies with isolated polyps (phaceloid) to colonies with polyps joined on the surface of the corallum (cerioid), then to colonies in which polyps have no walls (astreoid) or no walls and connected septa (thamnasterioid), to colonies with septa limited to the calice area and corallites connected by dissepiments only (aphroid), to corallites separated by coenosteum (coenosteoid) or chains of polyps closely integrated (meandroid, in Scleractinia only). Different

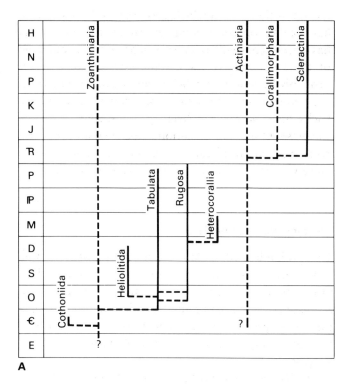

A

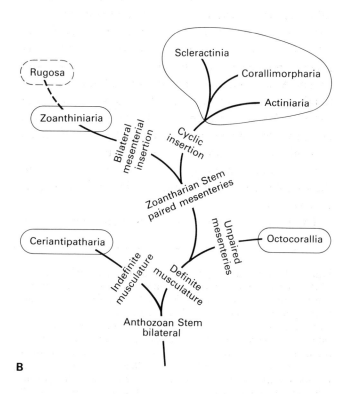

B

Figure 11.38. A, Stratigraphic ranges and possible relationships of the coral and anemone orders (Subclass Zoantharia). **B,** Possible relationships of major living groups within the Anthozoa; principal characters of stem groups are shown and the Rugosa are tentatively added as the representative Paleozoic group.

family groups went varying distances along these routes. For example, half of the Rugosa families that include colonial corals developed no further than to the phaceloid or cerioid stage, and only a small percentage of rugosan genera are astreoid, thamnasterioid, or aphroid. The sequences of colony form represent increasing biologic integration and cooperation within the colonies. In general, scleractinians achieved higher levels of integration than did the rugosans or tabulates. Coenosteoid and meandroid colonies, representing these high levels, are common in the Scleractinia but almost absent in Paleozoic corals except for the coenosteoid Heliolitida.

Other cnidarian fossils

Many fossil Cnidaria are not corals. The records of hydrozoans with massive calcareous skeletons and of octocoral spicules are good and of significance in understanding these groups. The fossil record of soft-bodied medusae and polyps is spectacular in its way but is not complete enough to present a coherent picture of the relationships or evolution of any group or of cnidarians generally. These other cnidarians do, however, add significantly to our knowledge of the timing of some events in cnidarian history and of cnidarian diversity, as already discussed.

Many possible, probable, or reasonably certain fossil cnidarians are known that cannot be assigned to an extant class. Classes have been established for some of these, but many remain unassigned and poorly understood. Mostly, this final section on the Cnidaria is to document groups that are uncommon as fossils but that have nevertheless played a key part in our discussion and interpretation of the phylum.

The earliest undisputed fossil metazoans are found in the latest Proterozoic Ediacaran fauna. This was first described from Ediacara, South Australia, but elements of the fauna are now known from most continents. Fossils are molds and casts in fine sandstone, but excellent detail is preserved in many individuals (Figure 11.19). More than half of the known specimens and species of this fauna are medusoids, and many of these can be assigned to a living class and even order. Others preserve features that suggest that these specimens are neither hydrozoans nor scyphozoans but may represent extinct classes.

Polypoid colonies are represented in the Ediacaran by impressions of sea pen–like forms that are probably octocorals. These are more questionably assigned than some medusoids.

The Ediacaran fauna indicates that cnidarians were common and diverse at an early stage in metazoan history, but it provides no real answers to basic questions about the origin of the phylum or the relationship of the classes. One would like to know what is missing: What was the nature of the polyps that presumably were in the life cycles of the species known only from medusae? Were there contemporaneous zoantharian polyps? After all, the Ediacaran fauna represents a shallow near-shore environment; other environments presumably had different faunas, including different but possibly even more varied metazoans. When these are known, a more balanced picture of the early Cnidaria will emerge.

In the Cambrian and Lower Ordovician, other fossils are known that are thought to be cnidarians. The possible and probable corals have already been discussed; others are the Protomedusae and Hydroconozoa described in the succeeding sections; still others are too poorly understood for inclusion here.

Later Phanerozoic noncoral cnidarians are mostly assignable to an extant group and are described accordingly.

CLASS PROTOMEDUSAE (Figure 11.39)

Medusa-like forms with variable number of radial pouches separated by grooves (sulci) commonly with smaller supplementary lobes. Tentacles unknown. Aboral central stomach with canals radiating to each lobe and to ventral cavity. (Late Precambrian(?), Cambrian to Ordovician; one genus.)

This class is known from one genus (*Brooksella*). The unique body plan, especially in the peculiar and variable lobes, clearly separates the class from any other cnidarian. Fossils are abundant locally.

CLASS HYDROCONOZOA (Figure 11.40)

Small, solitary, conical to cylindrical attached organisms with external skeletons; some with septa, tabulae, and axial canals. (Early Cambrian; five genera.)

All described forms have unusual features difficult to interpret but possibly analogous to features known in scyphozoans and rugose corals. Hydroconozoans seem to represent metazoans at a cnidarian level of development and are logically placed in the phylum even though they are poorly understood.

Figure 11.39. Protomedusae. **A, E,** and **F** are upper views and **B, C,** and **D** are under views of *Brooksella* (Middle Cambrian, Alabama, ×1). **A** and **B** are of the same specimen; top and bottom are conventional. These star cobbles with their variable number of lobes are interpreted as the remains of very simple medusae.

CLASS HYDROZOA

The class has been diagnosed in Part I as having an alternation of generations in which the polyp stage predominates. Within this general framework are a variety of groups of varying evolutionary and geologic importance. The orders are listed in Table 11.1; key ones are briefly characterized and discussed in this section.

ORDER TRACHYLINIDA

Small-to-medium-sized hydrozoan medusae with special marginal sensory cells (lithocytes) containing organic and carbonate grains (statoliths) that control orientation and equilibrium. Polypoid generation reduced or absent; tentacles commonly on upper bell surface. Planula develops into **actinula** (tentacle-bearing, short, stalkless, polyplike larva), which passes either directly or by budding into medusa. (Ediacaran(?), Cambrian(?), Jurassic to Holocene.)

Trachylines have a poor fossil record but are of interest to paleontologists because they may be the most primitive living cnidarians. The trachyline life cycle essentially passes from a planula to an actinula larva to a medusa (Figure 11.3), the tentacles of the actinula often becoming those of the medusa. If the earliest cnidarians are assumed to be medusae, by the development of an attached (actinula-like) form a polyp phase might be evolved. If the medusae were then released by budding of the polyp, a hydrozoan life cycle would result. This history would also explain why polyps generally reproduce asexually and medusae sexually. By emphasis on the medusoid phase, scyphozoans could be derived, and by neotenous suppression (youthful characteristics retained by adult) of the medusa to make an originally larval (actinuloid) polyp dominant, the anthozoans could be produced. In any case, the trachyline planula-actinula-medusa cycle may represent a primitive, prepolyp cnidarian arrangement.

ORDER HYDROIDA (Figures 11.11; 11.41)

Hydrozoans with well-developed polypoid stage; medusa present or absent. Polyps with or without chitinous skeleton (periderm); if present, skeleton may be limited to stalk or cover both stalk and hydranth. Benthic forms are solitary or colonial. Pelagic colonies (chondrophores) composed of variously modified polyps. (Ediacaran to Holocene.)

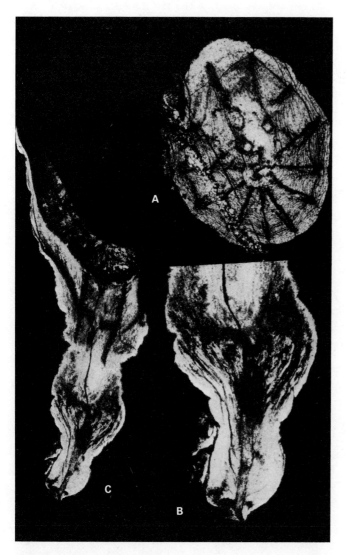

Figure 11.40. Hydroconozoa. **A,** Tranverse thin section of *Hydroconus* just below calice (×10). **B,** Longitudinal thin section of lower part of same specimen (×20). **C,** Longitudinal section of another specimen (×10), showing calice. (After Korde, K. B. *Paleontogicheskii Zhurnal* No. 2, pl. 1; 1963.)

The skeleton may be fossilized but is commonly broken up or flattened in the process. Such hydroid fossils have been reported to occur in every system from Cambrian to Holocene. They are seldom studied, however, and relatively few well-preserved specimens have been actually described (Figure 11.41).

Pelagic polypoid colonies (chondrophores) are now generally assigned to this order. Fossil chondrophores have been described from the Ediacaran and from various parts of the Paleozoic.

The vast majority of living hydrozoans belong to the Hydroida but most groups are not known as fossils. Fossilized polyps are rare.

ORDERS MILLEPORINA AND STYLASTERINA (Figure 11.42)

Hydrozoans with massive aragonitic skeletons that vary from encrusting to massive to bushlike in shape. Skeleton is coenosteum in which canals ramify. Polyps minute; two types rise from canals to exterior through distinct coenosteal pores. Feeding polyps (gastrozooids) project through gastropores; stinging and prey-gathering polyps (dactylozooids) occupy dactylopores. The two orders commonly united as hydrocorallines. (Cretaceous to Holocene; about 30 genera.)

The skeleton of hydrocorallines is deposited by a colony-wide tissue, the coenosarc, which also penetrates the skeleton as a ramifying system of canals or tubes either near the surface of the skeleton (millepores) or throughout it (stylasterines). In stylasterines, the dactylozooids lack tentacles and may form rings or cyclosystems around the gastrozooids; in millepores, dactylozooids are tentacular and cyclosystems occur but are less regularly developed than in stylasterines. Dactylozooids are generally armed with painful stinging nematocysts which give some hydrocorallines the familiar name fire coral. Hydrocorallines are important elements of both fossil and living reef faunas.

ORDER SIPHONOPHORIDA

Planktic or nektic colonies of polymorphic medusae and polyps attached to stem or disc. Polyps without oral tentacles; medusae rarely completely developed or free. (Holocene.)

The siphonophores include the most highly integrated colonies known in any phylum. Polymorphs of both polypoid and medusoid origin in the same colony are each adapted to a specific function. Medusoid polymorphs include swimming bells that have no mouth or tentacles, floats, gelatinous leaflike bodies that serve a protective function, and reproductive gonophores. Polyp polymorphs include feeding forms, reproductive forms, and tactile or prey-capturing polyps that have a basal tentacle but no mouth.

All polymorphs are budded from a stem of coenosarc that is either tubular or disc shaped. Repeated suites of polymorphs of both medusoid and polypoid types are termed cormidia, some of which separate and become independent units. Siphonophores have attained the most complex organization of any tissue level animal.

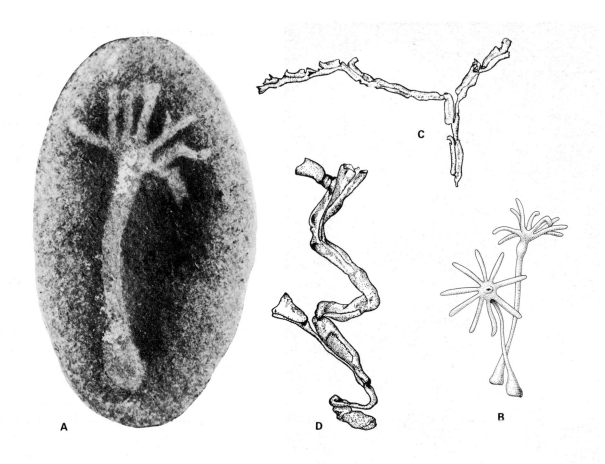

Figure 11.41. Hydroida. Photograph (**A**,×4) and restoration (**B**) of Pennsylvanian hydroid polyp (*Masohydra*) from Illinois. **C** and **D**, Sketches of chitinous sheaths (exoskeletons) of Ordovician colonies (*Desmohydra*, northern Europe, ×30 and ×60, respectively). (**A**, courtesy the Field Museum of Natural History. **B**, from Schram, F. R.; Nitecki, M. *Journal of Palaeontology* 49; 1975. **C**, **D**, from Kozlowski, R. *Acta Palaeontologica Polonica* 4: 228, 229; 1959.)

Polymorphic individuals are not only specialized, they are also highly integrated, so much so that in many taxa the colony looks and behaves like a solitary animal. In one sense, these colonies have reached an organ level of development, the polymorphs substituting for the organs of "higher" animals.

CLASS SCYPHOZOA

Scyphozoans were diagnosed in Part I as having an alternation of generations in which the medusoid stage predominates. They are rare as fossils because they generally lack skeletons. Impressions of medusoids that appear to be scyphozoans are known from the late Precambrian (Ediacaran) (Figure 11.19, *A* to *D*), Cambrian, and Jurassic to Holocene (Figure 11.12, *E*). Two important groups with chitinlike skeletons have been assigned to the class and are discussed here.

ORDER CONCHOPELTIDA (Figure 11.43)

Broad, low, conical, bilaterally symmetrical, tetra-radiate forms with lobate margins and chitinous(?) tests; specimens with marginal impressions apparently formed by tentacles are known. (Ediacaran to Ordovician; two genera.)

The conchopeltids have been assigned to the Cnidaria because of their medusoid form and apparent tentacles and to the Scyphozoa because of their squarish outline and fourfold radial symmetry. Recently, however, it has been shown that the anterior lobe is shorter than the other three and that the test is bilaterally symmetrical (radiobilateral). The shape and apparent tentacles are

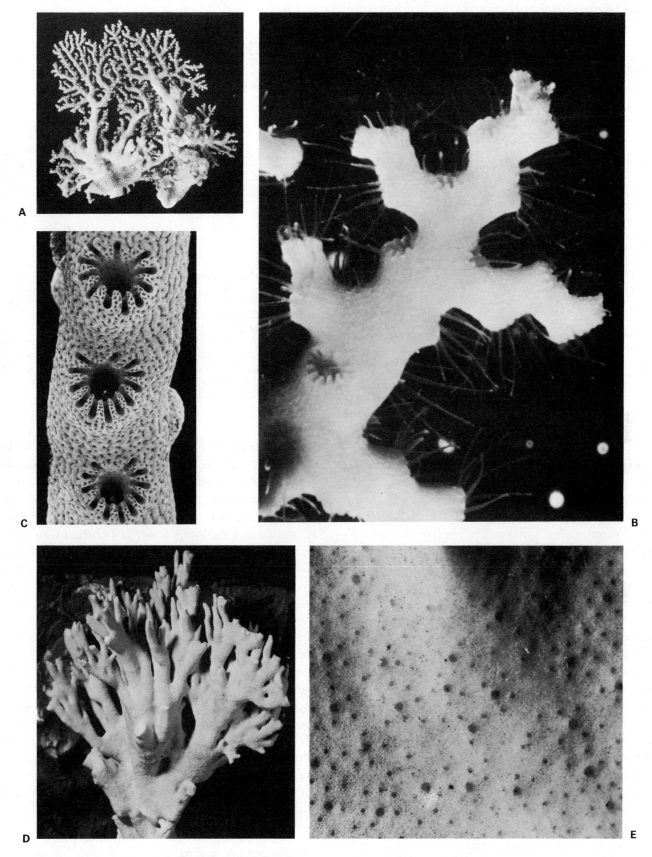

Figure 11.42. Hydrocorallines. **A** to **C**, *Stylaster (Holocene, Caribbean):* **A**, colony (×1); **B**, branch showing several cyclosystems with living polyps (seen by their tentacles) (×15); **C1**, Scanning electron micrograph of cyclosystem consisting of gastropore surrounded by dactylopores (×28). **D** and **E**, *Millepora* (Holocene, Caribbean: part of colony (**D**, ×0.5) and surface detail (**E**, ×10) showing pores. The largest openings are gastropores; dactylopores are smaller. (**A** and **C**, photos by S. D. Cairns; **B**, photo by R. Larson.)

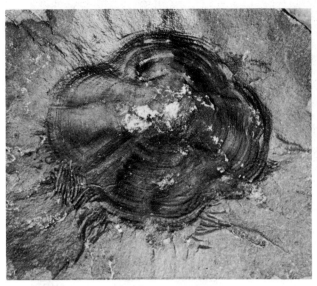

Figure 11.43. Conchopeltida. Unretouched photograph of upper surface of *Conchopeltis* (Middle Ordovician, New York, ×1). The four lobes (short one at top), indicating a fourfold radiobilateral symmetry, and the tentacles, apparently extending from underneath the thin test as though the specimen had been drifting (in the direction of the top of the photograph) when entrapped, are evidence that this is a cnidarian, and probably a medusoid (see text discussion). (Specimen in the New York State Museum.)

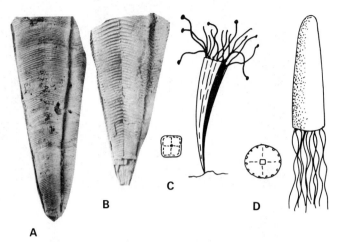

Figure 11.44. Conulariida. Two fossil specimens and reconstructions to show polypoid and medusoid interpretations (×1.5). The square cross-section shape and the external ornamentation of this group are characteristic. **A,** *Conularia* (Middle Ordovician, New York). **B,** *Paraconularia* (Pennsylvanian, Texas). **C** and **D,** Reconstructions showing inferred soft part morphology and way of life; only the conical tests with some intrnal thickenings, as suggested in the diagramatic cross sections, have actualy been observed. (**C** and **D** from Kiderlen, H. *Neues Jahrb. f. Mineralogie,* 77(B): 164; 1937.)

convincing evidence of cnidarian affinities but the chitin-like test is unique. Living scyphozoans apparently secrete chitin only for attachment to the substrate, and no living medusa has any kind of test or exoskeleton. Largely for this reason, some workers have interpreted conchopeltids as polyps, perhaps at an advanced hydrozoan level, and intermediate between hydrozoans and scyphozoans. In any case, recognition of the extinct order is justified even though fossils are rare.

The conchopeltids have formerly been included in the conulariids (discussed next) because of their pyramidal form and fourfold symmetry. They differ however, in lacking the internal septa, elongate form, and external markings, in being bilateral, and in having an organic, nonmineralized test. It has been suggested that the Conchopeltida were an early cnidarian offshoot ancestral to the Conulariida, and that the latter may have given rise to one or more of the modern scyphozoan groups. It is more likely that the conulariids are unrelated to either the conchopeltids or the Cnidaria.

ORDER CONULARIIDA (Figure 11.44)

Elongate, four-sided pyramidal tests (some with four inward projecting thickenings termed septa), and with characteristic external marking; chitinophosphatic. Soft parts not known. (Cambrian to Triassic; about 25 genera.)

Conulariids are important fossils in many parts of the Paleozoic. They tend to be associated with marine siltstones and fine sandstones apparently in a shallow, nearshore environment. They have been interpreted as either polyps or medusae by various workers, but there is good evidence that they were sessile (attached at their apex) during early ontogenetic stages at least. Four triangular flaps at the big end of the pyramid seem to have served as opercula, enabling the animal to enclose itself within the test. Attachment and the opercula fit the polyp theory, but are also compatible with the interpretation of the animal as a more advanced metazoan, possibly a lophophorate.

The fourfold radial symmetry of conulariids is a fundamental character of this group (Figure 11.44). This symmetry, plus the probably superficial similarity of the internal bifurcating septa, have suggested a possible relationship to the Scyphozoa, but this is not convincing. No good evidence exists that the conulariids are cnidarians or that they are related to the Conchopeltida. It seems best to treat them as a group of unknown affinities.

Supplementary reading

Bayer, F. M.; Owre, H. B. *The free-living lower invertebrates.* New York: Macmillan; 1968. An excellent medium-level textbook that emphasizes the Cnidaria; beautifully illustrated.

Coates, A. G.; Oliver, W. A., Jr. Coloniality in zoantharian corals, pp. 3–27. In: Boardman, R. S.; Cheetham, A. H.; Oliver, W. A., Jr., editors. *Animal colonies: development and function through time.* Stroudsburg, PA.: Dowden, Hutchinson, & Ross; 1973. Why coral colonies are, and something of their history, evolution, and function.

Heckel, P. H. Carbonate buildups in the geologic record: a review. In: Laporte, L. F., editor. Reefs in time and space: selected examples from the recent and ancient. *Society of Economic Paleontologists & Mineralogists, Special Publication* 18: 90–154; 1974. Defines and reviews the geologic history of reefs.

Hill, D. Rugosa and Tabulata. In: Teichert, C., editor. *Treatise on invertebrate paleontology, Part F, Coelenterata,* Supplement 1. Boulder, CO, and Lawrence, KS: Geological Society of America and University of Kansas Press; 1981. Comprehensive review of morphology and natural history of the principal Paleozoic groups of corals.

Hyman, L. H. *The invertebrates: Protozoa through Ctenophora.* New York: McGraw-Hill Book Co.; 1950. Compre-hensive volume on all aspects of the living Cnidaria.

Moore, R. C., editor. *Treatise on invertebrate paleontology, Part F, Coelenterata.* Boulder, CO, and Lawrence, KS: Geological Society of America and University of Kansas Press; 1956. The most comprehensive overview of fossil Cnidaria; morphology, function, evolution, history, and classification.

Oliver, W. A., Jr. The relationship of the scleractinian corals to the rugose corals. *Paleobiology* 6: 146–160, 1980. Recent discussion of this important problem in evolution.

Scrutton, C. T. Early fossil cnidarians, pp. 161–207. In: House, M. R., editor. *The origin of major invertebrate groups.* London: Academic Press, Inc.: 1979. Excellent review of the early record and its interpretation.

Sokolov, B. S., editor. *Fundamentals of paleontology,* Vol. 2, *Porifera, Archeocyatha, Coelenterata.* Moscow: Vermes; 1962. (English translation published by U.S. Department of Commerce.) The Soviet "Treatise" supplements the Moore *Treatise* in many important ways.

12

Annelida

Richard A. Robison

Various benthic, soft-bodied animals independently developed locomotion without limbs. The mechanical requirements for this type of locomotion have commonly produced a convergence in external form resulting in a bilaterally symmetrical, elongate body. Most animals of this form have been called worms, and, at one time, all worms were assigned to a single phylum, Vermes. Subsequent investigations have demonstrated several distinct grades of worms, causing repeated revisions in classification. Currently, as many as 14 worm phyla are recognized, which is about one-half of all the animal phyla (see Chapter 2, Table 2.1).

Considering the usual requirements for preservation of fossils, it should not be surprising that, except for hardened jaw apparatuses in a few groups, remains of fossil worms have seldom been discovered. Various kinds of worm burrows, tracks, castings, and tubes are much more commonly encountered; however, they rarely provide information useful for taxonomic identification of their makers. Because only one phylum, the Annelida, has an appreciable fossil record, it is the only worm phylum treated in this book.

The most familiar annelids are the earthworms (Figure 12.1, *D*), which are common in garden soil. Less familiar groups are the leeches and marine bristleworms (Figure 12.1, *A* to *C*). In addition to their elongate, bilaterally symmetrical bodies, annelids are at the triploblastic coelomate grade and are characterized by a metamerous coelom (see Chapter 2), which is divided into numerous segments, each with a repetition of body parts. A high degree of segmentation—the basis for the phylum name (from the Latin, *annelus*, meaning little ring)—is a characteristic that distinguishes the Annelida from all other phyla except the Arthropoda, which differs by the presence of an external skeleton, jointed appendages, and great reduction of the coelom. A secondary loss of segmentation is rare in specialized sessile annelids, but internal vestiges can still be recognized.

Annelids had almost certainly evolved by Proterozoic time. Although body fossils are rare, an increasing abundance and variety of tubes and trace fossils in Phanerozoic strata attest to a history of expanding annelid diversity during the last 600 million years. Today, annelids occupy many habitats extending from the deep oceans to high mountains and from equatorial to polar regions.

Fossils of these inconspicuous, soft-bodied creatures have had little use in biostratigraphy. However, because of the capacity of some annelids to ingest and modify great volumes of soil and sediment, their influence on the geologic record may surpass that of any other animal phylum. For example, densities of up to 5000 annelids per square meter are not unusual in marine sediments. On some sandy intertidal beaches, densities of 32,000 burrowing annelids per square meter have been reported. Repeated ingestion by these worms totals as much as 3 metric tons of sand per square meter per year. On land, densities of 50 to 500 earthworms per square meter are common, depending on soil type. These might move as much as 5 kg of castings per square meter to the surface each year. The castings are some of the world's finest fertilizer, being rich in nitrates, phosphates, and potash.

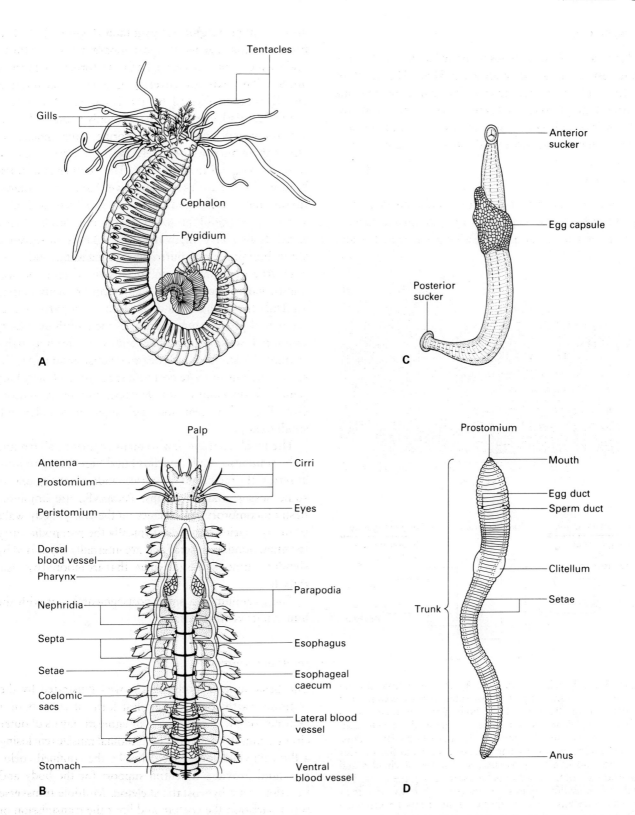

Figure 12.1. Morphology of representative modern annelids. **A,** *Amphitrite,* a tube-dwelling marine polychaete. **B,** General arrangement of internal organs in the anterior part of *Neanthes,* a crawling and swimming marine polychaete. **C,** *Hirudinaria,* a freshwater leech with an egg capsule being formed while the body is held in place by suckers. **D,** Ventral view of *Lumbricus,* the common earthworm. (**A, B,** and **D** modified from Brown, F. A. *Selected invertebrate types.* Copyright © 1950 by John Wiley & Sons, Inc. Reprinted by permission. **C** modified from Fretter, V.; Graham, A. *A functional anatomy of invertebrates.* New York: Academic Press, Inc.; 1976.)

Morphology

Adaptations of the annelids to many habitats and niches have resulted in a highly diverse phylum. The length of mature individuals ranges from less than 1 mm in some marine annelids that dwell between sand grains to almost 400 cm in some giant Australian earthworms. Representative annelid morphology is illustrated in Figure 12.1.

External features

The body is bilaterally symmetrical and is divided into a short head region or **cephalon** with a mouth and concentration of sensory apparatus, a long segmented **trunk**,

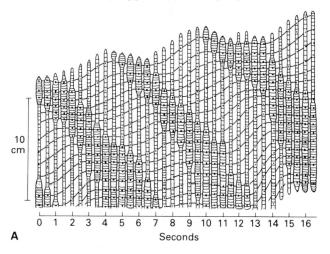

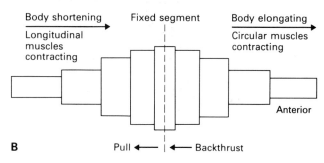

Figure 12.2. Annelid locomotion. **A,** Diagram prepared from motion pictures showing successive stages in the locomotion of an earthworm. Regions of the body with contracted longitudinal muscles are drawn twice as wide as those with contracted circular muscles and are also marked by larger dots. Progression begins as circular muscles contract in the anterior segments and continues as successive waves of circular and longitudinal contractions pass backward along the body. The track of individual points on the worm's body and their movements relative to each other are shown by the lines running obliquely forward from left to right on the diagram. **B,** Forces acting upon the stationary segment and adjacent moving segments of an earthworm performing peristalsis. (**A** from Gray, J.; Lissmann, H. W. *Journal of Experimental Biology* 1938 v. 15, p. 506–517. **B** modified from Clark, R. B. *Dynamics in metazoan evolution.* Oxford: Oxford University Press; 1964 p. 120.)

and a tiny postsegmental **pygidium** (Figure 12.1). This body division is generally most apparent in sessile forms and least conspicuous in vagile forms (forms free to move about). The exterior covering is a thin nonchitinous cuticle that is secreted by the epidermis. Muscularly movable, bristlelike **setae** are commonly implanted in the soft body wall and aid in anchoring some annelids in either their burrows or tubes, assist in surface locomotion, or even aid in swimming. The number and distribution of setae are useful definitive characters at higher taxonomic levels; and setal form, which is highly variable, is useful at lower taxonomic levels. Many annelids also have highly developed glands for secreting slime, building tubes, burrowing, or ingesting food.

Degree of cephalization, or fusion of anterior segments, varies according to adaptation. In many marine annelids the cephalon is divided into two parts. A presegmental prostomium in front of the mouth may carry unjointed sensory **antennae** and stout feeding **palps** (Figure 12.1). A mouth-bearing peristomium may be fused with one or more posterior segments and carry long slender feeding and sensory **tentacles** and shorter sensory **cirri**. Eyes, if present, may be simple or complex and small to large.

The trunk contains few to many segments, all tending to be similar in vagile annelids but differentiated to form anterior thoracic and posterior abdominal regions in some sessile marine groups. **Parapodia** are unjointed fleshy locomotory appendages on the lateral body walls of many marine annelids. Typically the parapodia carry terminal bundles of setae and are internally supported by slender chitinous rods (aciculae) that are moored by basal muscles.

The pygidium is a small postsegmental unit with the anus on its ventral surface.

Internal features

The basic annelid body plan is well illustrated by the common earthworm. In simplest form, it consists of a flexible body wall containing antagonistic layers of outer circular muscles and inner longitudinal muscles enclosing a fluid-filled coelom. Mechanically, the confined coelomic fluid provides a central support for the body and functions as a **hydrostatic skeleton**. Multiple transverse septa partition the coelom and limit the transmission of fluid pressure, thereby enabling localized changes in shape. Locomotion is accomplished by peristalsis, an action generated by successive waves of muscle contractions that pass backward along the body (Figure 12.2, A). In any part of the body, simultaneous contraction of circular muscles and relaxation of longitudinal muscles results in elongation, whereas reverse muscle activity results in shortening (Figure 12.2, *B*). Those parts of the body are fattest where longitudinal muscles are maximally contracted, and by wedging those fat parts against the wall of a burrow, thrust can be exerted. Simultaneous extension of setae provides additional resistance to slipping. Each segment in succession becomes stationary as a muscle wave passes backward.

This simple annelid body plan has advantages and limitations. It represents a significant functional advance over the unsegmented coelomate worms (such as sipunculids and echiuroids), in which individual muscle contractions influence the entire hydrostatic skeleton. In those animals, localized change in shape is difficult, and rapid movement is excessively wasteful of energy. Thus their mode of life is necessarily rather sessile. On the other hand, advanced lever systems are impossible for soft-bodied animals, and annelid locomotion over a flat substratum is much less efficient than that of arthropods and higher chordates.

Most other internal features also show a clear segmental arrangement of parts (Figure 12.1, *B*) and include the following:
1 A central nervous system consisting of paired cerebral ganglia merging posteriorly into a ventral cord with lateral nerve branches to each segment.
2 A closed circulatory system with median dorsal and ventral blood vessels joined in each segment by vessels to the gut and body wall and one to several pulsating heart-like organs.
3 An alimentary tract with pharynx, esophagus, stomach, and intestine.
4 An excretory system with simple to complex pairs of **nephridia** in most segments.
Like tubules in the human kidney, these nephridia extract wastes from the blood and coelomic fluid. Respiration is accomplished by gaseous exchange through the body wall or by gills in some tube dwellers.

Great variation has been superimposed on the basic, and presumably primitive, annelid body plan. In many marine annelids, parapodia project from the lateral body walls and are operated by a complex series of muscles. Secondary adaptation to a sessile life by some marine worms has been accompanied by a corresponding reduction in septa to become no more than suspensory muscles for the gut, and they no longer function as coelomic seals between segments. In the parasitic leeches, septa have disappeared, and the coelom is much reduced or almost absent, being largely replaced by connective tissue.

Reproduction

Reproduction in annelids may be sexual, asexual, or both. The sexes are separate in most marine annelids, but the reproductive system varies from simple to extremely complex. Gametes may be released directly into the sea or fertilized eggs may be deposited in gelatinous balls and strands. Among some tube dwellers, the eggs may be brooded within the tube. Earthworms and leeches are **hermaphroditic** (sexes united) and possess complex glands associated with reproduction. The **clitellum** (Figure 12.1, *D*) of earthworms and leeches is a swollen glandular region that secretes a mucous ring or capsule into which eggs and sperm are discharged. In what is thought to be a primitive means of reproduction, the fertilized eggs of some marine annelids develop into ciliated spheroidal larvae called **trochophores**. Other annelids reproduce asexually by budding or related processes.

Classification

Three major annelid classes, Polychaeta (bristleworms), Oligochaeta (earthworms), and Hirudinea (leeches), are generally recognized, but disagreement continues over the status of additional classes that have been proposed. Some of the more useful taxonomic characters of living annelids are: structure of the reproductive system; number, distribution, and form of setae; presence or absence of parapodia; structure and appendages of the cephalon; and structure and distribution of nephridia. Some of these characters are rarely, if ever, preserved in fossils. Hence, confident taxonomic assignment of some fossils to family, order, and even class may not be possible.

CLASS POLYCHAETA (Figure 12.1, *A* and *B*)

Morphology highly variable and strongly dependent on mode of life. Sexes usually separate with gonads in numerous segments; egg and sperm ducts simple. Parapodia common and bearing numerous setae in distinct bundles. Cephalic appendages common in sessile forms. (Possibly Proterozoic, Cambrian to Holocene; about 150 genera.)

Polychaetes usually occupy marine, rarely freshwater, and more rarely terrestrial habitats. Many are deposit and filter feeders, others are predators or scavengers, and a few are parasites.

During the first half of this century, the polychaetes were usually divided into two orders or subclasses of nearly equal size, Errantia and Sedentaria—the names reflecting their modes of life. More recently, depending on the author, that bipartite classification has been rejected in favor of 8 to 17 orders that include about 60 to 80 families.

Members of a few orders, and in particular the Eunicida, possess pharyngeal jaws composed of several hardened organic elements that paleontologists call **scolecodonts** (Figure 12.3). In the living animal these are joined by muscle tissue and cuticle but are usually found separated as fossils. The jaw apparatus typically includes a ventral pair of **mandibles** for chewing, a series of asymmetrically or incompletely paired **maxillae** for manipulating food, and a basal pair of **carriers** that support the posterior maxillae. Scolecodonts superficially resemble conodonts (see Chapter 20) but differ in composition and microstructure.

CLASS OLIGOCHAETA (Figure 12.1, D)

Hermaphroditic earthworms with gonads limited to a few anterior segments. Complex glands are associated wth specialized sperm and egg ducts; clitellum present at maturity. Cephalon indistinct and parapodia absent. Setae segmentally arranged, only rarely in bundles, and may be absent.

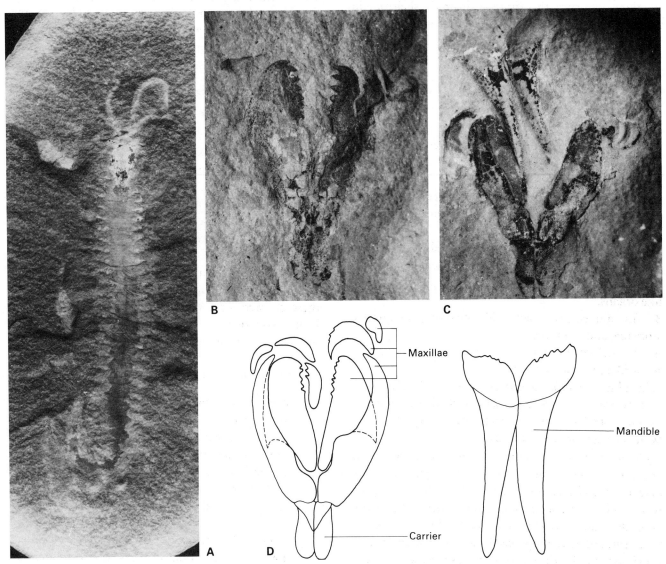

Figure 12.3. Jaw apparatus in the polychaete *Esconites* (order Eunicida) of middle Pennsylvanian age from Illinois. **A,** Complete specimen, about 60 mm long, with antennae, gills, and a poorly preserved jaw apparatus in place. **B,** Enlarged ventral view of associated jaw elements (scolecodonts) from another specimen with maxillae above and carriers below; entire apparatus about 6 mm long. **C,** Ventral view of another jaw apparatus with mandibles above and spread maxillae below. **D,** Dorsal reconstruction of jaw apparatus based on many specimens. (Modified from Thompson, I.; Johnson, R. G. *Fieldiana: Geology*; 1977.)

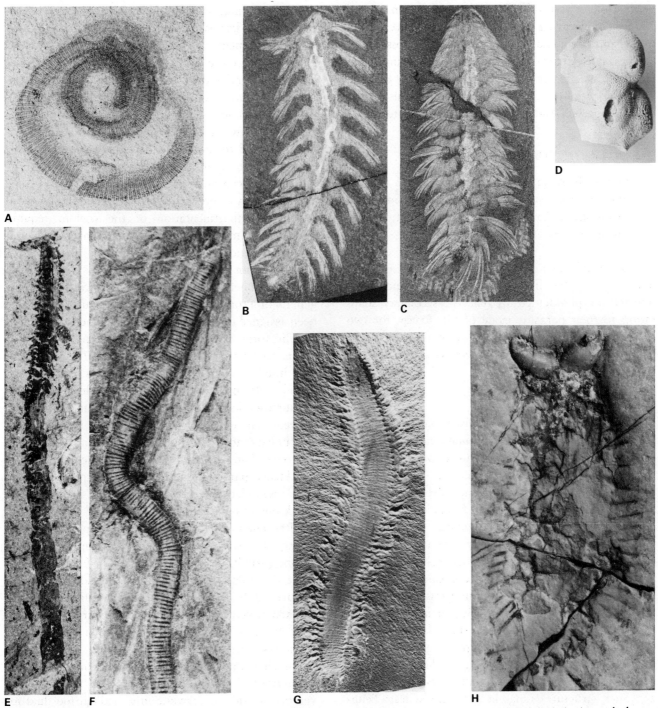

Figure 12.4. Examples of annelid body fossils and parasitic cysts. **A,** *Palaeoscolex* (Middle Cambrian, Utah), about 125 mm long if fully extended; coiling of this type in modern polychaetes is often induced by anaerobic conditions. **B** and **C,** *Burgessochaeta,* a possible burrower, and *Canadia,* a possible benthic swimmer; each with prominent but differently arranged setae, and each about 30 mm long (from Burgess Shale of Middle Cambrian age, British Columbia). **D,** Two perforated cysts formed on the arm of a Pennsylvanian crinoid in response to infestation of parasitic myzostomarian annelids. **E,** A polychaete, *Trentonia,* preserved at the end of its trace (Middle Ordovician, Quebec), illustrated portion about 60 mm long. **F,** Anterior part of *Protoscolex,* class undetermined (Upper Ordovician, Kentucky). **G,** A Late Mississippian polychaete, *Astreptoscolex* (Bear Gulch, Montana), about 100 mm long. **H,** Anterior part of *Eunicites* with jaw apparatus in place (top) and acicula preserved along trunk segments (Solenhofen Limestone of Middle Jurassic age, Germany); illustrated part about 40 mm long. (**A** from Robison, R. A. *Journal of Paleontology* 1969 v. 43, p. 1169–1173. **B** and **C** from Conway Morris, S. *Philosophical Transactions of the Royal Society of London* 1979 v. 285, p. 277–284. **D** from Welch, J. R. *Journal of Paleontology* 1976 v. 50, p. 218–225. **E** from Pickerill, R. K.; Forbes, W. H. *Canadian Journal of Earth Sciences* 1978 v. 15, p. 659–664. **G,** courtesy F. R. Schram; **H,** courtesy A. Seilacher.)

Most oligochaetes dwell in moist burrows in the soil and feed on decaying vegetable matter. Some live in fresh water, and a small number are marine. A few fossils have been assigned to the Oligochaeta, the oldest being Ordovician in age. However, the identifications of all such specimens are open to serious question because of inadequate preservation of critical taxonomic features.

CLASS HIRUDINEA (Figure 12.1, C)

Hermaphroditic and parasitic leeches with anterior and posterior suckers for attachment and aid in locomotion. Parapodia and septa absent, coelom much reduced, and setae usually absent.

Most hirudineans inhabit fresh water, but some are marine, and a few live in damp places on land. Their unusual morphology clearly reflects a secondary adaptation to their parasitic mode of life. Except for two genera of questionable identity from the Jurassic of Germany, hirudineans are unknown as fossils.

Other annelids

A small group of highly specialized, parasitic marine worms, the Myzostomaria, has an unusual discoid body with incomplete septa and five pairs of parapodia, each with a small chitinous hook on the end. Members resemble polychaetes in producing trochophore larvae but resemble oligochaetes and hirudineans in being hermaphroditic. Taxonomic assignment has been controversial, ranging from family or order under the Polychaeta to inclusion in a separate phylum, but recognition as an annelid class seems to be a developing consensus.

Myzostomarians infest echinoderms, particularly crinoids. Generally, groups of two or more myzostomarians invade tissue between skeletal plates of the crinoid arms, pinnules, and cups. The host secretes calcium carbonate to partially seal off the parasites and thereby produces a perforated gall or cyst (Figure 12.4, D). Such structures range in age from at least Pennsylvanian to Holocene. Superficially similar swellings are present on crinoid stems as old as Ordovician; however, these probably represent a different type of parasite of uncertain affinities.

Origin and evolution

The meager fossil record provides little knowledge regarding the origin of annelids and only a sparse and very fragmentary outline of their evolutionary history. Ad-

ditional information from comparative anatomy and embryology can be instructive.

Initial segmentation of the annelid coelom probably enabled peristaltic burrowing. Support for this conclusion is found in the fact that when its circular muscles contract, a simple annelid body must become circular in transverse section, thus making peristaltic locomotion ideal in burrows. In comparison, peristaltic locomotion is poorly suited to movement on flat surfaces where the amount of body-to-surface contact is minimal. The advance in locomotion over that possible in unsegmented coelomates was not achieved without cost, however, because compartmentation of the coelom requires repetition of most organ systems and makes specialization and concentration of function in those systems difficult, if not impossible.

When annelids first appeared is uncertain. A few genera of Proterozoic body fossils (Figure 12.5) have been assigned to the Annelida. None of these is of typical annelid form, however, and alternative phyletic assignments are preferred by some specialists. The attribution of a number of Proterozoic trace fossils to an annelid origin is also open to question. Nevertheless, the variety of burrows leaves little doubt that annelids were in existence at least by Proterozoic time. It is generally concluded that the ancestral annelid was a vagile marine burrower, had separate sexes that spawned into the sea, and lacked parapodia and anterior appendages. It may have resembled *Palaeoscolex* (Figure 12.4, A), which is first known from the Lower Cambrian.

Although fossils are rare, available specimens show enough diversity to indicate extensive adaptive radiation among early Paleozoic polychaetes. The famous Burgess Shale of Middle Cambrian age has yielded six genera (Figure 12.4, B and C) from one locality in western Canada. Several other genera from the Burgess Shale that were formerly assigned to the Annelida have recently been reassigned to other phyla.

Polychaete parapodia represent a further advance in locomotion. It has been suggested that they were developed as aids to crawling in a medium too fluid for effective peristaltic burrowing. A likely environment for such evolution is the nutrient-rich flocculent layer immediately above many shallow sea floors. Because parapodia provide little assistance in swimming, it seems improbable that they evolved as an adaptation to aid that type of locomotion. Parapodia are first definitely known from the Middle Cambrian (Figure 12.4, B and C), and they are present on many later fossil polychaetes (Figure 12.4, E, G and H).

Another important event in the evolution of poly-

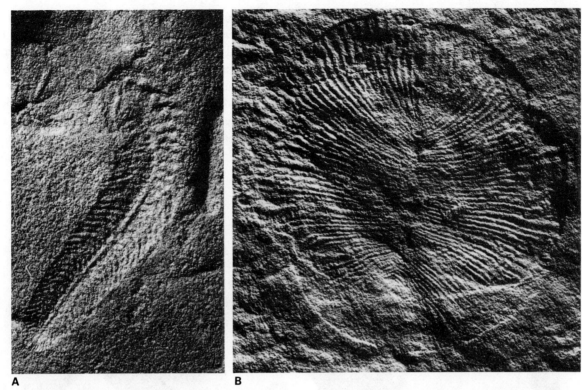

Figure 12.5. Examples of Proterozoic fossils from Australia that have been assigned to the Annelida by some specialists. **A,** *Spriggina,* an elongate genus that ranges up to 4 cm in length and resembles modern pelagic polychaete annelids of the family Tomopteridae; however, the absence of well-developed parapodia and the enlarged head region, among other features, also suggest an arthropod affinity. **B,** *Dickinsonia,* a flat ovoid genus, ranging up to 60 cm in length, resembles an unusual modern polychaete that is parasitic on sponges; possible affinities with the flatworms (phylum Platyhelminthes) is an alternative preferred by some workers. (Courtesy M. F. Glaessner.)

chaetes was the development of hardened jaw elements or scolecodonts. These elements are known to range from the Ordovician to the Holocene, and they are most abundant in shallow marine deposits. Although they hold potential for biostratigraphic zonation, scolecodonts have received relatively little study, largely because of taxonomic difficulties. The elements, which are parts of an assemblage or jaw apparatus, are usually found dissociated. Moreover, some element types are homeomorphic, being found in more than one kind of natural assemblage. In a few instances natural assemblages of scolecodonts have been found associated with other body remains (Figure 12.3), and sometimes by careful study the jaw apparatuses can be reconstructed from dissociated material (Figure 12.6); however, much more work will be necessary before the biostratigraphic potential of these fossils is fulfilled.

Trace fossils of probable annelid origin are abundant in many Phanerozoic marine deposits. A large number of the burrows and trails were probably made by polychaetes, but other animals also leave similar traces, and identification of the trace maker is seldom possible (see Chapter 21 for examples and further discussion).

For protection, some polychaetes secrete sturdy calcareous tubes in which to live, others construct leathery tubes of mucus, and still others build agglutinated tubes by cementing sand grains, shell fragments, or miscellaneous debris with mucus. One common calcareous form, *Spirorbis* (Figure 12.7, *A* and *B*), produces small, solitary, coiled tubes that are typically found attached to shells of other fossils. It has a geologic range from Ordovician to Holocene. Another long-ranging (Silurian to Holocene) and cosmopolitan genus is *Serpula* (Figure 12.7, *C*), which usually lives in groups and produces long, slender, coiled or contorted tubes. From the Jurassic onward, abundant serpulid tubes locally form the framework of extensive organic layers and mounds. Other polychaetes build closely packed tubes, both calcareous (Figure 12.7, *D*) and agglutinated (Figure 12.7, *E*), which may form wave-resistant reefs along tropical and subtropical coasts. One such modern example is essentially continuous along more than 300 km of the southeastern coast of Florida.

Although their numbers are great in modern soils, fossilized Oligochaeta have not been definitely identified. One genus (Figure 12.4, *F*) from Ordovician and Silurian marine deposits has previously been referred to this class; however, representatives show no evidence of a clitellum. Moreover each segment has two rings of small nipplelike structures that may have held setae, and no modern oligochaetes are known to possess double rings of setae. Affinities of this genus remain in doubt. One other possible representative of the class is an annulated burrow cast from terrestrial Paleocene deposits in Wyoming.

The oligochaetes exhibit a combination of primitive and derived characters. On one hand, they have mostly remained vagile burrowers with primitive segmentation and no appendages. On the other hand, oligochaete nephridia have become highly specialized for reducing the loss of body fluids. Hermaphroditism is a specialization advantageous where mates or water for transport of gametes are not easily available. The latter two

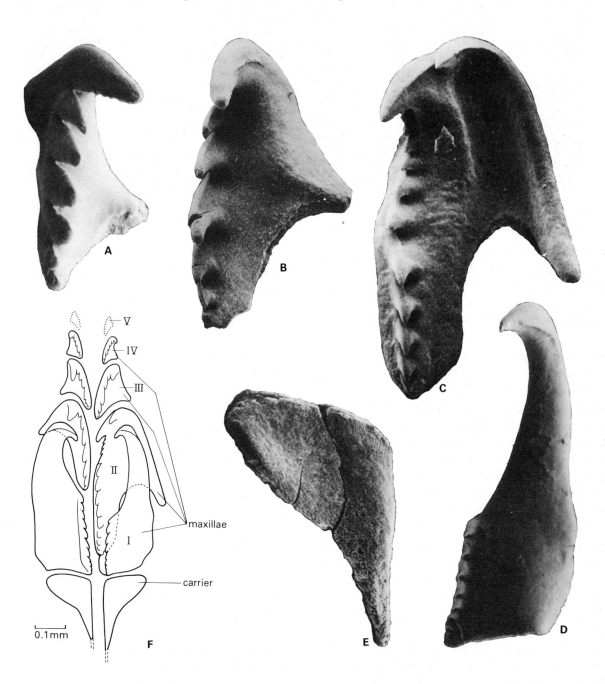

Figure 12.6. Jaw elements (scolecodonts) of the Jurassic polychaete *Arabella* from Poland. **A**, Right maxilla IV. **B**, Right maxilla III. **C**, Right maxilla II. **D**, Right maxilla I. **E**, Right carrier in ventral view. **F**, Reconstruction of jaw apparatus; maxillae V and posterior parts of carriers are unknown. (Modified from Szaniawski, H.; Gaździcki, A. *Acta Palaeontologica Polonica* 1978 v. 23, p. 3–29.)

Figure 12.7. Polychaete tubes. **A**, *Spirorbis* and bryozoans encrusting a Devonian brachiopod from Ontario. **B**, *Spirorbis* and bryozoans encrusting modern seaweed from Newfoundland. **C**, *Serpula* encrusting a Cretaceous oyster from Texas. **D**, *Dodecaceria*, modern calcareous tubes from California. **E**, *Phragmatopoma*, modern agglutinated tubes from Florida. (Photos by M. Frederick.)

features obviously correlate with the adaptation from marine to terrestrial habitats.

A close relationship between the oligochaetes and hirudineans is indicated by shared possession of several derived features, especially those of the reproductive and excretory systems. However, the abandonment of a burrowing habit for one of parasitism resulted in additional fundamental changes in the hirudineans. Peristaltic locomotion aided by setae was exchanged for crawling with a looping motion aided by suckers, and some are capable of swimming. These activities greatly reduce the need for a segmented hydrostatic skeleton. Consequently, the hirudinean coelom has been reduced, and the septa have been lost.

Supplementary reading

Clark, R. B. *Dynamics in metazoan evolution.* Oxford: Clarendon Press; 1964. Excellent functional analysis of the origin of the coelom and segments.

Clark, R. B. Systematics and phylogeny: Annelida, Echiura, Sipuncula. In: Florkin, M.; Scheer, B. T., editors. *Chemical zoology,* vol. 4. New York: Academic Press, Inc.; 1969. Good review of the classification and evolution of annelids and some related groups.

Conway Morris, S. Middle Cambrian polychaetes from the Burgess Shale of British Columbia. *Philosophical Transactions of the Royal Society of London* (B. Biological Sciences) 285: 227–274; 1979. Redescription and analysis of some extraordinarily well-preserved, early annelids.

Edwards, C. A.; Lofty, J. R. *Biology of earthworms,* 2nd ed. London: Chapman & Hall; 1977. General account of oligochaetes with a key to common genera.

Fauchald, K. The polychaete worms: Definitions and keys to the orders, families and genera. *Natural History Museum of Los Angeles County, Science Series* 28: 1–190; 1977. Reviews the classification of the Polychaeta and provides definitions and keys for all modern taxa down to generic level.

Glaessner, M. F. Early Phanerozoic annelid worms and their geological significance. *Journal of the Geological Society* (London) 132: 259–275; 1976. Analysis of early annelid evolution.

Howell, B. F. Worms. In: Moore, R. C., editor. *Treatise on invertebrate paleontology, Part W.* New York and Lawrence, KS: Geological Society of America and University of Kansas Press; 1962. Comprehensive summary of fossil annelids with descriptions of taxa down to generic level.

Kielan-Jaworowska, Z. Polychaete jaw apparatuses from the Ordovician and Silurian of Poland and a comparison with modern forms. *Palaeontologia Polonica* 16: 1–152; 1966. Monographic treatment of many articulated scolecodont assemblages.

Stephenson, J. *The Oligochaeta.* Oxford: Clarendon Press; 1930. Classic but somewhat outdated account of the biology and classification of oligochaetes.

Thompson, I. Errant polychaetes (Annelida) from the Pennsylvanian Essex fauna of northern Illinois. *Palaeontographica* (Stuttgart) 163: 169–199; 1979. Description of a well-preserved and diverse assemblage of late Paleozoic annelids.

13

Phylum Arthropoda

Richard A. Robison
Roger L. Kaesler

Part I   Phylum overview

The phylum Arthropoda comprises a vast array of diverse, often highly specialized invertebrates with a fossil record extending from the late Proterozoic to the present. Throughout their long history, members of the phylum have been particularly sensitive to environmental influences and have evolved rapidly. Included in the Arthropoda are such common living representatives as insects, crabs, shrimp, lobsters, ostracodes, barnacles, garden sowbugs, spiders, ticks, scorpions, limulids, centipedes, and millipedes (Figure 13.1), as well as numerous other organisms that are less well known. Also included are the extinct trilobites and eurypterids, which were common in early and middle Paleozoic seas.

As will be shown later, the grouping of all these kinds of animals into a single phylum may produce an artificial classification. The evolution of arthropod characters is thought by some investigators to have occurred more than once during the remote geologic past. Nevertheless, it is fairly easy to recognize an arthropod—either a living one or a fossil.

Arthropods are triploblastic and have a reduced coelom. Characteristically, they are bilaterally symmetrical with a segmented body divided into specialized regions, including a head with important sensory and nervous centers. A chitinous outer covering of the body, the **cuticle**, is jointed and functions as an **exoskeleton** to provide support and protection for the soft body parts. Arthropods also possess paired serial appendages and enlarged blood cavities. Although the primary function of the appendages is locomotion, evolution in all major groups has led to modification of some appendages for other functions such as feeding, respiration, sensory reception, and mating. This specialization has usually been accompanied by a reduction in the number of segments and appendages. Growth is confined to short periods when the exoskeleton is shed and a larger one is generated. Some adults are different from juveniles, and some parasitic species develop adult forms greatly different from those of close relatives.

The importance of arthropods in our world can scarcely be overstated. Some insects are invaluable as pollinators of plants, whereas others are among the worst pests known, destroying crops and spreading such diseases as malaria and yellow fever. Many arthropods, especially the lobsters, crabs, and shrimp, are among the most sought-after sources of food. By comparison, barnacles annually cause millions of dollars in damage by encrusting the hulls of ships and requiring removal of their skeletal accumulations in dry dock.

In the terrestrial environment, insects are the most abundant primary consumers by an overwhelming margin. They form major links in the food chain, competing with the large herbivorous vertebrates and supplying the food of many insectivores. Other arthropods form significant links in marine food chains.

The importance of the arthropods derives, in part, from their countless numbers of individuals and exceptional diversity, both of which far surpass those of any other animal phylum. The number of species of living insects alone has been estimated at from one to two million—more than all other animal species combined. Among such readily fossilized arthropods as trilobites

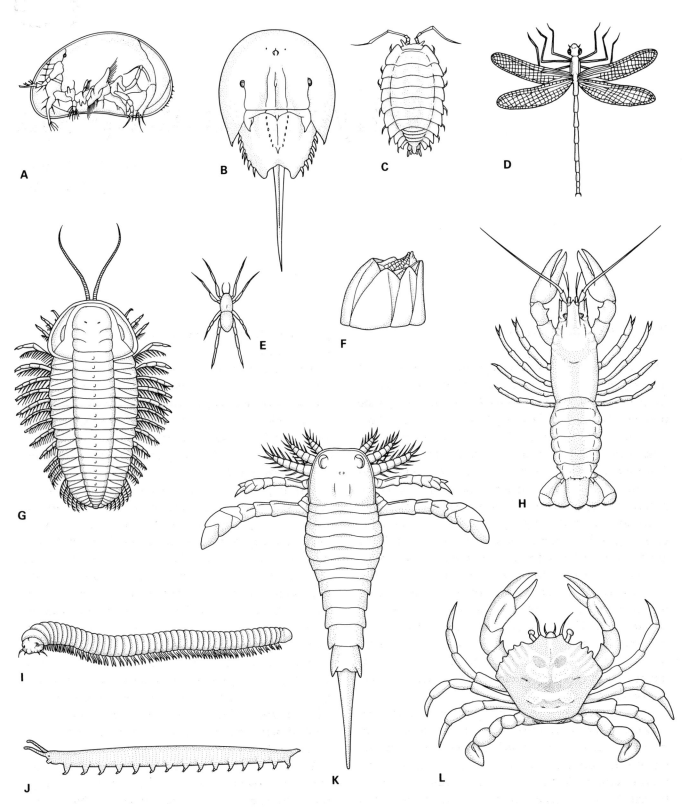

Figure 13.1. Representative arthropods. **A,** Ostracode (Crustacea) with left valve removed to show appendages. (×17.) **B,** Limulid or horseshoe crab (Chelicerata). (×0.2.) **C,** Garden sowbug (Crustacea). (×2.) **D,** Dragonfly (Hexapoda). (×1.) **E,** Spider (Chelicerata). (×1.) **F,** Barnacle (Crustacea). (×3.) **G,** Trilobite (Trilobitomorpha). (×7.) **H,** Crayfish (Crustacea). (×0.8.) **I,** Millipede (Myriapoda). (×1.) **J,** Onychophoran (Onychophora). (×1.) **K,** Eurypterid (Chelicerata). (×0.2.) **L,** Crab (Crustacea). (×0.5.)

and ostracodes, the number of described species is much lower, but nevertheless they make up an appreciable part of the fossil record.

Many different species of arthropods may be found living in one habitat or preserved as fossils in one kind of rock. This high local diversity presumably results from their highly specialized modes of life, which permit them to occupy many different ecologic niches. Members of some species, however, are gregarious or singularly adapted to unusual environments, resulting in strata with large numbers of a single species.

Arthropods are important to geologists in various ways. Rapid evolution is relatively easy to detect in their many distinctive morphologic features and makes some arthropods, especially trilobites and ostracodes, valuable for biostratigraphic correlation. Also, as complex, highly evolved organisms, some arthropods are particularly sensitive to the environments in which they live, and their fossils thus aid in reconstructing ancient environments.

The known record of fossil arthropods is significantly biased by the fact that members of most groups have a thin unmineralized exoskeleton that is not usually preserved. In a few groups, including the trilobites, ostracodes, and some crabs and barnacles, the exoskeleton is thickened and strengthened by the addition of calcium carbonate or calcium phosphate, and it is these groups that supply most of the arthropod fossils. Nevertheless, sufficient information is now available to indicate that arthropods probably originated in a marine environment during the late Proterozoic and were subsequently involved in at least four intervals of major adaptive radiation. The first extensive diversification was in the early Paleozoic as they rapidly adapted to most marine habitats. Further diversification became possible as members of at least three higher groups adapted to the land during the middle Paleozoic. Development of wings in the insects enabled a third major expansion during the late Paleozoic. A fourth and perhaps the greatest radiation was during the early Tertiary following the terrestrial appearance and rapid spread of the flowering plants (angiosperms).

Today, arthropods are known in abundance from nearly every habitat on earth, and they display an almost inconceivable variety of adaptations. They range from the deepest parts of the oceans to high mountains. Some are deposit or passive-suspension feeders, others are highly mobile predators, and still others are parasites of a wide variety of plants and other animals.

By almost any criterion, arthropods have been and continue to be a successful group. Compared with perhaps their chief competitors, the chordates, the arthropods have been many millions of years ahead in most major evolutionary advances. For example, arthropods first appeared during the late Proterozoic, whereas the oldest known chordate is Middle Cambrian in age. Some arthropods adapted to land environments at least by Early Devonian; chordates followed during the Late Devonian when they evolved legs. Arthropods possessed legs by the late Proterozoic. Some arthropods evolved wings for flight at least by Early Pennsylvanian time, whereas chordates first evolved wings during the Triassic.

In view of these and other successes by the arthropods, why have they not achieved total control of the earth? We suggest this question as a topic for student thought and discussion after the remainder of this chapter has been read.

Introductory morphology

Body plan

Conspicuous segmentation indicates a close relationship between the annelids (Chapter 12) and arthropods. Most major differences between the two phyla can be traced to a process of **sclerotization** in arthropods whereby the enclosing cuticle was differentially thickened and hardened to form a jointed exoskeleton. This initiated a new plan of structural organization in which control of body shape was transferred from the musculature in ancestral annelids to the exoskeleton in descendant arthropods. At the same time, sclerotized and jointed appendages were converted to a system of mechanical levers that enabled relatively small muscle contractions to be translated into relatively large movement. Motive force was shifted from musculature of the trunk to that of the limbs, resulting in considerably increased speed and efficiency of locomotion over firm substrates. Another change was the transformation of muscle fibers from the smooth, slow-acting type characteristic of annelids to a striated, fast-acting type found in all arthropods, except the relatively primitive onychophorans (Figure 13.1, *J*). Through later specialization, body segments and appendages were reduced in number as the remaining appendages were modified for several functions in addition to locomotion. Arthropods were the first animals to fly, but unlike vertebrates that evolved wings from the anterior limbs, arthropods evolved wings from lateral extensions of the dorsal exoskeleton.

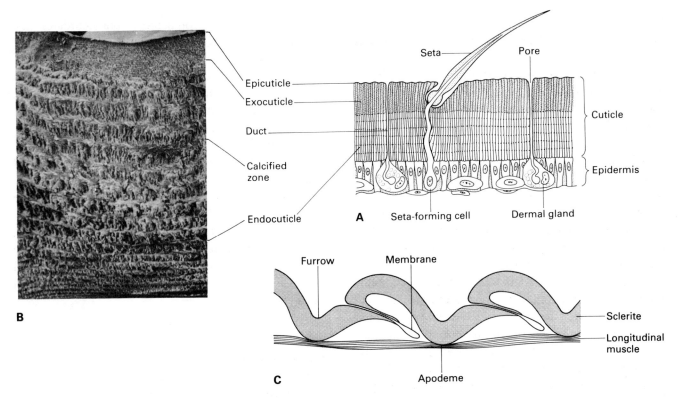

B

Figure 13.2. Structure of arthropod exoskeleton. **A**, Diagrammatic section of an insect cuticle overlying an epidermis of multiple cell types. **B**, Scanning electron micrograph of broken cuticle from a chela of the crayfish *Austropotamobius*. (×210.) **C**, Diagram showing relationship of sclerites and membranes. (**A** modified from Wigglesworth, V. B. *The principles of insect physiology,* 6th ed. New York: Methuen & Co., Ltd; 1960. **B** from Dalingwater, J. E.; Miller, J. *Palaeontology* 20: 1977; by permission of the Palaeontological Association. **C** modified from Hessler, R. R. *Journal of Paleontology* 36: 1305–1312; 1962.)

External features

The acellular arthropod cuticle functions as a supportive exoskeleton as well as a protective armor and is unlike either the soft skin or hard shell found as body coverings in most other invertebrate phyla. It is secreted at the outer surface of the epidermis and is layered (Figure 13.2, *A* and *B*). An innermost flexible layer, the **endocuticle**, is primarily composed of chitin, a horny material inert to most chemical agents. A rigid intermediate layer, the **exocuticle**, is high in protein, and in groups such as trilobites and ostracodes is further strengthened by impregnation with calcium carbonate or calcium phosphate. A thin extenal later, the **epicuticle**, contains a waxy material that prevents transmission of water and other substances. Glandular secretions and sensory information are transmitted through different kinds of ducts and pores.

The exoskeleton is constructed of thicker stiffened **sclerites** separated by thinner flexible **membranes** that form the articulating joints necessary for movement (Figure 13.2, *C*). Sclerotized cuticle may continue around the segment, but commonly dorsal and ventral plates are separated by lateral membranous areas. Dorsal sclerites are generally sturdier and thus are more apt to be preserved as fossils.

Other accessory features of the cuticle include hairlike sensory **setae** (Figure 13.2, *A*) and a variety of spines. The function of most spines is poorly understood; however, protection against predators and stabilization against water currents and turbulence are some of the more obvious possibilities. Infolds of the cuticle form internal **apodemes** (Figure 13.2, *C*) for muscle attachment, whereas other folds and surface ridges add structural strength.

Grouping of segments into regions specialized for particular functions occurs in all arthropods. These regions, called **tagmata**, may consist of either fused or mutually movable segments. Examples are illustrated in Figure 13.3, but related terms are not defined until later in this chapter. Despite other structural and functional changes, basic tagmatic patterns have persisted and are characteristic of major present-day groups; however, tagmata designated by the same terms do not necessarily include homologous segments. For example, the primitive segments that developed into the trilobitomorph thorax are not the same as those that developed into the hexapod thorax. In some arthropods an unsegmented tailpiece may bear a spikelike **telson** and associated with this may be a pair of bladelike or filamentous **furcae**.

Among primitive arthropods, most segments, whether separate or fused, bear a pair of ventrally directed appendages. As expressed in the phylum name (from the Greek *arthron,* meaning joint and *podos,* meaning foot), these appendages are characteristically jointed, each being composed of a series of sclerotized cylindrical **podomeres** that are connected at the joints by flexible membranes. Levered movement is usually effected by sets of opposed muscles, with some sets connecting podomeres and others running from a podomere to the inner

TRILOBITOMORPHA

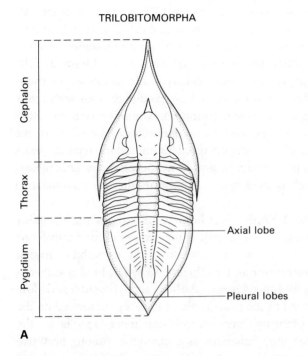

A

HEXAPODA

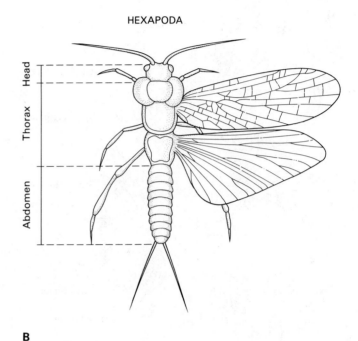

B

CRUSTACEA

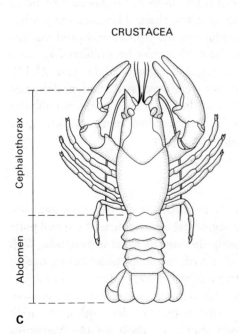

C

CHELICERATA

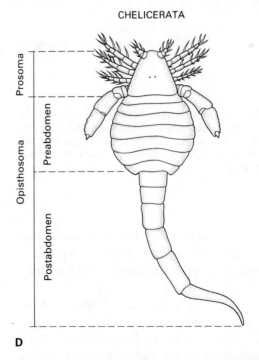

D

Figure 13.3. Specialized divisions (tagmata) of the body in arthropods. The trilobitomorph in **A** is also divided lengthwise into one axial and two pleural lobes. (Modified from Størmer, L.

In: Moore, R. C., editor. *Treatise on invertebrate paleontology, Part O.* Boulder, CO, and Lawrence, KS: Geological Society of America and University of Kansas Press; 1959.)

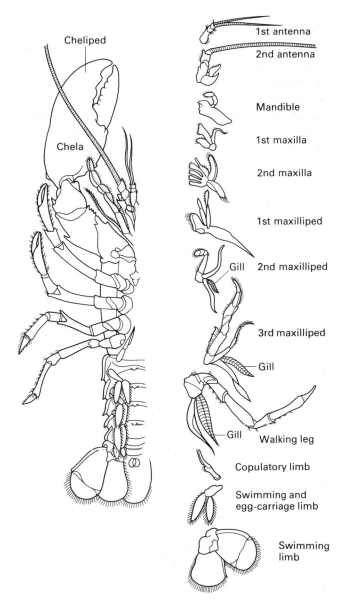

Figure 13.4. Specialization of arthropod limbs as exemplified by a lobster. A partial ventral view of the animal is shown on the left and examples of most individual appendages (not to scale) are shown on the right. (Modified from Buchsbaum, R. *Animals without backbones*. Chicago: University of Chicago Press; 1948.)

body wall. In a few groups (for example, onychophorans, centipedes, spiders) the appendages are extended by blood pressure, and muscles are necessary only for retraction and to control movement.

Sclerotization has enabled the development of functionally diverse appendages for division of labor, and a single advanced arthropod may possess several kinds. Common limbs of the head (Figure 13.4) include **antennae** for sensory purposes and locomotion, **mandibles** for chewing, **maxillae** and **maxillipeds** for

manipulating food, and **chelipeds**, each with a distal pincer or **chela** for grasping and cutting food and for use in offense and defense. Additional limbs may be used for walking, swimming, mating, egg care, and a variety of more specialized activities. Respiratory **gills** are also present on different types of appendages.

Comparative analysis of the structure and function of arthropod appendages has led to the recognition of at least two basic types (Figure 13.5, *A* and *B*), each extending from a proximal podomere called a **coxa**. One type has a single **ramus** or branch and is **uniramous**, whereas the other has two primary rami and is **biramous**. The primitive biramous limb typically possesses an **endopodite** for walking and an **exopodite** lined with many long filaments for respiration, swimming, or both. Other rami (exites) may be secondarily developed. Also, the coxa of each unspecialized biramous limb typically bears a spiny, medially projecting process, the **gnathobase**, which is used in one way or another for manipulating food.

Further differences between uniramous and biramous arthropods are evident in the manner of food transfer to the mouth and in evolution of the mandibles. Among primitive uniramians (Figure 13.5, *D*), food is gathered only by appendages on the head and is passed from below upward to the mouth. Biting surfaces are formed on the tips of the uniramous mandibular appendages; hence, the whole limb functions as a mandible. Among primitive biramians (Figure 13.5, *C*), particulate food is gathered by a long series of trunk limbs and is passed forward to the mouth along a ventral channel between the gnathobases. In some sediments the limb activity produces distinctive feeding trails that may be preserved as trace fossils (see Chapter 21, Figures 21.16 and 21.17). Biramian mandibles were developed from modification of gnathobases, and the distal part of the mandibular limbs may retain their original motive function, be further modified, or disappear.

Internal features

Arthropods are thought to have evolved from annelids. However, clear segmental arrangement of internal parts has been variously diminished in all arthropods. With abandonment of a hydrostatic skeleton during one or more transitions to an arthropod body plan, the coelom became considerably reduced, appearing in the embryo as a series of cavities in the mesoderm but in the adult being represented mainly by small cavities containing sexual and excretory organs. Most intersegmental septa were lost, and parts of the blood-vascular system were

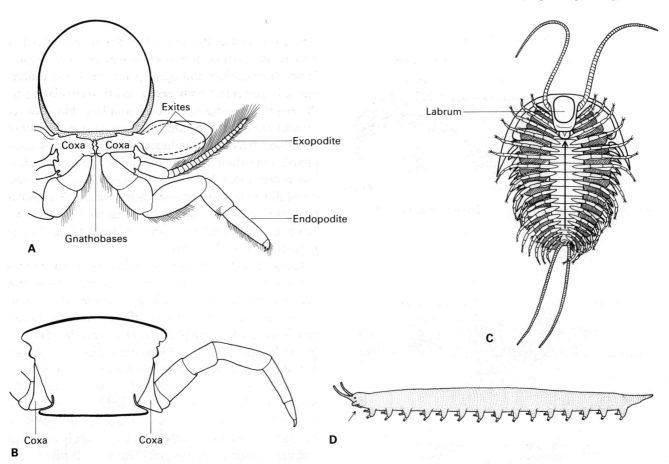

Figure 13.5. Basic arthropod limb types and primitive methods of food transfer. **A,** Biramous limb of the crustacean *Anaspides*; exites are secondary additions. **B,** Uniramous limb of the myriapod *Lithobius*. **C** and **D,** Comparison of the direction (arrows) of food movement to the mouth in the biramous trilobitomorph, *Olenoides,* and the uniramous onychophoran, *Peripatoides.* (**A** and **B** modified from Manton, S. M. In: Moore, R. C., editor. *Treatise on invertebrate paleontology, Part R.* Boulder, CO, and Lawrence, KS; Geological Society of America and University of Kansas Press; 1969. **C** and **D** modified from Hessler, R. R.; Newman, W. A. *Fossils and Strata*; 1975.)

expanded as cavities in tissue and as open spaces around internal organs, which collectively form a **hemocoel.** Some arthropod organ systems retain vestiges of segmental divisions, but most function as single large organs that serve the entire body. Internal features of the common lobster (Figure 13.6) are representative of the phylum.

Arthropods have a well-developed digestive system consisting mainly of a short tubular esophagus, a large stomach, and a long slender intestine (Figure 13.6, *A*). The anterior and posterior parts of the digestive tract are lined with chitinous cuticle, which in the esophagus may be studded with calcified "teeth" for further grinding of food. One or more pairs of liverlike digestive glands manufacture enzymes to aid digestion.

Principal components of the highly organized arthropod nervous system are an anterior brain and a ventral nerve cord that connect around the esophagus by a ring of nerve ganglia (Figure 13.6, *A* and *C*). Like some annelids, primitive arthropods have a segmented ladder-like ventral nerve cord consisting of two interconnected strands, but in advanced arthropods the paired strands may be partially or entirely fused. Concentration of neural tissue increased with specialization of the limbs as well as specialization of other organ systems and reached an extreme in such forms as the ticks where even the brain is fused with the ventral nerve cord and all traces of segmentation are lost. Closely coordinated with the nervous system are several kinds of highly specialized sensory organs for seeing, hearing, feeling, balancing, and receiving chemical stimuli.

The characteristic arthropod hemocoel functions as an open circulatory system for the transfer of nutrients and oxygen to all parts of the body (Figure 13.6, *A*, *B* and *D*). A dorsal heart pumps blood and usually consists of a pulsating tube. Arteries conduct the blood *not* into capil-

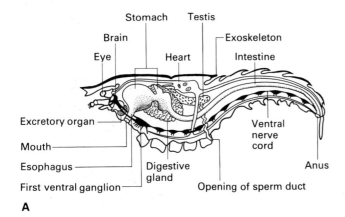

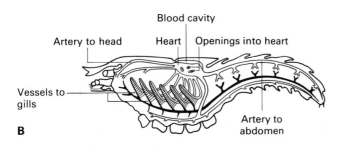

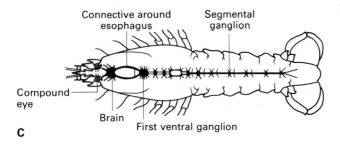

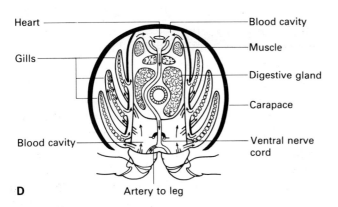

Figure 13.6. Internal features of a lobster. **A,** Digestive system and its relation to other organs. **B,** Main blood channels in the circulatory system. **C,** Anatomy of the segmented nervous system. **D,** Cross section of thorax showing relations of gill chambers to other organs and path of blood flow through main channels. (Modified from Buchsbaum, R. *Animals without backbones.* Chicago: University of Chicago Press; 1948.)

laries, as in vertebrates and some other invertebrates, but into tissue cavities and open spaces between organs. From these cavities and open spaces, the blood collects into a large ventral body cavity and then circulates past the respiratory organs and back to a large blood cavity around the heart. Openings with one-way valves admit the blood back into the heart where the cycle is completed. Important consequences of this system are low blood pressure, which limits maximum size, and a susceptibility to severe bleeding when wounded. To combat the latter problem, rapid clotting agents have been developed in the blood, providing a rewarding field of study for recent medical research.

Larger aquatic arthropods breathe by means of gills carried by the exopodites or biramous limbs. These gills are delicate filamentous or leaflike structures that tend to be less protected in primitive arthropods but are commonly covered by folds of the body wall in advanced forms (Figure 13.6, *D*). Additional diffusion of oxygen across the body surface may supplement that taken in by the gills. Some land arthropods possess **lung books** thought to have developed by further enclosure and modification of gills, but most land arthropods breathe by means of tubular **tracheae** that are usually branched and are formed from ingrowths of cuticle. Differences in structure are sufficient from group to group to indicate that tracheae probably evolved independently within different arthropod groups. In small arthropods, respiration may be entirely accomplished by diffusion of gases directly through the body wall, and special respiratory organs are absent.

The sexes are separate in most arthropods, but some are hermaphroditic (sexes combined) and a few are parthenogenetic (embryo develops from unfertilized egg). Sexual dimorphism is common in advanced arthropods, and even trimorphism is encountered in some species in which sexually dimorphic generations alternate with parthenogenetic generations. The gonads, which occupy the vestigial coelom, open ventrally to the exterior (Figure 13.6, *A*) in different segments in different groups. Also, in males and females of a single species the openings are not always on the same segment.

Classification

The higher classification of arthropods is complex and not firmly established. Characters commonly used to define higher taxa include the structure of tagmata, especially of the head region, as well as the structure, number, and arrangement of appendages. Many taxonomists recognize five major superclasses (Trilobitomorpha, Crustacea, Chelicerata, Myriapoda, and

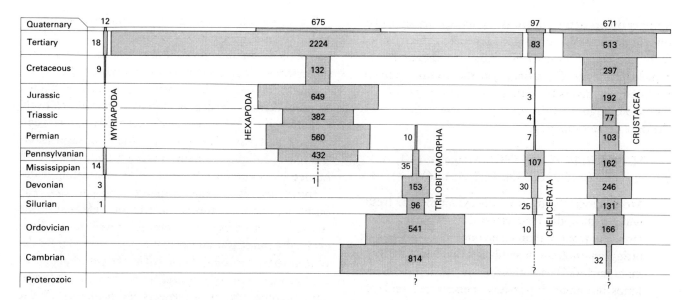

Figure 13.7. Geologic distribution of the major arthropod superclasses based mostly on data compiled from the *Treatise on Invertebrate Paleontology*. Numbers indicate the fossil genera in each superclass. (Data on generic diversity of the Hexapoda was supplied by F. M. Carpenter.)

Hexapoda) and at least four minor superclasses (Onychophora, Pentastomida, Tardigrada, and Pycnogonida). Based primarily on detailed functional and developmental studies, other taxonomists have recently proposed that the Crustacea, Chelicerata, and Uniramia (including the Onychophora, Myriapoda, Hexapoda, and possibly Tardigrada) be recognized as three separate phyla, each of which developed an arthropod body plan independently from different origins. It also has been suggested that the class Trilobita may constitute a fourth phylum; however, inadequate information makes the status of that extinct group less certain. We accept the possibility that the arthropods may be polyphyletic, but we also believe that recognition of multiple arthropod phyla deserves a thorough evaluation prior to possible general acceptance. In the meantime, we present a more traditional classification, similar to that listed in the *Treatise on Invertebrate Paleontology (Part R)*. Grouping is virtually identical in both classifications, the major differences being in the ranks assigned to certain groups.

SUPERCLASS TRILOBITOMORPHA
(Figures 13.1, *G*; 13.3, *A*; 13.5, *C*)

Primitive marine arthropods, typically elongate oval in dorsal view, with prominent elevation lengthwise along middle of back producing three-lobed appearance. Tagmata normally consist of cephalon, thorax, and pygidium. Appendages characteristically include one pair of jointed uniramous antennae in front of mouth and long series of similar, paired biramous limbs with weak gnathobases behind mouth. Mandibles and chelae absent. (Upper Proterozoic (?), Lower Cambrian to Upper Permian.)

The extinct superclass is named for the three-lobed appearance of its members. Most of the vital organs were housed beneath the central **axial lobe**, and mainly glandular tissue was present beneath the lateral **pleural lobes**, which also provided a cover for the appendages. The **cephalon** (Figure 13.3, *A*) contains all fused segments of the anterior body, the **pygidium** contains all fused segments of the posterior, and the intermediate **thorax** includes all movable segments. Those limbs behind the mouth were multifunctional, each being used for locomotion, food gathering, and respiration, but all are primitive (that is, relatively unspecialized) in structure. By mechanical manipulation, probably aided by water currents, food gathered by the trunk limbs was passed forward to the mouth along a midventral channel between the gnathobases (Figure 13.5, *C*). The mouth was directed backward and was covered by a platelike **labrum** that helped to prevent food from escaping ingestion. The chief source of food was probably organic detritus from the sea floor, but some members may have captured small soft-bodied organisms or secondarily become pelagic suspension feeders.

By far the most common representatives are the trilobites, whose more than 1500 described genera outnumber others of this superclass by approximately 100 to one. Four questionably assigned genera have been named from the upper Proterozoic. Adaptive radiation was rapid during the Cambrian, and the group gradually declined in diversity during the remainder of the Paleozoic (Figure 13.7). Among the arthropods, this is the only extinct superclass.

SUPERCLASS CRUSTACEA (Figures 13.1, *A, C, F, H* and *L*; 13.3, *C*; 13.4; 13.5, *A*; 13.6)

Morphologically diverse, mostly marine and fresh-water arthropods characterized by two pairs of antennae in front of mouth and one pair of grinding or biting gnathobasic mandibles behind mouth. Tagmata include head, thorax, and abdomen. Some limbs biramous in primitive members; secondarily uniramous and highly specialized in others. (Upper Proterozoic (?), Lower Cambrian to Holocene.)

Familiar crustaceans are crayfish, crabs, lobsters, shrimp, barnacles, and garden sowbugs. A host of other forms are common but are usually not as readily observed. Because of the great diversification, including nine classes, it is difficult to characterize a typical crustacean. Nevertheless, the anterior **head** contains the principal sensory organs and six fused segments, each of which, except the first, usually bear a pair of appendages. The middle tagma, the thorax, and the posterior tagma, the **abdomen**, vary considerably in form as well as number and function of appendages. Generally, the thorax extends backward to include the most posterior segment with a genital pore. The head and thorax may be united to form a **cephalothorax** (Figure 13.3, *C*).

Many crustacean features can be related to evolutionary trends involving specialization of some limbs, loss of other limbs together with some body segments, and development of a special covering over parts of the body. In contrast to the primitive multipurpose limbs of trilobitomorphs, limbs of many crustaceans are highly specialized for particular uses (Figure 13.4). Moreover, limbs in one crustacean group may perform functions different from those of corresponding limbs in other groups. Improved efficiency from such specialization was accompanied by reduction in the number of limbs as well as segments and shortening of the body. In some crustaceans, cuticle at the back of the head has been extended into a covering, the **carapace**, that protects more of the body and forms chambers to aid respiration and feeding (Figure 13.6, *D*). Such cuticular expansion has reached a peak in the ostracodes (Figure 13.1, *A*), with the entire body enclosed in a calcified **bivalve carapace** (one valve on each side).

Crustaceans had a marine origin and have become so abundant and diverse in that environment that they have been called "the insects of the sea." Several groups have successfully adapted to freshwater habitats, and a few are common in moist land areas. Some have remained benthic detritus feeders and have retained relatively primitive limbs. Others have become swimmers and floaters and have developed a variety of suspension-feeding techniques, usually involving limbs lined with numerous, sometimes large setae for generating water currents and for straining minute food particles. The barnacles (class Cirripedia) have become efficient sessile suspension feeders, and several other groups have evolved commensal and parasitic modes of life. By development of grasping chelae on limbs of the oral region, such groups as the shrimp and crabs (class Malacostraca) have become efficient scavengers and predators. Chelae gave rise to a new feeding style convergent on that shown by uniramous groups, where food particles are lifted upward to the mouth rather than being forwarded along a ventral food channel. Chelae are also used to cut food.

Biostratigraphically, the most important crustaceans are the minute ostracodes, characterized by an unusually expanded, relatively well-calcified, bivalve carapace (Figure 13.1, *A*). They flourish today in both fresh and marine waters and occur in abundance in many formations, ranging in age from Cambrian to Holocene. Their small size, rapid evolution, and abundance make them particularly useful for temporal correlation of rocks drilled for oil and gas.

Figure 13.8. Possible crustacean, *Parvancorina*, from the Upper Proterozoic of Australia. (×6.5) Note its superficial resemblance to the juvenile trilobite instar in Figure 13.9. (Photo by M. F. Glaessner.)

One late Proterozoic genus (Figure 13.8) has been assigned to the Crustacea, but its appendages are unknown and other morphologic features are equivocal. Other taxa range from the Cambrian to the present. Only one of nine classes is extinct. In general, crustacean history is a story of continued adaptive radiation and taxonomic diversification (Figure 13.7). More than 2000 genera of fossil crustaceans have been described.

SUPERCLASS CHELICERATA (Figures 13.1, *B*, *E*, and *K*; 13.3, *D*)

Mostly predaceous or parasitic, aquatic or terrestrial arthropods with single pair of chelate limbs in front of mouth. Except in parasitic ticks, body divided into two tagmata. Antennae and mandibles absent. (Cambrian (?), Ordovician to Holocene.)

Among the common living chelicerates are the spiders, scorpions, ticks, mites, and limulids. Two classes are recognized—class Merostomata (Figure 13.1, *B*) being relatively primitive and primarily aquatic, and class Arachnida (Figure 13.1, *E*) being specialized and primarily confined to the land. The extinct eurypterids (Figure 13.1, *K*) were distinctive inhabitants of marginal marine environments during middle Paleozoic time.

Morphologically, the anterior tagma, called a **prosoma** (Figure 13.3, *D*), bears six pairs of major appendages. A single pair of preoral chelate limbs, the **chelicerae**, are characteristic and give the superclass its name. Five pairs of postoral limbs on the prosoma may be used for feeding, locomotion, or both. The posterior tagma, the **opisthosoma**, commonly contains 12 segments. Opisthosomal appendages, if present, mainly carry gills or silk glands. Genital openings are usually found on the second opisthosomal segment.

The cuticle is uncalcified in the chelicerates. Although nearly 400 genera of fossils have been described, in actual number of specimens the group has a comparatively poor fossil record. Possible representatives are present in Cambrian rocks, and typical forms range from the Ordovician to the present (Figure 13.7).

SUPERCLASS MYRIAPODA (Figure 13.1, *I*)

Terrestrial uniramous arthropods with a long slender body and many legs. Tagmata poorly differentiated; consisting of short six-segmented head, long multisegmented trunk, and short legless tailpiece of one to a few segments. Head appendages typically including one pair of antennae in front of mouth and one pair of jointed whole-limb mandibles and one or two pairs of maxillipeds behind mouth. Respiration by tracheae. (Silurian to Holocene.)

Centipedes (class Chilopoda) with one pair of legs per trunk segment and millipedes (class Diplopoda) with two pairs of legs per trunk segment are well-known representatives. Myriapods have thin cuticle with little chitin, which, in combination with a terrestrial mode of life, has resulted in a poor fossil record. Only about 60 genera of fossils have been described (Figure 13.7).

SUPERCLASS HEXAPODA (Figures 13.1, *D*; 13.3, *B*)

Mostly terrestrial uniramous arthropods with well-developed tagmata, including head, thorax, and abdomen. Head of six fused segments, generally bearing single pair of antennae in front of mouth together with one pair of unjointed whole-limb mandibles and two pairs of other specialized limbs behind mouth. Thorax composed of three segments, each with one pair of walking legs for which group is named; dorsal wings present or absent. Abdomen usually containing 11 segments, number varying slightly among primitive groups. Respiration by tracheae. (Middle Devonian to Holocene.)

Members of this superclass, overwhelmingly represented by the insects, are familiar to everyone, but their classification is still unsettled. The scheme followed here is to assign those primitively wingless forms having covered mouthparts to three small classes (Collembola, Protura, and Diplura). The vast majority of hexapods have exposed mouth parts and are assigned to the class Insecta. Most insects possess wings, but a few (such as silverfish) are primitively wingless and others (lice, fleas, some ants) have become secondarily wingless.

Hexapods abound in most terrestrial habitats, and some have even adapted to marine habitats. Nevertheless, their soft bodies and unmineralized exoskeletons have resulted in a generally poor fossil record. The wings of insects are the parts most likely to be fossilized, which is fortunate for taxonomic reasons because characteristics of wings are important in classifying living insects. In spite of their poor overall record, more than 7000 species of fossil hexapods belonging to about 5000 genera have been described (Figure 13.7), attesting to a considerable diversity since the late Paleozoic.

Minor superclasses

The minor superclasses include the following:

1 Onychophora, represented by primitive caterpillar-like forms (Figure 13.1, *J*)
2 Pentastomida, containing aberrant highly specialized parasites, unknown as fossils
3 Tardigrada, represented by minute unsegmented eight-legged creatures, known in the fossil record from a single Cretaceous species and only doubtfully included with the arthropods
4 Pycnogonida, an unusual, spiderlike marine taxon sometimes included with the Chelicerata and with fossils known only from the Devonian of Germany.

Only the onychophorans, which some have described as a connecting link between the annelids and arthropods, are of further interest here. Actually, they do not fit well into either phylum but are usually included with the arthropods. Onychophorans are characterized by a short, poorly differentiated head and a long trunk with 14 to 43 pairs of similar, unjointed, uniramous legs. The head contains three segments and bears a single pair of antennae, one pair of jaws, and one pair of special oral appendages. The cuticle is thin, and the body wall is composed of circular and longitudinal muscles like the annelids, but the coelom is much reduced. A hemocoel is well developed, and respiration is by tracheae as in many arthropods. External segmentation is lacking except for the serial repetition of appendages. Living onychophorans are rare, being found primarily in moist tropical forests. One genus of Cambrian age can be assigned to the Onychophora; however, it was adapted to marine environments and differs in several characters from extant terrestrial genera. Fossilized terrestrial onychophorans are known from a single genus of Pennsylvanian age.

Convergent evolution

Convergent evolution has been commonplace among arthropods. In such a large group, it should not be surprising that basic mechanical requirements and natural selection have repeatedly produced **homeomorphy** (superficial resemblances) among arthropods of different origins. These similarities make the recognition of phyletic relationships difficult, and convergence cannot be ignored if some measure of stability is to be achieved at the higher levels of arthropod classification. Following are a few examples of convergent features in arthropods:
1 *Sclerotization*, a definitive character of the Arthropoda, which almost certainly evolved independently in ancestral lineages of marine biramians and terrestrial uniramians
2 *Unsegmented mandibles*, which in hexapods were derived from segmented uniramous limbs and in crustaceans were derived from segmented biramous limbs
3 *Chelae*, which occur on appendages that are not homologous in the Chelicerata and Crustacea
4 *Compound eyes*, in which some parts are not homologous as shown by different embryonic development (for example, Crustacea versus Hexapoda)
5 *Tracheae*, as indicated by histologic and embryologic differences in onychophorans, myriapods, arachnids, and certain crustaceans
6 *Hexapody*, or the special use of six legs for walking, which is found in all Hexapoda but also has been independently developed in rare Chelicerata and Crustacea.

Minor convergent features are numerous, and representative examples are discussed in Part II of this chapter.

Features of the phylum

Growth, autotomy, and regeneration

A rigid, boxlike exoskeleton provides good protection and support of internal organs, but it also presents a special problem for growth. Arthropods partially overcome this problem by periodically molting the exoskeleton. The animal then undergoes rapid growth, approximately doubling its former volume. Afterward, a new larger cuticle is formed, and the animal remains at a constant size until it molts again.

Advantages and disadvantages are inherent in this type of growth. On one hand, repeated disposal of juvenile exoskeletons enables the animal to change substantially in form during development. Displacement in developmental timing can be an important factor in causing some of these changes (see discussion of trilobite ontogeny in Part II of this chapter). On the other hand, the animal must expend considerable metabolic energy in periodically replacing its exoskeleton, it is subject to perils during times when it lacks the support and protection of an exoskeleton, and without continuous skeletal support its maximum size is restricted. Although some fossil eurypterids approach 2 m in maximum length, the great majority of arthropods do not exceed 10 cm.

Arthropod molting is under hormonal control. The stiff outer layer of the cuticle is separated from the animal after an enzyme dissolves the inner layers. Usually the entire exoskeleton is shed at about the same time, and all of the animal's growth occurs in discrete stages.

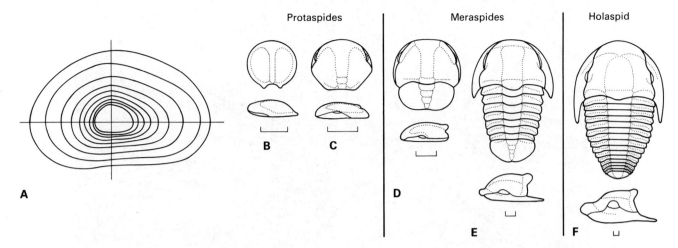

Figure 13.9. Instars of arthropods. **A,** Nested outlines of the extant ostracode *Cypridopsis* following successive growth increments. (×4.) **B** to **F,** Five of several growth stages in the Lower Cambrian trilobite *Crassifimbra* from Nevada; bar scales are 0.25 mm long. (**A** from Kesling, R. V. In: Moore, R. C., editor, *Treatise on invertebrate paleontology, Part Q.* Boulder, CO, and Lawrence, KS: Geological Society of America and University of Kansas Press; 1961. **B** to **F** from Palmer, A. R. *Journal of Paleontology* 32: 154–170; 1958.)

The stage between successive molts is referred to as an **instar** (Figure 13.9). Some arthropods (such as ostracodes and more advanced insects) have specific numbers of instars and usually terminate growth after reaching an adult stage. The remainder (such as crabs, lobsters) continue to molt and increase in size throughout their lives, but the frequency of molting decreases with advanced age.

Some arthropods have developed escape and repair mechanisms called **autotomy** and **regeneration**. They are able to separate a damaged or seized appendage along a preformed suture by means of a muscular reflex. The limb is regenerated during the next molt, but not necessarily in its original form. For example, some crabs have unequal chelae, a large one for crushing and a small one for cutting. If the crushing chela is removed autotomously, the small one may grow large to be used for crushing, whereas the regenerated one will be small and used for cutting. In a more extreme example, if a lobster loses an eye, it is regenerated as an antenna.

Discontinuous growth, molting, and autotomy and regeneration have important implications for interpretation of the arthropod fossil record. First, each arthropod can potentially contribute several molted exoskeletons or **exuviae** to the fossil record. Some exuviae are fragile and easily destroyed by scavengers and wave or current action. On the other hand, dead bodies are more likely to be attacked by scavengers than are empty exuviae, which in some instances may contribute to preservation of a greater proportion of exuviae. Some arthropods molt in such a way that their exuviae are difficult to discriminate from the remains of dead individuals, whereas others molt according to a set procedure that an alert paleontologist can sometimes determine (Figure 13.10).

Because growth commonly occurs in discrete steps that are morphologically distinct from each other, it is often possible to determine the relative ages of members in a population of fossil arthropods. This permits population dynamics and survivorship to be studied without reference to size, as must be done with such continuously growing organisms as pelecypods and brachiopods. Such studies, of course, depend on the preservation of early instars, which are more susceptible to destruction than are the sturdier remains of later instars.

Adaptation to land

Although members of several metazoan phyla have adapted to freshwater habitats, only among the arthropods, chordates, and mollusks have some groups fully adapted to life on land. Arthropods were the first animals to make that important transition. Beginning early in the Devonian Period, it was made not once but several times by some chelicerates and crustaceans as well as ancestors of the onychophorans, myriapods, and hexapods. This compares to possibly one transition each by chordates (amphibians) and mollusks (pulmonate snails). If success on land is measured by either numbers of species or numbers of individuals, arthropods have been far more successful than members of other phyla. In large part, this success was due to adaptations already made by aquatic

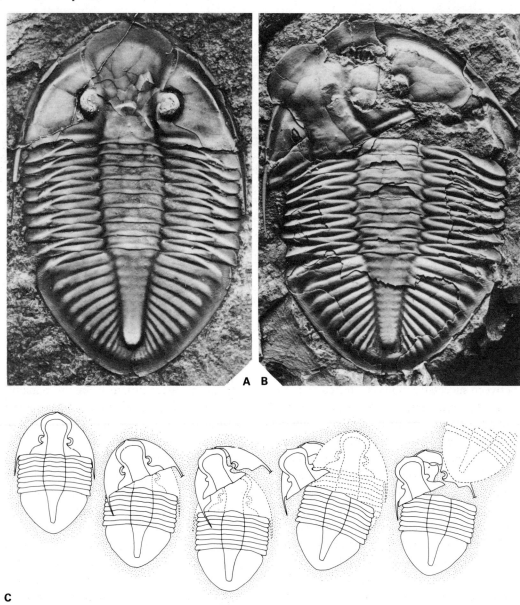

Figure 13.10. Molting in the Ordovician trilobite *Pseudogygites* from Ontario, Canada. **A,** Remains of a carcass. (×3.) **B,** Exuvia discarded during molting. (×2.) **C,** Interpretation of the sequence of events during molting (see Part II for explanation of terms). The trilobite first severed the connection between the thorax and cephalon by anchoring one genal spine in the mud and backing up. This caused the left side of the cephalon to rotate back over the front part of the thorax. With its head, the trilobite then nudged the cranidium free from the remainder of the cephalon and crawled obliquely forward through the gap left by the split facial suture. As the trilobite moved forward, it pushed aside the cranidium while the articulated thorax and pygidium became caught against the cephalon and slid off the back of the trilobite. Finally, when it was almost free of its dorsal exoskeleton, the trilobite dislodged and overturned the ventral labrum with its thrashing legs. (From Ludvigsen, R. *Fossils of Ontario, Part 1: The trilobites.* Toronto: Royal Ontario Museum; 1979.)

ancestors, not in response to any urge or strategy to live on land, but as an immediate aid to animals living in water. Structures that later serve other functions in different or changed environments are called preadaptive (Chapter 4). Preadaptations have evidently permitted the successful invasion of many new habitats. Because of the spectacular success of arthropods together with the extent of their land habitats, it is appropriate to consider briefly some of the major problems of life on land and some related facets of arthropod evolution.

Locomotion   Ocean currents carry nutrients to sessile animals and remove toxic body wastes. Air currents are virtually devoid of nutrients, however, and land animals must normally be mobile to feed and to escape being buried by their own body wastes. Aquatic arthropods had evolved efficient walking legs by the beginning of Cambrian time, but many millions of years passed before some forms ventured onto land. Some arthropods continued to improve locomotion on land by reducing the number of walking legs to six, increasing leg length, and concentrating leg attachment sites on the thorax or comparable region. These changes enabled stepping with two sets of three legs, the first and third legs on each side working with the opposite middle leg so that one set provides tripodal support while the other set steps. Other advantages were decreased leg interference, increased stride, and more powerful lever action. These are distinctive features of the Hexapoda, but they are by no means restricted to that superclass. Among the spiders (Chelicerata), one of the four pairs of walking legs is sometimes converted to other functions, making the animals effectively hexapodous. A similar trend is observed among aquatic crabs and other crustaceans.

Desiccation   Excessive loss of body fluids is a danger to land animals, especially small animals because the relative amount of surface area increases with decrease in size, and their total water content is low. Under these circumstances, it is difficult to imagine a better preadaptation for living on land than the waterproof arthropod cuticle. Additional water conservation has been achieved by a variety of modifications to the respiratory and excretory systems such as closing devices (spiracles) for the tracheae and special tubules that separate and resorb water from body wastes.

Respiration   Biramians have been most abundant in aquatic environments, whereas uniramians have been most abundant on land. Respiratory methods may have had an important influence on those distribution patterns. Diffusion of gases through the body wall is a primitive means of respiration in annelids, and it was apparently retained or redeveloped in some small arthropods. But as body size increases, relative surface area decreases, and as early marine arthropods increased in size, additional respiratory surface area was needed. One solution was to develop gills on exopodites of the biramous appendages. Except for some arachnids (such as spiders, scorpions) with modified gill structures, few biramians have successfully adapted to land. Even among the arachnids, many have developed supplementary tracheae. The apparent inability of uniramians to develop gills may have limited their diversity in aquatic habitats, but it also may have given them an adaptive

advantage on land because they retained a simple respiratory method. During the transition to land life, the respiratory surface of primitive uniramians, which was the body wall, was increased in area by infolding the cuticle to form tubules of a tracheal system. Further evolution is evident in comparison of the short, discrete tracheae of onychophorans with the much longer, anastomosing tracheae of many insects.

Gravity   Animals living on land are much more influenced by gravity than those living in water, which provides buoyancy. As arthropods adapted from water to land, their relatively small size, many closely spaced legs, and well-developed muscles virtually eliminated the potential problem of reduced bouyancy. In comparison, increased effects of gravity on the early amphibians, which were larger and quadrupedal, resulted in rapid evolution to strengthen the backbone and the more widely spaced legs.

Reproduction   Protection of the developing embryo against desiccation is a major problem of reproduction on land. Many advanced land arthropods have reduced this problem by evolving a special egg case with permeability finely balanced to permit sufficient diffusion of oxygen and carbon dioxide but little loss of water. Other land arthropods retain the eggs internally for protection, and some scorpions even have enclosing uterine sacs like the mammalian placenta.

Origin and taxonomic relationships

Until the mid-1900s, most taxonomists assumed a basic phyletic unity for all animals with arthropod features. Such animals were traditionally assigned to a single phylum, and a common ancestral arthropod was generally thought to have evolved from a parapodia-bearing polychaete annelid (Chapter 12). Since the mid-1900s, elaborate functional and developmental studies have produced increasing evidence that arthropods are polyphyletic. This evidence is thought to indicate that multiple annelid or annelid-like lineages, in response to similar selection pressures, independently evolved a sclerotized cuticle (exoskeleton), hemocoel, and related features. Therefore, some workers have suggested that the term "arthropod" is best used to define a grade of functional organization rather than a phylum. Others, however, still defend strongly the concept of a monophyletic origin for the Arthropoda. How many times arthropod characters have evolved is unknown.

If lack of specialization of segments is accepted as a

primitive character, arthropods can be divided into at least two groups whose early members differed fundamentally in structure of appendages and feeding methods. One group, including the Trilobitomorpha, Chelicerata, and Crustacea, is given the informal name "Biramia." Its primitive members were aquatic and possessed biramous appendages. Small food particles were collected by a long series of similar trunk limbs that passed the particles forward to the mouth along a ventromedial channel (Figure 13.5, C). Gnathobasic mandibles were later developed to aid in food processing, and most digestion probably took place in paired blind ducts branching from the gut. Such blind ducts are only rarely preserved in fossils, but are found in many living biramians. The other group, including the Onychophora, Myriapoda, and Hexapoda, has been given the informal name "Uniramia." Its primitive members lived mostly on land and possessed uniramous appendages. Food was gathered by oral limbs and was passed from below upward to the mouth (Figure 13.5, D). Whole-limb mandibles developed, and digestion may have been restricted to the gut as in all present-day uniramians.

The oldest fossils assigned to the Arthropoda are ovoid to subcircular forms (Figure 13.8) from the upper Proterozoic of Australia and the Soviet Union. Preservation of all specimens is poor, and interpretations of morphology, especially numbers of segments, tagmata, and appendages, as well as taxonomy are open to question. They do not closely resemble typical primitive uniramians or biramians and may represent either already specialized uniramians or biramians, another phyletic group of arthropod grade, or animals of undetermined affinities.

The oldest unequivocal arthropods are biramous trilobites near the base of the Cambrian system. Many trilobites have a relatively thick calcified exoskeleton, resulting in an excellent fossil record. In comparison, primitive uniramians have a thin, unmineralized exoskeleton and a poor fossil record. The oldest known uniramians are onychophorans from the Middle Cambrian. Uniramians and biramians probably evolved from different lineages of marine annelids during the late Proterozoic, but the known fossil record is deficient in details.

Part II   Additional concepts

Main divisions of the Arthropoda and their fossil records

Table 13.1 summarizes a provisional classification of the Arthropoda at higher taxonomic levels. It also includes numbers of genera of fossils taken mostly from the *Treatise on Invertebrate Paleontology* together with observed geologic ranges. For some groups, the number of described genera has significantly increased since publication of the respective *Treatise* volume. *Treatise* coverage of the Hexapoda is still in preparation, and the number of insect genera described from the fossil record is an estimate.

Table 13.1. Classification of Arthropoda

Taxa (number of genera)	Observed geologic ranges
Superclass Trilobitomorpha	Proterozoic(?); Cambrian–Permian
Class Trilobita (1401)	Cambrian–Permian
Class(es) uncertain (16)	Proterozoic(?); Cambrian–Devonian
Superclass Crustacea	Proterozoic(?); Cambrian–Holocene
Class Branchiopoda (119)	Proterozoic(?); Cambrian–Holocene
Class Malacostraca (586)	Cambrian–Holocene
Class Ostracoda (1900)	Cambrian–Holocene
Class Cirripedia (107)	Silurian–Holocene
Class Euthycarcinoidea (2)	Triassic
Class Copepoda (2)	Miocene–Holocene
Class Cephalocarida (1)	Cambrian–Holocene
Class Mystacocarida	Holocene
Class Branchiura	Holocene
Superclass Chelicerata	Cambrian–Holocene
Class Merostomata (89)	Cambrian–Holocene
Class Arachnida (289)	Silurian–Holocene
Superclass Pycnogonida	Devonian–Holocene
Class Pantopoda (1)	Devonian–Holocene
Superclass Onychophora	Cambrian–Holocene
Class(es) uncertain (2)	Cambrian–Holocene
Superclass Myriapoda	Silurian–Holocene
Class Archipolypoda (8)	Silurian–Pennsylvanian
Class Arthropleurida (1)	Pennsylvanian
Class Diplopoda (23)	Pennsylvanian–Holocene
Class Chilopoda (5)	Cretaceous–Holocene
Class Symphyla (1)	Oligocene–Holocene
Class Pauropoda	Holocene
Superclass Hexapoda	Devonian–Holocene
Class Collembola (25)	Devonian–Holocene
Class Insecta (±5000)	Pennsylvanian–Holocene
Class Protura	Holocene
Class Diplura (2)	Tertiary–Holocene
Superclass Pentastomida	Holocene
Class Linguatulida	Holocene
Superclass Tardigrada	Holocene
Class Eutardigrada	Holocene
Class Heterotardigrada	Holocene

Some classes of arthropods have excellent fossil records; some have none. The Trilobita and Ostracoda are represented by abundant fossils assigned to hundreds of genera. In comparison, ten classes are represented by fewer than ten genera each, and another eight classes lack known fossils. In the remainder of the chapter, the amount of information on each class is roughly proportional to the importance of its fossil record. Parts O to R of the *Treatise on Invertebrate Paleontology* are recommended for additional detail.

Superclass Trilobitomorpha

Richard A. Robison

Unassigned trilobitomorphs

More than a dozen unusual genera from Cambrian and Devonian rocks were formerly assigned to the class Trilobitoidea. Four other questionably related genera have been described from the upper Proterozoic. Most representatives are more or less trilobed but differ from trilobites in the structures of tagmata and appendages. At the posterior end, for example, some have a spikelike telson, whereas others are fanlike. Recent detailed studies of most of these genera have produced significant new information, particularly on appendages. Phyletic relationships appear to be diverse, and some are uncertain. Continued recognition of the class Trilobitoidea does not seem to be warranted. Thus, the superclass Trilobitomorpha includes only one widely accepted class, the Trilobita. For this reason the definitive character states for the superclass Trilobitomorpha and class Trilobita are nearly identical.

CLASS TRILOBITA (Figures 13.1, G; 13.3, A; 13.5, C)

Dorsal surface characteristically divided lengthwise into an axial and two pleural lobes (basis for class name). Tagmata normally including a cephalon, thorax, and pygidium. Appendages usually consisting of a single pair of preoral uniramous antennae and a long postoral series of similar, paired, biramous limbs. Adult length ranging from about 0.1 to 70 cm and averaging near 5 cm. (Lower Cambrian to Upper Permian.)

Wherever marine Cambrian rocks are found, trilobites are likely to be the most abundant fossils. Rocks of no other geologic system are so typified by a single invertebrate group, and for that reason, the Cambrian Period has sometimes been referred to as the "age of trilobites." Some stratigraphers have used the first appearance of trilobites to define the base of the Cambrian System, but formal international agreement on the position of that boundary is yet to be achieved.

Rapid adaptive radiation culminated in maximum trilobite diversity during the Late Cambrian. Diversity then declined sharply during the Ordovician, followed by more gradual decline until extinction during the Late Permian. Trilobites are among the most useful guide fossils in lower Paleozoic strata, and traditionally they have been the major basis for zonation and correlation of Cambrian rocks. With decreasing emphasis, they have been used in later systems up to and including the Devonian. Extinction of the trilobites is one of the biologic events that has traditionally been used to define the boundary between the Paleozoic and Mesozoic eras.

Complete trilobites may be found in quiet-water deposits, but more commonly, currents, scavengers, and molting have separated and scattered the skeletal parts. This scattering poses a special problem for taxonomists in reconstructing original body form, especially where two or more closely similar species are associated in the same collection.

An interest in trilobites has not been confined to paleontologists and stratigraphers. These fossils have aroused a sense of wonder and curiosity dating back at least 15,000 years to Paleolithic inhabitants of a rock shelter at Arcy-sur-Cure, France, where a layer with human artifacts has yielded a Silurian trilobite perforated for suspension, perhaps to adorn a necklace. It is noteworthy that the particular species of trilobite is unknown in the rocks of France and was probably imported. A Cambrian trilobite preserved in chert that was chipped to form an implement by an Australian aboriginal has been found far from the locality of its origin. Ute Indians of the western United States collected representatives of the species illustrated in Figure 13.11, *D*, which they fashioned into amulets—the trilobites being recognized by the Ute name *timpe khanitza pachavee* meaning "little water bug like stone house in." Trilobites have often been collected as curios in Bolivia, China, and England, sometimes being mounted as focal ornaments in elaborate pieces of jewelry. Over the years, interest in trilobites has increased, and today amateur as well as professional collectors continue eager searches for these ancient arthropods at many sites around the world.

Morphology

External features The dorsal cuticle is more heavily calcified in trilobites than in other arthropods except ostracodes and some malacostracans (such as crabs and lobsters). In most trilobites that have been analyzed, the cuticle consists of a principal layer composed primarily of calcite. Fine lamellae, formed by the distribution of

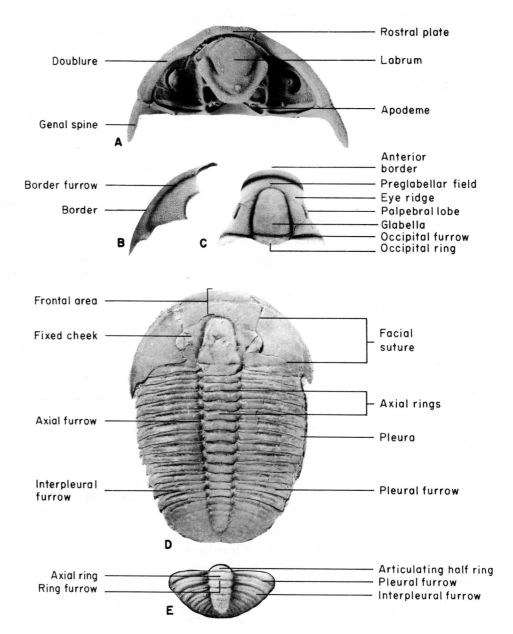

Figure 13.11. Morphologic features of trilobites. **A,** Ventral view of the cephalon of *Ceraurinella* from the Middle Ordovician of Northwest Territories, Canada. Dorsal views of disarticulated free cheek (**B**), cranidium (**C**), and pygidium (**E**) of *Elrathia* from the Middle Cambrian of Utah. **D,** Dorsal view of a complete specimen of another species of *Elrathia* from the Middle Cambrian of Utah. (**A,** photo by R. Ludvigsen.)

minor organic material, have been observed in the principal layer of some specimens. Rarely, an outer, much thinner layer of apatite is present. Microstructure suggests that these two layers correspond to the exocuticle of extant arthropods. Presumably, other primary layers of the cuticle have not been preserved. Cuticular calcification is clearly a major factor in the common preservation of trilobites.

Features of the trilobite cephalon (Figure 13.3, *A*) evolved rapidly, and variations in those features are therefore of much value in taxonomy. A transverse **occipital furrow** usually divides the cephalic axis into a longer anterior **glabella** and a shorter posterior **occipital ring** (Figure 13.11, *C*). Lateral glabellar furrows may partially subdivide the glabella and correspond to internal segmentation and apodemes for insertion of muscles. Pleural regions of the cephalon usually bear compound eyes situated along the outer margins of crescentic raised areas called **palpebral lobes.** Narrow **eye ridges** commonly extend from the anterior ends of the palpebral lobes to the axial furrow near the anterolateral corners of the glabella. To aid the animal in freeing its confined body during molting, the cephalon parted along symmetrical lines of weakness called **facial sutures** (Figure 13.11, *D*), which may be dorsal, marginal, ventral, or of mixed position. The large central portion of the cephalon between dorsal facial sutures is called a **cranidium** (Figure 13.11, *C*). A flangelike border is normally bounded at its inner edge by a **border furrow** (Figure 13.11, *B*) and continues onto the ventral surface as a rimlike **doublure** (Figure 13.11, *A*) of variable width. Together, the rounded border and doublure confer structural strength to the margin of the cephalon. If dorsal facial sutures are absent, areas at the sides of the glabella are called cheeks or genae. If dorsal facial sutures are present, the region between the glabella and each suture is a **fixed cheek** (Figure 13.11, *D*), and the area outside the suture, which fell away during molting, is a **free cheek** (Figure 13.11, *B*). The entire surface between the anterior end of the glabella and the anterior cephalic margin is the **frontal area** (Figure 13.11, *D*) and is subdivided by the border furrow into an **anterior border** and a **preglabellar field** (Figure 13.11, *C*). Cephalic spines may occur at many positions but are most common at the posterolateral corners of the cephalon, where they form **genal spines** (Figure 13.11, *A*), and on the occipital ring. In the absence of genal spines, the posterolateral corners of the cephalon are called **genal angles**.

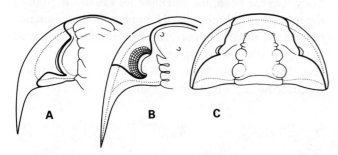

Figure 13.12. Types of facial sutures in trilobites. **A,** Opisthoparian. **B,** Proparian. **C,** Gonatoparian. (From Harrington, H. J. In: Moore, R. C., editor. *Treatise on invertebrate paleontology, Part O.* New York, and Lawrence, KS: Geological Society of America and University of Kansas Press; 1959.)

Where facial sutures are present on the dorsal surface, they normally lie between the eye and the palpebral lobe. A suture is defined as **opisthoparian, proparian,** or **gonatoparian** (Figure 13.12) according to whether the suture line from the posterior end of the eye intersects the posterior margin, lateral margin, or genal angle of the cephalon. In some early classifications of trilobites, great taxonomic significance was attached to the course of the facial sutures; however, it has subsequently been shown that the position of these sutures may change during ontogeny (Figure 13.9, *B* to *F*) and that similar adult suture types are polyphyletic in origin. Therefore, although terms for the suture types are useful descriptive adjectives, suture types alone may have little taxonomic value.

On the ventral cephalic surface, a labrum (Figure 13.11, *A*) of a variable size and shape covers the mouth area. A **rostral plate** may substitute for part of the anterior doublure and in some families adjoins or is fused to the labrum. The remainder of the ventral surface was presumably covered by uncalcified cuticle from which the appendages extended.

The thorax (Figure 13.3, *A*), between the cephalon and pygidium, is covered by freely articulating sclerites, which are usually similar in shape but generally decrease in size from anterior to posterior. One unusual trilobite genus (*Naraoia*) lacks a thorax, but otherwise the observed number of adult thoracic segments ranges from two to 61. The number of adult thoracic segments is fixed in some genera, usually where the total is less than about ten, but the number is variable in many other genera and even within some species.

Axial furrows divide each thoracic segment into a medial **axial ring** and two lateral **pleurae** (Figure 13.11, *D*) Along the anterior edge of the axial ring is a crescentic **articulating half-ring** that extends beneath the folded posterior margin of the next preceding sclerite, which together function as an overlapping hinge to allow longitudinal movement. Pleurae range from nearly flat to moderately convex, most are corrugated by an oblique **pleural furrow** that strengthens the straplike part against bending, and the distal ends may be rounded, truncated, or extended into spines of variable length.

Articulation along the transverse boundaries of the thoracic segments enabled most trilobites to enroll (Figure 13.23), presumably for protection of the appendages and other parts of the ventral surface that lacked calcified cuticle. Early trilobites show no more than minor overlap of contiguous pleurae, but later ones developed a variety of pleural mechanisms to aid articulation or to prevent overgliding of the free pleural tips during enrollment.

The tail region, or pygidium, varies greatly in size, shape, segmentation, and depth or effacement of furrows (compare specimens in Figures 13.20 and 13.24). If the pygidium is smaller, larger, or subequal in size as compared to the cephalon, a trilobite is said to be **micropygous, macropygous,** or **isopygous,** respectively.

In outline, the pygidium is commonly semi-elliptical or semicircular, but exceptions are numerous and some are complex in form. Convexity is highly variable, and the margin may be smooth or spinose. Segments number from one to more than 30 and, like those of the cephalon, are fused to form a rigid shield.

Important features of the pygidium (Figure 13.11, *E*) include the axis, which may extend to the posterior margin or terminate somewhere short of that point. It is divided into rings by **ring furrows**, and an anterior articulating half-ring provides connection with the posterior thoracic segment. Boundaries of fused segments may be indicated on the pleural regions by **interpleural furrows** which, like the pleural furrows and the border furrows, are variable in depth. In certain trilobites, especially some late Paleozoic genera that are macropygous, a disparity exists between the numbers of segments indicated on the axis versus those on the pleural regions (Figure 13.23, *D*).

The external surface of most trilobites is marked by sculpture related to various sensory and organ systems. These features commonly include pores and tubercles of one or more sizes (Figure 13.13, *A*). Most pores were probably occupied by sensory setae, but some led to sensory ducts. Various fine ridges may also be present. Minute, asymmetrical, scarplike **terrace ridges** (Figure 13.13, *B*), generally confined to the doublure and lateral body margins, are thought to have aided burrowing by increasing friction between the cuticle and substrate. Other fine ridges, which are symmetrical in transverse section, form an anastomosing network on genal areas of some cephala (Figure 13.13, *A*) and probably reflect part of the underlying circulatory system. Gaseous exchange across these same areas may have provided auxiliary respiration. On other trilobites, patterns of coarser radial ridges have been interpreted as external reflections of blind ducts branching from the gut (Figure 13.13, *C*). Paired, smooth, and usually somewhat depressed areas are **muscle scars** of various shapes, sizes, and positions, which overlie sites where muscles were inserted on the internal surface (Figure 13.13, *D*). These features commonly vary among species and hence are useful in taxonomy. Distinctive sculpture may be especially useful in the identification of disarticulated sclerites where several species are present in a single collection.

Appendages are fairly well known for only four genera of trilobites: two from the Cambrian (*Naraoia, Olenoides*), one from the Ordovician (*Triarthrus*), and one from the Devonian (*Phacops*). Less information is available for appendages of a few other genera. Although limited, this sample indicates uniformity (Figure 13.14) that differentiates the trilobites from all other major groups of arthropods. Also, as might be expected, it demonstrates minor variation in structure from genus to genus.

Trilobites possessed paired limbs on most, if not all, body segments (Figure 13.5, *C*). A single anterior pair of long, jointed, uniramous limbs probably functioned as sensory antennae like those found in most other arthropods. Typically, the remaining limbs are biramous and, except for a gradual anterior to posterior decrease in size, are closely similar to one another along the length of the body. On the head, the antennae are situated in front of the mouth; and three pairs of biramous limbs are situated behind the mouth. In one genus (*Olenoides*), the paired hindmost limbs are long and uniramous like the antennae. The biramous limbs each possess a relatively large but narrow and simple coxa with a spiny gnathobase on

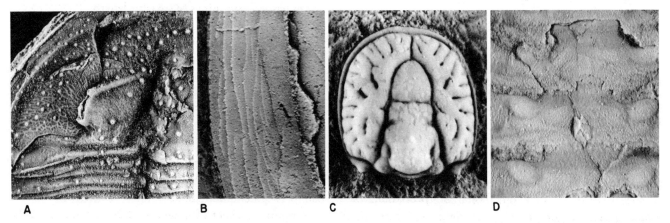

A **B** **C** **D**

Figure 13.13. Sculpture of the trilobite exoskeleton. **A**, Eye region of *Alokistocare* showing pore-bearing tubercles that were probably occupied by setae and weak anastomosing ridges that reflect parts of the underlying circulatory system. **B**, Terrace ridges on the doublure of *Asaphiscus*. **C**, Radial ridges on the cheek regions of *Ptychagnostus* that indicate the courses of underlying intestinal branches. **D**, Muscle scars on the axis of *Zacanthoides*.

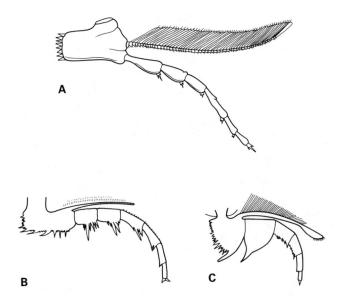

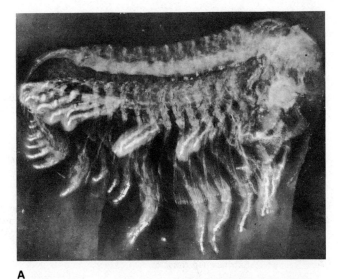

A

Figure 13.14. Biramous limbs of trilobites. **A**, Reconstruction of fifth thoracic limb of *Triarthrus*. As in many extant arthropods, grooves on the coxa are probably for strengthening rather than an indication of internal divisions. **B**, Reconstruction of first thoracic limb of *Olenoides*. **C**, Reconstruction of the second pygidial limb of *Naraoia*, which is the only trilobite known to lack a thorax. Because of its unusually large gnathobases, *Naraoia* has been interpreted as a predator and scavenger. (**A** modified from Cisne, J. L. *Fossils and Strata* 4; 1975. **B** modified from Whitington, H. B. *Fossils and Strata* 4; 1975. **C** modified from Whittington, H. B. *Philosophical Transactions of the Royal Society of London*: 280: 409–443; 1977.)

its medial surface. The endopodite, which is attached ventrolaterally to the coxa, is composed of either five or six podomeres and apparently had a dual walking and feeding function. At the end of the endopodite are variable numbers of short claws and setae that sometimes left distinctive traces (Figure 21.4). The expodite, which is attached dorsolaterally to the coxa, consists of a slender shaft bearing numerous, flat, respiratory filaments.

Internal anatomy Some internal features of trilobites are known from rare specimens preserved in dark shale where some soft parts with an originally high sulfur content have been replaced by fine-grained pyrite. Differences in absorption coefficients of the pyrite and shale enable photography of the specimens by exposure to x rays (Figure 13.15). Features mainly of the digestive system and parts of the muscle system have been revealed by this technique.

Based on radiographs of Ordovician specimens, a reconstruction of the trilobite digestive system (Figure 13.16, C) shows a reclining J-shaped gut commencing at a posteriorly directed mouth, which is situated dorsal of

B

Figure 13.15. Trilobites photographed with x rays. **A**, *Phacops* from the Devonian of Germany showing features of the appendages, compound eye, and intestinal tract. (×2.) **B**, *Triarthrus* from the Ordovician of New York showing features of the appendages. (×11.) (Photos by Wilhelm Stürmer.)

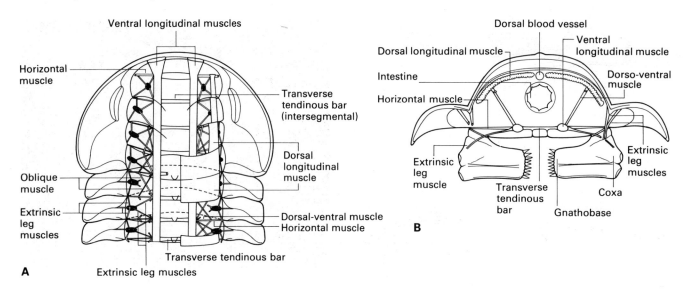

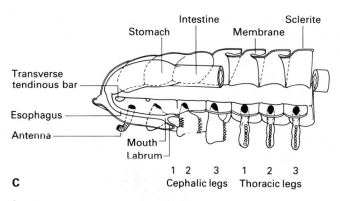

Figure 13.16. Reconstructed internal anatomy of *Triarthrus* from the Ordovician of New York. **A**, Dorsal view of head and first three thoracic segments showing some musculature. **B**, Transverse section of the fifth thoracic segment. **C**, Interior view of the head and first three thoracic segments showing the digestive tract and related musculature as well as the coxae of the right postoral limbs. (Modied from Cisne, J. L. *Fossils and Strata* 4: 45–63; 1975.)

the labrum. A narrow esophagus loops around the anterior-most transverse tendinous bar and connects above to a large stomach that lies beneath the glabella. The stomach passes posteriorly into a long narrower intestine that extends to an anus on the last segment of the pygidium. From comparison with living arthropods, certain cephalic muscle scars are thought to be related to suspension of the anterior digestive tract and movement of the labrum. For other trilobites, external patterns of ridges and grooves (Figure 13.13, *A* and *C*) indicate the presence of intestinal branches and parts of the circulatory system beneath cheek areas of the cephalon.

A reconstruction of basic trilobite musculature (Figure 13.16, *A* and *B*) most closely resembles musculature of primitive crustaceans such as cephalocarids. Paired sheets of parallel fibers along the upper body wall formed large **dorsal longitudinal muscles**. Two parallel bundles of fibers along the lower body wall comprised the **ventral longitudinal muscles**, which occurred in segmental blocks inserted end-to-end on a ladderlike series of transverse tendinous bars. Ends of the tendinous bars were anchored to the ventral exoskeleton by smaller horizontal muscles. Except in the first two limb-bearing segments of the cephalon, the large longitudinal muscles were linked by sets of smaller dorsal-ventral muscles in a box-truss pattern like that of primitive crustaceans. The association of unspecialized muscles in head segments with specialized muscles in the trunk has been cited as a primitive arthropod condition. Moreover, the musculature differed fundamentally from that of primitive uniramians by the absence of circular muscles and lateral longitudinal muscles.

Eyes   Trilobite eyes are of special interest because they are the most ancient visual system known, supplying some of the best direct evidence of eye evolution.

Scientists have suggested that compound eyes in most major groups of arthropods were independently derived from aggregates of simple eyes; however, early Cambrian trilobites had fully developed compound eyes, and they fail to provide evidence concerning the origin of this sensory system in arthropods.

Great variation in shape, structure, and relative size of eyes suggests differential use of vision among trilobites.

The eyes are usually crescentic in outline (Figure 13.11), but some are globose (Figure 13.17, *A*), conical, stalked, or anteriorly fused into a single band. Within some groups, the eyes progressively migrated laterally and then disappeared from the cephalon. Presumably the eyes were lost, but in some species the eyes may have only shifted to the ventral surface, as in some living arthropods.

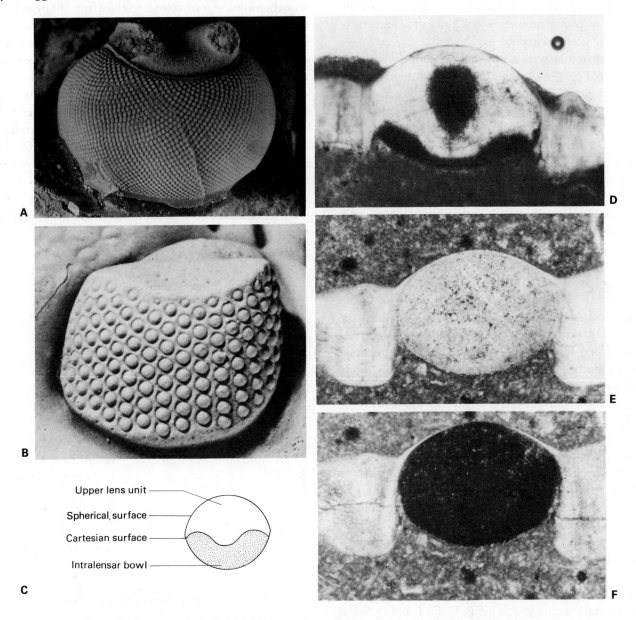

Figure 13.17. Trilobite eyes and their structure. **A,** Holochroal type with closely packed lenses; *Scutellum* from the Devonian of Bohemia. (×17.) **B,** Schizochroal type with separated lenses; *Phacops* from the Devonian of Ohio. (×7.) **C,** Reconstructed section of doublet lens in *Crozonaspis*; compare with lens in **D**. **D** to **F,** Thin sections of two schizochroal calcitic lenses from the same eye of *Dalmanites*. (×54.) **D,** one lens (in plane-polarized light) interpreted by some investigators to show a primary doublet structure and by others to show effects of secondary diagenetic alteration; **E** (in plane-polarized light) and **F** (in cross-polarized light), another representative lens for comparison with **C**. (**A** from Levi-Setti, R. *Trilobites*. Chicago: University of Chicago Press; 1975. **B,** specimen 211332, Smithsonian Institution. **C** modified from Clarkson, E. N. K.; Levi-Setti, R. *Nature* 234: 663–667; 1975. **D** to **F,** photos by K. M. Towe.)

Labels in figure C:
Upper lens unit
Spherical surface
Cartesian surface
Intralensar bowl

Preservation of eyes is unusual in the fossil record but is fairly common in trilobites due primarily to the composition of the lenses, each of which consists of a single crystal of calcite oriented with its principal optic axis (*c* axis) normal to the visual surface (Figure 13.17, *E* and *F*). Such precise crystal orientation is functionally desirable not only because it eliminates polarized rays but also because it is the only orientation of calcite that does not result in a double image. Similar structure and orientation are unknown among the few kinds of living arthropods with calcified lenses.

The earliest as well as most subsequent trilobites had **holochroal** eyes, each characterized by close packing of biconvex lenses beneath a single cornea (Figure 13.17, *A*). Generally, the lenses are hexagonal in outline, but some are either rhomboidal or quadrate, and they range in number from one to more than 15,000. This type of eye is poorly known in Cambrian trilobites because in most adults it was encircled by a functional suture, and upon molting or death, the lens-bearing unit normally dropped out and was lost. In some groups, beginning in the Late Cambrian, the suture fused between the holochroal eye and the free cheek, resulting in more common preservation of the adult visual surface.

An aggregated type, or **schizochroal eye**, is confined to certain trilobites (phacopid and dalmanitid) of Ordovician to Devonian age (Figure 13.17, *B*). Schizochroal eyes are characterized by a few to more than 700 relatively large thick lenses, each covered by a separate cornea. Each lens is positioned in a cylindrical mounting and is separated from its neighbors by material similar in composition and structure to that of the rest of the cuticle.

From the section of a single lens (Figure 13.17, *D*) and from indirect evidence provided by internal molds, each schizochroal lens has been interpreted to consist of a doublet structure (Figure 13.17, *C*). Here the surface of separation between the two calcitic parts approximates optically correcting 17th century designs by Descartes and Huygens. Such lens designs minimize spherical aberration and improve light collection in dim environments. Corroborative evidence for such a doublet structure is weak, however, and has been contradicted by further study of other lenses (Fig. 13.17, *E* and *F*). Some schizochroal lenses may have had this elegant doublet design, but available evidence is subject to different interpretation.

Ontogeny   Most trilobites show three fairly natural divisions in their developmental histories. The earliest, or **protaspid period**, includes juvenile instars characterized by an undivided dorsal exoskeleton (Figure 13.9, *B* and *C*). Protaspides are generally subcircular in outline, moderately convex, and less than 1 mm in length. Typically, the protaspid glabella is widest at its anterior end where it extends to and merges with the anterior border. An intermediate **meraspid period** began with separation of the dorsal exoskeleton into an articulated cephalon and pygidium (Figure 13.9, *D*). During subsequent molts, segments were released from the pygidium to form the thorax (Figure 13.9, *E*). Normally, one new segment was released during each meraspid molt, but exceptions are known. The adult, or **holaspid period**, includes those instars characterized by a full complement of thoracic segments (Figure 13.9, *F*). During the holaspid period, additional segments were sometimes added to the pygidium, and as much as a fortyfold increase in total exoskeletal length has been reported. In some trilobites, segments identified by distinctive spines show a progressive forward displacement with each instar and demonstrate that, as in annelids, new segments were generated at the anterior margin of the hindmost pygidial segment.

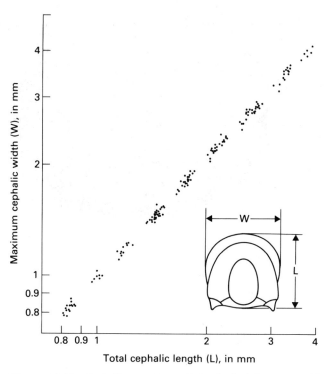

Figure 13.18. Cephalic ontogeny of the Ordovician agnostoid *Trinodus* shown by plotting measurements of length and width. The clusters represent instars. (Modified from Hunt, A. S. *Journal of Paleontology* 41: 203–208; 1967.)

When measurements of two morphologic dimensions are plotted against one another, the points may fall into clusters that represent instars (Figure 13.18). More

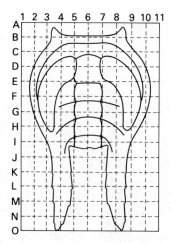

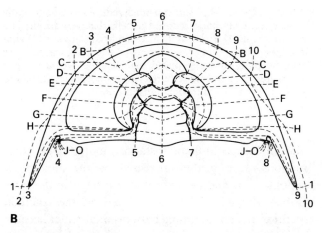

Figure 13.19. Ontogenetic change in the cephalon of the Early Cambrian trilobite *Olenellus* illustrated by coordinate transformation. **A** is an early meraspid instar and **B** is an intermediate holaspid instar. Intersections of the vertical and horizontal lines are at corresponding morphologic positions on each diagram. (From Palmer, A. R. *Journal of Paleontology* 31: 105–128; 1957.)

commonly, the points do not cluster but are scattered along a line, which indicates that size ranges of instars overlap.

Among trilobites, other ontogenetic modification varies from group to group, and a few examples have been illustrated by coordinate transformation. With this technique, the outlines of two different trilobite instars are reproduced, and a grid with equidistant coordinates is inscribed on the younger instar (Figure 13.19, *A*). Another grid is inscribed on the older instar (Figure 13.19, *B*), but the lines are drawn so that each coordinate passes through morphologic points corresponding to those on the younger instar. Details of transformation in morphology then become more evident. For example, comparison of the two instars in Figure 13.19 shows a

disproportionate increase in surface area of the cheeks, the disappearance of a pair of prominent posterior juvenile spines, and generation of new genal spines. An explanation for the relative increase in surface area is partly derived from elementary geometry, where it is known that the surface area of an object increases as the square, and volume as the cube, as linear dimensions increase. It follows that as an animal grows, a disproportionate increase in surface area is necessary to maintain a constant area-to-volume ratio. This rule of scaling inequality has been referred to as the **principle of similitude**. As already noted, trilobite cheeks were underlain by extensions of the digestive and circulatory systems, which both have surface-dependent functions. Therefore, as the trilobite grew, a disproportionate increase in surface area of the cheeks and their associated organ systems was necessary to avoid impairment of vital functions. Modifications of the spines (Figure 13.19), some of which were observed in intermediate instars, were probably related to changing hydrodynamic factors during the transition from a pelagic protaspid to a benthic holaspid mode of life.

During ontogeny, many trilobites changed from a proparian to a gonatoparian or opisthoparian type of facial suture. For example, the early protaspid in Figure 13.9, *B* is proparian. On the posterior fixed cheeks it possessed a single pair of short spines that disappeared before the late protaspid instar (Figure 13.9, *C*). By increase in length of the free cheeks, the early meraspid instar developed a gonatoparian suture (Figure 13.9, *D*). By subsequent addition and gradual enlargement of a new pair of spines on the posterior free cheeks, an opisthoparian suture was developed by the middle of the meraspid period. Thus, the type of holaspid facial suture is determined by ontogenetic changes in proportions of parts on either side of the suture.

Timing of the appearances of morphologic characters and their rates of development varied during trilobite ontogeny. Two trends in the displacement of developmental timing were particularly common. One was arrested development or **paedomorphosis**, which resulted in failure to develop some characters of the ancestor. This could have been achieved by early sexual maturity (progenesis) or delayed somatic development of certain parts or organs (neoteny). Separation of evidence of these processes may be difficult in the fossil record; however, progenesis tends to produce forms that are smaller than their immediate ancestors, whereas neoteny tends to produce forms as large or larger than their immediate ancestors. With little genetic change, paedomorphosis allowed rapid attainment of significantly

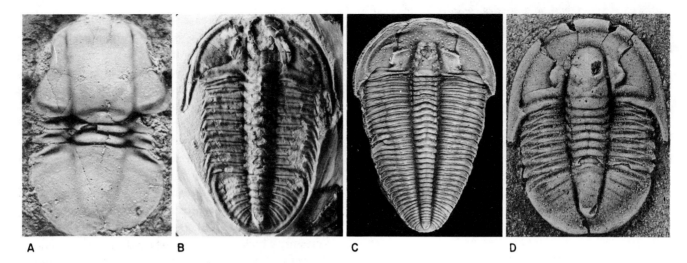

A   B   C   D

Figure 13.20. Trilobite holaspid forms resulting from differences in developmental timing of morphologic characters. **A** and **B**, *Thoracocare* (Middle Cambrian, Idaho, ×15) and *Bathyuriscus* (Middle Cambrian, Montana, ×1.3) show paedomorphic retention of a large pygidium, few thoracic segments, and a glabella that extends to the anterior margin. The exceptionally wide anterior axial lobe of *Thoracocare* was acquired by peramorphosis. **C**, *Alokistocare* (Middle Cambrian, Utah, ×1.3) shows a much reduced pygidium, many thoracic segments, and a reduced glabella that were acquired by peramorphosis. **D**, *Cedaria* (Upper Cambrian, Utah, ×12) with large pygidium and few thoracic segments, which are paedomorphic, and a reduced glabella and well-developed frontal area, which were acquired by peramorphosis. (D, photo by M. H. Lawson, Smithsonian Institution.)

different morphology, which may have further allowed escape from adult specializations or enabled the retention of juvenile capabilities (such as pelagic lifestyle, feeding techniques). Paedomorphosis usually gave rise to new trilobites characterized by a parallel-sided or anteriorly expanded glabella that extends to the anterior margin of the cephalon, few thoracic segments, and a relatively large pygidium (Figures 13.9, *E* and 13.20, *A* and *B*). In some trilobites it resulted in the holaspid retention of proparian facial sutures (Figure 13.12, *B*). Evidence also suggests that the schizochroal type of trilobite eye evolved by paedomorphosis from a holochroal eye.

An opposite trend was toward prolonged development or **peramorphosis**, which resulted in the progressive addition of characters to the last stages of ontogeny. This could have been achieved by delayed sexual maturity (hypermorphosis) or early somatic development (acceleration). Again, separation of evidence of these processes may be difficult in the fossil record. Peramorphosis contributes new adult capabilities, and trilobites that arose in this manner are typically characterized by a reduced and tapered glabella separated from the anterior margin of the cephalon by a well-developed preglabellar field, opisthoparian facial sutures, numerous thoracic segments, and a relatively small pygidium (Figure 13.20, *C*).

Different combinations in developmental timing seem to have been important in the generation of many novel holaspid forms. In some instances, morphologic combinations were further increased by the mixture of paedomorphic and peramorphic characters in the same trilobite (Figure 13.20, *D*). Moreover, the common repetition of certain timing patterns in different lineages resulted in numerous homeomorphs that have contributed to difficulties in classification. For example, in the *Treatise* classification of trilobites the order Corynexochida includes a polyphyletic assortment of Cambrian paedomorphs, whereas the order Phacopida includes many post-Cambrian paedomorphs.

Classification

During more than a century and a half, several classifications of trilobites have been proposed. Most species and genera can readily be grouped into families, but no scheme for grouping families into higher categories has received wide acceptance. At least two factors seem to have significantly contributed to this problem of higher classification. One is that undue importance has been placed on certain morphologic features. The other is a deficiency of knowledge on the origins and relationships of trilobite families. Both factors deserve brief review.

Most higher classifications of trilobites have been based on one or a few essential characters. Among those most commonly used are number of segments, type of facial suture, relative size of pygidium, and structure of the pleural lobes (especially whether the pleural lobes are expanded and merged anteriorly to form a preglabellar field). The preceding discussion of ontogeny showed that all these features in holaspid trilobites were strongly controlled by displacements in developmental timing. Moreover, repeated changes in timing appear to have resulted in numerous homeomorphs. The presence or absence of eyes was used in some early classifications, but scientists subsequently recognized that eyes are present in the oldest known trilobites and that secondary loss of eyes occurred independently within several groups. In addition, a character with a stable holaspid expression and corresponding taxonomic utility in one group of trilobites may be variable and have little or no taxonomic value in other groups.

It has gradually become evident, therefore, that phyletic classification of trilobites must be based on a complex of characters, and great care is necessary in their selection. Each character is nondefinitive by itself but acquires importance through its association with others.

The conventional pattern of phylogenetic classification has a treelike form in which species with connected evolutionary pathways are clustered to form genera, which in turn are clustered into families, families into orders, and so on back to the ancestral rootstock. This approach has traditionally been applied to the classification of trilobites, and it has worked reasonably well at lower taxonomic levels. But agreement has not been reached on how families should be arranged into orders.

The other major difficulty with the higher classification of trilobites arises from the fact that most families (140 are recognized in the *Treatise on Invertebrate Paleontology*) are of obscure origin, and most apparently became extinct without giving rise to new families. Many taxonomists seem to have tacitly assumed that large groups of families each shared a common ancestor and that all trilobites can be assigned to a few natural orders.

An alternative favored here is that many or most families arose independently from an unspecialized stock (ptychoparian) with features similar to those in Figure 13.11, *D*. This interpretation provides the options either of recognizing numerous orders, each with one or a few families, or of combining most families into a single large order. A few families are fairly well known, but the general problem of trilobite classification probably will not be satisfactorily resolved until much more information is obtained on origins and relationships for the many other families.

Based on all features, the trilobites are here provisionally divided into two disproportionate orders: the less specialized and more typical Polymerida with about 95 percent of described genera and the more specialized Agnostida with about 5 percent. For some of the better known families, grouping into small superfamilies is useful for expressing inferred relationships.

Some taxonomists prefer to assign a distinctive group of Lower Cambrian trilobites, the olenellids (Figure 13.21, *I*), to another small order. These micropygous forms lack facial sutures and known protaspides, the latter probably because of weak mineralization. Also, compared to typical polymeroids (Figure 13.9, *D*), the early meraspid cephalon of an olenellid (Figure 13.19, *A*) has unusually enlarged palpebral lobes and posterior border spines. Nevertheless, I do not consider these differences to be important enough to warrant discrimination of a separate order.

ORDER POLYMERIDA (Figures 13.20 to 13.23)

Small to large trilobites with variable pygidium. Eyes, facial sutures, and five or more thoracic segments usually present in holaspides. Labrum platelike with relatively small, short, inwardly directed processes. Anterior thoracic segment always with articulating half ring. Pleural regions of pygidium commonly segmented. (Lower Cambrian to Upper Permian.)

Included in the Polymerida is a great variety of trilobites that differ most noticeably in body outline, axial shape and extent, size and shape of pygidium, number of segments, depth of furrows, spinosity, and surface sculpture. Such characters document rapidly changing adaptations and are useful for discrimination of lower taxa. Although rare exceptions are known (such as those in Figures 13.20, *A* and 13.21, *D* and *E*), most polymeroids can be distinguished by the presence of five or more thoracic segments.

The majority of polymeroids were probably benthic deposit feeders, but some developed features indicative of a pelagic existence. Some may have preyed on small, soft-bodied organisms. Most were probably epifaunal but a few, as evidenced by extensive dorsal terrace ridges, may have been infaunal.

ORDER AGNOSTIDA (Figure 13.24)

Small, eyeless, isopygous trilobites that lack facial sutures and possess only two thoracic segments in holaspid stage. Cephalic axis always well separated from anterior border furrow. Labrum saddlelike with large, long, inwardly directed processes.

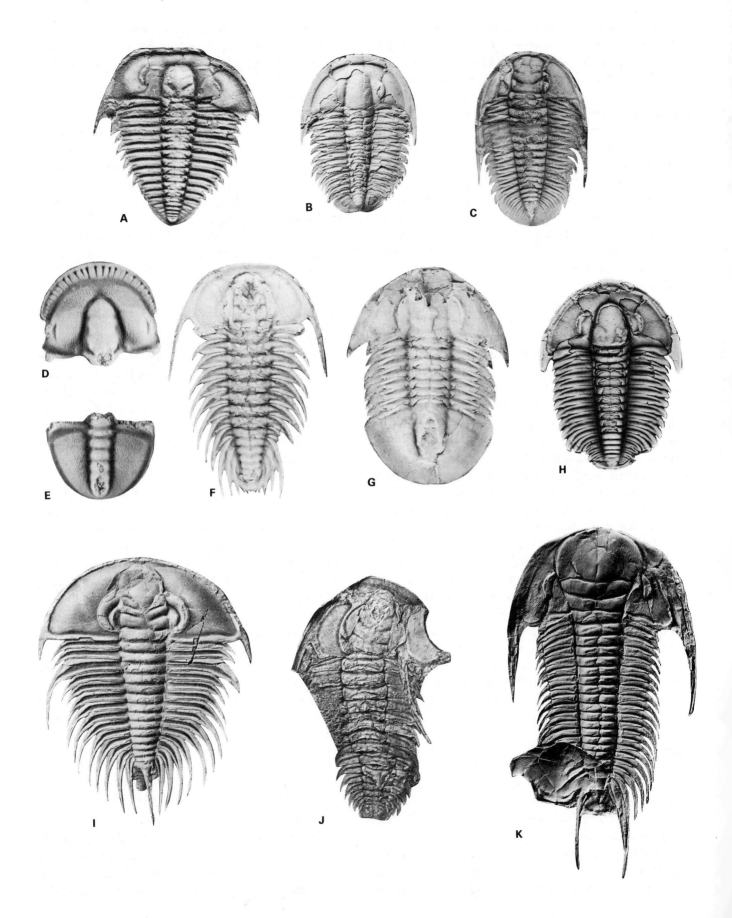

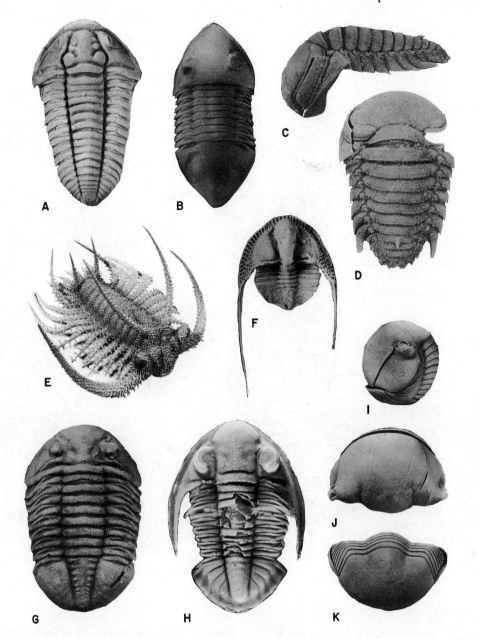

Figure 13.22. Representative Ordovician trilobites of the order Polymerida. **A,** *Flexicalymene* (Upper Ordovician, Ohio, ×1.3). **B,** *Isotelus* (Upper Ordovician, New York, ×0.6). **C** and **D,** Lateral and dorsal views of *Remopleurides* (Middle Ordovician, Virginia, ×2.3). **E,** Model of *Apianurus* showing extreme spinosity (Middle Ordovician, Virginia). **F,** *Cryptolithus* (Middle Ordovician, Pennsylvania, ×1.7).**G,** *Asaphus* (Lower Ordovician, Soviet Union, ×0.5). **H,** *Bathyurus* (Middle Ordovician, New York, ×1.3). **I** to **K,** Enrolled *Illaenus* (Middle Ordovician, Minnesota, ×1.1). (**A, E** and **F,** courtesy Smithsonian Institution. **B, G** and **I** to **K,** photos by M. H. Lawson. **C** and **D,** photos by H. B. Whittington. **H,** photo by R. Ludvigsen.)

Figure 13.21. Representative Cambrian trilobites of the order Polymerida together with *Pagetia* of doubtful assignment. **A,** *Olenus* (Upper Cambrian, Norway, ×4). **B,** *Aphelaspis* (Upper Cambrian, Nevada, ×2.7). **C,** *Saukia* (Upper Cambrian, Wisconsin, ×0.7). **D** and **E,** Disarticulated cephalon and pygidium of *Pagetia* (Middle Cambrian, Idaho, ×10). **F,** *Zacanthoides* (Middle Cambrian, Utah, ×1). **G,** *Glossopleura* (Middle Cambrian, Utah, ×1). **H,** *Modocia* (Middle Cambrian, Utah, ×4.7). **I,** *Olenellus* (Lower Cambrian, Vermont, ×0.2). **J,** *Redlichia* (Lower Cambrian, Australia, ×1.2). **K,** *Paradoxides* (Middle Cambrian, Newfoundland, ×0.4). (**A,** photo by V. Jaanusson. **D** and **E,** photos by D. P. Campbell. **F** and **G** courtesy of L. Gunther. **I** from Walcott, C. D. *Smithsonian Miscellaneous Collections*; 1910. **J,** photo by J. H. Shergold. **K,** photo by R. Levi-Setti.)

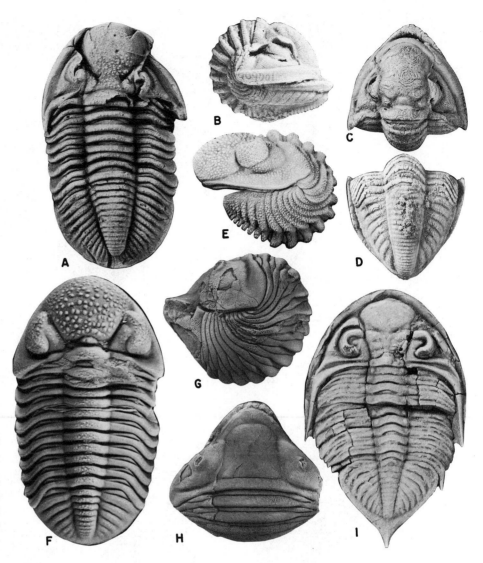

Figure 13.23. Representative Silurian to Permian trilobites of the order Polymerida. **A**, *Griffithides* (Mississippian, Illinois, ×1.6). **B** to **D**, Enrolled *Anisopyge* (Permian, Texas, ×1). **E** and **F**, Two species of *Phacops* (Devonian, Ontario and Ohio, ×1 and ×2.5). **G** and **H**, Enrolled *Dipleura* (Devonian, New York, ×0.5). **I**, *Dalmanites* (Silurian, England, ×1.2). (**A** and **G** to **I**, photos by M. H. Lawson. **B** to **D**, courtesy Smithsonian Institution. **E** and **F**, photos by Niles Eldredge.)

Anterior thoracic segment lacks articulating half ring. Pleural regions of pygidium unsegmented. (Lower Cambrian to Upper Ordovician.)

Included in the Agnostida are trilobites that were seemingly adapted to a specialized pelagic lifestyle. Worldwide distribution of many species in open-ocean deposits is consistent with this interpretation. Some Middle and Upper Cambrian species also are abundant and have short stratigraphic ranges, which give them particular importance for intercontinental correlation.

Several specialized agnostoid features are related to a general habit of spheroidal enrollment (Figure 13.24, *A, J* to *L*). Because of the extreme reduction in number of thoracic segments, this enrollment required a constant isopygous condition during ontogeny. A corresponding lack of change in size ratios of body parts (Figure 13.18), called **isometric growth**, is rare among all types of organisms. During enrollment, flexure of up to 90 degrees between the cephalon and thorax resulted in loss of the articulating half ring and development of an edge-to-edge hinge (Figure 13.24 *B* and *D*). An unusual aperture in the middle of the hinge line (Figure 13.24, *A*) was probably covered with an expansible membrane in the living animal. Possibly also related to agnostoid enrollment was the transformation of the occipital ring into a lateral pair of triangular **basal lobes** (Figure 13.24, *C*), which are found in no other trilobites.

Major differences in structure of the labrum suggest significant divergence in agnostoid and polymeroid nutrition or feeding techniques. Although variable in detail, all polymeroids possess an unperforated labrum

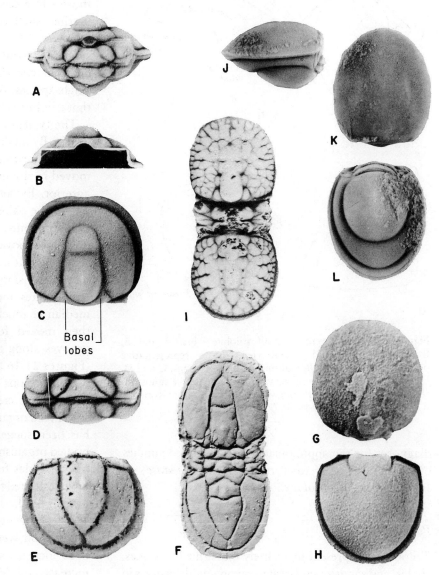

Figure 13.24. Representative trilobites of the order Agnostida. **A** to **E**, *Peronopsis* (Middle Cambrian, Utah): **A**, thoracic view of complete enrolled specimen showing median aperture of unknown function between cephalon and thorax (×8); **B** and **C**, posterior and dorsal views of disarticulated cephalon (×8); **D**, dorsal view of disarticulated thorax (×12); and **E**, dorsal view of disarticulated pygidium (×8). **F**, *Ptychagnostus* (Middle Cambrian, Utah, ×6). **G** and **H**, Disarticulated and exfoliated cephalon and pygidium of *Lejopyge* with axial features mostly effaced (Middle Cambrian, Utah, ×10). **I**, *Glyptagnostus* with strongly developed radial ridges on pleural regions of cephalon and pygidium (compare with Figure 13.13, *C*) (Upper Cambrian, Alabama, ×9). **J** to **L**, Lateral, cephalic, and pygidial views of enrolled *Sphaeragnostus* with axial features effaced only on cephalon (Upper Ordovician, Quebec, ×6.4).

Basal lobes

with a typically large, convex, central part, and all have two pairs of short **anterior** and **posterior wings** projecting inward from a generally narrow border (Figure 13.25, *A* and *B*). In comparison, the few known agnostoid labra typically have a small anterior boss with a broad perforated platform behind, and all have large, long, straplike wings (Figure 13.25, *C* and *D*).

One small superfamily, the Eodiscacea (Figure 13.21, *D* and *E*), is characterized by a mixture of typical polymeroid and agnostoid features. By possessing only two or three thoracic segments, a short tapered glabella, and an isopygous condition, eodiscaceans resemble agnostoids. Eyes, proparian facial sutures, and pleural segmentation on the pygidium may be present or absent. The eodiscacean labrum, occipital ring, and cephalothoracic hinge line are of polymeroid type. Phyletic

affinities and taxonomic assignment of the eodiscaceans have been subjects of disagreement between specialists. Preference here is to include them in the Polymerida. The agnostoids may have arisen from the polymeroids via the eodiscaceans, but suitable connecting links are unknown. In both the eodiscaceans and the agnostoids the small size, few thoracic segments, and isopygous condition are paedomorphic features probably related to adult retention of a pelagic larval lifestyle. A short tapered glabella is a peramorphic feature of unknown value.

Characters of the axial region, especially shape of the glabella together with the development and direction of its lateral furrows, are important in the definition of many trilobite families. Additional representatives of some of the more common or distinctive families are shown in Figures 13.21 to 13.24. (For information and

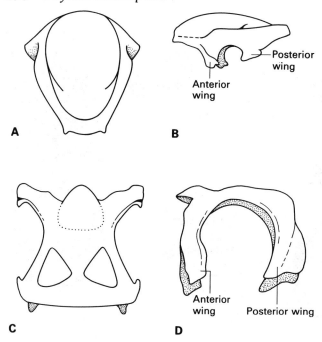

Figure 13.25. Different types of trilobite labra. **A** and **B**, Exterior ventral and lateral views of a representative polymeroid (*Ceraurinella*) labrum with short wings. **C** and **D**, Exterior ventral and lateral views of a representative agnostoid (*Peronopsis*) labrum with long wings. (Modified from Robison, R. A. *Lethaia* 5: 239–248; 1972.)

diagnoses for the approximately 140 described families of trilobites, refer to Part O of the *Treatise on Invertebrate Paleontology*.)

Paleoecology

Trilobites are present in a variety of marine limestones, shales, and sandstones. Greatest taxonomic diversity is in those strata representing deposition in shallow, normal marine, shelf environments with open circulation. As many as 29 trilobite species have been observed in a single such stratum. In comparison, some restricted-shelf deposits contain massive accumulations with only a single species. A scarcity in stromatolitic facies suggests that trilobites avoided elevated or fluctuating salinities.

The typical trilobite was probably a benthic detritus feeder. By analogy with such primitive living biramians as cephalocarids, the trilobite limbs are thought to have moved in rhythmic waves that passed from posterior to anterior. Evidence from trace fossils indicates that the tips of the paired endopodites dug into soft organically rich sediments, leaving two parallel series of distinctive oblique grooves arranged in a herringbone pattern (Figure 21.16). Dislodged detritus was probably swept into suspension by a rhythmic broomlike action of the endopodites and with the aid of setae was drawn into a median channel between the gnathobases where it was then moved forward to the mouth. Narrow linear grooves along the sides of many trilobite feeding trails (Figures 21.16 and 21.17) indicate that the edges of the pleural regions formed a seal with the sediment surface and thereby created enclosed chambers to contain the suspended organic detritus beneath the pleural regions. It has been suggested that some trilobites ate small soft-bodied organisms and other types of food; however, lack of mandibles for efficient shredding and chewing must have restricted their predatory activities as well as the size of their food.

Environmental influences, which are still poorly understood, were probably responsible for trends toward both increase and loss of surface sculpture as well as increase and decrease in depth of furrows. The smooth

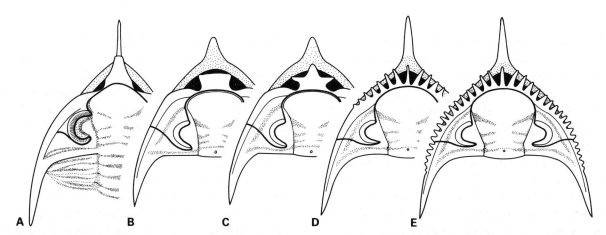

Figure 13.26. Modifications of the cephalic and pygidial margins in Devonian dalmanitid trilobites for possible water filtration. Reconstructions show the cephalon and pygidium juxtaposed as during enrollment. A to C, Three species of *Huntonia*. D and E, Two species of *Phalangocephalus*. (Modified from Campbell, K. S. W. *Oklahoma Geological Survey Bulletin* 123: 227; 1977.)

surfaces of some trilobites (Figure 13.22, *B, I* to *K*) probably minimized frictional drag during burrowing or in shallow water where current velocity or turbulence was moderate to strong. Various kinds of spines developed in many lineages (Figure 13.22, *E*) and may have aided in protection, hydrodynamic stabilization, sensory reception, or other needs. An unusual development of spines on the anterior cephalic margin of some Silurian and Devonian species (Figure 13.26) has been interpreted as an aid in filtering water exchanged through gaps between the juxtaposed cephalon and pygidium when the trilobite was enrolled.

Much evolution in the size, shape, structure, and position of the eyes, together with the number and arrangement of lenses, must have been controlled by changing visual needs. By analogy with present-day marine arthropods, reduction or loss of eyes is most common in species living in dark environments at depths greater than about 600 m, but exceptions are known from shallow water. Therefore, lack of eyes has limited application in paleoecologic interpretations.

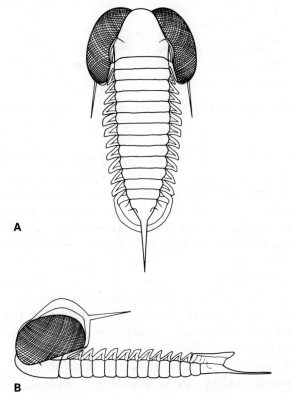

Figure 13.27. The pelagic Ordovician trilobite *Opipeuter*, which is known from Ireland, Spitsbergen, and Utah. (×4.) **A,** Dorsal view. **B,** Lateral view of its inferred swimming position in which the pygidial spine may have functioned as a rudder and the genal spines as stabilizers. As the animal swam on its back, probably the only area not visible to the tumid holochroal eyes was a small region above the ventral surface where the genal spines may also have provided protection. (Modified from Fortey, R. A. *Palaeontology* 17: 117–124; 1974.)

Members of several trilobite lineages show evidence of a range in adaptation from benthic to pelagic modes of life and presumably adaptation to additional feeding habits. One common trend was toward reduction in relative width of the pleural regions (Figure 13.27, *A*), which were no longer needed as chamber covers when benthic deposit feeding was abandoned for pelagic life. Also, large gaps commonly developed between the pleural tips of thoracic segments, which further indicate a change in function of the pleurae. Benthic feeders can easily retrieve food particles dropped during transfer to the mouth, but retrieval of particles dropped by similarly oriented pelagic feeders is a problem. Therefore, many living pelagic arthropods swim or float on their backs, which positions the food channel above the body and makes the dropping of particles less likely. Perhaps pelagic trilobites assumed a similar orientation (Figure 13.27, *B*).

Evolution and distribution

Although slightly older shelly invertebrates have been discovered, trilobites were among the first to develop a mineralized framework to support the body. Among the oldest known trilobites are polymeroids from near the base of the Cambrian in Morocco, northern Europe, and Siberia. At least four genera are represented, indicating an even earlier but still obscure origin. Presence of a stiff exoskeleton and jointed appendages endowed early trilobites with a significant advantage in benthic mobility and an efficient new method of deposit feeding. Their Early Cambrian entry into previously empty or poorly defended ecologic space resulted in rapid adaptive radiation. Maximum diversity of genera and families was reached during the Late Cambrian (Figure 13.28). A sharp decline in diversity followed during the Ordovician, which probably resulted from interactions with newly developed competitors for food and living space as well as the radiation of such predators as cephalopods and fish. Much more gradual post-Ordovician decline in diversity continued until extinction of the class during the Late Permian. A few trilobite lineages continued to fill restricted ecologic niches after the Ordovician, but most innovations were confined to species and genera. Of the 140 families recognized in the *Treatise on Invertebrate Paleontology*, only three originated after the Ordovician. An apparent mild fluctuation in Silurian and Devonian generic diversity (Figure 13.28) may partly reflect fewer studies of Silurian trilobite faunas.

Many evolutionary trends are evident among trilobites. The tempo of morphologic change was more rapid

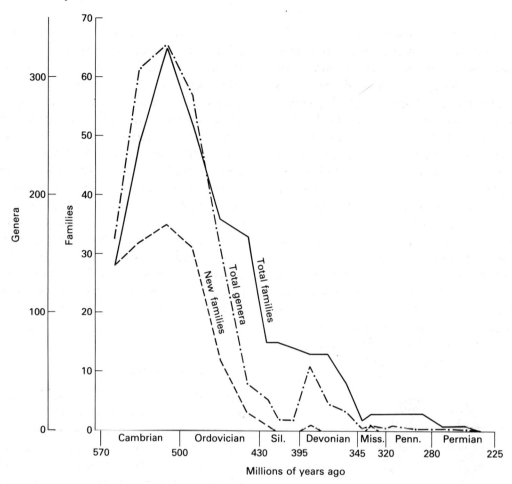

Figure 13.28. Diversity of trilobite genera and families. Appearances of new families are plotted for comparison. (Data from *Treatise on Invertebrate Paleontology, Part O.*)

for some trilobites than in many nonarthropod groups because juvenile exoskeletons were molted and new holaspid characters could be produced rather abruptly through paedomorphosis and peramorphosis. As already mentioned, changes in relative sizes of the cephalic frontal area and pygidium as well as in the number of thoracic segments were probably controlled by displacements in developmental timing, which allowed escape from overspecialization and for the addition of new capabilities. Decrease in relative width of the pleural lobes seems to have accompanied repeated and varied adaptations to pelagic modes of life; however, other factors may also have initiated such change.

The scant information available indicates that evolution of trilobite limbs was slight and the limbs did not provide great speed or maneuverability. Hence, development of various styles of protective enrollment may have helped later trilobites to compensate for such evolutionary conservatism.

Since early in their history, many trilobites show evidence of being strongly influenced by such environ-

mental factors as temperature, salinity, and sediment type. As an example, Cambrian lithofacies patterns in western North America indicate the existence of a broad marine shelf with shallow-water carbonate platform deposits of variable extent. These were flanked by lagoonal muds and near-shore sands on one side and by deeper water, outer shelf muds on the other side (Figure 13.29). Paleomagnetic data indicate equatorial latitudes. Almost mutually exclusive trilobite biofacies on opposite sides of the platform deposits suggest that faunal exchange was usually limited, perhaps by shoal-water barriers and unfavorable temperature and salinity gradients. Inner restricted-shelf biofacies are generally characterized by sparse, low-diversity, endemic polymeroid faunas. Outer open-shelf biofacies contain common to abundant, high diversity, mixed endemic and cosmopolitan, polymeroid and agnostoid faunas. As the result of shifting environments during the Cambrian, unrelated trilobite faunas commonly succeed one another in stratigraphic sections. Only by tracing trilobite lineages within related facies will it be possible to

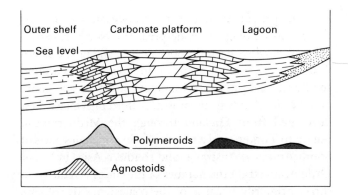

Figure 13.29. Generalized model for lithofacies of the Middle Cambrian shallow shelf in the western United States and relative abundance of polymeroid and agnostoid trilobites. (Not drawn to scale.) (From Robison, R. A. *Brigham Young University Geology Studies* 23: 93–109; 1976.)

determine many evolutionary histories, and much work remains. Moreover, these differences usually cause the biostratigraphic zonation for one biofacies to be a poor predictor of what will be found in other biofacies.

On a broader scale, trilobite provinces developed as the result of enduring genetic isolation. Boundaries between provinces may be gradational or rather abrupt, and their characteristics and positions are still being defined.

Cambrian trilobites   The first major adaptive radiation of trilobites occurred during the Early Cambrian when epeiric seas were relatively contracted. Primitive characters seen in many genera (Figures 13.21, *I* and *J*) include a long, segmented glabella; large, crescentic palpebral lobes, which are typically undifferentiated from the eye ridges and anteriorly abut against the glabella; usually more than 15 thoracic segments; and a tiny pygidium. In some genera the thorax is divided into two parts with segments in the posterior part being abnormally reduced in size (Figure 13.21, *I*). Also, one thoracic segment, variably positioned, may have much larger than average pleurae. Families with these characters have been used to define two partly overlapping Lower Cambrian trilobite provinces. One province, including present-day North America, South America, and northwestern Europe, is characterized by members of the family Olenellidae (Figure 13.21, *I*), which in addition to the listed characters, also lack dorsal facial sutures. The other province, including Antarctica, Australia, and much of southern Asia, is characterized by members of the family Redlichiidae (Figure 13.21, *J*), which are distinguished by opisthoparian facial sutures and usually an anteriorly tapered glabella. Mixed faunas are present in parts of northern Asia, southwestern

Europe, and northwestern Africa. Most olenellids and redlichiids were extinct by the end of the Early Cambrian.

A second major trilobite radiation took place during the Middle and Late Cambrian as shallow seas spread widely onto the cratonic areas of the world. With the great expansion of shallow-sea habitats, trilobite provinciality increased. Four main provinces, each typified by a different assemblage of endemic polymeroids, developed in restricted-shelf areas. These provinces include (1) the inner margins of the North American shelf, (2) a sandy facies in central Europe, (3) a broad carbonate platform in Siberia, and (4) a mixed carbonate and clastic facies in China and Australia. Open-ocean provinces of the same age are less well defined but contain a rich mixture of endemic and cosmopolitan faunas in shallower shelf areas and less diverse but more cosmopolitan faunas in deeper shelf and basinal areas.

Trilobite faunas from this second radiation are commonly dominated by relatively simple polymeroids belonging to many families (Figures 13.11, *D* and 13.21, *A*, *B*, and *H*). Typically, the frontal area of the cephalon is enlarged to include a prominent preglabellar field, and the relative size of the glabella is correspondingly reduced. Palpebral lobes are generally small and well separated from the glabella. Differentiated eye ridges connect the anterior palpebral lobes and the anterior glabella. Members of most genera have opisthoparian facial sutures, from ten to 14 thoracic segments, and a moderate-sized pygidium.

Although they first appeared well after the polymeroids, during the late Early Cambrian, the pelagic agnostoid trilobites also reached their maximum diversity during the Middle and Late Cambrian. They are abundant in some open-ocean lithofacies but rare in restricted-shelf lithofacies. Wide geographic distribution of many species (Figure 13.24, *F* and *I*) and rapid evolution place them among the best biostratigraphic indices for intercontinental correlation of Middle and Upper Cambrian strata.

Examples of biostratigraphically useful polymeroid genera in the Middle and Upper Cambrian of North America are illustrated in Figures 13.20, *B* to *D* and 13.21, *B* to *H*. In western Europe, micropygous paradoxidids with opisthoparian facial sutures and an anteriorly expanded glabella (Figure 13.21, *K*) are characteristic of Middle Cambrian strata, whereas olenids with a distinctive, generally subquadrate glabella (Figure 13.21, *A*) are abundant in Upper Cambrian strata.

Three significant extinction events—two within the Late Cambrian and one at the end of the Cambrian—saw the elimination of most polymeroid genera and several families in what were then low-latitude shelf areas of North America and Australia. Such events have been used to define the boundaries of a biostratigraphic unit called a **biomere**. After each of the three extinction events, the shelf areas were repopulated by low-diversity polymeroid faunas apparently immigrating from cooler and perhaps deeper habitats. Rapid intracontinental radiation followed each repopulation. Associated faunas and sediments were not noticeably affected by whatever caused elimination of the polymeroids. Modest temperature reduction is a likely agent, but the evidence is inconclusive.

Ordovician trilobites   In contrast to the great adaptive radiations during the Cambrian, the greatest decline in trilobite diversity occurred during the the Ordovician (Figure 13.28). During this period of decline, a few families became specialized along more easily recognized lines, and average family longevity approximately doubled.

Ordovician trilobite faunas are dominated by polymeroids (Figure 13.22), and these show many continuations of evolutionary trends that began in the Cambrian. For example, features related to enrollment are poorly developed in Cambrian polymeroids, which are rarely found enrolled. Articulating facets as well as a variety of structures that interlocked on enrollment became progressively more elaborate and more common in Ordovician trilobites. This change coincided with a progressively more common preservation of enrolled specimens. Fusion of the suture around the eye surface resulted in progressively more common preservation of lens-bearing eye units. Eyes increased in relative surface area, and some reached extraordinary proportions (Figure 13.27). On the ventral cephalic surface, reduction in width of the rostral plate ended in its loss from many Ordovician genera. On average, Ordovician polymeroids have fewer thoracic segments (commonly less than ten) and a larger pygidium than Cambrian polymeroids. Also, the number of holaspid thoracic segments is more stable in genera and families. In many lineages, especially some adapted to reef environments, members show either extraordinary spinosity (Figure 13.22, *E*) or the surface furrows and relief are greatly reduced (Figure 13.22, *B, I* to *K*). Members of three families have a distinctive pitted cephalic fringe (Figure 13.22, *F*), which may have enlarged the area of an auxiliary respiratory surface. Agnostoids are usually rare, and they disappeared near the end of the Ordovician.

Quantitative analysis has been used to outline four trilobite provinces in Lower Ordovician strata. One diverse assemblage is widespread in North America, parts of northwestern Europe, Siberia, and northeastern Asia. A more restricted fauna is found in regions around the Baltic Sea, the Ural Mountains of the Soviet Union, and Arctic islands to the north. A third province is best developed from England through the Mediterranean region to Turkey, and a fourth province includes parts of South America, Australia, and southern Asia. In Upper Ordovician strata the number of provinces is reduced to three, primarily because of the coalescence of trilobite faunas across North America and northern Eurasia.

Silurian and Devonian trilobites   From the Silurian onward, trilobites show little change in basic morphology, although details continued to change. With only three new families appearing after the Ordovician (Figure 13.28), much of the change in faunal aspect resulted from the continuing elimination of families.

Silurian and Devonian trilobite faunas are similar and may be difficult for the nonspecialist to distinguish. Two of the more distinctive families are the closely related Dalmanitidae (Figure 13.23, *I* and 13.26) and Phacopididae (Figure 13.23, *E, F*), each characterized by a forwardly expanding glabella, schizochroal eyes, proparian facial sutures, no rostral plate, and a thorax with 11 segments. The dalmanitids, which range from Ordovician to Devonian, have a large, generally spinose pygidium. The phacopids, which are restricted to the Silurian and Devonian, have a nonspinose pygidium of intermediate size.

Trilobites are present in many marine facies of the Silurian, but they have received relatively little study or emphasis in correlations. It has been estimated that trilobites average about 5 percent or less of Silurian invertebrate specimens. About 25 percent of Silurian trilobite genera are cosmopolitan, and of the 16 families represented, almost 90 percent are cosmopolitan.

At the generic level, trilobites had a slight resurgence during the Early Devonian (Figure 13.28). In North America a northeast-to-southwest land barrier (Canadian Shield and Transcontinental arch) separated an Appalachian province from a less-studied province of Old World type extending from Nevada through western Canada into the Arctic islands and Siberia. Eleven of 14 described trilobite families became extinct during the Middle and Late Devonian.

Mississippian-to-Permian trilobites   Late Paleozoic trilobites are generally scarce, being found mostly in reef deposits. The most commonly represented family is the

Phillipsiidae (Lower Mississippian to Upper Permian; Figure 13.23, *A* to *D*), which is characterized by a large glabella, holochroal eyes, opisthoparian facial sutures, nine thoracic segments, and a large pygidium that commonly has greater segmentation on the axis than on the pleural regions.

The cause of trilobite extinction near the end of the Permian remains conjectural. A eustatic fall in sea level and a corresponding major reduction in reef environments may have been significant factors.

Superclass Crustacea

Roger L. Kaesler

CLASS OSTRACODA (Figure 13.30)

Laterally compressed Crustacea; bivalve carapace hinged along dorsal margin and usually calcified, enclosing body. Head and thorax fused and segmentation reduced; five pairs of cephalic appendages, none to two pairs of thoracic appendages, and a pair of furcae; no abdominal appendages.

Most ostracodes can be distinguished from other crustaceans by their small size and bivalve carapace, both of which contribute to other morphologic features peculiar to the group (Figure 13.30). Although some are more than 1 cm in length, most ostracodes are much smaller, with adults usually being less than 1 mm long. Body and appendages are nearly always completely enclosed by the carapace, which consists of halves called **right** and **left valves**. The carapace is heavily calcified in all but a few highly specialized groups. The hinge along the dorsal margin of the carapace allows the valves to open along the ventral margin, so the animal may extend its appendages for feeding, moving about, and copulating. A strip of flexible, uncalcified material called a **ligament** connects the two valves at the hinge. Although not always well developed, the hinge may be complex and therefore useful in taxonomy. Like other arthropods, ostracodes grow in discrete steps after molting and before secretion of a new skeleton. As a result, their valves lack the growth lines found in pelecypods, a useful characteristic for discriminating large, clamlike ostracodes found in early Paleozoic faunas.

Ostracodes have been referred to as the accessory minerals of the biosphere. Although they are not often ecologically dominant in either modern marine communities or assemblages of fossils, they are, nevertheless, nearly always present. They range throughout the Phanerozoic and for much of their evolutionary history have occupied marine, brackish, and fresh water. A few modern terrestrial forms live in wet moss, some genera are parasitic, and a number of forms are found deep underground in the fresh water of caves. No ostracodes can fly, but in a sense, some are airborne, for eggs of many kinds of freshwater ostracodes are carried from place to place in mud stuck to the feet and feathers of waterfowl and may also be picked up by the wind from the bottom of dried ponds and lakes.

Ostracodes are characterized by small size, often great abundance, and excellent preservation. From a bulk sample of sedimentary rock, a micropaleontologist can commonly select hundreds of individual fossils for study. In contrast, larger fossils are often limited to a few individuals per sample, and among easily disarticulated forms complete, well-preserved specimens are rare.

The abundance of ostracodes in many kinds of rocks and their small size also contribute to a low sampling bias for two other reasons. First, with large samples the

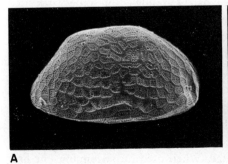 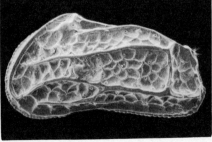 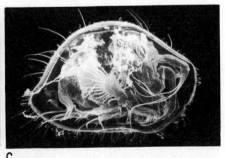

A **B** **C**

Figure 13.30. Representative Cenozoic Ostracoda of the order Podocopida from the Atlantic Coastal Plain and Gulf Coast area of North America. **A,** Exterior view of right valve of *Microcytherura* (Pleistocene, ×70). **B,** Exterior view of left valve of *Orionina* (Pliocene, ×58). **C,** Interior view of left valve of *Bairdia,* showing freeze-dried appendages and the manner in which body and appendages are enclosed in the carapace (Holocene, ×49). (**A** and **B,** courtesy J. E. Hazel and T. M. Cronin. **C,** courtesy P. A. Sandberg.)

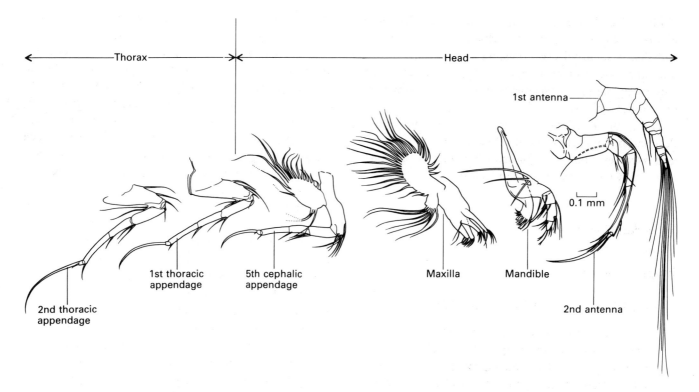

Figure 13.31. Appendage morphology of *Bairdia* (order Podocopida), an ostracode genus that ranges from Ordovician to Holocene. (Courtesy R. F. Maddocks.)

likelihood of failing to collect rare species is diminished. Second, the relative abundance of ostracodes at an outcrop is rarely biased by the outcrop's having been picked over by other collectors. Every paleontologist knows of once-productive outcrops from which the more obvious corals, crinoids, or trilobites have been depleted. Ostracodes and other microfossils, however, are collected in bulk samples that usually must be processed in the laboratory before the presence of fossils can be determined. At the end of a day's field work, micropaleontologists may have a few dozen samples, but they rarely know if they have found any microfossils. They can be certain, however, that they have collected the microfossils in the proportions in which they occur in the formations sampled.

Because of their small size, not all ostracodes are destroyed during exploration by rotary drilling, and they are therefore useful in dating and in correlation of many subsurface sedimentary deposits from a wide variety of environments. Ostracodes are commonly endemic, so they are of greater biostratigraphic value within basins than between biogeographic provinces. In the study of some nonmarine deposits, they are commonly the most biostratigraphically useful fossils encountered, although freshwater species have rather long geologic ranges.

Morphology of ostracodes

Appendage and body morphology   The most important and widely studied parts of the ostracode body are the appendages. Because the head and thorax are fused to form a cephalothorax, some thoracic appendages assist in feeding and some cephalic appendages assist in locomotion. Reduction of segmentation and lack of a well-differentiated head, presumably because the entire body is enclosed in the carapace (Figure 13.30, C), make it difficult to homologize segments of ostracodes with those of other crustaceans. Homologies of some associated appendages have also been difficult to establish but are now well understood. Ostracodes lack abdominal appendages, and in some taxa even the abdomen is lost.

Ostacodes usually have seven pairs of appendages, five of which are attached to the head. The first two pairs of cephalic appendages are the first and second antennae (Figure 13.31). In strongly swimming forms these bear long setae, whereas benthic ostracodes commonly have much shorter setae. When used for swimming, the first antennae (also called antennules) move back and forth above the carapace. Among crawling forms they are often spread in front of the animal to explore its path and maintain balance. The second antennae, which vary

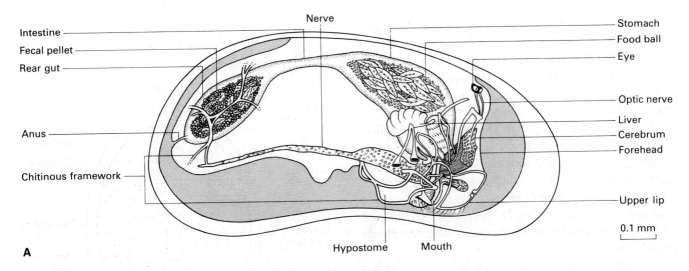

A

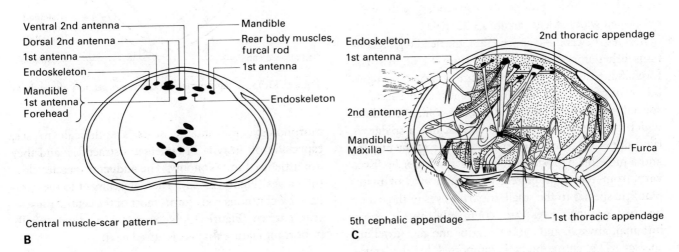

B

C

Figure 13.32. Body morphology of ostracodes. **A,** Morphology of head, digestive system, and nervous system of *Candona* (right valve removed), a typical freshwater ostracode (order Podocopida). **B,** Dorsal muscle-scar pattern on right valve of *Cypridopsis* (order Podocopida). (×100.) **C,** Position of muscles of *Cypridopsis*. (Courtesy R. V. Kesling.).

considerable among higher taxa, are the principal locomotory appendages and are used either for swimming or, along with the thoracic legs, for walking.

The third and fourth pairs of cephalic appendages are the mandibles and maxillae (Figure 13.31). The basal podomere of the mandible is elongate and heavily sclerotized at one end to form chewing teeth. In some, the other end terminates in a tiny ball that pivots against the inside of the valve. Both the mandibles and the maxillae also bear jointed, armlike, setose **palps** for sensing, capturing, and transferring food particles. Expodites of these appendages bear a respiratory plate that fans constantly to produce a current of water supplying oxygen and, among suspension feeders, food. The respiratory plates are not gills, however, and oxygen intake occurs over the entire surface of the body. The mandibular teeth

of one group of highly specialized ostracodes are sharply pointed and lie inside a tube used to penetrate plate tissue and to suck juices.

The fifth pair of cephalic appendages has been the subject of controversy because it differs in form and function among major groups. It is homologous with the second maxilla of other Crustacea. Among most benthic marine ostracodes, the fifth cephalic appendage is a walking leg that has often been erroneously regarded as a thoracic appendage because of its similarity to true thoracic walking legs (Figure 13.31). Among planktic marine, some benthic marine, and most freshwater ostracodes, this appendage is used only for feeding. It may also be sexually dimorphic, being modified in some males as a grasping device to help hold the female during copulation.

The first pair of thoracic appendages usually functions as walking legs (Figure 13.31). A movable terminal claw supports much of the weight of the posterior portion of the body. In some swimming and burrowing forms, these appendages are reduced in size, and in one benthic group both pairs of thoracic appendages are missing.

The second pair of thoracic appendages varies widely in function. In some groups it is simply a third pair of walking legs, but in others it is modified as a brush-tipped cleaning apparatus or a highly maneuverable pincer for removing parasites. Until recently the first and second thoracic legs were commonly regarded as second and third thoracic legs because homologies were imperfectly understood.

The furcae are not appendages but spinelike extensions of the abdomen. They are used for locomotion and are reduced in those groups of ostracodes with well-developed walking legs (Figure 13.32, *B*).

Besides the cephalic appendages, the head region contains other important features (Figure 13.32, *A*). The **forehead**, to which the antennae are attached, occupies the front of the body. In a few benthic marine species, a special gland in the forehead secretes a filament along which the ostracode moves somewhat like a spider. A rigid **upper lip** forms the anterior portion of the mouth, and a movable **hypostome** functions as a lower lip. Eyes vary from the large paired, compound eyes of many planktic species to the small, tripartite eyes of the major benthic groups. Eyes are commonly absent from infaunal, abyssal, and bathyal marine and cave-dwelling species. The entire body is supported by the endoskeleton, a framework of chitinous rods that stiffens the cuticle and provides a place of attachment for some muscles.

Some body systems of ostracodes are typical of arthropods in general, but in others the organs are either much reduced, presumably the result of small size and enclosure by a carapace, or modified for a special purpose. The digestive system (Figure 13.32, *A*), for example, is not unusual, but the mode of feeding varies widely.

The respiratory and circulatory systems are considerably reduced. Only large planktic ostracodes possess a heart, and only a few of these have gills, which develop from folds of posterodorsal cuticle.

The muscle system includes two major parts. Muscles that operate appendages are attached to the chitinous endoskeleton, the central part of the carapace, or the dorsal part of the carapace, where they form the **dorsal muscle-scar pattern** (Figure 13.32, *B* and *C*). Although these muscles perform important functions and knowledge of them is important in the study of functional

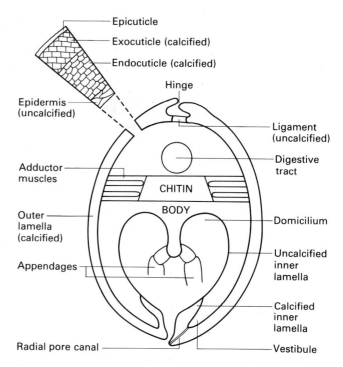

Figure 13.33. Diagrammatic transverse section through an ostracode.

morphology, the muscle scars are difficult to see, especially on heavily ornamented ostracodes, and they are little used in taxonomy. The **adductor muscles** close the valves (Figure 13.33). Their attachment to the carapace by chitinous rods forms most of the **central muscle-scar pattern** (Figure 13.34, *A* to *C*), a taxonomically important feature that is discussed next.

Carapace morphology The chitinous cuticle is calcified to form the bivalve carapace (Figure 13.33). Generally, the carapace is almost bilaterally symmetrical, although in some groups one valve overlaps the other. The valves join along a dorsal hinge that may be either straight or curved and ranges from short to being the longest part of the carapace. At the hinge, the two valves are connected by a narrow, three-layered, uncalcified ligament. Elsewhere the valves meet along a **free margin** that may differ in outline from the carapace where obscured by external features such as spines and ridges (Figure 13.34, *A*).

The carapace consists of a **calcified outer lamella** and a **calcified inner lamella** (Figure 13.33). A pocket called the **vestibule** sometimes forms between the two calcified lamellae. The shape and presence of the vestibule vary in taxonomically important ways. In some taxa the calcified inner lamella is developed only in the adult stage. Its absence can usually be taken as proof of immaturity, but in some genera it is narrow and easily overlooked. The calcified inner lamella commonly bears ridges that

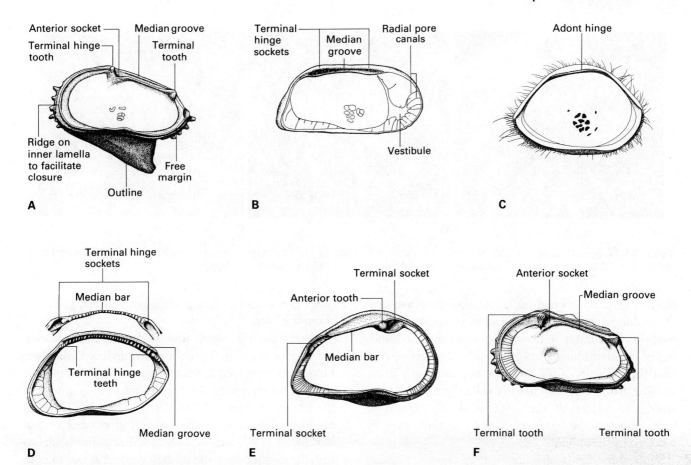

Figure 13.34. Morphologic features of interior of ostracode valves. **A**, Right valve of *Pterygocythere* showing amphidont hinge and difference between outline of the carapace and the free margin. (×40.) **B**, Left valve of *Krithe* showing the highly irregular anterior vestibule; long, curved radial pore canals; and merodont hinge. (×45.) **C**, Adont hinge, left valve of *Bairdia*. (×25.) **D**, Merodont hinge, right valve of *Loxoreticulatum*, with hinge of left valve shown above. (×65.) **E**, Amphidont hinge, left valve of *Brachycythere*. (×50.) **F**, Amphidont hinge, right valve of *Curfsina* with crenulated anterior terminal tooth. (×50.) (**A**, **E** and **F** from Benson, R. H.; Tatro, J. O. *University of Kansas Paleontological Contributions* Article 7, p. 1–32; 1964. **B** and **D** from Benson, R. H. *University of Kansas Paleontological Contributions* Article 6, p. 1–36; 1964. **C**, courtesy R. J. Maddocks.)

facilitate tight closure of the valves (Figure 13.34, *A*). In some living deposit-feeding forms, the margin of one valve strongly overlaps the other, to make a tighter closure, whereas suspension feeders have little valve overlap. It has been suggested that some large, extinct Paleozoic ostracodes with straight ventral margins and no valve overlap were also suspension feeders, although this hypothesis cannot be tested for groups without close living descendants.

Two kinds of canals penetrate the ostracode carapace. **Radial pore canals** are subparallel to the outer surface of the carapace, lie between the outer and inner lamellae, and exit near the free margin (Figures 13.33 and 13.34, *B*). **Normal pore canals** are simple or sievelike, usually scattered widely over the carapace, and perpendicular to the surface (Figure 13.35). Setae associated with these pores sense the environment outside the carapace. It has

been suggested that some sieve pores may be light sensors. The scanning electron mcroscope has made possible detailed study of normal pore canals, and a wide variety has been discovered that is taxonomically useful.

Like pore canals, the muscle scars of the ostracode carapace are points at which specific parts of the body are connected directly to the carapace. As a result, muscle-scar patterns, especially the central, adductor muscle scars, are important in taxonomy. The number and arrangement of muscle scars support classifications based on appendage morphology and are useful indicators of the superfamily or order to which an ostracode belongs (Figure 13.36).

Hinges of ostracodes vary greatly and are useful in taxonomy. A few decades ago, a classification of ostracode hinges was developed that expressed the degree of complexity of three parts of the hinge, the structures at

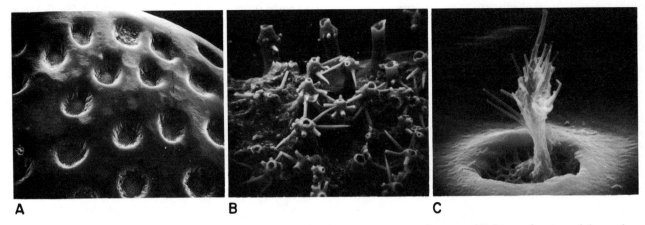

Figure 13.35. Normal pore canals. **A,** Sieve pores of *Haplocytheridea.* (×315.) **B,** Simple pores of *Echinocythereis,* each located on a spine. (×470.) **C,** Sieve pore of *Aurila* occupied by brushlike seta. (×3150.) (Courtesy P. A. Sandberg.)

the two ends of the hinge (called **terminal elements**), and the middle or **median element.** This classification is not as useful as once thought, however, because as understanding of ostracode hinges has grown it has become evident that the elements of the hinge are not necessarily homologous from one group of ostracodes to another. Moreover, the nature of the hinge is closely related to robustness of the carapace, which, in turn, may be affected by the habitat.

Only three kinds of hinges will be considered here: adont, merodont and amphidont hinges. Students should be aware, however, that all three types can be subdivided, that morphologic transition occurs between types, that

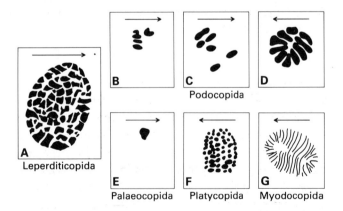

Figure 13.36. Central muscle-scar patterns of representative ostracodes; arrows point toward anterior. **A,** Order Leperditicopida, *Eoleperditia.* (×9.) **B** to **D,** Order Podocopida: **B,** *Trachyleberis.* (×50.) **C,** *Cypridopsis.* (×75.) **D,** *Darwinula* (×150). **E,** Order Palaeocopida, *Tvaerenella* (×20). **F,** Order Platycopida, *Cavellina* (×150). **G,** Order Myodocopida, *Entomoconchus.* (×90.) Muscle scars of the order Archaeocopida are unknown. (Modified from Scott, H. W. In: Moore R. C., editor. *Treatise on invertebrate paleontology, Part Q.* New York, and Lawrence, KS: Geological Society of America and University of Kansas Press; 1961.)

hinge types change during ontogeny, and that still more elaborate hinges are known.

Most Paleozoic, most freshwater, and many marine ostracodes with weakly calcified carapaces have **adont hinges** (without teeth). Adont hinges have a long median element on the right valve that fits into a long groove of the left valve (Figure 13.34, *C*). The terminal elements of adont hinges are either weakly developed or not developed at all, but on some modified adont hinges as broadly defined here the median element is not smooth but crenulated, consisting of a long row of tiny teeth and sockets. **Merodont hinges** (Figure 13.34, *B* and *D*) have elongated, strongly crenulated, terminal elements on the right valve (Figure 13.34, *D*) that fit into corresponding terminal sockets on the left valve. The median element may be smooth or crenulated and occurs as either a bar or groove on the right valve. **Amphidont hinges** (Figure 13.34, *A, E* and *F*) have short terminal elements that consist of well-developed teeth on the right valve that may be crenulated, lobate, divided, or smooth. The median element of the right valve consists of an anterior socket, which may be smooth or divided, and a median groove that is usually smooth but may be crenulated. In all three types, hinge structures of the left valve are usually counterparts of those of the right valve. Remember, however, that the two valves are held together by the uncalcified ligament and not by the hinge itself. In rare instances reversal of hinge structure is known in which hinge elements characteristic of the right valve appear on the left valve and vice versa.

The surface of the carapace is commonly marked with pronounced features usually referred to as ornamentation. The function of some ornamentation is easy to understand. An **eye tubercle,** for example, occurs on many valves as a transparent lens of calcite directly out-

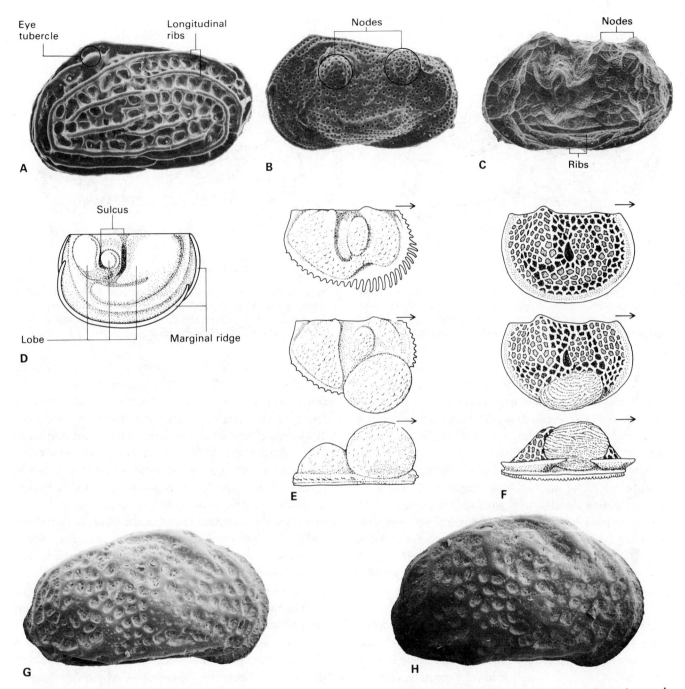

Figure 13.37. Surface ornamentation and sexual dimorphism of Ostracoda. **A,** *Climacoidea* (Pleistocene, order Podocopida), showing a pronounced eye tubercle near the anterior dorsal corner, reticulate ornamentation, and strong longitudinal ribs. (×94.) **B,** *Cluthia* (Pleistocene, order Podocopida), with nodes and a pitted surface. (×94.) **C,** *Palmanella* (Pleistocene, order Podocopida), with ribs and nodes. (×94.) **D,** Lobes, sulci, and marginal ridges of palaeocopid ostracodes. **E** and **F,** Examples of males and tecnomorphs in lateral view (upper) and females or heteromorphs in lateral and ventral view (lower and middle): *Beyrichia* (**E**), a palaeocopid from the Lower Devonian, and *Treposella* (**F**), palaeocopid from the Middle Devonian. (Both ×14.) **G** and **H,** Typical sexual dimorphism among Ostracoda of the superfamily Cytheracea (order Podocopida): male (**G**) and female (**H**) of *Hemicythere* (Holocene). (Both ×80.) (**A** to **C,** courtesy J. E. Hazel and T. M. Cronin. **D** to **F** from Benson, R. H. *et al.* In: Moore, R. C., editor. *Treatise on invertebrate paleontology, Part Q.* New York, and Lawrence, KS: Geological Society of America and University of Kansas Press; 1961. **G** and **H,** courtesy R. F. Maddocks.)

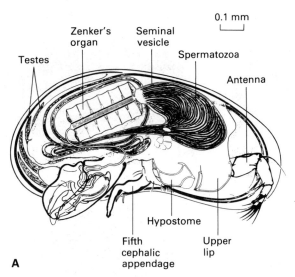

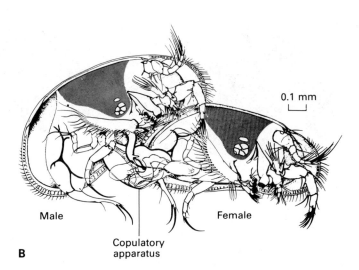

Figure 13.38. Reproductive system of ostracodes. A, Male *Candona* (superfamily Cypridacea, order Podocopida) with right valve and most appendages removed. B, Copulating ostracodes. Male, on the left, thrusts large copulatory apparatus well forward to clasp female's genital lobes and effect penetration of the female's two vaginas. (From McGregor, D. L.; Kesling, R. V. *University of Michigan, Contributions from the Museum of Paleontology*; 1969.)

side the eye (Figure 13.37, *A*). It is clearly associated with vision and allows light to penetrate the carapace. The functional morphology of most other surface ornamentation is not well understood, although much of it must serve to strengthen the carapace. Many ostracodes have a netlike **reticulate** ornamentation (Figure 13.37, *A*). Others are marked by ridges, ribs, spines, and nodes (Figure 13.37, *A* to *C*). Ornamentation is useful in discriminating species and genera. Ornamentation of some common Paleozoic ostracodes is different from that of living forms, being marked by the presence of lobes and sulci that probably reflect underlying morphology of the body and ridges on the margin that are nearly parallel to the free margin of the valve (Figure 13.37, *D*).

Reproduction and sexual dimorphism   The reproductive system of ostracodes is highly developed and specialized in some groups to the point of being bizarre (Figure 13.38). The sexes are always separate, and sex organs are large, often comprising about one-third of the male's volume and one-fifth of the female's. The complexity and size of the reproductive system result in part from the small size of the animal and the difficulty of copulating while enclosed in a bivalve carapace. A peculiarity not accounted for, however, is the fact that ostracode spermatozoa are the largest known in the animal kingdom, ranging from six to ten times longer than the male's body. In some groups a large pump (Zenker's organ) propels spermatozoa from the male into the female's seminal receptacles (Figure 13.38, *A*). Moreover, the ovary-uterus system of some females is not connected to the vaginal-seminal-receptacle system, requiring transfer of spermatozoa from one system to the other in a manner that is imperfectly understood. In nearly all species, each adult has duplicate, bilaterally symmetrical reproductive systems that are not connected and function separately. In most freshwater ostracodes, the ovaries and testes are situated immediately adjacent to the valves rather than in the main part of the body as in common benthic marine forms. Imprints of the sex organs are sometimes present on the inside of the valves, indicating both the sex and sexual maturity. Many ostracodes brood their eggs within the **domicilium**, that is, within the carapace but outside the female's body (Figure 13.33). Most benthic forms, however, lay eggs and abandon them. Eggs of freshwater ostracodes are able to withstand long periods of desiccation, at least 30 years for some species. With such an elaborate reproductive system, it is surprising that some ostracodes have given up sex altogether. Males of some freshwater species have never been found, and the females are entirely parthenogenetic. Among other species, parthenogenesis is common but not obligatory.

The size and importance of the reproductive system of ostracodes would be expected to lead to pronounced sexual dimorphism in the carapace as well. Among the common benthic ostracodes, sexual dimorphism of the carapace is usually expressed by males having a somewhat greater length-to-height ratio than females (Figure 13.37, *G* and *H*). The degree of sexual dimorphism varies widely, but separating members of a population into males, females, and different immature stages usually requires study of a large number of individuals. Among the dominant group of small, benthic ostracodes of the Paleozoic, however, sexual dimorphism is enhanced by the fact that females brooded their young in special pouches in the carapace wall (Figure 13.37, *E* and *F*). These females are sometimes referred to as **heteromorphs** to distinguish them from morphologically similar adult males and juveniles called **tecnomorphs**, which lack brood pouches. Several kinds of dimorphism have been recognized depending on the position of the brood pouch and the structure of the carapace from which it originated.

Classification of ostracodes

The classification of Ostracoda is in a state of flux at both generic and higher taxonomic levels. The abbreviated classification used here is similar to the one in the *Treatise on Invertebrate Paleontology*. Differences represent advances in knowledge of the morphology and phylogeny of ostracodes since the *Treatise* was published. Students should be aware that other classifications exist that will surely be encountered during further study of ostracodes. Some differences between classifications are

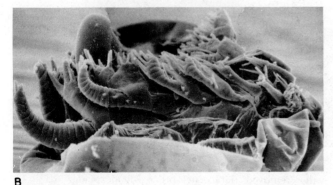

Figure 13.39. *Vestrogothia*, a Cambrian archaeocopid in which both the carapace and appendages are preserved as calcium phosphate, leading some workers to propose original secretion of phosphate and others to propose diagenetic phosphatization soon after death but before the soft parts had decayed. **A**, Exterior of left valve. (×54.) **B**, Oblique view of preadult from below showing six pairs of appendages. (×180.) (From Muller, K. J. *Lethaia* 12: 1–27; 1979.)

simply matters of choice between several available names.

The Ostracoda are usually divided into six orders. Two of these, the Archaeocopida and Leperditicopida, are restricted to the Paleozoic, and the order Palaeocopida is largely restricted to the Paleozoic. The orders Myodocopida, Platycopida, and Podocopida range from the Ordovician and have living species, some in genera and families that have changed little in carapace morphology since the Ordovician. Of course, study of most fossils must be based on carapace morphology alone, whereas appendages of living forms are available for study. Indeed, as with many other crustaceans, orders of living ostracodes are discriminated primarily on appendage morphology. As a result, not all specialists agree on the affinities of many taxa known only from the fossil record. Fortunately, several important characters of the carapace, such as the muscle-scar patterns, eye tubercles, and normal pore canals, are directly related to soft parts. Because many higher taxa of ostracodes with long geologic ranges have extant members, classification may be based on both soft parts and carapace morphology. The two sets of characters reinforce rather than contradict each other, giving ostracode specialists confidence that they are dealing with natural biologic groups rather than artifacts.

ORDER ARCHAEOCOPIDA (Figure 13.39)

Carapace weakly calcified or phosphatized, flexible; hinge long and straight or lacking; ventral margin convex; eye tubercle usually prominent; surface smooth, punctate, wrinkled, or ribbed, sometimes spinose; inner lamella not calcified; sexual dimorphism and muscle-scar pattern unknown. (Cambrian, Ordovician(?); 20 genera.)

The order Archaeocopida (Figure 13.39) is probably ancestral to all later Ostracoda, but recent discoveries have cast some doubt on its affinities. Phosphatic archaeocopids have recently been discovered in which even the appendages are preserved as calcium phosphate (Figure 13.39, *B*). Some specialists have assigned these to a separate suborder (Phosphatocopina), but others believe that they represent only a special kind of preservation. The appendages of these archaeocopids have many podomeres and are thus unlike those of any living ostracodes. Careful study of their homology with appendages of other Crustacea is needed to establish the relationship of the archaeocopids to other ostracodes and, indeed, to determine if they ought to be classed as ostracodes at all.

ORDER LEPERDITICOPIDA (Figures 13.36, A; 13.40, A)

Carapace large, thick, well calcified, and usually smooth with long, straight hinge and commonly prominent eye tubercle; inner lamella not calcified; large, complex muscle-scar pattern up to one-third height of valve with as many as 200 small, secondary scars; sexual dimorphism unknown. (Cambrian(?), Ordovician to Devonian; 30 genera.)

Some leperditicopids are nearly 3 cm long, and most of them are four or five times larger than the average ostracode. For this reason, some specialists doubt that they are true ostracodes. Others, however, regard them as closely related to archaeocopids and consider them to be ancestral to the myodocopids. Lack of valve overlap along the ventral margin suggests that they were suspension feeders, although commonly they lived in shallow-marine environments, often associated with algal stromatolites.

ORDER PALAEOCOPIDA (Figures 13.36, E; 13.37, D to F; 13.40, B)

Carapace well calcified, typically with long, straight hinge, broadly convex ventral margin, inner lamella not calcified, and either a deep pit occupied by the adductor muscle scar or prominent lobes, sulci, and ridges (Figures 13.37, D and 13. 40, B); usually with pronounced sexual dimorphism including development of brood pouch by female (Figures 13.37, E and F); no valve overlap; muscle scars poorly

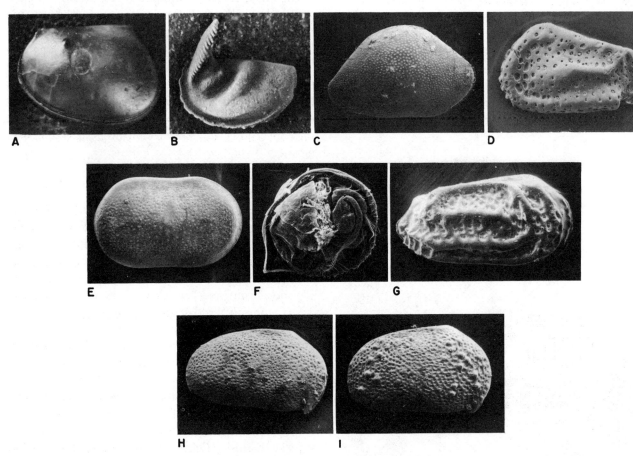

Figure 13.40. Ostracodes representative of higher taxa. **A,** Exterior view of left valve of *Leperditia* (Silurian, order Leperditicopida, ×1.9). **B,** Exterior view of left valve of tecnomorph of *Ceratopsis* (Ordovician, order Palaeocopida, ×12). **C,** Exterior view of right valve of *Bairdia* (Holocene, superfamily Bairdiacea, order Podocopida, ×30). **D,** Exterior view of left valve of *Hazelina* (Paleocene, superfamily Cytheracea, order Podocopida ×63). **E,** Exterior view of right valve of *Cytherella* (Holocene, order Platycopida, ×34). **F,** Interior view of freeze-dried myodocopid specimen with right valve removed (Holocene, order Myodocopida, ×37). **G,** *Lophocythere,* showing characteristic tripartite ornamentation of Mesozoic cytheraceans (Jurassic, superfamily Cytheracea, order Podocopida, ×53.) **H,** Lagoonal ostracode *Cyprideis,* which lacks nodes when living in water of normal-marine salinity (superfamily Cytheracea, order Podocopida, ×33.) **I,** *Cyprideis* from brackish water (×33). (**A** and **B,** courtesy S. M. Warshauer. **C** and **E,** courtesy R. F. Maddocks. **D,** courtesy J. E. Hazel and T. M. Cronin. **F,** courtesy P. A. Sandberg. **G,** courtesy R. C. Whatley. **H** and **I,** courtesy T. I. Kilenyi.)

known. (Ordovician to Triassic, Tertiary(?) to Holocene(?); 500 genera.)

Palaeocopids are confined to the Paleozoic and Lower Triassic except, perhaps, for one family whose members have been found in the Tertiary of Japan and in Holocene sediments in the eastern South Pacific. Unfortunately, these possible modern palaeocopids are rare and have received little study. Collections consist only of empty valves and carapaces, so morphology of appendages has not been studied. The apparent lack in these modern animals of the kind of sexual dimorphism that is typical of palaeocopids has led some specialists to classify them with other orders. The eventual discovery of live specimens with appendages and other soft parts may be extremely significant in advancing knowledge of Paleozoic ostracodes.

ORDER PODOCOPIDA (Figures 13.30; 13.31; 13.32; 13.34; 13.36, B to D; 13.37, A to C and G to H; 13.38; 13.40, C, D, and G to I)

Carapace well-calcified; valves of unequal size, one commonly overlapping the other; dorsal margin curved (Figure 13.32, B) or, if straight, shorter than total length of carapace (Figure 13.34, A); ventral margin commonly concave (Figure 13.32, A); inner lamella partly calcified, radial pore canals present (Figure 13.34, B), commonly numerous; hinge adont, merodont, or amphidont; muscle-scar pattern usually with few secondary scars homologous within superfamilies (Figure 13.36, B, C, D); males usually more elongate than females (Figure 13.37, G and H). (Ordovician to Holocene; 1200 genera.)

The order Podocopida includes most species of Mesozoic to Holocene ostracodes. Although podocopids first appeared in the Ordovician, they did not become diverse until late Paleozoic. As were the remaining two orders, this order was originally erected on the basis of appendage morphology; however, the podocopid carapace also possesses several distinguishing features that are listed in the diagnosis.

Podocopida are divided into two suborders. The second of these suborders comprises five major groups (superfamilies) each characterized by a distinctive appendage and carapace morphology. Some groups are good environmental indicators. One group (Figure 13.40, C) of entirely marine species ranges from Ordovician to Holocene with little morphologic change. These species are characteristically shaped like lemon seeds and have subdued ornamentation, usually adont hinges, and

wide calcified inner lamellae. Another group (Figure 13.40, D) predominates in the marine environment, although a few genera inhabit fresh water. They are benthic, well calcified, and commonly have strong ornamentation and merodont or amphidont hinges. A third group (Figure 13.32) predominates in fresh water, but numerous marine genera are known as well. They typically have less heavily calcified carapaces than genera of the first two groups, adont hinges, and adductor muscle-scar patterns consisting of a few secondary scars arranged in a tight cluster (Figure 13.36, C). The remaining two groups are represented by few species.

ORDER PLATYCOPIDA (Figures 13.36, F; 13.40, E)

Nearly smooth or ribbed, well-calcified carapace with oval outline, strongly unequal valves with larger, usually right valve overlapping smaller valve around entire carapace in uninterrupted marginal furrow; no calcified inner lamella or lobes and sulci; sexual dimorphism by inflation of posterior part of domicilium of female; no eyes; appendages modified for burrowing; muscle-scar pattern with numerous secondary scars among Paleozoic forms, fewer among Mesozoic and Cenozoic forms (Ordovician to Holocene; 30 genera.)

Although reduced in modern seas to two common genera, platycopids were abundant at times in the geologic past, both in marine and brackish-water environments. Some Paleozoic species superficially resemble palaeocopids, but they have a continuous marginal furrow and the kind of sexual dimorphism that is characteristic of all platycopids.

ORDER MYODOCOPIDA (Figures 13.36, G; 13.40, F)

Carapace commonly large, weakly calcified, with equal to subequal valves and no valve overlap, inner lamella partly calcified, and convex ventral margin; second antenna commonly extended through notch in anterior margin of carapace; muscle-scar pattern consisting of numerous elongate scars. (Ordovician to Holocene; 120 genera.)

The order Myodocopida (Figure 13.40, F) is another that was erected primarily on appendage morphology, notably second antennae with long setae for swimming and sometimes a notch in the carapace through which the second antennae are extended. The carapace is commonly larger than that of palaeocopids, podocopids, and platycopids, with some species reaching as much as 1 cm in length. Many myodocopids are strong swimmers, some are planktic, and others are benthic, including some that live infaunally. Some have compound eyes and a heart, and a few large species have gills. Although myodocopids are abundant in modern marine environments, they have a sparse fossil record. Their weakly calcified carapaces are seldom preserved, and most of their fossils come from Paleozoic rocks, where a few groups of presumed planktic forms are useful in biostratigraphy of deep-water sediments.

Paleoecology of ostracodes

For aid in interpreting paleoenvironments, one seeks taxa with long ranges and rather narrow environmental tolerances. Here some ostracodes are unsurpassed. One podocopid genus, for example, has occupied marine-shelf environments since the Ordovician with little change in morphology (Figures 13.30, *C*; 13.40, *C*). Another long-ranged podocopid species has occupied freshwater environments since the Pennsylvanian. Although individual species usually have narrow environmental tolerances, ostracodes inhabit virtually every aquatic environment from fresh to hypersaline, from alpine lakes to abyssal depths of the ocean, and from polar to equatorial regions. A few species are terrestrial, inhabiting moist substrates.

Ostracodes with robust carapaces have been regarded by some as indicators of high-energy environments, and those with delicate carapaces, as indicators of quiet water. Although this generalization is true in some instances, enough exceptions are known to make it too unreliable for application without good taxonomy to support interpretations. For example, many ostracodes from quiet-water, abyssal depths are heavily calcified, and some plant-dwelling forms from high-energy environments have delicate carapaces. Until the functional morphology of external features of the ostracode carapace is better understood, generalizations about carapace morphology and environment remain suspect.

Since the Jurassic, one group of podocopids has been the most abundant among benthic marine ostracodes. Most families range from the Jurassic or Cretaceous, and some genera have equally long ranges. Families and genera are often characteristic of specific environments the world over. Members of some families, for example, are found living on plants and are, hence, indicative of shallow, lighted benthic environments. Unfortunately,

some of these have fragile carapaces that rarely survive the rigors of the high-energy, nearshore environment, thus limiting their usefulness in the study of paleoenvironments. Members of other families live in shallow, high-energy, nearshore environments, and their carapaces, if robust, are often preserved in sediments deposited in that environment.

Even within a single genus, variation in morphology can be useful in studies of ancient environments. Populations of some species living in brackish water develop large, hollow nodes on the carapace, whereas populations from more nearly normal-marine salinity are smooth (Figure 13.40, *H* and *I*). The development of nodes has been ascribed both to individual response to environment and to genetic differences among populations, but experimental work to confirm either view remains to be done.

Evolution and biostratigraphy of ostracodes

Evolutionary trends   As is true for most classes of organisms, the early fossil record is inadequate to determine the origin of the Ostracoda. The oldest ostracodes are members of the order Archaeocopida, which appeared in the Early Cambrian and became extinct in the Late Cambrian or Early Ordovician. They are much larger than average ostracodes that lived from the middle Paleozoic to the present, a few species reaching more than 0.5 cm in length. They also possess rather different appendages (Figure 13.39, *B*) and some specialists have questioned whether they are, in fact, ostracodes.

Since the decline of the archaeocopids, most ostracodes have followed several evolutionary trends for which the functional significance is not well understood. First, throughout much of the Paleozoic, ostracodes generally evolved toward smaller size, a common trend among several groups of Crustacea. The leperditicopids of the Ordovician, some of which are more than 3 cm long, are the largest ostracodes that ever lived. During the Silurian and Devonian, the dominant forms were genera of the order Palaeocopida, which are smaller in size, rarely exceeding 2 mm in length. Late Paleozoic palaeocopids and podocopids are usually from 0.5 to 1 mm long, and most Mesozoic and Cenozoic ostracodes are about the same size. Notable exceptions to this trend occur among some modern myodocopids, which rival the leperditicopids in size. Some deep-sea, benthic podocopids are also large as ostracodes go, but none is as large as the giant myodocopids.

A second evolutionary trend has been for ostracodes to develop simpler muscle-scar patterns with fewer secondary scars (Figure 13.36). The Leperditicopida have a highly complex, aggregate muscle-scar pattern (Figure 13.36, *A*) whereas the Myodocopida, which probably evolved from them, have a simpler pattern with many fewer scars (Figure 13.36, *G*). The Paleozoic platycopids (Figure 13.36, *F*) have many more scars in their patterns than later marine platycopids thought to be closely related to them. Similarly, an extant freshwater genus that appeared in the late Paleozoic (Figure 13.36, *D*) has a more complex muscle-scar pattern than more recently evolved podocopids (Figure 13.36, *B* and *C*). Even within a single superfamily, Paleozoic species can have a more complex muscle-scar pattern than extant species. Some Cenozoic marine podocopid genera, however, have reversed the trend by evolving a pattern in which the individual muscle scars are split into pairs. The functional significance of this reversal, if any, is not understood. The muscle-scar pattern of palaeocopids is not well enough understood to determine evolutionary trends.

Evolution of the ostracode hinge shows several trends. In general, hinges have become shorter relative to the length of the carapace, and some families that originated in the Mesozoic have developed slightly curved hinges. In addition, hinges have tended to become more robust with more pronounced terminal elements that are less crenulated or subdivided, especially among genera of one group of Podocopida. The robustness of the hinge, however, is related to the robustness of the carapace as a whole. Because many ostracodes with delicate carapaces are still present in quiet-water environments, exceptions to the evolutionary trend toward robust hinges are common.

During the Mesozoic many podocopid ostracodes developed ornamentation consisting of three prominent longitudinal ribs that extend most of the length of the carapace (Figure 13.40, *G*). Although some modern representatives of this group also have this rib pattern, it is commonly modified so that the dorsal rib is arched and much shorter, the middle rib is less pronounced, and the ventral rib terminates in a posteroventral spine or hump (Figures 13.30, *B* and 13.40, *D*). In addition, the surface of modern species is often ornamented in much finer detail than that of their Mesozoic relatives (Figures 13.30, *A* and 13.37, *B* and *C*). Recent work has shown that some of these ornamental features can be homologized from genus to genus within a family, making them useful in determining phylogenies.

Biostratigraphy The usefulness of ostracodes in biostratigraphy may be considered at several taxonomic levels. Most biostratigraphic work, of course, is based on careful study of the stratigraphic ranges of species, but sometimes orders and suborders are useful, at least in a general way. Ostracodes of the order Archaeocopida are rare, but they are confined to Cambrian and Lower Ordovician rocks. The Leperditicopida are indicative of Ordovician to Devonian rocks, and most families of the order Palaeocopida were extinct by the end of the Permian. The orders Podocopida, Platycopida, and Myodocopida, however, range from Ordovician to the present and are useful biostratigraphically only at lower taxonomic levels.

Ostracodes have been used in biostratigraphy of rocks of virtually all Phanerozoic ages. The extent of their use depends in part on the biostratigraphic usefulness of other microfossils, notably the foraminifera and conodonts, in rocks of the same age and on the environment of deposition. For example, the use of ostracodes in biostratigraphy of lower and middle Paleozoic marine rocks has declined in recent years as intensive study has shown conodonts to be unsurpassed for global correlations. Moreover, during much of the late Paleozoic, some ostracode lineages evolved slowly, and many genera and species have long ranges. The fusulinid foraminifera, on the other hand, evolved rapidly and form the basis of biostratigraphy of most marine rocks deposited during the late Paleozoic. Similarly, the ostracodes are useful in biostratigraphy of the Jurassic but are little used for global correlations, especially for Cretaceous to Holocene rocks, in spite of rapid evolutionary turnover of many groups. A high degree of endemism among ostracodes and the appearance and rapid evolution of the planktic foraminifera have combined to limit the use of ostracodes in biostratigraphy of rocks of these ages.

In spite of their endemism and limited use for global correlation of marine rocks, ostracodes are extensively used in biostratigraphy within sedimentary basins. This is especially true for Cenozoic deposits of the Atlantic Coastal Plain of the United States, the Caribbean Sea, and the Aquitaine Basin of France, where rates of sedimentation were low and ostracodes are abundant. Individual ostracode species are sensitive to the environment, but the broad environmental tolerance of the group as a whole ensures that ostracodes are present in rocks from most of a basin's depositional environments.

The use of ostracodes in biostratigraphy of freshwater deposits deserves special mention. Although freshwater ostracode species typically have long geologic ranges, they are likely to be the most useful fossils for biostratigraphy of freshwater deposits because of the paucity of other forms. Their usefulness is enhanced by

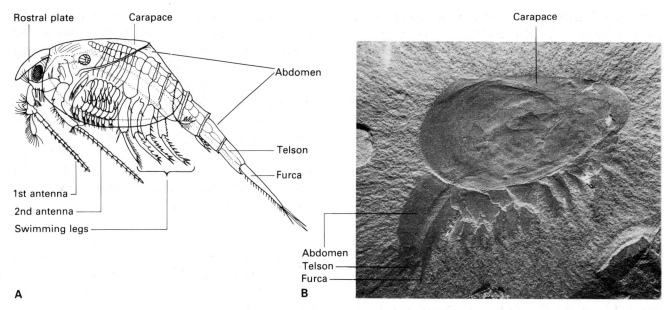

Figure 13.41 Modern and fossil phyllocarids. **A,** Modern *Nebalia* from the north Atlantic with left side of carapace shown as though transparent. (×4.) **B,** *Canadaspis,* the oldest fossil crustacean with well-preserved appendages (Middle Cambrian, Burgess Shale, British Columbia, ×2). (**A** from Moore, R. C.; McCormick, L. In: Moore, R.C., editor. *Treatise on invertebrate paleontology, Part R.* Boulder, CO, and Lawrence, KS: Geological Society of America and University of Kansas Press; 1969. **B** from Briggs, D. E. G. *Philosophical Transactions of the Royal Society of London* 281: 439–487; 1978.)

the ease with which their eggs are transported from basin to basin, ensuring that the same species are present over vast geographic areas. Cenozoic intermontane basins of western North America have been zoned with ostracodes, and some work has been done with late Paleozoic nonmarine deposits.

Other crustacean classes

CLASS MALACOSTRACA (Figures 13.41; 13.42)

Crustacea with carapace usually covering head and thorax; head with six segments, biramous first antennae and four other pairs of appendages; thorax with eight segments, and up to ten pairs of appendages. Abdomen with six (rarely seven) segments, commonly ventrally flexed, with telson. Compound eyes commonly on movable stalks.

Malacostraca inhabit a broad range of environments, being second among arthropods only to the Ostracoda in this respect. They occupy semiterrestrial, freshwater, and most marine habitats. Such forms as shrimp breed in coastal lagoons where food is abundant and later migrate onto the open shelf. Malacostracans include behaviorally sophisticated predators, scavengers, and deposit feeders that have left a good record of distinctive tracefossils.

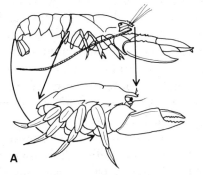

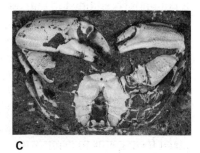

Figure 13.42. Decapod morphology. **A,** Position of the abdomen of crabs (below) compared to other decapods. **B,** Dorsal view of *Longusorbis,* a crab preserved in a Cretaceous concretion (×0.85). **C,** Ventral view of *Longusorbis.* (×0.85). (**A** from Moore, R. C. In: Moore, R. C., editor. *Treatise on invertebrate paleontology, Part R.* Boulder, CO, and Lawrence, KS: Geological Society of America and University of Kansas Press; 1969. **B** and **C,** courtesy B. C. Richards.)

SUBCLASS PHYLLOCARIDA (Figure 13.41)

Malacostraca with carapace bivalved, with or without dorsal hinge line; abdomen with seven segments and telson with pair of furcae.

Phyllocarids range from Cambrian to the present and are the most primitive malacostracans. The specimen shown in Figure 13.41, *B* from the Burgess Shale is the oldest undoubted crustacean, although some older specimens without preserved appendages have been assigned to the Crustacea on less solid evidence. Paleozoic genera are thought to have occupied a wide variety of ecological niches from which their descendants, the higher malacostracans, displaced them.

Phyllocarids resemble ostracodes in having a bivalve carapace, but the two groups show little resemblance beyond this. The phyllocarid carapace is light in structure and does not cover the posterior part of the abdomen (Figure 13.41). Some genera lack a hinge, and one group has a median dorsal plate separating the two valves. This latter structure provided double hinges, perhaps allowing the valves to extend laterally like wings. In addition to the two valves, the carapace also includes an anterior, movable rostral plate. The abdomen has a posterior pair of rodlike, bladelike, or filamentous furcae.

Modern phyllocarids (Figure 13.41, *A*), which some specialists place in a separate subclass, are usually a few millimeters long. The largest living individuals are only 4 cm in length. Paleozoic forms, however, were commonly much larger, reaching lengths of 75 cm.

SUBCLASS EUMALACOSTRACA (Figures 13.4; 13.42)

Malacostracans with carapace, if present, not bivalved; abdomen with six segments, and telson without furcae.

This relatively advanced group contains several orders; however, only one of these, the Decapoda, has a fossil record sufficient to warrant discussion here.

ORDER DECAPODA (Figures 13.4; 13.42)

Eumalacostraca with first three pairs of thoracic limbs modified as maxillipeds; at most five pairs of walking legs, one or more pairs with chelae; abdomen fully developed or variously reduced or incurved.

The decapods include the modern lobsters, crayfish, shrimp, and crabs and have a fossil record extending from the Late Devonian to the Holocene. The quality of the fossil record varies with the degree of calcification of carapaces and appendages. Identification of fossil decapods is often hampered by the fact that the classification of living decapods is based largely on appendages, features of the carapace, and information about life history that may not be available in the fossil record. For example, one important distinction between two suborders is that females of one lay eggs that hatch at an early stage of ontogenetic development, whereas females of the other carry on their abdominal appendages eggs that hatch at an advanced stage. Such information is rarely available from the fossil record.

As the name implies, Decapoda have ten legs, which are attached to the thorax (Figure 13.4). These may be modified in a variety of ways by loss of parts, fusion of podomeres, formation of chelae that flatten to a paddle shape, and even reduction and loss of those at the posterior end. Especially important developments of some decapods are chelae, the pincers used primarily for crushing shells and cutting food. These chelate appendages are often strongly calcified, especially among the crabs (Figure 13.42), accounting in part for the better fossil record of crabs compared to other decapods.

Decapods have numerous adaptations for protection, defense, and concealment that may strongly affect their morphology and the completeness of their fossil record. Relatively slow-moving forms, for example, are likely to be strongly calcified, whereas such swimming forms as shrimp are usually weakly calcified. Many decapods are able to swim backward rapidly by using their telson as a paddle. A pursuing predator thus encounters the chelae, which are not primarily for defense but are nevertheless useful in that function. Some decapods have spines and ridges that protect their stalked eyes. Crabs often cover themselves with sponges, shells, or plant debris, and some mimic the shape and color of corals or plants of irregular shape. Burrowing is often only a temporary means of concealment, but some shrimps make elaborate permanent burrows and are responsible for extensive bioturbation of the substrate. Perhaps the most elaborate means of concealment has taken place among the hermit crabs, which range from Jurassic to the present. Primitive forms are symmetrical and commonly occupy empty scaphopod shells or bamboo tubes. More advanced forms have lost their bilateral symmetry and live in abandoned gastropod shells, preferring dextrally coiled ones. One chela of the first walking leg serves as an operculum and may be modified in shape to give a tight closure to the snail aperture. Some hermit crabs are almost entirely terrestrial, returning to the sea only to mate. When on land, these crabs are very particular about the weight of the shells they occupy, usually discarding a heavy one in favor of a lighter one of the same size. In some nearshore marine environments, nearly every gastropod shell is occupied by a hermit crab, suggesting that the availability of shells may be the environmental factor that limits the size of the crab population. In such circumstances, the shells of gastropods may be transported by the hermit crabs into environments different from the ones in which the snails lived, complicating the interpretation of paleoenvironments.

Among the most significant evolutionary developments of the decapods is the evolution of the crabs, in which the abdomen is folded under the thorax and some segments are fused (Figure 13.42, *A*). This position of the abdomen protects the gills from clogging with sediment and enables crabs to live in muddy environments and to burrow more effectively. Crabs first appeared in the Jurassic, and their adaptive radiation is still proceeding. With about 635 modern genera, they outnumber all other decapods by a ratio of almost two to one and have a fair fossil record.

CLASS BRANCHIOPODA (Figure 13.43)

Carapace forming dorsal shield or bivalve shell or absent; number of segments highly variable; first antennae usually reduced and unsegmented; thoracic limbs generally uniform and leaflike; paired eyes usually present.

Two orders of branchiopods (Conchostraca and Cladocera) have useful fossil records. Because of their weak calcification, however, neither group has as complete a fossil record as the other groups of Crustacea that have been discussed.

The order Conchostraca includes crustaceans with leaflike (phyllopod) thoracic appendages (Figure 13.43, *A*). A short body is enclosed in a laterally compressed, bivalve carapace. During each molt, an inner chitinous layer called the duplicature is cast off, but usually the outer part of the carapace is retained, and a fine growth line is added peripherally to each valve. The shape of the valves and the growth lines give the carapace a superficial similarity to shells of some clams (Figure 13.43, *B*).

Many species of conchostracans show sexual dimorphism of the carapace, which can be important in taxonomy. Masses of eggs are attached to a pair of swimming appendages and are cast off with the exuvia during molting. Because as many as 120 eggs are produced at one time and molting may occur as often as every two or three days, a great number of eggs can be laid by a single female. In rare instances fossil eggs have been found associated with valves of female conchostracans.

Most conchostracans live in ponds or other small, temporary bodies of water, although a few species have been reported from brackish water and coastal salt flats. In recent environments, ponds rarely contain more than a single species of conchostracan at a time. In the fossil record, however, many species may be found together, presumably the result of time averaging. Because a typical geologic sample from an ancient pond deposit

A   **B**   **C**

Figure 13.43. Typical freshwater branchiopods. **A,** *Cyzicus,* a modern conchostracan with left valve removed to show body and appendages. **B,** Right valve of *Estherites* from the Tertiary of China. (×1.5.) **C,** *Ceriodaphnia* with eggs in special brood chamber. (×35.) (**A** from Moore, R. C.; McCormick, L. In: Moore, R. C., editor. *Treatise on invertebrate paleontology,*

Part R. Boulder, CO, and Lawrence, KS: Geological Society of America and University of Kansas Press; 1969. **B,** courtesy Chang Wen-tang. **C** modified from Tasch, P. In: Moore, R. C., editor. *Treatise on invertebrate paleontology, Part R.* Boulder, CO, and Lawrence, KS: Geological Society of America and University of Kansas Press; 1969.)

represents many years, conchostracans from such a sample may represent portions of many successive communities. By careful sampling some paleontologists have been able to detect successive assemblages of conchostracans, enabling them to determine the number of years during which a deposit formed.

The order Cladocera also includes phyllopod crustaceans (Figure 13.43, *C*). The body but not the head is enclosed by a one-piece uncalcified carapace that appears to be bivalved because it is compressed and has a distinct, lengthwise dorsal notch where it joins the body. Segmentation is only weakly developed, and unlike most other crustaceans, adult cladocerans lack maxillae. Many species are sexually dimorphic with the males being smaller than females and having larger first antennae and a modified abdomen. Male cladocerans are

rare, however, and many species reproduce largely asexually. Eggs and larvae are carried in a brood chamber, and young are released at the time of molting.

Although a few marine and brackish-water cladocerans are known, most live in fresh water. Their principal interest to paleontologists lies in the preservation of their skeletal remains in many postglacial lake deposits. Because most Holocene species have living representatives with well-understood ecology, study of fossil or subfossil cladocerans can aid in deciphering the depositional history of Holocene lakes.

CLASS CIRRIPEDIA (Figure 13.44)

Sessile marine crustaceans, initially attached by first antennae; other appendages and body typically surrounded by calcareous plates that are not molted; sexes separate or combined; males, when present, reduced, occurring with females or hermaphrodites.

Cirripeds or barnacles are highly unusual crustaceans that lose many of their typical arthropod features during ontogeny. All adults of the principal order are sessile, and all have a bivalve larval stage that bears a close superficial resemblance to a smooth-shelled ostracode. The carapace of the larva is discarded during metamorphosis and replaced by a fleshy mantle and a series of calcified plates that form a protective compartment for the body and appendages (Figure 13.44). Among nonparasitic forms, setose appendages are used to collect food. Parasitic forms feed with piercing mouth parts, by absorption through rootlike systems, or directly through the body wall.

Sexes of barnacles may be either separate or combined. Hermaphroditism occurs and was thought to be very common before the extreme sexual dimorphism was recognized. Cross fertilization is now known to be the rule.

The principal order, Thoracica, has a fossil record that begins in the Silurian, although fossils are rare except locally in Mesozoic and Cenozoic rocks. It is divided into four suborders, of which the acorn barnacles (suborder Balanomorpha), common worldwide in modern shallow-marine environments, have by far the best fossil record. They appear in Cretaceous rocks and today are widespread, occupying shorelines of all continents. They also commonly attach to such swimming and floating objects as whales and ships. The shell is bilaterally symmetrical and consists of four to eight calcified **compartmental plates** that are usually rigidly articulated to form a circular wall (Figure 13.44, *A* and *B*). The circular

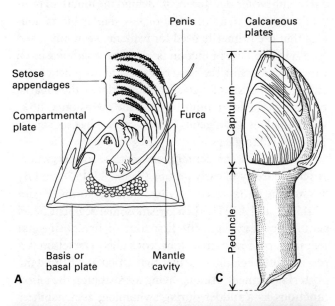

Figure 13.44. Morphology of Cirripedia. **A,** Cut-away view of a balanomorph barnacle showing relationship of body and skeletal features. (×4.) **B,** Modern balanomorph barnacles encrusting a gastropod shell. Only the compartmental plates of the circular wall remain. (×0.9.) **C,** External morphology of a lepadomorph barnacle. (×1.2.) (**A** and **C** from Newman, W. A. *et al.* In: Moore, R. C., editor. *Treatise on invertebrate paleontology, Part R.* Boulder, CO, and Lawrence, KS: Geological Society of America and University of Kansas Press; 1969. **B,** photo by Michael Frederick.)

wall is anchored to the substrate by either a membranous **basis** or a calcareous **basal plate**. The orifice is covered by four additional plates. Fossil balanomorphs are most often found with the circular wall intact. The goose-necked barnacles (suborder Lepadomorpha) range from Silurian to the present (Figure 13.44, C). The adult consists of a fleshy **peduncle** for attachment and a **capitulum** containing the body and appendages. The capitulum is commonly covered by calcareous plates, and in some forms even the peduncle bears plates. These plates are retained by the barnacle during molting and are marked by numerous growth lines. Two other suborders have very sparse fossil records.

Parasitism has developed in four of the five orders of barnacles. Two of these orders are unknown as fossils. Another order (Ascothoracica) has a sparse fossil record that extends back to the Cretaceous but is known only from cysts formed in octocorals and from small holes bored in echinoid tests. The free-living order Acrothoracica has also left a record of trace fossils by boring into the shells of pelecypods and gastropods, dead coral, and limestone.

Superclass Chelicerata

Richard A. Robison

CLASS MEROSTOMATA (Figures 13.45; 13.46)

Primitive aquatic chelicerates breathing by means of gills. Prosoma with seven pairs of locomotory and feeding appendages; opisthosoma with a variable number of covered appendages, most having a respiratory function.

The merostomes are represented by only three surviving genera (for example, Figure 13.45, D and E), all of which live in marginal marine environments. Although generally rare, sufficient fossil merostomes have been found to indicate a moderate diversity in middle Paleozoic seas. Members of one order (Eurypterida) reached lengths of almost 2 m, which places them among the largest known arthropods. They were possibly the largest invertebrates of their time.

Morphology of merostomes

The merostome body is composed of two discrete tagmata, an anterior prosoma and a posterior opisthosoma, the latter commonly terminating in a large styli-form telson (Figure 13.45, A). Functionally, the prosoma is primarily adapted for feeding, locomotion, and sensory reception, whereas the opisthosoma is specially adapted for respiration, reproduction, and excretion. Weak longitudinal trilobation of the dorsal surface is common.

Except for paired appendages, the fused prosoma shows no external evidence of segmentation and is generally semicircular or quadrate in dorsal view. Among the few features of the dorsal prosoma are two laterally situated compound eyes and a closely set pair of smaller, median eyes called **ocelli**. The mouth is centrally located on the ventral prosoma, and clustered nearby are seven pairs of limbs (Figure 13.45, E). The front pair, the chelicerae, are modified for biting or piercing. They originate with the second segment but become preoral as the result of posterior migration of the mouth during early embryonic development. Behind the mouth of most adults are five pairs of more or less similar, uniramous legs that are primarily used for walking, swimming, and feeding. The latter function is aided by gnathobases on the coxae. Usually the posterior prosomal segment bears either a pair of small specialized appendages, the **chilaria**, or a single highly modified plate, the **metastoma**; however, one Devonian genus (*Weinbergina*) has six similar pairs of walking legs.

The opisthosoma contains six to 12 segments, which articulate freely in early groups (Figure 13.45, A and B) but may be partly or entirely fused in later forms (Figure 13.45, D and E). The first opisthosomal segment bears genital openings, and the first four or five postgenital segments possess paired, foliaceous gills. The telson is a remarkably versatile structure that articulates behind the posterior segment. Among living merostomes, the telson functions as a rudder during swimming, as a stabilizer when the animal settles through the water, and as an aid in righting overturned individuals. It also contributes to propulsion during burrowing.

Classification of merostomes

During recent years, most classifications of the Merostomata have included four orders: the Aglaspida (Figure 13.45, C), Eurypterida, Xiphosurida, and Chasmataspida. In 1979, a detailed review was published of the single known aglaspid specimen with appendages approximately in place. This showed that the anterior tagma probably has only four pairs of appendages rather than six as previously reported, and the presence or absence of chelicerae could not be confirmed. Therefore, continued assignment of the Aglaspida to the Merostomata does not seem to be warranted, and other class

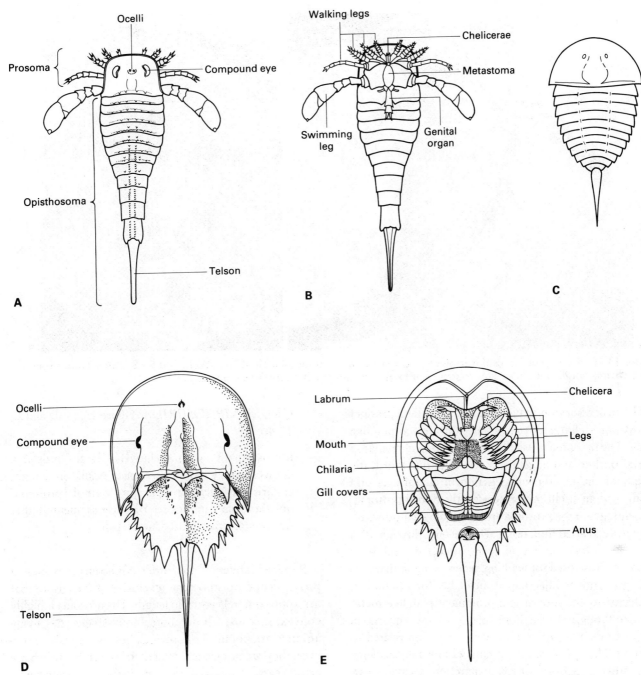

Figure 13.45. Morphology of representative merostomes and an aglaspid of uncertain class assignment. **A** and **B**, Dorsal and ventral views of *Eurypterus* (order Eurypterida) from the Silurian. **C**, *Aglaspis* (order Aglaspida) from the Cambrian. **D** and **E**, Dorsal and ventral views of *Limulus* (order Xiphosurida), a recent horseshoe crab. (Modified from Snodgrass, R. E. *Arthropod anatomy.* Ithaca, NY: Cornell University Press; 1952. Used by permission of Cornell University Press.)

assignment is uncertain at this time. The order Chasmataspida is based on a single unusual genus from Lower Ordovician strata in Tennessee, and because specimens with appendages are unknown, the higher taxonomic assignment of the Chasmataspida is open to question. This leaves only the Eurypterida and Xiphosurida as unequivocal orders of the Merostomata.

ORDER EURYPTERIDA (Figure 13.45, *A* and *B*; 13.46, *A*)

Merostomata with prosoma of moderate size, generally less than one-fourth the length of opisthosoma; metastoma on ventral prosoma; opisthosoma contains 12 movable segments and telson.

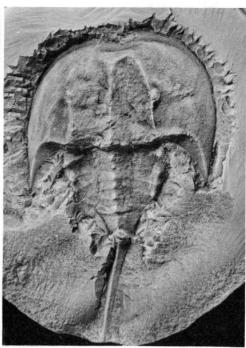

Figure 13.46. Representative fossil merostomes. **A,** *Eurypterus* from the Silurian of New York. (×0.5.) **B,** *Paleolimulus* from the Pennsylvanian of Illinois. (×1.5.) (**A,** courtesy Smithsonian Institution. **B,** courtesy D. C. Fisher.)

The order Eurypterida includes small-to-large, extinct merostomes characterized by an elongate scorpionlike body. The prosoma is usually quadrate to semicircular in dorsal outline and is generally less than one-fourth the length of the opisthosoma. Reniform lateral eyes and small median ocelli are present on the dorsal prosoma. Appendages of the prosoma are morphologically diverse, suggesting varied functions. The second through fifth pairs are often similar in form, however, and were apparently adapted for walking, whereas the sixth pair is enlarged and paddle shaped, probably for swimming. Posterior to the mouth is a prominent platelike metastoma, thought to be homologous to the chilaria in xiphosurids and presumably with a function related to feeding. The opisthosoma contains 12 movable segments and either a styliform or spatulate telson; segments two through six bear platelike appendages with attached, covered gills. In many specimens, a gradual to abrupt change in width of opisthosomal segments subdivides the tagma into a wider **preabdomen** with seven segments and a narrower **postabdomen** with five segments (Figure 13.3, *D*). Accentuation of the preabdomen tends to be greatest in large specimens and apparently was a modification to accommodate differentially increased surface requirements of the associated ventral gills. Like modern xiphosurids, most eurypterids probably had mixed benthic and nektic habits. The group is most common in marginal-marine, brackish-water, and freshwater facies.

ORDER XIPHOSURIDA (Figure 13.45, *D* and *E*; 13.46, *B*)

Merostomata with relatively large prosoma, approximately equal to opisthosoma in length; chilaria usually present on ventral prosoma; opisthosoma contains ten or fewer segments, either free or some or all fused, and telson.

Representatives of the order Xiphosurida possess a prosoma and opisthosoma (exclusive of the telson) that are approximately equal in length. The prosoma is highly convex. Because of its horse-hoof shape, the living animals are popularly known as horseshoe crabs; however, they are not closely related to true crabs, which are crustaceans. Contrary to some published statements, probably all xiphosurids possess both compound eyes and ocelli. In later and more advanced xiphosurids, the prosoma and opisthosoma articulate along a hinge line specialized to enable complete enrollment and rather precise occlusion of those two tagmata. The opisthosoma contains ten or fewer segments, and a common evolutionary trend was the progressive fusion of opisthosomal segments commencing at the posterior. In some instances, this fusion was accompanied by a reduction of segments. At least in later members, a small pair of platelike chilaria project downward from the posterior prosoma to block the back end of a median food channel located between gnathobases of the other pairs of prosomal appendages. Postgenital appendages of the opisthosoma are highly specialized for respiration. Although xiphosurids are capable of walking on substrates and swimming, much of their activity is spent burrowing near the surface in search of such prey as worms and mollusks.

Origin and distribution of merostomes

Evolution of the merostome body plan presumably resulted from adaptation to a carnivorous habit that required functional modification of some appendages. Small median structures on the gill-bearing limbs of xiphos-

urids, thought by some investigators to represent vestigial endopodites, suggest a biramous ancestry.

Because of their thin, uncalcified cuticle, the merostomes have left a fairly poor fossil record. Although the dorsal tagmata of several early Paleozoic genera possess merostomelike features, almost nothing is known of their appendages and cephalic segmentation. Therefore, the origin of the Merostomata, and hence the Chelicerata, remains obscure.

Typical eurypterids (Figure 13.46, *A*), known from about 25 genera, range from the Ordovician to the Permian, with maximum diversity in Silurian and Devonian rocks. Other less well-preserved fossils showing some eurypterid features date from the Early

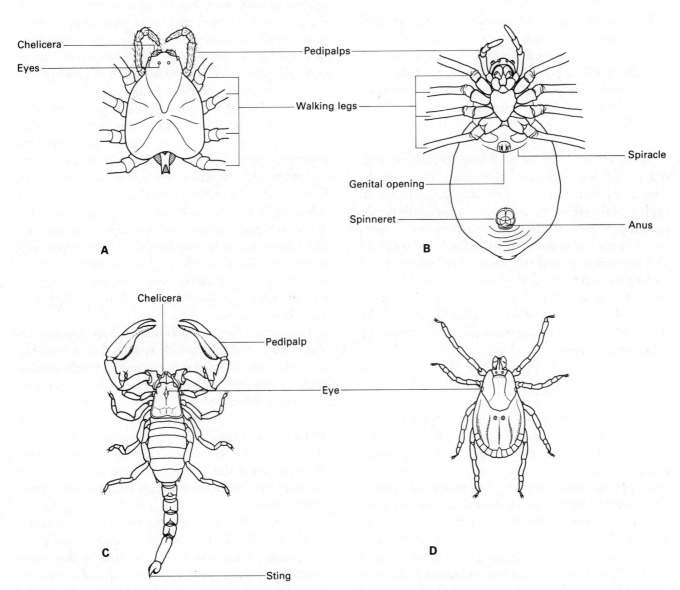

Figure 13.47. Morphology of representative living Arachnida. **A** and **B**, Dorsal view of prosoma and ventral view of entire female body of *Argiope* (order Araneida), a spider. **C**, *Chactas* (order Scorpionida), a scorpion. **D**, *Dermacentor* (order

Acarida), an unfed tick. (Modified from Snodgrass, R. E. *Arthropod anatomy.* Ithaca, NY: Cornell University Press; 1952. Used by permission of Cornell University Press.)

Cambrian. These may include the ancestors of the merostomes, but their classification is presently uncertain.

Typical xiphosurids, known from almost 30 genera (Figure 13.46, *B*), range from the Silurian to the present. A few rare fossils with some xiphosurid features range from the Early Cambrian; however, a lack of preserved appendages precludes confidence in their classification. Because the xiphosurids have maintained a remarkably stable morphology since the late Paleozoic (compare Figures 13.45, *D* and 13.46, *B*), extant representatives have sometimes been referred to as "living fossils."

CLASS ARACHNIDA (Figure 13.47)

Primarily terrestrial chelicerates usually breathing by means of lung books, tracheae, or both. Appendages generally restricted to prosoma, consisting of one pair of preoral chelicerae followed in order by one pair of postoral **pedipalps** (chiefly for feeding or sensory functions) and four pairs of walking legs.

Familiar arachnids are the abundant scorpions, spiders, ticks, and mites. Today this group comprises the dominant invertebrate carnivores on land. Scorpions and spiders eat great quantities of other arthropods, especially insects, and therefore are important to humans in the control of undesirable pests. Ticks and mites are also important because they are intermediate hosts of a variety of such serious diseases as Rocky Mountain spotted fever and African tick fever. Also, the common mange of several kinds of domestic animals is caused by mites, and plants as well as stored grains are attacked by other types of mites.

Morphology of arachnids

The prosoma contains six segments and the opisthosoma, 12. In more primitive groups (such as scorpions) (Figure 13.47, *C*), the opisthosoma joins the prosoma along a broad, fairly rigid juncture. In comparison, the first opisthosomal segment in spiders is greatly constricted to form a narrow waist that allows the opisthosoma to be easily flexed in all directions while constructing webs or binding prey with silk produced by special posterior glands. In the parasitic ticks (Figure 13.47, *D*) and mites, need for opisthosomal flexibility is negligible, and the prosoma and opisthosoma are fused in a short, flat body.

The arachnid prosoma bears a variable number of simple eyes. Unlike merostomes, most arachnids lack compound eyes; however, a few scorpions from the Silurian to Triassic Periods appear to possess compound eyes similar to those of the eurypterids. Six pairs of appendages are present on the ventral surface of the prosoma. Depending on the type of prey or host, considerable variation in the chelicerae and pedipalps has developed in response to different needs for grasping, piercing, tearing, and cutting. The four posterior pairs of prosomal appendages are primarily used for walking but in a few groups, one pair may be modified as tactile organs or for sperm transfer. Except for spiders with special **spinnerets** bearing silk glands (Figure 13.47, *B*) and scorpions with a pair of comblike structures, appendages are absent from the opisthosoma.

Most arachnids feed on liquids, whereas most merostomes probably took solid food. One common arachnid technique is to tear the prey open with the chelicerae, inject digestive juices, and then suck up the predigested, liquified contents, with the pedipalps providing a squeezing action. Finally, the empty hull of the prey is discarded. Another approach is to tear the prey into small pieces that are simultaneously bathed in digestive juices, and the resulting liquid is then ingested by sucking. Mouth parts and the foregut are extensively modified for intake of the liquid food and to filter out solids. Toxic substances for subduing the prey are produced by different glands found in the chelicerae of spiders, the pedipalps of one scorpion-like group, and in a sting on the posterior segment of scorpions. These arachnids have gained a sinister reputation because, when disturbed, they may inject humans with their painful poison; however, the results are seldom serious.

Respiration is accomplished directly through the cuticle in such tiny arachnids as mites. Larger arachnids require relatively larger surfaces, resulting in the development of internal lung books, tracheae, or a combination of the two. Where present, lung books occur on any of the second to sixth segments of the opisthosoma; the number of pairs varies from order to order but is fixed in each order. Lung books and tracheae open to the exterior through special slitlike closing devices (spiracles), which not only regulate intake of oxygen but also control loss of water vapor.

Excretion of nitrogenous compounds by merostomes and most orders of the arachnids is through coxal glands on prosomal legs. In scorpions, true spiders (order Araneida), and sun spiders (order Solpugida), the coxal glands are reduced in number and function. As a partial substitute, special tubules, usually restricted to the opisthosoma, collect waste from body fluids and void this waste into the hindgut. As an adaptation to dry terrestrial conditions, excess water and other materials are resorbed in the lower part of the tubules and in a connecting pouch of the hindgut.

Classification of arachnids

Arachnids have been assigned to orders primarily on the basis of such external features as general body form, structure of the prosoma and opisthosoma, degree of visible segmentation, and nature of appendages. Arachnid specialists generally recognize 16 orders, of which five are extinct (Table 13.2). Because of a thin, uncalcified cuticle and mostly terrestrial habitats, arachnids have a poor and unevenly known fossil record. For this reason, definitions of the several orders are not included here. Most fossil arachnids are from either Carboniferous swamp deposits or Tertiary amber.

Origin and distribution of arachnids

Many theories relating to the origin of arachnids have been proposed, but none has received wide acceptance. Earlier theories usually suggested a merostome origin for all arachnids by way of primitive scorpions. More recently, it has been recognized that although scorpions are ancient in origin and generalized in some characters, they are specialized in such characters as position of the mouth and structure of appendages. Alternative theories suggest that the class Arachnida is polyphyletic, resulting from transitions to terrestrial habitats by members of different groups of marine chelicerates.

The earliest known arachnids are scorpions from the middle Silurian, which also have commonly been cited as the earliest terrestrial animals. Additional recent discoveries and further studies have shown, however, that the earliest scorpions probably possessed gills (Figure 13.48, *B*) and were aquatic, living together with eurypterids in hypersaline to brackish marginal marine environments. Much like modern xiphosurids, early scorpions possessed cover plates that could have kept their gills moist during brief ventures out of water. The oldest known scorpions with ventral slitlike spiracles, which indicate the presence of lung books and hence complete adaptation to life on land, are from the Mazon Creek fauna of Pennsylvanian age in Illinois.

Exactly when the first arachnid became fully adapted to dwelling on land is still an open question. After their first appearance in aquatic habitats, it seems reasonable to assume that such groups as the scorpions passed a transitional stage in which progressive evolution enabled them to spend more and more time out of water. Because of its specialized leg structures, a representative of the order Trigonotarbida from Lower Devonian rocks in

Table 13.2. Numbers of known arachnid genera and species*

Order	Paleozoic Genera	Paleozoic Species	Mesozoic Genera	Mesozoic Species	Tertiary Genera	Tertiary Species	Quaternary† Genera	Quaternary† Species
Scorpionida (scorpions)	26	42	2	7	2	2	70	600
Pseudoscorpionida	—	—	—	—	13	18	234	1000
Phalangiida (harvestmen)	2	4	—	—	9	16	640	2350
Architarbida	14	25	—	—	—	—	—	—
Acarida (mites, ticks)	1	1	—	—	60	107	1389	6000
Haptopodida	1	1	—	—	—	—	—	—
Anthracomartida	11	20	—	—	—	—	—	—
Trigonotarbida	24	38	—	—	—	—	—	—
Palpigradida	—	—	1	1	—	—	4	21
Thelyphonida	2	5	—	—	1	1	10	70
Schizomida	—	—	—	—	2	2	3	28
Kustarachnida	1	3	—	—	—	—	—	—
Phrynichida	3	4	—	—	—	—	18	60
Araneida (spiders)	16	18	—	—	94	180	2735	25,000
Solpugida	1	1	—	—	—	—	134	600
Ricinuleida	2	9	—	—	—	—	2	16
Total	104	171	3	8	181	326	5239	35,745‡

*From Moore, R. C., editor. *Treatise on invertebrate paleontology, Part P*. Boulder, CO, and Lawrence, KS: Geological Society of America and University of Kansas Press; 1955, p. 48.
†Includes extant taxa.
‡In 1974, the estimated number of existing described species had increased to 65,000.

Figure 13.48. Fossil arachnids. **A,** Unnamed genus of spider (order Araneida) preserved in Eocene amber from Arkansas. (×20.) **B,** *Waeringoscorpio,* a primitive aquatic scorpion with probable gills at sides of the preabdomen (Lower Devonian, Germany. (×5.) **C,** *Alkenia,* the oldest known arachnid that appears to have been fully adapted to living on land. (×4.4.) (**A,** courtesy W. B. Saunders. **B** and **C** from Størmer, L. *Senckenbergiana Lethaea* 51: 335–369; 1970.)

Germany (Figure 13.48, C) has been cited as the oldest known terrestrial arachnid. The trigonotarbids have an observed range from Devonian to Carboniferous.

Whether as a result of one or several independent adaptations to terrestrial habitats, arachnids clearly underwent a rapid early radiation. Twelve of the 16 orders are represented in Carboniferous rocks, whereas only 11 orders are present in modern faunas (Table 13.2).

Among the arachnids, the carnivorous spiders comprise the most diverse order (Araneida in Table 13.2) among both fossil (Figure 13.48, A) and modern faunas. Their marked increase in Cenozoic deposits parallels the great adaptive radiations of flowering plants (angiosperms) and associated insects that began during the Cretaceous. The second most diverse group includes the predominantly parasitic mites and ticks of the order Acarida, whose marked Cenozoic increase at least partly reflects the radiation of mammalian hosts.

Superclass Hexapoda

Roger L. Kaesler

CLASS INSECTA (Figures 13.50 and 13.51)

Tracheate hexapods with 19 or 20 segments divided into three tagmata: head, thorax, and abdomen; head with six segments; thorax with three segments, each with pair of legs, second and third usually with

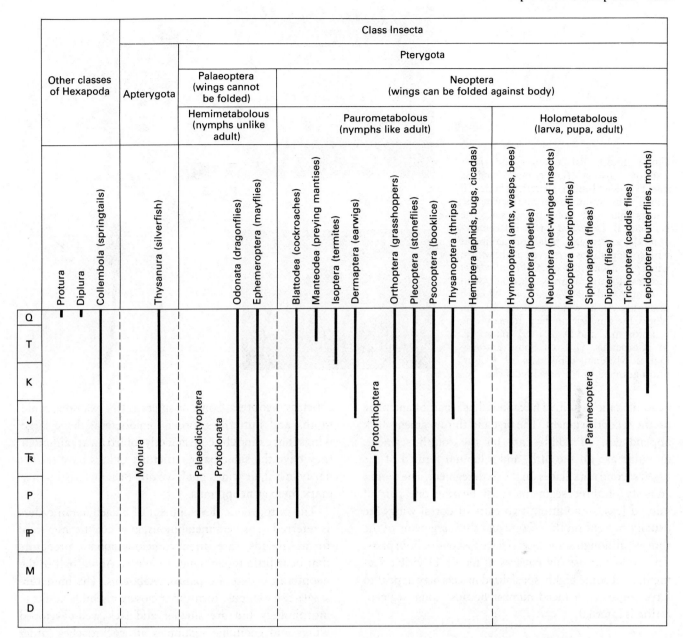

Figure 13.49. Geologic ranges of the minor hexapod classes and the common orders of insects. Major events in insect evolution shown at the top.

pair of wings, although some primitive forms wingless and some groups secondarily without wings; abdomen with ten or 11 segments, no legs.

Insects are familiar to all people, but the question of just what is and what is not an insect has not been settled. Most specialists prefer to separate such primitive, wingless hexapods as the proturans, diplurans, and springtails (Collembola) into classes separate from the insects. That procedure is followed here. A less common practice is to regard these groups as orders of the class Insecta together with 25 to 30 other orders of insects that are recognized

by nearly all specialists. Although the different approaches are of minor importance for most purposes, a discussion of the origin and early evolution of insects should specify which practice is being followed.

Morphology of insects

In spite of their great diversity, most insects adhere rather closely to a basic morphologic plan. The diversity usually arises from modifications of appendages rather than their loss or the loss of body segments.

The insect head contains six segments and, unlike the

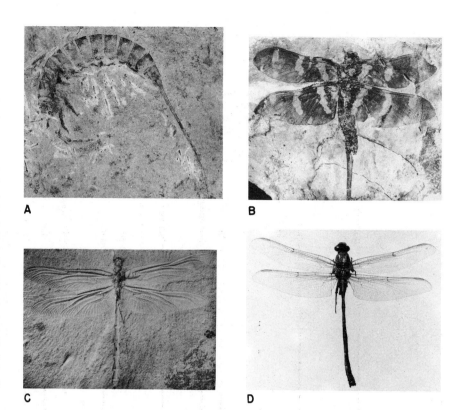

Figure 13.50. Palaeopteran and other primitive insects. **A,** Permian wingless insect (order Monura); length of body, 10 mm. The monurans are primitive, wingless insects that have been extinct since the Permian. **B,** *Dunbaria* (order Palaeodictyoptera), a Permian palaeopteran insect with color patterns preserved on the wings; wingspread is about 4 cm. **C,** *Protolindenia* (order Odonata), a Jurassic dragonfly with a wingspread of 12 cm. **D,** Modern dragonfly with wingspread of 10 cm. Notice the similarity in the pattern of wing venation in **C** and **D.** (**A** to **C,** courtesy F. M. Carpenter. **D** from Snow Entomological Museum, University of Kansas, photographed by Michael Frederick.)

head of some primitive hexapods, has a pair of antennae on the second segment. The first and third segments lack appendages. Mandibles are on the fourth segment, maxillae are on the fifth, and a labium formed of the fused second maxillae is on the sixth segment. The thorax consists of three segments, each bearing one pair of jointed legs. In addition, two pairs of dorsal wings are usually present on the second and third segments of the thorax, although some insects have lost one or both pairs. The abdomen usually consists of ten or 11 visible segments, but some highly specialized insects may appear to have a greatly reduced number because some segmentation is internal.

Classification of insects

The ability of people without specialized training to distinguish insects from most other kinds of arthropods attests to the fact that insects form a natural group. The many kinds of insects may be grouped rather easily into orders, the members of which have similar morphology and ontogeny (Figure 13.49). The fossil record reinforces this classification by showing that morphologic characters regarded as most primitive among living insects also appeared earliest. Orders of insects with wings that cannot be folded next to the body when the animal is at rest (for example, dragonflies, order Odonata), are grouped into the subclass Palaeoptera. Insects that can fold their wings against the body are assigned to the

subclass Neoptera. Some neopterans (for example, some moths and butterflies, order Lepidoptera), have large wings that cannot be conveniently folded away, although they have the same basic structure. Others have secondarily lost their wings but have other characteristics that mark them as neopterans.

The ontogenetic development of palaeopteran orders is referred to as **hemimetabolous,** in which the immature forms (naiads), are often voracious aquatic predators that bear little resemblance to adults. About half of the neopteran orders are **paurometabolous.** The immature stages are nymphs, forms that closely resemble adults in morphology but are smaller and lack well-developed wings and genitalia. Examples are cockroaches (order Blattodea) and grasshoppers (order Orthoptera), whose young may be easily recognized although rarely identified to species. The wings develop gradually outside the body as the nymphs grow. Other neopteran orders are said to be **holometabolous.** The larvae are quite different from the adults in most respects. Larval body shapes are wormlike, and they have mouth parts and hence feeding habits that are different from those of adults. Their larvae lack compound eyes, and their wings develop gradually within the body and do not appear externally until the end of larval life. Between the larval stages and the adult stage is an instar referred to as the **pupa,** during which the insect is transformed into an adult. The pupa has been called a resting stage, but it is actually a period of intense physiologic activity and morphologic change.

Figure 13.51. Examples of fossil neopteran insects, including some advanced Mesozoic and Cenozoic kinds. **A,** Pennsylvanian cockroach from Illinois (order Blattodea), length of forewing 2.6 cm. **B,** Permian scorpion fly (order Mecoptera), one of the oldest holometabolous orders. **C,** *Specomyrma* (order Hymenoptera), the oldest known ant, preserved in Cretaceous amber, a mode of fossilization that is responsible for its excellent preservation. **D,** *Apis* (order Hymenoptera), an Oligocene honeybee. **E,** Oligocene butterfly (order Lepidoptera) with unusual preservation of color markings on its wings. (**A,** courtesy F. M. Carpenter and R. Rock. **B** to **E,** courtesy F. M. Carpenter.)

Evolution of insects

Origin of insects and flight   Hexapods and myriapods are thought to have evolved from a common, unknown ancestor during the Devonian. The oldest known hexapods, a few fragments from Scotland, are primitive in form and appear to be springtails (class Collembola). Another fossil (*Eopterum*) was long interpreted as a wing and thought to represent the oldest known insect, but it is now interpreted as the telson of a primitive crustacean. Collections of fossil insects from Pennsylvanian and younger rocks contain numerous forms belonging to at least a dozen orders, about a third of which have living representatives. The evolution of wings, it seems, must have taken place during Mississippian time. Unfortunately, no fossil insects are known from Mississippian rocks, probably due to the widespread marine environments of the Mississippian and lack of deposits that were suitable for the preservation of insects. The eventual discovery of Mississippian insects promises to be an important event.

Development of wings was accompanied by a general increase in size. Wingless hexapods are nearly all small (Figure 13.50, *A*), whereas the Pennsylvanian insect fauna contains a number of large forms, including a primitive dragonfly with a wingspan of 75 cm, the largest insect known. Many other early palaeopteran insects are more usual in size, including other dragonflies, mayflies, and an extinct order, the Palaeodictyoptera (Figure 13.50, *B*). Early mayflies have wings of nearly equal size, whereas modern mayflies have posterior wings that are greatly reduced.

Although the oldest known fossil insects had wings, the use of them for flight may have evolved twice during the Mississippian or even earlier as indicated by two distinct methods of wing movement. Members of all orders of flying insects except the dragonflies (order Odonata) move all four wings more or less together. Dragonflies on the other hand, move their wings in pairs (Figure 13.50, C and D). When the front pair of wings is down, the rear pair is up, and vice versa.

Evolutionary history Four events were of primary importance in the history of insects (Figure 13.49). The first three—evolution of wings, the ability to fold the wings next to the body, and the holometabolus ontogeny—took place during the late Paleozoic. The fourth, coevolution of insects and flowering plants, occurred during Cretaceous and Cenozoic times.

Ability to fold the wings over the body, rather than leaving them extended outward as in dragonflies or upward as in mayflies, had evolved by Pennsylvanian time. Pennsylvanian faunas also include numerous kinds of neopteran insects. Especially noteworthy are cockroaches (Figure 13.51, A), some of which reached nearly 10 cm in length. Folded wings, of course, greatly enhance the ability of insects to move quickly over the ground, either in search of prey or to escape predators. Insects with folded wings are also able to escape into hiding places with less damage to their delicate wings. Dragonflies, which are typical palaeopteran insects, never use their legs for walking or running (Figure 13.50, C and D). In comparison, many neopterans, such as cockroaches and some beetles, use their wings only in unusual circumstances, and some species in most neopteran orders have lost the ability to fly.

Other Pennsylvanian neopteran insects of the order Protorthoptera are regarded as the ancestors of the grasshoppers, crickets, and related forms of the order Orthoptera. Several orders are found in the fossil record for the first time in the Permian: book lice (order Psocoptera); stone flies (order Plecoptera); and cicadas, leafhoppers, and aphids (order Hemiptera). Other hemipterans, the so-called true bugs, did not appear until the Triassic.

No early insect orders were holometabolous, but several orders of holometabolous insects first appeared during the Permian (Figure 13.51, B). The adaptive advantages of fundamentally different juvenile and adult forms are clear. Immature holometabolous insects are able to avoid competition with adults by using different food and occupying different habitats. The parasitic wasps, for example, are holometabolous. The larvae destroy their hosts and then emerge as adults that use different food and function primarily to sense new hosts and to disperse the population. In addition, in cold climates many holometabolous insects overwinter as pupae. Adults emerge in the early spring and seek out environments favorable to their larvae in which to lay eggs. Moreover, by getting an early start in the growing season, they may also produce two or more generations a year. Insects that lay their eggs in the fall, on the other hand, must rely on the chance that favorable environments for their larvae will be present during the following spring.

As for many marine invertebrates, the Late Permian was a time of extinction for several groups of insects, including most of the order Protorthoptera (Figure 13.49), which had declined during the Permian probably in response to increased competition from holometabolous insects. As a result of extinction of old orders and appearance of new ones, the orders of insects present at the end of the Permian were much the same as those found today.

Rise of the flowering plants during the Mesozoic was accompanied by coevolution of insect pollinators (Figure 13.51, D). The Mesozoic insect record is scanty, but it shows the evolution of a fauna of increasingly modern aspect. Of importance is the appearance of the parasitic wasps (order Hymenoptera), a group that has kept the plant-eating insects in check since the Cretaceous and is still far more efficient than chemical insecticides.

The Cretaceous marked the first appearance of the butterflies and moths (order Lepidoptera) (Figure 13.51, E), and the fleas (order Siphonaptera) appeared in the Tertiary. Lepidopterans are important pollinating insects, and their larvae also feed primarily on the tissues of flowering plants. Most fleas are parasites of mammals. It is not surprising, therefore, that both of these orders should appear first in the Tertiary along with the adaptive radiation of their hosts.

Supplementary reading

Clarke, K. U. *The Biology of the Arthropoda.* New York: American Elsevier Publishing Company; 1973. General introduction to arthropods written for beginning college students.

Eldridge, N. Trilobites and evolutionary patterns, pp. 305–332. In: Hallam, A., editor. *Patterns of evolution.* Amsterdam: Elsevier Publishing Company; 1977. Summary of several studies on the origin and evolution of trilobite taxa.

Fretter, V.; Graham, A. *A functional anatomy of invertebrates.* London: Academic Press, Inc.; 1976. Contains three chapters on arthropods; emphasis is on Crustacea and land arthropods.

Gould, S. J. *Ontogeny and phylogeny.* Cambridge: Harvard University Press; 1977. Chapter 7 is particularly recommended for its discussions of paedomorphosis and "recapitulation" or peramorphosis.

Gupta, A. P., editor. *Arthropod phylogeny.* New York: Van Nostrand Reinhold Co.; 1979. Collection of discussions and widely divergent conclusions by specialists in anatomy, embryology, paleontology, and physiology.

Hartmann, G., editor. Evolution of post-Paleozoic Ostracoda. Hamburg: *Abhandlungen und Verhandlungen des Naturwissenschaftlichen Vereins (NF)* 18/19: 7–336; 1976. Collection of 33 mostly specialized papers with references to much of the current literature on Ostracoda.

Loffler, H.; Danielopol, D., editors. *Aspects of ecology and zoogeography of recent and fossil Ostracoda.* Hague: Dr W. Junk b.v. Publishers; 1977. Collection of 43 papers.

Manton, S. M. *The Arthropoda.* Oxford: Oxford University Press; 1977. Excellent analysis of habits, functional morphology, and evolution. Uniramia, Crustacea, and Chelicerata are elevated to phylum rank and status of the Trilobita as a fourth phylum is left uncertain.

Martinsson, A., editor. Evolution and morphology of the Trilobita, Trilobitoidea, and Merostomata. *Fossils and Strata* 4: 1–467; 1975. Collection of papers summarizing recent research on primitive arthropods.

Moore, R.C., editor. *Treatise on invertebrate paleontology, Parts O–R: Arthropoda* 1–4. Boulder, CO, and Lawrence, KS: Geological Society of America and University of Kansas Press; 1955–1969. Comprehensive treatise on the morphology and systematics of fossil arthropods; two additional volumes on Hexapoda are in preparation.

Riek, E. F. Fossil history. In: Commonwealth Scientific and Industrial Research Organization. *The insects of Australia.* Melbourne: Melbourne University Press; 1970. Concise account of the evolution and fossil record of insects, which is not confined to Australia.

Schmitt, W. L. *Crustaceans.* Ann Arbor: University of Michigan Press; 1965. Brief, readable summary of the Crustacea.

14

Phylum Mollusca

John Pojeta, Jr.
Bruce Runnegar
John S. Peel
Mackenzie Gordon, Jr.

Part I Phylum overview

John Pojeta, Jr.

The concept Mollusca integrates much information about animals that at first glance appear to be radically different from one another (Figures 14.1 and 14.2). The concept is a triumph of comparative anatomy and was found to be readily usable by paleontologists because many mollusks have fossilizable shells surrounding some or all of the body. Furthermore, the shells bear a variety of marks caused by attachment of soft parts, which allow reconstruction of fossil species.

Included in Mollusca are snails, slugs, mussels, oysters, clams, squids, and octopuses. The diversity of the phylum is shown by the eight known classes. Mollusks range in size from microscopic snails and clams to squids 18 m long, and they live in most marine and freshwater environments; some slugs and snails live on land. In the sea, mollusks range from the intertidal zone to abyssal depths and are benthic, planktic, and nektic. There are even squids that glide for short distances through the air by means of enlarged lateral fins. The number of species of living mollusks is estimated to range from 50,000 to 130,000.

The word mollusca comes from Latin and is translated as "soft bodied." In the mid-18th century, Linnaeus presented the notion that most soft-bodied invertebrates are mollusks, but by the late 18th century Cuvier had developed a concept of Mollusca that approaches the one in use today. Because no one character is unique to every mollusk, it is not possible to frame a succinct morphologic definition of the phylum. The concept Mollusca is a phylogenetic one, and mollusks are unified by morphologic and anatomic similarities among the different forms, by embryologic similarities, and by fossil evidence of their evolutionary history.

Most mollusks can be described as free-living metazoans having a dorsal calcareous **exoskeleton** that provides structural support for a muscular **foot** and the **viscera**. Surrounding the soft parts is a space called the **mantle cavity**, which opens to the outside and whose outer wall is a thin flap of tissue called the **mantle** (Figure 14.1). The mantle secretes the shell. The mantle cavity serves as a passageway for incoming feeding and respiratory currents, as an exit for discharge of wastes, and sometimes as an organ of locomotion and sensory perception. Mollusks are unsegmented and have a small coelom. A few mollusks show serial repetition of some organs or structures, such as the gills of chitons (Figure 14.1, *B*), several structures in symmetrical univalves (Figure 14.1, *C*), and the valves of the shells of chitons (Figure 14.2, *B*), which is best termed **pseudosegmentation**. Mollusks have no septa subdividing the body into segments with which the various organ systems are numerically correlated by occurring in each segment as in annelid worms.

Mollusks have been diverse and abundant animals since earliest Cambrian time and are widespread as fossils. Early and Middle Cambrian mollusks include many unfamiliar forms, but by Late Cambrian time most extant classes are known as fossils. The first major radiation of mollusks occurred in the Ordovician, and at least 5000 species of mollusks are known from rocks of this age. In Mesozoic and Cenozoic time, mollusks

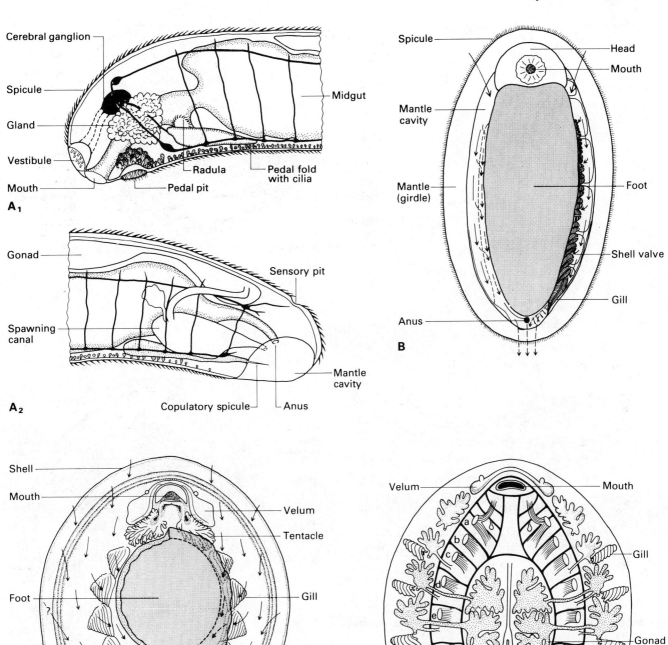

Figure 14.1. Major soft part features of each of the classes of mollusks. A_1 and A_2, Anterior and posterior ends of a generalized representative of the class Aplacophora (wormlike mollusks). Body wall on left side removed. **B**, Ventral view of a member of the class Polyplacophora (chitons). The arrows in the mantle cavity show the direction of the water currents. C_1 and C_2, Ventral exterior and interior views of a member of the class Monoplacophora (no common name). In C_1, the arrows show the direction of the water currents. C_2 shows pseudo-

segmentation; the muscles (*a-h*) are the foot retractors. **D**, Dorsal view of a generalized representative of the class Gastropoda (snails and slugs). **E**, Diagrammatic median section through the genus *Nautilus*, a living shelled representative of the class Cephalopoda (*Nautilus*, squids, octopuses). F_1 and F_2, Diagrammatic median section and right side view of a representative of the class Scaphopoda (tusk shells). **G**, Reconstruction of some of the soft parts on the left side of a member of the class Rostroconchia (no common name). The

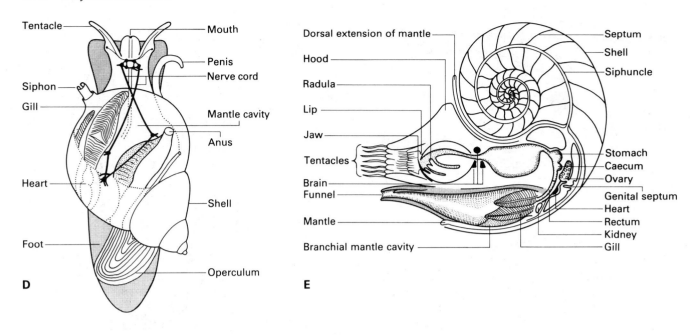

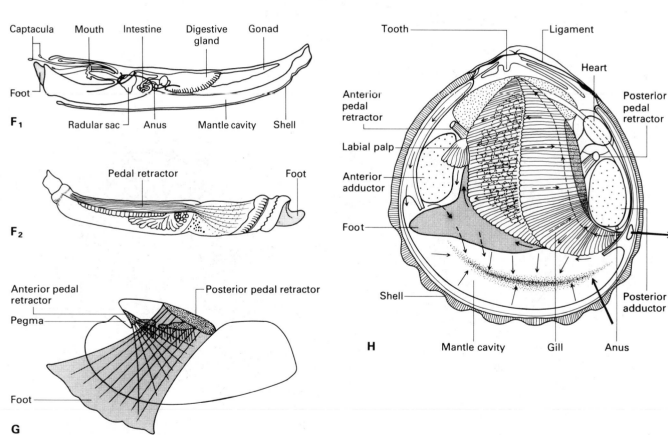

reconstruction is drawn on an internal mold. **H,** Left side view of a representative of the class Pelecypoda (clams, mussels, scallops, oysters). Left valve and mantle removed. Arrows on the gills, foot, and mantle cavity show the water currents in the body; the two long arrows above and below the anus show the main inhalant and exhalant currents of the gills. (**A** modified from Salvini-Plawen, L. *Malacologia* 9: 192; 1969. **B** modified from Yonge, C. M. *Quarterly Journal of Microscopical Science* 81: 371; 1939. **C** modified from Lemche, H.; Wingstrand, K. G. *Galathea Report.* Copenhagen, Denmark: Zoologisk Museum;

1959. **D** modified from Wenz 1938; in: Schindewolf, O. H. *Handbuch der Paläozoologie,* 1938, p. 2. **E** modified from Borradaile, *et al., The Invertebrata,* Cambridge University Press, 1958, p. 636, published with permission. **F** from Pelseneer, in: Lankester, E. R. *A treatise on zoology,* Adam & Charles Black, 1906, p. 198–199. **G** modified from Runnegar, B. *Royal Society of London Philosophical Transactions B* 284: 324; 1978. **H** from Yonge, C. M. *Proceedings of the Malacological Society of London* 38: 469; 1969.

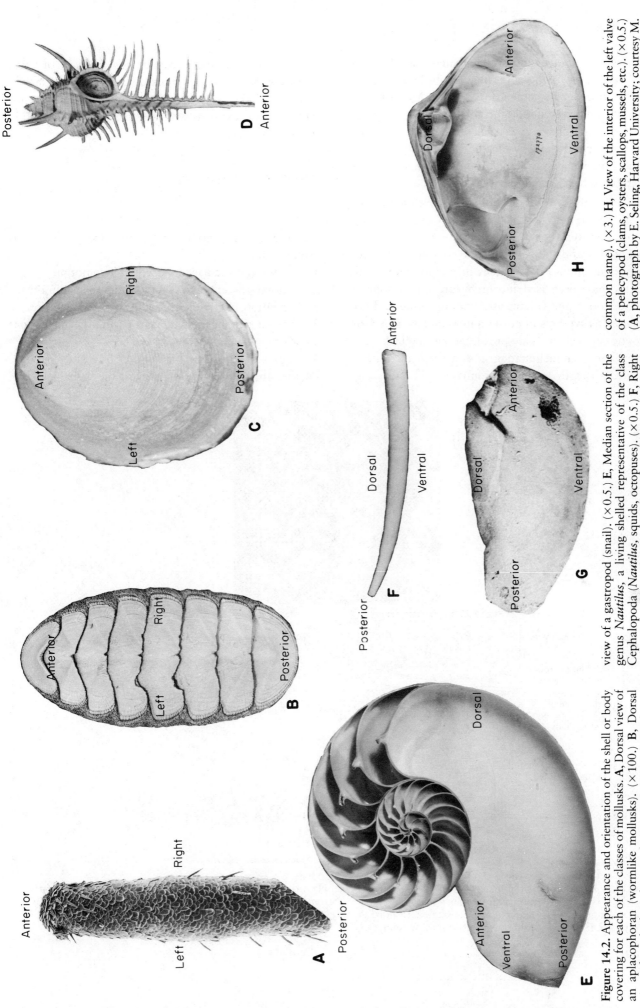

Figure 14.2. Appearance and orientation of the shell or body covering for each of the classes of mollusks. **A**, Dorsal view of an aplacophoran (wormlike mollusks). (×100.) **B**, Dorsal view of a polyplacophoran (chiton). (×0.75.) **C**, Dorsal view of a monoplacophoran (no common name). (×5.) **D**, Ventral view of a gastropod (snail). (×0.5.) **E**, Median section of the genus *Nautilus*, a living shelled representative of the class Cephalopoda (*Nautilus*, squids, octopuses). (×0.5.) **F**, Right side view of a scaphopod (tusk shell). (×0.5.) **G**, Right side view of an internal mold of the class Rostroconchia (no common name). (×3.) **H**, View of the interior of the left valve of a pelecypod (clams, oysters, scallops, mussels, etc.). (×0.5.) (A, photograph by E. Seling, Harvard University; courtesy M. P. Morse, Northeastern University, Nahant, MA.)

underwent great diversification, with many new lineages evolving and few dying out. Today mollusks are second to arthropods in the number of species. Mollusks are biostratigraphically important animals and are widely used for zonation in upper Paleozoic, Mesozoic, and Cenozoic rocks. In lower Paleozoic rocks, mollusk ranges are not yet well known, but some are important for biostratigraphic correlation.

Some bivalved mollusks (such as oysters, clams, scallops, and mussels), snails (periwinkles, escargot, and conchs), and squids occur in vast numbers. The estuaries of the world and the waters of the continental shelves supply many millions of tonnes of mollusks for food. Mollusks have been used since prehistoric time as sources of food, tools, and ornaments. Various mollusks have been used as symbols of power and as money, and their aesthetic appeal is widespread. Some mollusks are vectors for such debilitating and widespread human parasitic diseases as schistosomiasis. Virtually any modern seashore in any climate has a variety of living mollusks. Mollusks are most diverse and abundant in continental shelf and epicontinental sea environments, which in part explains why they are well represented in the fossil record.

Introductory morphology

Soft parts

The body of many mollusks can be divided into four parts (Figure 14.1):

1   A ventral muscular foot used for locomotion
2   A dorsal visceral mass that includes the major organs of circulation, excretion, reproduction, and digestion
3   An anterior head that has the mouth and various sense organs such as tentacles and eyes
4   A mantle, which is a fold of the body wall surrounding the other soft parts and secreting the shell.

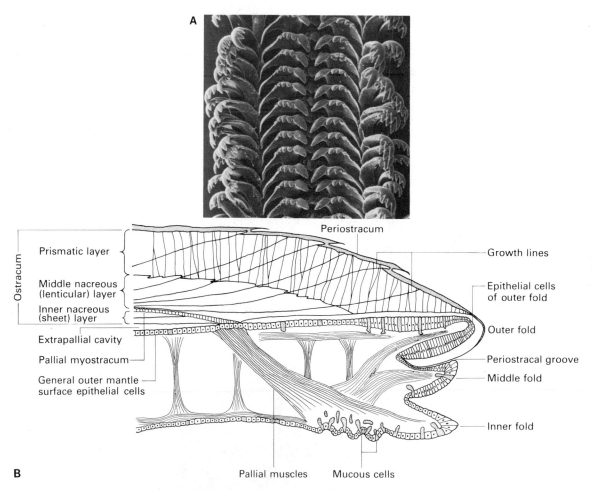

Figure 14.3. **A**, Radula of the living gastropod *Rhinoclavis*. (×75.) **B**, Diagram of shell-mantle relations in the marginal region of the pelecypod *Anodonta* as seen in radial section. (**A** courtesy R. S. Houbrick, Smithsonian Institution, Washington, D.C. **B** from Taylor, J. D.; Kennedy, W. J.; Hall, A. *Bulletin of the British Museum (Natural History) Zoology*, Supp. 3, p. 8; 1969.)

The space between the mantle and the rest of the body is the mantle cavity, which contains the gills and into which the digestive, reproductive, and excretory systems open. However, there are various exceptions to this arrangement of the body parts. For example, the worm-like mollusks (Figures 14.1, *A* and 14.2, *A*) lack a head and shell and the homolog of the foot is not muscular; clams (Figure 14.1, *H*) lack a head; and the homologue of the foot in *Nautilus*, squids, and octopuses (Figure 14.1, *E*) is divided into numerous grasping tentacles and a funnel used in swimming.

The gut is complete with an anterior mouth and a posterior anus. In *Nautilus*, squids, and octopuses the gut is bent into a U shape, and in snails it is twisted so that the anus opens into the mantle cavity above the mouth. Most living mollusks have a **radula** (Figures 14.1, A_1 and 14.3, *A*) at the back of the mouth. This elongate structure bears horny or mineralized teeth and can be protruded from the mouth to rasp the substrate while feeding. However, one class of mollusks that includes clams and oysters lacks a radula; these animals are presumed to have lost it because most of them filter microorganisms from water by cilia on the gills. The gut is subdivided and various organs empty into it. The stomach of some mollusks has an outpocketing termed the **style sac** that contains a rod known as the **crystaline style** whose function is not entirely clear. Feces are usually in pelletal form.

The nervous system has **ganglia** (Figure 14.1, A_1) and elongate or longitudinal **nerve cords**. Significant concentration of nervous tissue in the head occurs in some classes and is termed **cephalization**. Squids and octopuses have eyes comparable to those of vertebrates. The excretory organs are termed **nephridia** or kidneys (Figure 14.1, C_2), and their number, position, and openings are variable. Respiratory organs are the mantle and gills (Figure 14.1, *B*, C_1, *D*, *E*, and *H*), which are made up of thin partitions that surround spaces in which gaseous exchange occurs.

The circulatory system is open, although it approaches a closed condition in squids and octopuses. Most mollusks have a heart and veins and arteries between which circulation is completed in sinuses of the hemocoel in connective or muscular tissue. The oxygen-carrying substance in most mollusks is dissolved directly in the blood. In most mollusks the sexes are separate, but hermaphroditism occurs in some of the wormlike mollusks, clams, and snails. In many, gametes are shed directly into the sea where fertilization occurs. In others, fertilization is internal in the female.

Hard parts

The major calcareous part of mollusks is the shell or **conch**. It shows considerable variation in form (Figure 14.2) and ordinarily is bilaterally symmetrical except in snails. The shell is oriented with respect to the soft parts so that the **anterior** end is over the head, the **posterior** end is away from the head, the **dorsal** side is over the visceral hump, and the **ventral** side is toward the foot.

Most mollusks have a **univalved** (Figure 14.2, *C* to *F*) or a **bivalved** shell (Figure 14.2, *H*). Univalved shells are all in one piece. Bivalved shells have right and left halves held together on the dorsal side by an elastic organic **ligament** that is partly calcified; the line of junction along which the valves can be separated from each other is called the **commissure**. Some mollusks (Figure 14.2, *G*) have a shell that is univalved in the larval stage and bivalved in the adult, and the calcareous shell layers are continuous across the dorsal margin so there is no dorsal commissure along which the valves can be separated; these shells are **pseudobivalved**. In one class (Figure 14.2, *B*) the shell is **multivalved**, having eight dorsal pieces surrounded by a muscular **girdle** of mantle tissue in which calcareous spicules are embedded. Another class (Figure 14.2, *A*) has no shell, and the body is covered with calcareous spicules embedded in a cuticle. Various chitons, clams, snails, squids, and octopuses (whose ancestors had a shell) have lost that structure or have the conch embedded within the soft parts rather than surrounding them.

Univalved and bivalved shells are generally composed of aragonite, but some are calcite or of both minerals. Pseudobivalved shells were probably aragonitic. Living multivalved species have aragonitic shells. Calcareous spicules are all aragonitic.

Microstructure of the molluscan shell   Molluscan shells are multilayered. The outermost layer is organic and called the **periostracum**; it is secreted in the **periostracal groove** of the mantle edge in and around the periphery of the shell (Figure 14.3, *B*). Below the periostracum are two or more calcareous shell layers collectively called the **ostracum**. The calcareous crystals making up the ostracum are surrounded by thin sheets of organic material. All the organic material of the shell is collectively called **conchiolin**. Conchiolin is proteinaceous and thus differs from chitin, a carbohydrate found in the exoskeletons of arthropods and various other living organisms. Chitin occurs in the internal shells of a few squids.

The shell of mollusks is fabricated by the mantle. In some pelecypods the mantle edge is divided into three folds: an inner muscular one, a middle sensory one, and an outer secretory one (Figure 14.3, *B*). Shell secretion begins with the formation of the periostracum by cells in the periostracal groove. The outer calcareous layer is laid down by cells of the outer part of the outer fold. The rest of the shell is laid down by cells of the general outer

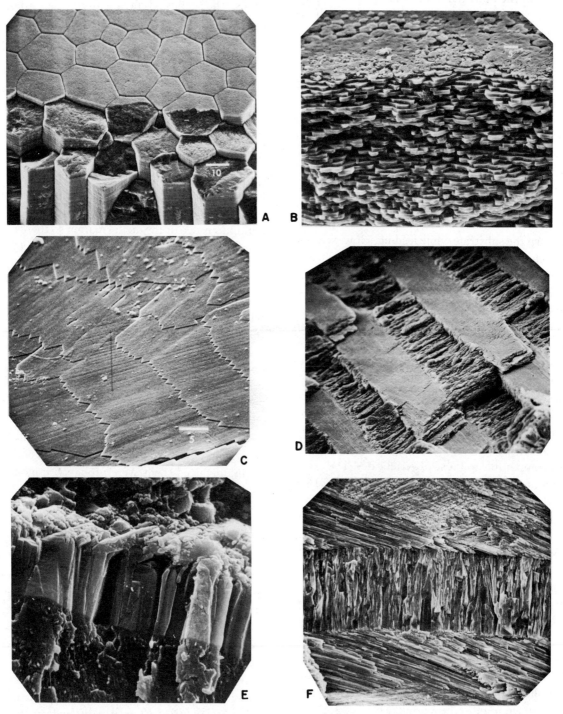

Figure 14.4. Examples of various types of microstructure of the molluscan shell. **A,** Prismatic microstructure in the pelecypod *Pinna.* (×375.) **B,** Nacreous microstructure in the pelecypod *Isognomon.* (×900.) **C,** Foliated microstructure in the pelecypod *Anomia.* (×1800.) **D,** Crossed lamellar microstructure in the pelecypod *Ctenoides.* (×600.) **E,** Myostracum (light layer) of the pelecypod *Mercenaria* attached to muscle tissue below (dark layer). (×4500.) **F,** Myostracum (center) of the pelecypod *Propeamussium* between crossed lamellar layers (above and below). (×2550.) (**A** to **D** and **F** courtesy T. R. Waller, Smithsonian Institution, Washington, D.C. **E,** courtesy P. W. Skelton, Open University, Milton Keynes, England.)

surface of the mantle. The **microstructure** of the ostracum shows considerable variation, with at least a dozen types having been identified to date. The most common microstructures are described next.

Prismatic structure (Figure 14.4, *A*) can be either aragonitic or calcitic. It consists of polygonal prisms of calcium carbonate, each surrounded by a thick wall of conchiolin. The axes of the prisms are commonly normal to the shell surface. Prismatic structure occurs in the outer layer of the ostracum of various members of almost all classes of mollusks.

Nacreous structure (Figure 14.4, *B*) is popularly called mother-of-pearl because of its pearly luster. It is always aragonitic and usually forms parts of the shell other than the outer layer of the ostracum. This microstructure consists of numerous thin lamellae that are composed of equidimensional, tabular, euhedral, or rounded crystals. The sheets of crystals alternate with sheets of conchiolin and are parallel to the shell surface. Nacreous structure is known only in various members of several molluscan classes.

Foliated structure (Figure 14.4, *C*) is always calcitic and occurs in oysters, scallops, and some snails. It is made up of elongate thin crystals, called laths, which join laterally to form sheets or folia. The laths and folia may be parallel, or oblique, to the surface of the shell. Foliated microstructure occurs in the outer or inner layers of the ostracum.

Crossed lamellar structure (Figure 14.4, *D*) is almost always aragonitic. It consists of a series of more or less parallel, rectangular, first-order lamellae, which on the whole are perpendicular to the surface of the layer and may be up to several millimeters long. The first-order lamellae seen in Figure 14.4, *D* run the length of the figure from the upper left to the lower right. They are themselves composed of transverse second-order lamellae, which are inclined to the sides of the first-order lamellae. Adjacent series of second-order lamellae slope in opposite directions and thus cross each other and form the apparent striping seen in Figure 14.4, *D*. Outer or inner shell layers can show crossed lamellar microstructure. This fabric is known in various members of several classes of mollusks.

Myostracum (Figure 14.4, *E* and *F*) is a general term for the shell that is laid down beneath the areas of muscle attachment, which are called **muscle scars**. Myostracal layers are always aragonitic and thin and usually have an irregular prismatic structure, which lacks the thick conchiolin wall around the crystals seen in the outer prismatic layer. Myostracum is found in all classes of mollusks that have shells.

Muscle scars and growth lines The principal muscles of shelled mollusks serve the foot and mantle edges; other smaller muscles, such as for the head, jaw, and gill, may be present. Many muscles are attached to the shell surface at muscle scars. Because the muscles move peripherally as the animal grows, the muscle scars leave trails of myostracum within the inner shell layer. These trails are elongate and enlarge in the direction of growth (Figures 14.3, *B* and 14.5).

The mollusk shell increases in size by the addition of calcium carbonate around its periphery. The shell increments added in this way produce **growth lines** on the outer surface of the shell that reflect the shape of the shell margin or aperture at the time that they were formed (Figure 14.3, *B*). Growth lines also reflect the shape of the leading (youngest) edge of any shell-inserted muscle. These lines are normally produced on the surface of a muscle scar (Figure 14.5, *A*). As the older edge of the muscle moves toward the shell margin during growth, new inner shell layers are deposited over the area where the muscle was previously inserted. Often these new shell layers are thinner across the old muscle insertion than they are in other parts of the interior of the shell. This thinning produces a smooth concave **muscle track** visible on the inside of the shell (Figure 14.5, *B*). The muscle scar is at the peripheral termination of the muscle track.

Growth lines on the interior of the shell identify muscle scars in fossils. The thin myostracum allows growth lines to show through it (Figure 14.5, *A*). Muscle scars of fossil mollusks are usually best observed on internal molds of the shell and are thus preserved as negative impressions of the original muscle insertions and tracks (Figure 14.5, *B* to *D*). Thus in extinct mollusks one can identify where the muscle scars are on the fossil from the trail of myostracum, interior growth lines, or the muscle track.

Embryology

The embryologic development of many mollusks is indirect: they pass through one or more free-swimming larval stages before metamorphosing into the adult form. The first larval stage is termed the **trochophore** (Figure 14.6, *A*), which then develops into the **veliger** larva (Figure 14.6, *B*). The veliger swims by using an organ known as the **velum**. However, there are numerous exceptions to this sequence. Various mollusks have direct development; the veliger stage is known only in some tusk shells, marine clams, and snails. Most freshwater clams have unique larvae that are parasitic on fish.

The larval shell of mollusks is often different from the adult shell in ornament, shape, or both; rarely, it is differ-

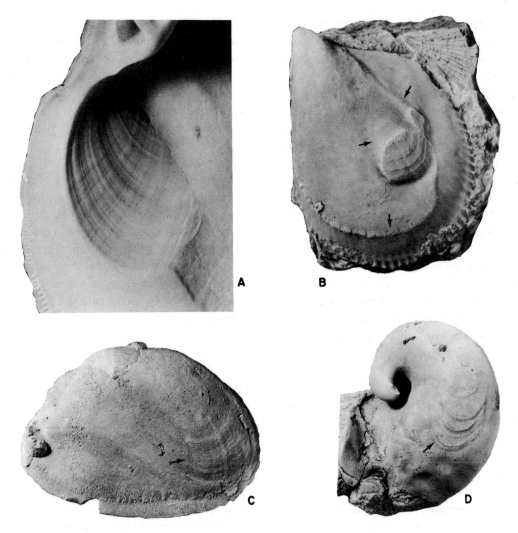

Figure 14.5. Muscle scars and tracks in mollusks. **A,** Muscle scar of the anterior shell-closing (adductor) muscle of the Holocene clam *Mercenaria* showing growth lines. (×2.5.) **B,** Fillings of muscle scars and tracks in an internal mold of the Ordovician pelecypod *Ambonychia*. Scars of the posterior closing muscle and foot-retracting muscle with growth lines are in the center rear of the specimen, and scars of the mantle-retracting (pallial) muscles are in the lower part. Muscle tracks of all muscles are indicated by arrows. (×3.) **C,** Fillings of muscle scars in an internal mold of the Silurian pelecypod *Megalomoidea.* Arrow points to growth lines on the posterior shell-closing muscle scar. (×1.) **D,** Fillings of muscle scars in an internal mold of the Devonian gastropod *Platyceras.* Arrow points to growth lines. (×1.5.)

ent in mineralogy. The larval shell is called the **proto-conch** in snails, pseudobivalved mollusks, and *Nautilus*. It is called a **prodissoconch** in most bivalves.

Classification

The classification of mollusks used here was developed in the middle and late 1970s. The phylum is divided into three subphyla and eight classes with the Rostroconchia being the only known extinct class. The various members of the phylum Mollusca, its subphyla, and its classes are placed together because of an enormously large number of observations and deductions based on fossils, comparative anatomy and morphology, comparative biochemistry, comparative embryology, and other studies. No morphologic or anatomic features are common to every member of the Mollusca exclusive of other phyla, and few such features are common to all members of any subphylum or class. Thus, extended morphologic and anatomic definitions can be confusing and misleading. For all the higher taxa presented in this chapter, the major morphologic and anatomic features are given in the diagnoses. Reasons other than those of morphology and anatomy for grouping of mollusks into higher level

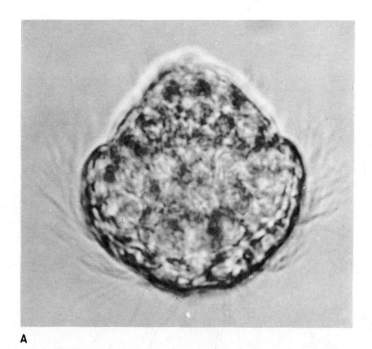

A

B

Figure 14.6. Molluscan larvae. **A**, Trochophore larva of the clam *Bankia*. **B**, Veliger larva of the European oyster *Ostrea*. (×450.) (**A** courtesy R. A. Turner. **B** from Waller, T. R. *Smithsonian Contributions to Zoology*, Figure 33; 1981.)

taxa are discussed in the section "Origin of Classes and Adaptive Diversity."

SUBPHYLUM AMPHINEURA

Mollusks with radula, gut with mouth and anus widely separated at opposite ends of body, and prominent primitive spicular skeleton. Veliger larval stage lacking. (Late Cambrian to Holocene.)

The subphylum includes both epifaunal and infaunal representatives. Two small classes are included that have other, more detailed anatomic similarities.

CLASS APLACOPHORA (Figures 14.1, *A*; 14.2, *A*; 14.7, *A*; 14.8)

Amphineurans with wormlike body form, and no head or shell; body covered with cuticle in which aragonitic spicules are embedded. (Holocene; about 70 genera.)

Aplacophorans have no unequivocal fossil record; in large part this may be because of the small size (microscopic to 4 mm), generalized shape (Figure 14.7, *A*), and aragonitic composition of their spicules. Aplacophorans are all part of the marine benthos; they range from the intertidal zone down to depths of 9000 m. Most are either coiled around various types of cnidarians or are burrowers in sediment. Some are interstitial in their mode of life, living between grains of sediment. The adaptive zone that the burrowing aplacophorans exploit is similar to that of various worms.

There are two main kinds of aplacophorans, which are best regarded as subclasses or orders. One kind has a midventral longitudinal groove running from the mouth to the anus (Figure 14.8, *A*). It is lined with cuticle, lacks spicules, and contains one or more longitudinal pedal folds. The space adjacent to the fold or folds is part of the mantle cavity. This type of aplacophoran moves on a mucous track, using cilia on the pedal fold(s); often they either crawl on the sea bottom or are entwined about cnidarians or seaweeds. The other kind of apalcophoran lacks a groove, and the cuticle and spicules encircle the body (Figure 14.8, *B*). This type of aplacophoran often constructs J-shaped tubes in the sediment in which the anterior end of the animal is lowermost.

The embryology of aplacophorans is not well known, but it does show that they are mollusks. Some late 19th century embryologic observations were interpreted as indicating that aplacophorans are degenerate polyplacophorans. More recent ontogenetic studies have not confirmed this interpretation.

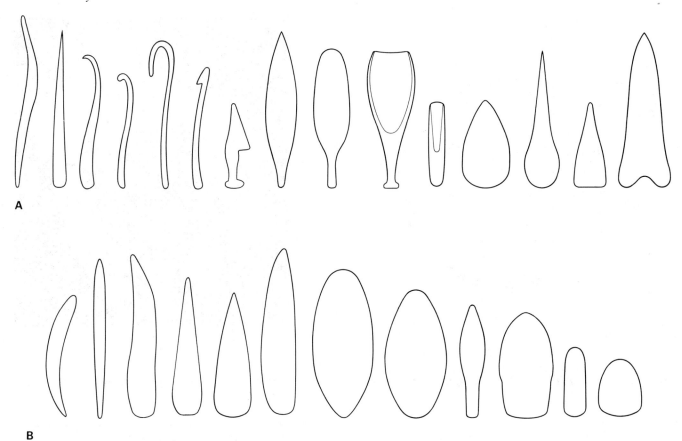

A

B

Figure 14.7. A, Outline drawings of various aplacophoran spicules. Acicular types to left; scalelike types to right. Size is variable ranging from microscopic to 3 or 4 mm long. **B,** Outline drawings of various Holocene polyplacophoran spicules. Acicular types to the left; scalelike types to the right. Note that the two types grade into each other. Size variable, ranging from microscopic up to 10 mm.

CLASS POLYPLACOPHORA (Figures 14.1, *B*; 14.2, *B*; 14.7, *B*)
Amphineurans with head, elongated body, and dorsal shell of eight articulated valves surrounded by muscular mantle girdle covered by cuticle with embedded spicules. (Late Cambrian to Holocene; about 100 genera.)

Polyplacophorans are also called chitons. The group is not well represented in the fossil record. All are part of the marine benthos. Most live in relatively shallow water, clinging to hard substrates with their foot and girdle. They are common inhabitants of the intertidal zone; a few occur as deep as 7000 m. Most crawl slowly over the substrate, rasping encrusting plant material with their radulae. As intertidal rock clingers, most chitons occupy an adaptive zone of relatively high wave energy along with limpet-shaped snails, marine mussels, and barnacles.

SUBPHYLUM CYRTOSOMA

Mollusks generally with conical univalved shell, spiral in many; single shell opening or aperture usually small. Anus usually close to mouth, gut usually bent or twisted into U-shape; radula in all classes. Body laterally compressed in few species. (Early Cambrian to Holocene.)

Cyrtosomes are an ancient phylogenetic entity, which is indicated by their high taxonomic level. The basis for the taxon Cyrtosoma is the likely common ancestry of the three included classes. Because the common ancestry was in the Early Cambrian and their morphology has become highly varied, it is necessary to study the early representatives to understand the phylogenetic origins of the classes. Most cyrtosomes are epifaunal or nektic; some have secondarily become infaunal.

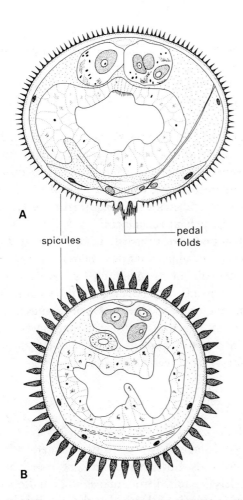

Figure. 14.8. Cross sections of the two types of aplacophorans. **A,** Those with a midventral longitudinal groove, which has cilia-covered pedal folds that lack spicules. **B,** Those lacking a midventral longitudinal groove and having spicules surrounding the body. (From Salvini-Plawen, L. *Malacologia* 9: 192, 194; 1969.)

CLASS MONOPLACOPHORA (Figures 14.1, C; 14.2, C; 14.9, *A* to *C*)

Pseudosegmented cyrtosome mollusks with a cap-shaped to helical shell. Soft parts not twisted (untorted) and anus not over the head (Figure 14.9, *A* to *C*), but some with gut bent into ⊂ shape. Most bilaterally symmetrical, with one to several pairs of muscles (pedal retractors) for clamping shell over soft parts. Apex of shell points anteriorly and over-hangs head (exogastric, Fig. 14.9, *A* and *B*). (Early Cambrian to Holocene; about 135 genera.)

Living monoplacophorans are part of the marine benthos, are found at water depths ranging from about 175 m to about 6500 m, and have a worldwide distribution in deeper waters. There are about eight named living species

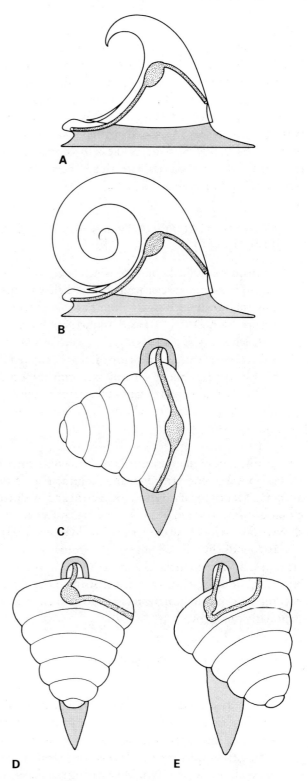

Figure 14.9. Diagram showing a way in which torsion may have originated in gastropods. The digestive tract is shown in coarse stippling. **A,** Monoplacophoran with shell curved over head; anterior to left. **B,** Monoplacophoran with shell coiled planispirally over head; anterior to left. **C,** Intermediate stage of a monoplacophoran with helically coiled shell projecting to one side; anterior up. **D,** Torted gastropod with shell resting on dorsal side of foot. **E,** Torted gastropod with shell carried in position that is common in living species.

and hundreds of extinct species. As interpreted here, monoplacophorans first appear as fossils in the Lower Cambrian and range into the Middle Triassic. They are unknown in the fossil record from the Middle Triassic on. Presumably they are missing from the younger rock record because their later evolution took place in deeper waters. Until the discovery of living monoplacophorans of the genus *Neopilina* in the middle of this century, fossil representatives of the class had usually been classified with the class Gastropoda.

CLASS GASTROPODA (Figures 14.1, *D*; 14.2, *D*; 14.9, *D* and *E*)

Cyrtosome mollusks most of which have body contained in asymmetric helically coiled shell with apex pointing posteriorly away from head (endogastric, Figure 14.9, *D* and *E*). Head distinct; can be moved independently of rest of body. Foot solelike and used for creeping. Visceral mass torted, that is, rotated up to 180 degrees about vertical axis with respect to foot so that anus and organs of the mantle cavity are above head. (Early Cambrian to Holocene; about 7800 genera.)

The rotation of the visceral mass known as **torsion** occurs in early ontogeny by rapid contraction of the asymmetrical right larval retractor muscle and by differential growth. The sequence of drawings in Figure 14.9 shows how torsion may have occurred phylogenetically.

Gastropods live in all marine environments at all latitudes, from the intertidal zone to 8000 m. Many are terrestrial or in fresh water. A few gastropods have even become parasitic and can be recognized as snails only in their early ontogeny.

CLASS CEPHALOPODA (Figures 14.1, *E*; 14.2, *E*)

Bilaterally symmetrical cyrtosome mollusks, usually with external shell with variety of shapes, rarely asymmetrical; shell with septa and siphuncle. In most geologically younger forms, shell reduced and internal; in few, shell lost. Most having radula. Head well developed, with brain and anterior sense organs including complex eyes. Foot modified into circlet of tentacles around mouth, and funnel for locomotion. Visceral mass untorted. Body dorsoventrally elongated so that the gut is bent into U shape, and mouth and anus near one another. (Late Cambrian to Holocene; about 3100 genera.)

Cephalopods are aggressive swimming carnivores and show a high degree of cephalization. Most living cephalopods have the shell embedded internally in the dorsal mantle. However, throughout most of their history, most cephalopods possessed a liquid- and gas-filled external shell (Figures 14.1, *E* and 14.2, *E*). Much of their evolution has been concerned with the development of structures to deal with regulating buoyancy and their position in the water column. The external shell is partitioned by **septa** (Figure 14.2, *E*), which are pierced by a buoyancy control organ called the **siphuncle** (Figure 14.1, *E*). Septa divide the conch into **chambers**.

Cephalopods are primarily nektic animals and, because they can adjust their postion in the water column, can be thought of as submarine-like. Because of their ability to swim, cephalopods are little controlled by bottom conditions, and they can achieve wide geographic distribution, which makes them useful biostratigraphic tools. In so far as known, all cephalopods throughout their history have been marine animals.

SUBPHYLUM DIASOMA

Mollusks with pseudobivalved, bivalved, or univalved shells that often gape at anterior and posterior ends. Gut relatively straight; mouth and anus widely separated from one another and opening at opposite ends of the shell so that the gut is not bent or twisted into U shape. Head poorly developed or absent, and radula present or absent. Body primitively laterally compressed. (Early Cambrian to Holocene.)

The Diasoma is a phylogenetic concept used to show the commonality of origin, ecology, and gross structure of the contained classes. The ancestral diasome class is the Rostroconchia which is inferred to have given rise to both the Pelecypoda and Scaphopoda. Probably all diasomes were primitively infaunal. An indication of this is that early representatives show lateral compression of the body, which allowed small ancestral diasomes to slice into the sediment and evolve into larger infaunal animals.

CLASS ROSTROCONCHIA (Figures 14.1, *G*; 14.2, *G*)

Diasome mollusks with pseudobivalved shell; radula probably present. In primitive rostroconchs an anterodorsal plate called the **pegma** (Figure 14.1, *G*) connects the two valves and was a site of muscle attachment. (Early Cambrian to Late Permian; about 35 genera.)

Rostroconchs are the only known extinct class of mollusks. They are always found associated with marine fossils and sediments. The vast majority of rostroconchs were infaunal benthos, although a few were probably epifaunal. Most rostroconchs were probably deposit feeders, although some were suspension feeders.

CLASS SCAPHOPODA (Figures 14.1, F; 14.2, F)

Diasome mollusks with tapering tubular univalved shell closed dorsally and ventrally and open at both ends; radula and unique anterior feeding structures

(captacula) present. (Middle Ordovician to Holocene; about 50 genera.)

Living scaphopods are selective burrowing predators and ciliary feeders. All are part of the marine benthos in water to 6200 m deep, and evidence from fossils suggests that they have always lived in the marine environment. Scaphopods are usually shallow burrowers in sand, often with the posterior end protruding above the substrate.

CLASS PELECYPODA (Figures 14.1, H; 14.2, H)

Diasome mollusks with bivalved shell. Dorsal margins of valves typically with crenulations known as teeth and sockets. Shell opened by dorsal ligament;

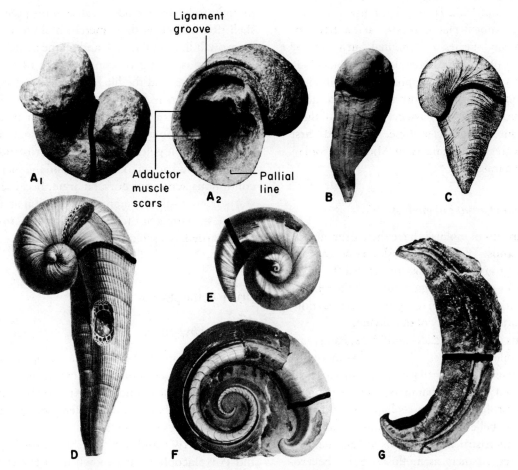

Figure 14.10. Variations in rudist shell form from the more or less typical pelecypod shell shape, in which it is easy to tell the positions of adductor muscle scars, pallial line, and ligament (**A**), to large horn-shaped bodies (**G**). **A**, *Diceras*: exterior front view (**A₁**) showing strongly coiled beaks and interior of valve (**A₂**). (×0.4.) **B**, *Coralliochama* with both valves attached. The lower valve is cone shaped. (×0.2.) **C**, Side view of *Plagioptychus* showing both valves and commissure. The shape is much like **B**. (×0.3.) **D**, *Caprinula* with coiled upper valve and conical lower valve. (×0.3.) **E**, *Caprina* with upper valve markedly coiled and larger than lower valve. (×0.1.) **F**, *Ichthyosarcolites* with upper valve strongly coiled and much larger than lower valve. (×0.1.) **G**, *Immanites* with straight commissure and each valve elongated into a large horn-shaped structure. (×0.1.) Heavy black lines mark commissure of the two valves. (From Moore, R. C.; Teichert, C., editors. *Treatise on invertebrate paleontology, Part N, Mollusca 6, Bivalvia,* vol. 2. Boulder, CO, and Lawrence, KS: Geological Society of America and University of Kansas Press; 1969.)

closed by adductor muscles (Figure 14.1, *H*). Head and radula lacking. (Early Cambrian to Holocene; about 3300 genera.)

From the earliest ontogenetic stage of its formation, the calcification of the shell of pelecypods occurs at two points; the uncalcified or poorly calcified portion between these two points becomes the ligament. There is never a calcified univalved stage in pelecypods, and the shell layers of one valve are not continuous across the dorsal margin with the shell layers of the other valve as occurs in the pseudobivalved rostroconchs. Several names besides Pelecypoda have been used for this class including Bivalvia and Lamellibranchiata.

Pelecypods are a morphologically diverse group of mollusks, with shell shapes ranging from typical bivalves (Figures 14.1, *H* and 14.2, *H*) to great horns and elongated cylindrical tubes (Figure 14.10); a few have internal shells. Many species are infaunal burrowers or borers, and many others are epifaunal; the infaunal mode of life is probably primitive. Pelecypods live in all modern marine environments from the intertidal zone to great depths in all modern seas and have been in most of these environments since Ordovician time. Pelecypods first invaded freshwater environments in Middle Devonian time. A few are parasitic.

Some major extinct groups of mollusks

Some extinct groups of mollusks have shell forms greatly different from those of living mollusks. How can one recognize these groups as mollusks given the difficulty of defining mollusks in morphologic terms and the reliance of any definition of mollusk on soft parts and deductions about the evolutionary history of the phylum?

Bellerophonts (Figures 14.22, *A* and 14.32, *I* and *J*) are planispirally coiled univalves; this shell form occurs widely in univalved mollusks and some protozoans. The bellerophonts lack sutures, chambers, and a siphuncle. Thus, they are probably neither cephalopods nor foraminifers. Some bellerophonts are known to have nacreous shell microstructure, and this feature makes it reasonable to regard them as mollusks. Some bellerophonts also have a reentrant in the shell aperture such as is found in some of the noncephalopod cyrtosomes but differs from any of the known reentrants in scaphopods. Thus bellerophonts are probably noncephalopod cyrtosome mollusks, even though they have been extinct since Triassic time. Whether they are monoplacophorans or gastropods is debated.

Rudists evolved some of the most unusual shell forms

in the Mollusca (Figure 14.10) and have been extinct since Paleocene time. One extreme of rudist variation is a 2 m long species. Some have each valve elongated into a large horn-shaped structure (Figure 14.10, *G*), and others have one cone-shaped valve, with the second valve forming a lid on the cone (Figure 14.10, *B* and *C*). As more information about Mesozoic mollusks was collected, it could be seen that there was a continuum of shell form and structure between the more bizarre rudists of the Late Cretaceous and those that showed standard pelecypod shell morphology (Figure 14.10, *A*). Once scientists understood that the extreme rudists had ligament insertion areas, hinge teeth, and adductor muscle scars, it was clear that they were pelecypods.

Rostroconchs became extinct in Late Permian time. Until very recently many members of this class were thought to be arthropods because of the pseudobivalved shell. Rostroconchs show incremental growth traceable by growth lines in the shell and on the muscle scars, and they also possess a protoconch (Figure 14.11, *A* and *B*). These features and the likelihood that rostroconchs are ancestral to pelecypods and scaphopods show that rostroconchs are mollusks. Because rostroconchs are symmetrical about the commissural plane, they superficially resemble pelecypods. However, rostroconchs lack a ligament and adductor muscles, and the shell layers are continuous across the dorsal margin. Because their shell starts from a single center of calcification and they have unique structures not found in pelecypods, rostroconchs are regarded as a separate class of mollusks.

Origin of the phylum

Flatworm hypothesis

All classes of mollusks were in existence by the end of Middle Ordovician time. The Rostroconchia is the only class that became extinct. The Aplacophora and Monoplacophora probably had a Proterozoic origin. The first Rostroconchia, Pelecypoda, and Gastropoda are known from the Early Cambrian. The first fossil Cephalopoda and Polyplacophora are known from Upper Cambrian rocks, but these classes probably originated earlier in the Cambrian. The first Scaphopoda are known from upper Middle Ordovician rocks.

The Proterozoic ancestor of mollusks was probably a small ciliated wormlike organism without a coelom. This organism probably showed pseudosegmentation by having a serial repetition of some organ systems along the length of the body, but was without transverse septa subdividing the body and was without numerical correlation of the various organ systems. Thus the number of digestive diverticula did not correspond with the number of lateral nerve branches nor with the number of excretory organs. This ancestor of the mollusks was probably much like the living turbellarian flatworms (Figure 14.12, *A*). Mollusks probably never possessed segmentation as occurs in annelid worms, which have a body composed of repeated segments separated from each other by septa and with most organ systems being numerically correlated and occurring in each segment (Figure 12.1, *B*). The embryologic and larval similarities between mollusks and annelids probably reflect their origin in a common ancestor, a pseudosegmented tur-

bellarianlike flatworm. Annelids developed segmentation, whereas in mollusks a septate coelom was not developed. The trochophore larva of mollusks and annelids was inherited from this common ancestor. Molluscan pseudosegmentation is best expressed in the Monoplacophora (Figure 14.1, *C*) and is significant in the Polyplacophora (Figure 14.2, *B*).

Most of the features that distinguish mollusks could have arisen in a turbellarian-like animal as the result of a single evolutionary accomplishment, the secretion of a mucoid cuticle over the dorsal body surface (Figure 14.13, *A*). The body wall surface secreting the cuticle would become the primitive mantle. Aragonitic spicules embedded in a cuticle are secreted in the primitive

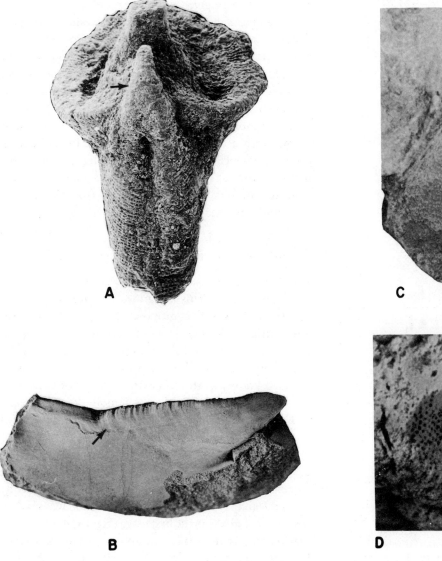

Figure 14.11. **A** and **B**, Molluscan morphologic features in rostroconchs. **A**, *Hippocardia*? showing protoconch (arrow) and growth lines. (×44.) **B**, *Ribeiria*, internal mold, showing growth lines on posterodorsal muscle scar (arrow). (×2.) **C**, The Late Cambrian chitons *Matthevia* (×7.5) and **D**, *Preacanthochiton* (×13) showing dorsal surfaces covered with tubercles, which suggest that the shell may have originally been secreted as spicules.

aplacophoran and polyplacophoran mollusks. Similar aragonitic spicules occur in the outer body wall of some living turbellarian flatworms, and it seems likely that the primitive molluscan skeleton was spicular and not layered calcium carbonate (Figure 14.12, B_1).

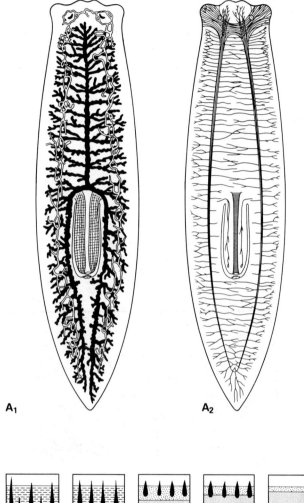

A_1 A_2

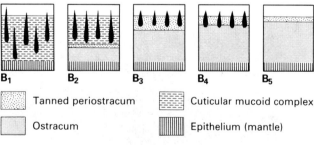

B_1 B_2 B_3 B_4 B_5

☐ Tanned periostracum ▦ Cuticular mucoid complex

☐ Ostracum ▥ Epithelium (mantle)

Figure 14.12. A, Pseudosegmentation of the flatworm *Dugesia* or *Planaria*. A_1, The digestive (black) and excretory (white) systems; A_2, the nervous system. **B,** Organization of cuticle and shell in some Holocene mollusks showing how a layered shell might be derived from a spiculose cuticle. *1,* Aplacophora; *2,* Polyplacophora; *3* and *4,* various Pelecypoda; *5,* most living Pelecypoda. Black tear drop-shaped structures are diagrammatic spicules. (**A** modified from Parker, T. J.; Haswell, W. A.; Lowenstein, O. *A textbook of zoology,* Hampshire, England: Macmillan Publishing Co., Inc.; 1961 p. 220. **B** from Carter, J. G.; Aller, R. C. *Lethaia* 8: 318; 1975.)

Cuticle spicules may have been subsequently cemented together by the outer layer of the primitive ostracum as occurs in some pelecypods (Figure 14.12, B_4), or the ostracum may have been secreted entirely below the spicule-bearing cuticle (Figure 14.12, B_2 and B_3). In whatever way the ostracum was formed, once it evolved, it raised the spiculose cuticle above the dorsal body wall (Figure 14.12, B_3 and B_4), and the cuticle became the periostracum of the shell (Figure 14.12, B_5). By the stage of evolution at which the ostracum had formed, the mantle and shell had become extended as eaves beyond the edge of the body, and a mantle cavity was formed in which the gills (Figure 14.13, *C*) could be protected. Gills also allowed size to increase. The gut of various living mollusks has horny shields. Such a horny shield in the foregut may have given rise to the radula.

The above scenario on the origin of mollusks is largely deduced from the zoology of the group, from the occurrence of spicules in turbellarians, and from current thought that the spiculose aplacophoran mollusks are anatomically primitive and not secondarily degenerate as had been previously believed. Using this approach, the ancestral mollusk was probably a spiculose wormlike animal with a cuticle. It was much like living aplacophorans, which have various molluscan anatomic features without the development of a shell. The shell-bearing mollusks would have evolved from this aplacophoranlike ancestor.

Hypothetical ancestral mollusk

Another approach is to think of the ancestral mollusk as having had a shell. This scenario is also largely a zoologic construct. It was first proposed about 100 years ago when little was known about the Aplacophora and when it was unclear that they are mollusks. Some version of this hypothetical ancestral mollusk or HAM (Figure 14.14, *A*), also known as the schematic mollusk or archetype, has appeared in most books about mollusks ever since.

The HAM is basically a least common denominator of all structures inferred to be primitive in the various classes of extant mollusks having a shell. Using this device to explain molluscan evolution, workers are forced to derive each class of mollusks independently from the archetype because this assumption was made in deriving the HAM in the first place (Figure 14.14, *B*). The HAM concept also makes the aplacophorans degenerate because they have to lose the shell of the ancestor, it means that the polyplacophorans have to multiply the number of shells of the ancestor, and it implies that the HAM existed essentially unchanged from the late

Proterozoic to the Middle Ordovician. A recent explanation of how the HAM may have looked (Figure 14.14, *A*) shows an animal with an univalved shell and multiple foot retractor muscles and with the mouth and anus at opposite ends of the body. Such an animal is effectively a monoplacophoran and if found would be classified as such.

Various zoologists have advised caution about the use

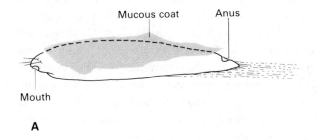

A

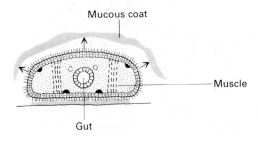

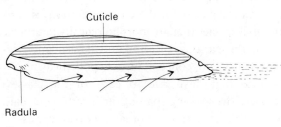

B

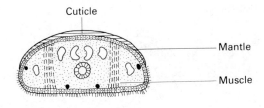

C

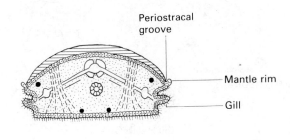

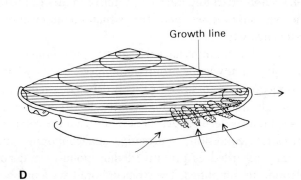

D

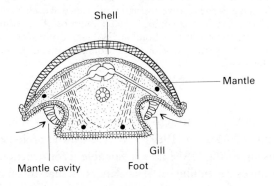

Figure 14.13. Possible sequence of events leading from a turbellarian-like flatworm to a shell-bearing mollusk. Left side is the lateral view, right side the cross-sectional view. **A,** Flatwormlike ancestor with a complete gut and the ability to secrete a protective mucous coat. **B,** Transitional stage with radula and cuticle. **C,** Transitional molluscan stage with the beginning of a mantle cavity. The cuticle is secreted in a shallow periostracal groove beyond which the mantle rim extends. **D,** Advanced molluscan stage with shell of calcified layers laid down under the uncalcified periostracum. (Modified from Stasek, C. R. in: Florkin, M.; Sheer, B. T., editors. *Chemical zoology, Mollusca.* New York: Academic Press, 7: 6; 1972; and Stasek, C. R.; McWilliams, R. *Veliger* 2; 1973.)

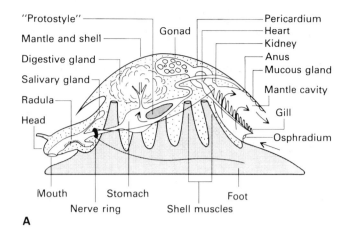

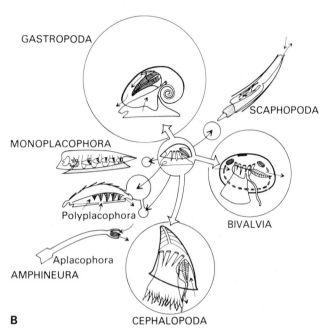

Figure 14.14. A, Reconstruction of the HAM. **B,** Suggested origins of the various classes of living mollusks from the HAM. Circles suggest the relative size and significance of each class. (**A** from Yonge, C. M. in: Yonge, C. M.; Thompson, T. E., editors. *Living marine mollusks.* London: William Collins & Sons; 1976, p. 25. **B** from Morton, J. E. *Molluscs,* 4th ed. London: Hutchinson & Co. Ltd.; 1967, p. 21.)

of the HAM as central to an understanding of molluscan evolution. Morton (1967) fully realized the problem of thinking of the HAM as a reasonable reconstruction of the ancestral mollusk: "The danger is that in mixing genealogical ideas with morphology our archetype may become like an heraldic animal—a lowest common multiple of incompatible organs." Although the HAM was a useful concept when little was known about early fossil mollusks, the HAM is probably no longer a necessary device for understanding molluscan evolution.

Origin of classes and adaptive diversity

The approach to molluscan evolution used here is to consider how one class of mollusks may have arisen from one or more species of another preexisting class. This approach concentrates on the study of the comparative morphology of molluscan shells from Cambrian and Ordovician rocks because all classes of mollusks had evolved by the end of the Ordovician. Vertebrate paleontologists working in younger rocks, have successfully used this method of study to develop insights into the origins of vertebrate classes. They have determined that certain fishes were ancestral to amphibians, these in turn are regarded as ancestral to reptiles, and reptiles are regarded as being ancestral to birds and mammals. Much of the empirical base for these interpretations is derived from fossil species that are morphologically intermediate between the various ancestor and descendant classes. There is now no concept of a hypothetical ancestral vertebrate that gave rise to each of the classes of Vertebrata over a long span of time; rather one class is derived from species of another preexisting class level taxon.

The morphologic diversity of mollusks is correlated with their ecologic diversity, which is greater than that of any of the other unsegmented invertebrates and rivals that of arthropods and vertebrates. Although the adaptive diversity of the large classes of living mollusks is great, empirical information from the fossil record and deductions based on comparative studies of living taxa suggest that the evolution of each class involved structural change that can be correlated with the development of a new mode of life or with a better way of carrying on an already established mode of life. Table 14.1 summarizes the inferred primitive modes of life of the various classes and the structural innovations associated with them.

The aplacophorans have the simplest morphology and anatomy of any living group of mollusks, but they have no unequivocal fossil record. It is debated whether the relatively simple morphology is a primitive condition or is secondarily derived (degenerate) from a more complex ancestral condition. Currently, aplacophorans are usually regarded as primitive living mollusks and not secondarily simplified. The Aplacophora have a number of anatomic features of mollusks that have evolved without the development of a shell. A shell cannot form because the cuticle in which the spicules are embedded does not contain the proper protein so it can act as a place for the deposition of calcium salts in crystalline layers. Aplacophorans evolved typically molluscan features—as

Table 14.1. Synopsis of mollusk evolution

Class	Distinguishing feature(s)	Inferred primitive mode of life	Structural innovation(s)
Aplacophora	Exoskeleton of spicules; no shell	Uncertain	Radula; complete gut
Polyplacophora	Shell of eight valves; spicules limited to mantle girdle	Benthic algal grazers in high-energy, shallow marine environment	Flexible multivalved shell, large foot, and mantle girdle allowing tight clinging to substrate
Monoplacophora	Univalved shell with single aperture; no operculum; untorted	Benthic algal grazers in shallow marine environment of varying energy	Univalved shell that can be pulled down to cover entire animal; large foot
Gastropoda	Univalved asymmetrical shell with single aperture and operculum; torted	Benthic algal grazers in shallow marine environments of varying energy	Asymmetrical, torted shell with columellar muscle for retraction of head and foot into shell and operculum for closing aperture
Cephalopoda	Shell of various shapes, univalved with single aperture and septa pierced by siphuncle; untorted	Swimming and/or crawling carnivores or carrion feeders	Siphuncle that controls gas and fluid levels in chambers to adjust buoyancy
Rostroconchia	Pseudobivalved, gaping shell with pegma	Infaunal ciliary deposit feeders	Pseudobivalved, laterally compressed gaping shell and probable anterior structures for ciliary deposit feeding
Scaphopoda	Univalved, tubular shell open at both ends; untorted	Infaunal ciliary deposit feeder including shelled protists	Tubular shell open at both ends
Pelecypoda	Bivalved shell with adductor muscles and ligament	Infaunal ciliary deposit and suspension feeders	Bivalved shell with complex musculature, ligament, and large foot allowing rapid burrowing

radula, restricted coelom, style, mantle cavity, and gills. The first mollusks reasonably can be regarded as being similar to aplacophorans in having a spiculose cuticle prior to the evolution of the shell. Small, spiculose animals are seldom fossilized.

What the ancestral mode of life of the shell-less aplacophorans may have been is uncertain because they have no unequivocal fossil record and because living species show a variety of feeding habits and may be infaunal or epifaunal. Many burrowing species are particle-size–dependent detritivores. Some burrowing species can keep the mouth open during rasping with the radula so that feeding is not particle-size dependent; these forms can break down and manipulate food and can rasp algae, organic detritus or prey on foraminifers, small worms, and small snails. Various aplacophorans are carnivorous on cnidarians; they have a suctorial mouth and, based on cladistic evaluation, some are regarded as having a primitive radula. In these primitive suctorial aplacophorans, the radula probably serves only to help move food backward toward the midgut. Cnidarians have been significant constituents of the Earth's fauna since late Proterozoic time, and the first radiation of shelled mollusks occurred in earliest Cambrian time.

Investigators have suggested that the polyplacophorans arose from a spiculose ancestor that evolved eight shell valves secreted under the cuticle, with the spicules being limited to the mantle girdle surrounding the shell. It is thought that the polyplacophoran ancestor diverged early from the rest of the mollusks and that the eight-part shell was probably developed independently from the shell of other mollusks. In part, these interpretations are based on the observations that chitons retain mucoid material in the cuticle, which is reminiscent of what the earliest molluscan covering may have been, and that chitons never achieved the threefold subdivision of the mantle margin (Figure 14.3, *B*) which is common among many other living shelled mollusks. The development of the exposed mantle rim to form the spiculose girdle, and the divided nature of the shell, point to a history of chitons long separate from that of other mollusks. The observation that the innermost shell layer of chitons was developed after they were already a differentiated group indicates that this shell layer is not homologous with the inner shell layer of other mollusks. The fact that some Late Cambrian chitons (Figure 14.11, *C* and *D*) have the surface of the shell covered with tubercles suggests that the shell may have originally been secreted as spicules, which were later cemented together, as occurs in some living pelecypods. Thus aplacophorans, polyplacophorans, and monoplacophorans have had long separate histories and are related to one another only near the base of the stem of the molluscan phylogenetic tree (Figure 14.15).

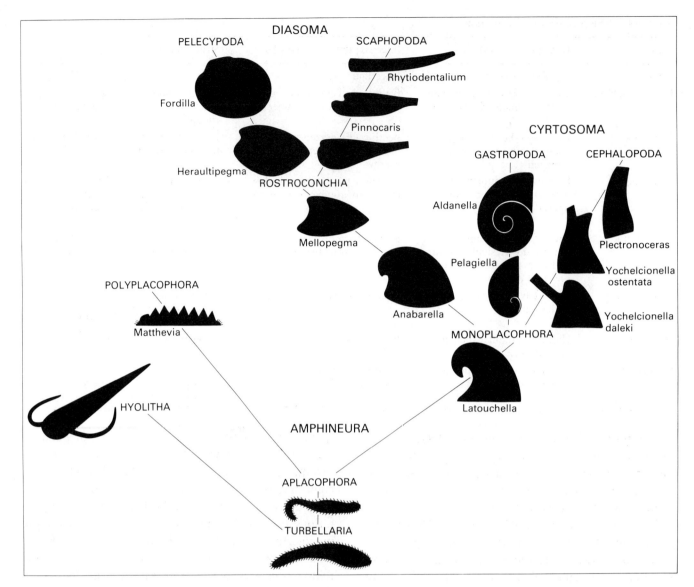

Figure 14.15. Proposed phylogeny of the Mollusca. The phylum Hyolitha probably had a common ancestry with the mollusks and is discussed elsewhere in this book. (From Pojeta, J., Jr. *Tulane Studies in Geology and Paleontology* 16: 78; 1980.)

Polyplacophorans evolved the multivalved shell, strong muscular foot, and mantle girdle that primitively enabled them to colonize high-energy, shallow marine environments where they sluggishly grazed algal mats. Most polyplacophorans have lived in this adaptive zone since the Late Cambrian, and the major feature of their evolution has been the development of flatter, better-articulated valves. In a few species, the valves have become internal, or nearly so, within the mantle, and some live as deep as 7000 m where they presumably feed as detritivores.

The remaining classes of mollusks are ultimately descended from a common ancestor that had a univalved dorsal exoskeleton. Thus the shells of all remaining classes contain homologous parts and structures. In this chapter these six classes are placed in two subphyla both of which appeared and began to diversify within a few millions of years in the Early Cambrian (Figure 14.15).

The Monoplacophora are probably descended from an aplacophoranlike spiculose ancestor that developed a conical shell below the cuticle. It is uncertain but seems likely that the first monoplacophorans maintained a spiculose mantle margin or even a spiculose periostracum. The development of the monoplacophoran shell was probably a Proterozoic event. The earliest known Cambrian monoplacophorans were probably epifaunal grazers on algae in contrast to living deeper-water forms.

The mode of life of living monoplacophorans is not well known and probably not significant for understanding the diverse shallow-water species that occur widely in

Paleozoic rocks. Living species seem to be unselective epifaunal detritus feeders. The morphology of many extinct Paleozoic species suggests that they were probably benthic grazers, and they are often found in near-shore facies. Monoplacophorans have paired dorsoventral muscles that can clamp the shell against the substrate, in effect, using it as a second valve (Figures 14.21, *E* and *F* and 14.22, *B*). They are untorted, very few are asymmetric, they lack an operculum, and enlargement of their shells did not provide the most compact configuration for the visceral mass.

Gastropods probably evolved from monoplacophorans (Figures 14.9 and 14.15) when their shells became helically coiled in the form of a corkscrew in the Early Cambrian. Helical coiling allows the animal to produce the most compact body mass in the least possible space. In monoplacophorans, the shell is carried on the dorsal side of the body as the organism moves over the substrate (Figure 14.9, *A*). If the planispiral shell (shell coiled in one plane; Figure 14.9, *B*) of some monoplacophorans became asymmetrical, the spire of the shell would stick out to the side of the body and the shell would tend to fall sideways (Figure 14.9, *C*). To keep the shell upright and creep along the substrate, the asymmetric monoplacophoran would have to carry the spire and the visceral mass above the foot, requiring continual muscular effort because of the asymmetry of the shell. If the shell could be rotated so that the spire did not protrude to the side but rather faced away from the head (endogastric) and was carried on the back of the foot, the shell could be dragged behind the animal rather than being carried above it (Figure 14.9, *D*). Some living monoplacophorans can rotate the shell 90 degrees with relation to the head-foot mass. This carrying of the helical shell on the back of the foot with the spire facing away from the head would cause torsion of the soft parts. Thus, torsion may have been a functional adaptation to having the shell coiled in an asymmetric spiral and may have existed in this state for a long time before it became a genetic trait. Figure 14.9, *E* shows the approximate position in which many living gastropods carry the shell. Originally torsion may have been much less than 180 degrees as some coiled Early Cambrian shells have a very flat spire, and it is not necessary to postulate a fully torted planispirally coiled intermediate form between monoplacophorans and gastropods.

The asymmetrical shell of gastropods combined with torsion allows them to house the visceral mass in the protective conch in the most compact way, to carry the shell so that it causes a minimum of drag and muscular exertion, and to have the gills and various sense organs in an anterior position. All gastropods have one or two columellar muscles that allow them to retract the head and foot into the protective shell, and all living species have an operculum or vestigial operculum-secreting cells on the foot; the operculum serves to close the shell aperture when the head and foot are retracted. These shell features, combined with the ability of gastropods to use the radula to manipulate and break down food, may be some of the reasons for the great diversification of this class. In their later evolution, various gastropods invaded freshwater and terrestrial environments (Pennsylvanian), some evolved into sophisticated predators that use the radula to bore into shells (Cretaceous) or use poisoned radular teeth to subdue their prey (Eocene), some became filter feeders (Ordovician), some swim (Cretaceous), some burrow, many have lost their shells, and a few are parasitic.

Cephalopods probably evolved from dorsoventrally elongated monoplacophorans, some of which are known to possess septa. Some of these monoplacophorans (Figure 14.24, center right) show a shell curvature that was probably away from the head (endogastric) and have a tube or snorkel as the highest point on the dorsal side of the shell. Such a snorkel could be sealed off and become the primitive siphuncle.

Cephalopod evolution is closely linked to the development of the siphuncle, which gave the class a method to control shell buoyancy when septa were formed. Septate monoplacophorans and gastropods cannot alter whatever fluid or gas fills the chambers because they lack a buoyancy control mechanism. Thus cephalopods were the first mollusks to live in the water column (Late Cambrian) and to become early swimming carnivores. Some cephalopods are carrion feeders, and some can swim as well as use the tentacles to crawl. Cephalopods with internal shells first occur in Devonian rocks; they balance the buoyancy of the shell with external deposits.

Among the Diasoma, the oldest known rostroconch (Figure 14.16, *D* to *F*) is from the Early Cambrian. It is a laterally compressed shell that gapes at the anterior, ventral, and posterior shell margins. This shell form can readily be related to laterally compressed monoplacophorans (Figure 14.16, *A* to *C*) from the Early and Middle Cambrian, which have a curved rather than straight apertural margin when viewed laterally and anterior, ventral, and posterior shell gapes. Thus, it is reasonable to argue that rostroconchs arose from monoplacophorans. The lateral compression of the shell in some monoplacophorans also provided the mechanism by which epifaunal grazers could become infaunal burrowers in the sediment and explains the origin of the

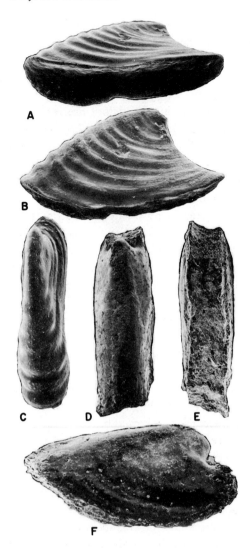

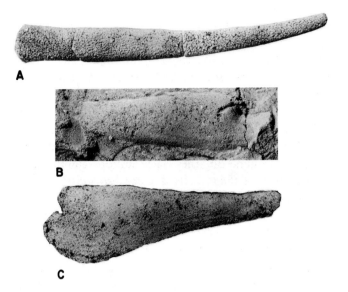

Figure 14.17. Suggested origin of scaphopods from rostroconchs. **A,** Oldest known scaphopod, *Rhytiodentalium,* from the Middle Ordovician rocks of Kentucky. (×3.) **B,** The rostroconch *Pinnocaris?* from Lower Ordovician rocks of Australia showing both the general body form and anterior elongation of scaphopods. (×2.) **C,** The rostroconch *Pinnocaris* from Upper Cambrian rocks of Australia. (×1.5.) (From Pojeta, J., Jr. *Tulane Studies in Geology and Paleontology* 16: 79; 1980.)

Figure 14.16. Suggested origin of rostroconchs from laterally compressed monoplacophorans in the Cambrian. Oldest known rostroconch *Heraultipegma* (**D** to **F**) with shell shape and gapes very similar to the monoplacophoran *Mellopegma* shown in **A** to **C**. (**A, B,** and **E** ×38; **C** ×75; **D** and **F** ×30.) (From Pojeta, J., Jr. *Tulane Studies in Geology and Paleontology* 16: 75; 1980.)

infaunal mode of life in the Diasoma. Most rostroconchs were probably infaunal deposit feeders, using specialized anterior protrusible structures to collect food by ciliary action; they were the first mollusks to exploit this mode of life. Some rostroconchs were probably suspension feeders (Early Ordovician), and a few almost certainly became epifaunal deposit feeders (Early Ordovician).

The scaphopods are all infaunal and also have specialized anterior ciliated feeding structures that originally may have been used for deposit feeding; the particles brought to the mouth included shelled protists, which could be crushed by the radula. The earliest known scaphopods (Figure 14.17, *A*) are Middle Ordovician in

age and probably had their origin from similarly shaped rostroconchs that first appeared in the Late Cambrian (Figure 14.17, *C*). These rostroconch shells are markedly elongated and almost tubular posteriorly. They gape anteriorly and posteriorly, and the valve edges touch ventrally. Dorsally the shell layers are continuous as in other rostroconchs. In shell form these elongated rostroconchs look remarkably like scaphopods and were probably burrowers. To derive a scaphopod from them, it is necessary to fuse the ventral margin and develop anterior elongation of the shell. Some species of rostroconchs show anterior elongation (Figure 14.17, *B*), and it is reasonable to consider that scaphopods arose from rostroconchs.

Pelecypods probably evolved from rostroconchs in the Early Cambrian by differential calcification of the dorsal margin in the latter group. This differential calcification could produce the elastic ligament of pelecypods. At about the same time that the ligament originated, adductor muscles would have had to develop. The change from a rostroconch to a pelecypod would be a biochemical one expressed morphologically as an elastic dorsal margin having more organic material than the shell.

Based on general morphology and comparison with younger species, the oldest known pelecypods may have been infaunal using the foot and coadapted complex musculature and ligament to burrow. The oldest known Early Cambrian species has muscle scars that suggest that it had anterior ciliated structures used for deposit feeding; another species that is only slightly younger was probably a suspension feeder. Suspension feeding has been the main method of food gathering in pelecypods ever since, although one group has maintained deposit feeding and another supplements suspension feeding with deposit feeding. Beginning in the Early Ordovician and accelerating in the Middle Ordovician, many non-burrowing semi-infaunal and epifaunal species of pelecypods evolved. These forms use special threadlike structures to attach to hard substrates. In Late Ordovician time, at least one species was a facultative borer, but this mode of life did not become widespread until the Mesozoic. The first freshwater pelecypods occur in upper Middle Devonian rocks. In Mississippian time some pelecypods became cemented epifauna much like Mesozoic and Cenozoic oysters. Swimming scallops first appear in the fossil record in Mississippian time. Beginning in the Mesozoic there was a great radiation of shallow- and deep-burrowing pelecypods in which large portions of the mantle edges fused and the posterior mantle was elongated into tubes called siphons. Siphons reach to the sediment-water interface for respiration and suspension feeding in deeply burrowing species. Shallow-burrowing species are known throughout the Paleozoic.

Supplementary reading

Broadhead, T. W., editor. *Mollusks, notes for a short course,* University of Tennessee Studies in Geology No. 13. Knoxville, TN: University of Tennessee; 1985. State-of-the-art summary of the paleontology of mollusks by 19 authors.

Florkin, M.; Scheer, B. T., editors. *Chemical zoology, Mollusca,* vol. 7. New York: Academic Press, Inc.; 1972. Introductory chapter has superior discussion of the molluscan framework.

House, M. R., editor. *The origin of major invertebrate groups,* Systematics Association Special Volume No. 12. New York: Academic Press, Inc.; 1979. Proceedings of a symposium providing summaries of molluscan classes on pages 323–414.

Morton, J. E., *Molluscs,* 4th ed. London: Hutchinson & Co., Ltd.; 1967. Highly readable general account of mollusks.

Pojeta, J., Jr. Molluscan phylogeny: *Tulane Studies in Geology and Paleontology,* 16: 55–80, 1980. Summary of paleontologic information useful for evaluating molluscan evolution; downplays the HAM concept.

Rhoads, D. C.; Lutz, R. A., editors. *Skeletal growth of aquatic organisms.* New York: Plenum Press; 1980. State-of-the-art summary of various aspects of growth, form, ultrastructure, and chemistry of the molluscan shell.

Runnegar, B. Molluscan phylogeny revisited: *Memoir Association of Australasian Palaeontologists* 1: 121–144; 1983. Discusses the most recent discoveries about the early fossil record of the phylum and integrates them into a new synthesis of molluscan phylogeny.

Runnegar, B.; Jell, P. A. Australian Middle Cambrian molluscs and their bearing on early molluscan evolution. *Alcheringa* 1: 109–138; 1976. Best summary presentation of Cambrian mollusks with full entry to the pertinent literature.

Salvini-Plawen, L. On the origin and evolution of the Mollusca. *Accademia Nazionale Dei Lincei* 49: 235–293, 1981 (in English). Zoologist's interpretation of molluscan evolution using various hypothetical forms and making little use of the fossil record.

Trueman, E. R.; Clarke, M. R., editors. In: Wilbur, K. M., editor in chief. *The Mollusca, Vol. 10, Evolution.* Orlando, FL: Academic Press, Inc.; 1985. Summary presentation of origin and evolution of mollusk classes.

Yochelson, E. L. An alternative approach to the interpretation of the phylogeny of ancient mollusks. *Malacologia* 17: 165–192; 1978. Paleontologic approach to the evolution of the phylum using the HAM concept.

Yonge, C. M.; Thompson, T. E. *Living marine mollusks.* London: William Collins Sons & Co., Ltd.; 1976. Highly readable account of the major groups of the phylum.

Part II   Additional concepts

Class Polyplacophora

John Pojeta, Jr.

Extant polyplacophorans or chitons live largely in the high subtidal to intertidal zones of rocky coasts; a few with considerably different morphology have been found as deep as 7000 m in the ocean. Chitons are usually flattened dorsoventrally and have a broad muscular foot and an expanded spiculose portion of the mantle termed the girdle (Figure 14.1, *B*). When they are disturbed, they clamp the foot and girdle against the substrate and are difficult to remove. Chitons cling to the substrate by a combination of vacuum and adhesive secretions. Detached chitons can roll up like a trilobite because they have eight articulated calcareous valves on the dorsal side (Figure 14.2, *B*) and four groups of muscles controlling the valves. Polyplacophorans have a distinct head lacking eyes and tentacles (Figure 14.1, *B*). Between the head and foot and the girdle is the deep groove of the mantle cavity with the gills (Figure 14.1, *B*). Anatomically, chitons show similarities to aplacophorans in spicules embedded in a cuticle, structure of the nervous system, and relationship between the space around the heart and the gonads.

Living chitons occur in all seas at all latitudes and are part of the marine benthos. Adults range in size from 8 mm to about 33 cm. Chitons are not well known from the fossil record, in part because their valves are readily disarticulated and they, or their molds, are then difficult to recognize. Recent work shows that the Ordovician chiton fauna was rich and varied and that the class is at least as old as Late Cambrian.

Shell morphology

The skeleton of modern chitons is a shell of eight articulated calcareous valves (Figure 14.2, *B*) and various calcareous spicules (Figure 14.7, *B*) embedded in the organic cuticle of the muscular girdle (Figure 14.19, *A*). In some, the tips of some of the radular teeth contain magnetite.

The valves of chitons (Figure 14.18) are embedded in the median part of the dorsal surface (Figure 14.2, *B*). In modern chitons, the **head** and **tail valves** are semicircular (Figures 14.2, *B* and 14.18, *I* and *J*). The dorsal side of each intermediate valve is often divided into a **median** and two **lateral areas**, which are roughly triangular in shape (Figure 14.18, *H* and *J*). Ornamentation of the valves varies; the central portion of a valve is sometimes ornamented differently from the rest and is termed the **jugum** (Figure 14.18, *J*). A slight elevation near the center or posterocentral portion of the tail valve is called the **mucro** (Figure 14.18, *J*).

Usually two anterior projections called **sutural laminae** are on the ventral part of each intermediate valve (Figure 14.18, *G* and *H*) and extend forward beneath the preceding valve to articulate the two valves together. The lateral portions of the posterior part of the ventral side of each intermediate valve usually have **insertion plates** (Figure 14.18, *G*) which attach the valves to the mantle. Insertion plates often have **insertion teeth**, as do the anterior margin of the head valve and the posterior margin of the tail valve (Fig. 14.18, *I*).

In modern polyplacophorans, the valves are composed of four major layers (Figure 14.19, *A*). The outermost is the thin organic periostracum secreted in the periostracal groove. This is underlain by the calcareous **tegmentum**, which has a high organic content and can be permeated by numerous canals that contain sense organs known as **esthetes** that are sometimes photoreceptors. The tegmentum passes over the apex of the valves and turns under onto the ventral side along the posterior margin of each valve (Figure 14.19, *C* and *E*). The turned-under tegmentum and periostracum are called the **apical area** (Figures 14.18, *C*, *F*, *G*, and *I* and 14.19, *B* to *E*). Beneath the tegmentum is the calcareous **articulamentum**, which forms the sutural laminae and insertion plates and teeth. The articulamentum is underlain by the **hypostracum**, a calcareous layer formed on the general inner surface of the valves (Figure 14.19).

Classification

Chitons are placed in two orders. The order Paleoloricata includes those forms that lack the articulamentum and thus do not have sutural laminae and insertion plates (Figures 14.18, *A* to *F* and 14.19, *B* and *C*). These animals have thick, massive valves with large apical areas and overlap of the valves brings tegmentum into contact with tegmentum. The valves are strongly arched, and in some geologically early forms (Figure 14.20) there seems to be no overlap of them. Paleoloricates have a known stratigraphic range of Upper Cambrian through Upper Cretaceous.

Members of the order Neoloricata possess the articulamentum. Primitive neoloricates have only sutural laminae (Figure 14.19, *D*), but most have both sutural laminae and insertion plates (Figure 14.18, *G* and *H*). Neoloricates have flat and thin valves (Figures 14.18, *G* to *J* and 14.19, *D* and *E*), and have only a small apical area (Figure 14.18, *G*). The known stratigraphic range is Mississippian through Holocene.

Primitive number of valves

Except for teratologic specimens, living and articulated fossil chitons are characterized by a shell made up of eight valves. However, some evidence suggests that the primitive number of valves may have been seven. During the ontogeny of chitons, seven of the valves are all formed at about the same time and the eighth valve is developed considerably later. In counting the three distinctive types of disarticulated valves of a Late Cambrian chiton, a ratio of 1:5:1 (head:intermediate:tail) was obtained. However, recent reexamination of an articulated Ordovician chiton that was thought to have only seven valves has revealed a small eighth valve.

Reconstruction of a Late Cambrian chiton

In Late Cambrian high-tidal and intertidal carbonate rocks of North America the remains of a fossil called *Matthevia* have been found (Figure 14.18, *D* to *F*). *Matthevia* is usually associated with algal stromatolites and is thought to have lived by rasping plant material with its radula as do most modern chitons. Several years ago *Matthevia* was interpreted to be unique and was placed in a class separate from all other mollusks. New evidence suggests that *Matthevia* is one of the oldest known polyplacophorans, and a reconstruction of *Matthevia* grazing on domal stromatolites is provided in

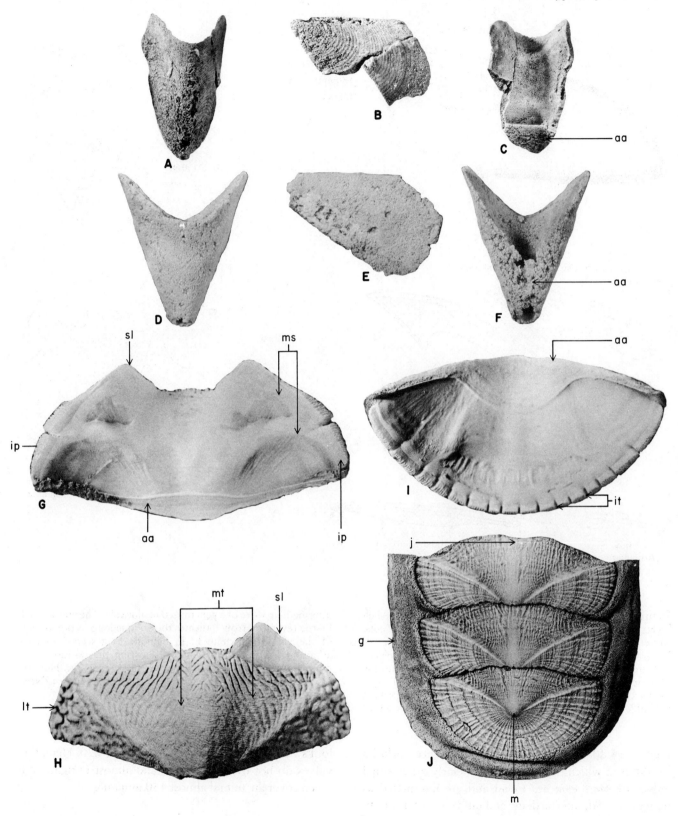

Figure 14.18. Paleoloricates (**A** to **F**) and Neoloricates (**G** to **J**). **A** to **C**, Silicified intermediate valve of *Chelodes* from the Early Ordovician of Australia. Dorsal, right lateral, and ventral views. (×2.) **D** to **F**, Silicified intermediate valve of *Matthevia* from the Late Cambrian of Texas. Anterior, right lateral, and posterior views. (×2.5.) **G** and **H**, Interior (ventral) and exterior (dorsal) views of intermediate valve of *Chiton*. (Holocene, ×3.5.) **I**, Interior (ventral) view of head valve of *Chiton*, posterior end up. (Holocene, ×6.) **J**, Specimen of *Stenoplax* showing dorsal side of valves 6, 7 (intermediate), and 8 (tail). Sutural laminae not shown on valve 6. (Holocene, ×3.5.) *aa*, Apical area; *g*, girdle; *ip*, insertion plate; *it*, insertion teeth; *j*, jugum; *lt*, lateral triangle; *m*, mucro; *ms*, muscle scars; *mt*, median triangle; *sl*, sutural lamina.

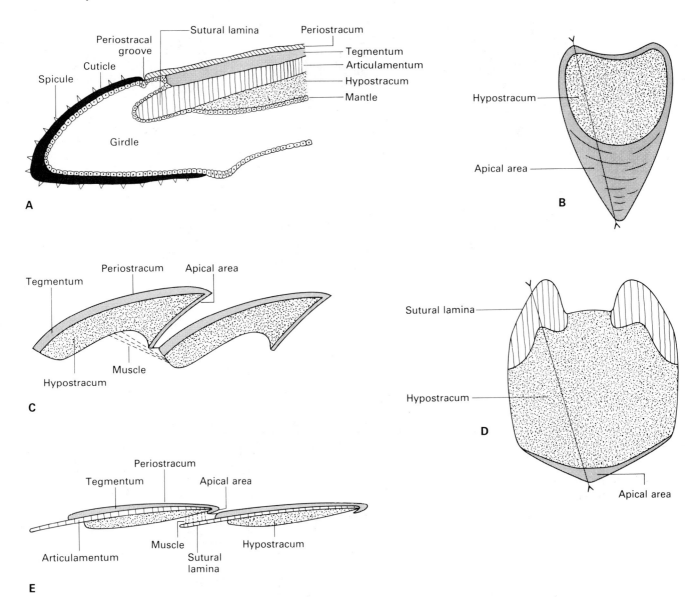

Figure 14.19. Morphology of the mantle edge and the shell layers of chitons. **A,** Generalized section of the mantle edge showing the relationship of the shell layers and mantle. **B,** Ventral view of intermediate valve of a generalized paleoloricate. The line marked by the reversed arrow indicates the oblique long section of one of the valves shown in **C. C,** Oblique long section of two adjacent valves of a paleoloricate showing probable position of muscle attachment. **D,** Ventral view of intermediate valve of a generalized neoloricate. The line marked by the reversed arrow indicates the oblique long section of one of the valves shown in **E. E,** Oblique long section of two adjacent valves of a neoloricate showing the position of a muscle attachment. (Modified from Stasek, C. R. In: Florkin, M.; Sheer, B. J., editors. *Chemical zoology, Mollusca.* New York: Academic Press, Inc.; 1972, p. 14–15.)

Figure 14.20. Some of this new evidence includes localities at which three shapes of similarly ornamented valves of *Matthevia* are found and are interpreted as being the head, intermediate, and tail valves of a chiton. In comparing the shapes of the intermediate valves of *Matthevia* with those of the intermediate valves of an acknowledged paleoloricate (Figure 14.18, *A* to *F*), there is obvious similarity. In placing *Matthevia* valves on a chiton-shaped body, one ends up with an animal that looks like the reconstruction in Figure 14.20. Here, the valves do not overlap. They stand almost upright, and seven cover an animal almost 130 mm long.

Supplementary reading

Baxter, J. M.; Jones, A. M. Valve structure and growth in the chiton *Lepidochitona cinereus* (Polyplacophora: Ischnochitonidae). *Journal of the Marine Biological Association, United Kingdom* 61: 65–78; 1981. Detailed study of the

Figure 14.20. Reconstruction of the Late Cambrian paleoloricate *Matthevia* grazing on stromatolites. (From Runnegar, B.; Pojeta, J.; Taylor, M. E.; Collins, D. *Journal of Paleontology* 53: 1375; 1979.)

ultrastructure, esthete development, and growth of polyplacophoran shells.

Beedham, G. E.; Trueman, E. R. The relationship of the mantle and shell of the Polyplacophora in comparison with that of other Mollusca. *Journal of the Zoological Society* 151: 215–231; 1967. Considers the polyplacophoran mantle edge and the origin of the molluscan shell.

Hunter, W. R.; Brown, S. C. Ctenidial number in relation to size in certain chitons, with a discussion of its phyletic significance. *Biological Bulletin* 128: 508–521; 1965. Demonstrates that both polyplacophorans and monoplacophorans are not metameric and that mollusks were probably derived from turbellarianlike animals.

Runnegar, B.; Pojeta, J., Jr.; Taylor, M. E.; Collins, D. New species of the Cambrian and Ordovician chitons *Matthevia* and *Chelodes* from Wisconsin and Queensland: evidence for the early history of polyplacophoran mollusks. *Journal of Paleontology* 53: 1374–1394; 1979. Interprets *Matthevia* as one of the oldest known chitons.

Yochelson, E. L. Matthevia, a proposed new class of mollusks. U.S. Geological Survey Professional Paper 523B; 1966. Interprets *Matthevia* as the only known member of an extinct class of mollusks.

Class Monoplacophora

Bruce Runnegar

The monoplacophorans are the only group of animals that has been described hypothetically before its first members were discovered, found as fossils (Figure 14.21, A to C) before being taken alive, and dredged from the depths of the ocean (Figure 14.21, D) before being collected from the shallower waters of the continental shelf. This unusual history of discovery, and a puzzling 350 million year break in the fossil record of primitive members of this class (Devonian to Holocene), are responsible for the continually changing ways zoologists and paleontologists have viewed these animals and their place in the Mollusca. At last, after almost a century of study of fossil forms and a quarter of a century of study of living forms, the first living monoplacophorans have been examined in aquaria. It seems that the ancestors of these primitive members of the class survived the events that extinguished their more highly evolved relatives at the beginning of the Mesozoic. The absence of these animals from much of the Phanerozoic fossil record might be the result of a retreat into deep water at the end of the Devonian.

Because the Monoplacophora probably gave rise to at least three other classes of mollusks, the phylogenetic limits of the Monoplacophora can only be defined precisely by specifying single events that distinguished the progeny from their ancestors. The following events are singled out for consideration:

1 Development of the pegma (Figure 14.1, G) separates the first rostroconch from its closest monoplacophoran relative.

2 The origin of a siphuncle (Figure 14.1, E) distinguishes the first cephalopod from its predecessors.

3 Torsion (Figure 14.9) changes a gastropod-like monoplacophoran to the first snail.

Even when these evolutionary steps are reversed subsequently (as they were in all three groups), the fact that one of its ancestors had made the step allows the newer form to be separated from the Monoplacophora. The reversed forms sometimes appear to be monoplacophoran-like but must be recognized as convergent homeomorphs in an evolutionary classification and placed in one of the other three classes. Thus, although the living naked opisthobranch snail *Adalaria* (a sea slug) shows no evidence of torsion at any stage in its life cycle, it is still classified as a gastropod because its ancestors were torted. This means that no all-embracing monothetic definition of the Monoplacophora can be stated, even if all living and fossil representatives were known. Phylogenetic connection, judged from comparative anatomy and morphology, embryology, and stratigraphy, is the only clear guide to whether a particular animal belongs to the Monoplacophora.

Morphology

Anatomy of living forms Viewed externally (Figure 14.21, D), living monoplacophorans have a thin, limpet-shaped shell that has an apex at or near the anterior margin. The shell may be smooth or have prominent growth lines or reticulate ornament. All have an outer prismatic and an inner nacreous shell layer; both layers are aragonitic. Shell layers formed at muscle scars and an

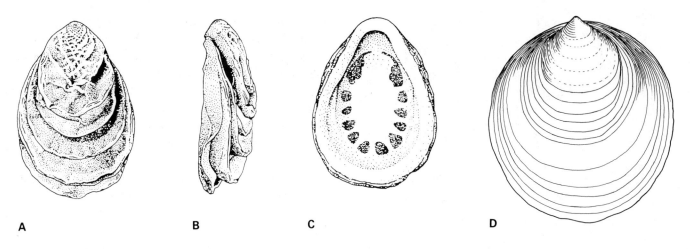

A   B   C   D

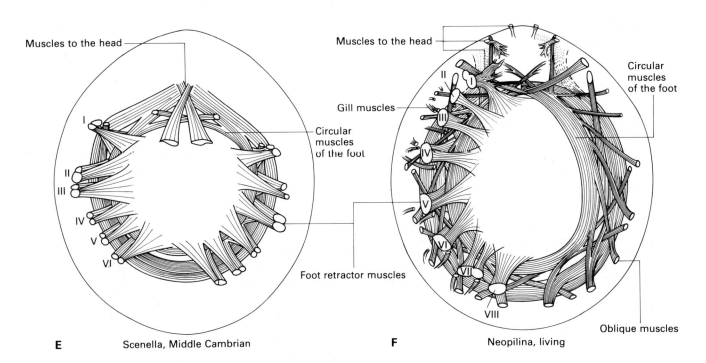

E   Scenella, Middle Cambrian

F   Neopilina, living

Figure 14.21. A to **C,** The limpet-shaped fossil monoplaco-phoran *Tryblidium* from the Silurian of Sweden has a thick shell and paired insertions of the pedal muscles. Exterior, left lateral, and interior views. (×1.8.) **D,** Exterior view of the living, limpet-shaped monoplacophoran *Neopilina,* which has a very thin shell up to 3 cm in length. Note that the apex is close to the anterior end of the shell. **E** and **F,** Musculature of primitive monoplacophorans. The shell attached muscles of *Scenella* (**E**) from the Middle Cambrian of Canada were reconstructed from the muscle scars (Figure 14.22, *B*) by using the known muscu-lature of the living *Neopilina* (**F**). (**A** to **C** from Knight, J. B., *et al.* in: Moore, R. C., editor. *Treatise on inverte-brate paleontology, Part I, Mollusca 1.* New York and Lawrence, KS: Geological Society of America and University of Kansas Press; 1960. **D** and **F** from Lemche, H.; Wingstrand, K. G. *Galathea Report.* Copenhagen, Denmark: Zoologisk Museum; 1959.)

organic periostracum probably exist, but they are diffi-cult to observe because the shells are so thin.

Looking at the underside (Figure 14.1, *C₁*), one sees a central subcircular foot bordered laterally by five or six pairs of gills. The gills are not elaborate structures, and they do not effectively divide the mantle cavity into two

chambers as in other mollusks. Water currents probably enter the mantle cavity on both sides near the front of the animal, pass through the gills, and exit posteriorly after joining up near the anus. The head lies in front of the foot and is attached to the shell by muscles. It is bordered by a pair of fleshy lobes and preoral and postoral tentacles.

The precise functions of these structures are not well understood, but they are probably used for sensing or feeding.

The internal anatomy of living monoplacophorans (Figure 14.1, C_2) is unique among living mollusks. The nervous system has a ladderlike ventral arrangement with weakly developed cerebral ganglia and is very primitive. The gut has a long coiled intestine, typical of an unspecialized herbivore or deposit feeder, and the body cavities are larger and more extensive than in most other mollusks. More than 24 pairs of pedal muscles are inserted on the shell, of which eight pairs are pedal retractors (Figure 14.21, F). There are six pairs of renal organs or nephridia, two pairs of gonads, and two pairs of auricles of the heart (Figure 14.1, C_2). The unexpected repetition of these structures has led some zoologists to suggest that monoplacophorans show evidence of metameric segmentation and a close relationship to segmented annelids and arthropods. The repetition of organ systems in monoplacophorans is best considered pseudo-metamerism and is analogous to the repetition of some structures in chitons. This pseudometamerism in some mollusks is a parallel development to the eumetamerism seen in annelids and arthropods and probably does not indicate a common, segmented ancestor of the three phyla.

Fossil forms Unlike living monoplacophorans, all of which have limpet-shaped shells, fossil monoplacophorans show great diversity of shell form (Figures 14.22 to 14.25). They are ordinarily bilaterally symmetrical (Figure 14.24) although a few are slightly asymmetrical (Figure 14.25). The shells of fossil monoplacophorans can be straight, curved, planispiral, laterally compressed, or limpet shaped. The known range of pedal retractor muscle scars in fossil forms is from one pair (Figure 14.26, E) to six to eight pairs (Figure 14.26, A). So far as is known, all monoplacophorans, both living and fossil, have an aragonitic shell that may have prismatic, nacreous, and/or crossed lamellar shell layers.

Classification

ORDER CYRTONELLIDA (Figures 14.22, B; 14.24)

Limpet-shaped, spiral, or laterally compressed Monoplacophora that lack a well-formed posterior shell aperture reentrant such as a sinus or slit. (Earliest Cambrian to Middle Devonian; about 50 genera.)

ORDER TRYBLIDIIDA (Figure 14.21, A to D and F)

Limpet-shaped Monoplacophora with an anteriorly placed larval shell. (Earliest Middle Cambrian to Holocene; about 20 genera and subgenera.)

ORDER BELLEROPHONTIDA (Figure 14.22, A)

Planispiral Monoplacophora in which all but the most primitive members have a well-formed posterior shell aperture reentrant in the form of a sinus or slit. (Earliest Middle Cambrian to Early Triassic; about 60 genera and subgenera.)

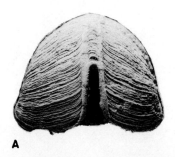

A

B

Figure 14.22. A, Posterior view of the Carboniferous bellerophont *Bellerophon* showing the slit and slit-band forming the posterior apertural reentrant. (×1.) **B,** Internal mold of the Middle Cambrian limpet-shaped cyrtonellid *Scenella* showing muscle scars used to reconstruct musculature in Figure 14.21, *E*. Note that the apex of the shell is subcentral and not near the anterior end as in tryblidiids. (×5.)

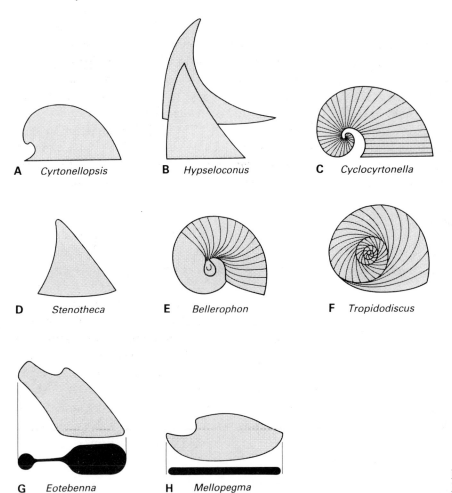

A *Cyrtonellopsis*   **B** *Hypseloconus*   **C** *Cyclocyrtonella*

D *Stenotheca*   **E** *Bellerophon*   **F** *Tropidodiscus*

G *Eotebenna*   **H** *Mellopegma*

Figure 14.23. Some shell forms in the Monoplacophora.

ORDER PELAGIELLIDA (Figure 14.25)

Asymmetrically coiled Monoplacophora with an expanded subtriangular or oval aperture. (Earliest Cambrian to Late Cambrian; about 5 genera and subgenera.)

The changing concept of the taxon Monoplacophora

All known living monoplacophorans have dorsoventrally flattened, limpet-shaped shells (Figure 14.21, D). Until fairly recently the concept Monoplacophora was largely limited to living and fossil, bilaterally symmetrical, univalved mollusks with a limpet-shaped shell and several pairs of pedal retractor muscle scars symmetrically placed about the midline (Figure 14.21). Included with these were a few fossil shells of similar shape, but with a continuous ring or horseshoe-shaped pedal muscle scar high in the shell, and some strongly curved symmetrical univalved fossils, some of which had two pairs of symmetrically placed pedal retractor muscle

scars (*Cyrtonellopsis*, Figure 14.23). At the time this concept of the Monoplacophora was being developed, some tall, cone-shaped, bilaterally symmetrical univalves, in which the muscle scars were poorly known, were tentatively placed in the class (*Hypseloconus*, Figure 14.23).

Subsequent to this attempt to define the concept Monoplacophora, some loosely (*Cyclocyrtonella*, Figure 14.23) and more tightly coiled (Figure 14.26, C) planispiral, bellerophontiform shells were shown to have several pairs of symmetrically placed muscle scars and were included in the Monoplacophora. Previously, the best known of these forms had been considered to be symmetrical gastropods and were placed in the bellerophonts. These first discovered planispiral forms with multiple, symmetrical muscle scars lacked an apertural reentrant in the form of a sinus or slit, which is present in almost all bellerophonts presumed to be gastropods (Figure 14.22, A). For a short time, the lack of a sinus or slit was used to separate bellerophontiform monoplacophorans from "true" bellerophonts that were then still

Figure 14.24. Silica replicas of the shells of tall, coiled cyrtonellids from the earliest Middle Cambrian of Australia. All specimens are 3 mm or less in length. The ones with anterior tubes (snorkels) are placed in the genus *Yochelcionella*. Those without snorkels are known as *Latouchella*. Young shells (upper center) show how the snorkel was formed.

placed in the Gastropoda. However, the concept Monoplacophora had been significantly expanded to accommodate a greater diversity of shell form.

About 15 years ago, a Devonian bellerophont was found that had both multiple symmetrically placed muscle scars and a posterior apertural sinus. Again the line of distinction between untorted bellerophontiform monoplacophorans and bellerophonts, which were thought to be symmetrical torted gastropods with a single pair of symmetrical muscle scars, had become blurred. The main criterion for distinguishing the two groups, which were thought to be homeomorphs, then became muscle insertions, with several symmetrical pairs in the Monoplacophora and only a single symmetrical pair in the planispiral gastropods. In many bellerophontiform shells, muscle scars are not known because no specimens showing them have been found.

The latest development in the expanding and changing concept of the Monoplacophora is to include all the forms classically considered to be symmetrical bellerophont gastropods in the Monoplacophora, regardless of the number of pairs of symmetrical muscle scars or whether these muscle scars are known. This approach does not admit to a homeomorphy between the two groups but rather admits a phylogenetic relationship, and it is followed in this section. Thus bellerophonts are considered to be untorted mollusks belonging to the class Monoplacophora. Following this line of reasoning, the living Monoplacophora may only reflect a small fraction of the morphologic and taxonomic diversity that once existed in this primitive class of mollusks. For an alternative treatment of bellerophonts, see the section "Class Gastropoda."

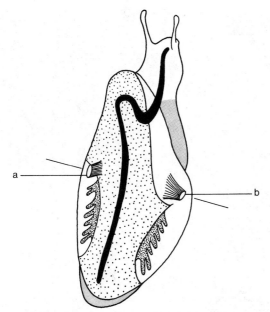

Figure 14.25. Reconstruction of the asymmetrical Cambrian monoplacophoran *Pelagiella*. The gut is shown in black, the gills in fine stipple, and the body mass in coarse stipple. *a* and *b* are the left and right pedal retractor muscles. The specimen from which the reconstruction was made is about 9 mm long.

Ecology and paleoecology

The two living genera of monoplacophorans (*Neopilina* and *Vema*) are dorsoventrally flattened or limpet-shaped forms that have been dredged from deep water of the Pacific, South Atlantic, and Indian Oceans and have usually been thought to have survived as relicts or living fossils in the cold stable environments of the ocean floor since their ancestors went there, perhaps as long ago as 350 million years (Figure 14.21, *D* and *F*). The great differences in temperature and pressure between the ocean floor and the surface kills these animals in transit, so that their habits and functions have to be inferred from anatomic studies and photographs of the collecting sites. Current thought is that these animals crawled on the sea floor feeding on organic detritus. Few, if any, of the living deep-sea monoplacophorans seem to have lived on a hard substrate, although they resemble rock-dwelling shallower water limpets in shell form and gross anatomy.

One small species of *Vema* has been found in shallower water (200 m) on the edge of the continental shelf off California. This species lives attached to rocks and feeds on organic detritus. Because these animals are blind, it has been speculated that they moved to the edge of the continental shelf from deeper water relatively recently.

Almost no work has been done on the paleoecology of the fossil Monoplacophora. One recent report indicates that the tall cone-shaped Late Cambrian monoplacophorans known as *Hypseloconus* (Figure 14.23) grazed digitate algal stromatolites in the intertidal zone. These and other primitive mollusks (Figure 14.20) may have been responsible for restricting most post-Cambrian stromatolites to hypersaline estuaries and geothermal pools.

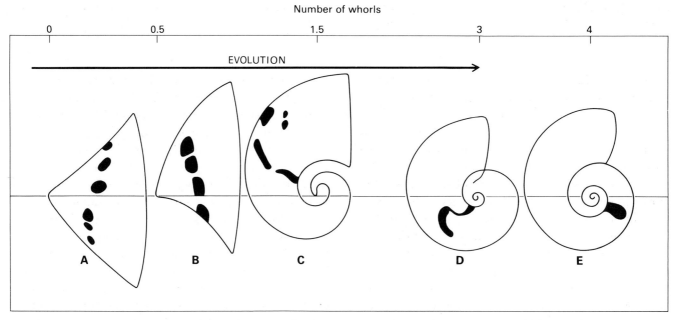

Figure 14.26. Pedal muscle scars on the shells of Ordovician monoplacophorans move further into the shell and decrease in number as the number of whorls increases. **A,** *Lenaella*, Soviet Union; **B,** *Nyuella*, Soviet Union; **C,** *Cyrtolites*, Canada; **D,** *Sinuites*, United States; E, *Bucania*?, Norway. **A** to **C** are cyrtonellids; **D** and **E** are bellerophonts.

The ancient environments of a few other monoplacophorans are reasonably well known. Hundreds of different-sized shells of a limpet-shaped form have been recovered from a horizon in the Middle Cambrian Burgess Shale of Canada, which also contains a spectacularly well-preserved soft-bodied fauna. The source of the common invertebrates in this deposit is thought to have been a local community inhabiting relatively deep-water muds that were piled up against the base of a submarine carbonate escarpment. Thus the Middle Cambrian form not only resembles living monoplacophorans in shell shape, but it also seems to have lived on soft bottoms.

By contrast, masses of shells of a late Paleozoic bellerophont are found in eastern Australia either washed on to Permian beaches or buried in lower shoreface sands. They are generally found with burrowing clams that have been disinterred and disarticulated by storm waves. Probably the Australian bellerophonts lived on subtidal algae in the cool temperate seas of eastern Gondwanaland. Their tropical North American counterparts appear to have inhabited Permian carbonate hard grounds along with a great variety of other mollusks. These few examples serve to show that much remains to be learned about the life habits of extinct Monoplacophora.

Evolution

Fossil shells that are not too different from those of living monoplacophorans are found in the oldest Cambrian fossiliferous horizons. Because these forms lack an anteriorly placed larval shell, they are classified as cyrtonellids. These limpet-shaped fossil shells (Figure 14.22, B) are thought to be primitive in the Monoplacophora because the ancestor of the class is likely to have had a dorsoventrally flattened limpet-shaped shell. Well-preserved muscle insertions on a Middle Cambrian species of the limpet-shaped monoplacophoran *Scenella* (Figures 14.21, E and 14.22, B) indicate that the anatomic organization of *Neopilina* has existed since that time (Figure 14.21, E and F). Thus, both morphologically and stratigraphically, living monoplacophorans can be viewed as members of a very primitive group of monoplacophoran mollusks.

The taller, partly coiled cyrtonellids vary in shape from smooth or wrinkled curved cones (*Latouchella*, Figure 14.24) to laterally compressed shells (*Stenotheca*, Figure 14.23, and *Mellopegma*, Figures 14.16, A to C and 14.23), to snorkel-bearing univalves (*Eotebenna*, Figure 14.23; *Yochelcionella*, Figure 14.24). These shell shapes reflect different evolutionary strategies within the group

and allow deductions about the soft anatomy of the various forms to be made. They were all strictly bilaterally symmetrical, the foot was probably controlled by a relatively few shell-inserted muscles, and the gills may have been reduced to a single pair, one on either side of the body. As in living monoplacophorans, water probably entered the mantle cavity anteriorly, either under the lip of the shell or through the snorkel. Microorganisms carried in with this water current were consumed as food, and spent water exited posteriorly. Laterally compressed genera probably lived partly within the substrate, but most taller, partly coiled cyrtonellids seem to have been epifaunal animals. The shell of the epifaunal forms represents a compromise between a limpet-shaped shell, which can be used on hard substrates that act as a second "valve," and the space-conserving, tightly coiled, small-apertured shells of most higher univalved mollusks. This morphologic compromise was not entirely successful, and the vast majority of the taller, partly coiled cyrtonellids were extinct by the Late Cambrian, although some similar shell forms of the Cyrtonellida range into the Devonian. The laterally compressed taxa are not known after the Cambrian.

However, the probable descendants of these taller, partly coiled cyrtonellids were eminently successful. By increasing the spiral curvature of their shells, some of the epifaunal forms developed planispiral skeletons; these were the first bellerophonts (*Bellerophon*, Figure 14.23). As the spiral curvature increased, geometric constraints forced the pair of gills into a more posterior position and reduced the pedal retractor muscles to one or two pairs (Figure 14.26). The pedal retractor muscles were attached to the shell near the axis of coiling (Figures 14.26, E and 14.44, B); in this position the muscle insertions move peripherally only a short distance as the shell grows. In these forms, water was taken in laterally through forward-opening lips and then passed through the gills to exit posteriorly. A posterior apertural re-entrant (Figure 14.22, A) in the form of a sinus, slit, or, in few cases, a series of holes evolved in all but the most primitive Cambrian forms to improve water circulation through the mantle cavity.

The more successful bellerophont shell design was exploited in a variety of ways. Some bellerophonts became globular, others evolved a discoidal shape, and a few reverted to a limpetlike shape similar to that of their distant ancestors. Many bellerophonts have a radial aperture (*Bellerophon*, Figure 14.23) in which the plane of the aperture lies along a radius from the axis of coiling to the shell periphery. These forms may have closed their shells with an uncalcified operculum, although no bellerophont opercula are known. Other bellerophonts developed a tangential aperture (*Tropidodiscus*, Figure 14.23) in which the plane of the aperture forms a tangent to the earliest formed whorls; such an aperture can be clamped against the substrate. A very few bellerophonts became slightly asymmetrically coiled, thus convergently developing the shell form of *Pelagiella* (Figure 14.25).

Pelagiella and a few other closely related genera are the only other asymmetrically coiled univalves currently considered to be monoplacophorans. They are found in Cambrian rocks and are thought to be transitional between symmetrical cyrtonellids such as *Latouchella* (Figure 14.24) and the first gastropods. *Pelagiella* (Figure 14.25) has a large subtriangular aperture and a coil that lies to the left of the head. There were two pedal retractor muscles, one inserted on the left side and the other attached in a more posterior postion on the right side. A large mantle cavity seems to have housed a pair of lateral gills.

The asymmetrical coiling of the shell of *Pelagiella* had probably caused about 10 degrees of torsion to occur. As the spiral curvature of such shells increased, the aperture became proportionally smaller, the spire less stable, and the angle of torsion greater. Such relatives of *Pelagiella* were the first gastropods.

Biostratigraphy

If the bellerophonts are monoplacophorans, the fossil record of the class spans the whole of the Paleozoic and a fraction of the Mesozoic; if not, it extends from the earliest Cambrian to the Devonian. In either case, there is a Holocene record of the group and a great gap in its fossil record. Within their fossil record, monoplacophorans were reasonably common inhabitants of shallow marine shelves and epicontinental seas. Rocks can be dated by identifying the various genera and species of monoplacophorans, which are discriminated mainly on differences in shape, size, and ornament. Usually, other fossils found with or near monoplacophorans provide more reliable age determinations. This may be due to the slower evolution of Monoplacophora than of other groups, but it probably reflects lack of detailed knowledge. For example, fossil monoplacophorans are potentially useful for dating Cambrian rocks, particularly those that contain few trilobites. In the last decade, Russian biostratigraphers have shown that monoplacophorans can be used with other fossils to zone the oldest part of the Cambrian System, known as the Tommotian. This discovery resulted from attempts to chemically extract microfossils from the oldest Cambrian rocks. It turns out

that most Early and Middle Cambrian mollusks are very small, 5 mm or less in size. Now that many different kinds of monoplacophoran mollusks can be chemically etched from Cambrian limestones, and there is a suitable tool for examining them (the scanning electron microscope), they probably will become increasingly important for studying Cambrian history.

Supplementary reading

Jenkins, M. M. *The curious mollusks*. New York: Holiday House; 1972. Describes the discovery of *Neopilina*.

Knight, J. B. Primitive fossil gastropods and their bearing on gastropod classification. *Smithsonian Miscellaneous Collections* 117: 1–56; 1952. Places class Monoplacophora in a historical perspective.

Lemche, H. A new deep-sea mollusc of the Cambro-Devonian class Monoplacophora. *Nature* 179: 413–416; 1957. The first report of the discovery of *Neopilina*.

Lemche, H.; Wingstrand, K. G. The anatomy of *Neopilina galatheae* Lemche, 1957 (Mollusca Tryblidiacea). *Galathea Report* 3: 9–71; 1959. Detailed description of the anatomy of *Neopilina*.

Lowenstam, H. A. Recovery, behaviour and evolutionary implications of live Monoplacophora. *Nature* 273: 231–232; 1978. Describes live monoplacophorans.

McLean, J. H. A new monoplacophoran limpet from the continental shelf off Southern California. *Contributions in Science, Natural History Museum of Los Angeles County* 307: 1–19; 1978. Detailed description of the anatomy of *Vema*.

Runnegar, B.; Jell, P. A. Australian Middle Cambrian molluscs and their bearing on early molluscan evolution. *Alcheringa* 1: 109–138; 1976. Reviews the early history and classification of the Monoplacophora.

Yochelson, E. L. Quo Vadis, Bellerophon?, pp. 141–161. In: Teichert, C.; Yochelson, E. L., editors. *Essays in paleontology and stratigraphy*. (R. C. Moore commemorative volume.) Lawrence, KS: University of Kansas Press; 1967. Summary of older views on the Monoplacophora.

Class Gastropoda

John S. Peel

Gastropods are the most diverse and abundant mollusks and occur in marine, freshwater, and terrestrial environments. Most are mobile benthos and have an anteroposteriorly elongated body with a well-developed head and a large, flattened foot, called the **head-foot** mass. However, some of different form have become adapted to pelagic, sedentary, and, in a few cases, parasitic modes of life. Most gastropods have an external shell that is usually coiled into a corkscrew helix (Figure 14.27, *A*). The shell contains the visceral mass, which is connected to the head-foot mass by a narrow neck, and provides a shelter into which the animal can withdraw if threatened or

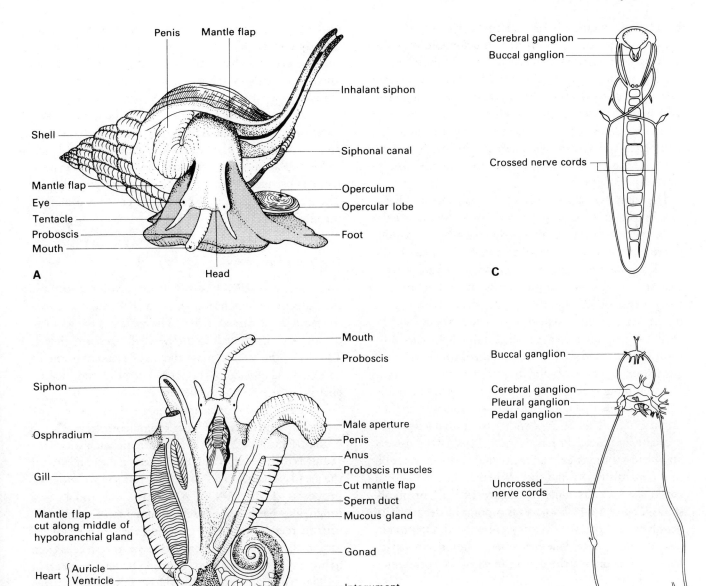

Figure 14.27. Soft parts of gastropods. **A**, *Buccinum*, neogastropod. Shell with a protruded head-foot mass and operculum. **B**, *Buccinum*. Shell removed, exposing the coiled visceral mass. The organs of the mantle cavity have been exposed by making a median longitudinal incision along the mantle and folding the two sides back. The proximal end of the proboscis and its muscle have been exposed by an additional incision. **C**, Streptoneurous (crossed) nervous system of a pleurotomariin archaeogastropod. **D**, Euthyneurous (uncrossed) nervous system of an opisthobranch. (**A, B** and **D** from Cox, L. R. in: Moore, R. C., editor. *Treatise on invertebrate paleontology, Part I, Mollusca 1*. New York and Lawrence, KS: Geological Society of America and University of Kansas Press; 1960, p. 89, 91, 103. **C** modified from Hyman, L. *The invertebrates, Mollusca I*, vol. 6. New York: McGraw-Hill Book Co.; 1967, p. 250.)

frightened. In many gastropods the shell is greatly reduced or lost, the helical form may be absent, and the animal may display a high degree of bilateral symmetry.

Gastropods are characterized by **torsion** (Figure 14.9, *D* and *E*), a process by which the posterior mantle cavity and anus are rotated laterally and anteriorly in a counter-clockwise direction so that they come to lie above the head. Various living gastropods undergo torsion during their early ontogeny by rapid contraction of the asymmetrical right larval retractor muscle and by differential growth. The majority of gastropods show evidence of having undergone torsion in their phylogeny by the asymmetry of their anatomy. However, many show evidence of a process termed **detorsion**, which to varying degrees modifies and even reverses the effects of torsion and forms a major trend in the later evolutionary history of gastropods. In a few sea slugs without shells, which are classified as gastropods for various reasons of anatomy, detorsion is so advanced that the animal is completely bilaterally symmetrical and shows no evidence of the torsion which its ancestors are presumed to have experienced.

The adaptive significance of torsion and its phylogenetic origin have been much debated. Studies of the embryology of some primitive living gastropods demonstrate that torsion is achieved by a slight difference in the time of development of the left and right larval retractor muscles, but this mechanism of torsion does not explain the process or its adaptive significance. There is little doubt that torsion resulted from life in a narrow tubelike shell. Gastropods such as sea slugs have phylogenetically lost their shell and show an accompanying high degree of detorsion. Torsion possibly occurred at the same time that the monoplacophoran ancestor of gastropods developed a coiled shell into which it was able to withdraw (Figure 14.9). The ability to withdraw into a narrow shell probably produced the narrow neck, between the visceral and head-foot masses, that facilitated torsional rotation of the mantle cavity and shell. Whether the pretorsional coiled shell was asymmetric (Figure 14.9, C) or planispiral (Figure 14.44 E and F) is a point of debate. It is possible that both shell forms were utilized. Opinion is also divided as to whether some relationship exists between torsion and the development of the corkscrew-coiled shell.

Once torsion occurred, the mantle cavity and anus opened anteriorly and the shell coiled backwards away from the head in the **endogastric** position (Figures 14.9, D and E and 14.44, C and D). In monoplacophorans the shell coils over the head and is **exogastric** (Figures 14.9, A and B and 14.44, E and F). The need for a narrow neck between the visceral and head-foot masses before torsion could occur suggests that it is unlikely that torsion occurred in an unshelled mollusk or in a monoplacophoran with a low cap-shaped shell such as *Neopilina* (Figure 14.21, D).

The abundance of gastropods in number of species and individuals suggests that torsion is of great adaptive significance. In rotating the mantle cavity to the front of the animal, torsion brought various organs from an originally posterior position to lie anteriorly above the head. In the anterior position these organs can better respond to environments into which the animal is moving, rather than sampling areas through which it has already passed. The anterior position is particularly advantageous to a chemical and sediment sensitive organ known as the **osphradium**, located at the entrance to the mantle cavity (Figure 14.27, B). Sediment fouling of the gills can be prevented more readily if these are located anteriorly. There is no obvious advantage in having the anus located above the head, and a tendency for the anus to be displaced toward the rear of the animal is evident from subsequent gastropod evolution (Figure 14.31). The ejection of feces in pelletal form prevents fouling of the gills.

Gastropod anatomy

Soft parts of shelled gastropods are divided into those that ordinarily extend outside the shell and those that normally are retained within. The head and the foot can usually be protruded from the shell aperture (Figure 14.27, A); the visceral mass and mantle cannot normally be extended outside of the shell. The head and foot are clearly separate but are called the head-foot mass.

The gastropod head usually bears eyes associated with tentacles, which are the most prominent external sense organs (Figure 14.27, A). The mouth, through which the radula is protruded, is usually located on the underside of the head or at the end of a long retractable **snout** or **proboscis** (Figure 14.27, A and B). The radula is a tonguelike band armed with thousands of tiny horny organic teeth in primitive gastropods where it is used to scrape algae from sediment surfaces or to rasp cells from living prey. In advanced gastropods, the number of radular teeth is often greatly reduced. In some specialized hunters, only a single row of harpoonlike teeth filled with poison is retained.

The foot may be adapted to a variety of functions. Primitively it was probably flat and used to creep over relatively hard substrates, as is still the case in many species, although it may be adapted to a variety of substrates. Usually, the foot has a mucous gland that produces a trail of slime to facilitate progress. Some small snails use ciliary locomotion, but in most, waves of muscular contraction pass over the foot from anterior to posterior and provide the power for locomotion. A few gastropods use the foot in inch-worm fashion, by periodically extending and anchoring the anterior portion and then advancing the posterior part by contraction. Others have the foot divided into right and left halves, which are advanced alternately or have a modified foot for burrowing or swimming.

The gastropod head and foot are withdrawn into the

shell by means of one (usually) or two (rarely) retractor muscles called **columellar muscles** because they are usually attached to the **columella** of the shell (Figure 14.33, *C*). The columella is a solid or perforated pillar of shell material formed by the inner walls of a helical shell. Extension of the head-foot mass after withdrawal is accomplished by contractions of intrinsic muscles and movement of the blood into the hemocoel spaces.

The mantle secretes and lines the inside of the shell. Toward the shell apex, the mantle adheres closely to the visceral mass; this part of the mantle is called the **integument** (Figure 14.27, *B*), and there is no mantle cavity in this region. Toward the aperture there is a space between the mantle and the other organs, called the mantle cavity (Figures 14.1, *D* and 14.27, *B*), which opens to the outside at the aperture. In a number of gastropods, the left side of the dorsal mantle is drawn out into an **inhalant siphon** (Figure 14.27, *A* and *B*), which may or may not secrete a shelly **siphonal canal** (Figure 14.28, *A*). The function of this siphon is to bring water to the gill and osphradium (Figure 14.27, *B*). Some gastropods, such as cowries (Figure 14.36, *B* and *E*), have the mantle reflected over the outside of the shell.

Most of the early whorls of the shell are filled by the gonad, digestive gland, heart, and excretory organs that form the visceral mass (Figure 14.27, *B*). The mantle cavity contains the gill(s), osphradium(ia), hypobranchial gland(s), and the openings of the digestive, excretory, and reproductive systems (Figure 14.27, *B*). The **hypobranchial glands** secrete mucus, which is used to trap particles that enter the mantle cavity with the respiratory current. These mucous-trapped particles are then removed from the mantle cavity by the exhalant current, together with the products of digestion, excretion, and reproduction. In a few shelled gastropods, the gills, osphradia, and hypobranchial organs are paired, but in most the ones on the right side have been lost.

Most gastropods have one internal **gill** or **ctenidium** (Figure 14.27, *B*). However, in land snails and slugs, the mantle cavity has evolved into a vascularized air-breathing lung and the gills have been lost. In aquatic gastropods with gills, there may be comblike surfaces on both sides of the gill; such gills are termed **bipectinate**. In many gastropods, the gills have comblike surfaces only on one side (Figure 14.27, *B*) and are termed **monopectinate**. Various living marine gastropods without shells have lost the mantle cavity and its internal gills; these forms respire through a series of fleshy external gills on the outside of the animal.

The gastropod nervous system has ganglia connected by nerve cords. The nerve cords form a loop, which in many gastropods has been twisted into a figure eight as a result of torsion; this condition is called **streptoneury** (Figure 14.27, *C*). In other gastropods, the nerve cords are detorted, and this condition is termed **euthyneury** (Figure 14.27, *D*).

Shell morphology

Exterior features The gastropod shell (Figure 14.28) is secreted by the mantle and forms a hollow cone regularly increasing in width from its first-formed point, the **apex** of the shell, to the **aperture** where the head-foot mass protrudes (Figure 14.28, *A*). A platelike cover known as the **operculum** commonly closes the aperture when the soft parts are withdrawn into the shell (Figure 14.28, *B*). The shell may have a **limpet** or **cap shape** (Figure 14.34, *A* and *E*), with a wide open aperture, but is usually coiled around an **axis of coiling** (Figure 14.28, *A*) that passes through the apex. Each coil of 360 degrees is called a **whorl** (Figure 14.28, *A* and *C*). The shell usually has the form of the corkscrew helix in which case it is called **helical** or **conispiral** (Figure 14.28, *E*). This type of coiling has the apex of the shell drawn out to one side and is thus three dimensional. In a few forms the shell coils in one plane and the coil, which is in two dimensions, is **planispiral** (Figure 14.28, *E*). Planispiral shells that are bilaterally symmetrical about the plane in which they coil are termed **isostrophic** (Figure 14.28, *E*); other planispiral shells and all helical shells are asymmetric and termed **anisostrophic** (Figure 14.28, *E*).

Various terms used to describe the shape of gastropod shells are derived from the names of common genera. Thus **turbiniform** is derived from the genus *Turbo* (Figure 14.32, *H*) and means *Turbo*like in shell shape. **Patelliform** is derived from the genus *Patella* (Figure 14.34, *E*) and is used to refer to all limpet- or cap-shaped gastropods such as those shown in Figure 14.32, *C* to *E*. **Trochiform** is derived from the genus *Trochus* and is applied to a variety of genera including those shown in Figure 14.32, *A* and *K*. This method of describing shell shapes provides a convenient substitute for long descriptions by formulating an initial, concise concept of shape. Many "–form" terms are possible.

The first formed whorls in conispiral gastropods are often different in style of coiling or ornamentation from the rest of the shell. These oldest whorls are termed the **protoconch**, and the rest of the shell is the **teleoconch**. Usually the axes of coiling of the protoconch and the teleoconch are parallel; but in a few forms they are not, and such shells are called **heterostrophic**.

A complete whorl generally covers the previous whorl

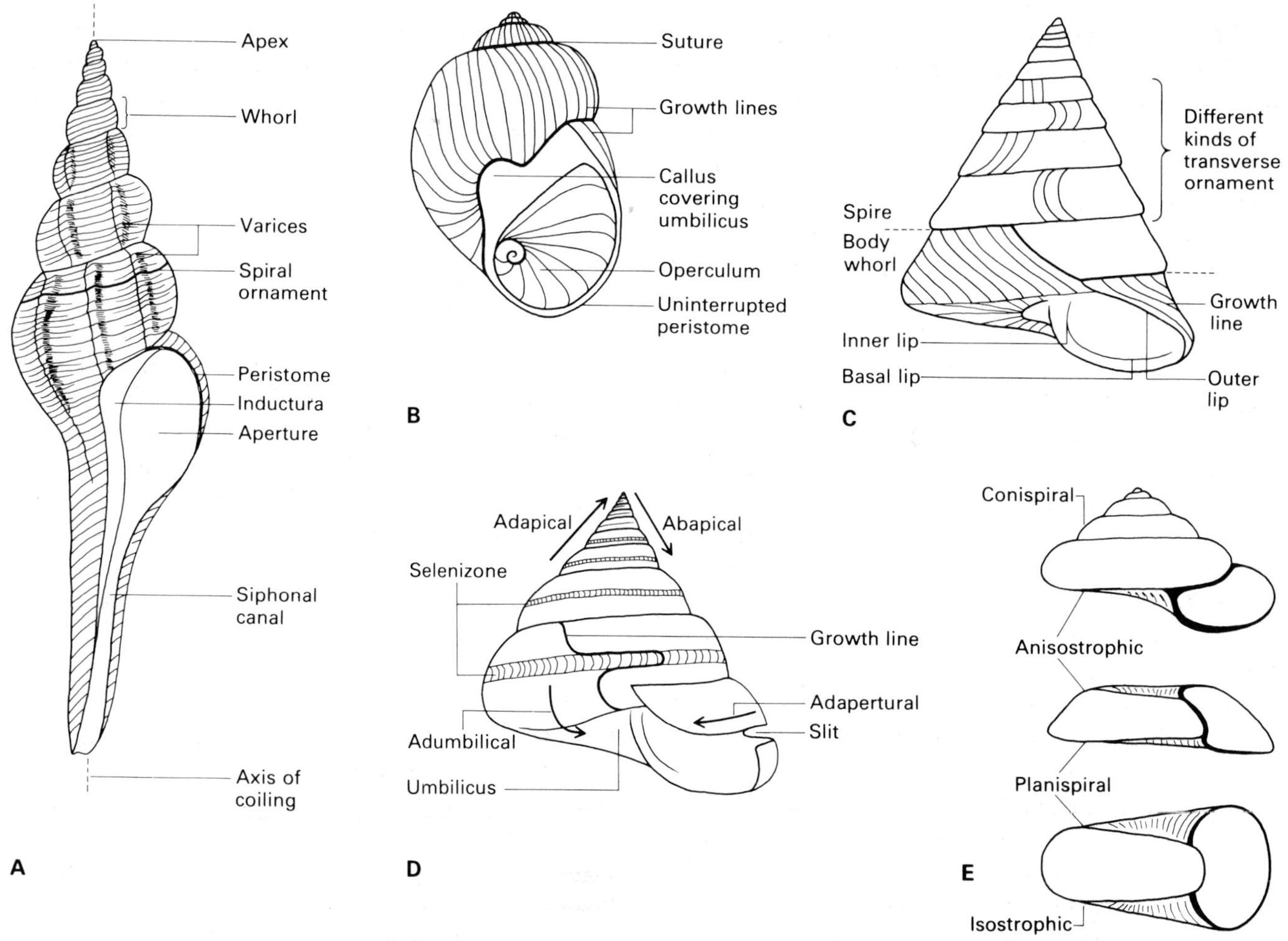

Figure 14.28. The gastropod shell: basic terminology. (See text for explanations of **A** to **E**.)

up to a line of contact known as the **suture** (Figure 14.28, B). The apical part of a conispiral shell, exclusive of the last complete whorl or **body whorl**, is referred to as the spire (Figure 14.28, C). Spires are loosely characterized as high (Figure 14.35, F) or low (Figure 14.35, C and E). The axial margins of successive whorls may be closely in contact with each other, producing a columella (Figure 14.33, C), or they may delimit a conical cavity termed the **umbilicus** (Figures 14.28, D and 14.35, D). On the inner side of the aperture is often a smooth shelly layer secreted by the surface of the mantle and known as the **inductura** (Figure 14.28, A). The inductura may partly or wholly cover the umbilicus (Figure 14.28, B) and is then called a **callus**; occasionally the inductura covers all the shell exterior as in cowries (Figure 14.36, B and E).

Successive increments of growth are added to the shell at the aperture, which may vary in shape from nearly circular (Figure 14.32, H) to strongly elongated (Figure 14.37, I). The margin of the aperture is termed the **peristome** and is often a simple uninterrupted curve (Figure 14.28, B). However, gastropods commonly have

an interrupted peristome where water enters the mantle cavity through the inhalant siphon. At this point the base of the peristome may be elongated into a siphonal canal (Figures 14.27, A and 14.28, A), or the position of the inhalant siphon may be marked by a smaller **inhalant notch** (Figure 14.31, C). The exhalant current from the mantle cavity may be marked by the presence of special structures on the outer lip of the aperture or near the contact of the outer lip with the spire. These structures form reentrants. A shallow reentrant is termed a **sinus** and can usually be seen in successive growth lines (Figure 14.29, B) as well as in the aperture. A deep parallel-sided reentrant is termed a **slit** (Figures 14.28, D and 14.32, A and B). With continuing growth the slit generates a spiral band on the surface of the whorls called a **selenizone** (Figure 14.28, D). In a few gastropods the exhalant current exits through a series of **perforations** (Figure 14.32, F).

The surface of the shell is marked by growth lines that are parallel to the margin of the aperture (Figure 14.28, B and C). Each growth line marks a successive growth stage. Periodic halts in growth can produce prominent **varices** (singular, varix) or ribs (Figure 14.28, *A*), which are regular features in some genera. **Transverse ornament** crosses the whorl from suture to suture, and is usually parallel to the apertural margin. The strength of transverse ornament varies from fine growth lines, to raised threads, to prominent varices. The inclination of transverse ornament may vary considerably (Figure 14.28, *C*), but most commonly it is almost perpendicular to the sutures or sloping backward from one suture to the next. **Spiral ornament** (Figure 14.28, *A*) parallels the suture and passes continuously around the shell on the surface of the whorls.

Use of the prefixes "ad–" and "ab–" provides a directional terminology for describing gastropod shells; ad– means toward, and ab– means away from (Figure 14.28, *D*). Thus **adapical** means toward the apex, and **abapical** means away from the apex. **Adapertural** means toward the aperture, generally in the sense of spiral movement around a whorl. **Adumbilical** means toward the umbilicus.

Terminology of coiling In most helical gastropods, which are positioned with the axis of coiling vertical, the aperture toward the observer, and the apex uppermost, the aperture lies to the right of the axis of coiling (Figures 14.28, *A* to *D*; 14.29, *A*; and 14.42, *A*). Such shells are termed **dextral** or **right-handed**; when viewed from the apex they coil in a clockwise direction (Figure 14.42, *B*). When the soft parts of dextral gastropods are examined, the genitalia lie on the right side when the animal is viewed from above with the head being anterior (Figures 14.1, *D* and 14.27, *B*). In the much rarer **sinistral** or **left-handed** gastropods the aperture lies to the left of the axis of coiling (Figures 14.29, *A* and 14.40, *E* and *H*). When viewed from the apex, sinistral gastropods coil in a counterclockwise direction and have genitalia on the left side of the body. Sinistral gastropods are mirror images of dextral ones. Left-handed coiling can occur at various taxonomic levels. Occasionally, sinistrality occurs as an individual variation in an otherwise dextral species.

Rarely, a gastropod whose soft parts are organized in a dextral fashion, with the genitalia on the right, has an apparently sinistral conispiral shell (or vice versa). Such snails are **hyperstrophic**. Normally organized helical shells are called **orthostrophic** and can be conveniently regarded as coiling down the axis of the shell or as having a positive coil (Figure 14.29, *A*). Hyperstrophic shells can be regarded as coiling up the axis of the shell and having a negative coil (Figure 14.29, *A*). Planispiral coiling is thus morphologically intermediate between depressed hyper-

strophic coils and raised orthostrophic coils. The exceptions to simple dextral or sinistral orthostrophic coiling are few.

Hyperstrophy is difficult to recognize in fossil material where only the shell is preserved. In living gastropods, whether orthostrophic or hyperstrophic, the direction of

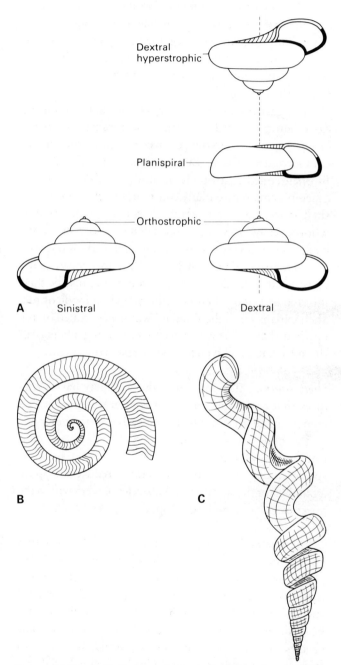

Figure 14.29. A, Coiling terminology. Equivalent areas of the apertural margin are shaded. **B** and **C,** Open coiling and uncoiling. **B,** Apical view of an open coiled Paleozoic euomphalin archaeogastropod that probably lived as a sedentary ciliary feeder. **C,** *Vermicularia*, a Holocene mesogastropod in life position. The mobile juvenile has a regularly coiled shell, but the whorls uncoil after the ciliary-feeding adult cements itself to the substrate with the apex lowermost.

coiling of the operculum, when viewed externally, is opposite to the direction of coiling of the shell (Figure 14.28, *B*). Thus the operculum shown in Figure 14.30, *D*, which coils in a counterclockwise direction, must have fitted a shell that was dextral. If *Maclurites* (Figure 14.33, *A*) were placed in the standard orthostrophic orientation (upside down from the figured position), both the shell and the operculum would coil sinistrally (counterclockwise). Thus *Maclurites* is regarded as being dextral and hyperstrophic. In this orientation, the coil of the operculum is still counterclockwise and therefore opposite to that of the shell.

Some gastropods are ordinary conispiral forms during early ontogeny but later change the parameters of shell growth so that the youngest whorls cease to maintain contact. Such shells are called **uncoiled** (Figure 14.29, *C*). In **open coiled** gastropods (Figure 14.29, *B*), shell parameters remain essentially constant, but the cone of the shell is too narrow to establish contact with previous whorls. Many uncoiled or open coiled, apparently sinistral shells occur as fossils and are flattened on one side like *Maclurites* (Figure 14.33, *A*). These forms are usually considered to be dextral hyperstrophic because of their inferred sedentary, benthic, ciliary-feeding mode of life. Shell expansion in these species was concentrated on the upper surface of the whorl due to the limiting influence of the substrate upon which the snail rested during life.

Shell interior   The interior of the shell may have structures that are visible only in broken or sectioned specimens or in internal molds. The **inner, basal,** and **outer** lips of the body whorl (Figure. 14.28, *C*) may have continuous **spiral folds** (Figure 14.33, *C*), and in some species these folds may considerably reduce the space available for the soft tissues. **Tubercles** or **teeth** within the aperture occur in many gastropods (Figures 14.34, *D* and 14.40, *B* and *E*).

Internal molds can contribute valuable information about internal structures of the shell (Figure 14.5, *D*). However, the shape of internal molds can differ greatly from the shell exterior, particularly in thick-shelled species (Figure 14.33, *C*). Internal whorl walls can also be reduced, thickened, or totally resorbed in some species. In some snails the apical whorls are closed off from the rest of the shell by a plug of shell material or a thin partition called a **septum.** Such apical obstructions produce internal molds with a bluntly rounded apical termination.

Most modern gastropods have the soft parts attached to the shell by a single columellar retractor muscle, which leaves a muscle scar on the columella. Some primitive snails have two columellar muscles. In the patelliform limpet shells, the retractor muscle is an incomplete ring (Figure 14.40, *A*). Muscle scars are poorly known in most fossil gastropods; they are reasonably well known in **bellerophontiform** shells (Figure 14.44, *A* and *B*), where the interpretation of their significance in relation to torsion and the origin of gastropods is a source of debate (Figures 14.21, *E* and *F*; 14.22, *B*; 14.26; and 14.44).

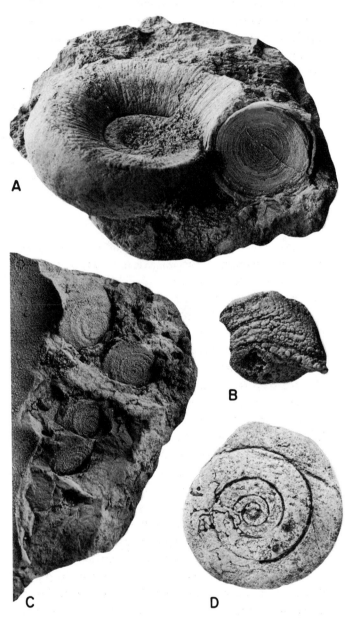

Figure 14.30. Opercula. **A,** Devonian euomphalin archaeogastropod with the multispiral operculum in place within the aperture. (×2.) **B,** Operculum of *Ceratopea* oriented with the internal muscle attachment surface downward. (Early Ordovician, ×2.) **C,** Silurian dolomite from Wisconsin with isolated paucispiral opercula. (×2.) **D,** Operculum of a Devonian trochin archaeogastropod, external surface; the operculum is sinistrally coiled in this orientation, indicating that the shell itself is dextral. (×2.)

Operculum The operculum is either a thin horny organic structure or a calcareous plate that closes the aperture of the shell when the soft parts are retracted. Opercula grow spirally (Figure 14.30), although their shape may be greatly modified (Figure 14.30, *B*). The operculum is carried dorsally on the posterior of the foot (Figure 14.27, *A*), where it acts as a pad upon which the shell rests during locomotion. An operculum is widespread in the adults of most Holocene species of one of the three subclasses of gastropods currently recognized (Prosobranchia) but is absent in adults of most species of the other two subclasses. When present, the operculum of most living gastropods is not calcified, and this situation was probably also true in the geologic past. Thus opercula are generally unsuited for preservation as fossils, and most occurrences of fossilized opercula are of calcified forms. Isolated opercula occur as fossils in rocks as old as the Early Ordovician, but finds of fossil gastropods with the operculum in place (Figure 14.30, *A*) are extremely rare. The unusual horn-shaped operculum of *Ceratopea* (Figure 14.30, *B*) is the oldest known fossil operculum and frequently is so common in Lower Ordovician rocks that it is an useful biostratigraphic tool. However, only one association of the shell of *Ceratopea* with the operculum in place has been reported. Opercula with a few whorls are **paucispiral** (Figure 14.30, *D*); those with many whorls are **multispiral** (Figure 14.30, *A*). Circular multispiral organic opercula are considered to be primitive.

Classification

The subclass taxa of gastropod classification are largely defined on the basis of the anatomy of living forms, and their names are derived from structures of the radular, respiratory, or nervous systems. Fossils are placed within these subclasses mainly by comparison of shell morphologies with those of living forms.

The names of the three subclasses of gastropods used herein are derived from respiratory systems. In the Prosobranchia the mantle cavity and associated gill or gills are located anteriorly. In the Opisthobranchia the mantle cavity and gill are progressively displaced posteriorly and are ultimately lost in some genera. The Pulmonata includes mostly freshwater and terrestrial gastropods in which the gill is lost and the mantle cavity is converted into an air breathing lung by vascularization of its inner surface.

An alternative classification is based on the state of torsion of the nervous system. In the Streptoneura (same as Prosobranchia), the visceral nerve cords become twisted into a figure eight during torsion. In the Euthyneura (same as Opisthobranchia and Pulmonata), inferred progressive detorsion has resulted in uncrossing of the nerve cords.

SUBCLASS PROSOBRANCHIA

Fully torted gastropods with anterior mantle cavity. In a few, mantle cavity with two gills; right gill lost in most (Figure 14.31); gill(s) bipectinate in primitive forms; monopectinate in advanced genera. Operculum commonly present. (Early Cambrian to Holocene; about 4500 genera.)

Prosobranchs are principally marine, although some are terrestrial or in freshwater. Most are benthic, epifaunal, or infaunal, but pelagic groups occur. Primitively, prosobranchs are browsing herbivores, but most advanced members of the group employ a variety of ciliary, carnivorous, and active predatory feeding mechanisms. More than half of all gastropods are prosobranchs.

ORDER ARCHAEOGASTROPODA (Figures 14.32; 14.33, *A* and *B*; 14.34)

Prosobranchs with bipectinate gill(s). Heart usually with two auricles. Inhalant siphon absent. Shells usually equidimensional and turbiniform; high-spired shells rarely developed. Sexes separate; fertilization usually external. (Early Cambrian to Holocene; about 1300 genera.)

Most living archaeogastropods are grazing herbivores. A few are scavengers or sluggish, browsing carnivores. Archaeogastropods are dominant in the Paleozoic and are usually divided into several suborders. The tendency in gastropod classification has been to increase the number of taxa at the ordinal level, and some of the suborders may eventually be elevated to order rank. In classifications that place bellerophonts in the Gastropoda, they are usually treated as a suborder of Archaeogastropoda.

SUBORDER PLEUROTOMARIINA (Figures 14.31, *A*; 14.32, *A* to *F*)

Archaeogastropods characterized by a pair of bipectinate gills. Shell shape variable. (Late Cambrian to Holocene.)

The primitive conispiral slit-bearing pleurotomariins

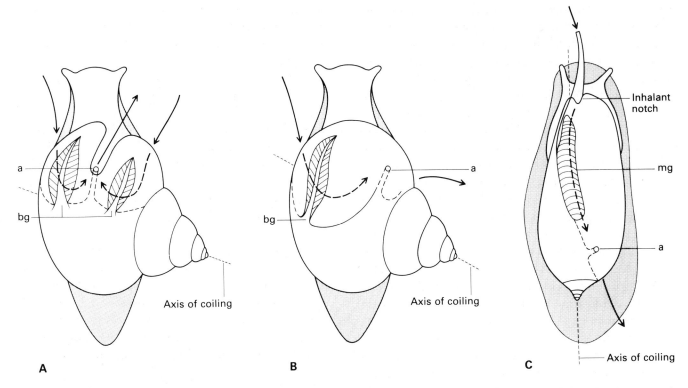

Figure 14.31. Prosobranch mantle cavity structure and shell orientation. **A,** Slit-bearing pleurotomariin archaeogastropod. **B,** Trochin archaeogastropod. **C,** Neogastropod. Heavy lines (broken within the mantle cavity) indicate the direction of respiratory water currents. *a,* Anus; *bg,* bipectinate gill; *mg,* monopectinate gill.

are today represented by rare deep-water species (Figure 14.32, *A* and *B*). Morphologically similar ancestors were widespread and abundant in shallow-water environments of the Paleozoic. Living shallow-water pleurotomariins either have a low spire and a series of exhalant perforations (Figure 14.32, *F*), or they have a patelliform shell with a slit (Figure 14.32, *E*) or an apical perforation (Figure 14.32, *C*). Most extant species are microherbivores, but some eat sponges.

SUBORDER TROCHINA (Figures 14.31, *B*; 14.32, *G, H, K* and *L*)

Archaeogastropods with single bipectinate gill and conispiral shell with simple aperture. (Early Ordovician to Holocene.)

Trochins abound on rocky shores, feeding on encrusting algae, but some Paleozoic forms were scavengers (Figure 14.32, *G*) or feeders on crinoid feces (Figures 14.5, *D* and 14.32, *L*).

SUBORDER MACLURITINA (Figure 14.33, *A*)

Archaeogastropods usually with large, flat or low-spired, umbilicate, hyperstrophic shell, often with probable exhalant channel on upper whorl surface; heavy operculum known in some. (Early Ordovician to Devonian.)

Most were sedentary and lay flat on the seafloor living by ciliary feeding.

SUBORDER ONYCHOCHILINA (Figure 14.33, *B*)

Archaeogastropods with small, umbilicate, hyperstrophic shell with higher spire than macluritins. Coiling apparently sinistral, but small sinus on abapical margin of aperture suggests coiling was hyperstrophic. (Early Cambrian to Devonian.)

SUBORDER EUOMPHALINA (Figures 14.29, *B*; 14.30, *A*)

Orthostrophic archaeogastropods that are similar to the macluritins. (Early Ordovician to Late Cretaceous, Holocene(?).)

Both suborders contain forms that have wide, flat,

Figure 14.32. Subclass Prosobranchia, order Archaeogastropoda, suborders Pleurotomariina (**A** to **F**) and Trochina (**G, H, K,** and **L**), with two bellerophontiform mollusks (**I** and **J**). **A** and **B,** *Mikadotrochus.* (Pliocene–Holocene, ×1.) **C** to **E,** The limpetlike shells *Diodora* (**C** and **D**) (Cretaceus–Holocene, ×1) and *Emarginula* (**E**) (Jurassic–Holocene, ×3). **F,** *Haliotis.* (?Cretaceous–Holocene, ×1.) **G,** *Cyclonema.* (Ordovician–Silurian, ×2.) **H,** *Turbo.* (Cretaceous–Holocene, ×2.) **I,** *Bellerophon.* (Silurian–Triassic, ×2.) **J,** *Euphemites.* (Devonian–Permian, ×2.) **K,** *Calliostoma.* (Miocene–Holocene, ×1.) **L,** *Platyceras.* Note the irregular growth lines. (Silurian–Permian, ×1.)

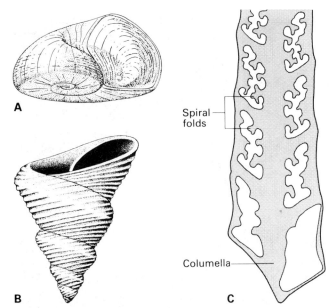

Figure 14.33. A and **B,** Subclass Prosobranchia: hyperstrophic Archaeogastropoda. **A,** *Maclurites,* an Ordovician macluritin, with the operculum in place. (×1.) **B,** *Mimospira,* an Ordovician to Silurian onychochilin. (×12.) **C,** Polished axial section of high-spired Jurassic opisthobranch *Bactroptyxis,* with the apical whorls at the top of the figure not preserved, showing the development of complex spiral folds within the shell interior. (×4.) Only the unshaded areas were occupied by the animal. An internal mold of this snail would therefore have the shape of the unshaded areas and bear little resemblance to the external shape of the shell. (**A** from Knight, J. B. *et al.* in: Moore, R. C., editor. *Treatise on invertebrate paleontology, Part I, Mollusca 1.* New York and Lawrence, KS: Geological Society of America and University of Kansas Press; 1960, p. 188. **B,** drawing by B. Blucher, after K. Wangberg-Eriksson. C courtesy M. J. Barker.)

sometimes open coiled shells, and euomphalins were largely adapted to the same sedentary mode of life as macluritins. However, most euomphalins are orthostrophic and were probably derived from a different stock than macluritins.

SUBORDER PATELLINA
(Figure 14.34, *A* and *E*)

Archaeogastropods with shell generally patelliform but lacking slit or apical perforation seen in patelliform pleurotomariins. Ordinarily only one gill present, but in some, several secondary gills evolved within shell margin. (Silurian to Holocene.)

Patellins are algal rasping limpets found clinging to rocks in the littoral zone.

SUBORDER NERITOPSINA (Figures 14.34, *B* and *D*)

Archaeogastropods with shell usually globose,

rounded and compact in response to life on exposed rocky coasts, usually with teeth on thickened inner lip. Most advanced archaeogastropods; developed internal fertilization and radiated into freshwater and terrestrial environments. (Middle Devonian to Holocene.)

SUBORDER MURCHISONIINA
(Figure 14.34, *C*)

Archaeogastropods commonly with high-spired shell with sinus and slit at midwhorl in outer lip; some with shallow siphonal notch. (Late Cambrian(?), Ordovician to Triassic.)

The slit suggests the presence of two gills as in pleurotomariins, but the high-spired shell and siphonal notch are features characteristic of the order Mesogastropoda.

ORDER MESOGASTROPODA (Figures 14.35; 14.36)

Prosobranchs with single monopectinate left gill. Heart with one auricle. Sexes separate; fertilization internal. Shells usually conispiral, often with a siphonal notch. (Middle Ordovician to Holocene; about 2000 genera.)

Although mesogastropods are present in Paleozoic rocks (Figure 14.35, *A*), they are not common and diverse until Mesozoic and Cenozoic time. Included in the order are such common living snails as the following: carrier shells (Figure 14.35, *G*), helmet shells (Figure 14.35, *E*), sundial shells (Figure 14.35, *C* and *D*), periwinkles (Figure 14.36, *C*), cowries (Figure 14.36, *B* and *E*), conchs (Figure 14.35, *B*), wentletraps (Figure 14.36, *J*), ceriths (Figure 14.36, *F*), turritellas (Figure 14.35, *F*), and moon shells (Figure 14.36, *I*). Mesogastropods live in various marine, freshwater, and terrestrial environments. They employ a variety of feeding patterns; many browse on algae, cnidarians, or ascideans or are active hunters of echinoids, pelecypods, or marine worms. Moon shells (Figure 14.36, *I*) bore holes through pelecypod shells to eat the soft parts (Figure 14.41, *B*). Some delicate shelled mesogastropods are pelagic marine carnivores (Figure 14.36, *G* and *H*). Terrestrial and freshwater mesogastropods (Figure 14.36, *A*) are known from rocks as old as Pennsylvanian.

ORDER NEOGASTROPODA (Figure 14.37)

Prosobranchs with single monopectinate left gill

Spiral folds

Columella

Figure 14.34. Subclass Prosobranchia, order Archaeogastropoda, suborders Patellina (**A** and **E**), Neritopsina (**B** and **D**), and Murchisoniina (**C**). **A**, *Acmaea*. (Oligocene–Holocene, ×3.) **B** and **D**, *Nerita*. (Cretaceous–Holocene, ×3.) **C**, *Murchisonia*. (Silurian–Permian, ×3.) **E**, *Patella*. (Cretaceous–Holocene, ×1.5.)

(Figure 14.31, *C*). Heart with single auricle. Sexes separate. Shells conispiral, with siphonal notch or canal. Mouth and radula at tip of retractile proboscis. (Cretaceous to Holocene; about 1200 genera.)

Neogastropods are less diverse than mesogastropods. However, they are common Cenozoic shells and include such widely distributed living snails as muricids (Figure 14.37, *A, B* and *E*), whelks (Figure 14.37, *F*), volutes (Figure 14.37, *D*), olives (Figure 14.37, *G*), and cones (Figure 14.37, *I*). Almost all are marine and are omnivorous scavengers or are carnivorous. The proboscis allows them to eat carrion or it can be thrust between the valves of pelecypods. Some neogastropods use the edge of the aperture to wedge open living pelecypods, whereas others can use the radula to bore a hole through the shells of barnacles or pelecypods. Some cones (Figure 14.37, *H* to *J*) hunt moving prey such as annelid worms, other gastropods, and fish. Cones kill their victims by means of a poisoned radular tooth located at the end of the proboscis; the poison can have a debilitating effect on humans.

SUBCLASS OPISTHOBRANCHIA
(Figures 14.33, *C*; 14.38; 14.39)

Usually strongly detorted gastropods with shell commonly concealed in mantle or absent. Mantle cavity usually located posteriorly or absent. With one internal gill or, in shell-less forms, with external gills. Operculum usually absent. (Mississippian to Holocene; about 800 genera.)

Opisthobranchs are less common than prosobranchs but show greater morphologic variation. Opistho-

Figure 14.35. Subclass Prosobranchia, order Mesogastropoda. **A,** *Cyrtospira.* (Silurian, ×4.5.) **B,** *Aporrhaias.* (Cretaceous–Holocene, ×3.) **C** and **D,** *Architectonica.* (Cretaceous–Holocene, ×3.) **E,** *Cassis.* (Eocene–Holocene, ×1.5.) **F,** *Turritella.* (Cretaceous–Holocene, ×3.) **G,** *Xenophora.* Apical view showing camouflage of cemented pelecypods and corals. (See also Figure 14.36, *D*). (Cretaceous–Holocene, ×1.5.)

branchs have been called emancipated gastropods because loss of the protective shell in many forms has accelerated the process of detorsion, opening the way for a host of morphologic and respiratory modifications. Some shell-less opisthobranchs, known as nudibranchs, retain little or no evidence of torsion. The mantle cavity and single gill are often displaced to the rear and are commonly lost. Respiration is accomplished by secondary external gills.

Various opisthobranchs that have lost their shell have

Figure 14.36. Subclass Prosobranchia, order Mesogastropoda. **A,** *Viviparus,* a freshwater mesogastropod with the operculum in place. (Cretaceous–Holocene, ×2.) **B,** *Cypraea.* Early whorls are completely enclosed by the final whorl. (Cretaceous–Holocene, ×1.) **C,** *Littorina.* (Eocene–Holocene, ×2.) **D,** *Xenophora,* with the cemented shells removed (see also Figure 14.35, G). (×1.) **E,** *Trivia,* a ribbed cowrie. (Eocene–Holocene, ×3.) **F,** *Cerithium.* (Pleistocene–Holocene, ×1.) **G** and **H,** *Carinaria.* The apex of the cap-shaped adult shell consists of a tightly coiled protoconch (to the left in **H**). (Eocene–Holocene, ×2.) **I,** *Polinices.* (Cretaceous–Holocene, ×2.) **J,** *Epitonium.* (Eocene–Holocene, ×2.)

become externally bilaterally symmetrical. Many of these sea slugs have developed obnoxious or acidic secretions as a form of protection. Some store the nematocysts from the cnidarians on which they feed in the tips of external outgrowths as a means of protection. Bright coloration is typical of some of the shell-less opisthobranchs and makes them blend into the background provided by the colorful tropical cnidarians and other animals on which they feed.

Opisthobranchs are mostly marine, but a few inhabit freshwater. Most opisthobranchs live as benthic carnivores or herbivores, but some are pelagic ciliary feeders or predators. Nearly all orders of opisthobranchs have swimming species.

Most classifications of opisthobranchs have at last 12 orders. Six of these are shell-less and have no known fossil record. Typical conispiral shells are largely restricted to the most primitive opisthobranch stocks, although cap-shaped and a variety of modified and reduced internal shells occur in many orders (Figure 14.38). The protoconchs of opisthobranchs are commonly heterostrophic, with the sinistrally coiled larval shell succeeded by a dextrally coiled adult shell; heterostrophic coiling is rare in other gastropod subclasses. One of the most unusual of gastropod shell forms is found in the **bivalved opisthobranchs,** which have a conispiral larval shell forming the apex of the left valve of the adult (Figure 14.39).

Figure 14.37. Subclass Prosobranchia, order Neogastropoda. **A, B** and **E,** Two species of *Murex* with contrasting styles of ornamentation. (Cretaceous–Holocene, ×3.) **C,** Early whorls of a Holocene volutacean showing changes in coiling and ornamentation from the protoconch to the teleoconch. (×3.) **D,** *Athleta* (Cretaceous–Holocene, ×3.) **F,** *Busycon,* a sinistral Holocene species. (×1.5.) **G,** *Oliva* (Eocene–Holocene, ×3.) **H,** *Terebra.* (Eocene–Holocene, ×3.) **I,** *Conus.* (Cretaceous–Holocene, ×3.) **J,** *Thatcheria.* (Holocene, ×3.)

Figure 14.38. Subclass Opisthobranchia. **A** and **B**, Reduced, platelike shells of *Umbraculum* (Eocene–Holocene, ×1.5) and *Aplysia* (Holocene, ×3). **C** and **G**, *Bulla*, the bubble shell, with **G** showing the sunken apex. (Jurassic–Holocene, ×3.) **D**, *Acteon* (Cretaceous–Holocene, ×3.) **E** and **F**, Highly modified shells of the pelagic ciliary-feeding opisthobranchs *Diacria* (**E**) and *Cavolinia* (**F**). (Miocene–Holocene, ×18.) The mouthlike aperture is at the bottom in both illustrations.

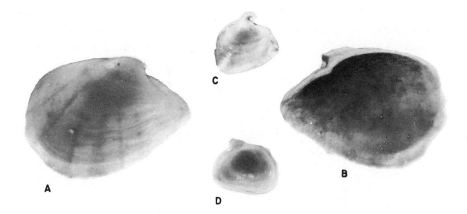

Figure 14.39. *Berthelinia,* a bivalved opisthobranch. **A** and **B**, Exterior and interior views of an adult left valve. **C** and **D**, Left valve of a juvenile showing the coiled protoconch. (Miocene, Poland, ×20.) (Courtesy W. Baluk.)

Opisthobranchs are known from rocks as old as Mississippian but have a poor record before Jurassic time. The most primitive types have typical gastropod shells (Figure 14.38, *C, D* and *G*) and are adapted to burrowing in sandy mud; some are microherbivores, whereas others feed on polychaetes, pelecypods, and foraminifers. However, even in these primitive forms the shell is commonly reduced and is progressively enclosed by the mantle. Some Jurassic and Cretaceous opisthobranchs have a high-spired shell in which the shell interior is greatly reduced by prominent spiral folds (Figure 14.33, *C*). Such folds are thought to have provided a means of disposal of excess calcium carbonate ingested during infaunal deposit feeding in lime-rich muds. Secretion of excess calcium carbonate in the form of external spines or varices, as is common in tropical prosobranchs (Figure 14.37, *A* and *B*), would have reduced feeding efficiency by hindering movement of these high-spired infaunal forms through the sediment.

Some opisthobranchs have a reduced cap-shaped or platelike shell (Figure 14.38, *A* and *B*) and live by feeding on sponges or as macroherbivores browsing on seaweed in the littoral zone. Some pelagic opisthobranchs with a delicate shell (Figure 14.38, *E* and *F*) are ciliary feeders, while others, in which the shell is lost, are voracious predators. These pelagic groups swim with the aid of winglike lateral extensions of the foot. One group of opisthobranchs lives as ectoparasites, sucking fluids from other mollusks or annelids with a long proboscis. A few opisthobranchs are highly modified internal parasites of holothurians and are recognized as gastropods only from their shelled embryos.

SUBCLASS PULMONATA (Figure 14.40)

Presumably detorted gastropods with conispiral shell commonly present, sometimes reduced and concealed in mantle. Gills absent; mantle cavity vascularized and altered into lung and fused with the neck except for small air hole. Operculum rarely present. (Pennsylvanian to Holocene; about 2500 genera.)

Pulmonates are highly successful terrestrial and freshwater animals; they are exceedingly abundant; the land snails and slugs make up about 30 per cent of all known gastropod genera. This successful colonization of land was the result of the evolution of the mantle cavity into a lung and the development of internal fertilization. A few archaeogastropods and mesogastropods have independently acquired these features and invaded land, but primarily the pulmonates have exploited land environments. Opisthobranchs have internal fertilization but lack a lung.

The euthyneurous nervous system suggests that pulmonates are detorted, but the mantle cavity (lung) is still anterior in shelled forms. A conispiral shell is commonly present but is reduced and internal in terrestrial slugs, although it is rarely lost. Prominent lamellae, folds, or teeth within the aperture are characteristic (Figure 14.40, *B* and *E*). An operculum is rarely present in adult pulmonates, although a few forms have a structure known as the **clausilium**, which is movable and attached by an elastic stalk to the columella rather than to the foot. During hibernation and estivation some advanced pulmonates close the aperture with a plate known as the **epiphragm**, which is usually dried mucus; it is sometimes calcified.

Two orders of pulmonates are generally recognized and are externally delimited on the basis of the number of tentacles and the position of the eyes on the head. The order Basommatophora has a single pair of tentacles with an eye at the base of each tentacle; this same condition is present in prosobranchs. In the order Stylommatophora, two pairs of tentacles are present on the head, and the eyes are located at the tips of the posterior pair. Fossil material is classified on the basis of similarity in shell form and microsculpture to living species. Turbiniform shells occur in both orders, but patelliform shells are known only in basommatophorans (Figure 14.40, *A* and *D*). Pupaeiform shells (Figure 14.40, *E*) and disklike shells (Figure 14.40, *I*) are more typically stylommatophoran. The oldest stylommatophorans are Early Pennsylvanian, and the oldest known basommatophorans are Late Jurassic.

Basommatophorans are the dominant freshwater snails of the world (Figure 14.40, *C, F* and *H*), although a few occur in shallow marine environments (Figure 14.40, *A, B* and *D*). They are all regarded as having evolved from terrestrial ancestors. Most feed by scraping plant material; they surface and take in air at the small air hole. Oxygen is also obtained under water by gaseous exchange between the bubble of air in the mantle cavity and the water. Some species can survive freezing, and a very few thrive in hot springs. When temporary bodies of water dry up, desiccation is avoided by burrowing into the mud, secretion of an epiphragm, and estivation.

Stylommatophorans are the dominant land snails and slugs of the world (Figure 14.40, *E, G* and *I*). The group is most abundant in modern humid tropical and subtropical environments but is also widespread in temperate environments. Most stylommatophorans are

Figure 14.40. Subclass Pulmonata. Marine and freshwater basommatophorans (**A** to **D**, **F** and **H**) and terrestrial stylommatophorans (**E**, **G**, and **I**). **A** and **D**, *Siphonaria.* Interior with incomplete ring-shaped, muscle scar and exterior views. (Cretaceous–Holocene, ×2.) **B**, *Melampus.* (Cretaceous–Holocene, ×3.) **C**, *Galba.* (Jurassic–Holocene, ×3.) **E**, *Clausilia.* (Pliocene–Holocene, ×6.) **F**, *Planorbis.* (Oligocene–Holocene, ×1.) **G**, *Helix.* (Pliocene–Holocene, ×1.) **H**, *Physa.* (Jurassic–Holocene, ×3.) **I**, *Discus.* (Paleocene–Holocene, ×8.)

herbivores, but some are omnivores or active carnivores hunting other gastropods and worms. Probably the best known stylommatophoran is the edible escargot (Figure 14.40, *G*).

Fossil traces of gastropods

Evidence of the occurrence of gastropods in the fossil record can be obtained from various traces in addition to preservation of their shells (Figure 14.41). Radular teeth are extremely rare as fossils, but traces of their use on algal coated surfaces have been recognized (Figure 14.41, *A*). Some land snails lay eggs with hard shells, and possible marine gastropod eggs have been reported from

Jurassic rocks (Figure 14.41, *C* and *D*). Probably the most widespread gastropod trace fossils are produced by some predatory mesogastropods and neogastropods, which, since the Cretaceous, have used the radula to bore holes in pelecypods and scaphopods through which they extract the edible soft parts (Figure 14.41, *B*). Many fossil tracks and burrows have been attributed to gastropods, although evidence of association is scarce.

Geometry of coiling

Shell measurements Traditional methods of quantifying gastropod shell shape combine simple linear measurements, such as height and width, with angular

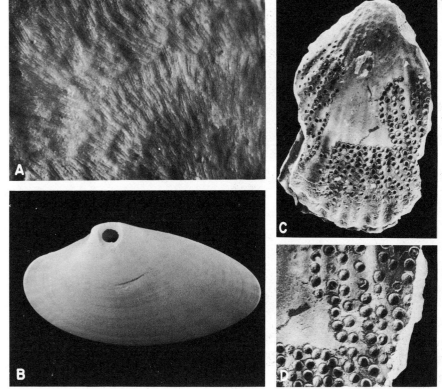

Figure 14.41. Fossilized traces of gastropods. **A,** *Radulichnus.* Scrape marks produced by the radula during feeding. (Cretaceous, ×20.) **B,** Tertiary pelecypod with gastropod boring near the umbo. (×2.) **C** and **D,** Fossilized gastropod spawn on the inside of a pelecypod shell (Jurassic, ×1 and ×2 respectively.) (**A** from Crimes, T. P.; Harper, J. C., editors. *Trace Fossils 2.* Copyright © 1977, Seel House Press, Liverpool, England. **C** and **D** from Kaiser, P.; Voigt, E. *Palaeontolgische Zeitschrift* 51: 5–11; 1977.)

measurements, such as the apical and incremental angles and the sutural slope (Figure 14.42, *A*).

The logarithmic spiral   It has long been recognized that coiling in the shell of most mollusks approximates to a logarithmic (equiangular) spiral. In such a spiral, the angle between any tangent to the spiral and the corresponding radius at the point of tangency is constant (Figure 14.42, *C* and *D*). The logarithmic spiral permits increase in size without change in shape, as demonstrated by the many regular whorls in the Cenozoic genus *Turritella* (Figure 14.35, *F*). Differences in coiling between gastropod species can be attributed to differences in the angle of the logarithmic spiral (Figure 14.42, *C* and *D*). Using this and other measurements, D'Arcy Thompson long ago devised a scheme that permitted description of the basic form of any gastropod shell in quantitative terms. His scheme is now little used, but it was a landmark in quantifying shell shapes.

Raupian parameters   More recently, David M. Raup has proposed four parameters to quantify the geometry of gastropod coiling. In a simplified model, with any particular shell form represented by an unvarying logarithmic spiral, the four parameters retain constant values during ontogeny. In reality, even the apparently regularly

coiled *Turritella* (Figure 14.35, *F*) shows slight allometry; an initial increase and subsequent decrease in shell width is more clearly evident in *Clausilia* (Figure 14.40, *E*). Significant coiling changes also characterize the passage from protoconch to teleoconch in most gastropods. This allometry causes the Raupian parameters to vary ontogenetically. Raup's studies mainly explored theoretical aspects of coiling in mollusks. However, periodic calculation of Raupian parameters during successive growth stages also provides a most useful tool for quantifying the allometry of coiled mollusks or evaluating variation in shell form produced by environmental change (Figure 14.43).

The four Raupian parameters are denoted by the letters S, W, D, and T (Figure 14.42, *E*). S is a representation of the shape of the generating curve of the shell, which is the cross-sectional shape of the whorl in any plane passing through the axis of coiling. The curve is usually complex but can be represented by a simple ratio, for example, height (h) : width (b) (Figure 14.42, *E*) or a series of ratios. W is the rate of expansion of the shell per whorl, measured as the ratio of linear dimensions (for example, whorl width, b, b^1, b^2, etc.) in successive whorls. D measures the distance of the generating curve from the axis of coiling, that is, the ratio of the distance from the axis to the adaxial margin of the curve (a^1, a^2,

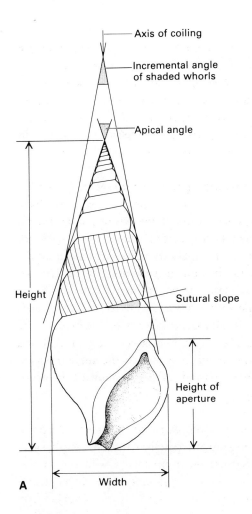

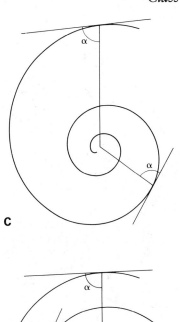

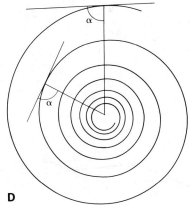

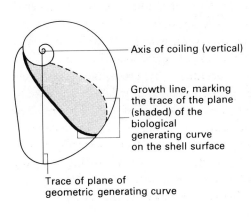

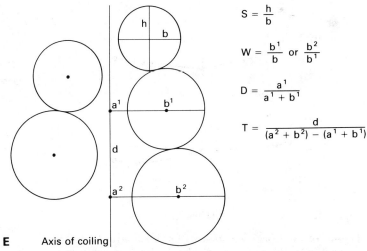

Figure 14.42. A, Gastropod shell, traditional measurements on a dextral shell in standard orientation. **B,** Biologic and geometric generating curves in *Nerita* (Figure 14.34, *B*). In this apical view, the plane of the geometric generating curve is vertical to the flat surface of the page; it passes through the shell apex and includes the axis of coiling, which is also vertical. The biologic generating curve is represented by a growth line (full line on upper whorl surface and broken line on base of whorl)

that lies in a plane approximately tangential to the previous whorl. This plane is shaded in the figure and slopes down toward the upper right. **C** and **D,** Variation in logarithmic spirals. **C,** *Turbo.* The spiral angle (α) equals 83 degrees. **D,** *Conus.* $\alpha = 88$ degrees. **E,** Raupian parameters. Measurements used for the determination of *S, W, D,* and *T,* with reference to a cross section through two consecutive whorls of a schematic high-spired shell.

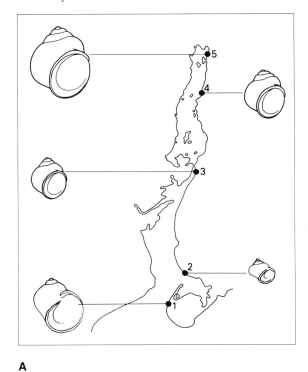

A

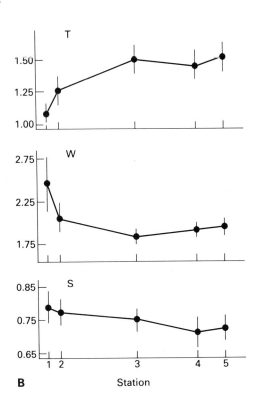

B   Station

Figure 14.43. Clinal variation in *Littorina saxatilis* in Pet-peswick Inlet, Nova Scotia. **A,** Sketches of *L. saxatilis* from sampling stations 1 to 5. **B,** Variation in Raupian parameters *T*, *W,* and *S* (mean values ± 1 standard deviation) from stations *1* to *5*. Stations *1* and *2* were on more exposed shores, and shells are rounder and have lower spires than shells from stations *3* to *5*, which are from more sheltered areas. *T* is calculated as the cotangent of half the apical angle; *S* is the ratio of apertural width to apertural height. (From Newkirk, G. F.; Doyle, R. W. *Marine biology.* 30: 229; 1975. Copyright © Springer-Verlag.)

etc.) to the distance from the axis of the abaxial margin $(a^1 + b^1, a^2 + b^2, \text{etc.})$. T is the rate of whorl translation and expresses the helicoid component of growth; for any whorl it is calculated as the displacement of the center of the generating curve along the axis of coiling (d) relative to the displacement perpendicular to the axis $(a^2 + b^2)$ in terms of the previous whorl $(a^1 + b^1)$.

Geometric and biologic generating curves   The generating curve (S) of Raup's model is the cross-sectional shape of the whorl in a plane through the axis of coiling and can be called the geometric generating curve; it is usually not the same as the plane of the true growing margin of the aperture, which is the biologic generating curve. The two may be approximately equal. In many gastropods the plane of the aperture forms a tangent to previous whorls (Figure 14.42, *B*) and lies at a considerable angle to the axis of coiling. This angle (E) between the geometric and biologic generating curves is termed the **angle of elevation** of the coiling axis and varies systematically in prosobranch gastropods.

Clinal variation of shell shape in Littorina   Morphologic variation in gastropod shells in response to physical changes in intertidal environments is well known, as shown by analysis of shape variation in *Littorina* using Raupian parameters (Figure 14.43). It was observed that shells of this species on more exposed shores are in general rounder and with a lower spire (Figure 14.43, stations 1 and 2) than shells found on more sheltered shores (Figure 14.43, stations 3 to 5). T values increase with decreasing exposure, whereas W and S decrease. Selection in higher energy environments decreased the effects of high turbulence by producing a relatively compact, semispherical shell with reduced resistance to water flow. On more protected shores, with reduced turbulence and wave splashing, desiccation becomes a problem when the tide is out. Here, selection acts to reduce size of the aperture relative to the shell, principally by reducing W. Concomitant increase in T ensures continued contact between the aperture and the axis.

Functional morphology

Gastropod higher classification is based on the soft parts that almost never occur as fossils. Most fossils are placed in this scheme by comparison of their shells with living forms. All fossil prosobranchs are ordinarily assigned to the three extant orders and are usually presumed (but probably incorrectly for many Paleozoic archaeogastropods) to have lived in similar ways to extant species.

Analysis of functional morphology permits this assumption to be tested and provides information about the anatomy of extinct lineages.

Absolute size is important in any functional interpretation. Discrete faunas of 1 to 2 mm Holocene gastropods live interstitially or upon algal and sea grass foliage. Many of these microscopic snails show great morphologic resemblance to epifaunal gastropods more than ten times larger (hundreds of times in terms of volume). However, the minute forms have different modes of life than the larger forms, which are discussed later, and are subject to other environmental pressures largely because of their size.

Gastropod shell form   Five laws of shell form have been proposed by R. M. Linsley as a foundation for functional morphologic interpretation in prosobranch gastropods:

1 *Law of radial apertures.* Gastropods of more than one whorl with radial apertures do not live with the plane of the aperture parallel to the substrate. The plane of a radial aperture lies along a radius from the axis of coiling to the shell periphery (Figure 14.35, *D*). Typically the plane of the aperture is approximately perpendicular to the substrate. Few Holocene shells have radial apertures, but this type of aperture was common in the Paleozoic. Living gastropods with radial apertures are typically sedentary, with the shell lying flat on the substrate for long periods. Many Paleozoic examples have flat shells that may be hyperstrophic or open coiled (Figure 14.29, *B*), and in view of their assumed sedentary behavior, were probably ciliary feeders.

2 *Law of tangential apertures.* Gastropods of more than one whorl with tangential apertures live with the plane of the aperture approximately parallel to the substrate. The plane of a tangential aperture forms a tangent to the earlier formed whorls (Figure 14.34, *B*). Most living prosobranchs have tangential apertures that can be clamped down over the body for protection; hence the apertural plane and the substrate are parallel.

3 *Law of shell balance.* If the shell is supported above the body, it will be positioned so that the center of mass of the shell and its contents is over the midline of the head-foot mass. Most living species carry the shell with its center of gravity above the center line of the body mass (Figure 14.31). Exceptions to this rule are cemented forms (Figure 14.29, *C*) and species with radial apertures which can rest the shell on the substrate.

4 *Law of reentrants.* Angulations or reentrants on the aperture usually indicate inhalant or exhalant areas; inhalant areas are directed as far anteriorly as possible.

Many gastropod shells show reentrants that represent the locations of inhalant or exhalant water currents. In advanced archaeogastropods and all mesogastropods and neogastropods, a single current enters the mantle cavity anteriorly, passes over the single gill, and leaves the cavity posteriorly carrying the excretory products with it (Figure 14.31, *C*). An inhalant siphonal notch or long siphonal canal is often present in mesogastropods and neogastropods but is absent in all archaeogastropods. In primitive extant archaeogastropods with a pair of gills, two currents enter the mantle cavity antero-laterally and unite prior to expulsion dorsally (Figure 14.31, *A*). The exhalant current in these archaeogastropods is marked by a sinus, slit, or one or more perforations (Figure 14.32, *B*, *C*, and *F*).

Not all reentrants are inhalant or exhalant in function. The Paleozoic archaeogastropod *Platyceras* (Figure 14.32, *L*) can show several indentations in the apertural margin, which correspond to the shape of the echinoderm calyx on which this feces-eating gastropod lived.

5 *Law of elongation.* Gastropods having elongated apertures possess only a single gill and develop a water flow through the mantle cavity from anterior to posterior along the long axis of the aperture; this axis is subparallel to the anterior-posterior axis of the foot.

Apertural elongation is a common feature in mesogastropods (Figure 14.36, *B*) and neogastropods (Figure 14.37, *G*). In primitive archaeogastropods, with two gills, the shape of the aperture remains equidimensional and allows two independent inhalant currents to operate (Figure 14.31, *A*). When only one gill is present, the development of a simple anterior-posterior water flow permits accompanying elongation of the aperture (Figure 14.31, *C*). On this basis, it can be inferred that the Paleozoic genus *Cyrtospira* (Figure 14.35, *A*) had a single gill. However, some gastropods with equidimensional apertures and without a siphonal notch or canal have only one gill.

Linsley's laws allow interpretation of the way in which many extinct gastropods carried their shells, how many gills they may have had, where inhalant and exhalant streams of water entered and left the shell, the approximate location of the anus, and the general shape of the mantle cavity. This information provides ways to test how extinct gastropods fit into the zoologically based higher taxa. An example is provided by the isostrophic bellerophontiform mollusks (Figures 14.22, *A* and 14.32, *I* and *J*). A debate exists about the class to which this group of Cambrian to Triassic mollusks belongs; a bellerophontiform mollusk reconstructed as a gastropod is shown in Figure 14.44, *C* and *D*, and one reconstructed as a monoplacophoran is shown in Figure 14.44, *E*

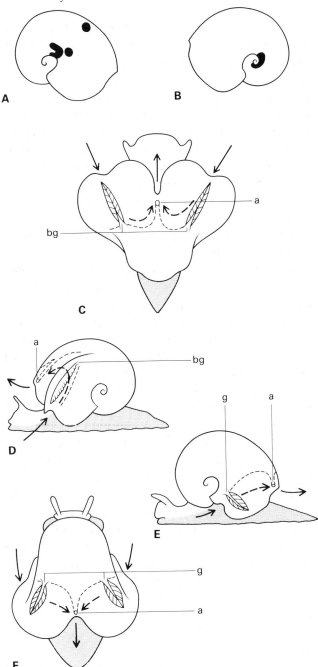

Figure 14.44. A and **B**, Muscle scars in bellerophontiform mollusks. **A**, Untorted monoplacophoran with three muscle scars on each side of the shell. **B**, Bellerophontiform mollusk considered to be a gastropod because the two symmetrical muscle scars, one on each side of the shell, near the umbilicus are considered to be comparable to the two asymmetrical muscle scars of primitive archaeogastropods. Muscle scars are in black; both shells are shown in lateral view with the anterior to the left. **C** to **F**, Inhalant and exhalant reentrants in bellerophontiform mollusks and inferred water circulation patterns (arrows). **C** and **D**, Bellerophontiform mollusk reconstructed as a torted gastropod (*Plectonotus*) in dorsal and lateral views showing dorsolateral inhalant reentrants to either side of the median exhalant reentrant. *a*, Anus; *bg*, bipectinate gill. **E** and **F**, Untorted monoplacophoran with inhalant reentrants located near the umbilici and a median dorsal exhalant reentrant. *a*, Anus; *g*, gill.

and *F*. The term bellerophontiform mollusk is used for these isostrophic shells to avoid any implications of zoologic affinity. The word bellerophont has taxonomic implications and is usually used for genera thought to be gastropods (for a contrasting assignment see the section "Class Monoplacophora").

Bellerophontiform mollusks—a question of torsion The bellerophontiform mollusks (Cambrian to Triassic) are a group of isostrophic univalved mollusks. In the section on Monoplacophora (p. 300), all members of this group are considered to be untorted and hence monoplacophorans. However, many specialists regard them as an incompletely understood undivided mixture of gastropods and monoplacophorans, and they are treated as such here.

Historically three major morphologic features have been relevant to the problem of the class level position of bellerophontiform shells. First, many genera in the group have a well-developed median dorsal reentrant which sometimes forms a slit generating a selenizone (Figures 14.22, *A*; 14.32, *I* and *J*; and 14.44, *C*). A median exhalant reentrant was once thought to be a diagnostic feature of primitive fossil and living gastropods (Figure 14.32, *A* and *B*); however, some fossil monoplacophorans are now known to have a median dorsal sinus (Figure 14.44, *F*).

Second, various bellerophontiform mollusks possess symmetrical multiple pedal retractor muscles scars (Figure 14.44, *A*). These scars indicate a lack of torsion by comparison with the symmetrical distribution of muscles around the aperture in the living monoplacophoran *Neopilina* (Figure 14.21, *F*) and fossil forms similar to *Neopilina* (Figures 14.21, *C*; 14.21, *E*; and 14.26, *A* to *C*). *Sinuites* (Figure 14.26, *D*), has a prominent median dorsal sinus and is now regarded as a monoplacophoran because of symmetrically disposed muscle scars, although it was long considered to be a typical gastropod. However, in other bellerophontiform genera, notably *Bellerophon* itself (Figure 14.32, *I*), a single pair of symmetrical pedal retractor muscle scars is present on the umbilical walls (Figure 14.44, *B*). Some living primitive gastropods also possess a single pair of retractor muscles, although these are usually not symmetrical because of the asymmetry of the animal. On the basis of having only a single pair of muscle scars, some bellerophontiform shells can be regarded as torted, even though the single pair of scars is symmetrical in bellerophontiform shells and asymmetrical in conispiral shells.

Third, analysis of water circulation patterns inferred from the study of reentrants suggests the presence of both gastropods and monoplacophorans among bellerophontiform mollusks. Unfortunately, information about the musculature and inferred inhalant reentrants is unknown for the majority of bellerophontiform mollusks.

Origin and evolution of gastropods

Origin Undoubted gastropods are known from rocks as old as early Late Cambrian. Bellerophontiform mollusks occur in lower Middle Cambrian rocks and several problematic conispiral univalved mollusks are

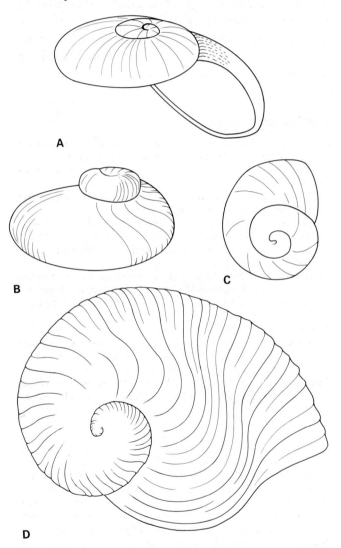

Figure 14.45. A to **C**, *Aldanella*, from the earliest Cambrian of the Soviet Union. (×40, 15, and 20, respectively.) **D**, *Costipelagiella*, Middle Cambrian of Bohemia. Note the bulbous protoconch of the 1.25 mm wide shell. (**A** to **C** modified from Missarzhevsky, V. V. In: Rozanov, A. Yu., *et al. The Tommotian stage and the Cambrian lower boundary problem.* Moscow: Nauka Publishers; 1969. **D** modified from Horny, R. J. *Sbornik* (National Museum of Prague) 20: 233–240; 1964.)

known from the Lower Cambrian. The latter group includes shells with a flat spire, the pelagiellaceans (Figure 14.25), which also range up into younger Cambrian rocks (Figure 14.45, *D*), and *Aldanella* (Figure 14.45, *A* to *C*), a Lower Cambrian genus in which the spire is distinct but low.

There is little agreement about the way in which these early gastropodlike mollusks are related to each other, with the result that there is no generally accepted theory concerning the origin of the Gastropoda. It should be remembered from p. 306 that torsion probably occurred in a monoplacophoran that had the ability to withdraw deeply into a narrow shell. Thus, in searching for an ancestor, we are looking for a monplacophoran in which the shell had several slowly expanding whorls rather than a cap-shaped shell or one with only a few rapidly expanding whorls.

It was suggested (Figure 14.15) that conispiral gastropods originated directly from pelagiellaceans. If true, *Aldanella* is an early dextral orthostrophic conispiral gastropod derived from a conispiral monoplacophoran and from which all other gastropods were presumably derived. The traditional theory suggests that the first gastropods were isostrophic bellerophontiform mollusks that originated from equally isostrophic but untorted monoplacophorans by torsion through 180 degrees. Such a sequence can be seen by going from Figure 14.44, *E* and *F* to Figure 14.44, *C* and *D*. This theory requires that the bellerophontiform ancestral gastropod, with its full 180 degrees of torsion (Figure 14.44, *C*) gave rise to the conispiral gastropods by progressive detorsion; Figure 14.44, *C* would be stage one; Figure 14.31, *A*, stage two; Figure 14.31, *B*, stage three; and Figure 14.31, *C*, the fourth and morphologically most advanced stage.

The two theories differ fundamentally in their evaluation of the relationship between torsion and conispiral coiling. The pelagiellacean–*Aldanella* theory described earlier (Figure 14.15) suggests that torsion took place as a result of the development of a conispiral shell in the monoplacophoran ancestor. Acquisition of this shell form caused a twisting of the original monoplacophoran body plan into the torted condition of the gastropods. The isostrophic monoplacophoran–isostrophic gastropod theory argues that torsion was a single-step process unrelated to conispiral coiling. In this theory, the development of the conispiral shell form reflects detorsion, namely the post torsional evolution of the gastropod respiratory system and shell reorientation (Figure 14.31).

A further consequence of the pelagiellacean–*Aldanella* theory of gastropod origin is that no torted isostrophic shell form is required in the evolution from monoplacophoran to gastropod. Supporters of this theory therefore propose that all bellerophontiform mollusks were untorted, hence monoplacophorans (see the section "Class Monoplacophora"). The alternative theory requires the presence of a torted isostrophic ancestral gastropod as an intermediate stage in the evolutionary process from untorted isostrophic monoplacophoran to conispiral gastropod. This theory permits some bellerophontiform mollusks to be interpreted as monoplacophorans (Figure 14.44, *E* and *F*) and some as gastropods (Figure 14.44, *C* and *D*). Analysis of functional morphology, discussed in a previous section, suggests that the bellerophontiform mollusk group includes both gastropods and monoplacophorans, although this in itself is no proof that the isostrophic monoplacophoran–isostrophic gastropod theory is the correct one of gastropod origin. It does point out a weakness in the pelagiellacean–*Aldanella* theory because that theory otherwise fails to explain the apparent occurrence of torted bellerophontiform mollusks.

Both theories assume that torsion was a unique event and that the gastropods were derived from a single ancestor. Little concrete evidence is available to support or deny this assumption. Torsion provides such great morphologic improvements and opportunities for evolutionary experimentation that it is possible, perhaps probable, that the process occurred on several independent occasions in different lineages. The ability to withdraw into a narrow shell is a simple protective adaptation.

Thus the gastropods may represent a grade of evolution rather than a single evolutionary event, and both theories may be partly correct. Unfortunately, too little is known about the Early Cambrian *Aldanella* and many other Cambrian univalves to be certain of their position in molluscan phylogeny. It is also possible that some gastropods may have arisen by other means, and various hypotheses have been suggested about the separate origins of pleurotomariins, onychochilins, and macluritins. Most recently, R. M. Linsley has placed pelagiellaceans, *Aldanella*, and the hyperstrophic onychochilins and macluritins in a new class Paragastropoda. Based on complex functional analyses, he concluded that these taxa were not torted.

The concept Gastropoda is currently under close scrutiny. Thus, it is not surprising that the classification of the Gastropoda is in a state of flux. It has been an accepted classification for some time, but new information and interpretations currently being assembled will undoubtedly lead to change over the next few years; this demonstrates both the dynamic nature and artificial character of classification. A nonphylogenetic approach to gastropod classification is unavoidable at this time because the relationships of various gastropod higher taxa are at present poorly understood. At the same time that the concept Archaeogastropoda is being reevaluated, various specialists also suggest that the mesogastropods, opisthobranchs, and pulmonates represent grades of organization attained in parallel phylogenetic lineages rather than single evolutionary events springing from one ancestral stock. On the other hand, the neogastropods are widely regarded as a monophyletic group that originated from advanced archaeogastropods or primitive mesogastropods during the Early Cretaceous.

Supplementary reading

Harper, J. A.; Rollins, H. S. Recognition of Monoplacophora and Gastropoda in the fossil record: a functional morphological look at the bellerophont controversy. *Third North American Paleontological Convention* 1: 227–232; 1982. Using features other than muscle scars, authors argue that bellerophontiform shells are gastropods.

Linsley, R. M. Some "Laws" of gastropod shell form. *Paleobiology* 3: 196–206; 1977. Applies the study of shell form and behavior of living prosobranchs to interpretations of Paleozoic gastropods.

Linsley, R. M. Shell form and evolution of the gastropods. *American Scientist* 66: 432–441; 1978. Discusses functional analysis of shell form as a tool for understanding the early history of gastropods.

McLean, J. H. The Galapogos rift limpet *Neomphalus*: relevance to understanding the evolution of a major Paleozoic–Mesozoic radiation. *Malacologia* 21: 291–336; 1981. Discusses a possible living euomphaline.

Moore, R. C.; Pitrat, C. W., editors. *Treatise on invertebrate paleontology, Part I, Mollusca 1.* New York and Lawrence, KS: Geological Society of America and University of Kansas Press; 1960. Extensive discussions of anatomy, hard part morphology, and taxonomy.

Peel, J. S. A new Silurian retractile monoplacophoran and the origin of the gastropods. *Geologists' Association Proceedings* 91: 91–97; 1980. Argues that the depth of insertion of muscles in the shell gives clues to the origin of torsion.

Raup, D. M. Geometric analysis of shell coiling: general problems. *Journal of Paleontology* 40: 1178–1190; 1966. Classical study of computer modeling of shell form.

Runnegar, B. Muscle scars, shell form, and torsion in Cambrian and Ordovician univalved molluscs. *Lethaia* 14: 311–322; 1981. Using muscle scars, postulates how and when torsion originated.

Sohl, N. F. Utility of gastropods in biostratigraphy.: in Kauffman, E. G.; Hazel, J. E., editors. *Concepts and methods of biostratigraphy.* Stroudsburg, PA: Dowden, Hutchinson, and Ross, Inc.; 1977. State-of-the-art summary of the usefulness of gastropods in correlation.

Vermeij, G. J. The Mesozoic marine revolution: Gastropods,

predators, and grazers. *Paleobiology* 3: 245–258; 1977. Relates changes in gastropod conchology and assemblages to the rise in importance of predators.

Yochelson, E. L. Quo Vadis Bellerophon?, pp. 141–161 in Teichert, C.; Yochelson, E. L., editors. *Essays in paleontology and stratigraphy.* (R. C. Moore commemorative volume.) Lawrence, KS: University of Kansas Press; 1967. Delightful discussion of the bellerophont problem to the date of publication.

Class Cephalopoda

John Pojeta, Jr.,
Mackenzie Gordon, Jr.

Cephalopods are the largest, most intelligent and agile mollusks and have a structural complexity greater than that of any other unsegmented invertebrate. As is typical of agressive carnivores, living cephalopods have a well-developed nervous system with marked cephalization. Some are rapid swimmers and compete with fishes. The eyes of some cephalopods are comparable to those of many vertebrates in structure and function (Figure 14.46, A and B); however, the two groups evolved their eyes independently. Many have a higher metabolic rate than most invertebrates and often have lively courtship displays. Study of modern octopuses shows that their brains are sufficiently well organized that they can learn from experience and that females can guard the young.

Shelled cephalopods have the related problems of the positive buoyancy of the gas-filled chambered shell and maintenance of their position in the water column. They have attained an approximately neutral buoyancy in sea water in various ways. Differences in distribution of the mass of the soft and hard parts, termed **shell equilibrium** or **poise adaptations,** balance the positive buoyancy of the gas-filled chambered shell (Figure 14.47). Poise adaptations keep the head-foot mass horizontal, and in straight-shelled forms they also keep the shell horizontal. **Hydrostatic adaptations** adjust buoyancy so that the animal does not float to the surface in an uncontrolled way. Hydrostatic changes are principally accomplished by liquid being added to or subtracted from the chambers of the shell by alterations in the osmotic pressure of the blood in the siphuncle. These adaptations avoid energy expenditure by the animal by keeping it at certain depths in the ocean; they also allow it to adjust to living at different depths. Cephalopods can be thought of as the molluscan equivalent of slowly adjusting submarines.

The foot of cephalopods has been modified into a **funnel** or **hyponome** and possibly the **tentacles** or **arms** (Figure 14.46, C and D). The funnel is controlled by powerful muscles and in some species can be pointed in a number of directions. Water is taken into the mantle cavity and then expelled out of the funnel under muscular pressure and the cephalopod is propelled in a direction opposite to that in which the funnel is pointing. Cephalopods move by jet propulsion through the water. They are streamlined to varying degrees; the most streamlined are some living squids that can reach speeds of 70 km/hr. The tentacles are grasping organs that surround the head, and they generally have suckers (Figure 14.46, C and D) and may also have hooks (Figure 14.65, E). Some have lateral fins that allow them to move without the aid of the funnel. In the center of the tentacles is the mouth that has a horny, organic, parrotlike beak ordinarily composed largely of conchiolin. The beak consists of upper and lower jaws. The radula is in the floor of the mouth, but is missing in some deep sea octopuses.

Most living cephalopods have an **ink sac** (Figure 14.46, C and D) that ejects a cloud of dark fluid through the funnel. The ink provides a screen behind which the cephalopod can elude predators, and it may anesthetize the chemoreceptors of some predators. The living shelled cephalopod *Nautilus* lacks an ink sac. The pigment sepia was named after the cuttlefish from whose ink it was first made. Fossilized ink sacs are known as far back as the Jurassic. Various squids, cuttlefishes, and octopuses are capable of rapid changes in color and color patterns because the mantle has pigment-bearing cells that are controlled by muscles. Muscular control of pigment-bearing cells is unique to cephalopods in the invertebrates. Some deeper water squids and octopuses possess special luminescent organs comparable to those of deep-water fishes. The luminescence is caused either by a secretion or by symbiotic bacteria.

Shelled cephalopods are widely used biostratigraphically in Paleozoic and Mesozoic rocks. These animals evolved rapidly and show great diversity that can be readily recognized in their hard parts, particularly the suture patterns and internal deposits. The **suture** is the line along which the outer margin of a septum meets the outer wall of the conch (Figures 14.49, A and 14.50). By these characters, cephalopods can often be identified from fragments of the shell or internal mold. Homeomorphy in shell form and ornament can make identification difficult if the sutures or internal deposits are not well preserved.

Many fossil cephalopods were probably nektic, like most modern ones, which indicates why some of them had similar cosmopolitan distributions. Also, dead individuals of the Holocene genus *Nautilus* float to the surface because the buoyancy of the largely gas-filled shell is

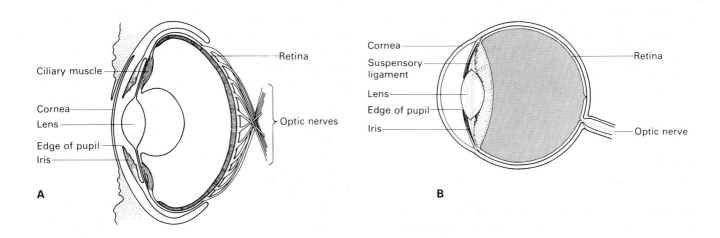

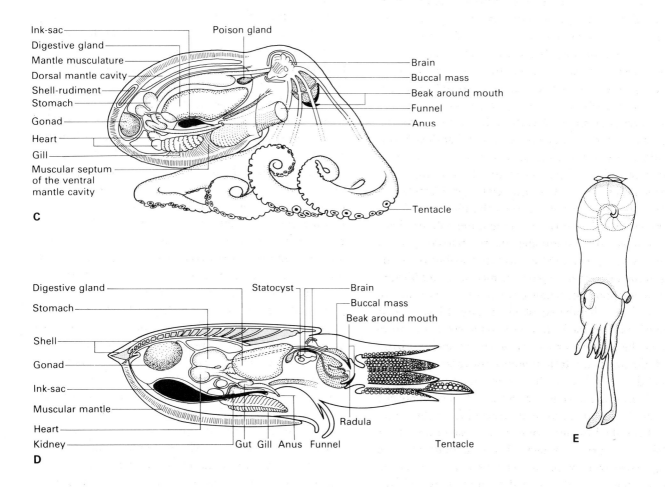

Figure 14.46. Soft parts of cephalopods. Comparison of cephalopod eye (A) with the human eye (B). C, Diagrammatic partial median section of *Octopus* showing major anatomic features. D and E, Ten-armed living cephalopods. D, Diagrammatic median section of the cuttlefish *Sepia* showing major anatomic features. E, *Spirula*, a genus with an internal planispiral shell. (A modified from Wells, M. J. *Brain and behavior in cephalopods*. Palo Alto, CA: Stanford University Press; 1962, p. 50. B modified, and with publisher's permission, from Weichert, C. K. *Anatomy of the chordates*. New York: McGraw-Hill Book Co.; 1958, p. 623. C to E modified from Lane, F. W. *Kingdom of the octopus*. New York: Sheridan House; 1960, p. 230, 232.)

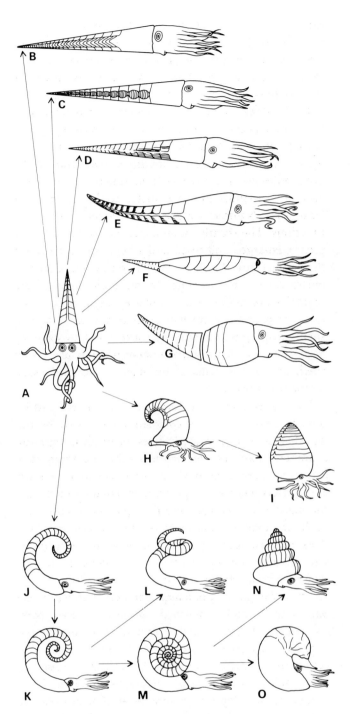

increased by the decay of the soft tissue, loss of the contents of the body chamber, and loss of the hydrostatic function of the siphuncle. Shells of dead *Nautilus* individuals are distributed by currents and winds in the Indian and western Pacific Oceans over a much greater range than where the animals live. Many ancient shelled cephalopods also probably floated after death. All living and fossil cephalopods are limited to marine conditions.

Living cephalopods can be placed in three informal groups:

1   Those having external shells and a thin internal mantle are represented only by the genus *Nautilus*, which has as many as 94 tentacles and about five species;

2   Those having internal shells, a thick external mantle, and ten tentacles are represented by more than 450 living species, most of which are *Spirula,* squids, and cuttlefishes;

3   Those having internal or no shells, a thick external mantle, and eight tentacles are represented by at least 150 living species including octopuses and the paper nautilus (argonaut). Living cephalopods are most abundant in the shallow coastal areas of the oceans.

Living *Nautilus* (Figure 14.49, *B*) is found in the southwestern Pacific and occurs in tropical waters at depths of about 5 to 550 m. It may occur along the coast of East Africa. It is an active swimmer, but may anchor itself to the substrate by its tentacles, a way of life termed nektobenthic. Most *Nautilus* average 15 to 18 cm in diameter as adults, are nocturnal, and are both carnivores and scavengers. The thickness of the shell of adults is about 1.5 mm. Known predators of *Nautilus* are sea turtles and fish such as sea perch, which have powerful jaws.

Ten-armed living cephalopods occur in all the oceans of the world (Figure 14.46, *D* and *E*). They range in size from 2 cm to 18 m and include the largest and most predatory of all invertebrates. Squids range from surface waters to depths of at least 3000 m, and some of them migrate from surface waters where they spend the night

Figure 14.47. Cartoon illustrating poise adaptations of shelled cephalopods. Mass redistribution giving poise adjustment was a major factor in early cephalopod radiation. The functional explanations of shell equilibrium adaptations are crudely summarized as follows, assuming a straight shell as a starting point (**A**).

1   Extra shell weighting may be added at the apical ends of the chambered straight shell. This weighting was achieved by endocones (Endoceratoidea, **B**), complex annular deposits (Actinoceratoidea, **C**), or cameral deposits (various Nautiloidea, **E**).

2   The chambered part may be reduced and may still lie above the soft parts with the body chamber and aperture adapted to the poise (various Nautiloidea, **H** and **I**).

3   Liquid may be retained in the chambers to negate the buoyant effects of the chambered shell (probably some straight Nautiloidea, **D**).

4   Chambers may extend over the body tissues so that poise problems are reduced (some Nautiloidea, **F**).

5   The early gas-filled chambers may be shed so that they do not affect adult poise (some Nautiloidea, **G**).

6   The chambered part may be coiled so that it lies above the body tissues (various Nautiloidea and most Ammonoidea, **J** to **O**).

7   Gas chambers may be lost (most Coleoidea, Figures 14.46, *C* and *D*; and 14.64, *C* to *E*).

(From House, M. R. In: House, M. R.; Senior, J. R., editors. *The Ammonoidea.* New York: Academic Press, Inc.; 1981, p. 6.)

to deeper waters where they spend the day. Many modern squids swim in large schools in surface waters both in the open ocean and near shore and are used for food. Various species undergo mass mortality after reproducing.

Eight-armed cephalopods (Figures 14.46, *C* and 14.64, *E*) live in nearly all the seas of the world and range in size from about 5 cm to about 10 m. Most species are found in shallow waters, are often used for food, and live in lairs that can be natural, manmade, or constructed by the cephalopod. A variety of them can autotomize arms and regenerate them. The deeper water forms occur to depths of 5000 m. Many have extensive webbing between the arms and probably swim by rhythmic movement of the webbing. The eight-armed female paper nautilus uses the first pair of arms, not mantle, to secrete an external shell (Figure 14.64, *E*). This is a brood chamber in which she also lives and is not homologous with the external shell of any other cephalopod. Fossil shelled cephalopods are interpreted to have lived in a variety of ways, including active swimming, and crawling and as nektobenthos and benthos, at varying depths but usually not below 500 m. Most are regarded as having been nektic or nektobenthic.

Anatomy

Living *Nautilus* is used as a guide to the arrangement of soft parts in fossil species having an external shell. *Nautilus* has three principal types of soft parts (Figures 14.1, *E* and 14.48, *A*):
1 At the aperture of the shell is the head-foot mass.
2 Behind this, in the body chamber of the shell, is the visceral mass, which extends adapically to the last formed septum.
3 The mantle surrounds the visceral mass and extends adapically to the first chamber as the siphuncular cord within the rest of the siphuncle.

In cephalopods having an internal shell, the head-foot and visceral masses are readily differentiated (Figure 14.46, *C* and *D*). The siphuncle, if present, is contained within a shell that is completely enveloped by the mantle.

As the name Cephalopoda (from the Greek words *kephalus* and *poda*, meaning head and foot, respectively) implies, the head and the foot are closely related in this class. These basic molluscan features have become so interrelated and modified in cephalopods that researchers cannot agree about how much is head and how much is foot. The **ring of tentacles** that surrounds the mouth (Figures 14.1, *E*; 14.46, *C* and *D*; and 14.48, *A*), is now often regarded as an anterior outgrowth of the head

rather than as part of the foot. The homology of the tentacles remains uncertain for several reasons, one of which is that the tentacles are innervated from the pedal (foot) ganglia (Figure 14.48, *A*). Workers generally agree that the hyponome is part of a modified molluscan foot.

In *Nautilus*, none of the numerous tentacles have hooks or suckers. Typically the tentacles have a partly or wholly retractile distal fingerlike extension that retracts into a sheath. The tentacle tips are grooved and may be both adhesive and prehensile in function, which apparently is to grasp prey and anchor to the substrate. At least some *Nautilus* tentacles have tactile or olfactory functions. The sheaths of an upper pair of tentacles are greatly enlarged and fused into a horny covering called the **hood** (Figure 14.48, *A*) that, on retraction of the soft parts, covers the aperture. In the male some of the tentacles are modified as copulatory organs. Tentacles grasp prey, and in many living species having an internal shell, grasping is facilitated by suckers and, in some, by hooks. In various living cephalopods without an external shell, one of the arms of the male is modified as an intromittent organ.

The visceral mass of *Nautilus* is short and saclike (Figure 14.48, *A*). It is completely enveloped by the weakly muscular mantle that lines the inside of the body chamber. Adapically the mantle is a thin membrane that forms the integument of the visceral mass and is known as the **visceral sac**. At the posterior end of the visceral sac, the mantle is prolonged into the siphuncular cord. Adorally the mantle is extended as a fold next to the shell, does not form an integument, and is separated from the visceral mass by the mantle cavity (Figure 14.48, *A*), which encircles the body. The lower part of the mantle cavity is the **gill** or **branchial chamber**; it contains one pair (eight- and ten-armed species) or two pairs (*Nautilus*) of gills and the terminal openings of the digestive, excretory, and reproductive systems. In cephalopods having internal shells, the body is elongated as in squids and cuttlefishes (Figure 14.46, *D* and *E*) or saclike as in octopuses (Figure 14.46, *C*). In these the mantle is strongly muscled and forms the outermost covering of the animal. In some of the more elongate forms, the mantle edges are modified into a pair of lateral fins.

Orientation of the cephalopod shell

In most of the literature dealing with fossil cephalopods having an external shell, orientation of the conch has been as follows (Figure 14.48, *D*):
1 Anterior is toward the aperture.

2 Posterior is toward the apex.

3 Ventral is the side of the shell with the hyponome, which commonly leaves a distinct embayment in the aperture known as the **hyponomic sinus**.

4 Dorsal is the side opposite ventral.

Cephalopods probably evolved from dorsoventrally elongated Cambrian monoplacophorans, some of which are known to have possessed septa and some of which had structures that could have evolved into a siphuncle (*Hypseloconus* and *Yochelcionella*, Figures 14.15 and 14.23). These ancestral forms indicate that the aperture would be ventral (Figure 14.48, *E*) because the aperture is where the foot protrudes in monoplacophorans. The apex is dorsal, the head is anterior, and the opposite end of the shell is posterior. Dorsal elongation has effectively bent cephalopods in half, so that they have a U-shaped gut, with the mouth and anus on the same side of the body and near to one another (Figures 14.46, *D* and 14.48, *A*).

Because the conventional orientation terminology of

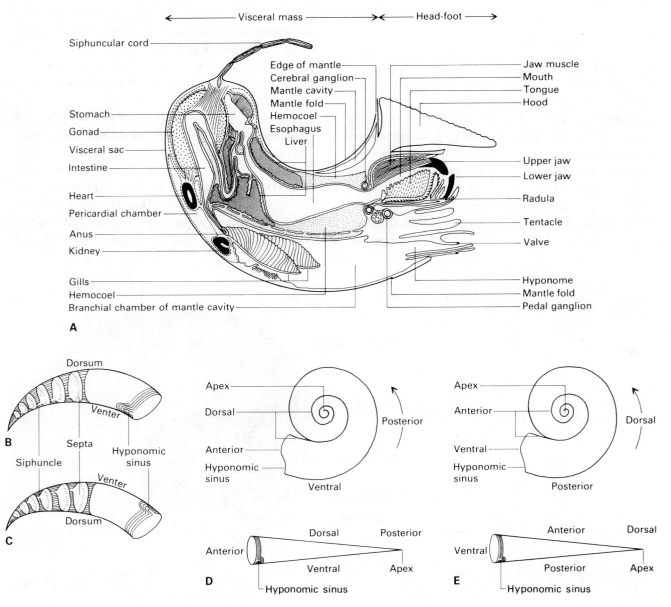

Figure 14.48 **A**, Diagrammatic median section of the soft parts of *Nautilus*. **B** and **C**, Difference between endogastric (**B**) and exogastric (**C**) shell curvature. The body chamber is shown in an external view; the phragmocone is cut in half lengthwise. **D** and **E**, Traditional (**D**) and phylogenetic (**E**) orientations of straight and coiled conch. (**A** modified from Stenzel, H. B. in Moore, R. C., editor. *Treatise on invertebrate paleontology,* *Part K.* New York and Lawrence, KS: Geological Society of America and the University of Kansas; 1964, p. 60. **B** to **E** modified from Teichert, C. in Moore, R. C., editor. *Treatise on invertebrate paleontology, Part K.* New York and Lawrence, KS: Geological Society of America and the University of Kansas; 1964, p. 22, 20.)

cephalopods has been widely used for many decades and continues to be used, it is used in this chapter. Furthermore, the aperture of shelled cephalopods is anterior in a functional sense because this is the direction in which the animal ordinarily meets the external environment and where the major sense organs are concentrated. However, it is important to understand the shortcomings of the conventional and functional terminology of orientation when considering homologies of the various parts of the external cephalopod shell in relation to the shells of other mollusks.

Shell morphology

Most fossil cephalopods had an external shell, but most modern ones have an internal shell. As in gastropods, the external shell, whether straight or coiled, is regarded as a cone (Figures 14.47; 14.48, *D* and *E;* and 14.49). The narrow end of the conch is divided by septa into a series of **chambers** or **camerae** and is termed the **phragmocone**. Anterior to this is the undivided **body chamber** at the wide end of the cone (Figure 14.49). As it grows, the animal moves forward in the shell and builds septa at the back of the body chamber. Connecting all the chambers is the siphuncle (Figure 14.1, *E*). The phragmocone

begins with an apical initial chamber, which in two subclasses is regarded as a protoconch. The exterior of the shell is ornamented by growth lines, which ordinarily record the configuration of the aperture and hyponomic sinus at earlier growth stages. The edge of the aperture is termed the **peristome**. On the inside of the shell of various cephalopods is a finely ridged structure called the **wrinkle layer** (Figure 14.58, *B*), which may have served for attachment of the mantle to the body chamber.

Curved or coiled shells in which the ventral side or **venter** is on the convex or outer side are termed **exogastric** (Figure 14.48, *C*); those in which the dorsal side or **dorsum** is on the convex side are called **endogastric** (Figure 14.48, *B*). Ventral can be recognized in a variety of ways including the presence of the hyponomic sinus. The siphuncle is commonly but not invariably ventral in position.

Because the cephalopod external shell can be regarded as a cone modified in shape by coiling, terms used to describe the various shapes end in the suffix "–cone" (noun) or "–conic" (adjective). Shells are termed **orthocones** if straight (Figure 14.47, *B* to *D*), **longicones** if long and tapering (Figure 14.47, *B* to *D*), and **brevicones** if short and stubby (Figure 14.47, *I*). Curved shells that complete less than one full circle are described as **cyrto-**

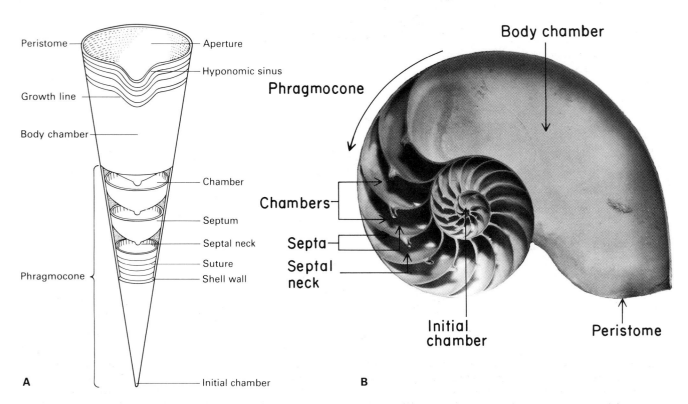

Figure 14.49. General morphology of straight (**A**) and coiled conchs (*Nautilus*, **B**). (**A** modified from Teichert, C. in: Moore, R. C., editor. *Treatise on invertebrate paleontology, Part K.* New York and Lawrence, KS: Geological Society of America and the University of Kansas; 1964, p. 14.)

cones (Figure 14.47, *E*, *G*, and *H*); they may be either longiconic (Figure 14.47, *E*) or breviconic (Figure 14.47, *H*).

Most cephalopod shells are coiled planispirally, and they are bilaterally symmetrical (Figure 14.47, *J*, *K*, *M*, and *O*). Each coil of the conch through 360 degrees is called a **whorl** or **volution**. The center of a planispirally coiled shell is the **umbilicus**. Shells having wide umbilici are termed **evolute** (Figure 14.47, *M*), and those with narrow ones are **involute** (Figure 14.47, *O*). Much variation exists in the overlapping of the whorls so that a gradation exists between widely evolute forms and nar-

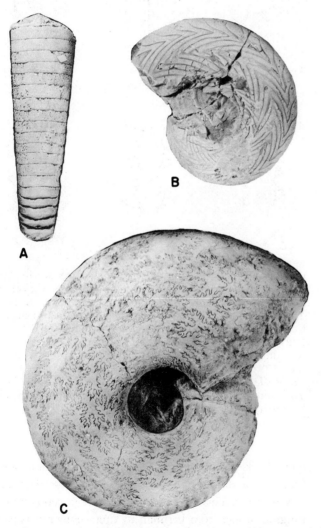

Figure 14.50. General morphology of sutures in shelled cephalopods. **A**, Simple suture, lacking lobes and saddles, of an orthoconic longicone from the Silurian of Ohio. (×0.5.) **B**, Simple sinuous suture, showing prominent lobes and saddles, of a planispiral shell from the Pennsylvanian of New Mexico. (×0.75.) **C**, Completely fluted sinuous suture with subsidiary lobes and saddles subdividing the major lobes and saddles of a planispiral shell from the Cretaceous of France. (×0.5.) All specimens are internal molds.

rowly involute ones (Figure 14.61). A few cephalopods of Silurian, Devonian, and Mesozoic age are coiled asymmetrically and are termed **torticones** (Figure 14.47, *L* and *N*). Homeomorphy in shell form is widespread and shell form is usually not a reliable guide to taxonomy or phylogenetic relationship.

Septa are curved calcareous walls that ordinarily are concave adorally and are pierced by the siphuncle (Figures 14.48, *B* and *C* and 14.49). Each septum meets the wall of the shell along a suture that is best seen on internal molds (Figure 14.50). The suture may be simple (Figure 14.50, *A*), or it can have complex flutings (Figure 14.50, *C*), which may have served as buttresses for strengthening the gas-filled phragmocone against hydrostatic pressure. However, the function of complex sutures is debated. In many cephalopods, the sutures are sinuous; the deflections of the suture directed adapically are termed **lobes** and those directed adorally are called **saddles** (Figure 14.50, *B* and *C*). Flutings subdivide the major lobes and saddles and form subsidiary lobes and saddles (Figure 14.50, *C*). Because sutures are highly diverse and complex, they are significant for classifying cephalopods. Although sutures go around the shell, when illustrated they are projected as a line on a plane (Figure 14.51, *A* to *F*). Ordinarily, only half of a suture, from midventer to middorsum, is illustrated as sutures are usually bilaterally symmetrical; the midventral point is marked by an arrow pointing adaperturally.

Five principal types of sutures are frequently recognized:

1 **Orthoceratitic sutures** (Figures 14.50, *A* and 14.51, *B*) either lack lobes and saddles or have broadly undulating and gently rounded lobes and saddles. This generalized type of suture is found in cephalopods ranging in age from Late Cambrian to Holocene.

2 **Agoniatitic sutures** (Figure 14.51, *C*) are composed of a few simple undivided lobes and saddles. They always have a narrow midventral lobe and a broad lateral lobe and may have additional lobes and saddles. Cephalopods having this type of suture are typically found in rocks of Early and Middle Devonian age.

3 **Goniatitic sutures** (Figures 14.50, *B* and 14.51, *D*) have more numerous undivided lobes and saddles than do agoniatitic sutures. Typically, goniatitic sutures have eight lobes around the conch. Lobes are narrowly rounded to pointed; the ventral one is commonly divided into two prongs by a median saddle. Saddles are typically but not invariably rounded. Cephalopods having this type of suture are common in rocks of Late Devonian to Permian age and occur rarely in rocks of Triassic and

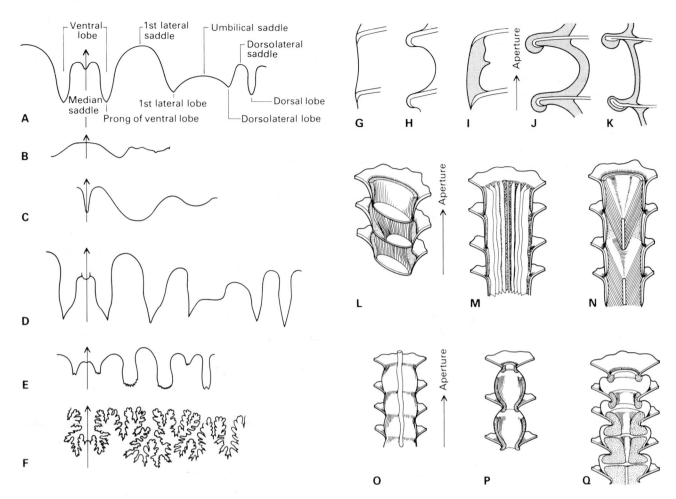

Figure 14.51. A to **F**, Major suture types of shelled cephalopods; arrows point adaperturally. **A**, General terminology of suture. **B**, Orthoceratitic suture of *Nautilus*. **C**, Agoniatitic suture of *Agoniatites*. **D**, Goniatitic suture of *Diaboloceras*. **E**, Ceratitic suture of *Xenodiscus*. **F**, Ammonitic suture of *Exiteloceras*. **G** to **K**, Various kinds of thin (**G** and **H**) and thick (**I** to **K**) connecting rings and their mode of attachment to septa and septal necks in nautiloids, endoceroids, and actinoceroids. **L** to **Q**, Types of endosiphuncular deposits. **L**, Diaphragms; **M**, longitudinal lamellae; **N**, endocones; **O**, central cylindrical tube; **P**, parietal deposits; **Q**, annulosiphonate deposits. In **L** and **M** connecting rings are stippled; in **P** and **Q** endosiphuncular deposits are stippled. (**A** to **D** and **F** modified from Moore, R. C., editor. *Treatise on invertebrate paleontology, Parts K and L*. New York and Lawrence, KS: Geological Society of America and the University of Kansas; 1957, 1964, pp. L13, K450, L30, L64, L225. **E** modified from Tozer, E. T. In: House, M. R.; Senior, J. R., editors. *The Ammonoidea*. New York: Academic Press, Inc.; 1981, p. 81. **G** to **Q** from Teichert, C. in: Moore, R. C., editor. *Treatise on invertebrate paleontology, Part K*. New York and Lawrence, KS: Geological Society of America and the University of Kansas Press; 1964, p. 40, 41.)

Cretaceous age. Ammonoids whose sutures have numerous undivided lobes and saddles are called **goniatites**.

4   **Ceratitic sutures** (Figure 14.51, *E*) have lobes, in which most of the tips are subdivided giving them a saw-toothed appearance, and rounded undivided saddles. This type of suture first appeared in cephalopods of Early Mississippian time, occurs in a few Pennsylvanian and Permian genera, is found in most Triassic forms, and occurs in a few Cretaceous species. Ammonoids whose sutures have subdivided lobes and entire saddles are called **ceratites**.

5   **Ammonitic sutures** (Figures 14.50, *C* and 14.51, *F*) have both the lobes and saddles much subdivided (fluted); the subdivisions are usually rounded rather than saw-toothed. This type of suture occurs in cephalopods ranging in age from Permian to Cretaceous but is particularly characteristic of Jurassic and Cretaceous species. Ammonoids whose sutures have both the lobes and saddles subdivided are called **ammonites**.

The terms goniatites, ceratites, and ammonites are useful in a broad way; goniatites are characteristic of the Paleozoic, ceratites of the Triassic, and ammonites of the Jurassic and Cretaceous. However, the suture types that are derived from these terms have arisen more than once

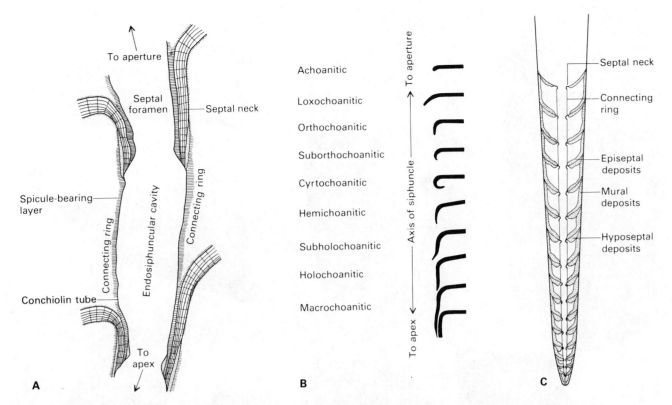

Figure 14.52. A, Longitudinal section of siphuncle of *Nautilus*. (×9.) **B**, Terminology applied to the shape of retrochoanitic septal necks. **C**, Reconstructed longitudinal section of the Devonian orthoconic nautiloid *Geisonoceras* showing regular deposition of cameral deposits in the phragmocone. (Modified from Stenzel, H. B.; Teichert, C. in: Moore, R. C., editor. *Treatise on invertebrate paleontology, Part K*. New York and Lawrence, KS: Geological Society of America and University of Kansas Press; 1964, p. 33, 38, 76.)

in cephalopod evolution. Thus, the Cretaceous forms with goniatitic and ceratitic sutures are homeomorphic to the Paleozoic and Triassic groups that have closely similar sutures, and they are classified separately. Likewise, the Permian forms having ammonitic sutures are regarded as homeomorphic to those having similar sutures in the Jurassic and Cretaceous and are not classified as being closely related to the younger group. Although the Mississippian forms having ceratitic sutures are regarded as ancestral to the Late Permian and Triassic forms that have the same suture type, the two groups are placed in different orders.

The siphuncle is unique to cephalopods and is composed of both soft and hard parts. In *Nautilus*, the soft tissue making up the siphuncular cord includes blood vessels, nerves, and mantle. Some extinct externally shelled cephalopods that had extremely large siphuncles are thought to have had some of the visceral mass in the cavity of the siphuncle. The nonliving tissue making up the outer sheath of the siphuncle is termed the **ectosiphuncle**. In *Nautilus* (Figure 14.52, *A*) the ectosiphuncle is composed of shelly material, conchiolin, and spicules

of calcium carbonate. The space inside this ectosiphuncle is termed the **endosiphuncle**, and any calcareous deposits (Figure 14.51, *L* to *Q*) in this space are **endosiphuncular deposits**. *Nautilus* has no shelly deposits in the endosiphuncle (Figure 14.52, *A*).

The ectosiphuncle forms a continuous tube from the apex to the living chamber. It is in two main parts: the **septal necks** where the septa are pierced and the edges bent (Figures 14.49 and 14.52, *A*), and the **connecting rings**, which are tubes that extend between two adjacent septa or septal necks (Figure 14.52, *A*). There is an elaborate terminology for the shapes of connecting rings and septal necks (Figures 14.51, *G* to *K* and 14.52, *B*). Septal necks that bend adapically are **retrochoanitic**; those that bend adorally are **prochoanitic**.

Various fossil cephalopods also deposited shell on the walls of the chambers. Such structures are called **cameral deposits** (Figure 14.52, *C*). These are named according to their position on the wall of the camera. **Episeptal deposits** are on the anterior wall of a septum, **hyposeptal deposits** are on the posterior wall, and **mural deposits** line the outer wall of the chamber. The shapes and forms of

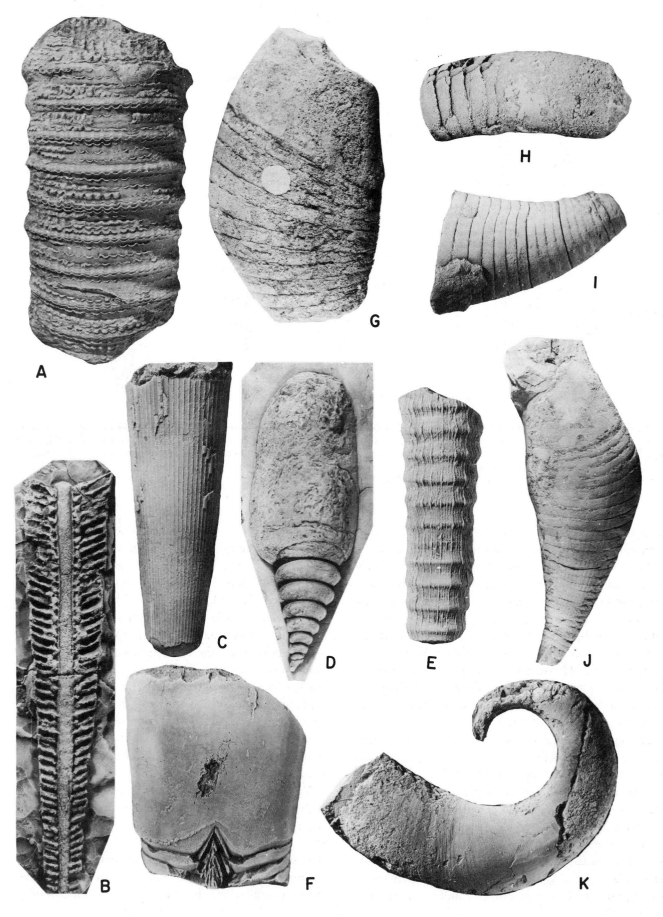

cameral deposits are important in the classification of some cephalopods. Functionally, they aided in counteracting the positive buoyancy of the gas-filled phragmocones of orthoconic or cyrtoconic longicones (Figure 14.47, *D* and *E*). Sometimes distinguishing cameral deposits from diagenetically deposited calcite is difficult. It is uncertain how cameral deposits formed, but various authors have postulated the presence of a cameral mantle. *Nautilus* lacks cameral deposits and does not have a cameral mantle.

Classification

The classification of cephalopods is complex; the six often recognized first-order groupings are usually called subclasses. At least 26 orders are recognized. Five of the subclasses have an external shell; the sixth has an internal shell or the shell is absent.

SUBCLASS NAUTILOIDEA (Figures 14.53; 14.54)

Small to large cephalopods with orthoconic to tightly coiled planispiral external shells (or much less commonly torticonic external shells). Sutures orthoceratitic, generally simple, less commonly with prominent lobes and saddles. Siphuncle not large in diameter, commonly marginal and ventral to subdorsal in position. Septal necks retrochoanitic; connecting rings thin to thick, with or without endosiphuncular deposits. Cameral deposits common in noncoiled forms. (Late Cambrian to Holocene; at least 700 genera.)

The diversity of species placed in this subclass is great, and the taxon is difficult to define in morphologic terms. The Nautiloidea is based on numerous observations that suggest that all species in the subclass are more closely related to one another than they are to any other subclass.

SUBCLASS ENDOCERATOIDEA (Figure 14.55)

Medium-sized to very large cephalopods usually having orthoconic or rarely cyrtoconic external shells. Sutures orthoceratitic, usually simple, but less commonly having midventral lobe or saddle. Siphuncle characteristically large and usually occupying at least one-fourth of the diameter of the shell; marginal to submarginal and ventral in most species, may be subcentral. Septal necks retrochoanitic, connecting rings thin to thick. Endosiphuncular deposits always present, characteristically consisting of a series of nested conical sheaths known as endocones down the center of which is an endosiphuncular tube. Cameral deposits absent. (Late Cambrian?, Early Ordovician to Silurian; at least 80 genera.)

Endoceratoids are common Ordovician megafossils. They show a marked decrease in diversity in Upper Ordovician and Silurian rocks. The largest known Paleozoic invertebrate is a Middle Ordovician endoceratoid whose shell reached a length of 9 to 10 m.

SUBCLASS ACTINOCERATOIDEA (Figure 14.56)

Medium-sized to very large cephalopods, most having orthoconic external shells, a few cyrtoconic. Sutures orthoceratitic. Cameral deposits characteristically present. Septal necks retrochoanitic, connecting rings complex. Body chamber contracted near aperture. Siphuncle typically large, as much as half the diameter of the body chamber, marginal to subcentral in position, ventral. Segments of siphuncle inflated between septa. Siphuncle contains characteristic complex endosiphuncular deposits laid down in such a manner that a system of endo-

Figure 14.53. Orthoconic (A to F) and cyrtoconic (G to K) Nautiloidea. A, *Dawsonoceras*. Part of phragmocone showing external ornament. (Silurian, England, × 1.) B, *Murrayoceras*. Most of weathered phragmocone showing septa, chambers, and siphuncle. (Ordovician, New York, × 1.) C, *Parakionoceras*. Part of phragmocone showing external ornament. (Devonian, Czechoslovakia, × 1.5.) D, *Discosorus*. Internal mold showing broadly expanded siphuncular segments and large living chamber. (Silurian, Ohio, × 0.67.) E, *Gorbyoceras?*. Part of phragmocone showing external ornament. (Ordovician, Canada, × 1.5.) F, *Bathmoceras*. Internal mold of distal part of phragmocone and proximal part of living chamber. Note blade-like endosiphuncular deposits, which actually extend inward from the connecting rings. (Ordovician, Czechoslovakia, × 1.5.). G, *Cyrtogomphoceras*. Internal mold of large part of the phragmocone and living chamber that becomes constricted adaperturally. (Ordovician, Wyoming, × 0.75.) H, *Byronoceras*. Internal mold of proximal part of phragmocone and most of living chamber. Note position of the siphuncle, which shows that the specimen is exogastric. (Silurian, Illinois, × 1.) I, *Augustoceras?*. Internal mold of part of the phragmocone. (Ordovician, Ohio, × 2.) J, *Westonoceras*. Internal mold of most of the conch. Note constriction of living chamber adaperturally. (Ordovician, Oklahoma, × 0.5.) K, *Richardsonoceras?*. Exterior of most of shell. (Ordovician, New York, × 1.)

siphuncular canals is created. (Late Cambrian?, Middle Ordovician to Late Mississippian; at least 40 genera.)

In this subclass, orthoconic shells commonly reach lengths of 0.9 m, and one of the late Paleozoic genera assigned here is thought to have reached a shell length of 6 m.

SUBCLASS BACTRITOIDEA (Figures 14.57; 14.58)

Small cephalopods having thin orthoconic or cyrtoconic and exogastric external shells. Sutures orthoceratitic, simple, with a midventral lobe in most genera. Septal necks retrochoanitic, straight or inflated. Siphuncular deposits absent; cameral de-

Figure 14.54. Coiled Nautiloidea: **A** to **D**, planispiral; **E**, torticone. **A**, *Halloceras*. Open coiled shell. (Devonian, New York, ×1.) **B**, *Metacoceras*. Silicified specimen showing ornament of nodes. (Permian, Texas, ×1.) **C**, *Cooperoceras*. Silicified specimen showing ornament of spines. (Permian, Texas, ×0.6.) **D**, *Curtoceras*. Specimen showing uncoiling. (Ordovician, Vermont, ×1.) **E**, *Lechritrochoceras*. Internal mold showing most of phragmocone. (Silurian, Ohio, ×2.5.)

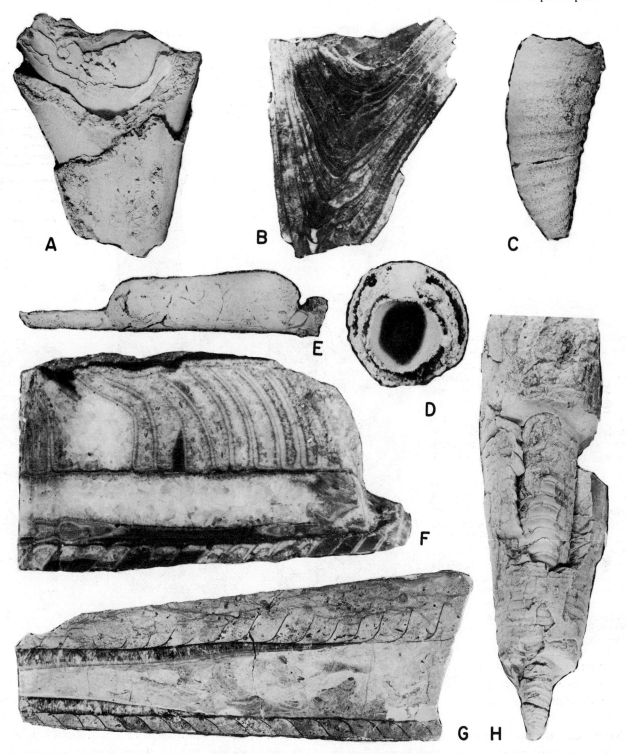

Figure 14.55. Endoceratoidea. **A** and **B**, *Cassinoceras*. Exterior and longitudinal polished section views of siphuncle showing nested endocones that fill the siphuncular space (compare to Figure 14.51, *N*). (Ordovician, Vermont, ×1.) **C**, *Piloceras*. Silicified siphuncle. (Ordovician, Tennessee, ×1.) **D**, *Allopiloceras*. Adapertural view of silicified siphuncle showing nested endocones. (Ordovician, Texas, ×2.) **E**, *Cotteroceras*. Silicified internal mold showing living chamber and proximal part of siphuncle. (Ordovician, Missouri, ×0.67.) **F**, *Cameroceras*. Midsagittal polished section of part of phragmocone showing septa, camerae, and siphuncle, which on the ventral side is partly filled with endosiphuncular deposits. Note that septa can be seen both above and below the siphuncle, showing that it is submarginal. (Ordovician, Vermont, ×1.) **G**, *Endoceras*. Midsagittal polished section of part of phragmocone showing septa, camerae, and endocone filling of siphuncle. (Ordovician, Indiana, ×1.) **H**, *Cassinoceras*. Ventral view of conch and siphuncle. Posteriorly the conch is weathered away and the siphuncle protrudes. (Ordovician, Vermont, ×0.5.)

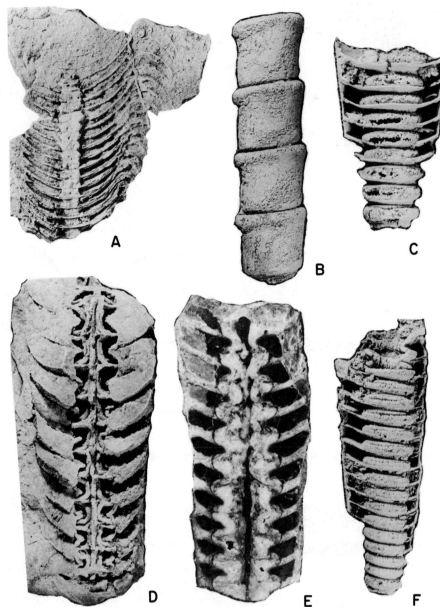

Figure 14.56. Actinoceratoidea. **A,** *Gonioceras.* Part of weathered phragmocone showing septa, chambers, and annular siphuncular deposits. (Ordovician, Virginia, ×1.) **B,** *Huronia.* Silicified siphuncular deposits. (Silurian, Iowa, ×0.67.) **C,** *Actinoceras.* Part of silicified phragmocone showing septa, camerae, and annular siphuncular deposits. (Ordovician, Kentucky, ×2.) **D,** *Armenoceras.* Internal mold showing the filling of the complex of endosiphuncular canals. (Silurian, Wisconsin, ×0.8.). **E,** *Actinoceras.* Polished longitudinal section showing annular siphuncular deposits (compare with Figure 14.51, *Q*), episeptal cameral deposits, septa, and camerae. (Ordovician, New York, ×0.75.) **F,** *Actinoceras.* Part of silicified phragmocone showing camerae, septa, and annular siphuncular deposits. (Ordovician, Kentucky, ×1.)

posits absent in most. Siphuncle small, marginal or slightly submarginal, and ventral. Protoconch calcareous, small, commonly globular, and constricted at attachment to conch.

Bactritoids are well represented in Devonian rocks where at least six genera are known. Upper Paleozoic rocks contain at least 20 known genera, and the upper limit of their range is the Upper Triassic where one genus occurs.

The taxonomic position and hierarchical level of bactritoids are debated, and not all texts treat them as a subclass. Some authors ally the bactritoids to the nautiloids and others place them in the next subclass, the Ammonoidea. Various genera in the Nautiloidea have straight shells and small marginal and ventral siphuncles suggesting bactritoid affinities, but their apical parts are poorly known. The bactritoids probably evolved from orthoconic nautiloids in latest Silurian or earliest Devonian time. The thin shell, bulbous protoconch, and small marginal siphuncle with a midventral lobe of bactritoids are also typical of primitive ammonoids. Most authors regard bactritoids as morphologically and phylogenetically intermediate between nautiloids and ammonoids. It is generally accepted that bactritoids gave rise to ammonoids by evolving planispiral coiling. Planispiral coiling (Early Ordovician to Holocene) is regarded as a homeomorphic feature in cephalopods, having been independently developed in various nautiloids as well as in the ammonoids.

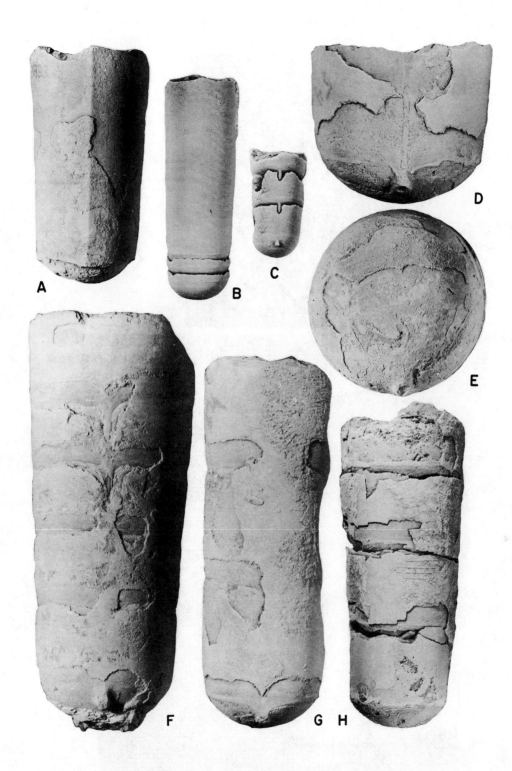

Figure 14.57. Bactritoidea. **A** and **D** to **H**, *Bactrites* from the Pennsylvanian of Texas; **B** and **C**, *Lobobactrites* from the Devonian of Morocco. **A**, Dorsal view of part of phragmocone showing carina. (×2.) **B**, Lateral view of internal mold showing distal part of phragmocone and much of living chamber. (×2.) **C**, Ventral view of internal mold of short section of phragmocone showing marginal siphuncle and midventral lobe in suture. (×3.5.) **D** and **E**, Ventral and posterior views of internal mold of one camera (with some shell preserved laterally) showing marginal siphuncle. (×3.) **F**, Ventral view of part of a partially decorticated phragmocone showing siphuncle and midventral lobe of suture. (×3.) **G**, Ventral view of part of a partially decorticated phragmocone showing siphuncle and midventral lobe of suture. (×3.) **H**, Lateral view of part of a partially decorticated phragmocone. (×1.5.)

Figure 14.58. Bactritoidea. **A,** *Bactrites*. Numerous small specimens showing small, globular, calcareous protoconchs. (Mississippian, Nevada, ×5.) **B,** *Bactrites*. Enlargement of specimen shown in Figure 14.57, *H,* showing wrinkle layer below the outer shell layer, some of which is still seen in the few growth lines to the left of center (×6.)

SUBCLASS AMMONOIDEA (Figures 14.59 to 14.62)

Small to large cephalopods with external shells usually planispirally coiled and exogastric. Sutures agoniatitic, goniatitic, ceratitic, or ammonitic. Septal necks most commonly prochoanitic in adult stages. Siphuncular deposits rare; cameral deposits absent. Siphuncle small in diameter, usually marginal and ventral in mature stages; in most, ventral throughout ontogeny; in few, dorsal. Protoconch bulbous and calcareous. (Early Devonian to Late Cretaceous; at least 2000 genera.)

As a group, ammonoids show great variety in shell form, although most of them are planispirally coiled. A general term for ammonoid shells that are not planispiral is **heteromorphs**; some of the more unusual heteromorphs are shown in Figure 14.62. Heteromorphs date back to the Devonian, but first appear in significant numbers in the Late Triassic. Shells of adult ammonoids range from 10 mm to 3 m in diameter. The great variety in size, shape, style of coiling, thickness of shell, external ornament, and sutural complexity suggests that am-

monoids were adapted to many modes of life. Nonammonoid shelled cephalopods also show great variety in shell form; they do not show the variation in morphology of the suture seen in ammonoids. Nonammonoid shelled cephalopods do show great variation in the position, width, and deposits in the siphuncle and in cameral deposits. Some ammonoids served as food for larger vertebrates. In the Carboniferous, ammonoids with shark bite marks are known; in the Jurassic, specimens are known with bony fish bite marks; and in the Cretaceous, specimens are known with mosasaur bite marks. Plesiosaurs also may have preyed on ammonoids.

SUBCLASS COLEOIDEA (Figures 14.63, E; 14.64)

Small to very large cephalopods in which mantle forms outer covering of body. Shell either internal (most) or absent (few). Body projectile- or sacshaped. Shell orthoconic, cyrtoconic, or rarely coiled. Head has 8 or 10 tentacles. (Early Devonian to Holocene; about 250 genera.)

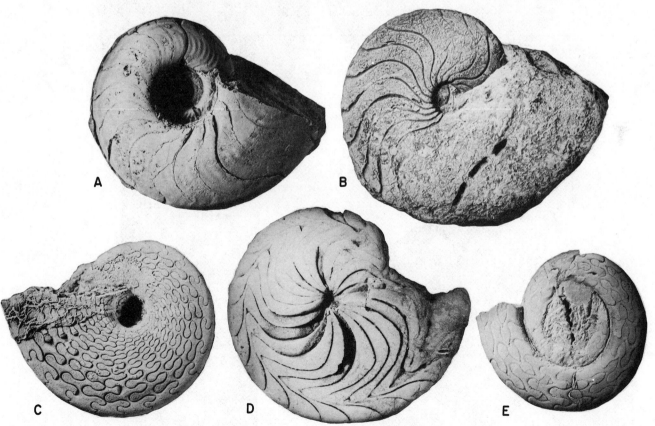

Figure 14.59. Paleozoic ammonoids showing agoniatitic (**A** and **B**) and goniatitic (**C** to **E**) sutures. All specimens are internal molds. **A**, *Agoniatites*. (Devonian, New York, ×1.5.) **B**, *Agon-* *iatites*. (Devonian, Morocco, ×1.) **C**, *Uddenites*. (Permian, Texas, ×4.) **D**, *Imitoceras*. (Mississippian, Indiana, ×1.) **E**, *Protocanites*. (Mississippian, Indiana, ×1.)

Figure 14.61. Variation in ornament and size of umbilicus in ammonoids. **A** is an evolute form, and **I** is an involute form. Note the gradation in width of umbilici between these end members. **A**, *Gastrioceras*. (Pennsylvanian, Arkansas, ×1.5.) **B**, *Columbites*. (Triassic, Idaho, ×1.5.) **C**, *Homoceras*. Pennsylvanian, Belgium, ×4.5.) **D**, *Muensteroceras*. (Mississippian, Kentucky, ×1.) **E**, *Tragodesmoceras*. (Cretaceous, South Dakota, ×2.) **F**, *Altudoceras*. (Permian, Mexico, ×1.) **G**, *Neogastroplites*. (Cretaceous, Montana, ×2.) **H**, *Pseudolioceras*. (Jurassic, Alaska, ×1.) **I**, *Sagenites*. (Triassic, California, ×3.) **A**, **C**, **D**, and **F**, Goniatites; **B** and **I**, ceratites; **E**, **G**, and **H**, ammonites.

Figure 14.60. A to **C**, Homeomorphy of ceratitic sutures in ammonoids. **A**, *Tissotia*. (Cretaceous, Peru, ×1.) **B**, *Meekoceras*. (Triassic, Idaho, ×2.) **C**, *Prodromites*. (Mississippian, Indiana, ×0.5.) These three genera are placed in different orders. **D** to **F**, Homeomorphy of ammonitic sutures in ammonoids. **D**, *Waagenoceras*. (Permian, Texas, ×2.) **E**, *Clioscaphites*. (Cretaceous, Montana, ×1.5.) **F**, *Baculites*. (Cretaceous, Montana, ×1.5.) The Permian genus is placed in a different order from the Cretaceous genera.

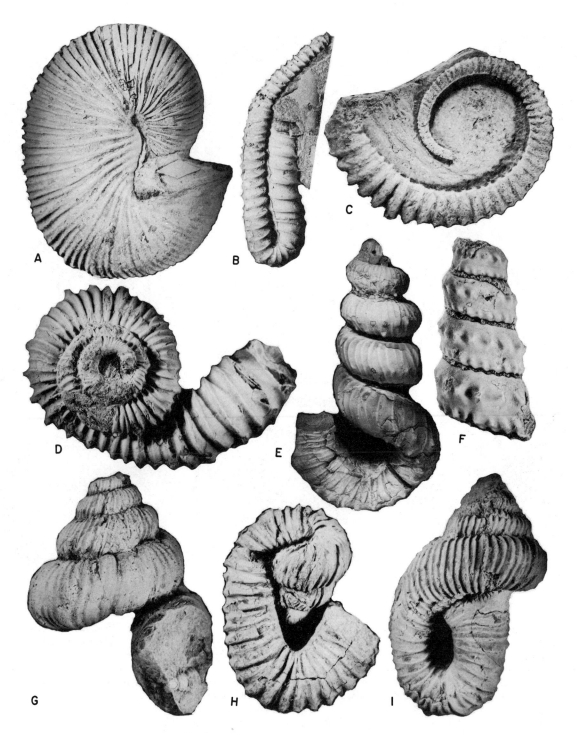

Figure 14.62. Heteromorphic ammonoids. **A, B,** and **D** to **I** are Cretaceous in age and **C** is Jurassic in age. **A,** *Scaphites.* (Montana, ×1.5.) **B,** *Solenoceras?.* (South Dakota, ×1.5.) **C,** *Leptoceras.* (Cuba, ×1.5.) **D,** *Nostroceras.* (Texas, ×1.5.) **E,** *Cirroceras.* (Wyoming, ×0.5.) **F,** *Neostlingoceras.* (New Mexico, ×1.5.) **G,** *Nostoceras.* (Texas, ×1.5.) **H,** *Anaklinoceras.* (Texas, ×2.) **I,** *Nostoceras.* (Texas, ×1.)

The stratigraphic range is Lower Devonian to Holocene, which is also the known stratigraphic range of ten-armed forms; eight-armed forms range from the Middle Jurassic to Holocene.

Paleontologically, the most important group of coleoids is the order Belemnitida, which ranges from the Upper Mississippian to Upper Cretaceous and is particularly well represented in Mesozoic rocks. Like the ammonoids, coleoids may have evolved from bactritoids. The internal skeleton of a representative belemnite is

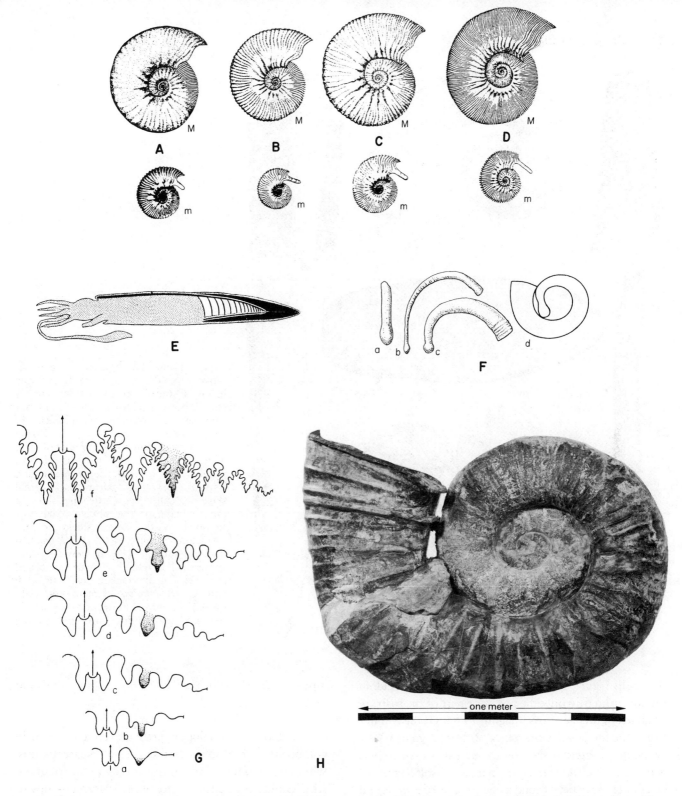

Figure 14.63. A to D, Comparison of macroconchs (*M*) and microconchs (*m*) having lappets of various Upper Jurassic Kosmoceratidae. (×0.3.) **E**, Belemnite skeleton (black) with soft parts (stippled) restored. Note that mantle surrounds the skeleton and that a counterweight of shell, called the rostrum, has been secreted around the phragmocone. Extending forward dorsally from the phragmocone is the thin proostracum. **F**, Protoconch and initial parts of conch of Lower Devonian bactritoids (*a*) and ammonoids (*b* to *d*). **G**, Ontogenetic development of the suture of *Perrinites hilli* (Smith) from the Permian of Texas, from earlier (*a*) to later (*f*) growth stages. Note the change from goniatitic to ammonitic suture types; shaded lobe is a homologous point. Arrows point adaperturally.

H, Giant Late Jurassic ammonoid *Lytoceras* from New Zealand; maximum diameter 1.5 m. (A to D from Callomon, J. H. *Leicester Literary and Philosophical Society Transactions* 62: plate I; 1963. E from Moore, R. C.; Laliker, C. G.; Fischer, A. G. *Invertebrate fossils.* New York: McGraw-Hill Book Co.; 1953, p. 344. F, from Teichert, C.; Yochelson, E. L. *Essays in paleontology and stratigraphy.* Department of Geology, University of Kansas, Special Publication 2; 1967, p. 183. G, from Miller, A. K.; et al. in: Moore, R. C., editor. *Treatise on invertebrate paleontology, Part L.* New York and Lawrence, KS: Geological Society of America and University of Kansas Press; 1957, p.21. H, courtesy of Graeme Stevens, New Zealand Geological Survey.)

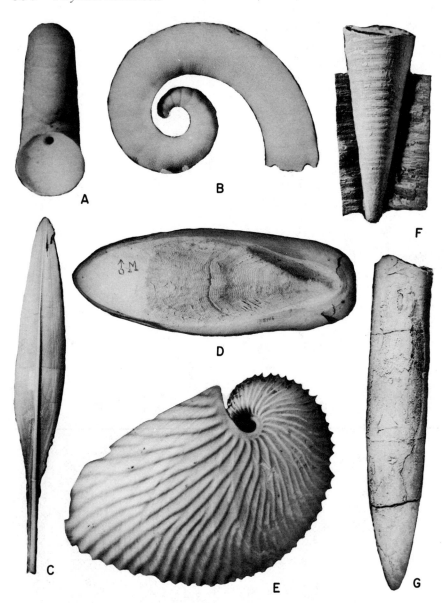

Figure 14.64. Hard parts of representative coleoids. **A** to **E** are from living forms; **F** and **G** are fossil belemnites. **A** and **B**, *Spirula* (Florida, ×4): **A**, anterior view showing opening of siphuncle; **B**, lateral view. **C**, *Loligo* (squid). Dorsal view of paper thin horny remnant of shell, which is called the pen or gladius. (England, ×1.2.) **D**, *Sepia* (cuttlefish). Ventral view of calcareous internal skeleton, called the cuttlebone, showing the remnant of the phragmocone surrounded by the dorsal shield or pro-ostracum. (France, ×1.) **E**, *Argonauta* (paper nautilus). Lateral view of planispiral shell secreted by female as a brood chamber (Mediterranean Sea, ×0.75.) **F** and **G**, *Belemnites* (Cretaceous, California): **F**, part of phragmocone partially surrounded by a portion of the rostrum (×2); **G**, rostrum (×1).

shown in Figure 14.63, *E*. Whole skeletons are seldom found; ordinarily only the posterior part of the **rostrum** or **guard** occurs as a fossil (Figure 14.64, *G*). Belemnites overcame the positive buoyancy of the phragmocone by evolving an internal shell on the outside of which was deposited the rostrum. Remnants of soft parts of belemnites have been found in the Jurassic of Europe and the Cretaceous of Syria; they show arm hooks and the contents of the ink sac. Mesozoic reptiles known as ichthyosaurs fed on belemnites.

Other cephalopod hard parts occasionally found as fossils

Radulae   Preservation of the horny radula as a fossil is rare, but fossil radulae have been found in a very few

ammonoids. They are much like those of some squids and octopuses.

Rhyncholites   Cephalopod jaws are largely horny in composition, but some of them have calcareous tips, which occur as fossils and have a worldwide distribution. Rhyncholite is a term for fossilized calcareous tips of cephalopod beaks (Figure 14.65, *A* to *C*). Strictly applied, the term refers only to the calcareous part of the upper jaw; remains of the calcareous part of the lower jaw are called conchorhynchs. Rhyncholites have a range of Lower Carboniferous to Eocene.

Aptychi   Aptychus is a general term for structures sometimes interpreted to have been a cephalopod operculum (Figure 14.65, *D*, *F* and *G*) serving the same function as

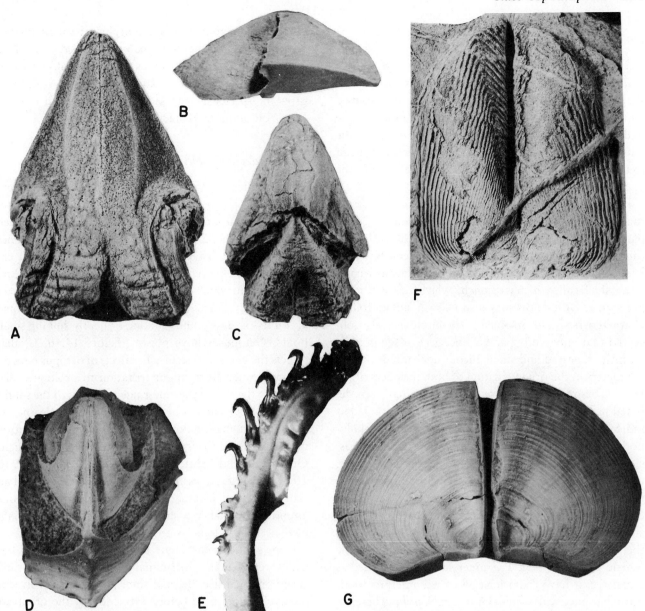

Figure 14.65. Hard parts occasionally found as fossils. **A** to **C**, Rhyncholites (Cretaceous, France, ×4): **A**, ventral view; **B**, lateral view; **C**, dorsal view. **D**, Probable aptychus within the shell (Cretaceous, South Dakota, ×2.) **E**, Distal part of one of the long arms of a modern squid showing hooks. (×2.) **F** and **G**, Paired aptychi from the Jurassic. **F** is from Cuba (×2), **G** is from Germany (×1).

the horny hood of *Nautilus*. A few aptychi have been found in place in the aperture. Some paleontologists believe aptychi are better interpreted as cephalopod jaws. There is an extensive terminology for aptychi depending on their number of parts. The total known range of aptychi is Lower Devonian to Cretaceous.

Onychites Onychites (Figure 14.65, *E*) are fossilized horny cephalopod arm hooks. They are ordinarily rare fossils, and range in age from Pennsylvanian to Cretaceous.

Statoliths Statoliths are aragonitic bodies usually less than 2 mm formed in the statocyst (Figure 14.46, *D*), an organ of equilibrium below the brain and comparable in function to the inner ear of vertebrates. Most classes of mollusks have statocysts, but not all of them have statoliths. Statolith form is reasonably constant in adults of living species of cephalopods and has been used as a tool to help understand the evolution of coleoids that lack other calcareous parts. Statoliths are not common fossils. The stratigraphic range is Eocene to Holocene.

Dimorphism in cephalopods

The sexes of living cephalopods are separate and in *Nautilus* have slightly different shell form in the size and shape of the aperture. The male genitalia are modified tentacles situated near the mouth and are bulky; thus the male has a larger hood and a broader aperture than the female. However, individual variation makes it difficult to tell a male from a female *Nautilus* by the shape of the aperture alone.

For more than 100 years paleontologists have noticed that the adults of many species of ammonoids occur in paired groups in the same strata. One of each pair (dimorph) has a small shell called a **microconch**, which has a more ornate aperture, and the other member has a large shell called a **macroconch**, which has a simpler peristome. Differences may also exist in shape, coiling characteristics, and ornament. Paleontologists also observed that apparently parallel lineages evolved in pairs with the repeated and simultaneous appearance of new characters in the ornamentation of both members of the pairs. The meaning of these observations has been debated at length; they are now most often interpreted as indicating sexual dimorphism in some extinct cephalopods. Examples of dimorphism are seen in various Jurassic species (Figure 14.63, *A* to *D*) where small, ornate, tightly coiled microconchs occur with larger, more loosely coiled, less ornate macroconchs. The microconchs have anterior projections called lappets on either side of the peristome and are interpreted as being males; the lappets are regarded as structures for clasping the female during mating. The larger shells are regarded as females that have room for eggs. These interpretations are at variance with what is known in *Nautilus* where the female shell is smaller, but are consistent with what is known in the eight-armed paper nautilus where the female is ten to 20 times larger than the male.

One of the major arguments advanced against the idea of sexual dimorphism in ammonoids has centered on the relative numbers of the dimorphs found together. They occur together in a variety of proportions, and either dimorph can occur almost to the total exclusion of the other. Sometimes this separation can be attributed to various sedimentologic sorting conditions. At times, species of living cephalopods segregate themselves sexually, and perhaps some extinct species did likewise.

Because we do not have the soft parts of extinct ammonoids, we can never prove absolutely that they exhibited sexual dimorphism, nor can we be entirely sure which dimorph was male and which female. Nonetheless, examples of dimorphism in ammonoids date back as far as the Devonian, so that dimorphism is an established fact in some fossil shelled cephalopods. It is reasonable to interpret this dimorphism as sexual because various living cephalopods show sexual dimorphism. Also, structures interpreted as egg sacs have been found in some Mesozoic ammonoid macroconchs.

Kraken, devilfish, and gigantism in cephalopods

Written tales of sea monsters are as old as the ancient Greek epic poems. The source of many of these tales seems to be the larger squids, which are sometimes called kraken, and the larger octopuses, which are sometimes called devilfish. There are innumerable tales about encounters with kraken, which can reach a length of 22 m, and devilfish, which may span 25 m and reach a diameter of 45 cm. There are reasonably well-documented accounts of large squids sinking or nearly sinking small boats. Octopuses have poison (Figure 14.46, *C*) that affects the nervous system. The effects of octopus bites on humans range from minor irritation to prolonged discomfort; the bite of the blue-ringed octopus of the southwest Pacific Ocean has been reported to cause death.

Gigantism in cephalopods has been a recurrent feature in their evolution (Figure 14.63, *H*). The largest straight Ordovician endoceratoids reached a shell length of 9 to 10 m, straight Mississippian actinoceratoids reached an estimated shell length of 6 m, and coiled Cretaceous ammonoids reached a shell diameter of 3 m and would have been 18 m long if unrolled.

Cope's rule is the name customarily applied to the widespread tendency of animal groups to evolve toward larger physical size. Because some giant forms evolved a relatively short time before extinction of the group to which they belong, some paleontologists have regarded gigantism as a sign of impending extinction. Various lineages have occasionally produced unusually large species that became extinct, but the lineage continued to exist and subsequently produced smaller species.

In a recent discussion of Cope's Rule it was noted that most animal taxa arise at a small body size and that early in their history most species are small. Niches requiring larger body size tend to be filled subsequently so that vacant regions of a group's maximum potential adaptive zone are progressively invaded. Cope's Rule cannot be explained by the presumed intrinsic advantages of large size. Rather it is the tendency of groups to arise at small body size that produced the widely observed pattern of net size increase.

The specialized nature of large species of a given body plan renders these forms unlikely potential ancestors for major new descendant taxa. The adaptive discontinuity that must be crossed for invasion of a new adaptive zone at large body size exists because of the need for descendant taxa to be specialized along new lines. Structural specialization at large body size for a given taxon gradually limits the range of potential morphologies and hence diversity. These factors tend to restrict large-scale adaptive breakthroughs to small body size.

Patterns of diversification in shelled cephalopods

The reasons for diversification and for the extinction of once diverse and abundant cephalopods are debated. However, 150 years of descriptive paleontology have given a good picture of the times of diversification and extinction. Different lineages show different patterns of diversification which are briefly discussed.

The first members of the Endoceratoidea appear near the Cambrian-Ordovician boundary. They diversified rapidly into at least 40 recognized genera before the end of Early Ordovician time (Figure 14.66). From this high point, endoceratoids show a dramatic decline to only about five genera at the end of the Ordovician. Although the proper taxonomic placement of the Silurian forms included in the Endoceratoidea is debated, the most genera known to be represented at any particular point in Silurian time is three. All endoceratoids were extinct before the end of the Silurian. Thus they show a pattern of rapid diversification, less rapid but dramatic decline, and a short-term relict existence. Their entire evolution, from origin to extinction, was 90 million years.

The first actinoceratoids (Figure 14.66) probably occur near the Cambrian–Ordovician boundary, but they did not begin to diversify until early Middle Ordovician time. By later Middle Ordovician time at least 20 genera existed. After this high point, actinoceratoids declined to about half that number of genera by the end of the Ordovician. Decline in the number of genera continued in the Early Silurian, but in Middle Silurian time there were again about 10 known genera. After this time, the subclass declined to about five known genera. It did not exceed this level of diversification for the rest of its known stratigraphic range and became extinct in Late Mississippian time. Actinoceratoids show a diversification pattern characterized by a slow start-up period, more than one period of diversification, and a prolonged relict period. Their history was about 170 million years long.

Most of the order level taxa of the Nautiloidea show diversification patterns comparable to the patterns of either the Endoceratoidea or Actinoceratoidea. Various orders of nautiloids show several times of diversification separated by prolonged intervals of low diversity, and many of them show extended periods of relict existence (Figure 14.66). By the end of the Triassic, all orders of nautiloids, except the one to which *Nautilus* belongs were extinct.

Bactritoids show no periods of extensive family or generic diversification (Figure 14.66). The time of their greatest number of genera was the Pennsylvanian and Early Permian when seven or eight such taxa existed. Bactritoids are not usually common fossils, and much of what is known about their evolution and diversification has been learned in the past 20 years. Nonetheless, we know that bactritoids did not undergo the diversification seen in other cephalopod subclasses.

The history of the Ammonoidea differs from that of the other externally shelled subclasses in that ammonoids became nearly extinct three times before reaching extinction at the end of the Cretaceous (Figure 14.66). Also ammonoids never showed any prolonged periods of dramatically decreased diversity, and they did not have an extended period of relict existence. After each near extinction, ammonoids reradiated rapidly and dramatically (Figure 14.66), and their final decline to extinction was also abrupt.

Homeomorphy is widespread in ammonoid shell form, width of umbilicus, ornament, and whorl cross-section shape. In general shell characteristics, gross homeomorphy even exists between ammonoids and planispirally coiled nautiloids (Figures 14.54 and 14.61). Among ammonoids, homeomorphy in the general pattern of sutures can also exist (Figure 14.60). However, knowledge of the details of suture morphology can sort out the homeomorphs. Even elementary knowledge of suture morphology makes it possible to distinguish between the oldest known species having ceratitic sutures (Figure 14.60, C), in which the lobes have parallel sides, and a Cretaceous form having ceratitic sutures (Figure 14.60, A), in which the lobes have arcuate constricted sides and expanded ends.

The oldest known ammonoids are Early Devonian in age, and among these agoniatitic sutures are most common. Early Devonian ammonoids show a progression in shell form toward tighter coiling in a sequence that passes from the orthoconic to cyrtoconic shells of the ancestral bactritoids, to loosely or open coiled shells, to tightly coiled planispiral adult whorls (Figure 14.63, F). Ammonoids diversified rapidly through the nearly 50 million years of the Devonian to a total of about 30 families in Upper Devonian rocks (Figure 14.66); most of these families have short ranges and are useful biostratigraphic

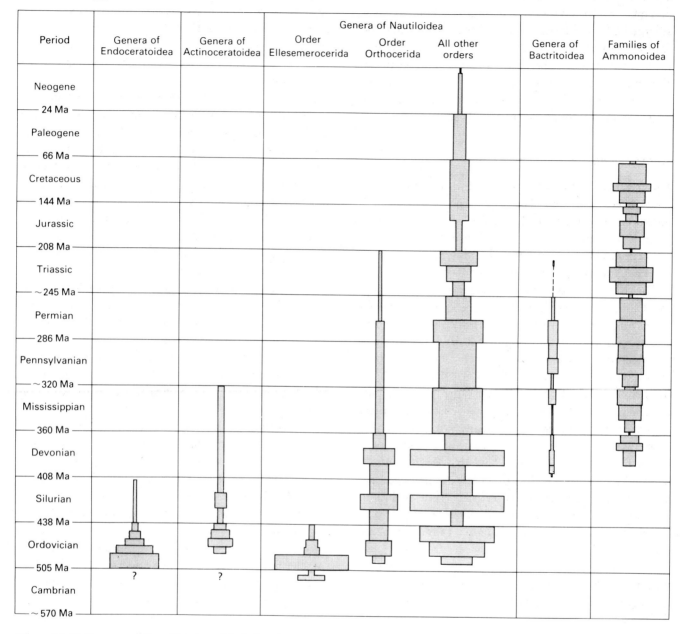

▢ = 10 Genera or families (width)

Period	Genera of Endoceratoidea	Genera of Actinoceratoidea	Genera of Nautiloidea			Genera of Bactritoidea	Families of Ammonoidea
			Order Ellesemerocerida	Order Orthocerida	All other orders		
Neogene							
— 24 Ma —							
Paleogene							
— 66 Ma —							
Cretaceous							
— 144 Ma —							
Jurassic							
— 208 Ma —							
Triassic							
— ~245 Ma —							
Permian							
— 286 Ma —							
Pennsylvanian							
— ~320 Ma —							
Mississippian							
— 360 Ma —							
Devonian							
— 408 Ma —							
Silurian							
— 438 Ma —							
Ordovician							
— 505 Ma —	?	?					
Cambrian							
— ~570 Ma —							

Figure 14.66. Patterns of diversification of shelled cephalopods. Families are diagrammed for the Ammonoidea; genera are used for all other subclasses. Coleoids are not included because of their generally poor fossil record.

tools. Thirteen families are known to have existed in latest Devonian time, and only one of these survived into the Mississippian Period.

The surviving ammonoids possessed goniatitic sutures, and they radiated into about 25 families during the Mississippian (Figure 14.66). Many Mississippian families have relatively long ranges. A decrease in the number of families occurs at the Mississippian–Pennsylvanian boundary, but it is not as dramatic as other declines in diversity of ammonoids. About nine families range across the boundary. About 30 families of Pennsylvanian am-

monoids exist, of which about 20 are present in latest Pennsylvanian time. Various Pennsylvanian families have restricted ranges. No decrease in family level diversity is found in Lower Permian rocks (Figure 14.66); 14 families cross the Pennsylvanian–Permian boundary.

In the Permian, about 27 families of ammonoids are present. However, many subfamilies are recognized in Permian ammonoids, resulting in a total of about 40 family-level taxa. During the later half of the Permian, forms having the goniatitic suture showed a marked decrease, and families having the ceratitic suture exhibited a modest diversification. Some Permian species, in their later ontogeny, developed an ammonitic suture (Figure 14.63, *G*). Ammonoids show a progressive decrease in the number of families in the Late Permian, and only three families survived into the Triassic. These surviving families include forms having highly specialized multi-lobate and digitate sutures and species having ceratitic sutures. All species having goniatitic sutures became extinct, although homeomorphs evolved in the Mesozoic.

The Triassic was the time of the great diversification of species having ceratitic sutures, and many Triassic families were short-lived. About 80 families of Triassic ammonoids are recognized. Almost all of them are ceratites and became extinct before the end of the period (Figure 14.66).

Species having the ammonitic suture are dominant in Jurassic and Cretaceous rocks, but which Triassic species might be ancestral to them is not certain. Only one small group of ammonoids that is present in Triassic rocks is also found in Jurassic rocks, and it is sometimes regarded as ancestral to Jurassic forms. Whatever their ancestry may have been, Jurassic ammonites radiated rapidly into about 46 known families, most of which have short ranges. The subfamily level is widely used in the classification of Jurassic ammonoids, and about 90 family level taxa are in the group. About 14 families of ammonoids were present in the latest Jurassic, and some nine of these also existed in the earliest Cretaceous. Cretaceous ammonoids underwent a dramatic radiation of their own, and about 85 family-level taxa are recognized; many of these have short ranges. At the end of Cretaceous time, all ammonoids became extinct, and of 11 families known at the beginning of the youngest Cretaceous stage, only five survived until the end of the period.

*Recapitulation and iterative evolution
in ammonoids*

In the latter part of the nineteenth century, naturalists developed the hypothesis known as the biogenetic law (Chapter 2). The most enthusiastic proponent of this concept was Ernst Haeckel who stated the generalization as "Ontogeny is the short and rapid recapitulation of phylogeny," which is often shortened to "Ontogeny recapitulates phylogeny." As usually promulgated, the biogenetic law held that ontogeny recapitulates a series of adult ancestors. If this were correct, one could readily construct phylogenies from the growth stages of the end members of a lineage, and because paleontology deals with biotas through time, the generalization could have had important applications in this field.

During the last third of the 19th century and the first third of the 20th century, influential American and British scholars of ammonoids attempted to study and classify this group in the light of the biogenetic law as they understood it. New characters added in the adult stage of an older species would appear at an earlier stage in a descendent species, and that descendent would add additional new characters in its adult form, and so on. Ammonoids with their complex shells and sutures seemed ideal candidates for the application of the biogenetic law because they show dramatic ontogenetic changes in suture pattern (Figure 14.63, *G*) and because sutural ontogeny can be readily studied by breaking back the shell around the coil.

Early in the 20th century, careful stratigraphic collecting showed that in various Jurassic ammonoids new characters first appeared in the young ontogenetic stages of stratigraphically older species and spread to the adult ontogenetic stages of descendants at stratigraphically higher levels. These observations are the reverse of the predictions of the biogenetic law. The biogenetic law was an important concept in the study of ammonoids from the late 1860s to about 1930, but it is now seldom used in the way that students of that time applied it.

Abundant zoologic and paleontologic information now available shows that related animals of different kinds resemble one another more closely in their young stages than in their adult stages. For example, the adult mammal bears little resemblance to the adult fish, and their embryos are quite different. Still, the embryonic mammal has gill arches much like those of the embryonic fish. The biogenetic law restated in this sense, that the young of related stocks show the relationship in their early ontogeny, is a useful generalization. It has usefulness for the study of ammonoid ontogeny and the implications that ontogenetic studies may have for determining ammonoid phylogenies.

Iterative evolution is the repeated occurrence of similar morphologic trends in successive offshoots of a group from a long-ranging conservative (and commonly morphologically simple) ancestral stock. As classically applied in Jurassic ammonoids, the concept of iterative evolution showed two long-ranging suborders, Lyto-ceratina and Phylloceratina (Figure 14.67, *A*), periodically giving rise to various superfamilies of the third suborder Ammonitina (in Figure 14.67, *A*, the Ammonoidea is treated as an order rather than as a subclass). This interpretation meant that the younger superfamilies of the Jurassic Ammonitina were not necessarily directly derived from the older ones, that similar morphologies were homeomorphic, and that the

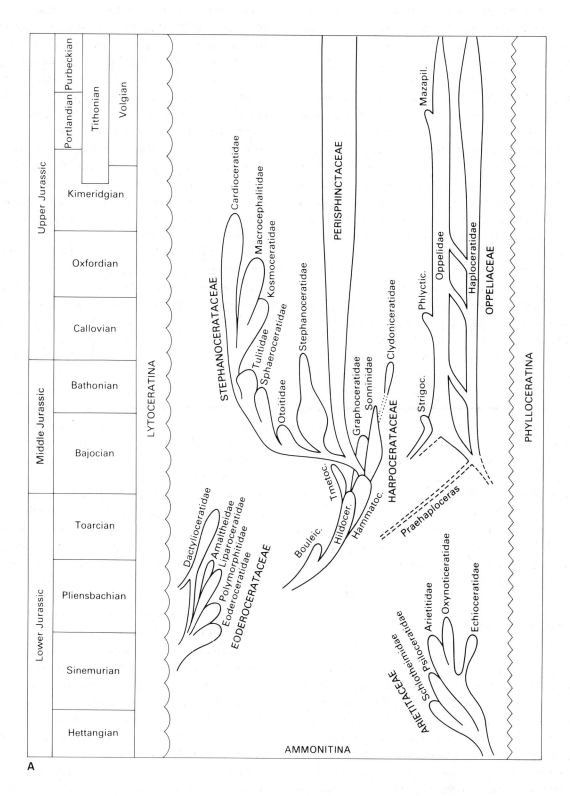

A

Figure 14.67. Comparison of phylogeny and classification of Jurassic ammonoids. **A,** Application of the concept of iterative evolution, where the Lytoceratina and Phylloceratina are regarded as long-ranging stem stocks periodically giving rise to the various superfamilies of the Ammonitina (center). **B,** Non-iterative approach in which the Phylloceratina and Lytoceratina are regarded as long-ranging stocks having no close relationship to the Ammonitina whose ancestry is regarded as unknown. (**A** from Arkell, W. J. *Journal of Paleontology* 24: p. 360; 1950. **B** from Donovan, D. T.; Callomon, J. H.; Howarth, M. K. In: House, M. R.; Senior, J. R., editors. *The Ammonoidea;* New York, Academic Press, Inc.; 1981, p. 125.)

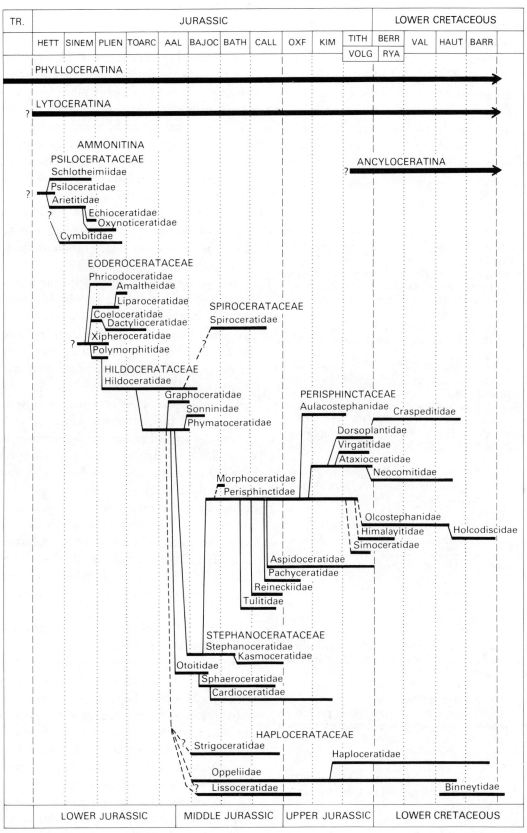

B

Ammonitina as then interpreted represented a grade of morphologic organization. In a 1981 classification of Jurassic ammonoids (Donovan and others in House and Senior, 1981) (Figure 14.67, *B*) the authors virtually abandon "the theory of Iterative Evolution . . . [because] new discoveries and studies . . . have not substantiated the idea of 'replenishment' of groups by successive homeomorphic waves, especially from the conservative suborders Phylloceratina and Lytoceratina. These suborders now stand in even more isolation than before, clearly distinguished in morphology, and probably in habitat, from the contemporary Ammonitina."

Supplementary reading

Arkell, W. J. A classification of the Jurassic ammonites. *Journal of Paleontology* 24: 354–364; 1950. Classical presentation of the concept of iterative evolution in ammonite studies.

Callomon, J. H. Sexual dimorphism in Jurassic ammonites. *Leicester Literary and Philosophical Society Transactions* 62: 21–56; 1963. Synthesis marshalling evidence of sexual dimorphism in extinct cephalopods.

Flower, R. H. Studies of the Actinoceratida. *New Mexico Bureau of Mines and Mineral Resources*, Memoir 2; 1957. Extended discussion of this subclass.

Flower, R. H. Some Chazyan and Mohawkian Endoceratida. *Journal of Paleontology* 32: 433–458; 1958. Extended discussion of this subclass.

House, M. R.; Senior, J. R., editors. *The Ammonoidea.* Systematics Association Special Volume No. 18. New York: Academic Press, Inc.; 1981. Proceedings of a symposium providing state-of-the-art summaries of the study of this subclass.

Jeletzky, J. A. Comparative morphology, phylogeny, and classification of fossil Coleoidea. *University of Kansas Paleontological Contributions*, Article 7; 1966. Monograph of fossils of this subclass.

Lehmann, U. *The ammonites: their life and their world.* Cambridge, England: Cambridge University Press (English Language Edition); 1981. Readable account of ammonoids with good introductory chapter on living cephalopods.

Mapes, R. H. Carboniferous and Permian Bactritoidea (Cephalopoda) in North America. *University of Kansas Paleontological Contributions*, Article 64; 1979. Monograph of representatives of this subclass in North America.

Moore, R. C., editor. *Treatise on invertebrate paleontology, Part K, Mollusca 3,* and *Part L, Mollusca 4.* New York and Lawrence, KS: Geological Society of America and University of Kansas Press; 1964 and 1957. Extensive treatment of all aspects of noncoleoid cephalopod studies.

Teichert, C. Major features of cephalopod evolution, pp. 162–210. *Essays in paleontology and stratigraphy.* Department of Geology, University of Kansas, Special Publication 2; 1967. Discussion of cephalopod evolution in terms of structural innovations in shell morphology.

Ward, P. D. *Nautilus*: have shell, will float. *Natural History* 91: 64–69; 1982. Easily readable account of the buoyancy system of *Nautilus.*

Class Rostroconchia

John Pojeta, Jr.

The idea that the taxon Rostroconchia is an extinct class of mollusks dates only from 1972. Prior to this interpretation the species now included in this class were placed in a variety of higher taxa. Most classifications placed primitive rostroconchs with certain bivalved arthropods, and advanced rostroconchs were treated as unusual pelecypods. The modern concept of Rostroconchia proved to be the key to the understanding of the phylogeny of diasome mollusks presented here and to the relationship of diasomes to cyrtosomes (Figure 14.15). This interpretation suggests that the HAM concept (Figure 14.14), which had long served to unify molluscan studies, is not now necessary for understanding molluscan evolution.

It now seems likely that molluscan evolution at the class level was much like vertebrate evolution, where species of one class-level taxon (for example, the reptiles) gave rise to species of another class-level taxon (for example, the mammals) through a series of morphologic intermediates. In both vertebrates and mollusks, this interpretation is based on known species of fossils.

An evolutionary classification is an attempt to divide and subdivide the continuum of life, and any continuum can only be divided by defining the points of division. In a sense, division is arbitrary and a matter of convenience. However, once the points of division are fixed, anyone can use them in an objective way, and the groupings defined by those points can then be treated as a series of axioms within the context of which various paleobiologic, biostratigraphic, and other deductions can be made. The whole effort of scientific classification of living things is based on the principle of organic evolution; all members of a subdivision are descended from a common ancestor.

All presently living species are the products of a long history of change and divergence. They can ordinarily be readily separated one from another at the class level because the great majority of classes had their origins long ago. However, knowledge of fossils and a few so-called living fossils shows that some species form morphologic transitions between some classes. These intermediate forms are difficult to classify in the Linnaean system because they do not have all the features of either the ancestral or descendant classes. Rather then erecting a new class for each intermediate form, many workers place them in already recognized classes and specify a single morphologic change that distinguishes an ancestral class from a descendant class. Examples of this procedure were discussed in the section on Monoplacophora. The morphologic event used to separate rostroconchs from monoplacophorans was the development of the pegma (Figures 14.69, *E, F,* and *H* to *J* and 14.77) by rostroconchs.

As an example, about ten years ago, studies of morphology and probable function suggested that some 400 known species of rostroconchs represented a separate evolutionary lineage that was best treated as a class of mollusks. Since then, some Cambrian cyrtonellid monoplacophorans have been found that differ little from the oldest known rostroconchs (Figure 14.16) in shell shape and ornament and are morphologic transitions between the two classes. Once these cyrtonellids evolved the pegma, they became rostroconchs by definition.

Rostroconchia, as a taxon, could be treated as a subclass or an order of either its ancestral class (Monoplacophora) or one of its descendant classes (Pelecypoda or Scaphopoda). Rostroconchia could probably not be treated at a level lower than order, because the known generic and species diversity could not be readily accommodated at such a low level. As a subclass or order, Rostroconchia would still contain all the species, genera, and families placed in it as a class, and the likely descendants of monoplacophorans as well as the likely ancestors of pelecypods and scaphopods could be indicated. Thus the concept Rostroconchia would remain intact at a lower hierarchical level than class, and it could be used to make various deductions.

Rostroconchs are treated as a class-level taxon because treating rostroconchs as a subordinate taxon of Pelecypoda, Scaphopoda, or Monoplacophora would be comparable to saying that a parent is more closely related to one of his or her children or one of his or her parents than to another. The taxon Rostroconchia is the stem group of the Diasoma and is placed at the same hierarchical level as Pelecypoda and Scaphopoda to point up its ancestral status.

In addition to giving rise to the other diasomes, rostroconchs underwent a dramatic diversification of their own (Figure 14.68). This evolution continued throughout the Paleozoic long after the scaphopods and pelecypods diverged from rostroconchs. Fully evolved rostroconchs are morphologically so different from both their monoplacophoran ancestors and pelecypod and scaphopod descendants, that to treat rostroconchs as a subordinate taxon of any of those classes would so skew those class concepts as to make them unmanageable. If mollusks having a pegma had not diversified dramatically in the Cambrian and Ordovician (Figure 14.68) and if they had

not given rise to pelecypods and scaphopods, it is extremely unlikely that two such morphologically similar shells as those shown in Figure 14.16 would be placed in different classes. Rather these first forms that had a pegma probably would have been classified as a family of cyrtonellids.

In producing a hierarchy for a particular taxon, systematists look for evolutionary groupings that can be used for the hierarchical levels below that taxon. From a practical point of view, putting Rostroconchia at the class level allows each of the two major lineages of rostroconchs to be dealt with as orders. Within each order are two lineages, each of which can be treated at the family level. Families contain from five to ten genera, and genera have from one to dozens of species.

Shell morphology

Orientation The problem of morphologic orientation exists in any major group of extinct metazoans because there is no direct access to soft parts. The orientation of rostroconchs was established by comparing them with other classes of mollusks and considering their probable phylogenetic relationships.

Rostroconchs are bilaterally symmetrical about a plane that passes between the two valves and separates them into right and left mirror images of one another (Figure 14.69, *A* and *C*). By analogy with bivalved mollusks, the hinged margin of the rostroconch shell is regarded as dorsal (Figure 14.69, *A* and *C*). In pelecypods, the dorsal margins of the two valves are joined by an elastic horny ligament, whereas in rostroconchs, shell layers are continuous across the dorsum. Phylogenetic conclusions indicate a relationship of rostroconchs to monoplacophorans and pelecypods and reinforce the interpretation that the hinged margin of rostroconchs is dorsal. The shell margin opposite the hinge is ventral (Figure 14.69). Primitive rostroconchs often have a ventral gape (Figure 14.71, *J*); advanced forms evolved ventral margins that touch along some or all of their length (Figure 14.71, *M* and *N*).

Various advanced rostroconchs have a relatively large **gape** at one end of the shell (Figure 14.69, *C*) and a narrow tubular **rostrum** at the other end (Figure 14.69, *B*). The large gape probably allowed for protrusion of the foot from the shell and was anterior or anteroventral, based on the existence of large anterior gapes for protrusion of the foot in various pelecypods and in scaphopods. By analogy with scaphopods and some pelecypods and gastropods, the rostrum probably allowed water and excretory products to enter or leave the shell. Because the

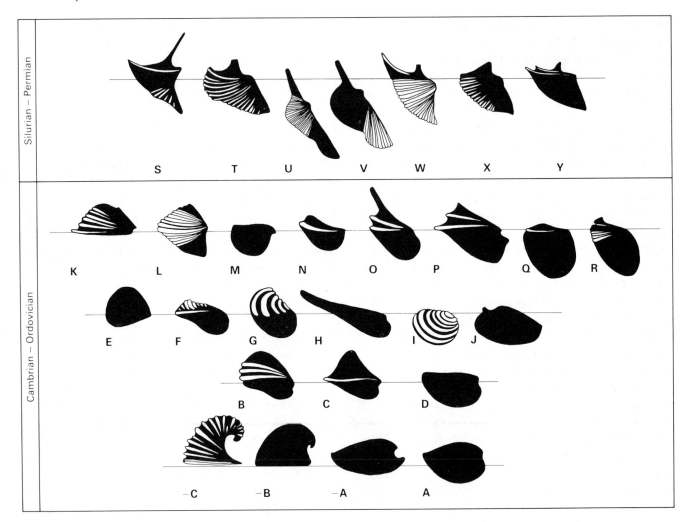

Figure 14.68. Diversification, probable life positions, and origin of rostroconchs. Silhouettes depict most known rostroconch genera (A to Y) and the probable monoplacophoran ancestors of the Rostroconchia (−C, −B, and −A). Horizontal lines connecting genera represent the sediment-water interface. −C, *Latouchella*; −B, *Anabarella*; −A, *Mellopegma*; A, *Heraultipegma*; B, *Pleuropegma*; C, *Oepikila*; D, *Ribeiria*; E, *Wanwania*; F, *Kimopegma*; G, *Cymatopegma*; H, *Pinnocaris*; I, *Ptychopegma*; J, *Pseudotechnophorus*; K, *Euchasma*; L, *Eopteria*; M, *Apoptopegma*; N, *Anisotechnophorus*; O, *Technophorus*; P, *Myocaris*; Q, *Pauropegma*; R, *Ischyrinia*; S and W, *Hippocardia*; T, *Pseudoconocardium*; U, *Conocardium*; V, *Arceodomus*; X, *Mulceodens*; Y, *Bigalea*. (Modified from Runnegar, B. *Royal Society of London Philosophical Transactions B.* 284: 320; 1978.)

rostrum is opposite the large gape, it was probably posterior (Figure 14.69, *B*).

Some primitive rostroconchs also have a rostrum at one end of the shell (Figure 14.69, *G* and *H*), and by analogy to advanced forms, this end is regarded as posterior. These primitive forms have a plate, called the **pegma**, connecting the two adult valves dorsally at the end of the shell opposite the rostrum (Figures 14.69, *E, F, H,* and *I* and 14.77); therefore the end with the pegma is considered to be anterior. Other primitive rostroconchs lack a rostrum and gape at both the anterior and posterior ends (Figure 14.71, *A, E, J,* and *O*). However, they have a pegma that marks the anterior end.

As a general rule, advanced rostroconchs are anteriorly expanded, a greater length of the shell being in front of the **beak** than behind it (Figures 14.69, *B* and 14.85, *G* and *H*); the beak is the place at which the protoconch occurs on the mature shell. Primitive rostroconchs are posteriorly expanded and have an anterior pegma at the opposite end (Figures 14.69, *H* and 14.85, *A* to *C*). Between these extremes are some rostroconchs that have an approximately equal length of shell anterior and posterior to the protoconch (Figure 14.69, *D*). These intermediates are ancestral to advanced rostroconchs, and they possess the anterior pegma (Figure 14.69, *E*).

External features   The shells of rostroconchs have a single beak (Figure 14.70) because growth originated

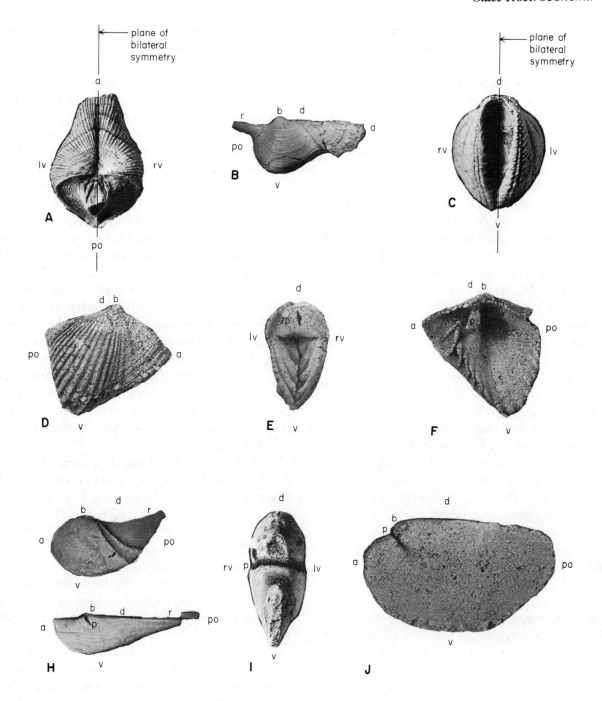

Figure 14.69. Orientation of rostroconchs: A to C, Advanced rostroconchs that are anteriorly elongated and have a greater length of shell in front of the beak than behind it; D to F, intermediate rostroconchs having about an equal length of shell in front of and behind the beak; G to J, primitive rostroconchs that are posteriorly elongated and have a greater length of shell behind the beak than in front of it. *a*, Anterior; *b*, beak; *d*, dorsal; *lv*, left valve; *r*, rostrum; *rv*, right valve; *p*, pegma; *po*, posterior; *v*, ventral. A and C, *Pseudoconocardium*. Dorsal and anterior views showing the plane of bilateral symmetry separating the two valves. C shows the large anterior gape. (Upper Pennsylvanian, Texas, ×1.5.) B, *Arceodomus*. Right lateral view. (Lower Permian, USSR, ×0.8.) D to F, *Eopteria*. D, Right lateral view. In this genus the rostrum projects a little or not at all past the posterior margin of the shell, but shell convexity and ornament marking the position of the rostrum change conspicuously. E, In this view, the posterior end of the shell has been broken away and we are looking anteriorly into the shell at the pegma and at the irregular anterior margins where the valves of the dissoconch touch below the pegma. F, Interior of right valve. (Lower Ordovician, Australia, ×3.) G, *Technophorus*. Left lateral view. (Upper Ordovician, Ohio, ×1.5.) H, *Pinnocaris*. Left lateral view of an artificial cast showing the notch made by the pegma and the posterior end of the rostrum broken and offset. (Upper Ordovician, Scotland, ×1.5.) I, *Ribeiria*. Anterior view of an internal mold showing transverse slit made by pegma and narrow anterior gape. (Ordovician, Czechoslovakia, ×3.) J, *Ribeiria*. Left lateral view of an internal mold showing the notch made by the pegma. (Upper Cambrian, Australia, ×4.5.)

from a univalved cap-shaped, hemispherical-shaped, or limpet-shaped protoconch or larval shell (Figure 14.70, *B, E, G,* and *K*). By subsequent growth the beak became situated between the **umbos** (areas of maximum convexity of the shell in dorsal profile) of the postlarval shell, or **dissoconch,** of advanced forms (Figure 14.70, *H, I,* and *K*). The protoconch or its internal mold is known in about half of the genera of the class (Figure 14.70, *C, D,* and *F* to *H*). Ordinarily no well-defined junction exists between the protoconch and dissoconch (Figure 14.70, *K*), and metamorphosis from the univalved to the bivalved stages took place by rapid growth of the right and left flanks of the protoconch, producing the two valves of the dissoconch. Such a shell is pseudobivalved because it arises from a single center of calcification and because one or more shell layers are continuous across the dorsum (Figure 14.71, *A* to *C*).

In primitive rostroconchs, the margins of the bivalved dissoconch do not touch (Figure 14.16, *D* and *E*), and there are anterior (Figures 14.70, *B* and 14.71, *O*), ventral (Figure 14.71, *J*), and posterior (Figure 14.71, *E*) gapes along the commissure. The commissure is defined as the growing edge of the shell because all rostroconchs have gapes in the shell margins (Figure 14.71) and because in all of them, one or more shell layers are continuous across the dorsum (Figure 14.71, *A* to *D*). The two valves of the dissoconch cannot be separated from one another without breaking the dorsum, and rostroconchs are usually found articulated.

In most rostroconchs, the shell gapes are closed, reduced in size, or partially occluded by other structures (Figure 14.71). Typically the ventral margins of the two valves touch along most of their length and there is no ventral gape (Figure 14.71, *F, L, M, N,* and *P*). The posterior gape (Figure 14.71, *E*) is reduced to one to three small openings (Figure 14.71, *F* to *I*). The dorsal-most of these is commonly drawn out into the tubular rostrum (Figure 14.72). In a few rostroconchs, the anterior gape is lost and the margins of the valves touch, but most often the anterior gape is retained, although in considerably modified form, from the primitively wide, straight-sided gape (Figure 14.71; *O*). The anterior gape may become a narrow slit (Figure 14.71, *P*); keyhole shaped (Figure 14.71, *R*); partly or largely occluded by **marginal denticles** (Figure 14.71, *Q* to *U*), some of which can become greatly enlarged into **apertural shelves** (Figure 14.71, *W*); or the anterior gape may become much expanded for a large foot (Figure 14.71, *V*).

In one family of advanced rostroconchs, the posterior part of the body of the shell lateral and ventral to the rostrum is surrounded by one or two **hoods** (Figure 14.73), each of which has a tube extending the length of the hood (Figure 14.73, *A* to *J*) and opening to the outside (Figure 14.73, *G, H,* and *J*). The hood(s) is highly variable in size (Figure 14.73, *A* and *G*) and can even be larger than the rest of the shell (Figure 14.73, *A*). The hood is formed of calcareous laminae concave toward the tube (Figure 14.73, *H* and *K*), which were originally separated by open spaces. Toward the tube the open spaces were filled with prismatic outer shell layer (Figures 14.73, *K* and 14.80, *C*). The tube connects internally with the body of the shell (Figure 14.73, *J*). The mantle that secreted the hood extended the length of the tube.

Shells of the most advanced rostroconchs lack a ventral gape (Figure 14.71, *N*), have the posterior gape reduced to the aperture of the rostrum (Figure 14.71, *W*), and have the anterior gape reduced to a circular opening that is largely occluded by apertural shelves (Figure 14.71, *W*). Shells of many advanced rostroconchs are divisible into an elongate posterior rostrum, a central globular body, and an elongate anterior **snout** (Figure 14.72, *I* to *N*). The most fully developed of these forms (Figures 14.71, *N* and *W* and 14.72, *M* and *N*) are homeomorphic to scaphopods because they have an elongate tubular shell in which openings are limited to the anterior and posterior ends.

Both primitive and advanced rostroconchs can have dorsal fissurelike external structures known as **clefts** (Figures 14.71, *D* and *I* and 14.72, *F, H,* and *I*). Clefts can be longitudinal (Figures 14.71, *D* and *I* and 14.72, *F, H,* and *I*), or transverse (Figure 14.71, *D*). Clefts associated with the rostrum are rostral clefts (Figure 14.71, *D*). Clefts in front of the beak are anterior clefts and posterior to the beak are posterior clefts. Explanations of the development and function of clefts have to do with interpretations of how rostroconchs grew and are considered later.

External ornament of rostroconchs is typically molluscan and much like that of pelecypods. It may consist of growth increments, markings, or bands that parallel the valve edges (Figure 14.74, *A* to *G*). These lines and coarse markings record interruptions in the secretion of the shell at its margins during growth or changes in the rate of shell secretion. Such growth markings form a series of conformable curves at the former position of the shell margin and are called **comarginal ornament** (also spelled commarginal). The strength (width) of comarginal ornament ranges from fine growth lines (Figure 14.74, *B* to *D*), to slightly coarser increments called growth **threads** or **lirae** (Figure 14.72, *J*, snout of shell; *M*, body of shell), to still coarser markings called growth **rugae** (Figure 14.74, *E* to *G* and *I*). All rostroconchs have

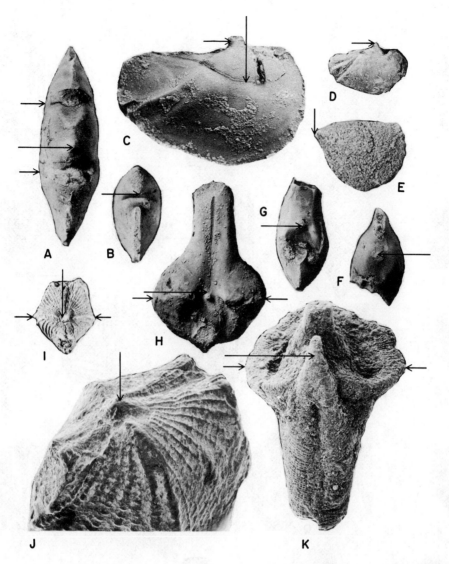

Figure 14.70. Protoconchs or beaks of various rostroconchs. **A** to **D** and **F** to **H** are natural internal molds, **E** is a silicified replica, and **I** to **K** preserve the shell. **A** to **E** are primitive rostroconchs (ribeirioids); **F** to **K** are advanced rostroconchs (conocardioids). All dorsal views have the anterior end up except **K**. **A**, *Tolmachovia*. Dorsal view showing the protoconch (long arrow) near the center and slits (short arrows) left by anterior and posterior pegmas. (Middle Ordovician, USSR, ×3.5.) **B**, *Ribeiria*. Anterior view showing a nearly terminal protoconch, (arrow) with a slit left by the pegma below the protoconch and an anterior gape. (Lower Ordovician, Texas, ×2.) **C** and **D**, *Technophorus*. Right lateral views showing an anteriorly placed protoconch (short arrows), notches left by pegmas. **C** shows side muscle scar (long arrow). (Upper Ordovician, Kentucky and Ohio; **C**, ×3.5; **D**, ×1.5.) **E**, *Apoptopegma*. Left lateral view showing a terminal protoconch (arrow). (Lower Ordovician, Australia, ×7.) **F**, *Eopteria*. Dorsal view showing a subcentral protoconch (arrow). (Lower Ordovician, Vermont, ×3.5.) **G**, *Pseudotechnophorus*. Dorsal view showing a subcentral protoconch (arrow). (Upper Cambrian, China, ×2.) **H**, *Bransonia*. Dorsal view showing a posteriorly placed protoconch (long arrow) between umbos (short arrows). (Permian, Australia, ×1.) **I** and **J**, *Pseudoconocardium*: **I**, dorsal view with the protoconch (long arrow) between umbos (short arrows) (×2); **J**, same specimen, oblique right lateral view (arrow points to protoconch), (×18). (Upper Pennsylvanian, Texas.) **K**, *Hippocardia*? Dorsal view with anterior end down showing the protoconch (long arrow) above the rostrum and between umbos (short arrows). (Pennsylvanian, Oklahoma, ×30.)

comarginal ornament, although a few have growth lines so fine that the shell appears to be smooth (Figure 14.74, *A*).

Radial or **ribbed ornament** diverges from the direction of the beak and cuts across the comarginal ornament. Patterns of radial ornament include one or two ribs in each valve (Figures 14.72, *B* to *E*; 14.74, *H, I, K,* and *N*; and 14.75, *B* and *F*), several to many ribs limited to part of the shell (Figures 14.72, *I, J,* and *M*; and 14.75, *A* and *I*), or ribs covering the entire shell (Figures 14.72, *K* and 14.75, *C* to *E, G,* and *H*). The strength of radial ornament ranges from fine (Figure 14.75, *C* to *H*) to

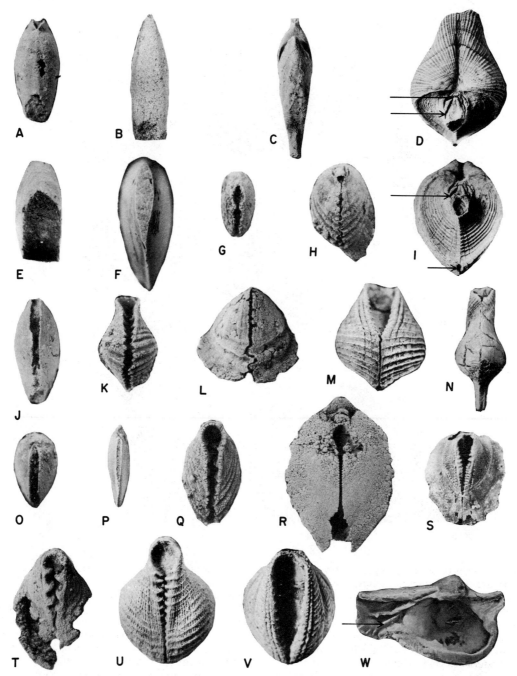

Figure 14.71. Views of valve margins of rostroconchs. A to D, Dorsal margin, anterior end up; E to I, posterior margin, dorsal side up; J to N, ventral margin, anterior end up (except L); O to V, anterior margin, dorsal side up; W, interior of the right valve, anterior end to left. Each series of views is arranged from primitive forms on the left to advanced forms on the right. B, C, E, F and P are natural internal molds; all others preserve the shell. A to C, E to G, J, O and P are primitive rostroconchs (ribeirioids); all others are advanced rostroconchs (conocardioids). **A,** *Ribeiria*. (Upper Cambrian, Australia, ×2.5.) **B,** *Ribeiria*. (Lower Ordovician, Australia, ×1.5.) **C,** *Ribeiria*. (Ordovician, locality unknown, ×1.5.) **D,** *Pseudoconocardium* showing transverse (long arrow) and longitudinal (short arrow) rostral clefts. (Upper Pennsylvanian, Texas, ×1.5.) **E,** *Heraultipegma*. (Lower Cambrian, France, ×25.) **F,** *Ribeiria*. (Ordovician, Czechoslovakia, ×2.5.) **G,** *Pauropegma*. (Lower Ordovician, Australia, ×2.) **H,** *Euchasma*. (Lower Ordovician, Australia, ×2.) **I,** *Pseudoconocardium* showing longitudinal rostral clefts (long arrow) and a small ventral aperture (short arrow). (Upper Pennsylvanian, Texas, ×1.5.) **J,** *Ribeiria*. (Upper Cambrian, Australia, ×2.5.) **K,** *Eopteria*. (Lower Ordovician, Australia, ×2.5.) **L,** *Euchasma*. (Lower Ordovician, Australia, ×2.) **M,** *Pseudoconocardium*. (Upper Pennsylvanian, Texas, ×1.5.) **N,** *Arceodomus*. (Upper Pennsylvanian, Texas, ×1.5.) **O,** *Ribeiria*. (Upper Cambrian, Australia, ×2.5.) **P,** *Pinnocaris*. (Upper Ordovician, Scotland, ×2.5.) **Q,** *Eopteria*. (Lower Ordovician, Australia, ×2.5.) **R,** *Euchasma*. (Lower Ordovician, Australia, ×2.) **S,** *Hippocardia*. (Middle Ordovician, Virginia, ×7.) **T,** *Eopteria*. (Lower Ordovician, Australia, ×2.5.) **U,** *Mulceodens*. (Upper Silurian, Sweden, ×5.) **V,** *Pseudoconocardium*. (Upper Pennsylvanian, Texas, ×1.5.) **W,** *Arceodomus* showing apertural shelves (arrow). (Upper Pennsylvanian, Texas, ×1.5.)

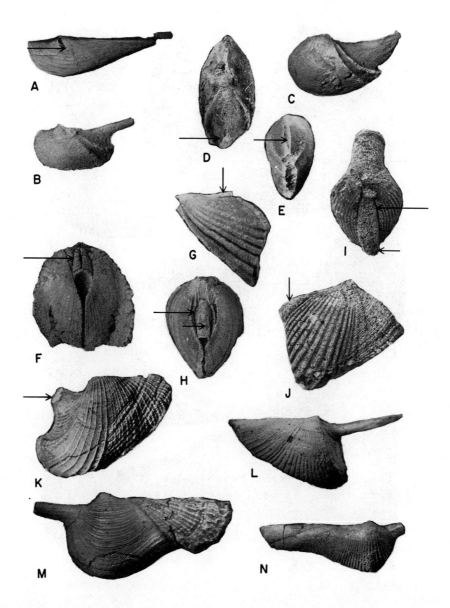

Figure 14.72. Clefts and variation in form of rostrum in rostroconchs. **A,** Artificial cast; **B** and **L,** natural molds; all others preserve shell or are silicified replicas. **A** to **E** are primitive rostroconchs (ribeirioids); **F** to **N** are advanced rostroconchs (conocardioids). **A,** *Pinnocaris.* Left lateral view showing the pegma (arrow) and rostrum. Posterior-most part of the rostrum is broken and offset above the rest of the shell. (Upper Ordovician, Scotland, ×1.5.) **B,** *Technophorus.* Left lateral view showing a long and narrow rostrum. (Upper Ordovician, Ohio, ×3.5.) **C,** *Technophorus.* Left lateral view showing a broad and long rostrum. (Upper Ordovician, Ohio, ×1.5) **D,** *Technophorus.* Dorsal view showing short ribs surrounding the rostrum (arrow). (Upper Ordovician, Kentucky, ×3.5.) **E,** *Technophorus.* Posterior view showing the narrow opening of a short rostrum (arrow) surrounded by a rib on each side. The shell is broken below the rostrum. (Middle Ordovician, Missouri, ×2.5.) **F,** *Euchasma.* Anterior view showing anterior longitudinal clefts (arrow). (Lower Ordovician, Australia, ×2.) **G** and **H,** *Euchasma.* Left lateral and dorsal views showing posterior longitudinal rostral clefts (long arrow) and rostrum (short arrows). (Lower Ordovician, Australia, ×2.) **I** and **J,**

Eopteria. Dorsal and right lateral views showing posterior longitudinal rostral clefts (long arrow), rostrum (short arrows), and division of shell into snout (comarginal ornament), body (radial ornament), and rostrum. (Lower Ordovician, Australia, ×2.5.) **K,** *Pseudoconocardium.* Right lateral view showing a short rostrum (arrow) that is below and at an angle to the dorsum of the rest of the shell. The body and snout of shell are both covered with radial ornament, but ribs on the snout are more numerous than on body. (Upper Pennsylvanian, Texas, ×1.5.) **L,** *Conocardium.* Left lateral view showing a greatly elongated rostrum that is below and at a slight angle to the rest of the dorsum. The snout and body of the shell are not readily separated from one another. (Middle Devonian, Michigan, ×2.) **M,** *Arceodomus.* Right lateral view showing a short rostrum that is below and at a slight angle to the rest of the dorsum. The shell is readily divisible into the snout (radial ornament), body (comarginal ornament), and rostrum. (Lower Permian, USSR, ×1.5.) **N,** *Conocardium.* Left lateral view with a short rostrum and greatly elongated snout. (Mississippian, England, ×1.5.)

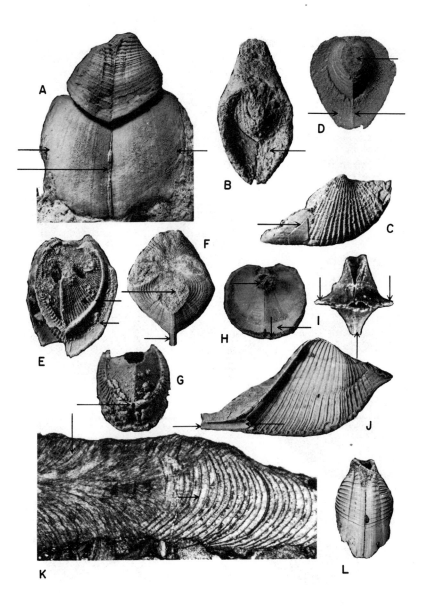

Figure 14.73. The hood of rostroconchs is only developed in some advanced forms (conocardioids). **A, D** and **F** are natural molds; all others preserve the shell or are silicified replicas. **A** to **D, F,** and **H** to **L,** *Hippocardia*; **E** and **G,** *Bigalea.* **A,** Specimen is flattened parallel to bedding. This ventral view shows the ribbed body of the shell and the greatly enlarged hood (short arrows). Between the two halves of the hood is the filling of the tube (long arrow). (Lower Devonian, New York, ×0.7.) **B,** Oblique dorsal view showing hood (arrow) surrounding lateral and ventral portions of the posterior part of the shell. (Middle Devonian, Kentucky, ×2.) **C,** Right lateral view showing unribbed hood (arrow) attached to posterior part of the ribbed body of the shell. (Middle Devonian, Kentucky, ×2.) **D,** Posterior view showing hood (right-facing short arrow) surrounding the ribbed body of the shell. Separating the two halves of the hood is the filling of the tube (long arrow). The rostral part of the posterior of the shell is broken away (left-facing short arrow). (Upper Ordovician, Estonia, ×2.) **E,** Posterior view showing two small unribbed hoods (arrows). (Middle Devonian, Michigan, ×4.) **F,** Oblique dorsal view of a specimen on which the hood is not preserved but that still has a small part of the filling of the tube preserved (short arrow). The rostrum is broken away (long arrow). (Middle Devonian, Kentucky, ×0.7.) **G,** Posterior view showing two small hoods, the inner one of which shows the opening of the tube to the outside (arrow). (Middle Devonian, Kentucky, ×4.) **H,** Posterior view showing the rostrum (short arrow) and hood (long arrow). The hood shows the aperture of tube and growth laminae (downward-facing arrow) concave toward the tube. (Middle Devonian, Kentucky, ×1.) **I,** Ventral view showing an anterior gape, right and left sides of the hood (short arrows), and the rostrum (long arrow). (Middle Ordovician, Virginia, ×7.) **J,** Interior of the left valve showing exposed internal ribs and marginal denticles on the body of the shell to the right and the hood and tube to the left; arrows mark internal and external apertures of tube. (Middle Devonian, Kentucky, ×1.5.) **K,** Cross section of part of right half of the hood showing concave calcareous laminae (short arrow) originally separated by open spaces to the right and a prismatic shell structure (long arrow) deposited close to the tube to the left. The tube (not shown) at the commissure of the hood would be toward the left. (Devonian, Ohio, ×20.) **L,** Ventral view showing the ribbed body of the shell (above) and the unribbed hood (below), which is marked only by growth lines. (Middle Devonian, Kentucky, ×1.5.)

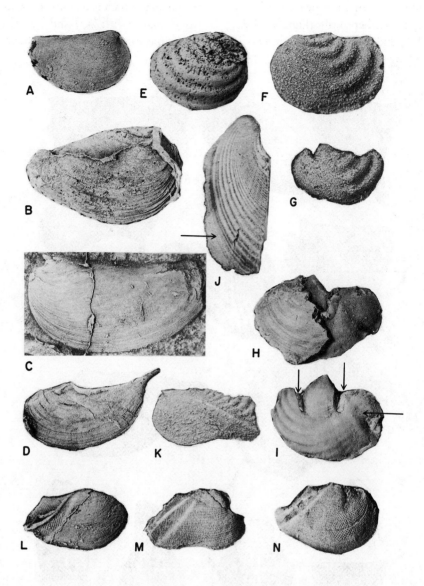

Figure 14.74. External ornament of rostroconchs. **A** to **G**, Comarginal ornament (all others show some form of radial ornament); **K** to **M**, divaricate ornament; **N**, quincunxial ornament. **B** to **D**, **F**, **G**, and **I** to **N** are either natural or artificial molds; all others preserve shell or are silicified replicas. **A** to **I** and **K** to **N** are primitive rostroconchs (ribeirioids); **J** is an advanced rostroconch (conocardioid). **A**, *Ribeiria*. Right lateral view showing smooth ornament. (Upper Cambrian, Australia, ×2.5.) **B**, *Ribeiria*. Right lateral view showing growth lines. (Ordovician, locality unknown, ×1.5.) **C**, *Ribeiria*. Left lateral view showing growth lines. (Ordovician, Czechoslovakia, ×1.5.) **D**, *Pinnocaris*. Left lateral view showing growth lines. (Upper Ordovician, Scotland, ×1.5.) **E**, *Ptychopegma*. Right lateral view showing growth rugae covering entire shell. (Lower Ordovician, Australia, ×1.5.) **F** and **G**, *Cymatopegma*. **F**, Left lateral view of shell exterior showing growth rugae limited to posterior part of shell. (×3.5.) **G**, Left lateral view of internal mold showing pegma and rugae limited to posterior part of specimen. (Upper Cambrian, Australia, ×2.5.) **H** and **I**, *Tolmachovia*. **H**, Left lateral view of a specimen preserving shell on the anterior end showing growth lines; the shell is broken away from posterior part revealing internal mold and the impression of one rib. (×3.5.) **I**, Internal mold of the left valve with two pegmas (short arrows), a single posterior rib (long arrow), and anterior rugae. (Middle Ordovician, USSR, ×3.5.) **J**, *Euchasma*. Anterior view of the right valve showing fine and coarse ribs, multiplication of ribs, and carina (arrow). (Lower Ordovician, Australia, ×2.) **K**, *Kimopegma*. Left lateral view showing a single rib with divaricate rugose ornament dorsal to the rib. (Upper Cambrian, Australia, ×2.) **L**, *Technophorus*. Right lateral view showing a combination of comarginal, radial, and fine lines of divaricate ornament to either side of the rib. (Middle Ordovician, Minnesota, ×3.5.) **M**, *Technophorus*. Right lateral view showing a combination of comarginal, radial, pitted, and fine lines of divaricate ornament between the ribs. (Middle Ordovician, New York, ×4.) **N**, *Technophorus*. Right lateral view showing a combination of comarginal, radial, and quincunxial ornament. (Upper Ordovician, Quebec, ×2.5.)

coarse (Figure 14.72, *G*, *J*, and *K*). Radial ornament in most rostroconchs is simple because during ontogeny, no increase takes place in the number of ribs (Figures 14.72, *G* to *N*; and 14.75, *C*, *E*, and *I*). A few show increase in the number of ribs during ontogeny (Figure 14.74, *J*).

Occasionally, there is a single sharply defined rib called the **carina**, which is usually associated with the umbonal ridge (Figures 14.74, *J* and 14.75, *B*). The development of carinas may have led to the formation of hoods (compare Figure 14.75, *A* and *B*).

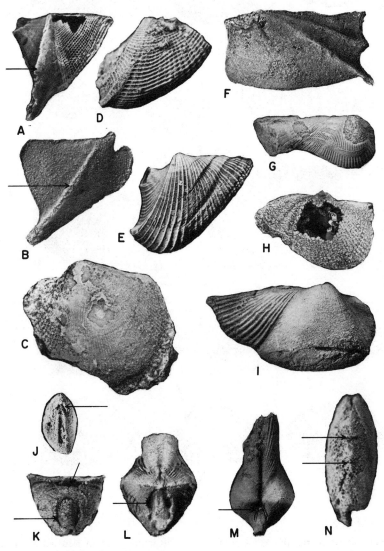

Figure 14.75. External ornament and clefts of rostroconchs. **A** to **I**, All have some form of radial ornament. **J** to **N**, All showing clefts. **B** and **F** are natural molds; all others preserve the shell or are silicified replicas. **B**, **F**, **J**, and **N** are primitive rostroconchs (riberioids); all others are advanced rostroconchs (conocardioids). **A**, *Hippocardia*. Right lateral view showing reticulate ornament on the snout and a small carina-like hood (arrow). Compare with **B**. (Middle Ordovician, Virginia, ×5.) **B**, *Oepikila*. Right lateral view showing carina (arrow). Compare with **A**. (Upper Cambrian, Australia, ×10.) **C**, *Eopteria*. Left lateral view showing fine ribs. (Lower Ordovician, Quebec, ×2.) **D**, *Bigalea*. Right lateral view showing reticulate ornament. (Devonian, Kentucky, ×4.) **E**, *Pseudoconocardium*. Right lateral view with coarse ribs covering the entire shell. A few broadly spaced, deeply incised growth lines outline broad rugae between them. (Upper Pennsylvanian, Texas, ×1.5.) **F**, *Myocaris*. Left lateral view showing two ribs. (Middle Ordovician, England, ×0.7.) **G**, *Conocardium*. Left lateral view showing coarse ribs. (Pennsylvanian, Missouri, ×1.5.) **H**, *Bransonia*. Left lateral view showing fine ribs. (Lower Ordovician, Australia, ×3.5.) **I**, *Arceodomus*, Left lateral view showing snout covered with coarse ribs. (Lower Permian, California, ×1.) **J**, *Ribeiria*. Anterior view showing anterior clefts (arrow). (Lower Ordovician, Ontario, ×1.5.) **K**, *Euchasma*. Dorsal view showing anterior transverse clefts (short arrow) and posterior longitudinal rostral clefts (long arrow). (Lower Ordovician, Australia, ×1.) **L**, *Mulceodens*. Dorsal view showing posterior longitudinal rostral clefts (arrow). (Middle Silurian, Sweden, ×5.) **M**, *Arceodomus*. Dorsal view showing posterior longitudinal rostral clefts (arrow). (Upper Mississippian, Montana, ×1.) **N**, *Pauropegma*. Dorsal view showing transverse clefts (arrows) anterior and posterior to the protoconch. (Lower Ordovician, Australia, ×1.)

The intersecting of comarginal and radial ornament of nearly equal strengths produces a pattern resembling a series of small boxes (Figure 14.75, *A* and *D*) termed **reticulate** or **cancellate ornament**. Ornament that is neither clearly comarginal nor radial is **divaricate ornament** (Figure 14.74, *K* to *M*); divaricate ornament is limited to only a part of the shell and is found associated with the ribs of species that have only one or two ribs. The strength of divaricate ornament can be described by the terms lines, threads, and rugae.

A few rostroconchs have much of the shell covered by small depressions in a quincunxial arrangement (Figure 14.74, *N*). Such depressions can be resolved into arrangements of five, with one depression being at each corner of a square and one in the center of the square as in the fives on dice. External ornament and shell shape are major criteria used to define genera in rostroconchs.

Internal features   Internal structures of rostroconchs are best seen on natural (Figure 14.76, *H* and *M*) and artificial (Figure 14.76, *C*) internal molds of the shell. Artificial molds are usually made by layering liquid latex on the fossil, each thin layer being allowed to dry before the next is added. Sufficient applications make the rubber thick enough to preserve the shape and size of the specimen as a reversed replica; the rubber can then be removed (Figure 14.76, *B*, *C*, *K*, *L*, *N*, and *O*). Sometimes internal structures can be observed on specimens that have been broken so that the inside of the shell is visible (Figure 14.76, *B*). Internal molds are negative impressions of the space inside the shell. Depending on where the shell is thickened internally, the internal mold can be considerably different from the shell exterior (Figure 14.76, *A* to *D*). External molds reverse the pattern of shell sculpture.

Sometimes both the internal and external molds of a single specimen are preserved together; the space between them is a measure of the minimum thickness of the shell (Figure 14.76, *P*). Such preservation is most common in sediments that undergo little compaction during diagenesis. Occasionally the external mold can be broken away from the internal mold; the two resulting negative impressions are referred to as **part** and **counterpart**. Most often, the part and counterpart are a combination of both internal and external features because diagenetic processes impress the external mold on the internal mold and obliterate the space between them (Figure 14.76, *J*); molds that combine a mixture of internal and external features are termed **composite molds**.

Paleozoic mollusks are usually preserved as molds because the aragonitic shell is readily dissolved. Most commonly, preservation is poor, and molds show little beyond general shape and gross internal features (Figure 14.76, *F*). Internal molds that show muscle scars are uncommon as are specimens in which the shell has been replaced by silica (Figure 14.76, *A* and *B*), phosphate, limonite, or other minerals that allow the acid extraction of fossils from carbonate rocks. Thus much knowledge of Paleozoic mollusks is derived from the relatively few specimens that were not preserved in the ordinary way.

The most prominent internal structure of primitive rostroconchs and older advanced rostroconchs is the pegma, a calcareous plate connecting the right and left valves in the upper umbonal cavity (Figures 14.69, *E*; 14.76, *L*; and 14.77, *A*, *C*, *E*, *F*, and *I*). Internal molds viewed anteriorly show the position of the pegma as a slit (Figures 14.69, *I*; 14.70, *B*; and 14.77, *G*). When viewed laterally, the pegma forms a conspicuous notch in the anterodorsal part of the shell (Figure 14.76, *F*, *J*, *K*, *M*, and *N*). The pegma can be at almost any angle from nearly horizontal to nearly vertical (Figures 14.76, *E*, *F*, and *J* to *M* and 14.77, *B*, *E*, *F*, and *H*). A few rostroconchs have two pegmas (Figures 14.70, *A*; 14.74, *I*; and 14.77, *J*); in these, the posterior end is indicated by the rostrum.

The upward- or inward-facing surface of pegmas is crossed by growth lines (Figures 14.77, *I* and 14.78, *B* to *F*), indicating the pegmas were insertion areas for muscles. These muscles probably served as **pedal retractors** (Figure 14.1, *G*) and were inserted across the plane of commissure. In rostroconchs having a single anterior pegma, a second probable pedal retractor was inserted across the dorsal midline posterior to the umbonal cavity (Figures 14.1, *G*; 14.11, *B*; 14.76, *M*; 14.78, *C* and *D*). A few rostroconchs have pegmas that are continuous with the walls of the umbonal cavity, and the anterior and posterior retractors of the foot were inserted on the walls of that cavity (Figure 14.76, *B*). Connecting the anterior and posterior pedal retractor muscle scars of primitive rostroconchs is a narrow **side muscle scar**, which may have continuous (Figures 14.70, *C*; 14.76, *C*, *D*, and *M*; and 14.78, *H*), broken (Figure 14.78, *I*), or several separate insertion areas (Figure 14.78, *G*). Side muscles may have served as accessory pedal retractors or may have attached the visceral mass to the shell.

The muscles of primitive rostroconchs are termed **median muscles** because the pedal retractors were inserted across the dorsal midline. Advanced rostroconchs do not have median muscle scars. Their muscle scars are paired, and most of them are placed laterally in each valve (Figures 14.76, *G* and *H*; 14.77, *K* and *L*; and 14.78, *K* and *O*). The larger muscle scars of advanced rostroconchs are difficult to interpret. Some may have

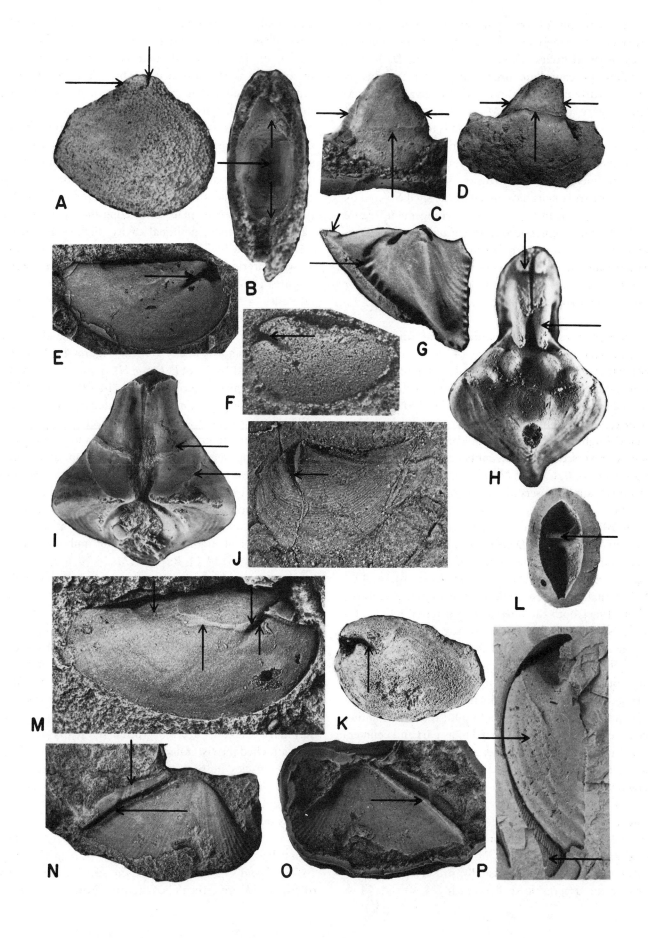

functioned as **protractors** to help pull the foot or the mantle out of the shell (Figure 14.76, *G* and *H*), some were probably pedal retractors (Figure 14.78, *J*, *K*, and *O*), and some were **mantle retractor muscles** used to withdraw the mantle from the shell margin (Figures 14.76, *I*; 14.77, *K* and *L*; and 14.78, *M* to *O*). The line of insertion of mantle retractor muscles is the **pallial line**. A reentrant in the pallial line is the **pallial sinus** (Figure 14.78, *M* to *O*), and it indicates that a structure could be protruded from and withdrawn into the shell in that region. In rostroconchs, pallial sinuses are at the anterior end.

Primitive and advanced rostroconchs are classified together for the variety of morphologic reasons just given. In addition, the ontogeny of *Pseudotechnophorus* demonstrates the relationship of the two groups in its musculature and is an apparent example of recapitulation. The larval shell of *Pseudotechnophorus* (Figures 14.70, *G* and 14.78, *L*) has the median and side muscle scars of primitive rostroconchs, whereas the adult shell has the laterally placed paired muscle scars of advanced rostroconchs (Figure 14.78, *K*).

Like other mollusks, rostroconchs had a shell composed of at least two main calcareous layers (Figure 14.80, *A*); the outer layer was probably prismatic and the inner probably nacreous. The marginal denticles on the commissure in some primitive (Figure 14.79, *A* to *D*) and most advanced (Figure 14.79, *E* to *G*) rostroconchs were formed from the outer shell layer secreted at the edge of the mantle. As rostroconchs grew, the denticles increased in length and formed internal ribs on the inside of the outer shell layer (Figures 14.76, *P* and 14.79, *K*). Internal ribs are independent of external ornament and occur in species having comarginal ornament (Figure 14.79, *B* to *D* and *K*) or external ribs (Figure 14.71, *Q* to *V* and 14.79, *J*, *N*, and *O*). In the living animals, most of the length of the internal ribs was submerged and covered by the inner shell layer secreted by the external surface of the mantle. Therefore only the youngest parts of the internal ribs projected at the shell margin as denticles (Figure

---

Figure 14.76. Modes of preservation of rostroconchs. **A** and **B**, Silicified replicas of original shell; **C** to **I**, **K**, and **M**, internal molds; **J**, composite mold; **N**, internal mold with submerged ribs of outer shell layer impressed on it; **P**, internal and external molds (part and counterpart) of the same shell; **C**, **J**, **L**, and **O**, latex replicas. **A** to **F** and **J** to **M** are primitive rostroconchs (ribeirioids); all others are advanced rostroconchs (conocardioids). **A** to **D**, *Pauropegma*. **A**, Right lateral view showing shell shape, protoconch (long arrow), and anterior transverse cleft (short arrow). **B**, Specimen with the ventral part of the shell broken away, exposing the umbonal cavity (long arrow), the anterior and posterior walls of which have growth lines marking the insertions of the anterior and posterior median muscle scars (short arrows). **C**, Right side of a latex internal mold of the umbonal cavity in **B** showing median (short arrows) and linear side (long arrow) muscle scars. **D**, Natural mold of the right valve showing impressions of median (short arrows) and linear side (long arrow) muscle scars for comparison with **B** and **C**. Note the difference in shape between **A** and **D** which are different modes of preservation of the same species. (Lower Ordovician, Australia, ×3.) **E** and **F**, *Ribeiria*. Right lateral and left lateral views of internal molds showing the shape and notch formed by the pegma (arrows); the space above the dorsum shows the approximate shape and thickness of the shell in this region. (**E**, Middle Ordovician, Czechoslovakia, ×3; **F**, Upper Cambrian, Australia, ×4.) **G** and **H**, *Hippocardia*?. Internal molds. **G**, Left lateral view showing marginal denticles (long arrow), which become enlarged anterodorsally forming apertural shelves, and probable protractor muscle scar (short arrow). (×3.) **H**, Dorsal view of **G** showing filling of the funnel-shaped spaces (see Figure 14.79, *M*) formed by the apertural shelves (long arrow) and probable protractor muscle scars at the anterodorsal tip of the specimen (short arrow). (Devonian, Germany, ×3.5.) **I**, *Bransonia*?. Dorsal view showing raised impressions of the anterior and posterior branches of the Y-shaped pallial line (arrows) (see Figure 14.77, *K*). (Middle Devonian, New York, ×3.) **J**, *Technophorus*. Composite mold of the left valve showing the notch formed by the internal pegma (arrow) and the external ornament. (Ordovician, Czechoslovakia, ×3.) **K** and **L**, *Ribeiria*. **K**, Internal mold of the left valve showing the notch formed by the pegma (arrow). **L**, Latex cast of the anterior end of **K** looking anteriorly showing the pegma (arrow) as a nearly horizontal plate connecting the two valves of the dissoconch. The umbonal cavity is above the pegma (see Figures 14.69, *I* and 14.70, *B*). (Lower Ordovician, Texas, ×2.) **M**, *Ribeiria*. Internal mold of the right valve showing the notch formed by the pegma (upward-facing short arrow), the lateral-most parts of the anterior and posterior median muscle scars (downward-facing short arrows), and the linear side muscle scar (long arrow). (Middle Ordovician, Czechoslovakia, ×4.) **N** and **O**, *Pseudotechnophorus*. **N**, Internal mold of the left valve, with submerged internal ribs of the outer shell layer impressed on it, showing a greatly elongated notch (long arrow) formed by a much elongated anterior pegma. About halfway down on the mold of the space above the pegma are a few growth lines (short arrow), suggesting a muscle insertion. The muscle may have been used for retracting the soft parts that filled this space. **O**, Latex cast of **N** reproducing structures as they would have looked on the inside of the valve. The arrow marks the pegma (Upper Cambrian, China, ×3.) **P**, *Euchasma*. Anterior view of a specimen that preserves both the internal mold to the right (right-facing arrow) and the external mold to the left (left-facing arrow). The internal mold is marked by the internal ribs formed as the marginal denticles elongated; dorsally the internal ribs become less distinct because they were being buried by the inner shell layer. The space between the internal and external molds shows the minimum thickness of the original shell, which was very thick in the umbonal region. (Lower Ordovician, Australia, ×3.)

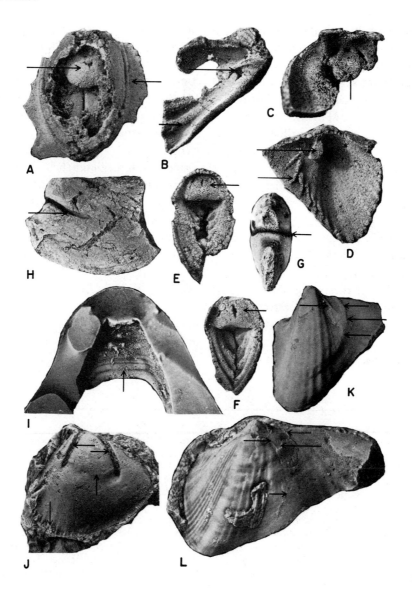

Figure 14.77. Pegmas and muscle scars. **A** to **F**, Silicified specimens; all others are internal molds. **G** to **J** are primitive rostroconchs (ribeirioids); all others are advanced rostroconchs (conocardioids). **A**, *Euchasma*. Posterior part of the shell is broken away. The large circular structure near the top is the pegma (long arrow) and covers the inside of the top of the keyhole-shaped anterior gape seen in Figure 14.71, *R*. Around the margin of the shell, much of the umbonal carina (short arrow) is preserved. (Lower Ordovician, Australia, ×2.) **B**, *Euchasma*. Interior of the left valve showing the subhorizontal pegma (long arrow) in anteroposterior section and the marginal denticles and internal ribs (short arrow), which disappear anterodorsally where they are covered by the inner shell layer. (Lower Ordovician, Australia, ×2.) **C**, *Euchasma*. Much of the shell is missing, but the left umbo and pegma (arrow) are preserved. This view shows the pegma from its posterior side, the same as in **A**, with only four point attachments rather than being attached along most of its length as in **E**, **F** and **I**. (Lower Ordovician, Malaysia, ×2.) **D**, *Eopteria*. Interior of the right valve showing the vertical pegma (long arrow) in dorsoventral section and the marginal denticles (short arrow). (Lower Ordovician, Australia, ×2.5.) **E** and **F**, *Eopteria*. Posterior parts of shells are broken away. The large semicircular structures near

the tops of the figures are the vertical pegmas (arrows). (Lower Ordovician, Australia, ×2.5.) **G**, *Ribeiria*. Anterior view of internal mold showing the slit (arrow) left by a nearly horizontal pegma. (Ordovician, Czechoslovakia, ×2.) **H** and **I**, *Ribeiria*. (Ordovician, Czechoslovakia.) **H**, Left lateral view of the anterior part of the internal mold showing the subhorizontal notch formed by the pegma (arrow). (×1.5.) **I**, Latex mold of the notch formed by the pegma shown in **H**. Note the growth lines on the posterodorsal face (arrow), indicating that the pegma was the site of a muscle insertion. (×2.5.) **J**, *Ischyrinia*. Right lateral view showing notches of two pegmas (short horizontal arrows), marginal denticles (downward-facing arrow), and side muscle scar (upward-facing arrow). (Upper Ordovician, Quebec, ×1.5.) **K**, *Bransonia*?. Right lateral view showing Y-shaped pallial line (short arrows). At the junction of the three branches is a muscle scar (long arrow), which may have been the site of a muscle used to retract structures that could be protruded through the anterior gape (compare with Figure 14.76, *I*) (Middle Devonian, New York, ×2.) **L**, *Bransonia*. Right lateral view for comparison with **K**. This specimen has a Y-shaped (short arrows) pallial line with a large muscle scar (long arrow) at the junction of the branches. (Permian, Australia, ×2.)

Figure 14.78. Legend on p. 374.

14.77, *D*). Where the inner shell layer does not completely submerge internal ribs, traces of them can extend well up toward the dorsal margin (Figures 14.76, *P* and 14.77, *B*, *D*, and *K*). During preservation of some specimens, differential solution of the inner shell layer took place before complete lithification of the internal mold (Figure 14.79, *F*), and the ribbed inner surface of the outer shell layer may be imprinted on the internal mold (Figure 14.79, *I* and *L*).

Classification

Although Rostroconchia is sometimes divided into three orders, it now seems best to recognize two. Primitive rostroconchs are placed in the Ribeirioida, and advanced rostroconchs are placed in the Conocardioida. Until recently, ribeirioids were usually classified as bivalved arthropods and conocardioids, as unusual pelecypods.

ORDER RIBEIRIOIDA (Figures 14.72, *A* to *E*; 14.74, *A* to *I* and *K* to *N*; 14.80, *D* to *F*)

Rostroconchs with all shell layers continuous across dorsum, one or two pegmas, and usually posteriorly elongated. Dorsal clefts present in some, mostly anterior to protoconch. Musculature high in shell; consists of anterior and posterior median pedal retractor muscles connected by right and left side muscles; pallial line in a few. (Earliest Cambrian to latest Ordovician; Early Silurian (?); about 20 known genera.)

ORDER CONOCARDIOIDA (Figures 14.72, *F* to *N*; 14.73; 14.80, *A* to *C*)

Rostroconchs with only one shell layer continuous across dorsum in most, pegma usually absent, and shell usually anteriorly elongated. Dorsal clefts present in most, usually posterior to protoconch, but may be both anterior and posterior. Musculature of adult stage consists of laterally placed paired pedal retractors and other muscles; pallial line in most. (Latest Cambrian to latest Permian; about 14 known genera.)

Functional morphology

Mode of life Because the valves of rostroconchs were rigidly joined dorsally and could not be opened at will, all rostroconchs have one or more permanent apertures in the shell margins. These openings allowed the animal to extend its foot to the outside and move or burrow, to bring in feeding and respiratory currents, to remove body wastes and foreign or unusable particles called **pseudofeces**, and to shed sex cells. In the primitive ribeirioids, the shell gapes along all the margins of the dissoconch (Figure 14.71, *E*, *J*, and *O*), but more highly evolved

---

Page 373.

Figure 14.78. Muscle scars of rostroconchs. **M** is silicified; all others are molds. **A** to **I** are primitive rostroconchs (ribeirioids); all others are advanced rostroconchs (conocardioids). **A** and **B**, *Pauropegma.* **A**, Dorsal view showing raised posterior median pedal retractor scar (arrow). **B**, Latex mold of the umbonal cavity of a shelled specimen showing growth lines marking the site of insertion of the anterior median pedal retractor muscle (arrow). (Lower Ordovician, Australia, ×3.) **C** and **H**, *Technophorus.* **C**, Dorsal view showing slit marking the site of the pegma (short left-facing arrow), protoconch (long left-facing arrow), growth lines marking the site of insertion of the small posterior median pedal retractor muscle just above the posterior gape (upward-facing arrow), and side muscle scar (right-facing arrow). **H**, Right lateral view of the same specimen showing the linear side muscle scar (arrow). (Middle Ordovician, USSR, ×3.) **D**, *Ribeiria.* Dorsal view showing slit marking the site of the pegma (short arrow) and growth lines marking the large posterior median pedal retractor muscle insertion (long arrow). (Middle Ordovician, Portugal, ×2.) **E** and **F**, *Tolmachovia.* Posterior views showing growth lines (arrows) marking the sites of insertions of the posterior median pedal retractor muscles. (Middle Ordovician, USSR, ×5.) **G**, *Ribeiria.* Right lateral view of an incomplete specimen showing the notch of the pegma (left-facing arrow) and separate side muscle insertions and tracks (upward-facing arrows). The muscle scars in this species were used to reconstruct the muscles in Figure 14.81. (Lower Ordovician, Alberta, Canada, ×4.) **I**, *Ischyrinia.* Left lateral view showing interrupted side muscle insertion (arrow). (Middle Ordovician, Norway, ×3.) **J**, *Conocardium.* Right lateral view of an incomplete specimen showing a probable pedal retractor muscle scar at the top of the umbo (arrow). (Mississippian, England, ×3.) **K** and **L**, *Pseudotechnophorus.* (Upper Cambrian, China.) **K**, Right lateral view of a postlarval shell showing several laterally placed, probable pedal retractor muscle scars (arrows). (×3.) **L**, Enlargement of the protoconch of the specimen illustrated in Figure 14.70, *G* showing the anterior median pedal retractor muscle scar (downward facing arrow) and the left side muscle scar (left-facing arrow). (×60.) **M** and **N**, *Euchasma.* **M**, Posterior part of the specimen broken away and right side raised up toward viewer. In the center of the figure is the denticulate lower part of the keyhole aperture (oblique arrow) shown in Figure 14.71, *R*. To the left of this is the pallial line (right-facing arrow), the dorsal part of which shows the reentrant of the pallial sinus (left-facing arrow). (Lower Ordovician, Australia, ×3.) **N**, Anterior view showing the pallial line (short arrows) and pallial sinus (long arrow). The fine lines to the outside of the pallial line insertions are the muscle tracks of the radial mantle muscles. (Lower Ordovician, Newfoundland, ×2.) **O**, *Eopteria.* Right lateral view showing a large probable pedal retractor muscle scar (right-facing short arrow), pallial line (right-facing long arrow), and pallial sinus (left-facing arrow). (Lower Ordovician, Vermont, ×5.)

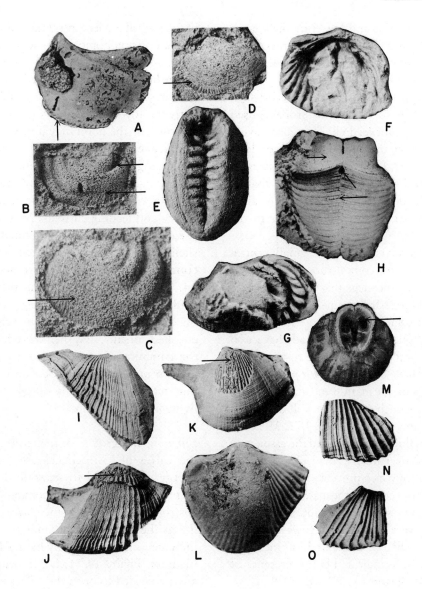

Figure 14.79. Marginal denticles and internal ribs. **A** to **D** are primitive rostroconchs (ribeirioids); all others are advanced rostroconchs (conocardioids). **A,** *Technophorus.* Internal mold of left valve showing marginal denticles (arrow) along ventral and anterior sides (Lower Ordovician, Australia, ×5.) **B** and **C,** *Cymatopegma.* **B,** Internal mold of the right valve showing the pegma (short arrow) and marginal denticles (long arrow) along anteroventral side. (×3.5.) **C,** Exterior of left valve showing ornament that has been decorticated anteroventrally, exposing marginal denticles and internal ribs (arrow). (×5.) (Upper Cambrian, Australia.) **D,** *Tolmachovia?* Left valve composite mold showing comarginal ornament and marginal denticles (arrow). (Lower Ordovician, Australia, ×1.5.) **E** to **G,** *Mulceodens.* (Middle Silurian, Sweden.) **E,** Anterior view showing marginal denticles blocking most of gape. (×5.) **F,** Interior view of right valve showing internal ribs. (×3.5.) **G,** Interior view of the left valve showing marginal denticles. (×3.5.) **H,** *Hippocardia.* Internal view of the ventral part of the dissoconch. The inner shell layer dissolved away, and only the outer shell layer structures are preserved and silicified. Extending the length of the center of the figure is the ventral commissure with interlocking denticles (left-facing arrow). The

small opening near the top (oblique arrow) is the entrance to the tube of the hood. The hood (right-facing arrow) lacks both radial ornament and internal ribs (see Figure 14.73, *J*). (Middle Devonian, Kentucky, ×2.) **I,** *Hippocardia.* Left lateral view of an internal mold on which have been impressed the internal ribs. (Middle Devonian, Michigan, ×1.5.) **J,** *Hippocardia.* Right lateral view of a specimen from which most of the shell (arrow) has broken away, revealing the impressions of the internal ribs. (Middle Devonian, Kentucky, ×1.5.) **K,** *Arceodomus.* Right lateral view of an incomplete specimen from which some of the comarginal ornament on the body of the shell has been weathered away, revealing the outer surfaces of the internal ribs (arrow). (Upper Mississippian, Montana, ×2.) **L,** *Eopteria.* Internal mold of the right valve showing the imprint of internal ribs. (Lower Ordovician, Texas, ×2.) **M,** *Arceodomus.* View of anterior gape showing the funnel-shaped spaces (arrow) formed by longitudinal shelves (see Figures 14.71, *W* and 14.76, *H*). (Upper Pennsylvanian, Texas, ×1.5.) **N** and **O,** *Hippocardia.* Fragment of silicified shell showing external radial ornament on the outside (**N**) and internal ribs on the inside (**O**). (Middle Devonian, Kentucky, ×1.5.)

ribeirioids have only anterior and posterior shell gapes (Figure 14.71, *F* and *P*). In a few ribeirioids there is only a posterior shell gape; presumably these animals were immobile once they acquired the adult shape.

Conocardioids ordinarily have a large anterior shell aperture (Figure 14.71, *Q* to *V*) that was probably used for moving and, in most, for feeding. They also have a small posterior orifice at the end of a short or long rostrum (Figures 14.72, *H, I* and *K* to *N*). The rostrum is analogous to the posterior end of scaphopods in that its opening was probably too small for suspension feeding. Probably the rostrum served to circulate water to the gills.

Some advanced rostroconchs have a small **ventral aperture** in the posteroventral commissure (Figure 14.71, *I*). In some advanced rostroconchs this ventral aperture occurs at the end of a long narrow tube on either side of which is the hood surrounding the rostral area (Figure 14.73). A few rostroconchs with two hoods have two ventral apertures below the opening of the rostrum. The ventral apertures may have been outlets for pseudofeces carried to them by cilia on the mantle and visceral mass. A few advanced rostroconchs that have well-developed apertural shelves (Figures 14.71, *W* and 14.79, *M*) lack a ventral aperture; probably the shelves and the associated ciliated mantle folded over them provided an effective food-sorting mechanism, which prevented sediment and other foreign matter from entering the mantle cavity. The shelves also block the only sizable shell gape, so it is likely that these animals were immobile. However, most rostroconchs probably had a muscular foot and were mobile or could burrow as indicated by the presence of probable pedal muscles (Figures 14.77 and 14.78).

Prominent pallial lines in some rostroconchs (Figure 14.78, *M* to *O*) indicate that they could retract the edges of the mantle. Anterior pallial sinuses in a few indicate the existence of protractible anterior structures, which may have been mantle tissue or tentacles used for deposit feeding. Species having small rostral openings were probably also deposit feeders. Forms having large posterior gapes and those with closed anterior margins were probably suspension feeders. Most rostroconchs were probably infaunal or semi-infaunal (Figure 14.68), but a few that had broadly flattened and anteriorly carinate shells were probably epifaunal (Figures 14.68, *K*; 14.71, *R*; and 14.77, *A* and *B*).

Function of the hood   The hood occurs in some conocardioids that range in age from Ordovician to Mississippian. The hood is a structure composed only of the outer shell layer. It contained living secretory mantle tissue only along its central tube (Figure 14.80, *C*), which is in a position homologous with the ventral aperture in some other conocardioids (Figure 14.71, *I*).

The hood could have evolved as a device to support the shell in soft substrates and/or could have deflected water currents toward or away from the aperture of the rostrum, or the hood may not be of primary functional significance. The significant structure may be the tube containing the secretory mantle that formed the hood. The elongated tube might have been used to eliminate pseudofeces brought into the mantle cavity by respiratory and feeding currents. Most pelecypods eliminate pseudofeces by sudden and rapid contractions of the shell-closing muscles, an option not available to rostroconchs. If the ventral aperture of rostroconchs (Figure 14.71, *I*) evolved into a long thin tube, rostroconchs could maintain the tube during growth only by producing a thin curved structure on each valve. As the valves enrolled dorsally, new shell was added along the length of the commissure of the tube during growth; continued growth produced the hood. Once the hood evolved, it may have preadapted those rostroconchs that had it to living in soft sediment.

Opening and growth of the rostroconch shell   Because the valves of rostroconchs are rigidly joined dorsally, energy was needed to separate them as growth increments were added at the commissure. Growth lines in rostroconchs show that the valves opened slowly during life as the shell enlarged, and the ventral edges of early increments may eventually have a spread of 180 degrees or more (Figure 14.71, *D*). In pelecypods, much of the energy needed to open the valves is stored in the dorsal elastic ligament, which, depending on its structure, is either compressed or stretched when the shell-closing muscles contract. In rostroconchs, this energy could be supplied in several not necessarily exclusive ways. Although applied in small amounts for long periods, the energy was sufficient to rupture parts of the dorsal margin in most rostroconchs (Figure 14.71, *D* and *I*), forming clefts.

The most obvious primary source of mechanical energy in any metazoan is its musculature. Energy produced by the contraction of the muscles could be transmitted hydrostatically to the shell either through: fluids in the hemocoel or by sea water in the mantle cavity. If the volume of the blood in the pedal hemocoel of the foot could be kept constant (as it can be in pelecypods), the foot could be protracted between the valve edges by contraction of its intrinsic muscles and then expanded by means of the pedal retractors inserted on the shell. This

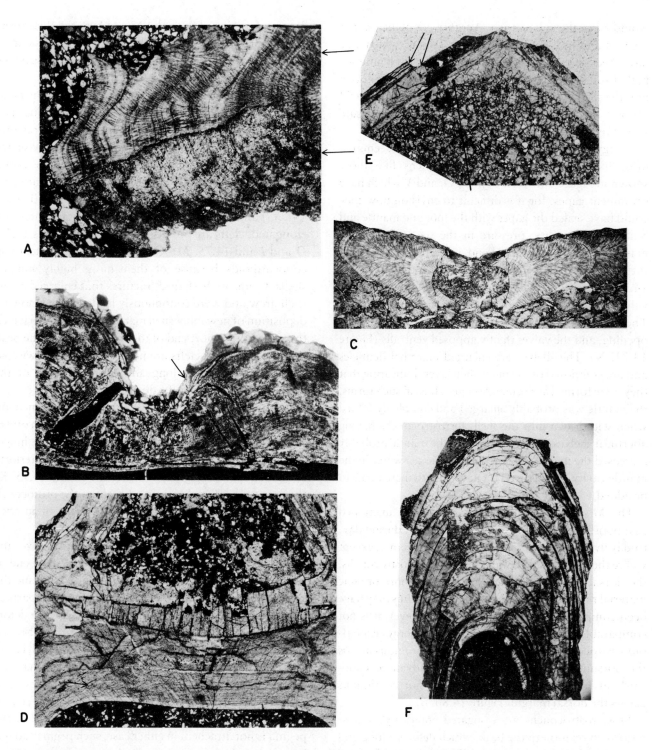

Figure 14.80. Microstructure of the rostroconch shell. **A** to **C** are concocardioids; **D** to **F** are ribeirioids. **A**, *Bransonia*. Section of shell edge showing outer prismatic layer (upper arrow) and recrystallized, probably originally nacreous inner layer (lower arrow); growth lines extend across both layers. (Permian, Australia, ×20.) **B**, *Pseudoconocardium*. Transverse section of anterior part of the hinge showing progressive deformation of the inner shell layers from the youngest (bottom) to oldest (top). Arrow points to outer discontinous shell layer. (Upper Pennsylvanian, Texas, ×20.) **C**, *Hippocardia*. Cross section of the hood showing the tube (center) and prismatic structure of the hood adjacent to the tube. (Devonian, Michigan, ×7.5.) **D** to **F**, *Ribeiria*. (Ordovician, Czechoslovakia.) **D**, Transverse section through the pegma showing an upper fractured layer (upper arrow) underlain by younger unfractured layers (lower arrow). The space above the pegma is the umbonal cavity; the space below the pegma is the mantle cavity. (×20.) **E**, Transverse section of posterodorsal margin of the shell showing three main layers (arrows). (×7.5.) **F**, Transverse section of the dorsal margin of the shell in front of the pegma showing shell layers continuous across the dorsal margin. (×7.5.)

would force the valves apart. Another possibility is that the foot and/or hypertrophied mantle tissue (which may have been present in rostroconchs that had anterior pallial sinuses or apertural shelves) could be withdrawn into the shell by appropriate muscles. Because this would prevent sea water from escaping, hydrostatic pressure would tend to open the valves.

Energy transmitted by fluids in the hemocoel seems the most likely explanation for rostroconchs like those shown in Figure 14.71, *A, E, J, M, O,* and *V* which have prominent gapes, for it is difficult to envision how they could have sealed the gapes with the foot and mantle and built up hydrostatic pressure in the sea water of the mantle cavity. On the other hand, it is difficult to see how forms that had apertural shelves (Figures 14.71, *W* and 14.79, *M*) could have had a foot large enough to open the valves, as the anterior gape is almost completely blocked. The forms that have apertural shelves have a small rostral opening, and the valves tightly apposed ventrally (Figure 14.71, *N*). The shelves are enlarged marginal denticles and are composed of the outer shell layer, indicating that they were formed at the mantle edge. Thus, in such forms, the mantle was probably enlarged and complexly folded when withdrawn into the shell. In rostroconchs having apertural shelves, the withdrawal of the mantle probably increased the hydrostatic pressure of the sea water in the mantle cavity and allowed for new growth increments to be added.

The dorsal shell margin of all rostroconchs functioned as a poorly elastic hinge during growth, but during day-to-day living, the valves were rigidly held together, except where they gaped. In a few of the advanced rostroconchs, the dorsal margin may have had slightly more organic material than in more primitive forms and thus may have been compressed a little more easily; however, it is not comparable to a ligament, which stores energy. Ribeirioid rostroconchs have all shell layers continuous across the dorsum (Figure 14.80, *F*), but in advanced conocardioids the thin outer shell layer is not continuous across the dorsal margin (Figure 14.80, *B*).

Most rostroconchs are elongated anteriorly or posteriorly in relation to the beak, which defines a long and short part of the dorsal margin (Figure 14.85). The long part of the dorsum is ordinarily straight. (Figure 14.69, *B, H,* and *J*). The short part of the dorsum is usually a little less elevated than the long part and is more or less parallel to the long part rather than being colinear with it. Sometimes the short part of the dorsum is at a slight to conspicuous angle to the long part (Figure 14.72, *K, M,* and *N*). During growth, the long straight part of the dorsum functioned as the hinge and here the shell layers are compressed (Figure 14.80, *B*). The long dorsal margin is bent inward as new shell is added at the growing margin, but it does not ordinarily fracture.

As the long part of the dorsum enrolls along the hinge axis, the short depressed part is not functioning as a hinge. It is pulled apart, and clefts or tensional fractures are formed (Figures 14.71, *D* and 14.75, *M*) between the hinge and the shorter less elevated parts of the dorsal margin (Figure 14.81). Thus in ribeirioids that are posteriorly elongated, clefts are anterior in position (Figures 14.75, *J* and 14.81); in conocardioids that are anteriorly elongated, clefts are posterior in position (Figures 14.71, *D* and *I* and 14.75, *M*). Clefts form more frequently in conocardioids because of their more highly inflated shells. Clefts are V-shaped fractures that enlarged as the shell grew and were continuously healed from below by deposition of new inner shell material. In species in which the gape of the short end of the shell extends to the beak (Figure 14.71, *A*), clefts are not developed. Likewise, in species in which the long and short portions of the dorsum are colinear, clefts do not form.

In some equidimensional species, clefts formed on either side of the protoconch (Figure 14.75, *N*). In others (Figure 14.72, *I*) the anterior end functioned as the hinge, and only rostral clefts were formed. In a few posteriorly elongated species (Figures 14.72, *F* and *H* and 14.75, *K*) the area of the dorsal margin adjacent to the protoconch functioned as the hinge, and clefts formed both anterior and posterior to the protoconch.

Ribeirioids and primitive conocardioids have the valves connected by a dorsal pegma on the short end of the dorsal margin. As the long end is compressed and the short end is pulled apart during growth, the older parts of the pegma are deformed by tensional stresses; the deformation in thin section resembles that of fault-block mountains (Figure 14.80, *D*). New horizontal shell layers are added below the deformed parts of the pegma. In a few primitive conocardioids (Figure 14.77, *A* to *C*) the pegma is not attached along its full length laterally; either there are point attachments or the posterior end of the pegma is not attached. In either case, such pegmas do not become cracked during growth. Apparently the areas or points of attachment could be resorbed dorsally and added to ventrally. Growth of forms having two pegmas is not well understood because of the lack of adequate study material. The tensional deformation shown in the clefts and pegma was probably a gradual process rather than a sudden event. Continuous inner shell probably always existed as a layer below the fractured parts.

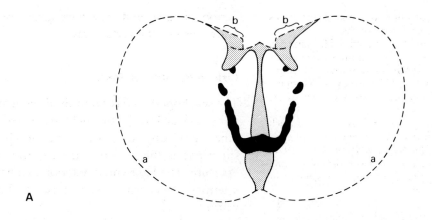

A

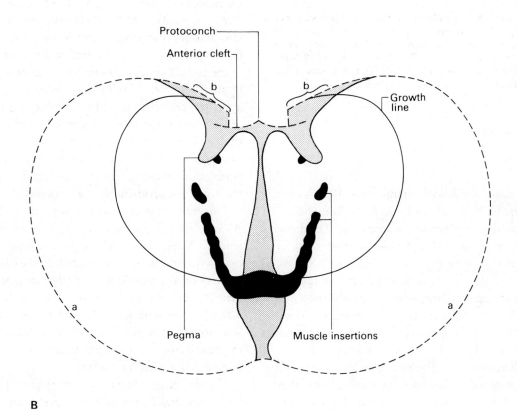

Protoconch

Anterior cleft

b   b

Growth line

a   a

Pegma   Muscle insertions

B

Figure 14.81. Paper cutout models to illustrate the growth of the ribeirioid shell. Drawings are based on internal molds of the Lower Ordovician species *Ribeiria lucan* (Figure 14.78, *G*). Black areas represent posterior median pedal retractor and side muscles; stippled areas show thickness of shells along the dorsal margins of completed models. Compare models to see how the anterior clefts enlarged mechanically during growth because the shell at the ventral edges of the clefts could not be resorbed. The ventral edges of each fracture enlarged as growth continued so that the long and short parts of the dorsum remained the same proportional distance apart, irrespective of the size of the shell.

Growth line on model **B** is the same size as the entire model **A**. *Instructions:* (1) Photocopy page ; (2) cut each model from photocopy, cut along *all* dashed lines; (3) construct each model separately; (4) in **A**, staple the points marked *a* together, fold model so the muscles are on the outside, do not crease dorsum; (5) in **A**, use transparent tape to fix edges marked *b* together; (6) flatten area below protoconch by gently pushing taped area inward; (7) repeat steps 4 to 6 for model **B**. (From Pojeta, J., Jr.; Runnegar, B. *U.S. Geological Survey Professional Paper.* 968: 6; 1976. This figure is not copyrighted and may be photocopied without permission.)

Supplementary reading

Pojeta, J., Jr. Geographic distribution of Cambrian and Ordovician rostroconch mollusks, pp. 27–36. In: Gray, J.;

Boucot, A. J., editors. *Historical biogeography, plate tectonics, and the changing environment.* Corvallis, OR: Oregon State University Press; 1979. Discusses the stratigraphic, lithologic, and paleobiogeographic distributions of Cambrian and Ordovician rostroconchs.

Pojeta, J., Jr.; Gilbert-Tomlinson, J.; Shergold, J. H. Cambrian and Ordovician rostroconch molluscs from Northern Australia. *Australian Bureau of Mineral Resources, Geology, and Geophysics, Bulletin* 171; 1977. Study of taxonomy and stratigraphic distribution of rostroconchs across the Cambrian—Ordovician boundary.

Pojeta, J., Jr.; Runnegar, B. The paleontology of rostroconch mollusks and the early history of the phylum Mollusca. *U.S. Geological Survey Professional Paper* 968; 1976. Monographic study of the functional morphology and taxonomy of the class and the significance of rostroconchs in early molluscan evolution.

Pojeta, J., Jr.; Runnegar, B.; Morris, N. J.; Newell, N. D. Rostroconchia: a new class of bivalved mollusks. *Science* 177: 264–267; 1972. Original proposal of this group as a class-level concept.

Runnegar, B. Origin and evolution of the class Rostroconchia. In: Yonge, M.; Thompson, T. E., editors. *Evolutionary systematics of bivalve molluscs. Royal Society of London Philosophical Transactions B* 284: 319–333; 1978. Summary of rostroconchs, their ecology, and their importance in understanding early molluscan phylogeny.

Class Scaphopoda

John Pojeta, Jr.

This small class of Mollusca contains about 1000 species ranging in age from Middle Ordovician to Holocene. All are marine benthos living with the anterior half to three-fourths of the tusklike shell embedded in mud, sand, or gravel in water as deep as 6200 m (Figure 14.82, A). Scaphopods are deposit feeders; most eat benthic foraminifers. Scaphopods about 40 mm long are known to eat foraminifers as much as 0.5 mm in diameter. Many scaphopods are 30 to 70 mm long, but total size range is 2 mm to 130 mm long. Holocene scaphopods have a worldwide distribution and occur in all oceans from cold to tropical water environments; in some places, they occur in vast numbers. Fossil scaphopods are major constituents of occasional shell beds in rocks as old as Permian.

The most recently proposed classification of scaphopods recognizes about 50 genus-level taxa. From the Ordovician through the Mississippian, known diversity was at its lowest level, and only one or two genera occur per system. No unequivocal record of the class is known in Silurian rocks. In both the Pennsylvanian and Permian, three to four genera are recognized. Five or six genera are known from both Triassic and Jurassic rocks. Nine genera have been recognized in the Cretaceous. Paleogene rocks have so far yielded about 13 genera. About 29 genera have been identified from the Neogene. Scaphopods show a modest diversification through a long period of time, and they have given rise to no other known group of invertebrates.

Orientation and anatomy

The scaphopod shell is a univalved, elongated cone open at both ends. Usually the shells are curved and regularly tapering (Figure 14.82, A), but they may be straight. The small end is the posterior **apex**, bearing the **posterior aperture**. The large end is anterior, and has the **anterior aperture**; the concave side is dorsal, and convex side is ventral.

A mantle secretes and lines the entire shell and is conical in form (Figure 14.82, A). The first formed mantle is dorsal in position, but early in ontogeny it develops two lateral lobes that grow ventrally, fuse at the ventral midline (Figure 14.82, C), and secrete the tusklike shell open at the ends. Virtually all subsequent growth of the shell is by the addition of tubular increments at the anterior aperture; the shell may also be thickened internally. At the posterior aperture, resorption or dissolution of the shell enlarges the opening as the animal grows to maintain the circulation of water for respiration and the elimination of feces. This enlargement destroys the larval shell, and the protoconch is seldom seen.

The mantle cavity is ventral in position (Figure 14.82, A) and has a slight constriction about midway along its length. Water does not flow continuously through the mantle cavity. Water is brought into the cavity by ciliary activity on the mantle through the small posterior aperture. Plungerlike contractions of the foot then drive spent water and feces back out through the posterior opening. Scaphopods lack gills, and gas exchange occurs through the vascularized walls of the mantle cavity. Feces are not compacted into pelletal form.

The mouth is a horizontal slit (Figure 14.82, D) and lies at the end of the muscular **proboscis**, which is the anterior part of the gut and is located on the posterodorsal side of the foot (Figures 14.82, E and 14.83, E to H). To either side of the proboscis, there is a large lobe from which arise numerous prehensile feeding tentacles known as **captacula** (singular, captaculum) (Figures 14.82, D and 14.83, E). Some species have as many as 150 captacula on each side of the proboscis (Figure 14.82, D), and they can break off or be autotomized and new ones grown. Captacula can be greatly extended and protruded by ciliary action and hydrostatic inflation; they are withdrawn by longitudinal muscle fibers. The gut is complete, and the anus opens into the mantle cavity within the shell about halfway back on the ventral side (Figure 14.82, A).

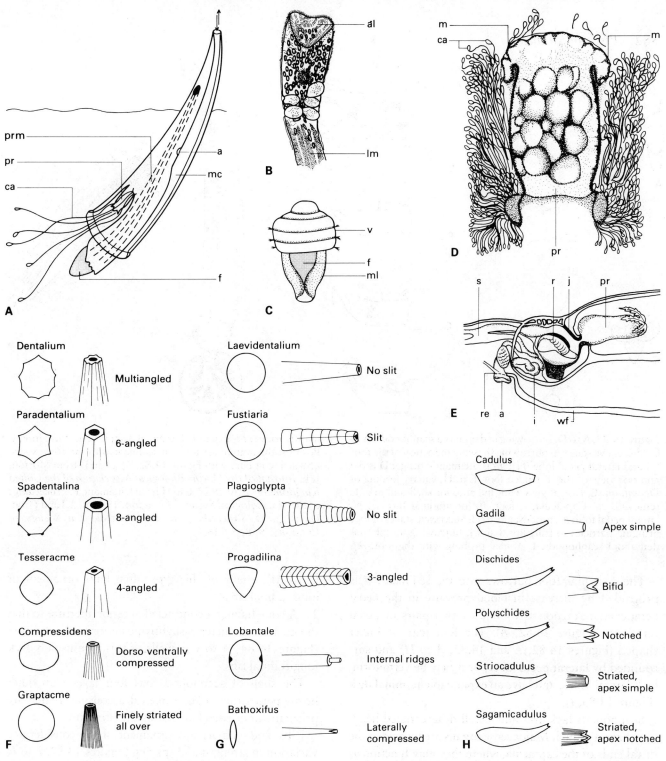

Figure 14.82. A, Major internal features of a living dentalioid scaphopod in life position drawn as though the shell were transparent. **B**, Tip of captaculum of *Dentalium*. **C**, Veliger larva of *Dentalium* showing paired lateral mantle folds touching at midventral line. **D**, Anterior portion of gut showing the captacula and proboscis filled with foraminifers; dorsal mantle and shell removed. **E**, Diagram of the base of the foot and digestive tract of *Dentalium*. **F** to **H**, Variation in cross section, ornament, and apical structures of scaphopods. **F**, Ribbed dentalioids; **G**, dentalioids ornamented with comarginal growth increments; **H**, siphonodentalioids. *a*, Anus; *al*, alveolus; *ca*, captacula; *f*, foot; *i*, intestine; *j*, jaw; *lm*, longitudinal muscles; *m*, mouth; *mc*, mantle cavity; *ml*, mantle lobe; *prm*, pedal retractor muscles; *pr*, proboscis; *r*, radula; *re*, rec-

tum; *s*, stomach; *v*, velum; *wf*, ventral wall of foot. (**A** from Stasek, C. R. In: Florkin, M.; Scheer, B. T., editors. *Chemical zoology, Mollusca*, vol. 7. New York: Academic Press, Inc.; 1972, p. 34. **B** modified from Morton, J. E. *Journal of the Marine Biological Association of the United Kingdom* 38: 230; 1959. **C** from Pelseneer, P. In: Lankester, E. R., editor. *Treatise on zoology, Mollusca*, vol. 5. London: Adam and Charles Black; 1906, p. 203. **D** from Morton, J. E. *Journal of the Marine Biological Association of the United Kingdom* 38: 233; 1959. **E** modified from Fischer-Piette, E.; Franc, A. In: Grasse, P.-P., editor. *Traite de zoology*, tome V, fascicule III. Paris: Libraires de l'Academie de Medicine; 1968, p. 1000. **F** to **H** modified from Palmer, C. P. *Veliger* 17: 116; 1974.)

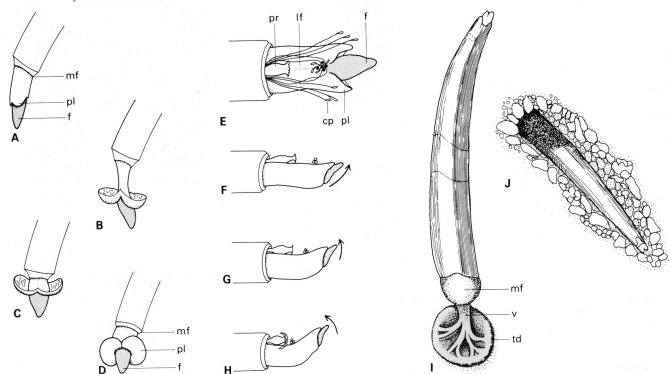

Figure 14.83. A to D, Burrowing in dentalioid scaphopods. **A** to **C,** Successive stages in burrowing showing the action of the foot (*f*) and lateral pedal lobe (*pl*); *mf* is the mantle fringe; **D** is the anterior view of the distended foot. **E to H,** Ciliary feeding of *Dentalium.* **E,** Dorsal view with the anterior shell and mantle removed. *cp,* Captacula; *f,* foot; *lf,* longitudinal furrow; *pl,* lateral pedal fold; *pr,* proboscis. **F to H,** Successive stages of foot bringing particles to frilly lips. **I** and **J,** Burrowing in siphono-dentalioid scaphopods. **I,** Animal probing with the foot. *mf,* mantle fringe; *td,* expanded terminal disk; *v,* vermiform foot. **J,** Position the animal assumed in sediment with the convex side upward (compare with Figure 14.82, *A*). (**A** to **D** from Morton, J. E. *Journal of the Marine Biological Association of the United Kingdom* 38: 226; 1959. **E** to **H** from Dinamani, P. *Proceedings of the Malacological Society of London*; 1964, p. 2. **I** and **J** from Davis, J. D. *Proceedings of the Malacological Society of London*; 1968, p. 136.)

The muscular foot burrows into the sediment; it is protruded by increase in blood pressure in the pedal hemocoel; retraction is by one or two pairs of pedal retractors (Figure 14.82, *A*). The foot may be bullet shaped (Figures 14.82, *A* and 14.83, *A* to *H*) and surrounded by **lateral pedal lobes**; or it may be vermiform, lack lateral lobes, and have an expandable **terminal disk** (Figure 14.83, *I*).

Scaphopods lack eyes and a well-differentiated head; they have a radula. Major sense organs are located at the distal ends of the captacula, where they may function as chemoreceptors for the recognition of food. Also statocysts, which contain statoliths, are embedded in the foot. The vascular system is simple; there is no heart. A large hemocoel has extensions into the mantle, foot, and captacula.

Shell morphology

The scaphopod shell has two basic shapes:

1   A continually expanding cone with the greatest diameter at the anterior aperture (Figure 14.84, *D, E,* and *G* to *M*); shells of this type often have ornament of longitudinal ribs.

2   A cone having a contracted anterior aperture so that the greatest diameter is slightly posterior to the aperture (Figure 14.84, *A* to *C*); shells of this type usually lack longitudinal ribs.

The shell of scaphopods has four layers: an outer horny periostracum and three calcareous layers that are prismatic or crossed lamellar in ultrastructure.

Ribs and comarginal ornament show considerable variation in strength and spacing (Figures 14.82, *F* to *H* and 14.84, *A* to *M*); these features are used to separate families and genera. Ornament type may change ontogenetically (Figure 14.84, *I*). A major feature used to distinguish genera is the cross-section shape of the shell (Figure 14.82, *F* and *G*).

Variation of the posterior end is sometimes used to define genera and species; however, sometimes this variation is an individual feature. The posterior opening may be simple and have no projections (Figure 14.82, *F*); however, sometimes the inner shell layers may project posteriorly past the outer shell layers as a structure called

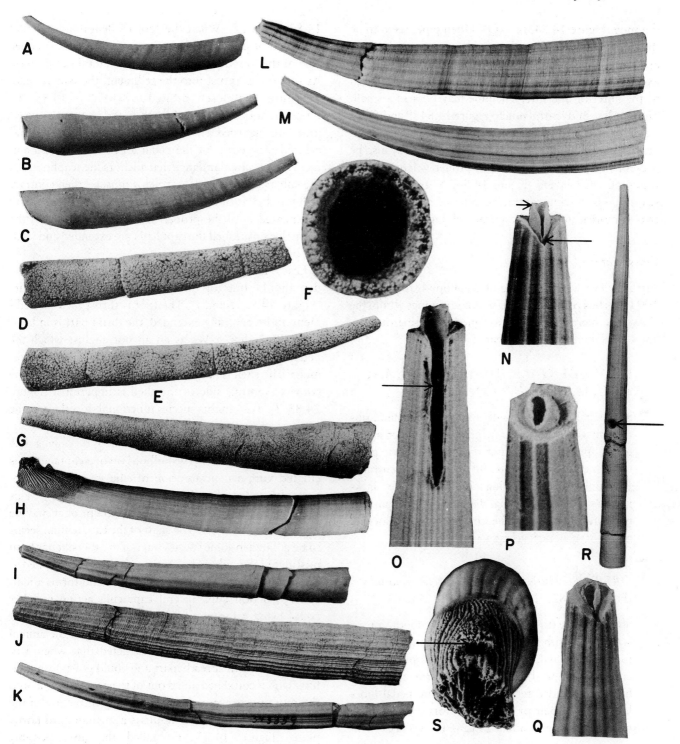

Figure 14.84. Shell morphology and diversity of scaphopods. **A** to **C** are siphonodentalioids; all others are dentalioids. **D** to **F** are Ordovician; **G** is Permian; **J** is Cretaceous; all others are Holocene. **A** to **C**, *Cadulus*. (Philippines, ×4.5.) **D** to **F**, *Rhytiodentalium*. (Kentucky, ×4.) **G**, *Prodentalium*. (Texas, ×2.5.) **H** and **S**, *Fissidentalium* showing coral overgrowth of posterior end. **S** is a view of the posterior end showing that the scaphopod was alive when the coral overgrew the shell, because the posterior aperture of the scaphopod remained open (arrow). (Taiwan. **H**, ×1.2; **S**, ×4.) **I**, *Dentalium*. (Florida, ×2.) **J**, *Dentalium*. (Tennessee, ×2.) **K**, *Paradentalium*. (Japan, ×1.5.) **L**, *Dentalium*. (Moluccas Islands, ×2.) **M**, *Dentalium*. (Philippines, ×1.5.) **N**, *Dentalium*. View of apical end showing pipe (short arrow) with a slit projecting above notch (long arrow) of the shell. (Moluccas Islands, ×5.) **O**, *Fissidentalium*. View of apical end showing a pipe and slit (arrow) extending from the pipe down the shell. (Japan, ×5.) **P**, *Dentalium*. Apical view showing a pipe with a small notch. (Mediterranean, ×11.) **Q**, *Dentalium*. View of apical end showing a small pipe with an eccentric slit and the shell with a notch. (Moluccas Islands, ×4.) **R**, *Spadentalina*. Ventral view showing a circular gastropod boring (arrow). (Mediterranean, ×2.)

the **pipe** (Figure 14.84, *N* to *Q*). Often pipes seem to be the result of resorption, but in some species they are apparently built after maturity is reached and thus are an exception to the general rule that shell increase is limited to the anterior aperture. The pipe may possess a small, more or less midventral **notch** (Figure 14.84, *P* and *Q*) or an elongated, parallel-sided reentrant termed the midventral **slit** (Figure 14.84, *N*). A slit or notch may also be present midventrally in the shell proper where it penetrates all shell layers (Figure 14.84, *N* and *O*). Some scaphopods, having contracted anterior apertures, have two or more notches at the apical end (Figure 14.82, *H*).

Classification

Virtually every classification of scaphopods for the past 100 years has recognized two major groupings within the class. The hierarchical level assigned to these groupings has varied; herein they are treated as orders.

ORDER DENTALIOIDA (Figures 14.82, *A*, *F*, and *G*; 14.83, *A* to *H*; 14.84, *D* to *R*)

Shell small to large, usually having longitudinal ribs or transverse annulations, greatest diameter at anterior aperture; pipes limited to this order; no more than one notch or slit present, located midventrally. Foot bullet shaped, having lateral pedal lobes, lacking an expanded terminal disk, having two pairs of pedal retractors. (Middle Ordovician to Holocene; about 30 generic level taxa.)

ORDER SIPHONODENTALIOIDA (Gadilida) (Figures 14.82, *H*; 14.83, *I*, *J*; 14.84, *A* to *C*)

Shell minute to medium, usually smooth, rarely sculptured with ribs or annulations, anterior aperture constricted; pipes absent, multiple notches common at apical end. Foot vermiform, commonly having expanded terminal disk, lateral pedal lobes lacking; one pair of pedal retractors present. (Permian to Holocene; about 21 generic level taxa.)

Mode of life

Living scaphopods all burrow in sediment and live with the anterior end lowermost and the posterior end above the sediment-water interface (Figures 14.82, *A* and 14.83, *J*). Dentalioids position themselves with the concave side up. When the foot is withdrawn, the anterior opening of dentalioids is closed by a **mantle fringe** (Figure 14.83, *A* to *D*). When the foot of dentalioids is fully extended, it is encircled by the mantle fringe. On either side of the foot dentalioids have two lateral pedal lobes. As the foot is thrust into the sediment, the lateral pedal lobes are held closely pressed to it (Figure 14.83, *A*). Once the foot is extended, both it and the lateral pedal lobes are engorged with blood (Figure 14.83, *B* to *D*) and erected. The edges of the pedal lobes are recurved and form a large circular flange that anchors the scaphopod in the substrate. The pedal retractor muscles then contract and the shell is drawn into the sediment. This procedure is repeated until the proper depth is reached; once the animal is positioned the captacula are extended and feeding begins.

Less is known about the burrowing habits of siphonodentalioids, but for one species the following occurs (Figure 14.83, *I* and *J*). The foot is thrust into the sediment and when fully extended, the distal part is inflated into a disk that serves as an anchor against which the pedal retractors contract and pull the shell into the sediment. The shell is at a small angle to the surface, and the convex (ventral) side of the shell is uppermost (Figure 14.83, *J*). Thus in this species functional dorsal is not the same as anatomic dorsal.

Observations of feeding are limited to only a few species of dentalioids. The bulbous tip of a captaculum is ciliated with a concentration of cilia in a terminal depression called the **alveolus** (Figure 14.82, *B*). This alveolus can be contracted to a deep cup, or it may be opened out shallowly. The shaft of the captaculum seems to be ciliated in some species but not in others. It has been postulated that when the captaculum is closely applied to a foraminifer, the alveolus flattens out and forms a temporary suction cup. Several captacula acting together could bring a foraminifer to the mouth. The radula of scaphopods is extremely large for the size of the animal (Figure 14.82, *E*); it is so constructed that when protruded, the teeth spread apart and could grasp foraminifers, which could be hauled out of the proboscis and into the more posterior parts of the gut by retraction of the radula. Lying above the radula is a median hard horny plate (Figure 14.83, *E*) called the **jaw**; possibly foraminifer shells are broken by being pressed against the jaw by the radula. Ordinarily only broken foraminifer shells are found in the gut posterior to the radula.

In some other dentalioids, feeding seems to be accomplished largely by ciliary action, and in these the captacula are ciliated along their entire length. Particles are moved by cilia to a longitudinal groove on the dorsal side of the foot (Figure 14.83, *E*). The foot is then thrust

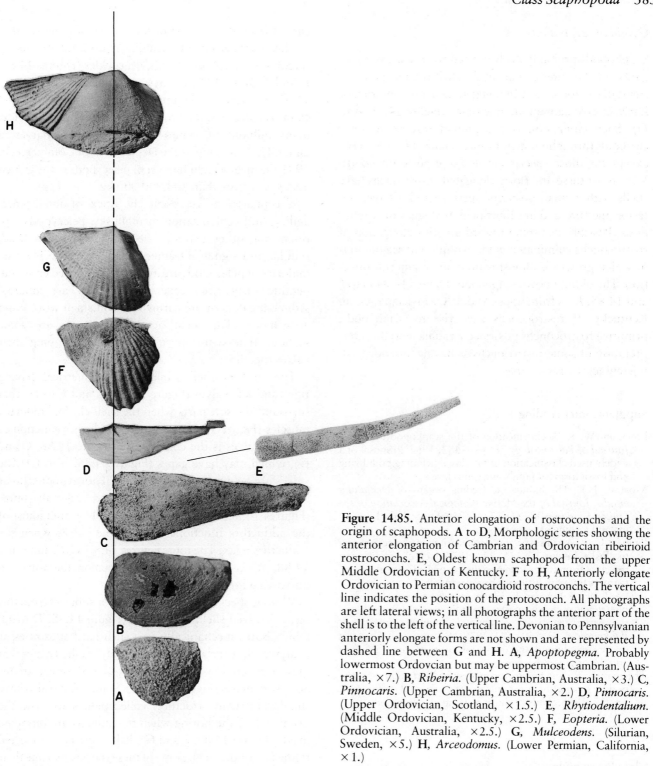

Figure 14.85. Anterior elongation of rostroconchs and the origin of scaphopods. **A** to **D,** Morphologic series showing the anterior elongation of Cambrian and Ordovician ribeirioid rostroconchs. **E,** Oldest known scaphopod from the upper Middle Ordovician of Kentucky. **F** to **H,** Anteriorly elongate Ordovician to Permian conocardioid rostroconchs. The vertical line indicates the position of the protoconch. All photographs are left lateral views; in all photographs the anterior part of the shell is to the left of the vertical line. Devonian to Pennsylvanian anteriorly elongate forms are not shown and are represented by dashed line between **G** and **H. A,** *Apoptopegma.* Probably lowermost Ordovician but may be uppermost Cambrian. (Australia, ×7.) **B,** *Ribeiria.* (Upper Cambrian, Australia, ×3.) **C,** *Pinnocaris.* (Upper Cambrian, Australia, ×2.) **D,** *Pinnocaris.* (Upper Ordovician, Scotland, ×1.5.) **E,** *Rhytiodentalium.* (Middle Ordovician, Kentucky, ×2.5.) **F,** *Eopteria.* (Lower Ordovician, Australia, ×2.5.) **G,** *Mulceodens.* (Silurian, Sweden, ×5.) **H,** *Arceodomus.* (Lower Permian, California, ×1.)

upward and the particles are moved toward the mouth (Figure 14.83, *F* to *H*).

The anterior aperture of scaphopods commonly is damaged or there are prolonged intervals of little growth (Figure 14.84, *D, E, H,* and *I* to *L*). These breaks or times of little growth leave pronounced transverse markings on the shell. Whether the broken apertures are the result of the animals being removed from the sediment by storms or are caused by annelid and arthropod predators is uncertain. Predatory gastropods are known to feed on scaphopods, but they leave distinctive circular holes in the shell (Figure 14.84, *R*).

Origin of scaphopods

Scaphopods probably evolved from rostroconchs. Elongation of the anterior part of the shell is a morphologic trend in rostroconchs beginning in the Late Cambrian or Early Ordovician among ribeirioids (Figure 14.85, A to D). Such elongation is a dominant feature in conocardioids throughout their history (Figure 14.85, F to H), except for those species having large posterior hoods. Various of these anteriorly elongated rostroconchs have shells with a small posterior aperture and a larger anterior aperture and are functional analogues of scaphopods. Possibly this trend toward anterior elongation of rostroconchs culminated in the evolution of scaphopods in which growth is almost entirely in an anterior direction. The oldest known scaphopod (Figures 14.84 D to F and 14.85, E) is from upper Middle Ordovician rocks in Kentucky. If rostroconchs gave rise to scaphopods, primitive rostroconchs possessed a radula, and the interpretation of some rostroconchs as having anterior feeding tentacles is reasonable.

Supplementary reading

Emerson, W. K. A classification of the scaphopod mollusks. *Journal of Paleontology* 26: 461–482; 1962. Proposal of a widely used classification of the class, including both living and fossil forms at family and genus levels.

Morton, J. E. The habits and feeding organs of *Dentalium entalis*. *Journal of the Marine Biological Association of the United Kingdom* 38: 225–238; 1959. Study of the burrowing and feeding of scaphopods.

Palmer, C. P. A supraspecific classification of the scaphopod mollusks. *Veliger* 17: 115–123; 1974. New classification of the class from order to genus, illustrating all critical generic features but with little emphasis on fossils.

Pojeta, J., Jr.; Runnegar, B. *Rhytiodentalium kentuckyensis*, a new genus and new species of Ordovician scaphopod, and the early history of scaphopod mollusks. *Journal of Paleontology* 53: 530–541; 1979. Description of the oldest known scaphopod and discussion of its zoologic affinities and the origin of the class.

Trueman, E. R. The burrowing process of *Dentalium* (Scaphopoda). *Journal of the Zoological Society of London* 154: 19–27; 1968. Study of the burrowing of scaphopods.

Class Pelecypoda

John Pojeta, Jr.

Shells of pelecypods litter the beaches of the earth from the polar seas to the equator and are also common along the shores of rivers and lakes. Most valuable pearls are produced by these animals. Until recently, buttons were made from their shells and much Indian wampum was carved from them. Most of the species of mollusks that are eaten are pelecypods including many kinds of mussels (Figures 14.86, A and 14.87, B), cockles (Figures 14.86, B and C and 14.87, F), oysters (Figures 14.86, D and 14.87, D), scallops (Figures 14.86, E and 14.87, A), and clams (Figures 14.86, F to H and 14.87, C). Worldwide, many millions of tonnes of pelecypods are harvested annually. According to the Department of Commerce, in 1981 the meat weight harvest of pelecypods in the United States was more than 100,900,000 kg.

If cephalopods represent the apex of intelligence, agility, and cephalization in mollusks, pelecypods represent the other extreme, because they lack a head, radula, and significant anterior sense organs. The vast majority of pelecypods are infaunal or attached epifaunal benthic suspension feeders; a minority are infaunal sediment eaters or are carnivorous on small worms and arthropods. The most active pelecypods are some scallops that swim short distances by clapping their valves together.

Typically the shell is bilaterally symmetrical, having right and left valves (Figure 14.86, C and F to H) that surround the soft parts when the shell-closing **adductor muscles** (Figure 14.87) contract. The line of junction of the two valves is the commissure (Figure 14.86, C and G), which may have gapes (Figure 14.88, A to C). The valves are held together by the dorsal horny, noncellular ligament composed of conchiolin and aragonitic fibres (Figures 14.86, C and G and 14.87). This antagonist of the adductors functions to open the valves when the adductors relax. The foot (Figures 14.87, C, E, and F and 14.88, A) is also important in opening the valves in burrowing forms.

Various specialized species add to or subtract from this basic bivalved shell. One group (Figure 14.88, D to F), which bores mechanically into such hard substrates as peat, wood, stiff clay, shell, and coral, has shell gapes that may be covered by horny (Figure 14.88, B) or calcareous **accessory plates** (Figure 14.88, F). Some species also have small calcareous structures called **pallets** to close the distal end of the boring when the animals are retracted inside (Figure 14.89, F and G). Pallets are not connected to the rest of the shell. Some of these species also line their borings with a calcareous tube that can reach a length of 1.2 m (Figures 14.88, D and E and 14.89, F and G).

The other extreme in the number of hard parts is represented by hydraulically burrowing species that live vertically in muddy sand in shallow water (Figure 14.90, A to E). These pelecypods are bivalved as juveniles. In the adult, one or both valves become embedded in an elongate calcareous tube that is open at the posterior end and

Figure 14.86. Shells of some species of living edible pelecypods. **A,** *Mytilus edulis,* right lateral view. (Long Island, New York, ×2.) **B** and **C,** *Cerastoderma edule,* right lateral and dorsal views. (Scotland, ×2.) **D,** *Crassostrea virginica,* left lateral view. (Chesapeake Bay, Maryland, ×1.) **E,** *Argopecten irradians,* right lateral view. (Florida, ×1.5.) **F** to **H,** *Mercenaria mercenaria,* right lateral, dorsal, and left lateral views. (Long Island, New York, ×1.2.) *c,* Commissure; *l* ligament; *ls,* ligament space; *lv,* left valve; *rv,* right valve.

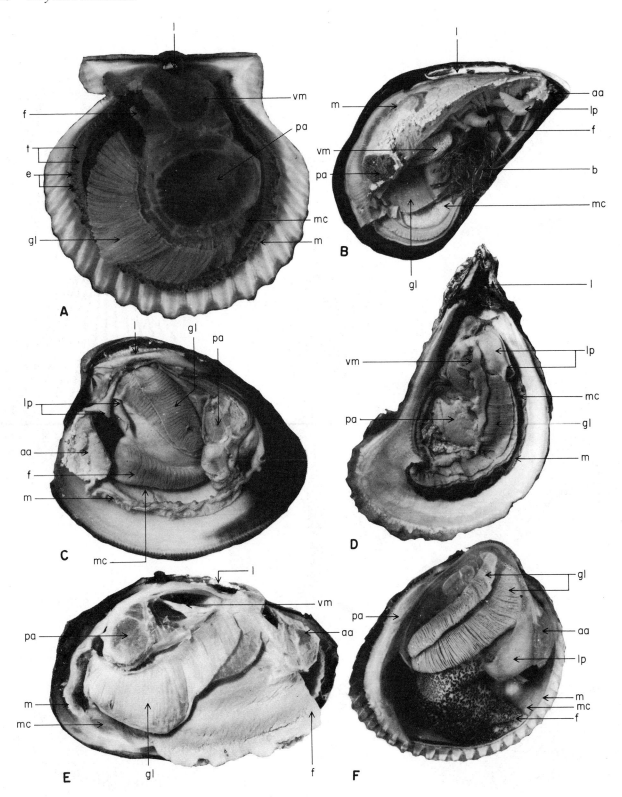

Figure 14.87. Gross anatomy of pelecypods. **A**, *Argopecten.* The left valve and mantle are removed. (North Carolina, ×2.) **B**, *Mytilus.* The right valve and most of the mantle are removed. (East Coast, United States, ×1.3.) **C**, *Mercenaria.* The left valve and mantle are removed, and the visceral mass is largely covered by the gill. (East Coast, United States, ×1.1.) **D**, *Crassostrea.* The right valve and mantle are removed. (Chesapeake Bay, Maryland, ×1.) **E**, *Leptodea.* The right valve and mantle are removed. (Ohio River, Cincinnati, Ohio, ×1.5.) **F**, *Dinocardium.* The right valve and mantle are removed, and the visceral mass largely covered by the gill. (Florida, ×1.5.) *aa*, Anterior adductor; *b*, byssus; *e*, eye; *f*, foot; *gl*, gill; *l*, ligament; *lp*, labial palp; *m*, mantle; *mc*, mantle cavity; *pa*, posterior adductor; *t*, tentacles; *vm*, visceral mass.

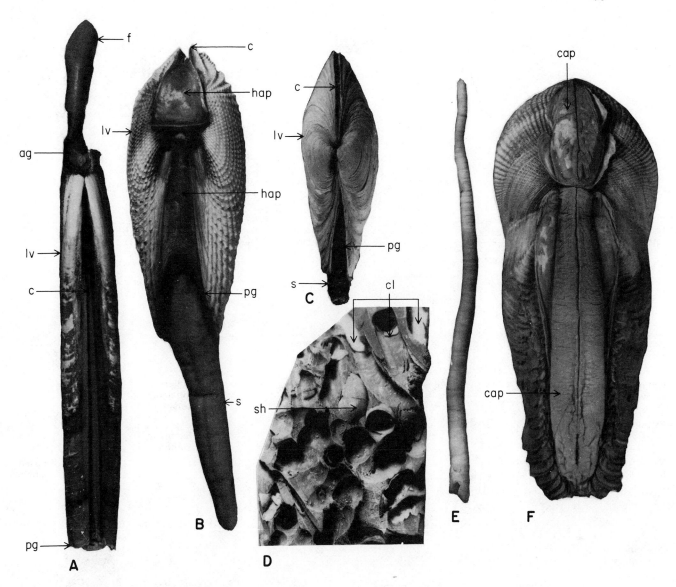

Figure 14.88. A to C, Dorsal views of living burrowing pelecypods. The anterior end is up. **A**, *Ensis*. (East Coast, United States, ×1.) **B**, *Cyrtopleura*. (Florida, ×1.) **C**, *Mya*. (Long Island, New York, ×2.) *ag*, Anterior (pedal) gape; *c*, commissure; *f*, foot; *hap*, horny accessory plate; *lv*, left valve; *pg*, posterior (siphonal) gape; *s*, fused siphons. **D** to **F**, Shelly structures of living piddocks and shipworms. **D**, *Teredo*. A piece of wood is broken to show how thoroughly this shipworm's boring penetrates and ramifies. Most of the calcareous tubular linings of the borings break away when the wood is split. (Atlantic Ocean, ×1.5.) **E**, *Teredo*. The calcareous tubular lining of boring removed from wood. (Atlantic Ocean, ×1.) **F**, *Parapholas*. Dorsal view showing paired calcareous accessory plates. (California, ×1.5.) *cap*, Calcareous accessory plate; *cl*, calcareous tubular lining of boring; *sh*, bivalved shell.

that can be up to 1 m long. The anterior end of the tube is connected to the outside environment in various ways (Figure 14.90, *B* to *E*).

Living pelecypods range in adult size from less than 1 mm to over 1 m long. No Cambrian pelecypods longer than 5 mm are known, and most are less than 2 mm long. Equally small sizes occur throughout the stratigraphic range of pelecypods. By the Early Ordovician, species larger than 1 cm evolved, and throughout most of the Phanerozoic, most species are megafauna. By Late Ordovician time, one species reached a length of 140 mm. The largest Paleozoic pelecypods occur in the Silurian of North America (250 mm) and the Permian of Malaysia (600 mm). Distant relatives of the marine mussels became more than 1 m in length in Cretaceous time and were contemporaneous with some of the horn-shaped rudists (Figure 14.10, *G*), which can be well over 1 m in diameter.

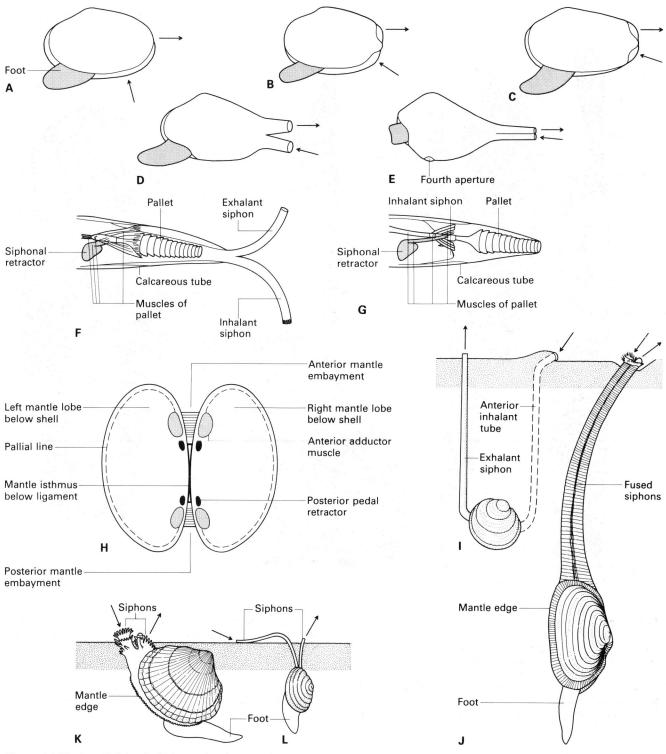

Figure 14.89. A to **E,** Mantle fusion and siphons in dimyarian pelecypods. Inhalant and exhalant currents are marked by appropriate arrows. The foot is stippled, and the shell is removed. All views are shown from the left side. The mantle margins are confluent where fused. **A,** The mantle margins are completely open and unfused. **B,** The mantle margins are fused at one posterior place between the exhalant and main mantle apertures. **C,** The mantle margins are fused at two posterior points, forming inhalant, exhalant, and large pedal apertures. **D,** The mantle margins are fused except at openings for posterior siphons and a small anterior pedal opening. **E,** The mantle margins are fused all around and much as in **D,** except for the presence of a fourth aperture. **F** and **G,** *Xylotrea* showing pallet in retracted **F** and protracted **G** conditions. **H,** Diagrammatic representation of pelecypod shell, mantle features, and major muscles. **I** to **L,** Life positions of various burrowing pelecypods.

I, *Loripes* is a suspension feeder with a posterior exhalant siphon and a mucous-cemented, anterior inhalant tube formed by the foot. **J,** *Mya* is a deep-burrowing suspension feeder with long fused siphons covered by periostracum. **K,** *Cerastoderma* is a shallow-burrowing suspension feeder with short siphons not covered with periostracum. **L,** *Tellina* is a shallow-burrowing deposit feeder with long unfused siphons. Inhalant and exhalant currents are indicated by appropriate arrows. (**A** to **G** and **I** to **L** modified from Cox, L. R. in: Moore, R. C., editor. *Treatise on invertebrate paleontology, Part N, Mollusca 6, Bivalvia 1.* Boulder, CO, and Lawrence, KS: Geological Society of America and University of Kansas Press; 1969, p. 4, 34, 65. **H** modified from Morton, J. E.; Yonge, C. M. In: Wilbur, K. M.; Yonge, C. M., editors. *Physiology of Mollusca,* vol. 1. New York: Academic Press, Inc.; 1964, p. 6.)

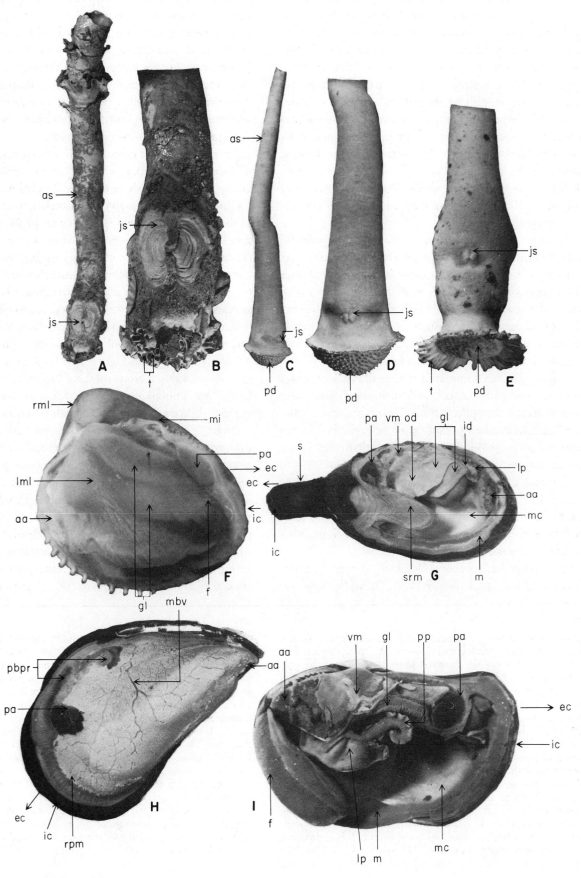

Figure 14.90. Legend on p. 392.

Cambrian and Early Ordovician pelecypods are few in known species but can be abundant in number of individuals. From the Middle Ordovician onward, pelecypods are highly diverse in species and abundant in the fossil record. After the Early Ordovician, Paleozoic pelecypods can be numerically dominant in benthic faunas at scattered stratigraphic levels, particularly in near-shore clastic and cratonic carbonate facies. In the Mesozoic and Cenozoic, pelecypods are a common to numerically dominant element of shallow-water benthic faunas. Estimates of the number of living species range from 6000 to 15,000; there are about 7000 named Paleozoic species, 15,000 Mesozoic species, and 20,000 pre-Holocene Cenozoic species.

Throughout their stratigraphic range, pelecypods are aquatic benthic animals, although a few living freshwater clams have been reported to crawl about in wet leaf litter on damp days. Debate exists as to whether some Paleozoic and Mesozoic species were pelagic as adults, but widespread geographic distributions are now usually attributed to larval dispersal; in living species, larvae can be planktic for a few days to a few weeks.

Pelecypods are a dominantly marine to brackish-water group (Figure 14.87, *A* to *D* and *F*), although some have lived in freshwater since Middle Devonian time (Figure 14.87, *E*). In modern oceans, some pelecypods occur in virtually every major benthic environment that is not anoxic and are known at all depths from the intertidal to the abyssal zones. Some pelecypods are known to live attached to or bored into turtles and floating or rooted plants; a few are commensal with echinoderms, crustaceans, tunicates, or sponges; and one genus is parasitic in the gut of holothurian echinoderms.

Pelecypods have been ecologically diverse benthos since Ordovician time (Figure 14.91, *D* and *J*). They have been able to burrow into soft substrates since at least the earliest Ordovician (Figure 14.91, *F* and *J*). Some infaunal pelecypods bore into hard substrates, including rock, coral, and wood (Figure 14.91, *G* to I). In most pelecypods, boring is mechanical through the use of the two valves; a few bore by use of nonacidic chemical secretions. Boring pelecypods have a range of Jurassic to Holocene, but one known Ordovician species was a facultative borer in stromatoporoids.

Large numbers of pelecypods live as epibenthos or are semi-infaunal and partly buried in the sediment. Both groups are known since the Middle Ordovician. Semi-infaunal species are most often attached to coarse debris in the sediment by the **byssus** (Figure 14.91, *E*), a structure composed of numerous hairlike threads. Most epifaunal species attach to hard substrates either by the byssus (Figure 14.91, *J*) or by cementation; a few recline on the sea floor and are not attached. Some byssate species nestle in preexisting cavities. Scallops may attach byssally or recline; both kinds can swim for short distances by the rapid expulsion of water from the hinged margin of the shell (Figure 14.91, *A* to *C*). Nestling species are known from the Late Ordovician, cemented forms are known in the Late Mississippian, and the earliest scallops occur in Middle Silurian rocks.

Most pelecypods are suspension feeders, eating diatoms, dinoflagellates, other algae, and bacteria. Fewer pelecypods are deposit feeders. One group of deposit feeders uses specialized ciliated anterior structures that are thrust into the sediment and connect internally near the mouth. The other group of deposit feeders uses a posterior structure to suck organic debris and microorganisms from the sediment-water interface. One small group of pelecypods is partly carnivorous and uses a specialized muscular septum to create a current by which

---

Page 391.

Figure 14.90. A to **E**, Shells of living salt and pepper shaker pelecypods. The anterior end is down. **A**, *Humphreyia.* The entire shell showing the bivalved juvenile stage and the univalved tubular adult stage. (Japan, ×0.5.) **B**, Enlargement of the anterior end of the specimen in **A** showing anterior tubules. (×1.25.) **C**, *Brechites.* The entire shell showing the small bivalved stage, the tubular univalved adult stage, and the perforated disc. (Borneo, ×0.67.) **D**, Enlargement of the anterior end of the specimen in **C**. (×1.3.) **E**, *Brechites.* Enlarged anterior end showing the juvenile bivalved stage, the lower part of the univalved adult stage, the perforated disc, and tubules. (Torres Strait between New Guinea and Australia, ×2.) *as*, Tubular adult shell; *js*, bivalved juvenile shell; *pd*, perforated disc; *t*, tubules. **F** to **I**, Mantle and internal anatomy of living pelecypods. **F**, *Dinocardium* having the shell removed, showing the outer surface of both mantle lobes. Internal structures are vaguely discernible. (Florida, ×1.5.) **G**, *Mya.* The right valve and mantle are removed, and the siphonal retractor muscles can be seen. The fused siphons are dark because they are covered with periostracum. (Long Island, New York, ×1.5.) **H**, *Mytilus.* The right valve is removed, and the outer surface of the right mantle lobe is exposed, showing major blood vessels and, posteriorly and ventrally, the radial pallial muscles. The vascularization of the mantle is clearly shown because the mantle is filled with gonadal tissue. (East Coast, United States, ×1.5.) **I**, *Yoldia (Megayoldia).* The left valve and mantle are removed. The retracted palp proboscis can be seen. (Northwest Pacific Ocean, ×1.5.) *aa*, Anterior adductor; *ec*, exhalant current; *f*, foot; *gl*, gill; *ic*, inhalant current; *id*, inner demibranch; *lml*, left mantle lobe; *lp*, labial palp; *m*, mantle; *mbv*, mantle blood vessel; *mc* mantle cavity; *mi*, mantle isthmus; *od*, outer demibranch; *pa*, posterior adductor; *pbpr*, posterior byssal-pedal retractor muscle; *pp*, palp proboscis; *rml*, right mantle lobe; *rpm*, radial pallial muscles; *s*, fused siphons; *srm*, siphonal retractor muscle, *vm*, visceral mass.

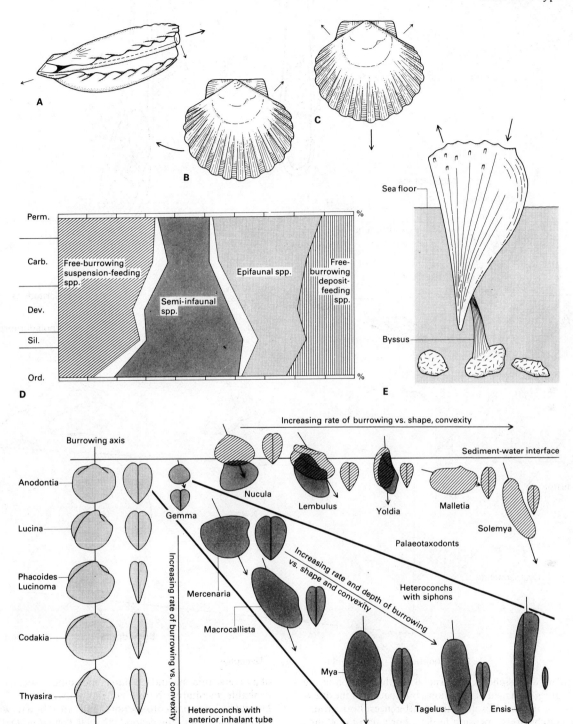

Figure 14.91. A to **C,** *Pecten.* A free-swimming scallop, edge and lateral views. The direction of movement is indicated by larger arrows; the direction of propelling water currents is indicated by smaller arrows. **D,** Gross changes in the life-habit spectrum of the Pelecypoda during the Paleozoic. Mean percentage of species in each life-habit group for faunas of each period is randomly chosen from the literature. Blank spaces mean percentages of species whose life habits could not be assigned to either neighboring category with a high degree of certainty. **E,**

Pinna, semi-infaunal species in living position partly buried in the sediment and attached by byssus to coarse debris. **F,** General relationships in living pelecypods of shell outline and convexity to depth and rate of burrowing. Dark arrows indicate approximate axes of burrowing. All shells are shown in normal living position but not necessarily at uniform scale. For heteroconchs with an inhalant anterior tube, a consistent depth zonation is implied; although some studies are in accord with depth relations here depicted, others indicate a different ordering. For

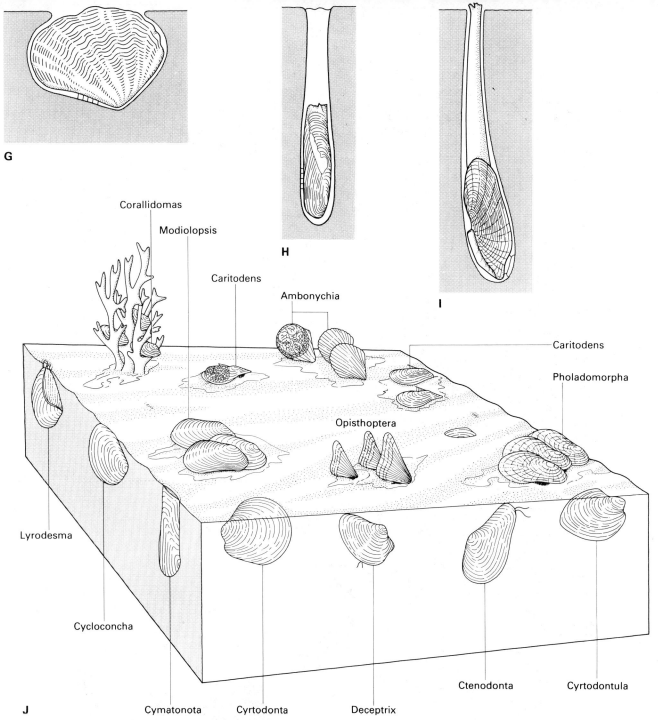

G

Corallidomas

Modiolopsis

Caritodens

Ambonychia

H

I

Caritodens

Pholadomorpha

Opisthoptera

Lyrodesma

Cycloconcha

Ctenodonta

Cyrtodontula

J

Cymatonota   Cyrtodonta   Deceptrix

heteroconchs with siphons, the general depth zonation in the diagram agrees with field observations, but spacing of the drawings does not indicate absolute depth differences. Both groups of heteroconchs are suspension feeders. For palaeotaxodonts, indicated depth distribution is approximately to scale; the dual images of the first three genera represent ultimate depth and position attained by burrowing (dark background) and progress toward these (light background) in the time it would take *Malletia* and *Solemya* to attain their normal burrowing depth as shown; most palaeotaxodonts have palp proboscides and are deposit feeders. **G** to **I**, Rock-boring species; the surface of rock is directed upward. **G**, *Tridacna*. Lateral view, the animal held in place by the byssus. **H**, *Lithophaga*. Longitudinal section of boring, the animal attached to its wall by byssal threads. **I**, *Pholas*. Longitudinal section, animal attached by suckerlike foot. **J**, Composite diagram of inferred ecology of Upper Ordovician pelecypods based largely on material from the tristate area of Kentucky, Ohio, and Indiana. There is some debate as to the interpretation of two or three genera, but the general picture

of a diverse infauna and epifauna, including nestlers, is correct. Probable facultative borers are not shown. *Ctenodonta* and *Deceptrix* are palaeotoxodonts and are shown with palp proboscides extending beyond the shell. *Cymatonota, Pholadomorpha,* and *Modiolopsis* are isofilibranchs; the nestling *Corallidomas* is attached to a trepostome bryozoan. *Cycloconcha* and *Lyrodesma* are heteroconchs. *Cyrtodontula* and *Cyrtodonta* are semi-infaunal pteriomorphs, although the byssus is not shown. *Opisthoptera, Ambonychia,* and *Caritodens* are pteriomorphs, and the epibionts are edrioasteroids. No anomalodesmatans are shown. (**A** to **C**, **E**, and **G** to **I** from Cox, L. R. and **F** from Kauffman, E. G., in: Moore, R. C., editor. *Treatise on invertebrate paleontology, Part N, Mollusca 6, Bivalvia, 1.* Boulder, CO, and Lawrence, KS: Geologic Society of America and University of Kansas Press; 1969, p. 7, 10, 164. **D** from Stanley, S. M. *Journal of Paleontology.* 46: 202; 1972. **J** From Pojeta, J. Jr. U.S. Geological Survey, Professional Paper 695:33 1971.)

small invertebrates are brought into the mantle cavity.

At the lower taxonomic levels, pelecypod diversity increases from intertidal and shallow-water marine habitats, across the shelf, and into the deep sea. However, diversity at the higher taxonomic levels, and in terms of adaptive strategies is greatest in shallow shelf habitats.

Because pelecypods had an Early Cambrian origin, are readily preserved, show great diversity, and are found in most aquatic sedimentary facies, they are important for biostratigraphic zonation and correlation and the interpretation of paleoenvironments. Their geologic applications are enhanced by the existence of well-studied, taxonomic, functional, or adaptive counterparts in the modern world for most fossil groups.

In the past 200 years at least 14 different names have been proposed for this large class. In the 20th century, the most widely used names have been **Pelecypoda, Bivalvia,** and **Lamellibranchiata.** Pelecypoda is used here because it is less cumbersome than Lamellibranchiata and because bivalve is a useful nontaxonomic term for all bivalved invertebrates.

Anatomy

The soft parts of pelecypods are divided into five groups: mantle or pallium, gills, foot and byssus, muscles, and visceral mass (Figures 14.87 and 14.90, *F* to *I*).

The **mantle** secretes the periostracum, ligament, and all shell layers and may contain extensions of the gonads (Figure 14.90, *H*). It consists of two thin lateral sheets of tissue called **mantle lobes** (Figures 14.89, *H* and 14.90, *F*), one corresponding to each valve. Anteriorly and posteriorly along the dorsum are midline **mantle embayments** (Figure 14.89, H) that can extend toward the beaks past the adductor muscles. Between the mantle embayments, the mantle lobes are joined by the **mantle isthmus** (Figures 14.89, H and 14.90, F). Various muscles attach to the shell (Figure 14.90, *F* and *H*), and the mantle between shell and muscle secretes the myostracum.

Dorsally the mantle lobes are fused to the visceral mass and form its integument; however, ventrally the lobes surround the open space of the mantle cavity (Figures 14.87 and 14.90, G and I) into which project the gills, foot, and anterior feeding structures known as **labial palps** (Figures 14.87, *B* to *D* and *F* and 14.90, *G* and *I*). In most sediment-eating pelecypods, anterior, elongate, extensible, ciliated feeding structures known as **palp proboscides** (singular, proboscis) are attached to the labial palps (Figure 14.90, *I*). Inhalant currents created by cilia on gills and mantle bring water into the mantle cavity;

spent water is removed by the exhalant current. The exhalant current is always posterior in position (Figure 14.89, *A* to *E*). In most pelecypods the inhalant current is also posterior and ventral to the exhalant current (Figure 14.89, *B* to *E* and *J* to *L*). However, the inhalant current may be anterior (Figure 14.89, *I*) or ventral (Figure 14.89, *A*).

The idealized edge of the mantle has three folds (Figure 14.3, *B*), each of which has a different function; this condition is present in many, but not all, pelecypods. The outer fold is secretory, the middle fold is sensory, and the inner fold is muscular. Between the outer and middle folds is the periostracal groove. Shell secretion begins with the formation of the periostracum by the cells in this groove. The outermost part of the calcareous shell is laid down by cells of the outer part of the outer fold, and the rest of the shell is laid down by cells of the general outer surface of the mantle.

The middle fold bears various sense organs and sometimes **mantle eyes** and/or **tentacles** for sorting and screening sediment (Figure 14.87, *A*). The mantle contains **radial pallial muscles** (Figure 14.90, *H*), which retract its edge and control the flow of water into and out of the mantle cavity. The pallial muscles may insert along a well-defined arc on the inside of the shell known as the **pallial line** (Figure 14.92). Often the thin, vascularized mantle functions as an accessory respiratory organ (Figure 14.90, *F* and *H*).

Recent work indicates that the idea of a homologous three-part mantle edge in all pelecypods needs to be modified. In some species that have a two-part mantle edge, **compound eyes** (Figure 14.93, *C* and *D*) are on the outer of the two folds, and this outer fold assumes the secretory, sensory, and water-control functions. The only homologous mantle-edge structure between major groups of pelecypods is the periostracal groove.

In many species, the mantle cavity is open to the outside along the entire anterior, ventral, and posterior margins (Figure 14.89, *A*). However, in most species, there is fusion of the mantle margins. Some species have the mantle lobes fused at a single place at the posterior end of the mantle cavity (Figure 14.89, *B*), which separates a small posterodorsal exhalant aperture from the main mantle aperture through which the inhalant current enters posteriorly and the foot protrudes anteroventrally. Various species have the mantle fused in two places (Figure 14.89, *C*), resulting in three apertures; the posterior two of these apertures are often extended into inhalant and exhalant **siphons** (Figure 14.89, *D*). The third opening is the **pedal aperture** for the foot. Siphons are posterior, tubular, and muscular extensions of the

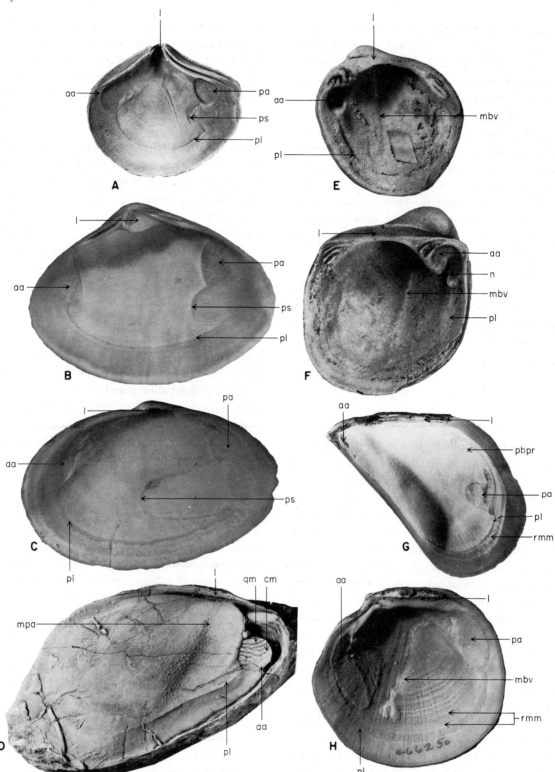

Figure 14.92. Soft-part impressions on the inside of the shell and position of the ligament. **A,** *Myadora*. Right valve. (Holocene, New Zealand, ×1.5.) **B,** *Spisula*. Right valve. (Holocene, Florida, ×1.5.) **C,** *Mya*. Right valve. (Holocene, Chesapeake Bay, Maryland, ×1.5.) **D,** *Modiomorpha*. Right valve internal mold. (Devonian, West Virginia, ×1.5.) **E,** *Cyrtodonta*. Silicified right valve. (Ordovician, Tennessee, ×1.) **F,** *Vanuxemia*. Silicified left valve. (Ordovician, Tennessee, ×2.) **G,** *Mytilus*. Right valve. (Holocene, New York, ×2.) **H,** *Megaxinus*. Right valve. (Holocene, Florida, ×2.) *aa,* Anterior adductor muscle scar; *cm,* catch muscle scar; *l,* ligament or ligament attachment area; *mbv,* mantle blood vessel impression; *mpa,* point attachment scars of mantle; *n,* notch in umbonal septum; *pa,* posterior adductor muscle scar; *pbpr,* posterior byssal-pedal retractor muscle scar; *pl,* pallial line; *ps,* pallial sinus; *qm,* quick muscle scar; *rmm,* radial mantle muscle impressions. (**A** courtesy H. E. Vokes, Tulane University.)

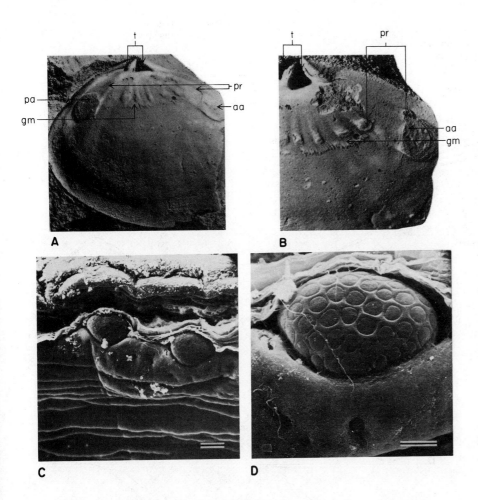

Figure 14.93. A and **B**, *Babinka*. Right valve internal mold showing probable small, numerous gill muscle scars (*gm*), pedal muscle scars (*pr*), and adductor muscle scars (*aa, pa*). (*t*, cardinal teeth). (**A**, ×2; **B**, ×3.5; Middle Ordovician, Czechoslovakia.) **C** and **D**, Compound eyes in the mantle margin of *Arca*. **C**, Two compound eyes beneath remnant of periostracum. Bar = 50 μm. **D**, Enlargement of the right compound eye in **C**. The simple eye is at bottom center. Bar = 20 μm. (From Waller, T. R. *Smithsonian Contributions to Zoology* No. 313; 1980.)

mantle (Figure 14.88, *B* and *C*) that allow burrowing species to maintain communication with the water above the burrow (Figure 14.89, *I* to *L*). A few pelecypods have three places of mantle fusion, forming four apertures (Figure 14.89, *E*).

Siphons are attached to the shell by **siphonal retractor muscles** (Figure 14.90, *G*), which are enlarged parts of the pallial retractor muscles. The insertion of the siphonal retractors often causes an embayment in the pallial line known as the **pallial sinus** (Figure 14.92, *A* to *C*). There may also be **point attachments** of the outer surface of the mantle to the shell (Figure 14.92, *D*). Some of the larger blood vessels may also leave impressions (Figure 14.92, *E*, *F*, and *H*) on the shell. All the soft-part impressions on the interior of the shell make it possible to reconstruct detailed mantle features in extinct species.

The length of siphons varies, and the depth of the pallial sinus is some measure of this length. Ordinarily the more deeply a pelecypod burrows, the longer the siphons, and the larger and deeper the pallial sinus (Figure 14.92, *A* to *C*). In deeply burrowing species, the siphons cannot be fully retracted into the shell (Figures 14.88, *B* and *C* and 14.90, *G*). Some pelecypods have the

inhalant and exhalant siphons separated lengthwise (Figure 14.89, *L*); most are deposit feeders using the inhalant siphon to suck up food from the sediment-water interface. Other species have the siphons fused lengthwise (Figures 14.88, *B*; 14.89, *J*; and 14.90, *G*); these are suspension feeders. Long fused siphons may be naked (Figure 14.88, *B*) or covered with periostracum (Figure 14.90, *G*). A few pelecypods have an eversible posterior siphon and use the foot to construct an anterior mucus-lined inhalant tube (Figure 14.89, *I*).

When pelecypods feed, the valves are slightly parted and the mantle edge is extended a little bit beyond the shell margins (Figure 14.89, *K*). In one superfamily the mantle is commonly reflected over the exterior of the shell; in this respect these pelecypods are similar to cowrie gastropods. In a few pelecypods the shell has become internal and is embedded in reflected mantle.

Pearls are calcareous bodies that are made in tissue sacs of the mantle or that form as blisterlike objects on the interior of the shell. The cause of pearl formation is irritation of the outer layer of the mantle by a parasite or small inorganic grains. The irritant is a nucleus around which the mantle secretes shell. Where the mantle

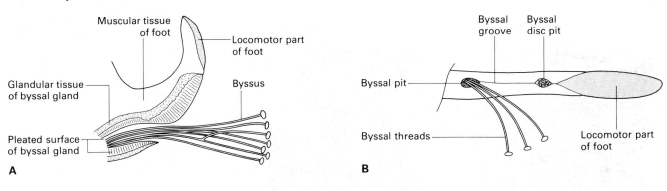

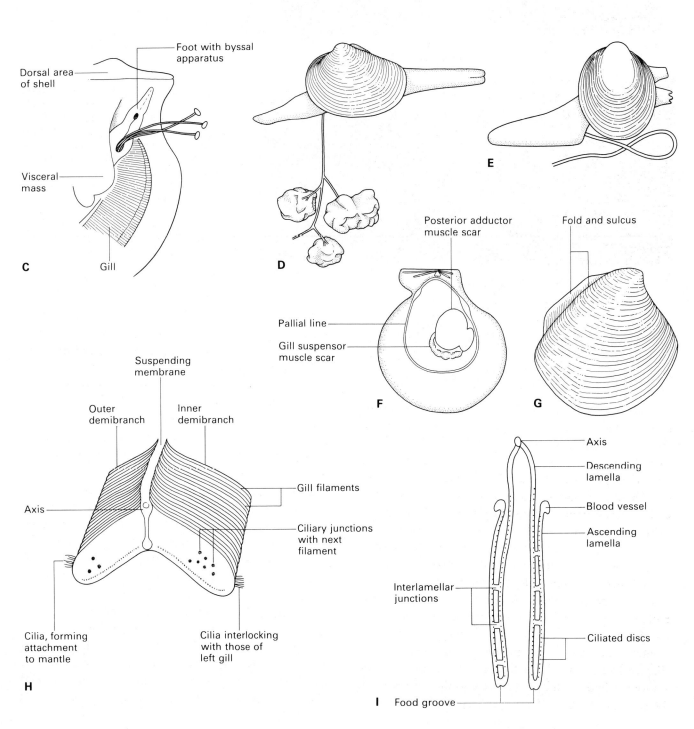

A

Muscular tissue of foot
Locomotor part of foot
Glandular tissue of byssal gland
Byssus
Pleated surface of byssal gland

B

Byssal groove
Byssal disc pit
Byssal pit
Byssal threads
Locomotor part of foot

C

Dorsal area of shell
Foot with byssal apparatus
Visceral mass
Gill

D

E

F

Posterior adductor muscle scar
Pallial line
Gill suspensor muscle scar

G

Fold and sulcus

H

Suspending membrane
Outer demibranch
Inner demibranch
Axis
Gill filaments
Ciliary junctions with next filament
Cilia, forming attachment to mantle
Cilia interlocking with those of left gill

I

Axis
Descending lamella
Blood vessel
Ascending lamella
Interlamellar junctions
Ciliated discs
Food groove

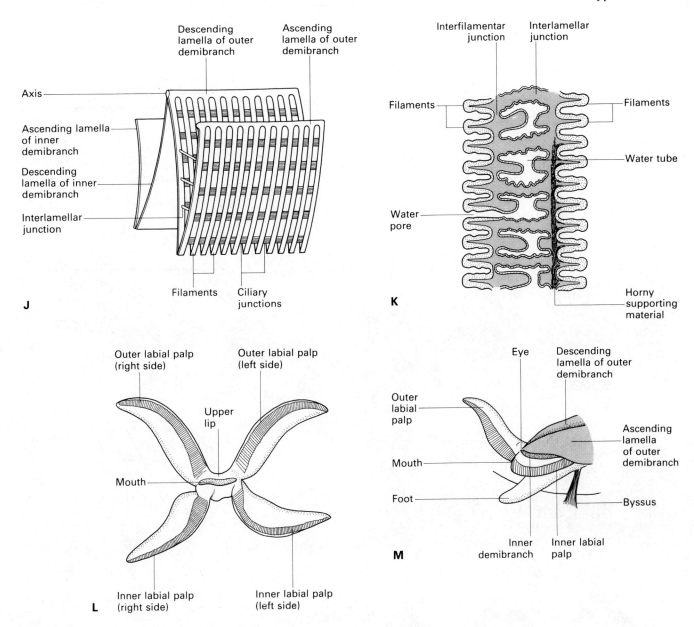

Figure 14.94. A to **C**, Foot and byssus of the Holocene pearl oyster *Pinctada*. **A**, Sagittal section. **B**, Ventral surface of the foot. **C**, Anterior part of the animal showing relationship of the foot, byssus, and gill. **D** and **E**, Early postlarval stages of *Mya* (**D**, ×25) and *Bankia* (**E**, ×74) showing single byssal threads in a juvenile; the adult lacks byssus. The byssal thread of *Mya* is divided and attached to sand grains. **F** to **M**, Gills and palps of pelecypods. **F**, *Chlamys*. Interior of the right valve showing the position of the gill muscle scar. (Holocene, ×0.6.) Compare with Figure 14.87, *A*. **G**, *Thyasira*. Exterior of the right valve showing the fold and sulcus marking approximate position of gills. (Holocene, ×1.) **H**, *Nucula*. Gross anatomy of portion of the protobranch gill looking anteriorly to posteriorly. **I** and **J**, *Mytilus*. Cross section (**I**) and oblique lateral view (**J**) of filibranch gill. **K**, *Tellina*. Horizontal longitudinal section of eulamellibranch gill. **L** and **M**, *Mytilus*. Labial palps. Anterior view (**L**) with ridged inner surfaces separated from each other and left lateral view (**M**) with outer palp cut at base and turned up. In life the anterior end of the gill is between the pair of palps. (**A** to **C** and **G** to **M** modified from Cox, L. R.; Kauffman, E. G., in: Moore, R. C., editor. *Treatise on invertebrate paleontology, Part N, Mollusca 6, Bivalvia 1,* Boulder, CO, and Lawrence, KS: Geological Society of America and University of Kansas Press; 1969, p. 14, 17–20, 31; **D** and **E** from Yonge, C. M. *Journal of Marine Biological Society* 42: 115; 1962. **F** modified from Newell, N. D. *Late Paleozoic pelecypods: Pectinacea.* Lawrence, KS: State Geological Survey of Kansas; 1937, p. 28.)

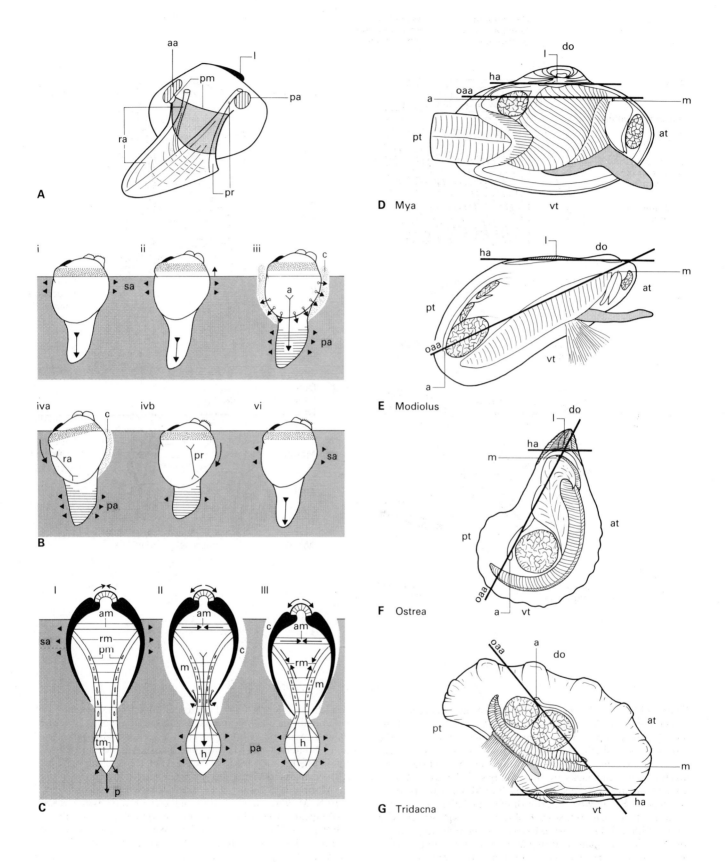

A

B

C

D Mya

E Modiolus

F Ostrea

G Tridacna

secretes nacreous aragonite, lustrous and sometimes valuable pearls are formed. The oldest reported traces of pearls are from the Late Silurian.

The *ciliated gills* or ctenidia have a complex anatomy. Because most pelecypods are suspension feeders, the gills function both for respiration and food gathering. In deposit feeders, the primary function of the gills is respiration. A few pelecypods, mostly deep-sea carnivorous species, lack gills. Some species brood eggs in the gills.

The gills are usually flat, elongated thin sheets of tissue that attach to the two sides of the visceral mass and/or to the proximal part of the foot (Figures 14.87 and 14.90, *G* and *I*). In some species that have only a posterior adductor muscle, the gills wrap around this structure (Figure 14.87, *A* and *D*). Gills may attach by membranes (Figure 14.94, *H*) or muscles along their upper sides to the foot or visceral mass medially and to the mantle laterally. Sometimes, the **lateral gill muscles** attach to the shell and leave muscle scars (Figures 14.93, *A* and *B* and 14.94, *F*). In a few species, the position of the lateral insertion of the gills is marked by an umbonal fold or sulcus on the exterior of the shell (Figure 14.94, *G*). However, gills ordinarily leave little indication of their position or form on the shell.

Each of the two gills consists of a hollow, longitudinal **axis** to each side of which are attached numerous **gill filaments** (Figure 14.94, *H* to *J*). All the gill filaments on a side form a **demibranch** (Figure 14.94, *H* and *J*). The inner demibranch is closest to the visceral mass, and the outer demibranch is closest to the mantle (Figures 14.90, *G* and 14.94, *H*).

The anatomically simplest type of gill has broad leaf-like gill filaments (Figure 14.94, *H*) that are either unconnected or attached by ciliary junctions. Pelecypods having this **protobranch gill** probably occur in rocks as old as the Early Ordovician, and this gill type was probably in the oldest known Cambrian pelecypod. In living species, the protobranch gill is largely limited to deposit feeders that have palp proboscides.

In most pelecypods, the gill filaments are narrow, elongate, lathlike bodies that ventrally are bent back on themselves (Figure 14.94, *I* and *J*). A typical gill consists of four lamellae, a descending and an ascending one in each demibranch (Figure 14.94, *I* and *J*). Adjacent filaments have small patches of elongate cilia on their anterior and posterior sides that interlock (Figure 14.94, *J*). Pelecypods with narrow gill filaments having ciliary junctions are **filibranchs**. Some filibranchs also have **interlamellar junctions** between the ascending and descending lamellae of each demibranch (Figure 14.94, *I* and *J*). Many epifaunal pelecypods have filibranch gills. **Eulamellibranch** pelecypods have gills, on which the interfilamentar and interlamellar junctions are ordinarily composed of vascularized connective tissue (Figure 14.94, *K*). Most burrowing suspension-feeding pelecypods are eulamellibranchs.

Each gill lamella has numerous cilia along its long surface whose beating generates the inhalant current. Upon entering the mantle cavity, the current is slowed, and the heavier particles sink to the bottom of the cavity; here cilia on the inside of the mantle carry the particles to the commissure. These pseudofeces are periodically

---

Figure 14.95. A to C, Burrowing in pelecypods. **A,** Generalized diagram of a pelecypod with the foot extended showing the principal muscles involved in burrowing. *aa,* Anterior adductor muscle; *l,* ligament; *pa,* posterior adductor muscle; *pm,* protractor muscle; *pr,* posterior retractor muscle; *ra,* anterior retractor muscle. **B,** Series of diagrams showing the exterior of a generalized pelecypod from the left side burrowing at the six steps (*i* to *vi*) of the digging cycle. The stippled band across the left valve indicates movement of the animal with reference to the sediment-water interface. Horizontal lines on the foot show the region of the pedal anchor (*pa*). ◄———, Movement of the shell; ◄———O, water ejection from the mantle cavity loosening the sediment around the shell; ◄———◄, probing and extention of the foot; ◄—<, hydrostatic pressure produced by adduction (*a*) of the valves filling the pedal hemocoel and causing pedal dilation; >———<, contraction of first the anterior (*ra*) and, second, the posterior (*pr*) retractor muscles; ◄, indicates pedal (*pa*) and shell (*sa*) anchors. **C,** Cross-section diagrams of successive stages in the burrowing of a generalized pelecypod, showing shell (*sa*) and pedal (*pa*) anchors (◄). *I* represents the steps i, ii, or vi of the digging cycle with the valves pressed against the sediment by means of the opening thrust to the ligament and the foot extending and probing (*p* ◄———). *II* represents step iii, where contraction of the adductor muscles (*am*) ejects water (◄———O) from the mantle cavity (*m*), loosening the sediment (*c*) around the valves; high pressure simultaneously produced in the hemocoel (*h*) gives rise to pedal dilation (◄———<). *III,* represents step iv, contraction of retractor muscles (*rm*) pulls the shell down into the loosened sediment. *pm,* Protractor muscle fibers; *tm,* transverse intrinsic pedal muscles; ►◄, tension in ligament, adductor, or retractor muscles. **D to G,** Orientation and relationship of hinge and mouth–anus axes in selected pelecypods. In all species, the rectum leading to the anus is dorsal to the posterior adductor muscle. **D,** *Mya.* Isomyarian with the hinge axis (*ha*) parallel to the mouth–anus axis (*oaa*). This burrowing form has large foot and siphons. **E,** *Modiolus.* Anisomyarian with the hinge axis at a low angle to the mouth–anus axis. It attaches byssally with a small foot and no siphons. **F,** *Ostrea.* Monomyarian with the hinge axis at a high angle to the mouth–anus axis. This cemented form lacks both the foot and siphons. **G,** *Tridacna.* Monomyarian with the hinge forming an oblique angle with the mouth–anus axis. It byssally attaches with a small foot and siphons (not shown; see Figure 14.96, *A*). In *Tridacna* the umbo is regarded as ventral. *a,* Anus; *at,* anterior; *do,* dorsal; *l,* ligament; *m,* mouth; *pt,* posterior; *vt,* ventral. (**A to C** modified from Trueman, E. R. *Symposium of Zoological Society of London* 22: 167–186; 1968. **D to G** modified from Cox, L. R. *Treatise on invertebrate paleontology, Part N, Mollusca 6, Bivalvia,* vol. 1. Boulder, CO, and Lawrence, KS: Geological Society of America and University of Kansas Press; 1969, p. N80.)

ejected by sudden contraction of the adductor muscles. The remaining suspended material is strained off by the gills, entangled in mucus, and carried by ciliary action along the ventral food groove (Figure 14.94, *I*). The anterior end of each gill terminates at the labial palps (Figures 14.87, *C*, *D*, and *F*; 14.90, *G*; and 14.94, *L* and *M*) on either side of the mouth. Additional sorting occurs on the ciliated labial palps. The accepted material then goes into the mouth. In deposit-feeding species with palp proboscides, these structures are attached to the outer pair of labial palps (Figure 14.90, *I*).

The third group of soft parts is the *foot and byssus*. In burrowing pelecypods, the foot projects into the mantle cavity and is large, highly inflatable and muscular (Figures 14.87, *C*, *E*, and *F*; 14.88, *A*; and 14.90, *I*). It is located along the midline and is directed ventrally, anteroventrally, or anteriorly from the visceral mass. Primitively the foot is an organ of burrowing and locomotion in relatively soft substrates. In species that have a small anterior adductor muscle (Figure 14.87, *B*) or only a posterior adductor muscle (Figure 14.87, *A*), the foot of the adult is small and not used for burrowing. In these species, the foot makes the byssus, a structure composed of hairlike threads that allows pelecypods to attach to hard substrates (Figure 14.87, *B*). A few species with only a posterior adductor muscle lose the foot entirely in early ontogeny and lack a byssus (Figure 14.87, *D*). In a few specialized species, the foot has evolved into an organ for creeping over hard substrates.

Burrowing consists of a series of movements involving the integrated action of the muscular system. The movements that occur during each downward thrust into the sediment (Figure 14.95, *B*) are repeated until the desired position in the substrate is achieved.

The hemocoel extends into the ventral part of the foot (Figure 14.95, *CII* and *III*). Before inflation, this region is often compressed like the blade of an ax (Figure 14.95,

CI) and is adapted for penetration of relatively soft substrates. The musculature of the foot is complex and consists of the following:

1 Two or more pairs of shell-inserted **pedal retractors** (Figures 14.93, *A* and *B* and 14.95, *A*), which function to withdraw the foot into the shell or to pull the shell over the foot (Figure 14.95, *C*).

2 **Intrinsic transverse pedal muscles** (Figure 14.95, *CI*), which oppose the pedal retractors.

3 Often a pair of shell-inserted **pedal protractors** (Figure 14.95, *A*), which also oppose the pedal retractors and help pull the foot outside of the shell.

The fibers of the pedal retractors extend throughout the ventral region of the foot (Figure 14.95, *A*), forming a basketlike stress grid (geodetic network). The transverse muscles insert into connective tissue in the epithelium of the foot (Figure 14.95, *C*). The blood in the pedal hemocoel functions as the fluid in a hydrostatic skeleton and is prevented from flowing out by a valve. Differential tension in the retractor muscles often imparts a rocking motion to the shell during burrowing, as first the anterior and then posterior pedal retractors contract (Figure 14.95, *Biv*).

Digging cycles consist of repeated adduction and opening of the valves integrated with protraction and retraction of the foot. Adduction creates a **pedal anchor** by causing maximum swelling of the foot (Figure 14.95, *Biii*, *Biv*, *CII*, and *CIII*). It also reduces the width of the shell (Figure 14.95, *CIII*) and ejects water from the mantle cavity (Figure 14.95, *Biii* and *CII*), which loosens the sediment around the shell so that during retraction (Figure 14.95, *Biv* and *CIII*) the shell moves downward more easily. Loss of pedal anchorage occurs at the end of retraction, and the distal end of the foot shrinks (Figure 14.95, *Bivb*). When the adductors are relaxed, the valves are opened by the ligament. This presses the valves against the sediment, forming a **shell anchor** (Figure

---

Figure 14.96. A to D, Byssus and foot in living pelecypods. **A,** *Tridacna*. Epifaunal form with the byssus near the beak and small foot. The left valve is removed. For ease of comparison the ligament of this form is at the top of the figure. Detailed functional studies indicate that *Tridacna* differs from all the other pelecypods in having both the ligament and byssus in the ventral position. Compare with Figure 14.95, *G*. (Holocene, Eniwetok Atoll, Marshall Islands, central Pacific Ocean, ×1.5.) **B,** *Arca*. Epifaunal form with a large byssus and foot near the midventral length. Byssus fibers are short. The right valve is removed. (Holocene, Eniwetok Atoll, Marshall Islands, central Pacific Ocean, ×2.) **C** and **D,** *Pinna*. (Holocene, Florida.) Semi-infaunal form with a large byssus near the anterior end. This pelecypod is strongly anisomyarian. **C,** Entire soft parts, with right valve removed. (×1.) **D,** Enlargement of the anterior end showing the byssus, byssal pit, byssal groove, and foot. (×3.)

aa, Anterior adductor; *b,* byssus; *bg,* byssal groove; *bp,* byssal pit; *c,* catch part of posterior adductor; *f,* foot; *g,* gill; *l,* ligament; *lp,* labial palp; *mc,* mantle cavity; *pa,* posterior adductor; *ppr,* posterior pedal retractor; *q,* quick part of posterior adductor. **E to H,** Comparison of gross forms of living species of byssate pelecypods (**E** and **F**) with similarly shaped Ordovician species (**G** and **H**). All four species are in the same superfamily. **E** and **F,** *Pteria*. **E,** Species with prominent posterior wing (*w*), lobe (*lo*), and small byssal sinus (*bs*). (Eniwetok Atoll, Marshall Islands, central Pacific Ocean, ×2.) **F,** Species with a less prominent wing, more prominent lobe, and larger byssal sinus. (Florida, ×2.) **G** and **H,** *Caritodens*. **G,** Species with a prominent wing. The anterior part of the shell is not preserved. (Upper Ordovician, Ohio, ×2.5.) **H,** This species has a small wing, small lobe, and prominent byssal sinus. (Upper Ordovician, Ohio, ×4.)

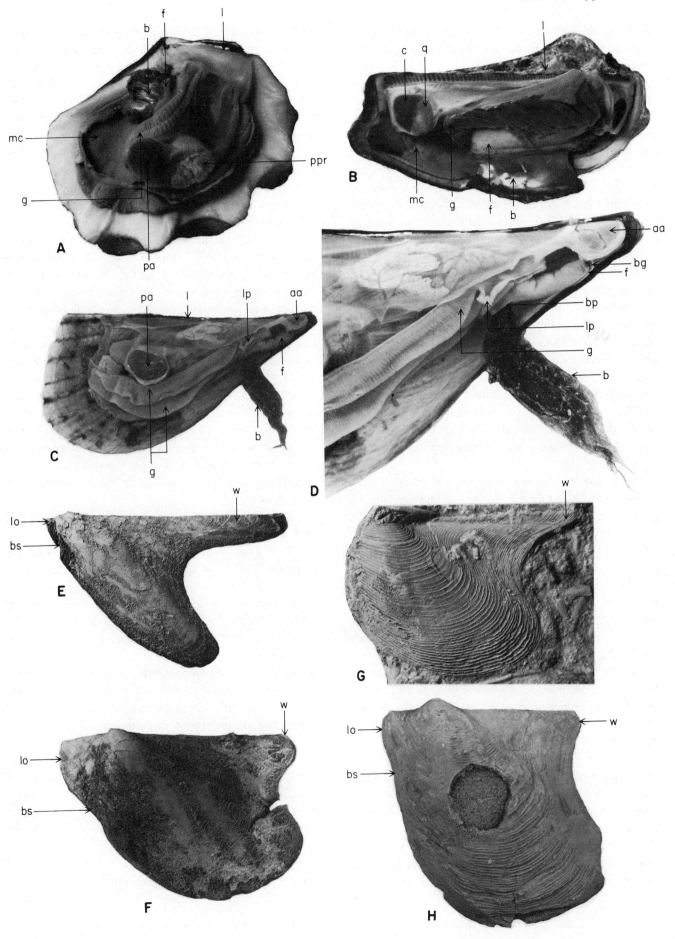

14.95, *Bi*, *Bii*, and *CI*), which prevents the animal from being pushed upward in the burrow during the next pedal protraction and probing.

Pelecypods are among the most successful inhabitants of soft substrates. They have been adapted for this since at least Early Ordovician time, when fairly large species first appear in the fossil record. The small size of known Cambrian pelecypods suggests that terms like epifaunal and infaunal may not be applicable to them. Some Cambrian species may have been interstitial in their mode of life, living between grains of sediment because they were often of the same size range as medium to coarse sand, or they may have lived in the slurry at the sediment-water interface.

The ability to burrow, combined with increase in size, may in part explain the diversification of pelecypods in the Ordovician. In the Cambrian, known rostroconch diversity is greater than that of known pelecypods. However, in the Ordovician, pelecypods diversified rapidly and rostroconchs declined. Most rostroconchs were also burrowing animals, but they had a pseudobivalved shell that lacked a ligament and adductor muscles. In post-Ordovician time, rostroconchs remained a minor element of Paleozoic faunas, whereas pelecypods are represented by hundreds of genera and thousands of species.

The byssus is secreted by the **byssal gland** (Figure 14.94, *A*) located at the base of the foot and typically is a group of hairlike threads (Figures 14.87, *B*; 14.94, *C*; and 14.96, *D*) that attach to rocks, shells, plants, or other hard substrates. Attachment may be throughout postlarval ontogeny or only in the early postlarval stage.

Byssal glands secrete a fluid that accumulates in the **byssal pit, byssal groove**, and **byssal disc pit** (Figures 14.94, *B* and 14.96, *D*). This fluid solidifies instantaneously in water and becomes a **byssal thread** (Figure 14.94, *B*). All the structures associated with the formation of the byssus are the **byssal apparatus**. The end of the foot places the distal end of the byssal thread on the attachment surface, and each byssal thread ends in a small disc-shaped sucker. The proximal ends of the byssal threads remain attached in the foot (Figures 14.94, *B* and *C* and 14.96, *D*). Byssal threads are made and attached one at a time and are chemically a conchiolin-like protein.

Adults that have a byssus usually have a highly reduced foot (Figures 14.87, *B*; 14.94, *A* to *C*; and 14.96, *A*, *C*, and *D*), the main function of which is to make and attach byssal threads. The ark shells (Figure 14.96, *B*) are an exception in that some of them have a byssus and a

well-developed foot. Byssate pelecypods can adjust themselves on the **byssal anchor** so as to maximize feeding currents, they can clamp against it for protection, and they can break the attachment and reestablish it elsewhere. These movements on the byssus are accomplished by modified pedal retractor muscles, which are termed **byssal–pedal retractor muscles** (Figure 14.90, *H*). The scar of the posterior byssal–pedal retractor muscles may be markedly elongated (Figure 14.97, *D*, *H*, and *J*).

A well-developed byssus is present in the adults of many species (Figure 14.96, *A* to *D*), but adults of many other species have only a vestigial byssal gland that does not secrete byssal threads. A small byssus is widespread in the early postlarval stage of many distantly related pelecypods (Figure 14.94, *D* and *E*). In *Mya*, at the 2–3 mm long stage, a single branching byssal thread enables it to anchor to sand grains and other small solid objects. In this way, the animal lives on the substrate while it is developing the necessary structures and proportions to assume the vertical burrowing habits of the adult (Figure 14.89, *J*). Because the presence of a byssus or a vestigial byssal gland is widespread taxonomically in distantly related pelecypods, and because of the importance of the byssus in various pelecypods in the early postlarval stage, the byssal apparatus in adult pelecypods is interpreted to represent the persistence of an early postlarval structure, and these pelecypods are thought to be neotenous in this regard.

The following shell features are generally associated with the presence of an adult byssus:

1   Reduction of the anterior end so that the beaks are **terminal** (Figure 14.97, *B*, *D*, *F*, and *G*) or nearly so (Figure 14.97, *C*).

2   Reduction in the size (Figure 14.97, *D*) or loss (Figure 14.97, *H*) of the anterior adductor muscle.

3   Presence of a **byssal gape** (Figure 14.97, *A* and *I*) in the commissure.

4   Occurrence of a **byssal sinus**, an anterodorsal embayment in the lateral shell profile (Figures 14.96, *E*, *F*, and *H* and 14.97, *F* and *G*) or, in scallops, a small **byssal notch** that sharply sets off the anterodorsal portion from the rest of the shell (Figure 14.97, *E*).

5   Occurrence of an **umbonal septum**, which is a thin shelly plate near the beak on which the anterior adductor muscle inserts (Figure 14.97, *B* and *C*).

6   Comparison of shell shapes of fossil species with similar living forms that have a byssus (Figure 14.96, *E* to *H*).

Byssally attached pelecypods are known in the fossil record from at least the late Early Ordovician.

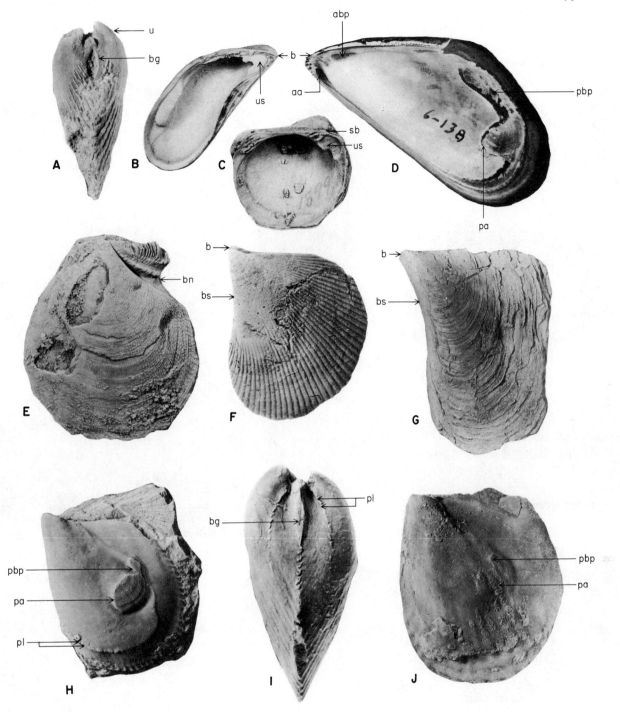

Figure 14.97. Structures associated with the presence of a byssus. **A,** *Ambonychia.* Anterior view of a composite mold showing the fusiform byssal gape (*bg*). *u,* Umbo. (Upper Ordovician, Ohio, ×2.) **B,** *Septifer.* Interior of the left valve showing the umbonal septum (*us*), with a notch made by the passage of the anterior byssal-pedal retractor muscle, and the terminal beak (*b*). (Holocene, Japan, ×1.5.) **C,** *Vanuxemia.* Interior of silicified left valve, showing umbonal septum (*us*) with a notch and subterminal beak (*sb*). (Middle Ordovician, Tennessee, ×1.5.) **D,** *Mytilus.* Interior of right valve, showing scars of anterior (*abp*) and posterior (*pbp*) byssal-pedal retractor muscles, anisomyarian adductor muscles (*aa, pa*), and the terminal beak (*b*). (Holocene, Long Island, New York, ×2.) **E,** *Streblochondria.* Right valve showing the byssal notch (*bn*) and groove. (Pennsylvanian, Texas, ×2.5.) **F,** *Ambonychia.* Left lateral view of a composite mold, showing the terminal beak (*b*) and byssal sinus (*bs*). (Upper Ordovician, Ohio, ×2.) **G,** *Myalina (Orthomyalina).* Left lateral view, showing the terminal beak (*b*) and byssal sinus (*bs*). (Pennsylvanian, Texas, ×0.8.) **H,** *Ambonychia.* Internal mold of the left valve, showing the subcentral posterior adductor scar (*pa*), scar of posterior byssal-pedal retractor (*pbp*), and pitted pallial line (*pl*). (Upper Ordovician, Kentucky, ×3.) **I,** *Ambonychia.* Anterior view of an internal mold, showing a slitlike byssal gape (*bg*). The anterior part of the pitted pallial line (*pl*) with no expanded portion indicates the lack of an anterior adductor muscle. (Upper Ordovician, Ohio, ×2.) **J,** *Anomalodonta.* Internal mold of the left valve showing the subcentral posterior adductor scar (*pa*) and scar of posterior byssal-pedal retractor (*pbp*). (Upper Ordovician, Ohio?, ×1.)

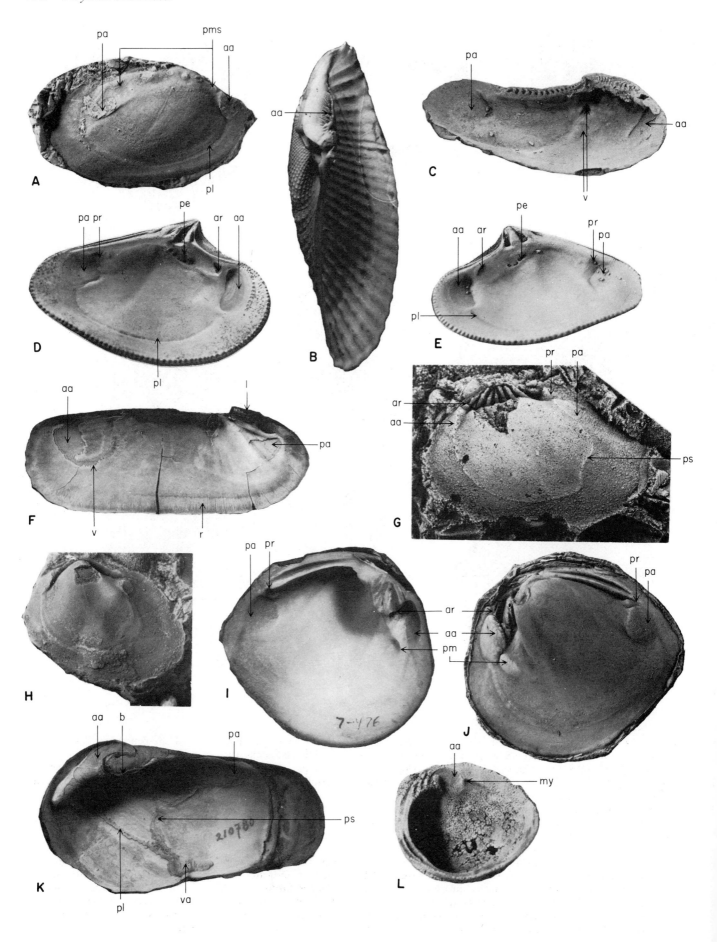

Muscles are the fourth group of soft parts. Although muscle scars are found on the shell, they are discussed here along with the muscles that make them. Muscle scars are widely used by paleontologists for the interpretation of the soft-part anatomy, behavior, and habitats of extinct pelecypods. In pelecypods, as in other shelled mollusks, muscle scars are often best seen on well-preserved internal or composite molds (Figures 14.92, *D*; 14.93, *A* and *B*; and 14.97, *H* to *J*). Pelecypod muscles are divided into five groups: intrinsic, adductor, pallial, shell, and miscellaneous muscles.

Intrinsic muscles are entirely within the soft parts. The intrinsic transverse muscles of the foot were discussed earlier (Figure 14.95, *C*). Adductor muscles are used for closing the shell. The evolutionary origin of these is debated; one school of thought regards them as enlarged pallial muscles formed in the mantle embayments (Figure 14.89, *H*). Pallial muscles are associated with the mantle and can insert as the pallial line and pallial sinus (Figure 14.92) and perhaps as point attachments (Figure 14.92, *D*). The term **shell muscles** is sometimes applied to muscles that attach to the shell and are associated with movements of the foot and/or byssus (Figures 14.89, *H*; 14.90, *H*; 14.93, *A* and *B*; and 14.98, *A*, *D*, *E*, *G*, *I*, and *J*). Miscellaneous muscles may attach such structures as the visceral mass (Figure 14.98, *C* and *F*) and gills (Figure 14.93, *A* and *B*) to the shell.

Adductors are the largest muscles in the class. They pass from valve to valve and close the two halves of the shell (Figures 14.87 and 14.89, *H*). Most pelecypods have both anterior and posterior adductor muscles and are **dimyarian** (Figure 14.92, *A* to *C*, *G*, and *H*). If the two adductors are subequal in size, they are **isomyarian** (homomyarian) (Figure 14.92, *A*). If they differ significantly in size, they are **anisomyarian** (heteromyarian) (Figure 14.97, *D*). In the vast majority of anisomyarian

species, the anterior adductor is smaller than the posterior one (Figure 14.97, *D*); however, in a few, the posterior adductor is reduced (Figure 14.98, *F*). Species with a single adductor are **monomyarian** (Figures 14.87, *A* and *D* and 14.97, *H* and *J*). In most of these, it is the anterior adductor that has been lost, although in a few, the posterior adductor is absent.

The adductors are most often within the dorsal half of the shell (Figures 14.92, *A* to *C*, *E*, *F*, and *H* and 14.98, *A*, and *C* to *E*). In anisomyarian species, the posterior adductor is lower than the anterior one (Figures 14.96, *C* and 14.97, *D*). In monomyarian species, the posterior adductor is subcentral in position (Figures 14.96, *A* and 14.97, *H* and *J*). In most, the adductor scars are flush within the inner surface or form shallow depressions in it (Figure 14.92). A few species, the oldest of which is Middle Ordovician, have the anterior adductor mounted on an umbonal septum (Figure 14.92, *F* and 14.97, *B* and *C*), which often has a prominent notch formed by the anterior byssal–pedal retractor muscle. In some species, most of which are Mesozoic rudists but some of which are as old as Middle Ordovician (Figure 14.98, *L*), adductors were mounted on shell thickenings called **myophores**. A few forms have the anterodorsal part of the shell reflected over the umbo, and the anterior adductor is attached to this reflected part (Figure 14.98, *K*). In such species, contraction of the anterior adductor opens the valves, and contraction of the posterior adductor closes them. However, in most dimyarian pelecypods, the two adductors close the shell and contract or relax at the same time.

Often the adductors are formed of two distinct bundles of muscle fibers, each attached to its own part of the muscle scar (Figures 14.92, *D* and 14.96, *B*). **Quick muscle fibers** are striated, located to the inside of the catch muscle fibers, and function for rapid closing of

---

Figure 14.98. Major muscle scars. **A**, *Cycloconcha*. Oblique lateral view of the internal mold of the right valve, showing several pedal muscle scars (*pms*), adductor scars (*aa, pa*), and pallial line (*pl*). (Upper Ordovician, Kentucky, ×3.) **B**, *Cyrtopleura*. Dorsal view showing anterodorsal part of the shell reflected over the umbo. This is the site of the anterior adductor muscle scar (*aa*), which is above the hinge. (Holocene, Florida, ×1.) **C**, *Phestia*. Oblique internal view of a silicified specimen showing teeth, adductor scars (*aa, pa*) and visceral retractor scars (*v*). (Permian, Texas, ×1.5.) **D** and **E**, *Astarte*. Oblique internal views of left (**D**) and right (**E**) valves showing adductor (*aa, pa*), pedal retractor (*ar, pr*), and pedal elevator (*pe*) muscle scars and integripalliate pallial line (*pl*). (Miocene, Chesapeake Bay, Maryland, ×2.) **F**, *Solemya*. Interior of the right valve showing large anterior adductor (*aa*), smaller posterior adductor (*pa*), visceral (*v*), and radial mantle (*r*) muscle scars. *l*, Ligament. (Holocene, New Zealand, ×1.5.) **G**, *Lyrodesma*. Internal mold of the left valve, showing adductor (*aa, pa*) and

pedal retractor (*ar, pr*) muscle scars and a small pallial sinus (*ps*) in the pallial line. (Upper Ordovician, New York, ×2.5.) **H**, *Lyrodesma*. Internal mold, showing the same features as in Figure 14.98, *G*. In this specimen, the pallial insertions are doubled. This is the oldest known species with a pallial sinus. (Lower Upper Ordovician, Kentucky, ×2.5.) **I** and **J**, *Fuscanaia*. Interior of the left valve (**I**) and latex mold of same (**J**) showing adductor (*aa, pa*), pedal retractor (*ar, pr*), and pedal protractor (*pm*) muscle scars. (Holocene, Ohio River, Ohio, ×1.) **K**, *Pholadidea*. Interior of the right valve, showing reflected part of shell, which is the site of the anterior adductor muscle scar (*aa*) wrapping around the beak (*b*), posterior adductor muscle scar (*pa*), ventral adductor muscle scar (*va*), pallial line (*pl*), and pallial sinus (*ps*). (Holocene, Pacific Ocean, Alaska, ×2.) **L**, *Vanuxemia*. Oblique interior view with anterior end up showing teeth and myophore (*my*), on which is the anterior adductor muscle scar (*aa*). (Middle Ordovician, Ontario, Canada, ×2.)

the valves. **Catch muscle fibers** are smooth, located to the outside of the quick muscle fibers (Figure 14.96, *B*), fatigue less quickly than quick muscle fibers, and serve to hold the two valves together in a fixed position for prolonged periods of time.

Pallial muscles are in the mantle and consist of both radially directed and comarginal fibers (Figure 14.92, *H*). The radially directed fibers are strongest near the periphery of the shell (Figures 14.90, *H* and 14.92, *H*), where they may attach at the pallial line. The radial mantle muscles withdraw the mantle edge before the valves close. The pallial line may have a continuous attachment ventrally from one adductor muscle scar to the other (Figure 14.92, *A* to *C*), or the fibers may be bundled and form a discontinuous attachment (Figure 14.97, *H* and *I*). Some species have a line of pallial attachment dorsal to the adductors as well as ventral to them. Most isomyarian pelecypods have a distinctly impressed pallial line and are burrowers; byssate anisomyarian or monomyarian pelecypods often lack a pallial line. Occasionally, some or all of the pallial line may be doubled (Figure 14.98, *H*). This doubling can occur in at least two ways: (1) earlier insertions of the radial mantle muscles may not be completely covered by inner shell layer, or (2) the numerous radial mantle muscles may have more than one line of insertion.

In siphonate species, the posterior radial mantle muscles may form siphonal retractors that may attach in the pallial sinus of the pallial line (Figure 14.92, *A* to *C*). Pelecypods having such an embayed pallial line are **sinupalliate**; those having an entire pallial line are termed **integripalliate**. The oldest known sinupalliate forms are Late Ordovician in age (Figure 14.98, *G* and *H*). A few pelecypods have other modifications of the pallial muscles. The most widespread of these is an accessory or **ventral adductor muscle** (Figure 14.98, *K*).

Shell muscles associated with the foot or byssus can range from one (Figure 14.97, *H*) to several pairs (Figures 14.93, *A* and *B* and 14.98, *A*) and are divided into pedal retractors (Figure 14.98, *I* and *J*), **pedal protractors** (Figure 14.98, *I* and *J*), and **pedal elevators** (Figure 14.98, *D* and *E*). In isomyarian species, the pedal retractors are most often positioned above or near the top of the adductor muscles (Figure 14.98, *D* and *E*). There are usually two pairs of pedal retractors, and most species having this arrangement are infaunal. In byssate species, the position of the byssal–pedal retractor(s) is variable with relation to the unequal-sized or single adductor muscle. When present, pedal protractors are always anterior in position, and ordinarily one pair is placed low with relation to the anterior adductor (Figures 14.95, *A* and

14.98, *I* and *J*). Pedal elevators serve to raise the foot into the shell; there may be one to several pairs. When present, they insert in the umbonal cavity between the pedal retractor muscles (Figure 14.98, *D* and *E*).

Three kinds of miscellaneous muscles are widespread. Several families have **visceral retractor muscles** that attach to the shell (Figure 14.98, *C* and *F*); such muscles were present in some Cambrian species. Some species lack gills and have a **muscular septum** that is used to pump water and small worms and crustaceans into the mantle cavity. The insertion of the muscular septum can be continuous or pitted (Figure 14.105, *E*), and species possessing it are known from Cretaceous time. The third kind of miscellaneous muscles, the gill muscles (Figure 14.94, *F*), insert on the shell in several living families and were probably present in an Ordovician species (Figure 14.93, *A* and *B*).

The last soft-part group, the *visceral mass* is a complex of organs in the umbonal cavity (Figure 14.105, *F*) and contains the major organs of digestion, excretion, circulation, and the nervous system. Except for species that have visceral retractor muscle scars (Figure 14.98, *C* and *F*), the visceral mass leaves little evidence of its complexity in fossils. The size and convexity of the umbos give a rough estimate of the size and position of the visceral mass.

Shell morphology

External features and orientation The shell of pelecypods has two beaks, one in each valve (Figures 14.99, *B* and *E* and 14.100, *B*, *C*, *E*, and *F*), because calcification of the larval shell begins to the right and left of an uncalcified cuticle, the medial part of which remains uncalcified and becomes the hinge line. The bivalved larval shell is formed prior to metamorphosis and is the **prodissoconch**. Prodissoconch I is the shell secreted by the cuticle and early mantle (Figure 14.99, *B*, *D*, *E*, and *G*); it is often D-shaped with a straight dorsal margin. In planktotrophic larvae, prodissoconch I ranges from 70 to 150 μm in length. Prodissoconch II is secreted entirely by the mantle (Figure 14.99, *B*, *D* and *E*). It ranges in size from 200 to 600 μm in length. The junction between prodissoconchs I and II is the point in ontogeny at which the larval shell surrounds the entire organism. At metamorphosis, the velum is lost, and secretion of the **dissoconch** or adult shell begins. The ornament, surface texture, and shape of the dissoconch may differ markedly from those of the prodissoconch (Figure 14.99, *A*, *C*, and *G*). Often a distinct line (Figure 14.99, *A*) marks the prodissoconch–dissoconch boundary.

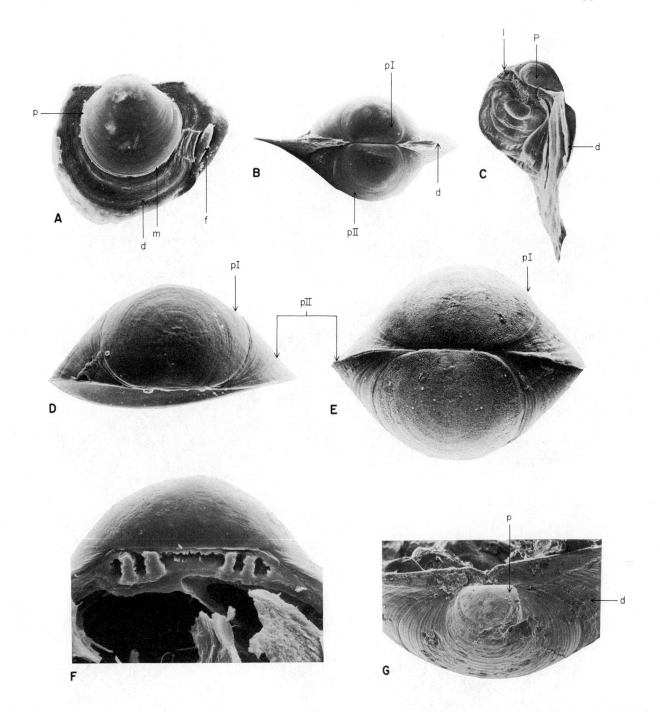

Figure 14.99. Larval shells. **A** and **B**, *Argopecten irradians.* (Holocene, Virginia.) Right lateral and dorsal views of a 9-day-old scallop. **A** shows the prodissoconch, dissoconch, the boundary between the two, and the byssal notch with the foot protruding. (×250.) **B** shows the D-shaped prodissoconch I, prodissoconch II, and dissoconch. (×300.) **C** to **E**, *Ostrea edulis.* **C,** Oblique dorsal view of a 20-day-old oyster showing the prodissoconch and dissoconch. The valves of prodissoconch are separated by growth of the dissoconch ligament. **D,** Dorsal view of the prodissoconch left valve shows the D-shaped prodissoconch I, prodissoconch II, anal notch, and track of anal notch. **E,** Dorsal view of the prodissoconch of an articulated specimen showing the same features as in **D.** The left valve is lowermost.

(Holocene, Long Island, New York; **C** ×38; **D** ×406; **E** ×350.) **F,** *Crassostrea gigas* showing the hinge of the left valve with taxodont-like teeth. (Holocene, Long Island, New York, ×1260.) **G,** *Striarca webbervillensis.* Dorsal view of one valve, showing the protoconch and dissoconch. (Upper Cretaceous, Maryland, ×280.) *d,* Dissoconch; *f,* foot; *l,* ligament; *m,* protoconch–dissoconch boundary line; *p,* prodissoconch; *pI,* prodissoconch I; *pII,* prodissoconch II (**A** to **F** courtesy of T. R. Waller, Smithsonian Institution, Washington D.C. **C** to **F** from Waller, T. R. *Smithsonian Contributions to Zoology No. 328;* 1981, pp. 51, 55, 58, 60. **D** courtesy David Jablonski, identified by N. F. Sohl.)

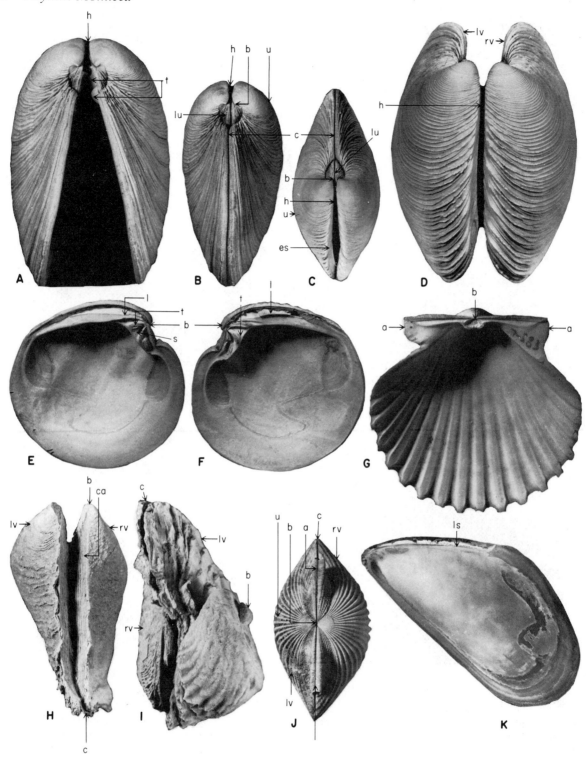

Figure 14.100. General shell features; all interior views have the dorsal margin up. **A** to **F**, *Dosinia*. (Holocene, New Zealand.) Equivalved inequilateral shell with prosogyrate beaks. **A** and **B**, Anterior views of articulated valves open (**A**, ×2) and closed (**B**, ×1.5) showing pivoting on hinge axis. **C** and **D**, Oblique dorsal and dorsal views of articulated valves closed (**C**, ×1) and open (**D**, ×2). **E** and **F**, Interior views of left (**E**) and right (**F**) valves. (×1). **G** and **J**, *Argopecten*. Inequivalved, equilateral shell with orthogyrate beaks. **G**, Interior of right valve showing straight dorsal margin. **J**, Dorsal view of closed articulated valves. (Holocene, eastern North America, ×1.5.) **H**, *Myalina*

(Orthomyalina). Dorsal view of a shelled specimen showing terminal beaks, interumbonal growth, and valve inequality. (Upper Pennsylvanian, Texas, ×2.) **I**, *Exogyra*. Dorsal view of a shelled specimen showing the coiled beak of the left valve and valve inequality. The anterior end is down. (Upper Cretaceous, Texas, ×2.) **K**, *Mytilus*. Interior of right valve showing muscle scars and elongate ligament support. (Holocene, Long Island, New York, ×2.) *a*, Auricle; *b*, beak; *c*, commissure; *ca*, cardinal area; *es*, escutcheon; *h*, approximate position of hinge axis; *l*, ligament; *ls*, ligament support; *lv*, left valve; *rv*, right valve; *s*, socket; *t*, teeth; *u*, umbo.

In most pelecypods, the valves are **hinged** and open and close along one margin (Figure 14.100, *A* to *F*). Passing through this margin is a hypothetical straight line termed the **hinge axis** that goes through the ligament (Figure 14.95, *D* to *G*) and joins the points about which the valves rotate. All structures along the hinged margin that function during the opening and closing of the valves are collectively termed the **hinge** (Figure 14.100, *E* and *F*).

The hinge and mantle isthmus are regarded as dorsal to the soft parts in most pelecypods. Ventral is the part of the shell opposite to the hinge, where the valves open most widely (Figure 14.100, *A*). Anterior is towards the mouth, and posterior is toward the anus (Figure 14.95, *D*). In typical isomyarian species, the hinge axis is parallel to a line connecting the mouth and anus (Figure 14.95, *D*); however, in many pelecypods, the line connecting the mouth and anus is at a considerable angle to the hinge axis (Figure 14.95, *E* to *G*). Even in most of these species, the hinge is regarded as dorsal and the other body coordinates are established with relation to this margin as mutually orthogonal directions in the commissural plane (Figure 14.95, *E* and *F*). In the giant clam, *Tridacna*, the hinge axis extends anteroposteriorly, but based on complex anatomic and phylogenetic interpretations, it is regarded as being anteroventral in position (Figure 14.95, *G*).

The dorsal margin of the shell is divided into an anterodorsal part in front of the beaks and a posterodorsal part behind them (Figure 14.100, *E* and *F*). These margins may form compressed often triangular areas called **auricles** or **ears** (Figure 14.100, *G* and *J*), or they may be extended beyond the body of the shell as **wings** or **lobes** (Figure 14.96, *E* to *H*); wings are usually at the posterior and lobes at the anterior end. Various dorsal features are helpful in determining the direction of the hinge axis, including an elongate ligament (Figure 14.100, *E* and *F*), a long straight dorsal margin (Figure 14.100, *G*), and elongate ligament supports (Figure 14.100, *K*). However, even in some living species with short ligaments, the direction of the hinge axis must be established by opening and closing the valves manually.

When describing the lateral profile of genera of pelecypods, shapes of new genera are compared with those of already well-known genera. The shape terms are based on the well-known genera. **Mytiliform** means a shape similar to *Mytilus* (Figure 14.100, *K*); **pteriaform** means a shape similar to that of *Pteria* (Figure 14.96, *E* and *F*); **pectiniform** means a shape similar to *Pecten* (Figure 14.100, *G*); and so on. Typically the right and left valves of pelecypods are bilaterally symmetrical, and the commissure coincides with the midsagittal plane (Figures 14.86, *C* and *G*; 14.88, *A*, *B*, and *F*; and 14.100, *B* and *C*); such species are **equivalved**. However, the valves of many species are not symmetrical about the commissure and are **inequivalved** (Figure 14.100, *H* to *J*).

Debate exists about the conventions to be used when measuring pelecypod shells. Here it is suggested that measurements be made with the hinge axis horizontal. The **length** of the shell is the distance between two planes perpendicular to the hinge axis (Figure 14.101, *A* and *B*) and just touching the anterior and posterior extremities of the shell. **Height** is the greatest dimension perpendicular to length (Figure 14.101, *A* and *B*). **Width** is the distance between two planes parallel to the commissural plane that are tangent to the outermost parts of the two valves (Figure 14.101, *C*).

If in the adult shell, the beaks occupy a position close to the middle of the length, the shell is **equilateral** (Figure 14.100, *G* and *J*); in reality, few pelecypods are truly equilateral. Even in scallops, most species show slight differences in the shape of the anterior and posterior auricles (Figure 14.100, *G*), and thus the anterior and posterior parts of the shell are not exact mirror images. If the beaks are closer to one end of the shell than to the other, the species is **inequilateral** (Figures 14.100, *E* and *F* and 14.101, *A* to *C*). If the beaks form the anteriormost point of the dorsal margin, they are **terminal** (Figure 14.100, *H*).

In transverse profile, each valve of many pelecypods rises from the beak to a prominence that usually extends above the hinge line and laterally becomes the point of maximum curvature of each valve (Figures 14.97, *A* and *I* and 14.100, *B*, *C*, and *J*). The term **umbo** (plural, umbos or umbones) is used for this region of each valve, which is continuous with the beak and with the body of the shell. The umbos are **prosogyrate** if they curve in such a way that the beak points in an anterior direction (Figure 14.100, *B* and *C*); this is the condition in most pelecypods. If the umbos curve so that the beaks point in a posterior direction, they are **opisthogyrate** (Figure 14.102, *A* and *B*). If the beaks point directly at one another, the umbos usually do not curve anteriorly or posteriorly; this **orthogyrate** condition is common in scallops (Figure 14.100, *J*). The umbos are **coiled** if they are incurved in such a way that their outline forms a spiral of a complete whorl or more (Figure 14.102, *E*). In most pelecypods, the beaks and umbos are incurved toward the commissure (Figure 14.100, *B* and *C*); however, in some they point away from the sagittal plane (Figure 14.102, *C*). Sometimes they are helically coiled away from the midline (Figure 14.102, *D*) and are **spirogyrate**.

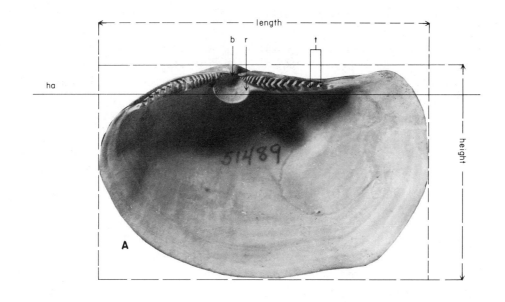

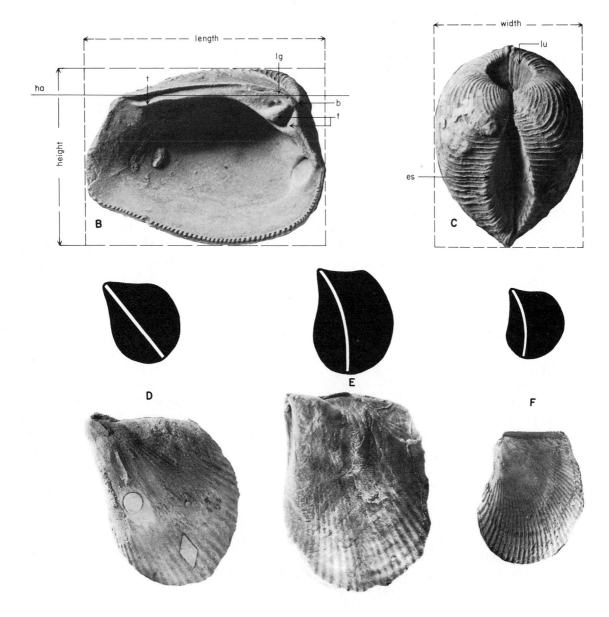

The **obliquity** of a shell is measured by the angle formed between the midumbonal line and the dorsal margin or the hinge axis (Figure 14.101, *D* to *F*). If the angle is acute, the shell is **prosocline** (forward obliquity, Figure 14.101, *D*); if it approaches 90 degrees, the shell is **acline** (erect obliquity, Figure 14.101, *E*). In a relatively few species, the ventral part of the mature shell grows forward and forms an angle of more than 90 degrees with the dorsal margin; such shells are **opisthocline** (backward obliquity, Figure 14.101, *F*). Obliquity can change ontogenetically (Figure 14.101, *E* and *F*).

In some cemented (Figure 14.102, *F*) or byssally attached (Figures 14.100, *H* and 14.102, *G*) species, the dorsal margin of one or both valves forms the base of a flat to slightly concave triangular **cardinal area** that is at a marked angle to the commissure. Cardinal areas are the result of interumbonal growth, which produces considerable separation of the beaks. Gapes in the valve margins are named for the soft parts that protrude through them. The **siphonal gape** is posterior (Figure 14.88, *A* to *C*); a **pedal gape** may be present anteriorly or anteroventrally (Figure 14.88, *A*). The byssal gape is highly variable. It may be a narrow ventral slit (Figure 14.102, *H*); small, symmetrical, and situated anterodorsally (Figure 14.97, *A* and *I*); large, symmetrical, and placed midventrally (Figure 14.102, *I*); or a small foramen limited to one valve (Figure 14.102, *J*). All three types of gape are known in fossils as old as the Late Ordovician.

Some pelecypods have a heart-shaped structure called the **lunule** located anteroventral to the beaks (Figures 14.100, *B* and *C* and 14.101, *C*) which functions to help maintain stability while the animal is burrowing. A large broader area posterior to the beaks is termed the **escutcheon** (Figures 14.100, *C* and 14.101, *C*); it surrounds the ligament and may be bounded by sharp or rounded ridges. The external ornament of pelecypods is similar to that of rostroconchs (Figures 14.86; 14.88; 14.90, *A* to *E*; 14.96, *E* to *H*; 14.97; 14.100; 14.102; and 14.108 to 14.111). A few pelecypods have prominent, overlapping growth lamellae (Figure 14.102, *E*), and some have spines (Figure 14.102, *F*). Although the ligament can be seen externally in many pelecypods, it is discussed here as part of the hinge.

Internal features The major internal features are the hinge and muscle scars; muscle scars were discussed earlier. The hinge includes the teeth and ligament and all structures that support and attach these to the rest of the shell (Figures 14.100, *E* and *F* and 14.101, *A* and *B*).

The hinge **teeth** usually lie below the hinge axis (Figure 14.100, *A*). Teeth function to: (1) guide the two valves into alignment when the adductors close the shell; (2) interlock the valves and prevent shearing or rotational movements from separating or opening the valves; and (3) maintain contact between the valves dorsally during burrowing, which is important during retraction when the effect of one valve meeting with some greater resistance in the substrate might misalign the valves. The teeth of one valve fit into corresponding depressions, called **sockets** (Figure 14.100, *E*), in the other valve. Between the two sets of teeth and sockets is a thin film of mantle. All the teeth and sockets of the shell are the **dentition**. Not all adult pelecypods have hinge teeth. The oldest known Cambrian pelecypods have teeth. The hinge teeth of the prodissoconch (Figure 14.99, *F*) can differ markedly from the adult teeth. In many pelecypods, some or all of the teeth in each valve are on the shelly vertical **hinge plate** (Figure 14.103, *C*, *F*, *I*, and *K*), which is situated below the beak and attached to the adjacent parts of the dorsal margin. The **umbonal cavity** (Figure 14.103, *A* to *D*) is the space within the umbo on the inside of the shell.

In many distantly related pelecypods, two groups of teeth are recognized: **cardinal teeth** are situated below the beaks (Figure 14.103, *A*, *C*, *G*, and *L*), and **lateral teeth** are located at some distance from the beak. Lateral teeth are separated by an **edentulous** (toothless) **space** from the cardinal teeth (Figure 14.103, *A*, *C*, *G*, and *L*), so that the lateral and cardinal teeth do not overlap. A full array of hinge teeth would include anterior and posterior laterals and cardinals (Figure 14.103, *G* and *L*). Most pelecypods do not have the full array of teeth, often having only cardinal and posterior lateral teeth (Figure 14.103, *A* and

Figure 14.101. A to **C**, Measurements of pelecypod shells. **A**, *Yoldia* (*Megayoldia*), subclass Palaeotaxodonta, with internal ligament and taxodont dentition. Internal view of the right valve. (Holocene, Massachusetts, ×2.) **B** and **C**, *Astartella*, subclass Heteroconchia, with ligament groove and heterodont dentition (shelled specimens). **B**, Internal view of the left valve. (×3.) **C**, External dorsal view. (Upper Pennsylvanian, Texas, ×3.) *b*, Beak; *es*, escutcheon; *ha*, hinge axis; *lg*, ligament groove; *lu*, lunule; *r*, resilifer; *t*, teeth. **D** to **F**, Variation in shell obliquity of Upper Ordovician species of the genus *Ambonychia*, subclass Pteriomorphia from the tristate area of Ohio, Indiana, and Kentucky. **D**, *A. ulrichi* having prosocline obliquity throughout ontogeny. (×1.5.) **E**, *A. suberecta* having obliquity varying from prosocline to orthocline when mature. (×1.5.) **F**, *A. cultrata* having obliquity varying from prosocline to opisthocline throughout ontogeny. (×0.5.) All views are left valve composite molds. (Modified from Pojeta, J., Jr. *Palaeontographica Americana* 4 (30): 173; 1962.)

C), cardinal teeth (Figure 14.103, *I*), or anterior and posterior lateral teeth (Figure 14.103, *H*). Cardinal teeth are usually short and radiate ventrally from the beak (Figures 14.100, *E* and *F* and 14.103, *A* and *I*), whereas lateral teeth tend to be elongate and subparallel to the adjacent shell margin (Figure 14.103, *A, C, H,* and *L*). The presence of cardinal and lateral teeth is known in species as old as Middle Ordovician. Any or all of the teeth may have **denticles** (Figure 14.103, *C, G,* and *I*).

Some species, ranging in age from Early Ordovician to Holocene, have elongate anterior and/or posterior teeth that overlap shorter teeth that radiate outward from beneath the beak (Figure 14.103, *B, D* to *F,* and *K*); the elongate teeth may reach the beak (Figure 14.103, *D* and *E*), or they may overlap other elongate teeth (Figure 14.103, *B* and *F*). In such species without an edentulous space, the short teeth radiating from below the beak are **pseudocardinals**, and the elongate teeth that are subparallel to the adjacent shell margin are **pseudolaterals**.

The terms cardinal, lateral, pseudocardinal and pseudolateral teeth are descriptive. They indicate the position of the teeth, to some degree their shape, and the presence or absence of an edentulous space; they do not indicate homologies or phylogenetic relationships between taxa.

Several terms group more or less similar patterns of dentition. Some of these group the variable dentition of large taxonomic units, whereas others are applicable to only one or two superfamilies. Four of these terms are useful for large numbers of species dating back to the Early Ordovician: edentulous, taxodont, heterodont, and actinodont.

Edentulous refers to the simplest hinge of pelecypods, which has a ligament but lacks teeth (Figures 14.98, *F* and 14.103, *J*). **Taxodont** describes dentition made up of many small short teeth of variable shape, which often occupy most of the length of the dorsal margin (Figures 14.101, *A* and 14.102, *K* and *L*). Taxodont teeth occur in three subclasses, but the condition is most widespread in species with protobranch gills (Figure 14.101, *A*). Homeomorphic taxodont dentition also occurs in a few brachiopods and ostracodes. **Heterodont** refers to dentition in which the types of teeth are separated by an edentulous space (Figure 14.103, *A, C, H,* and *L*). Most often heterodont dentition consists of cardinal and posterior lateral teeth (Figure 14.103, *A* and *C*), but anterior lateral teeth may be present (Figure 14.103, *G* and *L*), and cardinal teeth may be absent (Figure 14.103, *H*). Heterodont dentition is known in two subclasses. **Actinodont** refers to dentition composed of pseudocardinal and pseudolateral teeth. This dentition is most widespread in early Paleozoic forms (Figure 14.103, *B, E,* and *F*) but is also present in some living freshwater species (Figure 14.103, *K*).

Several schemes of tooth and socket notation have been proposed that result in dental formulas. Most of these schemes have tried to show homologies between the various dentitions of advanced pelecypods that are sometimes grouped in a high-level taxon called Heterodonta. However, detailed studies of a single group by different workers have produced different interpretations and have resulted in different dental formulas for the same species. Furthermore, the methods proposed have not been applicable to less-advanced pelecypods. An objective method of tooth and socket notation, which is equally applicable to advanced and primitive pelecypods, represents each tooth by the number 1 and each socket by a 0.

The **ligament** connects the two valves and acts as a spring to help open them when the adductors relax. It is composed of horny elastic conchiolin and aragonitic fibers. In a general way, ligaments that are readily visible

Figure 14.102. A to J, Beaks and byssal gapes. **A,** *Lyrodesma.* Dorsal view of a silicified specimen, showing opisthogyrate beaks. (Upper Ordovician, Kentucky, ×2.) **B,** *Phestia.* Dorsal view of a silicified specimen, showing opisthogyrate beaks. (Permian, Texas, ×2.) **C,** *Johnmartinia.* Dorsal view of internal mold, showing beaks and umbos pointing away from the midline. (Middle Ordovician, Northern Territory, Australia, ×3.) **D,** *Requienia.* 'Posterior' view of a shelled specimen, showing the spirogyrate umbos of this rudist. (Upper Cretaceous, Texas, ×2.) **E,** *Exogyra.* Exterior of the left valve of a shelled specimen, showing the coiled umbo. (Upper Cretaceous, Texas, ×2.) **F,** *Spondylus.* Dorsal view, showing cardinal area and cemented part of the right valve, with the anterior end down. (Holocene, Japan, ×2.) **G** and **I,** *Arca.* **G,** Dorsal view, showing cardinal areas and interumbonal growth of both valves. **I,** Ventral view, showing the byssal gape. (Holocene, Eniwetok Atoll, Marshall Islands, central Pacific Ocean, ×1.5.) **H,** *Geukensia.* Ventral view, showing the place where the byssus exits from the shell with no apparent byssal gape. (Holocene, Long Island, New York, ×2.) **J,** *Argopecten.* Oblique anterior view showing the byssal foramen in the right valve. (Holocene, North Carolina, ×2.) *b,* Beak; *bg,* byssal gape; *c,* commissure; *ca,* cardinal area; *cm,* cemented part of shell; *u,* umbo. **K** and **L,** Left valves showing the variation in taxodont dentition in two subclasses of Pelecypoda; a third subclass with taxodont dentition is shown in Figure 14.101, *A.* **K,** *Iridina,* subclass Heteroconchia, with external opisthodetic ligament (*l*). Most of the teeth are peglike; some posterior ones are chevron shaped. (Holocene, freshwater, Senegal, Africa, ×1.) **L,** *Arca,* subclass Pteriomorphia, with the cardinal area (*ca*) and duplivincular external ligament. Teeth are erect to oblique peglike structures. (Holocene, Eniwetok Atoll, Marshall Islands, central Pacific Ocean, ×2.) (**J** courtesy of T. R. Waller, Smithsonian Institution, Washington D.C.)

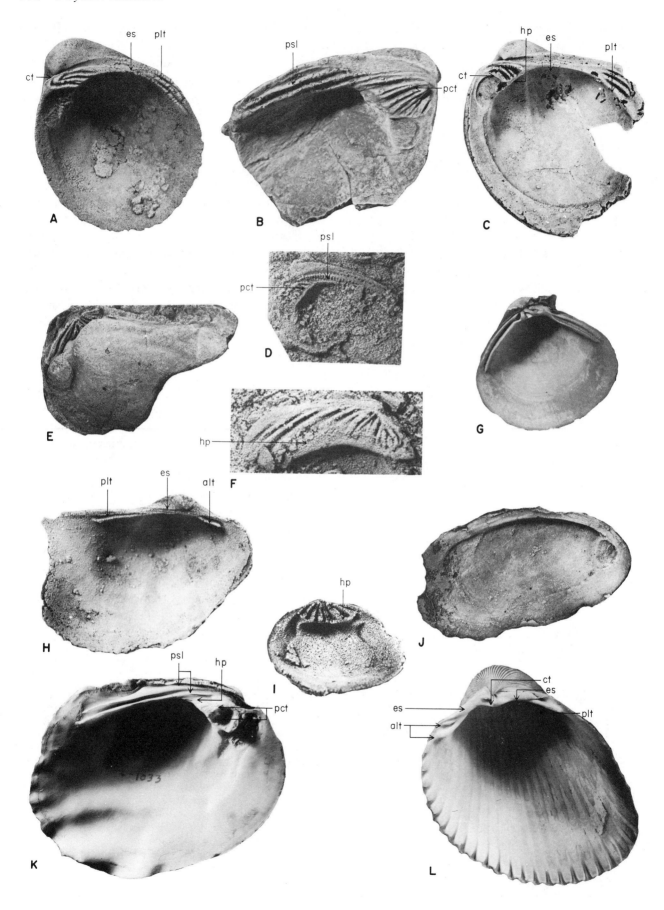

on the exterior of the shell are termed **external ligaments** (Figure 14.86, C) and those that are easily seen only when the valves are separated are called **internal ligaments** (Figure 14.100, G and J).

Elongate external ligaments may attach to narrow internal platforms called **nymphs** (Figure 14.104, G and I), or they may attach in **ligament grooves** above the teeth (Figure 14.104, J). An internal ligament that occupies a more-or-less triangular depression in the hinge (Figure 14.100, G) or occurs between the anterior and posterior tooth rows (Figures 14.101, A and 14.104, F) is a **resilium**. The calcareous depression that houses the resilium is a **resilifer**. The internal ligament may also occupy the spoon-shaped **chondrophore** (Figure 14.104, E) in one or both valves. The resilium may be reinforced by a small calcareous plate called the **lithodesma**.

The idealized pelecypod ligament consists of three layers (Figure 14.105, Da): an outer periostracum, a middle **lamellar layer**, and an inner **fibrous layer**. The periostracal and lamellar layers contain no calcareous material and are elastic to both compressional and tensional forces. The fibrous layer is impregnated with fibers of aragonite and is elastic only to compressional stresses. The pelecypod ligament is strained when the valves close. Above the hinge axis, the strain is tensional, and below it, the strain is compressional (Figure 14.105, C). Normally the ligament is constructed so that the junction between the functional parts of the lamellar and fibrous layers coincides with the hinge axis. Some authorities compare the three ligament layers with the three layers of the shell, the shell and ligament differing only in the amount of calcification. Recent histologic and ontogenetic studies of the ligament indicate that this interpretation is unlikely.

Often a secondary extension is at one or both ends of the ligament. The so-called secondary ligament may consist either of periostracum or fusion layer (Figure 14.105, Db and c), which is secreted by the outer surface of the outer folds of the mantle edge after these folds have fused in early postlarval life. In general, the fusion layer functions only as a cover between the dorsal margins of the valves.

The ligament has a variety of forms, each of which leaves a characteristic impression on the dorsal margin. The position of the ligament is **amphidetic** if it extends on both sides of the beaks (Figures 14.101, A; 14.104, B, F, and H; and 14.105, Aa) and **opisthodetic** if it is mainly posterior to the beaks (Figures 14.86, C; 14.100, E and F; and 14.105, Ac and d). The form of the ligament is described as follows: An **alivincular** ligament is a flattened central structure with the lamellar layer both anterior and posterior to a central fibrous layer (Figures 14.104, H; 14.105, Aa, and b); ordinarily, the central part occupies a resilifer (Figures 14.100 G; 14.101 A; and 14.104 F and H). Alivincular ligaments date from the Silurian. A **parivincular** ligament is largely opisthodetic and elongate and hemispherical in shape (Figure 14.105, Ad and e). This is the classical C-spring ligament, and it is often attached to nymphs and/or elongate ligament grooves. The oldest known species having a parivincular ligament are Early Cambrian in age. Some elongate opisthodetic ligaments are deep set and do not form a C-spring (Figure 14.105, Ac). The **multivincular** ligament has several fibrous resiliums inserted along the dorsal margin completely isolated from each other by lamellar ligament (Figures 14.104, D and 14.105, Af). This ligament type is known with certainty in pelecypods

Figure 14.103. Heterodont and actinodont dentition. **A,** *Vanuxemia.* Right valve interior of a silicified specimen showing heterodont dentition of cardinal and lateral teeth separated by edentulous space. The shadowed area is the umbonal cavity, (Middle Ordovician, Ontario, Canada, ×2.) **B,** *Tanaodon.* Left valve interior of the umbonal region of a shelled specimen, showing actinodont dentition of overlapping pseudocardinal and pseudolateral teeth. The shadowed area is the umbonal cavity. (Middle Devonian, Sichuan, China, ×1.5.) **C,** *Vanuxemia.* Right valve interior of a silicified specimen, showing heterodont dentition of cardinal and lateral teeth separated by edentulous space. The teeth have denticles. The shadowed area is the umbonal cavity. (Middle Ordovician, Kentucky, ×2.) **D,** *Noradonta.* Latex replica of the right valve, showing actinodont dentition of overlapping posterior pseudolateral teeth with denticles and pseudocardinal teeth on hinge plate. The shadowed area is the umbonal cavity. (Lower? Ordovician, Queensland, Australia, ×2.5.) **E,** *Actinodonta.* Internal mold of the left valve, showing overlapping actinodont teeth. (Lower Silurian, Great Britain, ×2.) **F,** *Copidens.* Latex replica of the hinge of the left valve, showing actinodont dentition and the hinge plate. (Lower? Ordovician, Queensland, Australia, ×2.5.) **G,** *Corbicula.* Right valve interior of a shelled specimen, showing heterodont dentition with small edentulous spaces separating the cardinal and anterior and posterior lateral teeth. (Holocene, Ohio River, Cincinnati, Ohio, ×1.5.) **H,** *Palaeopteria.* Left valve of a silicified specimen, showing heterodont dentition of anterior and posterior lateral teeth separated by an edentulous space. (Middle Ordovician, Kentucky, ×4.) **I,** *Lyrodesma.* Oblique view of the interior of a silicified right valve, showing cardinal teeth with denticles and mounted on a hinge plate. (Middle Ordovician, Kentucky, ×3.5.) **J,** *Modiolopsis.* Latex replica of the left valve, showing edentulous hinge. (Upper Ordovician, New York, ×1.) **K,** *Amblema?.* Interior of the left valve, showing actinodont dentition with overlapping pseudolateral and pseudocardinal teeth. (Holocene, Little Miami River, Ohio, ×0.7.) **L,** *Dinocardium.* Interior of the right valve, showing heterodont teeth with large edentulous spaces separating anterior and posterior lateral teeth from cardinal teeth. (Holocene, Florida, ×1.) *alt,* Anterior lateral teeth; *ct,* cardinal teeth; *es,* edentulous space; *hp,* hinge plate; *pct,* pseudocardinal teeth; *plt,* posterior lateral teeth; *psl,* pseudolateral teeth.

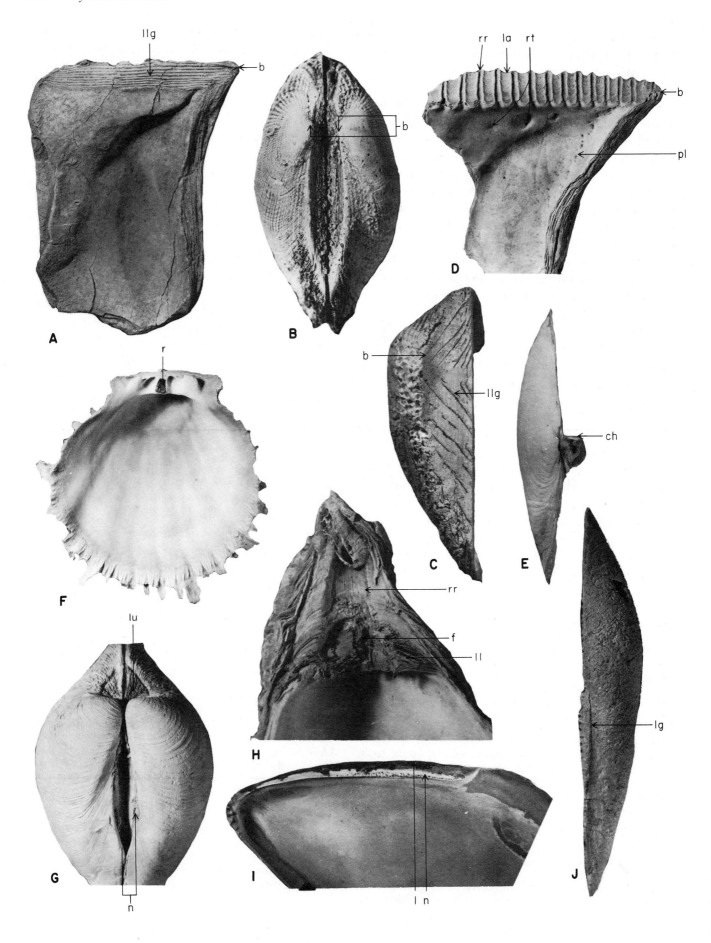

as old as Permian; it may occur in the Mississippian. The **duplivincular** ligament is similar to the multivincular one in that lamellar and fibrous components are separated and repeated. However, in the duplivincular ligament, the ligament-attachment areas of the shell are crossed by a series of grooves and ridges (Figures 14.104, *C* and 14.105, *Ag, Ah,* and *C*). The grooves represent the insertions of separate parallel sheets of lamellar material, and the ridges as well as the junction between the two valves are crossed by fibrous ligament. In living species that have a duplivincular ligament, this ligament is amphidetic (Figure 14.104, *B* and *C*), and the lamellar insertion grooves form a series of chevrons. In various fossil species, one arm of the chevron may be considerably smaller than the other, or the ligament may be entirely opisthodetic and composed only of half chevrons (Figure 14.104, *A*). The duplivincular ligament type first occurs in Middle Ordovician species.

During growth of the pelecypod shell, the ligament enlarges and elongates. It is a functional necessity for pelecypods to maintain a straight hinge axis. To lengthen a straight line on the border of a rounded shell, either the axis of the ligament must move ventrally, or the outline of the shell must be elongated. In some pelecypods, such as oysters, the hinge axis elongates largely by ventral movement as new ligamental material is added at the ventral surface of the ligament (Figure 14.104, *H*). Marine mussels show posterior extension of the dorsal valve margin as they grow (Figure 14.104, *I*). Little ventral movement of the ligament takes place; rather, posterior extension forms an opisthodetic deep set ligament (Figure 14.105, *Ac*).

The ligament can be described as a spring because it stores energy supplied to it by contraction of the adductor muscles. In many pelecypods, the anterior part of the ligament is split and torn apart as each valve grows in a separate helical spiral (Figure 14.105, *B* and *C*). Thus in the marine mussel *Mytilus edulis,* the whole of the original ligament of a shell 16 mm long is destroyed when the mussel has grown to a length of 70 mm. In oysters, the alivincular ligament grows ventrally as the size of the shell increases (Figure 14.104, *H*), and the earlier secreted fibrous portions of the ligament end up above the hinge axis, where tensional stresses fracture them and sea water corrodes the dorsal part along the cracks. In ark shells, which have prominent interumbonal growth, the lamellar sheets of earlier formed parts of the duplivincular ligament rupture as the shell grows (Figure 14.105, *C*) because they come to be farther and farther apart. When pelecypods die, the valves ordinarily gape because of the opening moment of the ligament.

Classification

The diversity of pelecypods is shown by the fact that more than 4500 generic names are available for species of this class. Of these, slightly more than 2600 are based on fossil type species. Almost 3300 of the available names are at present taxonomically useful. About 50 superfamilies of pelecypods are recognized.

Above the level of superfamily, the taxonomy of pelecypods has a complex history, but it is now generally agreed that the subclass is a useful category. How many subclasses should be recognized, and on what bases, is

---

Figure 14.104. Ligament structures and supports. **A,** *Myalina (Orthomyalina).* Interior view of the left valve of a shelled specimen with a terminal beak showing opisthodetic external insertion grooves of the lamellar part of the duplivincular ligament as half-chevrons on the dorsal margin. The most recently formed half-chevrons are the shortest, at a low angle to the dorsum, and on the most ventral part of the ligament insertion area. (Upper Pennsylvanian, Texas, ×1.) **B,** *Arca.* Dorsal view showing the external amphidetic duplivincular ligament with unequal chevrons. The long part is behind beaks, and the short part anterior to beaks. (Holocene, Eniwetok Atoll, Marshall Islands, central Pacific Ocean, ×1.5.) **C,** *Arca.* Dorsal view of the left valve, showing the external chevron-shaped insertion grooves of the lamellar part of the amphidetic duplivincular ligament anterior and posterior to the beak. (Holocene, Florida, ×2.) **D,** *Isognomen.* Interior view of much of the left valve of a shelled specimen, showing several external erect resilifer grooves of the multivincular ligament separated by areas for attachment of lamellar parts of the ligament, terminal beak, pitted pallial line, and byssal-pedal retractor muscle scars. (Miocene, Chesapeake Bay, Maryland, ×0.7.) **E,** *Mya.* Dorsal view of the left valve, showing the dried internal ligament in the spoon-shaped chondrophore. (Holocene, Long Island, New York, ×2.) **F,** *Spondylus.* Interior of the right valve with a dried internal amphidetic resilium in the resilifer between the two sockets and below the beak. (Holocene, Japan, ×0.7.) **G,** *Mercenaria.* Dorsal view, showing nymphs at the bottom of the opisthodetic ligament space and the lunule. (Holocene, Long Island, New York, ×1.5.) **H,** *Crassostrea.* Dorsal part of the left valve, showing an amphidetic resilifer, the dorsal part of which no longer contains a functional ligament. The ventral part of the resilifer has the fibrous ligament in place and a lamellar ligament to either side of the fibrous part. (Holocene, Chesapeake Bay, Maryland, ×3.) **I,** *Mytilus.* Dorsal part of the right valve, showing an elongate opisthodetic external ligament and nymph. (Holocene, Long Island, New York, ×3.5.) **J,** *Ctenodonta.* Dorsal view of the right valve, showing the ligament groove of the external opisthodetic parivincular ligament. (Middle Ordovician, Ontario, Canada, ×3.) *b,* Beak; *ch,* chondrophore; *f,* fibrous ligament; *l,* ligament; *la,* lamellar ligament attachment area; *lg,* ligament groove; *ll,* lamellar ligament; *llg,* lamellar ligament groove; *lu,* lunule; *n,* nymph; *pl,* pallial line; *r,* resilium; *rr,* resilifer; *rt,* byssal-pedal retractor muscle scar.

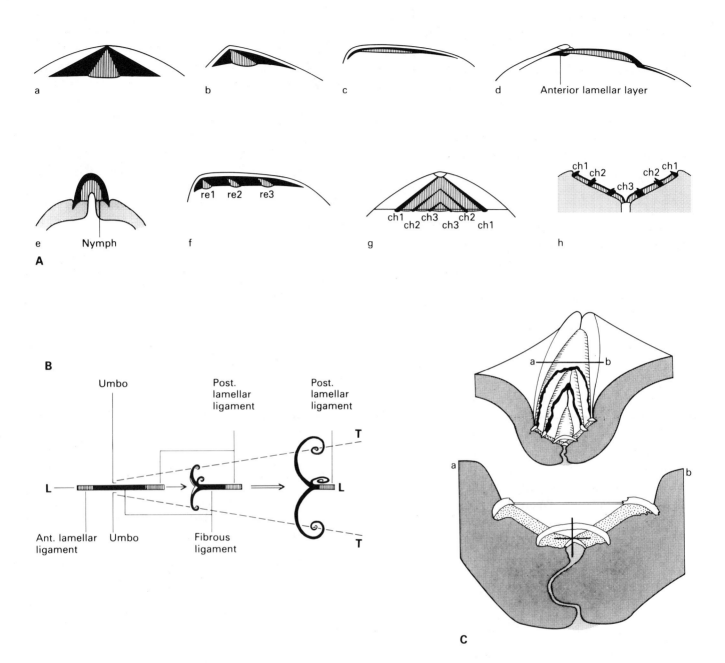

Figure 14.105. Diagrammatic sections of pelecypod ligaments. **A** shows the location of the ligament in relation to the right valve. Black indicates the lamellar layer (and, where present, fusion layer) of the ligament, the vertical lines indicate the fibrous layer of the ligament, and the stippled pattern represents the calcareous shell of the valves in cross section. *a,* Alivincular, amphidetic, erect, symmetrical ligament (compare with Figure 14.104, *H*). The fibrous part is in the center with a lamellar ligament on either side; *b,* alivincular, amphidetic, oblique, asymmetrical ligament with the fibrous part and large lamellar part posterior to the beak and the small lamellar part anterior to the beak; *c,* opisthodetic ligament placed directly between valve margins and projecting little or not at all above them (compare with Figure 14.104, *I*); *d,* opisthodetic parivincular ligament (compare with Figure 14.86, *C*); *e,* cross section of *d* showing C-spring shape of the ligament and nymphs; *f,* multivincular ligament showing the sequence of resilium formation (*re1, re2, re3*) (compare with Figure 14.104, *D*); *g,* duplivincular ligament with successive chevrons (*ch1, ch2, ch3*) (compare with Figure 14.104, *A* to *C*); *h,* cross section of *g.* **B** shows the effect of helical and interumbonal growth of the two valves on the ligament. *L–L,* longitudinal axis; *T–T,* transverse axis; broken lines indicate divergence of umbos. C, Camera lucida drawing of the duplivincular ligament of *Arca* showing the transverse section through the hinge posterior to the beaks. The upper figure (×6) shows three lamellar ligament sheets and the fibrous ligament, which covers the cardinal areas between sheets; note that closest to the beaks, the outer two lamellar sheets are fractured by interumbonal growth. The lower figure (×20) (along line *a–b* of the upper figure) shows two lamellar ligament sheets with fibrous ligament lining the rest of the cardinal area; cross shows the position of the hinge axis; lightly shaded part is

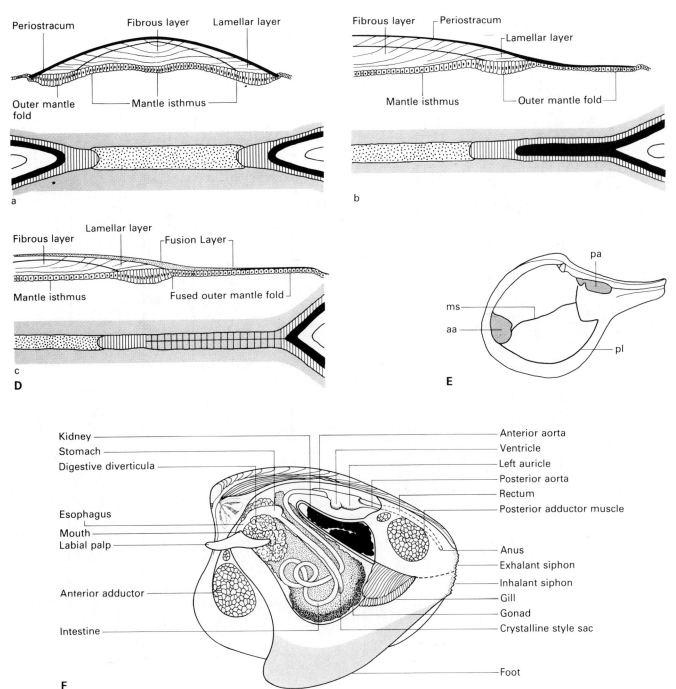

the dorsal extension of the mantle, which secretes both the teeth and ligament; darkly shaded part is the shell. **D**, Diagrammatic representations of pelecypod ligament. The upper diagram of each pair is a median cross section of ligament, and the lower diagram of each pair shows the dorsal aspect of mantle tissues involved in secretion of ligament. Fine stipple indicates general outer surface mantle; coarse stipple, mantle isthmus secreting **fibrous layer of ligament; closely spaced vertical rules, outer surface of outer mantle fold secreting periostracum; widely spaced vertical rules, fused outer mantle folds secreting fusion layer.** *a*, Idealized pelecypod ligament with the fibrous layer complete under the lamellar layer; *b*, idealized ligament with secondary periostracal extension; *c*, idealized ligament with secondary extension produced by fusion layer. **E**, *Cuspidaria*. Major muscle scars of Holocene species from Florida. *aa*,

Anterior adductor; *ms*, muscular septum; *pa*, posterior adductor; *pl*, pallial line. **F**, Diagram of major soft parts of *Mercenaria* (compare with Figure 14.87, C). (**A** modified from and **D** from Trueman, E. R. in: Moore, R. C., editor. *Treatise on invertebrate paleontology, Part N, Mollusca 6, Bivalvia, 1.* Boulder, CO, and Lawrence, KS: Geological Society of America and University of Kansas Press; 1969, p. 61, 59. **B** modified from Perkins, B. F. in: Moore, R. C., editor. *Treatise on invertebrate paleontology, Part N, Mollusca 6, Bivalvia, 2.* Boulder, CO, and Lawrence, KS: Geological Society of America and University of Kansas Press; 1969, p. 756. **C** from Newell, N. D. *Geological Survey of Kansas* 10: 28; 1937. **E** modified from Runnegar, B. *Journal of Paleontology* 48: 924; 1974. **F** modified from Barnes, R. D. *Invertebrate zoology*, 4th ed. Philadelphia: W. B. Saunders Co.; 1980, p. 425.)

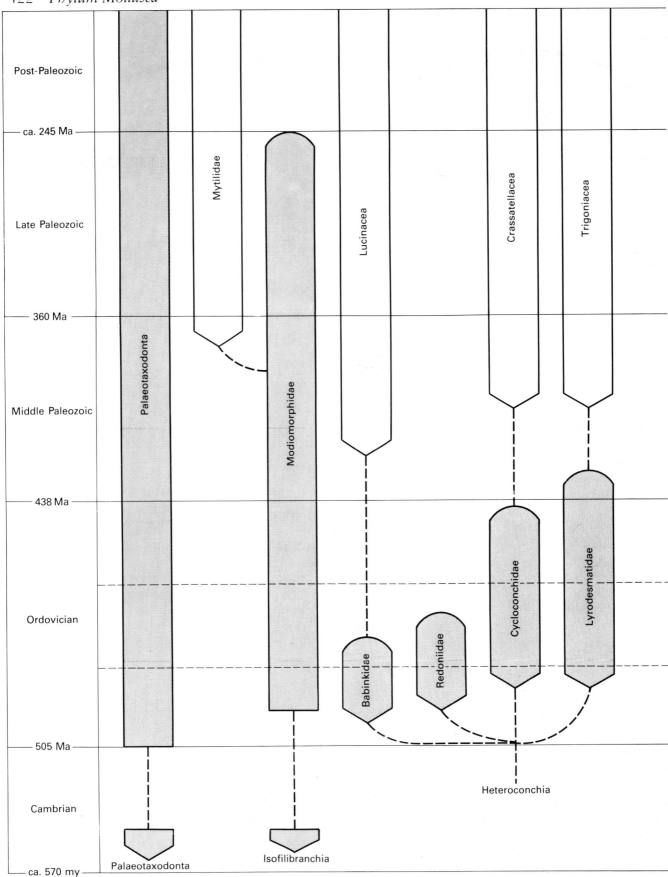

Figure 14.106. Diagram of the phylogenetic relationships of Paleozoic pelecypods showing that all major lineages (subclasses) had evolved by Middle Ordovician time. Lineages in dark color originated in the Cambrian or Ordovician; light-colored lineages have post-Ordovician origins. (Modified from Pojeta, J., Jr.; Runnegar, B. *American Scientist* 62: 707; 1974.)

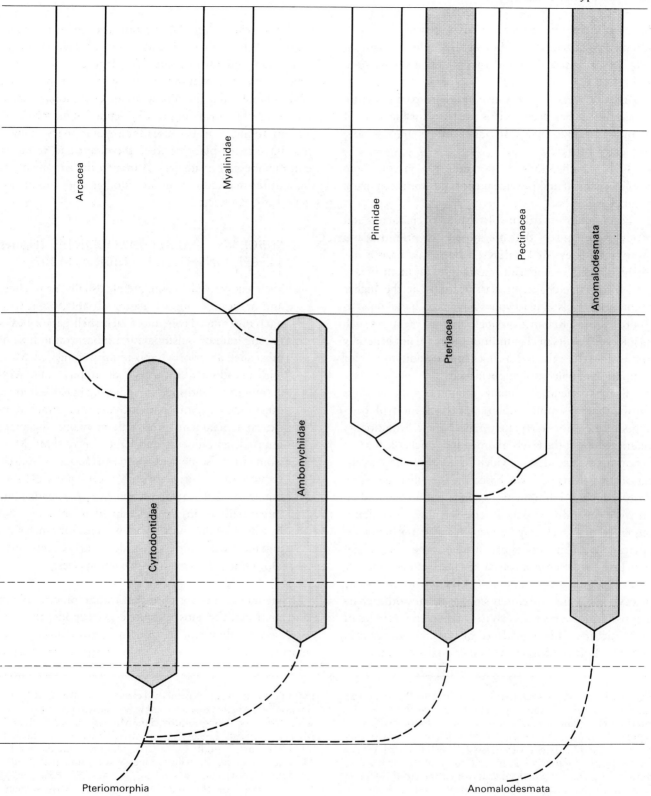

debated. In two classifications advocated in 1978, one author recognized seven subclasses, the other recognized two. The ordinal level of pelecypod taxonomy is in a state of flux.

Virtually every organ system and hard-part structure of pelecypods has been used as the basis for higher level groupings. Single-character classifications are not now advocated because:

1 Such classifications do not take into account the whole organism and provide no grounds for recognizing parallel evolution.

2 Evolution ordinarily involves suites of characters and produces organisms that are mosaics of primitive and more recently evolved features; a single taxobasis is unlikely to be phylogenetically meaningful for many taxa.

3 Most single-character classifications at the higher taxonomic levels in pelecypods have been based on anatomic features. Reliance on single anatomic features such as gill type results in classifications that are not generally applicable to fossils, and such schemes do not provide for evaluation of paleontologic evidence.

4 The experience of many generations of workers shows that a simple utilitarian classification of pelecypods does not adequately reflect their evolutionary relationships at the levels of order and subclass.

For many decades, a widely used zoologic classification of pelecypods was based on the gills; the terms Protobranchia, Filibranchia, Eulamellibranchia, and so on were generally used as orders. We now know that a sequence of stages can be recognized in the union of adjacent gill filaments and that identical grades of gill structure have been evolved in pelecypods that are otherwise unlike one another. On the other hand, some pelecypods, which are closely similar in many other characteristics, have been shown to have different grades of gill structure. This is evidence of parallel evolution in pelecypods through a series of functional gill grades.

The subclass-level classification used here is largely based on the primary radiation of the Pelecypoda in Cambrian and Ordovician time (Figure 14.106). The scheme groups primitive forms with their presumed descendants. Thus the subclasses are evolutionary units, and some of them are not readily defined in morphologic and anatomic terms because of the evolutionary changes that have taken place between the oldest and youngest end members of a subclass. Although the end members may differ markedly from one another, they are phylogenetically connected.

SUBCLASS PALAEOTAXODONTA (Figures 14.90, *I*; 14.98, *C* and *F*; 14.101, *A*; 14.107)

Shell equivalved, inequilateral, usually prosocline, and aragonitic; ultrastructure variable, often with nacreous inner layer; most lack shell gapes; beaks prosogyrate or opisthogyrate, subcentral in few; in most shell anteriorly or posteriorly elongated. Most with taxodont teeth, a few edentulous, some with actinodont dentition. In some, ligament external, opisthodetic, short parivincular, and attached in external ligament grooves; in many, ligament internal, amphidetic, and largely in resilifer. Most isomyarian, some with posterior adductor muscle reduced or absent; often with visceral muscles forming prominent scars in umbonal cavities; pallial line integripalliate in most, sinupalliate in some. No byssus in adult. Living forms with protobranch gills, and most with palp proboscides. (Early Cambrian to Holocene; 150 genera; 500 living species.)

Palaeotaxodonts are both the stratigraphically oldest and anatomically most primitive pelecypods; they are diverse and abundant in modern deep oceans. All are marine and infaunal, and most living species are detritus

Figure 14.107. Subclass Palaeotaxodonta. **A** to **D**, *Deceptrix*. Silicified specimens, showing right lateral view (**A**, ×2.5); interior of the right valve with taxodont teeth, most of which are peglike (**B**, ×2.5); umbonal cavity view showing muscle scars (**C**, ×2.5); and left lateral view (**D**, ×2.5). (Middle Ordovician, Kentucky.) **E**, *Johnmartinia*. Internal mold of the right valve, showing peglike, sigmoidal, and chevron-shaped teeth. (Middle Ordovician, Northern Territory, Australia, ×1.5.) **F** and **G**, *Ctenodonta*. Silicified specimens, showing right and left valves, dentition, and various muscle scars. (Middle Ordovician, Ontario, Canada, ×0.7.) **H**, *Ctenodonta*. Silicified left valve, showing reversed chevron-shaped teeth. (Middle Ordovician, Tennessee, ×2.) **I**, *Cardiolaria*. Internal mold of left valve, showing some muscle scars, taxodont dentition, and two large modified cardinal teeth below the beak. (Middle Ordovician, Finistere, France, ×6.) **J**, *Alococoncha*. Internal mold of the left valve, showing a few of the teeth and adductor muscle scars. (Middle Ordovician, Northern Territory, Australia, ×2.) **K**, *Similodonta*. Shelled specimen left valve, showing dentition and adductor scars. (Upper Ordovician, Minnesota, ×3.5.) **L** and **M**, *Tancrediopsis*. **L**, Silicified left valve, showing dentition and adductor scars. (Middle Ordovician, Ontario, Canada, ×2.5.) **M**, Internal mold, showing adductor scars and pallial line. (Middle Ordovician, Wisconsin, ×2.5.) **N**, *Palaeoneilo*. Internal mold of the right valve. (Lower Mississippian, Michigan, ×1.5.) **O**, *Nuculites*. Internal mold of the right valve. The large slit behind the anterior adductor is formed by a shell thickening similar to that seen in Figure 14.107, *G*. (Middle Devonian, Maryland, ×1.5.) **P**, *Phestia*. Silicified left valve, showing chevron-shaped teeth, adductor scars, pallial line, and resilifer. (Lower Permian, Texas, ×1.5.) **Q**, *Nucula*. Shelled right valve, showing muscle scars, dentition, and resilifer. (Upper Cretaceous, Maryland, ×2.) (**I** courtesy Claude Babin.)

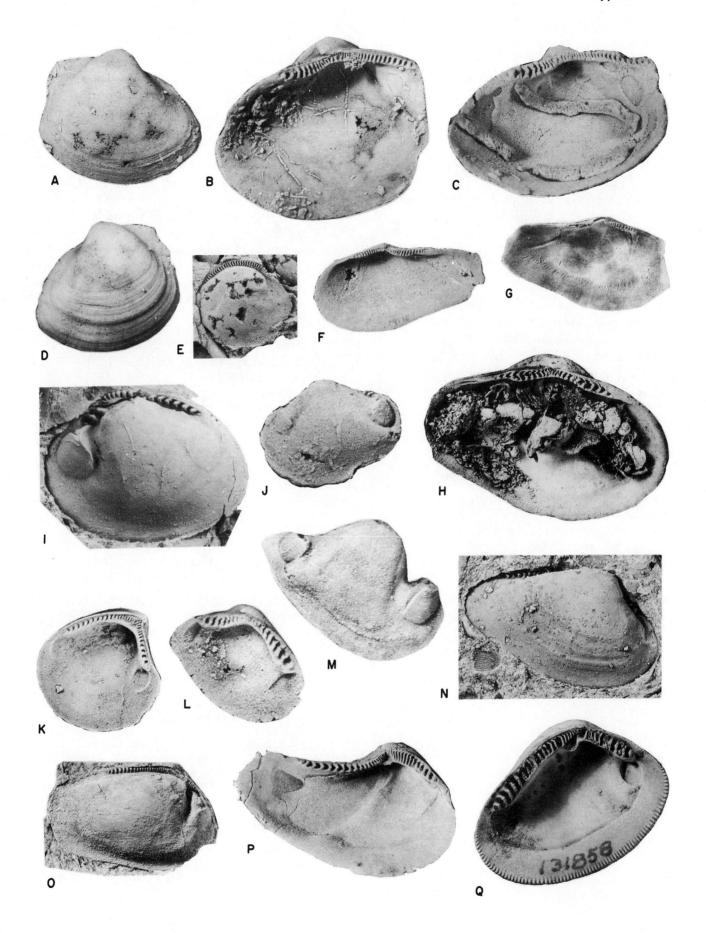

feeders. The depth range of living species is shallow subtidal to more than 5000 m. The subclass shows considerable variation, but the lack of an adult byssus and presence of an equivalved shell, taxodont teeth, protobranch gills, and palp proboscides are characteristic of the group.

SUBCLASS ISOFILIBRANCHIA (Figures 14.87, B; 14.92, D and G; 14.102, H; 14.103, J; 14.108)

Shell equivalved, strongly inequilateral, usually prosocline, calcitic and aragonitic or entirely aragonitic, inner shell layer commonly nacreous; ordinarily lack shell gapes, may have narrow slit where adult byssus exits; beaks prosogyrate, near anterior end or terminal; shell posteriorly elongate and mytiliform, modioliform, or soleniform. Hinge often edentulous; may have one or more small cardinal teeth or anterior and/or posterior small, taxodont-like teeth. Ligament usually external, deeply set, and opisthodetic, typically not a C-spring, in geologically younger forms supported by nymphs. Most anisomyarian, anterior adductor absent in few, posterior adductor typically confluent with posterior byssal–pedal retractors; pallial line integripalliate, siphons absent. Living species with small vermiform foot and filibranch gills. (Early Cambrian to Holocene; 140 genera; 250 living species.)

Most isofilibranchs are marine or brackish, but some live in fresh water. Mainly epibenthic, sedentary, and attached by a well-developed byssus, some are byssally attached semi-infauna, some nestle and may make byssal nests, some bore into rock, and a few are commensal with tunicates. Most live in the shallow water of the photic zone, and some are intertidal. One species is part of the Galapagos Rift fauna and occurs at 2500 m.

The oldest known isofilibranchs are only slightly younger than the oldest known palaeotaxodonts. The Paleozoic soleniform shells are included here in the Isofilibranchia (Figure 14.108, K, L, and N) because there seem to be species that have shapes intermediate (Figure 14.108, J) between these soleniform species and typical modioliform isofilibranchs (Figure 14.108, H and I). Soleniform isofilibranchs are minor constituents of Paleozoic faunas through the Late Permian; no soleniform pelecypods are known in Triassic or Jurassic rocks, but the shell shape reappears in the Early Cretaceous (Figure 14.108, M) and continues to the present as the razor clams (Figure 14.88, A). Some Late Permian soleniform shells have musculature and dentition much like those of Cretaceous and younger species.

SUBCLASS HETEROCONCHIA (Figures 14.87, C, E, and F; 14.88; 14.93, A and B; 14.102, A, D, and K; 14.109)

Shell variable, usually aragonitic, ultrastructure variable, with or without gapes, internal in some; typically shell external, equivalved, inequilateral, and without adult byssus; beaks often prosogyrate and not terminal. Well-developed teeth in most, actinodont in many, heterodont in most, taxodont in few; hinge plate commonly present. Ligament mainly external, opisthodetic, and parivincular, sometimes with resilium and chondrophore; lithodesma absent. Usually isomyarian, some anisomyarian, ventral adductor in some, internal shells lacking adductors in few; pallial line with or without sinus, many with mantle fusion and siphons. Well-

Figure 14.108. Subclass Isofilibranchia. A to I, O, and P, Modioliform. K to N, soleniform; Q and R, mytiliform. A, *Modiolopsis*. Internal mold of the right valve showing adductor scars, pallial line, and posterior byssal-pedal retractor. (Upper Ordovician, Kentucky, ×1.5.) B, *Corallidomus*. Right valve exterior. (Upper Ordovician, Ohio, ×1.5.) C, *Volsellina*. Right valve exterior. (Upper Pennsylvanian Texas, ×3.) D, *Promytilus*. Right valve exterior. (Upper Pennsylvanian Texas, ×3.) E, *Modiolus*. Right valve exterior. (Middle Jurassic, Colorado, ×2.) F, *Liromytilus*. Latex replica of the right valve exterior. (Middle Devonian, Indiana, ×0.66.) G, *Modiolopsis*. Left valve exterior. (Upper Ordovician, Kentucky, ×1.) H, *Modiolopsis*. Composite mold of the left valve. (Upper Ordovician, New York, ×1.) I, *Geukensia*. Left valve exterior. (Holocene, Long Island, New York, ×1.5.) J, *Cymatonota*?. Right valve composite mold. This specimen is similar to the soleniform shells shown in Figure 14.108, K to N in being elongated anteroposteriorly, but the dorsal and ventral margins are not subparallel because the shell is higher posteriorly than anteriorly. In this feature, the specimen is modioloform (Figure 14.108, G to I). (Upper Ordovician, New York, ×3.) K, *Cymatonota*. Dorsal view of a mold of spread valves of this soleniform genus. (Upper Ordovician, New York, ×3.) L, *Cymatonota*. Mold of the left valve. (Upper Ordovician, Ohio, ×1.5.) M, *Leptosolen*. Left valve of a shelled specimen. This Cretaceous specimen is similar in shape to the Paleozoic forms shown in Figure 14.108, K, L, and N. (Upper Cretaceous, Mississippi, ×1.5.) N, *Orthonota*. Dorsal view of a mold of spread valves. Compare with Figure 14.108, K and M. (Middle Devonian, New York, ×1.3.) O and P, *Modiolodon*. Silicified specimens, showing interior of the left valve and exterior of the right valve. (Middle Ordovician, Kentucky, ×2.) Q, *Brachidontes*. Shelled specimen, showing left valve exterior. (Upper Cretaceous, Texas, ×1.75.) R, *Mytilus*. Shelled specimen, showing left valve exterior. (Jurassic, Black Hills, South Dakota, ×2.)

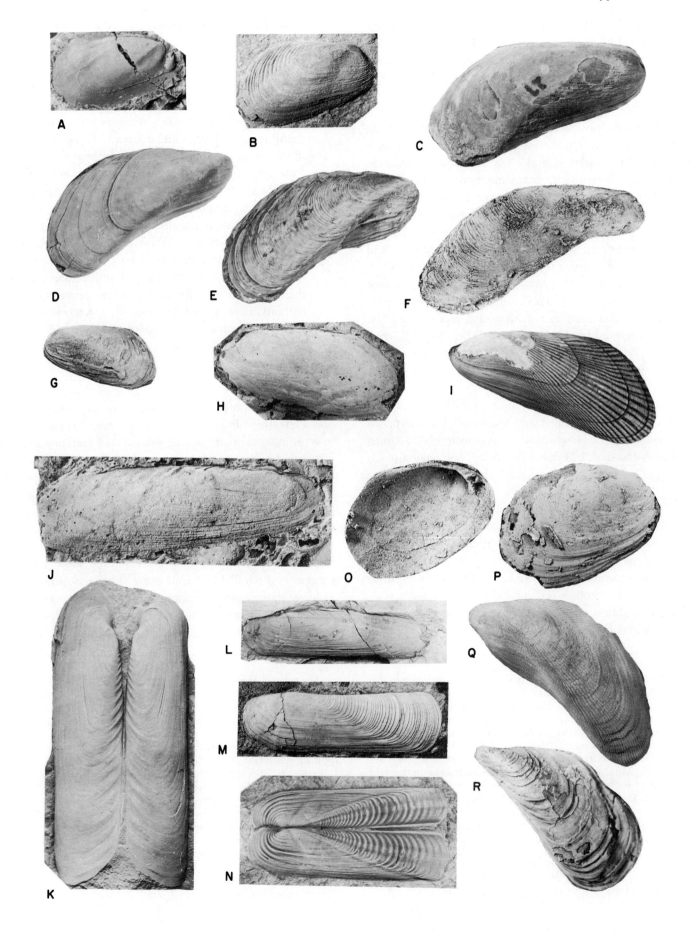

developed burrowing foot in most, crawling gastro-pod-like foot in few, reduced foot in few; gills filibranch in some, eulamellibranch in most. (Early Early Ordovician to Holocene; 2000 genera; 5000 living species.)

Although most heteroconchs are marine, most living freshwater species are in this subclass, some hetero-conchs are estuarine, and a few are almost terrestrial. Most are burrowers, but some bore, nestle, or are cemented or byssally attached; some are commensal, and a few are parasitic. Most are suspension feeders, but some are detritus feeders using the inhalant siphon. Many live in the shallow waters of the photic zone, and some are intertidal, a few range to depths of 4800 m.

In the Paleozoic, heteroconchs were relatively minor constitutents of pelecypod faunas. They underwent a dramatic radiation in the Mesozoic, and they are the most diverse of living pelecypods. Their post-Paleozoic radiation is thought to be a consequence of mantle fusion and siphon formation, which opened the way for many new infaunal modes of life.

The highly qualified definition of the Heteroconchia points out that the subclass is a phylogenetic concept. To understand the Heteroconchia, its constituent taxa must be appreciated. Living forms include freshwater mussels (Figure 14.103, *K*), fingernail clams, cockles (Figure 14.103, *L*), piddocks (Figure 14.88, *F*), giant clams (Figure 14.96, *A*), hard shell clams (Figure 14.87, *C*), false mussels, jewel box clams, coquina clams, and surf clams (Figure 14.92 *B*). Of the various extinct taxa, the most unusual are rudists (Figures 14.10 and 14.109, *E*). All heteroconchs seem to be descended from three Ordovician families (Figure 14.106) that have actino-dont dentition or only cardinal teeth (Figures 14.93, *A* and *B* and 14.109, *M* and *N*).

SUBCLASS PTERIOMORPHIA (Figures 14.87, *A* and *D*; 14.96, *B* to *H*; 14.97, *A*, *C*, and *E* to *J*; 14.101, *D* to *F*; 14.102, *E* to *J* and *L*; 14.110)

Shell variable, most with reduced anterior end and development of lobes, auricles, and/or wings, aragonitic, calcitic, or both, ultrastructure variable, many inequivalved, some equivalved; beaks vari-able, sometimes terminal; adult byssus in many, some cemented, byssal gape in many; dentition vari-able, heterodont, actinodont, taxodont, or edentu-lous; hinge plate in some. Ligament amphidetic or elongate opisthodetic, alivincular, multivincular, or duplivincular, with or without resilium. Commonly anisomyarian but may be isomyarian or mono-myarian. Mantle mostly unfused and lacking well-differentiated inhalant and exhalant apertures or siphons. Foot commonly reduced, absent in some. Gills filibranch in most; eulamellibranch in some. (Late Early Ordovician to Holocene; 900 genera, 1200 living species.)

Most pteriomorphs are marine, some are estuarine, and a few live in fresh water. Most are attached epifauna or semi-infauna, many attach by the byssus, and some are cemented. Some nestle, some bore, some are shallow burrowers, a few recline, and some swim. All are suspension feeders. Some are intertidal, most live in shallow marine situations, and a few occur at abyssal depths.

The Pteriomorphia, like the Heteroconchia, cannot be succinctly defined. Various superfamilies show con-vergent and parallel trends in many characters, and the range of morphologic variation in most features is great. Thus to understand the Pteriomorphia one must appreci-ate its constituent taxa. Living forms include the ark

---

Figure 14.109. Subclass Heteroconchia. **A**, *Megalomoidea.* Right lateral view of an internal mold, showing some muscle scars, pitted pallial line, and teeth. (Middle Silurian, Ontario, Canada, ×1.3.) **B**, *Cyprimeria.* Interior of the left valve of a shelled specimen, showing muscle scars, teeth, and pallial line. (Upper Cretaceous, Tennessee, ×0.8.) **C**, *Granocardium.* Exterior of the right valve of a shelled specimen. (Upper Cre-taceous, Mississippi, ×1.5.) **D**, *Crassatella.* Interior of the right valve of a shelled specimen. (Upper Cretaceous, Mississippi, ×1.5.) **E**, *Monopleura* (rudist). Exterior of five specimens cemented to each other. The opercular (left) valve is in place in some (short arrows). Two of the largest specimens are attached to a separated opercular valve at the base (long arrow). (Cre-taceous, Texas, ×1.7.) **F**, *Actinodonta.* Latex replica of the right valve showing actinodont dentition. (Lower Silurian, Great Britain, ×1.) **G**, *Schizodus.* Shelled specimen showing the exterior of the right valve. (Upper Pennsylvanian, Texas, ×1.5.)

H, *Paracyclas.* Shelled specimen, showing the exterior of the right valve. (Middle Devonian, Indiana, ×3.) **I**, *Astartella.* Shelled specimen, showing the exterior of the right valve. (Upper Pennsylvanian, Texas, ×3.) **J**, *Veniella.* Shelled speci-men, showing the right valve interior. (Upper Cretaceous, Tennessee, ×1.) **K**, *Aphrodina.* Shelled specimen, showing the right valve interior. (Upper Cretaceous, North Carolina, ×1.5.) **L**, *Redonia.* Latex replica of the right valve, showing actinodont dentition. (Middle Ordovician, Czechoslovakia, ×4.) **M**, *Cycloconcha.* Left valve interior of a shelled specimen, showing actinodont dentition. (Upper Ordovician, Kentucky, ×4.) **N**, *Lyrodesma.* Right valve interior of shelled specimen. (Upper Ordovician, Kentucky, ×5.) **O** and **P**. *Pterotrigonia (Scabro-trigonia).* Interior and exterior of the left valve. (Upper Cre-taceous, Tennessee, ×1.3.) **Q**, *Panopea.* Exterior of the right valve. (Upper Cretaceous, Texas, ×0.8.)

shells (Figure 14.110, *Q*), pearl oysters (Figure 14.96, *E* and *F*), pen shells (Figure 14.96, *C* and *D*), file shells (Figure 14.110, *G*), oysters (Figure 14.110, *I* and *M*), scallops (Figure 14.110, *A* and *B*), and jingle shells (Figure 14.110, *T*). Pteriomorphs have been abundant constituents of shallow-water marine faunas since the Middle Ordovician, and there are many extinct forms (Figures 14.92, *E* and *F*; 14.97, *A*, *C*, and *F* to *J*; 14.110, *C*, *E*, *F*, *J* to *L*, *N*, *P*, *R*, and *S*).

All pteriomorphs seem to be descended from three Ordovician families (Figure 14.106) that possessed the duplivincular ligament. The most primitive and stratigraphically oldest of these families is the Cyrtodontidae (Figures 14.92, *E* and *F* and 14.110, *P*), and shells of this type may have been ancestral to the rest of the pteriomorphs.

SUBCLASS ANOMALODESMATA (Figures 14.90, *A* to *E*; 14.111)

Shell short to elongate, primitively equivalved, most younger species subequivalved, a few markedly inequivalved, overlap of dorsal margins common in short shells, aragonitic, ultrastructure usually nacreous internally and prismatic externally; some with little interumbonal growth and beaks penetrate each other; deep burrowing species with pedal and siphonal gapes, a few species with adult byssus. Hinge often thickened, edentulous, or with one amorphous tooth, lateral teeth seldom present. Ligament parivincular, opisthodetic, primitively external, secondarily semi-internal and lying between the overlapping posterior dorsal edges of the valves, internal in some with lithodesma and chondro-phore, absent in a few. Usually isomyarian, rarely anisomyarian, never monomyarian. Siphons common, often with ventrally fused mantle lobes and sometimes with fourth aperture; foot usually well developed, gills eulamellibranch or absent, some with a muscular septum. Microornament of small granules or conical spines common on exterior of shell. (Late Middle Ordovician to Holocene; 100 genera; 400 living species.)

All species are marine. Most anomalodesmatans are burrowing, some are nestling, some are semi-infaunal, some are reclining, a few are cemented, and a few are commensal. Most are suspension feeders; some are partly carnivorous. Depth range is from littoral to abyssal.

The Anomalodesmata contains some of the rarest living pelecypod species that lack common names and are often found only in deep water or in remote parts of the world. Nonetheless the group shows considerable diversity occupying habitats from littoral to abyssal depths in equatorial to polar latitudes; these pelecypods are adapted for life as ecologic analogues of common shallow- and deep-burrowing clams, mussels, and oysters (heteroconchs and pteriomorphs).

Cambrian history, origin, and early evolution

In the past decade, a major happening in the study of pelecypods was the demonstration of the existence of the class in Cambrian rocks. *Pojetaia* (Figure 14.112, *A* to *D*) occurs in the Lower Cambrian of Australia, and *Fordilla* (Figure 14.112 *E* to *I*) has been found in Lower Cambrian rocks of North America, Denmark, and Siberia. Both genera have a bivalved shell, and the ligament,

Figure 14.110. Subclass Pteriomorphia. **A**, *Chesapecten*. Exterior of the right valve. (Pliocene, Chesapeake Bay, Maryland, ×0.5.) **B**. *Aviculopecten*. Exterior of the left valve. (Upper Pennsylvanian, Texas, ×2.) **C**, *Gryphaea*. Oblique view of a shelled specimen, showing an opercular right valve and an incurved convex left valve. (Upper Jurassic, Montana, ×1.1.) **D**, *Caritodens*. Exterior of the left valve, showing anterior and posterior wings. (Upper Ordovician, Indiana, × 1.) **E**, *Gervillia*. Exterior of the right valve with a posterodorsal auricle. (Upper Cretaceous, Tennessee, ×0.8.) **F**, *Opisthoptera*. Internal mold of the left valve, showing the posterior wing (Upper Ordovician, Indiana, ×1.) **G**, *Lima*. Exterior of the right valve. (Upper Cretaceous, North Carolina, ×2.5.) **H**, *Glycymeris*. Left valve interior, showing taxodont dentition and duplivincular ligament grooves. (Upper Cretaceous, Texas, ×1.5.) **I**, *Rastellum* (*Arctostrea*). Interior of the left valve, showing the posterior adductor muscle scar. (Upper Cretaceous, Alabama, ×0.5.) **J**, *Buchia*. Anterior view of a shelled specimen, showing a larger left valve. (Lower Cretaceous, California, ×2.) **K**, *Daonella*. Right valve exterior. (Middle Triassic, Nevada, ×2.) **L**, *Inoceramus*. Exterior of the left valve. (Upper Cretaceous, South Dakota, ×1.25.) **M**, *Agerostrea*. Interior of the left valve, showing an adductor muscle scar. (Upper Cretaceous, Netherlands, × 1.5.) **N**, *Halobia*. Exterior of left valve. (Upper Triassic, California, ×3.) **O**, *Aguileria*. Interior of the left valve, showing multivincular ligament grooves. (Upper Cretaceous, Texas, ×1.25.) **P**, *Cyrtodonta*. Interior of the left valve showing heterodont teeth, pallial line, and anisomyarian adductor scars. (Middle Ordovician, Kentucky, ×1.) **Q**, *Cucullaea* (*Idonearca*). Interior of left valve showing muscle scars, pallial line, anterior and posterior lateral teeth, taxodont teeth, and duplivincular ligament grooves. (Upper Cretaceous, Mississippi, ×1.25.) **R**, *Monotis*. Right valve exterior. (Upper Triassic, California, ×0.8.) **S**, *Exogyra*. Interior of the left valve, showing an adductor muscle scar. (Upper Cretaceous, Alabama, ×1.66.) **T**, *Anomia*. Articulated specimen, showing the right valve with the byssal foramen closed by pluglike, partially calcified byssus. (Upper Cretaceous, Colorado, ×2.)

muscle scars, and teeth are known. *Pojetaia* had a shell composed of a single layer of spherulitic aragonite prisms (Figure 14.112, *C*).

Fordilla and *Pojetaia* postdate the oldest ribeirioids by about 10 million years. The critical event in the evolution of pelecypods was the appearance of a completely divided shell that had an intervening ligament; pelecypods possess a ligament, ribeirioid rostroconchs do not. From what is known of the ultrastructure of *Pojetaia*, the shell and ligament were similar at the ultrastructural level. The only difference between the shell and ligament of this small pelecypod probably was in the amount of protein between aragonitic fibers. The conversion of the posterior dorsal margin of the ribeirioid rostroconch shell into the pelecypod ligament may have resulted from the production of additional protein at the hinge. The oldest known rostroconch, *Heraultipegma* (Figure 14.16, *D* to *F*), also had a prismatic and probably aragonitic shell.

The general morphology of *Fordilla* and *Pojetaia* indicates that they shared a common ancestor; there is no evidence for a polyphyletic origin of the Pelecypoda. Various morphologic features of *Pojetaia* suggest that it is closely related to the palaeotaxodonts, and it is very similar to Ordovician species of that subclass (Figure 14.107, *A* to *D*). The morphology of *Fordilla* is much like that of some Ordovician isofilibranchs (Figure 14.108, *O* and *P*), and it is probably the oldest member of that subclass.

The structure of Ordovician isofilibranchs suggests that they are the likely ancestors of the Anomalodesmata. The dentition of some Ordovician palaeotaxodonts indicates that they may be ancestral to the Heteroconchia. That most of the Ordovician heteroconchs have actinodont dentition suggests an ancestral relationship to the cyrtodontids, which have strong heterodont dentition; the cyrtodontids, in turn, appear to be ancestral to the other pteriomorphs (Figure 14.106).

Some geologic applications of the study of pelecypods

The study of the morphology, taxonomy, ecology, and phylogeny of living and fossil pelecypods, integrated with concepts of speciation, has provided the basic data necessary to use these animals for constructing geologic frameworks and syntheses and for solving various geologic problems.

The most widespread application of basic research on pelecypods is the production of regional biostratigraphic zonations, which are useful for correlating events and providing detailed geologic histories. Because pelecypods occur in diverse facies of widely different age, they have considerable utility for biostratigraphic correlation. In general, basic research on pelecypods of the Paleozoic is now reaching the stage where biostratigraphic syntheses will become possible in near-shore, shallow-water, and cratonic carbonate facies; these studies are more advanced for the late Paleozoic than for the early Paleozoic.

Basic studies are more advanced in post-Paleozoic rocks than in Paleozoic rocks, and various groups of pelecypods have been empirically established as useful for biostratigraphic zonations. In the Triassic, the pteriomorphs known as halobiids (Figure 14.110, *K* and *N*) and monotids (Figure 14.110, *R*) are found in quiet water, deeper basinal rocks where they provide stratigraphic resolution equivalent to that of ammonites. In the Jurassic, the pteriomorphs known as buchiids (Figure 14.110, *J*) are very useful for the correlation of continental-margin, offshore, fine, clastic sedimentary rocks. For various other Jurassic facies, the pteriomorphs known as pectinaceans (scallops) and inoceramids (Figure 14.110, *L*) are useful biostratigraphically. In shallow-water facies of the Jurassic and Cretaceous, the heteroconchs known as trigonioids (Figure 14.109, *O* and *P*) are biostratigraphically useful. In the Cretaceous, inoceramids are biostratigraphically useful fossils in chalks and other outer-shelf and basin center carbonate rocks, buchiids are still present, and scallops and various oysters (Figure 14.110, *C*, *I*, *M*, and *S*) are useful in facies ranging from lagoonal or estuarine sedimentary rocks to shore-face and shelf clastics and carbonate rocks. In both the Cretaceous and Cenozoic, freshwater mussels (unionids, Figure 14.98, *I* and *J*) are useful biostratigraphically. In the Cenozoic, various pteriomorphs, including pectinaceans (Figure 14.110, *A*), oysters (Figure 14.86, *D*), and arcids (Figure 14.102, *G*, *I*, and *L*), and the heteroconch venerids (Figure 14.86, *F* to *H*) are biostratigraphically useful in a variety of facies from estuarine to open marine. For the most part, Cenozoic pelecypods are most abundant in facies of mixed sediments such as argillaceous clastics and marls. They are generally less abundant in very coarse or very fine sediments and in high-energy environments.

Once sound species-level taxonomies and biostratigraphic zonations for correlating events have been established, pelecypods can be used for further geologic deductions including:

1 The measurements of fault movements.
2 The establishing of onlap—offlap sequences and the timing of sea-level events in general.
3 The plotting of distributions to determine paleogeography and paleoclimatology, which in turn can assist

Figure 14.111. Subclass Anomalodesmata. **A,** *Cimitaria.* Right valve exterior. (Middle Devonian, New York, ×1.) **B,** *Wilkingia.* Right valve exterior. (Pennsylvanian, Nebraska, ×0.8.) **C,** *Euciroa.* Right valve exterior, showing tuberculate ornament. (Holocene, Pacific Ocean, ×1.2.) **D,** *Pholadomya.* Right valve exterior. (Jurassic, Utah, ×2.) **E,** *Pholadomya.* Right valve exterior. (Jurassic, Montana, ×2.) **F,** *Anatimya.* Composite mold of the right valve. (Upper Cretaceous, Texas, ×1.33.) **G,** *Pandora.* Interior of the right valve. (Holocene, California, ×2.) **H,** *Rhytimya.* Right valve exterior. (Upper Ordovician, Ohio, ×2.5.) **I,** *Rhytimya.* Exterior of right valve, showing tuberculate ornament. (Ordovician, New York, ×4.) **J** to **L.** *Cuneamya.* Dorsal, left lateral, and anterior views. (Middle and Upper Ordovician, New York and Ohio, ×2.) **M,** *Grammysia.* Dorsal view of spread valves. (Middle Devonian, New York, ×1.) **N,** *Osteomya.* Dorsal view, showing a large siphonal gape. (Middle Jurassic, England, ×1.5.)

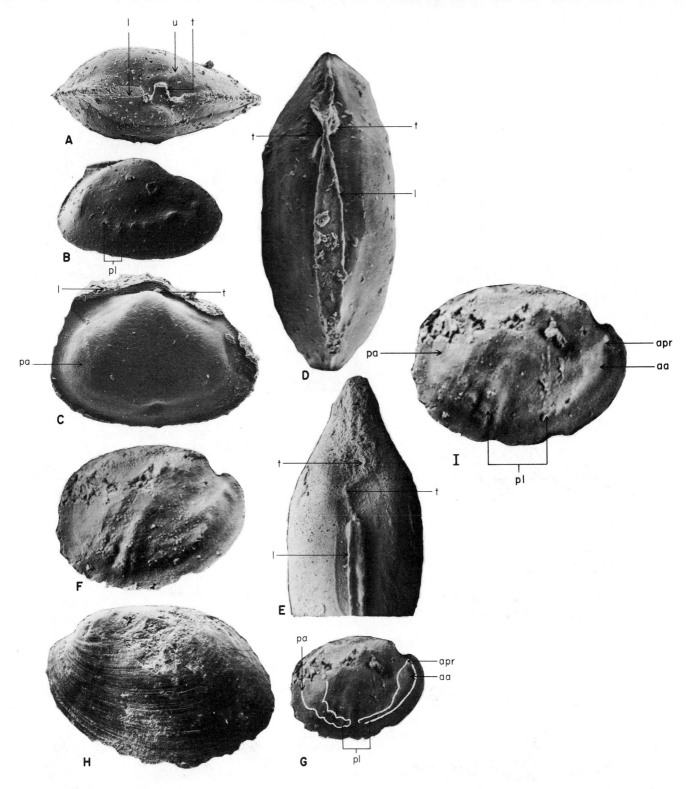

Figure 14.112. Cambrian pelecypods. **A** to **D**, *Pojetaia runnegari.* All views are phosphatic internal molds. (Lower Cambrian, South Australia.) **A,** Dorsal view showing ventral trace of teeth between the umbos and the filling of the ligament area. (×50.) **B,** Oblique ventrolateral view of the right valve, showing pitted pallial line. (×35.) **C,** Oblique dorsolateral view of the right valve, showing teeth, ligament area, posterior adductor muscle scar, and impressions of prisms of shell ultrastructure. (×55.) **D,** Dorsal view, showing the filling of the ligament space and molds of teeth. (×110.) **E** to **I,** *Fordilla.* **E,** *F. sibirica.* View of the anterior end of the dorsum of an internal mold showing the ligament filling and molds of teeth. (Lower Cambrian, Siberia, ×35.) **F** to **I,** *F. troyensis.* **F,** Right lateral view of an internal mold showing muscle scars (×10); **G,** same view of the same specimen with muscle scars outlined in white (×8); **H,** exterior of the left valve with shell (×15). (Lower Cambrian, New York.) **I,** Enlarged view of **G** with same features labeled, but muscle scars not outlined. (×13.) *aa,* Anterior adductor muscle scar; *apr,* anterior pedal retractor muscle scar; *l,* ligament filling; *pa,* posterior adductor muscle scar; *pl,* pallial line; *t,* teeth; *u,* umbo. (**A** to **D** courtesy Bruce Runnegar, University of New England, Armidale, Australia. **E** courtesy A. Yu. Rozanov, Palaeontological Institute, Akademia Nauk, USSR.)

in the study of continental margins, accreted tectono-stratigraphic terranes, and microplate movements.

In addition, the shells of pelecypods have been used to study amino acid racemization, to measure isotopes of oxygen, and study fission-track dating, to determine structural style through deformation fabrics of the shells, for salinity sensing and the determination of coal quality, and for paleobathymetry.

Supplementary reading

Addicott, W. O. Mid-Tertiary zoogeographic and paleo-geographic discontinuities across the San Andreas Fault, California. Proceedings of a Conference on the Geologic Problems of the San Andreas Fault System, Stanford University Geological Sciences Publications 9: 144–165; 1968. Study using pelecypods and other mollusks to measure large-scale movements of the San Andreas Fault.

Blackwelder, B. W. Late Cenozoic stages and molluscan zones of the U.S. Middle Atlantic Coastal Plain. *Paleontological Society Memoir 12*; 1981. Application of basic studies to the biostratigraphic zonation of Pliocene and Pleistocene rocks.

Moore, R. C.; Teichert, C., editors. *Treatise on invertebrate paleontology, Part N, Mollusca 6, Bivalvia.* Boulder, CO, and Lawrence, KS: Geological Society of America and Uni-versity of Kansas Press; 1969–1971. Extensive treatment of all aspects of pelecypod studies.

Pojeta, J. Jr. Review of Ordovician pelecypods. U.S. Geological Survey, Professional Paper 695; 1971. Summary treatment of all Ordovician pelecypods and their early evolution.

Pojeta, J., Jr. *Fordilla troyensis* Barrande and early pelecypod phylogeny. *Bulletins of American Paleontology 63:* 363–384; 1975. Monograph of the oldest known isofilibranch pelecypod.

Runnegar, B.; Bentley, C. Anatomy, ecology, and affinities of the Australian Early Cambrian bivalve *Pojetaia runnegari* Jell. *Journal of Paleontology 57:* 73–92; 1983. Monograph of the oldest known palaeotaxodont pelecypod.

Stanley, S. M. Relation of shell form to life habits of the Bivalvia (Mollusca). *Geological Society of America Memoir 125;* 1970. Monograph of the ecology of living pelecypods and morphologic features that indicate life habits.

Stanley, S. M. Functional morphology and evolution of byssally attached bivalve mollusks. *Journal of Paleontology 46:* 165–212; 1972. Monograph of the ecology of byssate pele-cypods.

Waller, T. R. Scanning electron microscopy of the shell and mantle in the order Arcoida (Mollusca: Bivalvia). *Smith-sonian Contributions to Zoology* No. 313; 1980. Modern summary of concepts of the mantle edge of pelecypods.

Yonge, M.; Thompson, T. E., editors. Evolutionary systemat-ics of bivalve molluscs. *Royal Society of London, Philo-sophical Transactions B 284:* 199–436; 1978. Proceedings of a symposium, with state-of-the-art summaries of the paleontology and zoology of pelecypods.

15

Phylum Hyolitha

John Pojeta, Jr.

The concept Hyolitha has changed dramatically over the past 20 years. Most recently hyoliths have been considered to be (1) a class of mollusks, (2) a specialized group of monoplacophorans, and (3) a separate phylum having a distant, common ancestry with mollusks (see Figure 14.15). This chapter follows the third approach. The consensus is that hyoliths and mollusks are related, but the closeness of that relationship and how to express it taxonomically are being debated.

More is known about hyoliths than about many other extinct groups of animals. For example, the gross morphology of the hyolith digestive tract is known as is the fact that Cambrian priapulid worms ate hyoliths. However, most information about hyolith soft parts must still be interpreted by comparing markings on their shells with similar markings made by soft parts of extant metazoans. The different interpretations of hyoliths result from emphases given to different structures and their presumed functions.

Hyolitha are marine, bilaterally symmetrical, solitary metazoans with a generally conical, probably aragonitic shell that has a single aperture and one or three other calcareous parts. Hyoliths had numerous muscles and a long, sinuous gut; they existed from earliest Cambrian to Late Permian; and at least 40 genera and hundreds of species are known. Many mature hyoliths are from 15 to 30 mm long, but they may be as short as 4 mm, and some Cambrian and Ordovician species are as much as 200 mm long. One incomplete specimen of a Permian species thought to be a hyolith is 400 mm long.

Morphology

All hyoliths have a bilaterally symmetrical exoskeleton with at least two pieces: a large, tapering conical piece called the **conch**, which has an aperture only at the bigger anterior end (Figure 15.1, *I*); and a smaller **operculum** that closes the aperture of the conch (Figure 15.1, *A, E,* and *F*). Hyoliths may have imperforate septa at the apical end of the conch (Figure 15.3, *M*). One major group of hyoliths has two additional skeletal pieces called **helens** (Figures 15.1, *A* and *C*; 15.3, *M*; and 15.4, *I*), which are thin, slightly asymmetrical, scimitar-shaped, lateral calcareous appendages placed between the operculum and the conch. Whether helens curved dorsally or ventrally during life is interpreted differently by various authors (Figure 15.1, *A* and C). The conchs of some hyoliths have crossed lamellar ultrastructure.

Soft parts

The structure of the hyolith digestive tract is known from sediment fillings of the gut in Cambrian and Ordovician species found in Antarctica, Czechoslovakia, and France. These sediment fillings of the gut show that the mouth was near the ventral side of the aperture and that a long sinuous intestine (Figure 15.4, *Li* and *Mi*) made a U-shaped dorsal turn near the apex of the shell and became a straight rectum (Figure 15.4, *Lr* and *Mr*) that terminated in a dorsal anus at the aperture. Functional analysis, epibionts, and the filling of the gut indicate that the less convex, flattened, or concave side of the conch was ventral (Figure 15.1, *J* to *W*) and that hyoliths were nearly sessile, epifaunal, benthic deposit feeders. Considerable knowledge exists about the muscle scars on both the conch and operculum (Figure 15.1, *B* and *D*), but there is no consensus about interpreting this information.

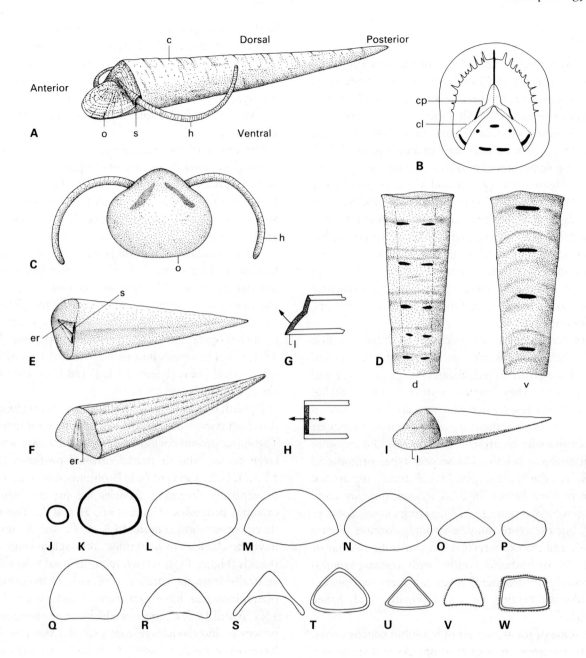

Figure 15.1. Morphology of hyoliths. **A,** Reconstruction of the hyolithid *Hyolithes* showing parts of the shell and orientation. In this reconstruction, the helens are concave dorsally. *c,* Conch; *h,* helen; *o,* operculum; *s,* slot between operculum and conch at the point where helens emerge. **B,** Interior of the operculum of the hyolithid *Maxilites,* showing muscle scars (black). *cl,* Clavicle; *cp,* cardinal process. (Cambrian, Czechoslovakia, ×2.5.) **C,** Reconstruction of the operculum showing the exterior with helens concave ventrally. **D,** Dorsal *(d)* and ventral *(v)* views of an internal mold of the conch of the hyolithid *Gompholites* showing multiple muscle scars (black). (Ordovician, Czechoslovakia, ×4.5.) **E,** Oblique ventral view of an orthothecid conch showing the operculum in place and slots *(s)* formed between external opercular ridges *(er)* and conch. **F,** Oblique left lateral view of an orthothecid conch showing the vertical operculum in place in the aperture and lack of a ligula. *er,* External ridge. **G** and **H,** Diagrammatic midsagittal sections of anterior ends showing the opercula (stippled) in place in conchs of hyolithids (**G**) and orthothecids (**H**); arrows show directions in which operculum could move. *l,* Ligula. **I,** Oblique left lateral view of a hyolithid conch showing the ligula *(l)* extending anteriorly from the ventral side of the aperture. **J** to **W,** Variation in cross section of conchs of Cambrian (**J, R,** and **S**) and Ordovician (**K** to **Q** and **T** to **W**) hyoliths of Europe. The ventral side is lowermost. **K** to **P, T,** and **U** are hyolithids, **J, Q** to **S, V,** and **W** are orthothecids. (**A** and **D** modified from, and **F** from Marek, L. *Sborník Geologických Věd, Paleontologie, Řada* P, 1: 65–67; 1963. **B** modified from Marek, L. *Časopis Národního Muzea,* odd. přírodovědný; 141: 71; 1972. **C** from Yochelson, E. L. U.S. Geological Survey, *Journal of Research* 2: 721; 1974. **E** and **I** from Marek, L. *Sborník Geologických Věd, Paleontologie, Řada* P, 9: 63, 57; 1967. **G** and **H** from Marek, L. *Časopis Národního Muzea,* odd. Přírodovědný; 1966, p. 61.)

Hard parts

In most hyoliths, the ventral side of the conch is flatter than the rest of the shell, and in many the mature cross section is roundly subtriangular (Figure 15.1, *N*) or sub-semicircular (Figure 15.1, *M*). In fewer species, the cross section is nearly circular (Figure 15.1, *J* and *K*), oval (Figure 15.1, *L*), or various other shapes (Figure 15.1, *O* to *W*). The venter may be convex (Figure 15.1, *J* to *P*), concave (Figure 15.1, *Q* to *S*, and *V*), or essentially flat (Figure 15.1, *T* and *W*). In many species, somewhat sharp lateral edges develop where the venter meets the dorsum (Figure 15.1, *M*, *O*, *T*, *U*, and *W*). The two sides of the dorsum may meet in a carina (Figure 15.1, *S* and *U*), but the dorsum ordinarily is rounded (Figure 15.1, *J* to *N*, *Q* and *R*). A few species have triangular or quadrangular cross sections (Figure 15.1, *T* to *W*). Cross-section shape can vary ontogenetically.

There are two types of conchs in hyoliths. The first type, **hyolithid conchs**, have an anterior shelflike extension of the aperture called the **ligula** (Figures 15.1, *I* and 15.2, *A* to *F*). They commonly have a convex to flat venter and a cross section in which the dorsum is rounded (Figure 15.5, *G*), but the cross section may be triangular and a carina may be present. Only hyolithid conchs are known to have helens. The second type, **orthothecid conchs**, lack the ligula (Figure 15.1, *E* and *F*) and are not known to have helens. In cross section, they are commonly reniform (Figure 15.1, *R*), having a gently concave venter, but the venter may be strongly concave (Figure 15.1, *S*), and the cross-section shape may be circular or square. In orthothecid conchs with concave ventral surfaces, the ventrolateral edges can form two projections like the runners on a sled (Figures 15.1, *S* and 15.4, *G*).

The slope of the peristome of hyolithid conchs varies. In most, the aperture is erect or upright and about perpendicular to the long axis of the conch (Figure 15.2, *D*), but it can slope forward (Figure 15.2, *E*) or backward (Figure 15.2. *F*). In orthothecid conchs, the apertural margin is erect and nearly perpendicular to the long axis (Figure 15.1, *F*). The ventral margin of the lateral profile of hyolithid conchs may be straight (Figure 15.2, *A*), but in most species it is curved upward adapically (Figures 15.2, *B* and 15.5, *D* and *I*). In a few forms, the apical part of the shell is curved downward (Figure 15.2, *C*).

Muscle scars are known in several species having hyolithid conchs. The conchs commonly have a pair of scars dorsally (Figure 15.2, *L*, *O*, and *Q*) and ventrally (Figure 15.2, *P*) just inside the aperture. In some conchs, the muscle tracks can be seen (Figure 15.2, *O* and *P*). On the

ligula, some specimens show a transversely elongated structure (Figure 15.2, *M*, *N*, *R*, and *U*), which is usually considered to be a muscle scar. At least two species of hyolithids show multiple paired dorsal muscle scars arranged in two rows (Figures 15.1, *Dd* and 15.2, *S*) and a single row of multiple ventral muscle scars (Figure 15.1, *Dv*).

Ornament of hyolith conchs is variable. It may be comarginal growth increments (Figures 15.4, *B* and *C* and 15.5, *B* and *F*) or longitudinal ribs (Figure 15.4, *F* and *J*); either of these can vary in strength. Ornament generally is better developed on the conch's dorsal side (Figure 15.2, *T* and *U*).

The operculum of hyolithid conchs fits on an aperture that is not in a single vertical plane but has a marked angulation (Figures 15.1, *A* and *G* and 15.2, *H*), and the operculum extends to the anterior margin of the ligula (Figure 15.1, *A* and *G*). The inner margins of the hyolithid operculum abut against the peristome (Figure 15.1, *G*). The operculum of orthothecid conchs lies in one vertical plane (Figure 15.1, *E* and *F*) and fits within the aperture (Figure 15.1, *H*).

Hyolith opercula are complex structures (Figure 15.3, *A* to *I*) that show considerable variation on the inner face. The most prominent opercular structures are small-to-large dorsal bifid or paired **cardinal processes** (Figure 15.3, *A*, *C*, *E*, and *G* to *I*). In hyolithids, one or more pairs of ventrally diverging **clavicles** are present below the cardinal processes (Figure 15.3, *H* and *I*). The orthothecid operculum (Figure 15.3, *A*, *C*, and *E*) does not have the clavicles of hyolithids, although in some orthothecids (Figure 15.3, *A*) two vague internal ridges diverge ventrally from the cardinal processes. In hyolithids, the operculum can have numerous muscle scars (Figures 15.1, *B* and 15.3, *I*), some of which can be on the cardinal processes and clavicles (Figure 15.1, *B*). Many hyolithids have two especially conspicuous opercular muscle scars placed ventrally to either side of the midline (Figures 15.2, *I* and 15.3, *I*). It is not known for certain if muscle scars are on the orthothecid operculum. The exterior of the hyolithid operculum is divided into a dorsal **cardinal shield** and a ventral **conical shield** (Figure 15.3, *F*). The operculum of hyoliths has well-defined comarginal growth lines (Figures 15.3, *G*; 15.4, *D* and 15.5, *J*).

The proximal ends of the helens lodge in shallow sockets on the left and right inner sides of the operculum above the clavicles (Figure 15.2, *I*). When the operculum is closed, the lateral edges of it and the conch form a nearly vertical slot (Figure 15.1, *A*) on each side, through which the helens exit to the outside. The larval shells of hyoliths are not known in many species, but two general

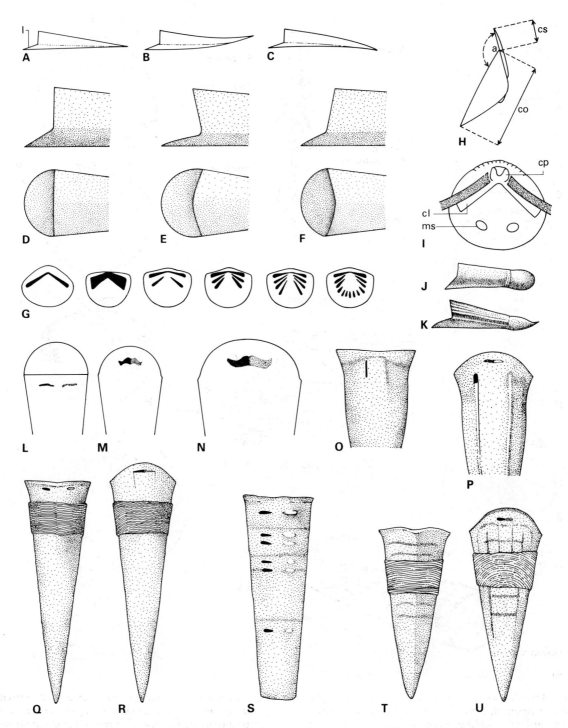

Figure 15.2. Morphology of hyoliths. **A** to **C**, Lateral profiles of left sides of conchs of hyolithids showing variation in ventral margin. **A**, Straight; **B**, curved upward; **C**, curved downward. **D** to **F**, Apertural variation in the conch of hyolithids: top row, left lateral views; bottom row, dorsal views. **D**, Erect aperture; **E**, forward sloping aperture; **F**, backward sloping aperture. **G**, Variation in number, position, and width of clavicles on inside of the opercula of hyolithids. **H**, Midsagittal section of a hyolithid operculum showing the angle (*a*) between the cardinal shield (*cs*) and conical shield (*co*). **I**, Diagram showing the position of the proximal part of the helens (stippled) on the inside of an operculum. In this reconstruction, the helens curve upward. *cp*, Cardinal process; *cl*, clavicle; *ms*, muscle scar. **J** and **K**, Protoconchs of hyoliths. **J**, Globose smooth protoconch. **K**, Pointed larval shell showing growth lines. (Ordovician, Poland.) **L** to **U**, Probable muscle scars and/or tracks on internal molds of conchs of hyolithids (black on left side). All from Czechoslovakia. **L** to **N**, Cambrian; **O** to **U**, Ordovician. **L** to **N**, *Maxilites*. Dorsal (**L**) and ventral (**M** and **N**) views. **O** and **P**, *Dilytes*. Dorsal and ventral views. **Q** and **R**, *Joachimilites*. Dorsal and ventral views. **S**, *Gompholites*. Dorsal view. **T** and **U**, *Eumorpholites*. Dorsal and ventral views. Dorsal view (**T**) does not show muscle scars, it is for comparison of ornament with ventral view (**U**). (**A** to **G**, **Q**, **R**, **T**, and **U** From Marek, L. *Sborník Geologických Věd, Paleontologie, Řada P*, 9: 60, 77, 81. 1967. **H**, **I**, and **R** from Marek, L. *Sborník Geologických Věd, Paleontologie, Řada P*, 1: 60, 63, 67; 1963. **J** and **K** from Dzik, J. *Lethaia* 11: 295; 1978. **L** to **N** from Marek, L. *Časopis Národního Muzea, odd. Přírodovědný* 1972, p.71. **O** and **P** from Marek, L. *Věstník Ústředního Ústavu Geologického* 49; 289, 1974.)

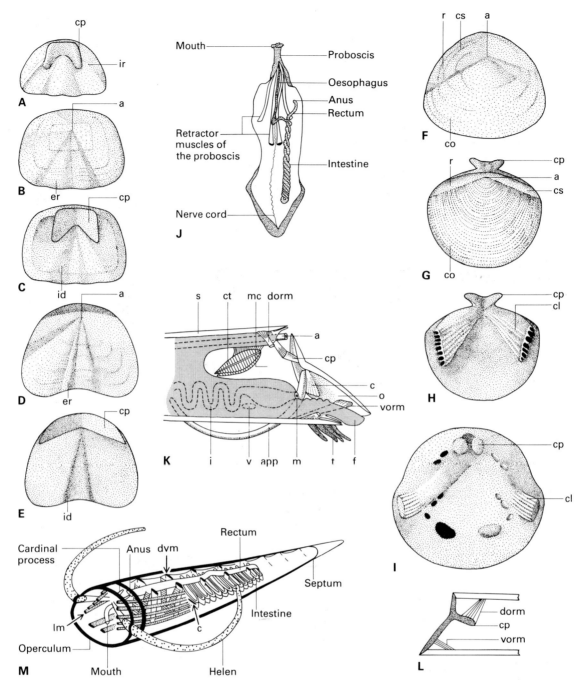

Figure 15.3. A to I, Views of opercula. **A to E,** Orthothecids; **F to I,** hyolithids. *a,* Apex; *cl,* clavicle; *co,* conical shield; *cp,* cardinal process; *cs,* cardinal shield; *er,* external ridge; *id,* internal depressions, *ir,* internal ridges; *r,* rooflet. **A,** *Orthotheca.* Interior view. (Devonian, Czechoslovakia). **B and C,** *Quadrotheca?.* Exterior and interior views. (Ordovician, Czechoslovakia). **D and E,** *Nephrotheca.* Exterior and interior views. (Ordovician, Czechoslovakia.) **F,** *Joachimilites.* Exterior view. (Ordovician, Czechoslovakia). **G and H,** *Nevadalites.* Exterior and interior views. (Cambrian, Nevada). **I,** *Gompholites.* Interior view. Muscle scars are black on the left and stippled on the right. (Ordovician, Czechoslovakia.) **J,** Diagrammatic representation of a dissected sipunculoid. **K,** Reconstruction of sagittal section of the anterior end of a hyolithid as a mollusk. Postulated tissue on the inside of the operculum and lateral and visceral muscles not shown. Mantle folds at the dorsal and ventral apertural margins are deliberately overemphasized and not stippled. Postulated tentacles are shown on only one side. Foot, tentacles, and main visceral mass are stippled. The operculum shows only the cardinal process and clavicle on the left side. The intestine contains many loops as might be expected in a detritus feeder. In several species of orthothecids, the loops of the intestine are in a horizontal plane; for ease of drawing, the hyolithid intestine is shown in a vertical plane. *a,* Anus; *app,* helen; *c,* clavicle; *ct,* paired gills; *cp,* cardinal process; *dorm,* dorsal opercular retractor muscle; *f,* foot; *i,* intestine; *m,* mouth; *o,* operculum; *t,* tentacles; *v,* vesicles; *vorm,* ventral opercular retractor muscles; *mc,* mantle cavity; *s,* shell wall. **L,** Reconstruction showing only the proposed retractor muscles of operculum seen in **K,** (abbreviations are defined in **K**). **M,** Reconstruction of *Gompholites.* Only parts of the musculature and gut are shown. *lm,* Insertions of longitudinal muscles; *dvm,* insertions of dorsally attached dorsoventral muscles; *c,* insertions of ventrally attached dorsoventral muscles. (**A to I** modified from Marek, L. *Sborník Geologických Věd, Paleontologie, Řada* P, 1: 61, 68; 1963; 9: 77, 85, 99, 105; 1967; and *Časopis pro Mineralogii a Geologii* 21: 278; 1976. **J and M** from Runnegar, B.; *et al. Lethaia* 8: 190; 1975. **K** from Marek, L.; Yochelson, E. L. *Lethaia* 9: 75; 1976. **L** modified from Marek, L. *Sborník Geologických Věd, Paleontologie, Řada* P, 1: 68; 1963.)

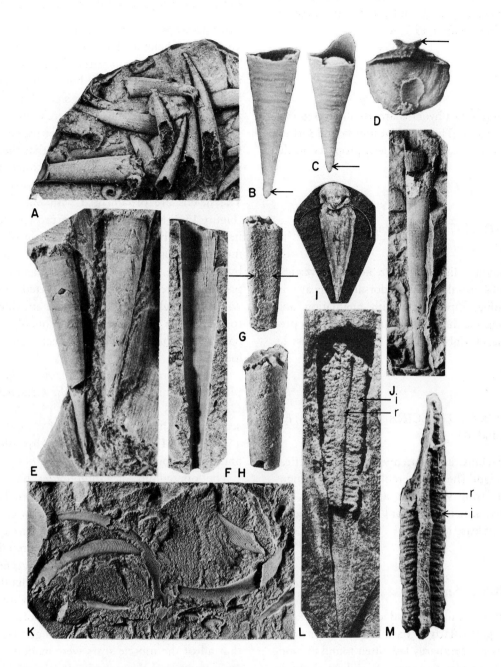

Figure 15.4. A, E to **H, J, L,** and **M,** orthothecids; **B** to **D, I,** and **K,** hyolithids. **A,** *Orthotheca.* Latex cast showing group of conchs. (Ordovician, France, ×3). **B** to **D,** *Nevadalites.* **B** and **C,** Silicified early growth stages showing the protoconch (arrows) separated from the mature conch by a slight constriction. Note the change in the apical angle between the larval shell and mature conch. (×20). **D,** Silicified operculum. Note the angulation between the cardinal shield and conical shield that allows one to see cardinal processes (arrow) in this view. (Cambrian, Nevada, ×20). **E,** *Orthotheca,* showing two conchs. (Cambrian, Mexico, ×5.5). **F,** *Bactrotheca.* External mold of conch. (Ordovician, Czechoslovakia, ×3.5). **G** and **H,** *Tcharatheca.* Internal molds. **G,** Ventral view with runners (arrows); **H,** dorsal view. (Cambrian, Antarctica, ×6.) **I,** *Hyolithes.* Dorsoventrally compressed articulated hyolithid showing operculum, helens, and conch. (Burgess Shale, Cambrian, British Columbia, ×2.) **J,** *Bactrotheca.* Latex cast showing the conch. (Ordovician, France, ×4). **K,** *Hyolithes.* Disarticulated helens showing possible growth lines. (Cambrian, Newfoundland, ×3). **L,** Dorsal view of in-place gut filling in conch of the orthothecid *Orthotheca* showing sinuous intestine *(i)* below straight rectum *(r).* (Ordovician, France, ×8). **M,** Silicified gut filling of a hyolith. (Cambrian Antarctica, ×15). (**L** from Thoral, M. *L'Etude Paleontologique de l'Ordovician Inferieur, Montpellier*; 1935, pl. 14.)

types of protoconchs seem to exist: smooth globose protoconchs (Figure 15.2, *J*) and pointed protoconchs having growth lines (Figure 15.2, *K*).

Classification

Two major groupings of hyoliths are usually recognized and treated as orders. They are Hyolithida and Orthothecida. Class level taxa have not been proposed in the Hyolitha.

ORDER HYOLITHIDA (Figures 15.4, *B* to *D*, *I*, and *K*; 15.5)

Hyoliths having a ligula; aperture not in one vertical plane. Shells usually gently, dorsally concave in lateral profile. Operculum with angulation and abutting against the apertural margin; helens present. (Earliest Cambrian to Late Permian; at least 25 genera.)

ORDER ORTHOTHECIDA (Figure 15.4, *A*, *E* to *H*, *J*, *L*, and *M*)

Hyoliths lacking a ligula; aperture in a vertical plane. Straight shells. Operculum without angulation and fitting within the aperture; helens not known to be present. (Earliest Cambrian to Middle Devonian; at least 15 genera.)

Stratigraphic distribution

Hyoliths are most abundant and diverse in rocks of Cambrian age, particularly in dark shales and in some limestones; they are ordinarily less often found in sandstones. More genera and species of hyoliths occur in Cambrian rocks than in any succeeding system. However, the number of genera is difficult to estimate because generic level taxonomy of hyoliths is in a state of flux. The number of specimens of hyoliths decreases markedly in Ordovician rocks, but the generic level diversity remains high. In post-Ordovician Paleozoic rocks, hyoliths are ordinarily uncommon. They may be represented by many or a few specimens at any one locality; diversity is markedly decreased. In Permian time, some hyoliths lived in the cold-water Gondwana environments of Australia.

Functional morphology and systematic position

Most investigators agree that the shell of hyoliths formed an external skeleton. The well-known Cambrian Burgess Shale preserves the soft parts of many animals. Hyoliths in the Burgess Shale fauna show well-preserved skeletal parts (Figure 15.4, *I*) but no sign of any external soft parts. Both the conch and the operculum of hyoliths have well-defined, regularly spaced, comarginal growth lines (Figure 15.5, *B* and *J*), indicating that these shell parts were secreted incrementally by an internal mantlelike epithelium similar to that of brachiopods and mollusks. The helens are ornamented with fine transverse lines (Figure 15.4, *K*). If these lines are growth increments, the helens probably grew by the addition of shell at their internal ends and their external surfaces were not covered by soft tissue. The helens lodge in shallow sockets and probably were attached to and secreted by mantle tissue within the shell. No trace of muscle attachment has been found on the helens, and they probably functioned as passive stabilizers and/or inefficient organs of locomotion.

The interpretation of hyoliths as a separate phylum places particular emphasis on the Ordovician hyolithid *Gompholites* (Figures 15.1, *D*; 15.2, *S*; and 15.3, *I*), sedimentary fillings of the gut of orthothecids (Figure 15.4, *L* and *M*), and comparisons of hyoliths to mollusks and sipunculoid worms.

The operculum of *Gompholites* has many paired muscle scars (Figure 15.3, *I*), and internal molds of the conch have an anterior to posterior row of linear muscle scars across the midventral part of the conch (Figure 15.1, *Dv*) and two rows of similar paired dorsal muscle scars (Figure 15.1, *Dd* and 15.2, *S*). The dorsal and ventral rows of muscle insertions are slightly offset from one another, and the muscle scars show traces of muscle tracks (Figure 15.1, *Dd*). This morphology suggests that the muscles of the conch were arranged segmentally and that all of the muscle scars were made by muscles that were functional throughout life (Figure 15.3, *M*). Two sets of muscles probably passed dorsoventrally and ventrodorsally to connective tissue on opposite body walls (Figure 15.3, *M*). The muscles could not attach to the shell at both ends because this would make them nonfunctional. Contraction of the dorsoventral and ventrodorsal muscles would compress a hydrostatic skeleton, lengthen the body, and move the operculum away from the aperture of the conch. Muscles that are functionally analogous to the ones postulated for hyoliths are present in the living, tube-dwelling annelid *Pectinaria*. Longitudinal body muscles apparently were attached to the numerous muscle scars of the operculum but not to the conch. Contraction of these muscles would close the operculum and possibly move the distal ends of the helens forward.

If each of the muscle scars in the conch of *Gompholites* marks the position of a muscle that was functional for life, an alternative to the previous interpretation is that the muscles of the conch ran directly to insertions on the operculum and served to close that structure. In this model, the shell-inserted muscles would have to be opposed by circular body muscles or by a structure analogous to the molluscan foot to open the operculum; the latter would require that hyoliths had pedal retractors. The extant sipunculoid worms (Figure 15.3, *J*)

have longitudinal muscles of the type required in this explanation. They are inserted in pairs on the dorsal and ventral body walls and are used to retract a feeding proboscis. Sipunculoids are sediment eaters, and the proboscis is extended by hydrostatic pressure. The transverse section of the muscles that retract the proboscis resembles the shape of the muscle insertions on hyolith conchs.

As in hyoliths, the mouth of sipunculoids opens into a recurved gut with a highly convoluted intestine that is followed by a straight rectum, which opens to the outside through an anus at the dorsoanterior end of the body (Figure 15.3, *J*). Some sipunculoids have a calcareous shield at the dorsal end of the trunk of the body behind

Figure 15.5. All hyolithids. **A,** *Hyolithes.* Dorsoventrally compressed conch showing the ligula (arrow). (Cambrian, Idaho, ×7.) **B,** *Hyolithes.* Right lateral view of a conch having a partially decorticated shell showing prominent growth increments (arrow) anteriorly. The apparent downward bending of the apical part was probably caused by distortion (Permian, Australia, ×3.) **C,** *Nevadalites.* Silicified early growth stage showing the protoconch (arrow). (Cambrian, Nevada, ×17.) **D,** *Hyolithes.* Right lateral view of an internal mold of a conch showing strong upward bending. (Ordovician, Estonia, ×2.) **E,** *Hyolithes.* Right lateral view of a sandstone internal mold of a conch showing the ligula. (Cambrian, Wisconsin, ×3.) **F,** *Hyolithes.* Ventral view of a conch with the shell preserved on the anterior end showing ornament. (Permian, Australia, ×3.) **G,** *Elegantilites?.* Cross section of an internal mold. (Silurian, Ohio, ×12.) **H** and **I,** *Hyolithes.* Left lateral views of sandstone internal molds of a conch showing a slight upward bending of the apex. (Cambrian, Wisconsin, ×3.) **J,** *Hyolithes.* Exterior view of a compressed operculum. (Cambrian, Idaho, ×7.)

the proboscis. Sipunculoids are at least functional analogues to hyoliths because they have a ventral mouth, a similar gross morphology of the gut, a dorsal anus, and both groups are detritivores. It is likely that hyoliths were at least at the sipunculoid grade of organization.

All classes of mollusks evolved from animals that had developed a dorsal exoskeleton of cuticle, spicular cuticle, or shell. The hyolith shell with its conch, operculum, and helens cannot be homologized with the dorsal shell of monoplacophorans from which virtually all shelled mollusks ultimately descended. Workers generally agree that the known specimens of the sediment filling of the hyolith gut (Figure 15.4, *L* and *M*) indicate the anus was above the mouth (Figure 15.3, *K* and *M*) and the opening of the gut on the ventral side was the mouth. In mollusks, this condition is known only in the torted gastropods and no sign of torsion exists in hyoliths. Untorted mollusks, such as cephalopods, which bend the gut into a U-shape, have the anus below the mouth.

The conclusion that the monoplacophoran shell cannot be homologized with the hyolith conch is supported by the serial arrangement of muscle scars in *Gompholites.* Some have argued that the scars were made by muscles that moved forward as the conch grew or that they are the result of stepwise growth. Stepwise growth results when septa are formed and the animal has to move the soft parts forward while the new septum is secreted. The muscle scars of *Gompholites* probably do not represent muscles that were always near the aperture as the animal grew either by stepwise growth or by secretion of small increments. *Gompholites* has only a single apical septum. The serially arranged muscle scars in *Gompholites* are all of about equal strength. If the more apically placed muscle scars represented older insertions of the same muscles that were near the anterior end of the youngest part of the conch and had moved forward with growth, the older scars should have been submerged by subsequently deposited inner shell layer. Because each set of muscle scars in *Gompholites* is equally strong and discrete, each was probably the site of a muscle insertion at the time of death of the animal.

Advocates of the molluscan nature of hyoliths include those who regard hyoliths as a separate class and those who consider them to be related to monoplacophorans. These individuals stress similarities between hyoliths and mollusks such as shell shape, composition, size, and microstructure, as well as protoconch similarities. Nearly all the observable structures of the hyolith shell are shared by other phyla as well as by Mollusca. An exception that has received much emphasis is crossed lamellar shell fabric, which is known only in mollusks and hyoliths. Crossed lamellar ultrastructure is widespread in mollusks, but no compelling reason is known for considering it to have been transmitted to all mollusks that have crossed-lamellar shell fabric from a single remote common ancestor. Close similarities of other shell fabrics between some pelecypods and some brachiopods and between tentaculites and some brachiopods are known, suggesting that similar shell microstructures can develop convergently in widely separated groups. A reconstruction of hyoliths as mollusks is shown in Figure 15.3, *K*; in this figure only the muscle scars closest to the aperture are regarded as having functional muscles (Figure 15.3, *L*).

Hyoliths were probably wormlike animals, having an exoskeleton that formed around the entire surface of the body, rather than dorsoventrally elongated mollusks. Hyoliths and mollusks seem to have shared a common ancestry in the late Proterozoic; their common ancestors may have been animals resembling living spiculose flatworms (see Figure 14.15). Hyoliths are not readily placed in the Mollusca or the Sipunculoidea and are here treated as a separate taxon of phylum rank. Gross similarities of hyoliths to shelled mollusks are attributed to convergence.

Supplementary reading

Fisher, D. W. Small conoidal shells of uncertain affinities—calyptoptomatids, pp. W116–W130. In: Moore, R. C. editor. *Treatise on invertebrate paleontology.* Boulder, CO, and Lawrence, KS: Geological Society of America and University of Kansas Press; 1962. State-of-the-art summary of Hyolitha to 1962.

Marek, L. New knowledge on the morphology of hyolithes. *Sborník Geologických Věd, Paleontologie, Řada* P, 1: 53–73; 1963. Discussions of hyolith operculum, ecology, systematics, and conch. (In English.)

Marek, L. The class Hyolitha in the Caradoc of Bohemia. *Sborník Geologických Věd, Paleontologie, Řada* P, 9: 51–113; 1967. Descriptions of Ordovician hyoliths and terminology of the shell. (In English.)

Marek, L.; Yochelson, E. L. Aspects of the biology of Hyolitha (Mollusca). *Lethaia* 9: 65–81; 1976. Discussion of functional morphology of hyoliths; treats them as mollusks.

Runnegar, B.; Pojeta, J. Jr.; Morris, N. J.; Taylor, J. D.; Taylor, M. E.; McClung, G. Biology of the Hyolitha. *Lethaia* 8: 181–191; 1975. Discussion of functional morphology of hyoliths and establishment of the phylum Hyolitha.

16

Phylum Brachiopoda

Albert J. Rowell
Richard E. Grant

Part I Phylum overview

Brachiopods are one of only seven phyla that have a geologic record spanning the entire Phanerozoic. Their fossils appear in rocks of earliest Cambrian age, and the distant descendants of these ancient forms are still living in the oceans and seas of the world. Brachiopods were particularly abundant during the Paleozoic and at many localities are among the most common fossils to be found in beds of this age.

Brachiopods are solitary animals in that they do not form colonies, although many cling together in clusters or banks. They secrete a shell consisting of two valves that enclose most of the animal (Figure 16.1). Because these shells consist of two valves they superficially resemble those of bivalved mollusks, especially clams. However, there is little similarity in detail. The two valves of a brachiopod are not equal unlike those in most clams. Moreover, each brachiopod valve is bilaterally symmetrical. Major differences also exist in the anatomy of the two groups because brachiopods and clams are not closely related. Brachiopods are of the triploblastic coelomate grade with an oligomerous body plan.

All brachiopods are marine, but some have wide salinity tolerances. The majority living today are epifaunal and attached by a fleshy stalk, the **pedicle,** to some object on the seafloor (Figure 16.2) such as boulders, other shells, reefs, or rock outcrops. The attachment is permanent, with few exceptions, and these animals are mobile only during their larval stage. A few species can attach directly to soft sediment, and others remain unattached. One group of brachiopods lives infaunally and excavates elongated vertical burrows in the bottom sediments (Figure 16.3).

Brachiopods are incapable of actively pursuing food, and their sessile mode of life imposes severe limits on potential feeding mechanisms and available trophic resources. All living brachiopods pump water through the cavity between the valves and extract nutrients from it. They are filter feeders in part, recovering particulate matter from the sea. Some species seem to be able to exist on dissolved nutrients that are absorbed directly into their body tissue.

Most brachiopods live in the relatively stable environment below low tide in seawater with salinity close to the normal value of 35‰. Some living species are very hardy, and the tolerance variations for salinity range from a low of 17‰ to a high of 45‰. These extreme values occur most commonly in extremely shallow water or in the intertidal zone. Only a few living brachiopods are able to withstand the rigors of life in the intertidal zone, and even these tend to live in protected positions in rock pools, on the undersides of boulders, or in burrows. During the Paleozoic a larger percentage of genera were probably euryhaline and able to colonize marginal marine environments.

The majority of present-day species occupy the shallow continental shelves and the upper part of the continental slope. The greatest number of living species are found at depths between 100 and 200 m, and the majority live in water shallower than 500 m (Figure 16.4). Brachiopods are not confined to shallow depths, however. Indeed the most geographically widespread species typically are found at depths greater than 2000 m,

445

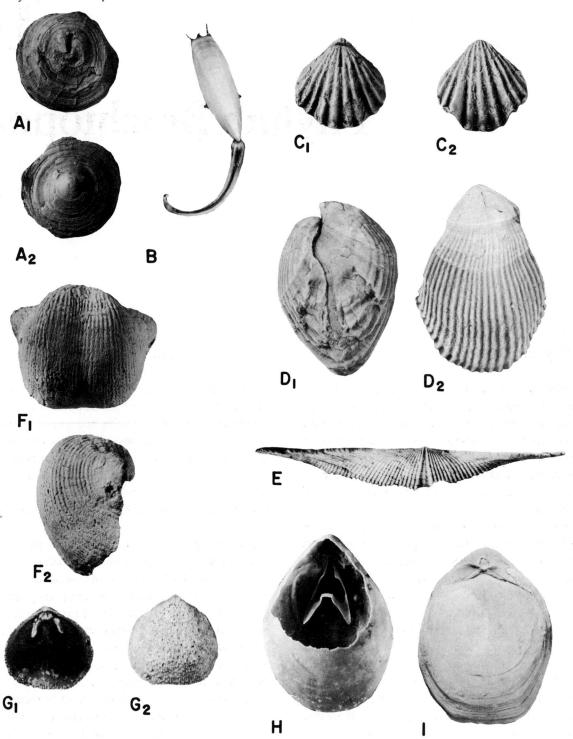

Figure 16.1. Representatives of the two classes of brachiopods. **A** and **B** are inarticulates; **C** to **I** are articulates. **A,** *Orbiculoidea.* Ventral and dorsal views. (Silurian, England, ×3.) **B,** *Glottidia* with pedicle and setae, (Holocene, west coast Panama, ×1.) **C,** *Rostricellula.* Dorsal and ventral views. (Ordovician, New York, ×3.) **D,** *Kirkidium.* Lateral and ventral views. (Silurian, Tennessee, ×1.) **E,** *Mucrospirifer.* Ventral view. (Devonian, Ontario, ×1.) **F,** *Reticulatia.* Ventral and lateral views of complete shell. (Mississippian, England, ×1.) **G,** *Rhipidomella.* Dorsal interior and exterior views. (Permian, Texas, ×2.) **H,** *Monsardithyris.* Ventral view through broken ventral valve showing loop. (Jurassic, France, ×1.) **I,** *Waconella.* Dorsal view. (Cretaceous, Texas, ×1.) (**A** from Rowell, A. J. In: Moore, R. C., editor. *Treatise on invertebrate paleontology. Part H* and Lawrence, KS: Geological Society of America and University of Kansas Press; 1965, p. 284. **B** and **H,** photo by G. A. Cooper. **C** from Cooper, G. A. *Smithsonian Miscellaneous Collections,* 127: pl. 132, Figures 16 and 18; 1956. **D** from Amsden, T. W. *Journal of Paleontology* 43: pl. 116, Figures 5 and 6; 1969. **G** from Cooper, G. A.; Grant, R. E. *Smithsonian Contributions to Paleobiology* No. 24, p. 2993, Figures 32 and 43; 1976.)

Figure 16.2. Attachment of brachiopods is normally by a fleshy pedicle visible in several specimens of this cluster of the modern *Liothyrella*. (Palmer Peninsula, Antarctica, ×1.) (Photo by G. A. Cooper.)

and one species has been dredged from 6179 m in the South Atlantic. Consequently, finding fossil brachiopods in a rock reveals that the beds were deposited in a marine or marginal marine environment, but says little about the water depth in which they accumulated. Analyses of the associated fauna and enclosing rock type are needed to estimate both original water depth and salinity.

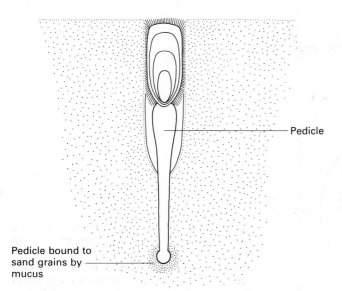

Pedicle

Pedicle bound to sand grains by mucus

Figure 16.3. Infaunal brachiopod (*Lingula*) in a feeding position within its burrow. (×1.)

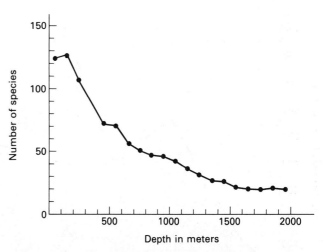

Figure 16.4. The number of species of brachiopods known at various water depths in modern oceans. (From Zezina, O.N. *Paleontologicheskii Zhurnal* 2: 5; 1970.)

Brachiopods are not only common as fossils but are also diverse, and many different kinds have been described. The phylum presently contains approximately 3400 genera, and judging from the rate of description of new forms, many others have yet to be recognized. Evolutionary change was relatively rapid in many lineages. The average brachiopod genus existed for about ten million years. Because they are abundant and easy to collect, they have proved stratigraphically useful, particularly in the Paleozoic.

Diversity, as measured by the number of genera, built up slowly in the Cambrian and has fluctuated since that time (Figure 16.5). It reached a maximum value in the Paleozoic, but even during that era three episodes of higher diversity and morphologic experimentation are recognizable in the Ordovician, Devonian, and Permian. The widespread extinction that affected many groups of organisms near the close of the Permian also took its toll of brachiopods. Many major groups of brachiopods became extinct at that time. Although the phylum partially recovered during the Jurassic, it has never regained either its relative dominance or former diversity.

Some 300 species of living brachiopods currently are known, belonging to about 100 genera. They therefore form only a minute proportion of the total biota of the sea and only a small fraction of the modern invertebrate fauna with preservable parts. In general, brachiopods are currently rare inhabitants of the oceans, but they may be locally abundant where they do occur because they tend to be gregarious. Only exceptionally are they the dominant element of the megafauna, as they are in some regions of the Antarctic. The phylum may be in decline, but fortunately for paleontologists it is by no means extinct.

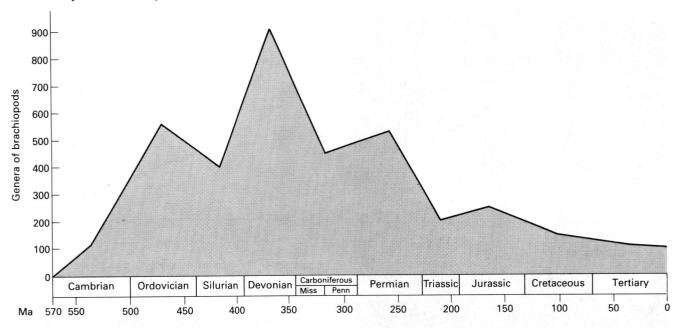

Figure 16.5. Variation of brachiopod diversity with time represented by number of genera recorded from each geologic period. (From Grant, R. E. *Journal of Paleontology* 54: 500; 1980.)

Introductory morphology

Valve orientation, primary function, and relationship

Unless secondarily distorted, the brachiopod shell is bilaterally symmetrical about a median longitudinal plane (Figure 16.6, *A*). A pedicle commonly anchors the shell at is **posterior** end, the opposing end being **anterior** (Figure 16.6, *B* and *C*). The pedicle emerges through an opening in the shell, the **foramen**, which is entirely or largely in one valve, called the **pedicle valve**. The pedicle

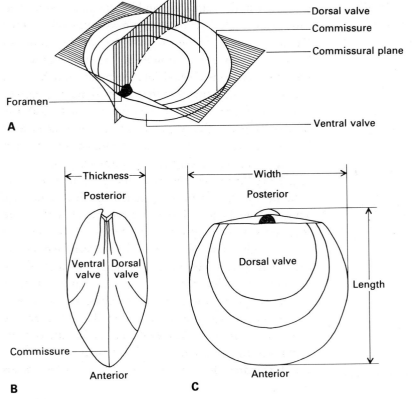

Figure 16.6. Major external features of a generalized brachiopod. (Modified from Williams, A.; Rowell, A. J. In: Moore, R. C., editor. *Treatise on invertebrate paleontology, Part H.* New York and Lawrence, KS: Geological Society of America and University of Kansas Press; 1965, p. 58.)

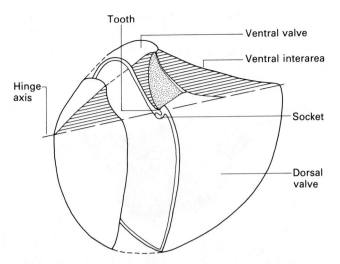

Figure 16.7. Hinge mechanism of a generalized brachiopod. Part of shell is drawn as if cut away to show one of the two teeth and sockets.

valve is ventral relative to the organs of the individual and is normally called the **ventral valve**. The terms pedicle valve and ventral valve are synonymous, and there is little agreement as to which should be preferred. In the majority of species the ventral valve is larger than the associated **dorsal valve** (sometimes called the **brachial valve**). Dimensions like length, width, and thickness are conventionally measured as in Figure 16.6, *B* and *C*.

The primary function of the shell is protection. The shell is mineralized in all brachiopods and consists predominantly either of calcite or apatite, a complex form of calcium phosphate. Brachiopods are virtually defenseless, and the strength, color, and location of the shell are the only forms of protection among epifaunal species. For effective protection, the shell must cover most of the tissue of the organism when the valves are closed. The closed valves of all living brachiopods and the majority of

fossil forms fit snugly along their **commissure** (Figure 16.6, *A*), and the only organ that protrudes is the pedicle. Even the pedicle is protected by a thick layer of organic cuticle.

Most brachiopods either possess a hinge mechanism at the posterior end of the valves or are descended from forms that possessed this mechanism. The hinge serves to restrict relative movement between valves that are free only to rotate about a hinge axis (Figure 16.7) and cannot slide past each other. In its simplest form the hinge mechanism consists of two peglike **teeth** in the ventral valve that articulate in two depressions, or **sockets**, in the dorsal valve (Figure 16.8).

Valve shape and ornament

The gross shapes of brachiopods vary considerably. The ventral valve typically is externally convex (Figure 16.1, C_2, D_2, and F_1). The dorsal valve may have a similar form (Figure 16.1, C_1, D_1, and *I*), but in many brachiopods it is externally concave or more rarely conical (Figure 16.1, A_2).

Small-scale features on the outer surface of the valves, usually arranged either radially from the posterior end of the shell or subparallel with the shell margin, are called **ornament** (the term is not meant to imply mere decorative lack of function).

Although the surfaces of many brachiopods appear to be smooth (see Figure 16.1, *I*) close inspection usually shows the presence of fine concentric **growth lines** (Figure 16.9, *A*) marking periods when the shell temporarily ceased to grow. Growth lines mark the configuration of the commissure of the individual brachiopod at particular stages in its ontogenetic development. More conspicuous concentric ornament may consist of well-

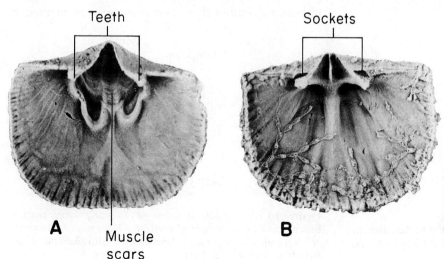

Figure 16.8. Location of teeth and sockets in an orthide brachiopod *Glyptorthis*. (Ordovician, Oklahoma.) **A**, Internal view of ventral valve. (×5). **B**, Internal view of dorsal valve. (×6.) (From Cooper, G. A. *Smithsonian Miscellaneous Collections* 127: pl. 44, Figures 12, 19; 1956.)

developed lamellae, or concentric wrinkles (rugae). **Ribs** are arranged radially (Figure 16.9, *B*); typically they are relatively low and triangular or semicircular in cross section. Major features with similar form and origin are termed **folds** and **sulci** (Figure 16.9, C). A fold is an elevation on the shell surface, and a sulcus is a major depression. The boundary between ribs and folds is rather subjective, but typically no more than three folds and associated sulci are on a shell; commonly there is only one and it is median.

Spines are conspicuous features of the external ornament of many brachiopods, and several types may be recognized. Some spines are only short triangular ex-

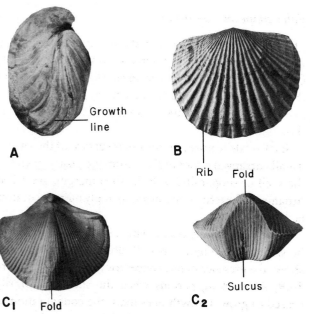

Figure 16.9. Examples of ornament and associated features of typical brachiopods. **A**, Growth lines in *Pseudoglossothyris*. (Jurassic, England, ×1.5.) **B**, Ribs in *Dinorthis*. (Ordovician, Minnesota, ×3.) **C**, Fold and sulcus in *Cyrtospirifer*. (Devonian, Iowa, ×2.) (**B**, photo by G. A. Cooper.)

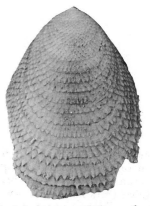

Figure 16.10. Spines as extensions of growth lamellae in *Spinilingula*. (Ordovician, Alabama, ×10.) (Photo by G. A. Cooper.)

tensions of growth lamellae (Figure 16.10). Others are slender, hollow cylindrical bodies whose length may approach or even exceed that of the shell itself (Figure 16.11).

Figure 16.11. Long hollow spines of the productidine *Yakovlevia*. (Permian, Texas, ×1.) (Photo by G. A. Cooper.)

Features of the posterior sector

Similar ornament usually marks the anterior and lateral sectors of a valve, but the posterior sector is commonly marked off from them, particularly in those brachiopods that have hinged valves. Both hinge mechanism and pedicle opening involve some modification of the posterior sector of the shell. The effects may be expressed in both valves, but they commonly are more marked in

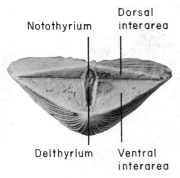

Figure 16.12. Typical features of an early hinge-bearing brachiopod, the orthide *Hesperorthis*. (Ordovician, Oklahoma, ×3.) (From Cooper, G. A. *Smithsonian Miscellaneous Contributions* 127: pl. 52; 1956.)

the ventral one, because the pedicle opening is either completely or largely in that valve. In its simplest form the pedicle opening is a triangular notch that extends from the apical region of the ventral valve (ventral beak) to the posterior valve margin. Such an opening is termed a **delthyrium** (Figure 16.12). A comparable opening in the dorsal valve is a **notothyrium** (Figure 16.12). Open delthyria and notothyria are common features of many early Paleozoic calcareous-shelled brachiopods. Later forms typically lack a notothyrium, and their delthyrium may be partially closed (Figure 16.1, *I*).

In hinged brachiopods, the posterior sector of the shell commonly consists of plane or curved triangular shelves, called **interareas**, between the beaks and the growing margin of the valves (Figure 16.12). The interareas are typically bisected by the delthyrium or notothyrium and lack the ornament of the remainder of the valve, although they may have distinct growth lines parallel to the posterior valve margin. They are conspicuous features of the valves of many early Paleozoic hinged brachiopods, but the dorsal interarea is commonly not developed in many younger species (Figure 16.1, *I*).

Internal features

Since the shell is an external skeleton, the tissue that secretes it and the majority of the animal's organs lie within the space enclosed by the valves. The inner surfaces of the valves may bear various projections and depressions that reflect the location and form of organs or muscle systems, and these vary from one type of brachiopod to the next. Consequently, knowledge of the internal features of the valves is needed to understand how the animal functioned and to aid in the animal's identification.

Some brachiopods are externally similar to each other yet differ conspicuously in the structures present in the interior of the shell. The external similarities are superficial and the totality of available characters implies that the organisms are not closely related, but that they had different ancestors. Such pairs of superficially similar organisms are **homeomorphs**.

Space enclosed by valves

The entire inner surface of a valve, including any of its projections, is lined with epithelium (Figure 16.13). The distribution of tissue readily divides the space enclosed by the valves into two parts: a posteriorly located **body cavity** and an anterior **mantle cavity**. These two cavities are separated from each other by the **anterior body wall**.

The shell in front of the anterior body wall is underlain by a double layer of epithelium, forming the dorsal and ventral **mantles**. Although very thin, the mantles are folds of the body wall (Figure 16.13). When the valves are open, the mantle cavity is in direct communication with the sea. The body cavity remains enclosed and is the main coelom of the animal. This coelom extends forward as thin, fingerlike extensions into the mantle called the **mantle canals**. All the coelomic spaces are lined with a membrane called a **peritoneum**, and the principal space, the body cavity, contains the major organs of the animal. The body cavity also encloses the musculature responsible for opening and closing the valves and for positioning the shell relative to the pedicle.

Organs within the body cavity

The alimentary canal and shell musculature are the most conspicuous features inside the body cavity. All brachiopods possess a well-developed alimentary system with a mouth opening on the midline of the lophophore into the mantle cavity. Behind the mouth is an esophagus, bulbous stomach, and intestine (Figure 16.14). In some brachiopods the intestine is relatively long and ends in an anus, opening on the right side of the anterior body wall. The intestine of the majority of living forms, however, is much shorter and the alimentary canal terminates blindly without an anus.

An open circulatory system is apparently developed in all brachiopods, although it is not well known. The main channel lies dorsal to the alimentary canal and is served by one or more contractile appendages that probably act as a heart. The respiratory function, however, is largely performed by the coelomic fluid in the main coelom rather than by the circulatory system. Oxygen uptake is believed to occur primarily in the lophophore and mantle.

A simple nervous system also is developed with main nerve ganglia lying near the junction of the esophagus and anterior body wall. All brachiopods have well-developed shadow reflexes and tactile responses; the valves snap shut when either reflex is triggered.

The gonads of brachiopods with calcareous shells are located in the mantle canals; those of the phosphate-shelled group are housed in the main body cavity. During the breeding season, sex cells escape from the body cavity through delicate funnel-shaped openings (nephridia) into the mantle cavity. The principal function of these openings is excretory, but they also act as gonoducts during reproduction. Fertilization normally is external. The sex cells are pumped from the mantle cavity into the sea

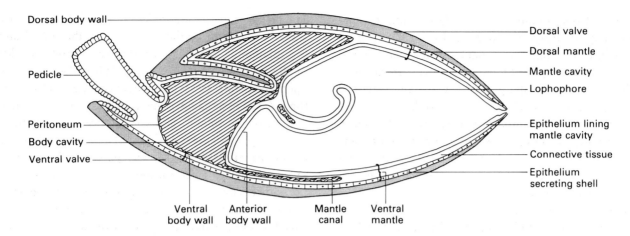

Figure 16.13. Relation between shell and epithelium together with principal internal features of a brachiopod. Longitudinal median section of an articulate; organs within body cavity omitted.

where they unite to produce zygotes that subsequently grow into larvae.

Musculature and valve movement

Unlike most of the organs considered previously, the location of the muscles frequently is reflected in the shell. The tissue at the base of the muscles may secrete shell material at a rate different from that of the tissue immediately adjacent to them. Consequently, the location of the muscles may be indicated by depressed or raised areas on the valve interior called **muscle scars** (Figure 16.8).

The principal muscles of brachiopods control opening and closing of the valves. Other large muscles also may be present that are used to position the valves relative to the pedicle. The number and disposition of the muscles vary, but two basic patterns are readily recognizable. The muscle pattern depends fundamentally on the relation between the valves.

In brachiopods with hinged valves, relative movement is controlled basically by two pairs of muscles: one pair is largely responsible for opening the shell, and the other pair is responsible for closing it. Contraction of the **adductor muscles** brings the valves together (Figure 16.15, *A*). These muscles are located near the midline of

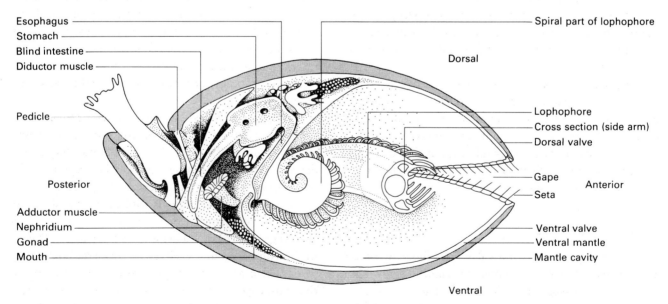

Figure 16.14. Principal organs of an articulate brachiopod viewed as if the animal were cut open along plane of symmetry. (From Williams, A.; Rowell, A. J. In: Moore, R. C., editor. *Treatise on invertebrate paleontology, Part H.* New York and Lawrence, KS: Geological Society of America and University of Kansas Press; 1965. p. 6.)

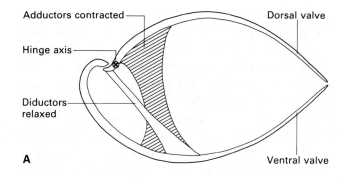

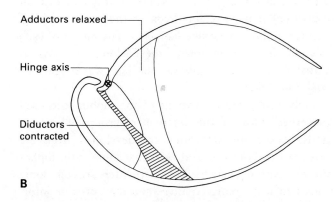

Figure 16.15. Closing and opening of valves of a typical hinged brachiopod. **A,** Contraction of adductor muscles closes valves. **B,** Contraction of diductor muscles opens valves using lever principle.

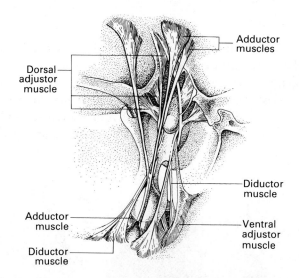

Figure 16.16. Musculature of a typical brachiopod possessing a hinge mechanism between the valves, viewed obliquely from the front. (*Macandrevia,* Atlantic Ocean.) (Drawn by L. B. Isham; from Cooper, G. A. *Journal of Paleontology* 49: 918; 1975.)

the shell and extend dorsoventrally between the two valves (Figure 16.16). Their insertion produces a pair of narrow muscle scars in the ventral valve; the muscles split dorsally into two pairs, and thus four scars are typically present in the dorsal valve (Figure 16.16). Relaxation of the adductor muscles combined with contraction of the **diductor muscles** opens the shell (Figure 16.15, *B*). The diductors work on a lever principle about the hinge. On the ventral valve they form large scars lateral to the adductors (Figure 16.16), and a second smaller pair may exist behind the main scars. The diductor muscles combine to form a single muscle pair that curves back and is inserted on the dorsal valve behind the hinge axis. Two other pairs of muscles are typically developed, the ventral and dorsal **pedicle adjustor muscles** (Figure 16.16). These allow the shell to rise, descend, and rotate on its pedicle.

Rigid hinge mechanisms are absent in the group of mainly phosphatic-shelled brachiopods, and the distribution and function of the principal muscles is markedly different (Figure 16.17). These brachiopods possess an anterior and posterior pair of adductor muscles that extend dorsoventrally and are inserted on the valves near

the margins of the body cavity. A variable number of **oblique muscles** (Figure 16.17) assist in shell closure and also control relative longitudinal and rotary movement of the valves. Because these brachiopods lack diductor muscles, the oblique muscles are also largely responsible for opening the valves. Some of the oblique muscles are attached to a valve at one end and to the anterior body wall at the other. Contraction of these muscles pulls the anterior body wall into the body cavity, and the relatively incompressible coelomic fluid causes the valves to separate. This mechanism is analogous to squeezing the equatorial region of a balloon, which produces increased separation of the poles.

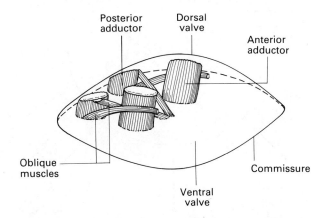

Figure 16.17. Musculature of a typical brachiopod that lacks a hinge mechanism between the valves, veiwed obliquely and dorsally. A generalized *Crania.*

Lophophore

The **lophophore** is a fleshy, hollow organ with ciliated filaments that occupies most of the space in the mantle cavity (Figure 16.18). The organ extends forward as two variously folded or coiled arms, or **brachia,** from the anterior body wall (Figure 16.14). A food groove extends along the length of the lophophore and the mouth opens into it.

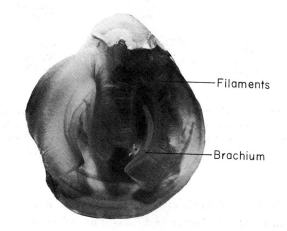

Figure 16.18. *Laqueus* with ventral valve cut away to show two filament-bearing brachia that form the lophophore. (Offshore Washington, ×2.)

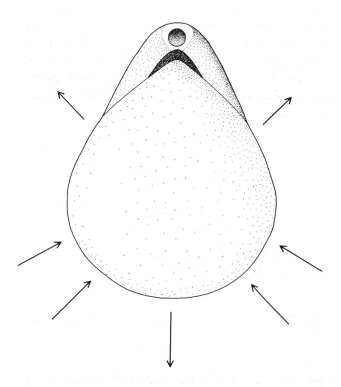

Figure 16.19. Location of principal inhalant and exhalant currents in a typical brachiopod, a generalized terebratulide.

The lophophore has three interrelated functions, acting as a pump, a sieve, and a respiratory organ. As a pump, the beat of the cilia on the filaments of the lophophore draws water with oxygen and nutrients into the mantle cavity in an ordered flow. In most adult brachiopods the water enters the mantle cavity anterolaterally and exhalant water leaves the cavity medially and posterolaterally (Figure 16.19). As a sieve, the lophophore's cilia trap particulate nutrients, which are conveyed by ciliary beat to the food groove and then to the mouth. As a respiratory organ, the lophophore's surface area is relatively enormous because the organ of an adult individual may have hundreds or thousands of long, slender, hollow fingerlike filaments. The lumen of each filament is an extension of the coelom, and gaseous transfer can occur between the sea water and the coelomic fluid.

Only rarely are indications of the lophophore itself preserved in fossil forms, but in many groups of brachiopods, calcareous structures are developed from the posterior end of the dorsal valve that support the lophophore to varying extents. These structures are commonly found in fossil representatives and may provide information on the shape of the organ.

Pedicle

The pedicle is the only part of the soft tissue that protrudes outside the shell. It is considered in this part of the text because of its intimate connection to the body cavity in some brachiopods.

Although most brachiopods possess a pedicle, it is not a unifying feature of the phylum. Ontogenetic studies and examination of the gross morphology of the pedicle of adult living brachiopods reveal that two different structures are described by the word "pedicle." The two structures are similar only in their function as a stalk that anchors the animal to the substrate. Among living brachiopods the distribution of these two pedicle types coincides with the grouping of brachiopods into those that possess a hinge mechanism and those that lack articulation. The pedicle of forms that lack a hinge mechanism is a specialized outgrowth of the ventral body wall. It is hollow, and its lumen connects with the main coelom of the body cavity (Figure 16.20). In contrast, the pedicle of living forms whose valves are hinged is solid, or a bundle of fibers, and separates dorsal and ventral body walls (Figure 16.21). The use of one word for these two dissimilar structures is unfortunate, but the term is well entrenched.

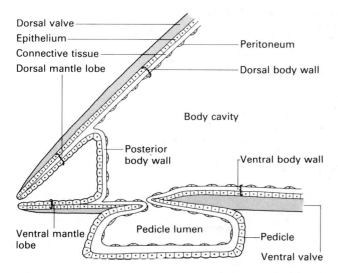

Figure 16.20. Diagrammatic longitudinal section through the posterior region of *Discina*, a hingeless brachiopod, showing pedicle as outgrowth from ventral body wall.

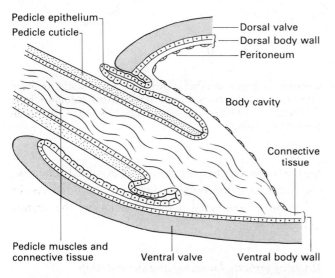

Figure 16.21. Diagrammatic longitudinal section through posterior region of *Terebratulina*, an articulate brachiopod, showing pedicle separating dorsal and ventral body walls.

Features of the phylum

Secretion and growth of the exoskeleton

The external surface of the shell of all living brachiopods is covered by a thin organic layer, the **periostracum** (Figure 16.22). It is degraded quickly after death and is not found in fossils. The mineralized shell and periostracum together form the exoskeleton.

The exoskeleton is secreted by epithelial cells of the dorsal and ventral body walls and both mantles (Figure 16.13). An increase in the area of a shell, as opposed to an increase in thickness, is controlled entirely by secretory activity of cells at the mantle margin. This process is best known in brachiopods with calcite shells, and these serve as our model.

New epithelial cells are proliferated at a generative zone (Figure 16.22, *A*) that lies on the inner side of the mantle and extends circumferentially parallel to the mantle margin. The generative zone moves distally with the advancing margin of the shell, but because of differences in vector velocities, differential movement of newly secreted cells is complex. These movements are understood most readily when viewed relative to the generative zone.

With continued growth, those cells that originated on the side of the generative zone away from the mantle margin migrate away from the zone and remain as part of the inner surface of the mantle. They have no role in the development of the shell or periostracum. The cells proliferated on the side of the generative zone facing the mantle margin, however, are intimately involved in exoskeletal secretion and initially play a complex and changing part in its development.

The position of any of these newly proliferated cells, relative to both the generative zone and the mantle margin, changes during growth of the animal (Figure 16.22, *B* to *D*). A cell initially lies adjacent to the generative zone on the inner side of the mantle, but with growth of the margin the cell rotates and comes to occupy a position on the outer surface of the mantle close to its edge. Further incremental growth of the mantle margin causes the cell to become increasingly distant from it. Thus, any particular cell (such as cell 1 in Figure 16.22, *B* to *D*) appears to move around the tip of the mantle as if on a conveyor belt propelled by new cells proliferating behind it at the generative zone. The secretory activities of the cells change in a regularly ordered manner during this migration.

Each cell in its lifetime is responsible for secreting part of the periostracum and part of each of the layers of shell material. The cell begins to assemble the organic periostracum (Figure 16.22, *A* and *B*) as it moves away from the generative zone and continues to do so until it reaches the mantle margin. The secretory role changes at this position, and the cell begins to deposit mineralized shell on the inner surface of the periostracum (Figure 16.22, *C*, cell 1). The initial calcite crystallites enlarge and fuse to form a continuous sheet that constitutes the **primary layer** of the shell (Figure 16.22, *A* and *C*). The thickness of the primary layer is controlled entirely by the few cells near the mantle margin that are involved in its growth. These cells secrete only calcite, but as the growing mantle

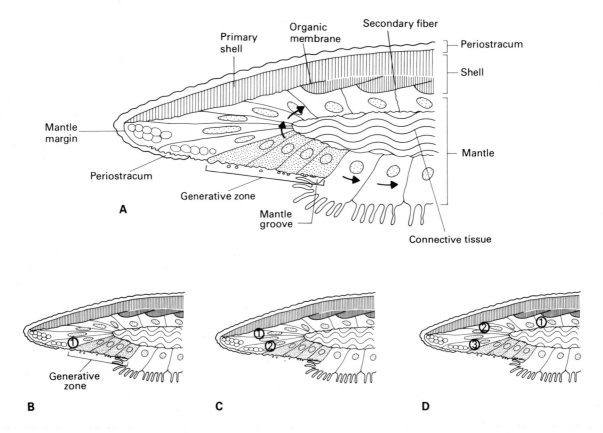

Figure 16.22. Section of exoskeleton of a typical brachiopod. **A,** Generalized cross section through margin of the mantle showing the three layers of the exoskeleton. Arrows indicate movement of cells relative to the generative zone. **B to D,** Three sequential stages in the growth of the mantle showing movement and secretory activity of cells 1 to 3. (Modified from Williams, A. *American Zoologist* 17: 118; 1977. American Society of Zoologists.)

margin moves away from a particular cell, the cell's secretory regime changes yet again. It now lays down an organic membrane and continues to deposit calcite (Figure 16.22, *D*, cell 1). The form of the organic membrane and the resultant shell ultrastructure vary among taxa of brachiopods, but the part of the shell that consists of both calcite and interleaved organic membranes is termed the **secondary layer.**

As the animal continues to grow, a given cell of the outer surface of the mantle becomes increasingly removed from the advancing mantle margin. The cell remains in contact with the shell and may continue to secrete secondary shell material, adding to valve thickness. It may, however, absorb shell material. Many of the more complex projections on the inner surface of a valve involve a delicate balance between absorption and secretion to produce them.

Thus the external shape of a valve is controlled by the secretory activity of a narrow band of cells at the mantle margin. Any particular cell occupies this marginal position only briefly. Radially directed components of growth of the two mantles must be closely coordinated

for the two valves to meet at the commissure. In contrast, the shape of the internal surface of a valve reflects the behavior of epithelial cells proximal to the mantle margin.

Rates of growth, longevity and survivorship

The brachiopod shell does not bear any clear indication of its precise age, so it is difficult to obtain accurate information on rates of growth or age of the individual at death based only on examination of the shell. Growth lines, for example, are seemingly not produced regularly with a monthly or annual periodicity. The only data that are above suspicion are those obtained by repeated observation and measurement of living forms over a period of many years. Such data are rare and obviously time-consuming to collect, but enough is known to permit tentative generalizations about living brachiopods.

Figure 16.23 shows the growth of some 120 individuals of a particular brachiopod species over a period of one year. Each specimen was measured at the beginning and end of the year. Comparable results have been

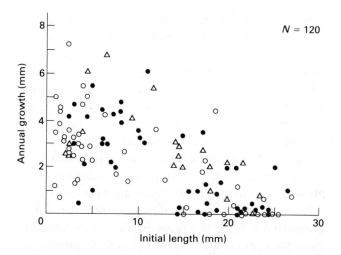

Figure 16.23. Rate of annual shell growth of 120 specimens of the living terebratulide *Terebratalia* from Washington. Symbols represent populations attached to separate rocks. (From Thayer, C. W. *Paleobiology* 3: 100; 1977. Paleontological Society).

obtained for a second species. This diagram shows that the average brachiopod initially grows rather rapidly, but its rate of growth is not constant and declines with increasing age. Many specimens between 20 and 25 mm in length at the beginning of the experiment scarcely grew during the year. Figure 16.23 also shows that there is considerable variation in the amount of annual growth, even for individuals of the same initial size. Specimens with an initial length of about 4 mm grew between 0.5 and 7 mm during the year; the average increment, however, was about 4 mm.

A few individual brachiopod specimens are known to have lived at least eight years because they have been observed repeatedly during that time. Reasonable estimates, based on growth rates, suggest that this value is of the right order of magnitude for many living species, but exceptional individuals may live 15 years or more. Sexual maturity occurs when the brachiopod is between one-third and one-half of the maximum size of the species.

Few figures are available on the number of eggs shed by a female. One infaunal female shed over 130,000 eggs in a year, but more typical estimates are in the order of 20,000 for epifaunal species. Even with the more conservative figure, reproduction consumes a great deal of the female brachiopod's energy. This problem is not unique to members of the phylum Brachiopoda but is encountered by most species of marine invertebrates. The production of large numbers of gametes is a typical adaptive response of an organism whose eggs, embryos, and larvae are preyed upon heavily. The greatest losses

occur prior to the settlement of the larvae. Some eggs are not fertilized; others fail to develop; some larvae cannot find suitable settling sites; and many will be eaten. The energy expended by the female in developing the eggs is not completely lost to the ecosystem, since the eggs and larvae can serve as a source of food to other creatures. This apparent extravagance is required to ensure perpetuation of the species and to maximize the possibility that the individual will reproduce itself successfully at least once.

The probability that a larva will grow into a sexually mature adult is increased once it has attached itself and undergone metamorphosis. Mortality rate, however, still remains high. Mortality rates are best shown graphically on **survivorship curves** (Figure 16.24), semilogarithmic plots that show the percentage of an initial population still alive with increasing age. The mortality rate at a given age is the slope of the survivorship curve at that age. The steeper the slope, the higher the mortality rate.

Three types of survivorship pattern are commonly recognized (Figure 16.24). In one, juvenile mortality is high, but mortality rate decreases with increasing age so the resulting survivorship curve (Figure 16.24, *A*) is concave upward. A second type of survivorship curve reveals uniform mortality; a constant percentage of the population dies during fixed time intervals so the survivorship curve plots as a straight line (Figure 16.24, *B*). The final type of survivorship curve (Figure 16.24, *C*) is

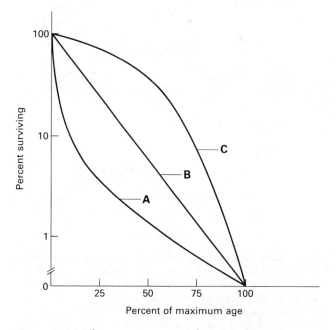

Figure 16.24. Three common types of survivorship curve: semilogarithmic plots of the percentage of a population surviving after elapse of a given percentage of its maximum life span.

concave downward because juvenile mortality is relatively low, but mortality rate increases with increasing age. This type of survivorship curve characterizes our own species in developed countries today.

Only one experiment has measured mortality rates directly in living brachiopods by repeated census of the same area. The results of this study are summarized in Figure 16.25. Mortality for individuals greater than 1 mm long was constant for both years in which the census was carried out because the size-frequency curves for living and dead forms were similar. On average, approximately 35 percent of a given size class died during a year, but mortality rate was not dependent on size. A survivorship curve would be like that of Figure 16.24, *B*.

Study of survivorship is informative in several paleobiologic contexts. Survivorship is a crude, generalized measure of the fitness of a population in a given environmental setting. Organisms differ in the form of their survivorship curves, and interpretations of the cause of such differences may shed light on the biology of the animals. Survivorship among fossil brachiopods has been actively studied for the past few years, and some of the possibilities and limitations of the study approaches used are becoming clear. The concepts and methodology employed are not limited to brachiopods but may be applied equally well to other organisms with comparable shell development.

Several basic conditions have to be met for such a study of fossil forms to have any biologic significance. It must be established that the animals lived in the region where they are now found, that their remains stayed in the area after death, and that the probability of being fossilized did not depend on their size. These are rather stringent requirements, but the validity of any resulting conclusions depends, to a considerable extent, on these conditions being satisfied. Survivorship data based on a transported assemblage whose small individuals were previously winnowed away clearly are meaningless.

Two distinct patterns of survivorship have been recognized in fossil brachiopods. There may have been more patterns in nature, however, because survivorship of fossils preserved in limestone can be studied only when they have been silicified, since too many juveniles would be lost or overlooked in trying to hammer them free from the enclosing rock. It is somewhat easier to study survivorship of populations preserved in shale. The most common pattern is shown in Figure 16.26. The mortality rate of small forms, those under 2 mm long, is exceedingly high. Subsequent mortality is lower and, within the limits of the method used, is approximately constant. The

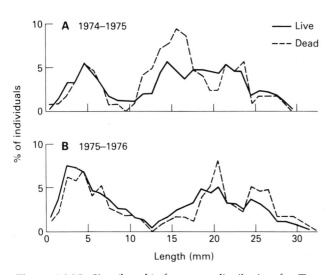

Figure 16.25. Size (length)–frequency distribution for *Terebratalia* from Washington. **A**, Solid line: live individuals, 1974; frequencies as percentages of total 1974 live population. Broken line: 1974–1975 deaths; frequencies as percentage of 1974–1975 deaths. **B**, Solid line: live individuals 1975; frequences as percentage of 1975 live population. Broken line: 1975–1976 deaths; frequencies as percentage of 1975–1976 deaths (From Thayer, C. W. *Paleobiology* 3: 103; 1977. Paleontological Society.)

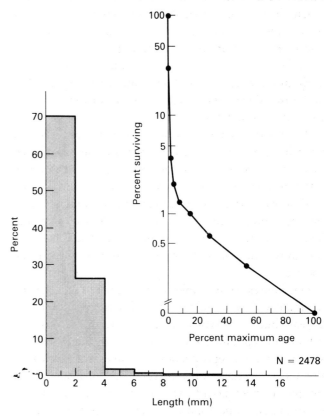

Figure 16.26. Typical size-frequency histogram and survivorship curve for fossil brachiopods that inhabited a soft bottom. (*Composita*, Pennsylvanian, Illinois.) (From Richards, R. P.; Bambach, R. K. *Journal of Paleontology* 49: 790; 1975.)

pattern is seemingly characteristic of brachiopods that inhabited soft-bottom environments. High juvenile mortality is usually attributed to clogging of the lophophore of small forms, but other explanations could also be invoked.

The second type of survivorship pattern that has been recognized is typical of brachiopods that inhabited firmer substrates and usually attached to pieces of shell. Juvenile mortality was relatively low in these forms (Figure 16.27), but mortality subsequently increased to a higher but constant level. The increase in mortality may be real and associated with the onset of sexual maturity. Alternatively, the apparent increase may be an artifact caused by reduction in growth rate. Additional studies are needed to explain the change. Survivorship obviously was not uniform throughout the phylum; mortality rates were influenced by environmental conditions. Study of survivorship provides another glimpse of the adaptive response of brachiopods to their surroundings.

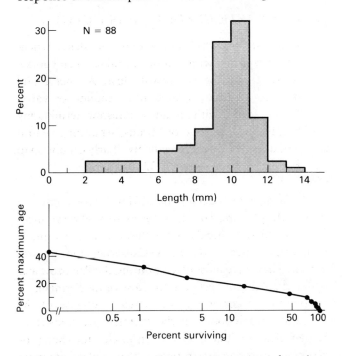

Figure 16.27. Typical size-frequency histogram and survivorship curve for fossil brachiopods that inhabited a hard bottom. (*Platystrophia*, Ordovician, Indiana.) (From Richards, R. P. Bambach, R. K. *Journal of Paleontology* 49: 793; 1975.)

Classification

The major classification of brachiopods has been remarkably stable for most of the past 150 years. The group is commonly regarded as a phylum that may readily be subdivided into two classes. The names of these two classes, the Inarticulata and the Articulata, are based on the absence or presence of a hinge mechanism that permits articulation between the valves. The classification is polythetic, but, because character states are strongly correlated, two or three features suffice to identify about 99.9 percent of all brachiopods to class level. If the shell is calcite and possesses teeth or lacks teeth but has a hinge line, it is an articulate. If the shell is phosphatic or is of calcium carbonate but lacks teeth or a hinge line, it is an inarticulate.

Paradoxically, although the classification has been stable, the cladistic relationship of the Articulata and

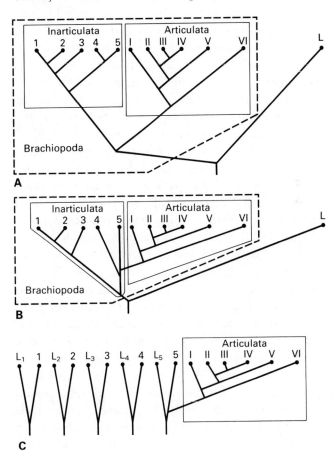

Figure 16.28. Cladograms showing alternative cladistic relationships of inarticulate and articulate brachiopods. Orders of inarticulate brachiopods numbered *1* to *5*; orders of articulate brachiopods numbered *I* to *VI*; ancestral lophophorate stock shown by *L*. **A,** Inarticulata and Articulata separately derived from a common ancestor that cladistically is a member of the phylum. Limits of phylum and classes are shown diagrammatically by boxes; phylum and both classes are holophyletic. **B,** Inarticulata derived from a common ancestor that cladistically is a member of the phylum; Articulata derived from an inarticulate stock. Limits of phylum and classes are shown diagrammatically by boxes; phylum and class Articulata are holophyletic, and class Inarticulata is paraphyletic. **C,** Principal orders of inarticulate brachiopods separately derived from lophophore-bearing, wormlike ancestral lophophorates (*L₁* to *L₅*). The class Articulata is a holophyletic taxon. The Brachiopoda and Inarticulata are polyphyletic groupings.

Inarticulata is not well understood. The traditional view is that the phylum is holophyletic; the most recent common ancestor is cladistically a member of the phylum, which includes all descendants. The classes are likewise traditionally viewed as holophyletic taxa (Figure 16.28 *A*).

It is conceivable that the Articulata were derived from some inarticulate stock and that the cladistic relationships are as in Figure 16.28, *B*. In these circumstances, both the phylum Brachiopoda and the class Articulata would remain holophyletic, but the Inarticulata would be paraphyletic because the class does not include all descendants. Some systematists argue that if this relationship is substantiated, the class Inarticulata would have to be abandoned and replaced by new holophyletic taxa (see Chapter 6, "Classification: Philosophies and Methods"). The more orthodox view is that paraphyletic taxa are permissible and cladistic aspects of evolutionary history should be depicted graphically as in Figure 16.28, *B*.

It has been claimed recently that brachiopods represent a grade of organization, not a clade. Proponents argue that the principal groups of inarticulate brachiopods arose independently from different groups of infaunal, lophophore-bearing, tube-dwelling, wormlike organisms (Figure 16.28, *C*). The articulate brachiopods are considered to have developed from either one of the groups of inarticulate brachiopods (as in Figure 16.28, *C*) or another group of infaunal, wormlike organisms. In either case, the Articulata would remain a holophyletic group and pose no problem. If the cladistic relationships of Figure 16.28, *C* were correct, however, both the class Inarticulata and the phylum Brachiopoda would be polyphyletic taxa. The majority of the resemblances between their members would be a measure of convergence and the taxa would be unsatisfactory groups in that they would be inconsistent with phyletic history.

Present evidence suggests that although marked differences between many of the early brachiopod stocks exist, they and their living descendants share a number of synapomorphies. These uniquely derived features, which are unknown outside the brachiopods, strongly support the argument that the phylum is monophyletic. It is not possible to assert confidently which of the two patterns of cladistic relationship depicted in Figure 16.28 (*A* and *B*) is nearer the truth. The following classification has the merits of accord with previous usage and of being utilitarian, but future studies may see it challenged.

CLASS INARTICULATA (Figure 16.1, *A* and *B*)

Brachiopods with mineralized shell usually of apatite; some of calcite; one group of aragonite. Articulation typically absent. Adductors and oblique muscles present; no diductors. Muscles inserted near margin of body cavity. Living forms with anus opening on right hand side of body; pedicle hollow, consisting of modified extension of ventral body wall. Posterior body wall separating ventral and dorsal mantles posteriorly. (Early Cambrian to Holocene; approximately 220 genera.)

The inarticulates are much the smaller of the two classes of Brachiopoda but range almost completely through the Phanerozoic. Their diversity built up quickly in the Cambrian and reached its acme in the Ordovician. By the end of the Devonian, most families were extinct, but three groups have survived into modern times and show only modest differences in shell morphology from their middle Paleozoic ancestors.

CLASS ARTICULATA (Figure 16.1, *C* to *I*)

Brachiopods with shells invariably of calcite. Articulation typically with pair of ventral teeth and dorsal sockets, may be secondarily lost. Adductor and diductor muscles located near midline of valve. Living forms without anus, intestine terminating blindly; pedicle solid or fibrous, separating ventral and dorsal body walls medially. (Early Cambrian to Holocene, approximately 3200 genera.)

The articulate brachiopods obviously constitute the bulk of the phylum. The oldest forms are slightly younger than the oldest inarticulates, but the articulates also made their appearance in the Early Cambrian and are still extant. They were most diverse during the Paleozoic and suffered an abrupt decline at the close of the Permian.

Hypotheses concerning the sudden collapse of the articulate brachiopods in the Permian are numerous. Earlier ideas attributed it to effects of climatic change or geologic uplift and diastrophism. Newer ideas cite more subtle ecologic changes associated with relative motion of continents and reduction of habitats, specifically the coming together of continental masses to form Pangea, the Triassic supercontinent. Despite the plethora of theories, no generally accepted explanation for the Permo–Triassic extinction has yet emerged. Research into this and other major extinctions of brachiopods (for example, in the Devonian and at the end of the Cretaceous) continues, along with study of probable causes of major extinctions of other animal groups (such as large mammals after the Pleistocene).

A similar and perhaps related problem is why articulate brachiopods suffered another decline in the Cretaceous after the sudden extinction of many groups in the Permian and their slow revival to another peak in the Jurassic. A possible cause may be in the concurrent evolution of several kinds of predators, more advanced than those of the Paleozoic or early Mesozoic, as well as more efficient gastropods and starfish adapted to the size of brachiopods.

Relationship of brachiopods to other phyla

Three phyla, the Bryozoa, Brachiopoda, and the Phoronida, are commonly grouped together as lophophorates. The implication is that in the distant past they may have shared a common evolutionary history that was not shared by members of other phyla.

The fossil record does not afford much help in trying to assess the reasonableness of this postulated relationship. The brachiopods have a much longer fossil record than either the phoronids or the bryozoans. That we know little about the early history of the phoronids is not surprising since they lack a durable skeleton. The same reasoning may explain the pre-Ordovician absence of bryozoans. The first appearance of these organisms in the fossil record may well mark the time at which they underwent a significant physiologic evolutionary step and developed the ability to secrete calcite, rather than the time at which the group developed.

The structure of the lophophore and the origin of the coelomic spaces within it seem to be true synapomorphies linking the three groups. If more evidence of the close phylogenetic relationship of the three phyla is to be provided, it will have to come from recognition of additional features that are uniquely derived evolutionary novelties found only in lophophorates. Such evidence will most likely be found by studying the development and morphology of living animals together with investigations of their protein chemistry.

Part II Additional concepts

Class Inarticulata

The inarticulates are a relatively small class, including only about five percent of the known genera of brachiopods, but they were dominant in the initial radiation of the phylum. Some 50 Cambrian genera are known and all four orders are represented in Lower Cambrian rocks. The earliest forms and most of their descendants had

phosphatic shells, but on at least four separate occasions inarticulate brachiopod lineages developed shells of calcium carbonate. Most of these lineages were short-lived, but one (the Craniidina) ranges from the Ordovician to the present day.

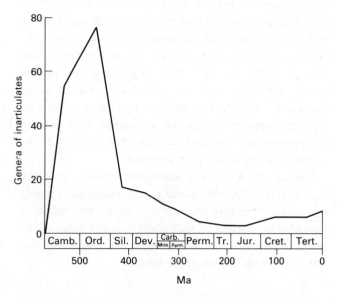

Figure 16.29. Variation of inarticulate brachiopod diversity with time represented by number of genera recorded from each geologic period.

The diversity of inarticulate brachiopods continued to increase during Early and Middle Ordovician time (Figure 16.29), but by the middle of the period they had lost their position of preeminence to the Articulata, which were diversifying even more rapidly. During the Late Ordovician, inarticulate diversity dropped precipitously, perhaps as a consequence of competition with the articulates. The latter typically were bigger animals that devoted a smaller percentage of space between the valves to the body cavity. Consequently, they had relatively larger lophophores and were probably more efficient feeders.

Diversity continued to decline during the remainder of the Paleozoic, and the class was represented by only three genera in the Triassic. These three genera are still extant, and the inarticulates have been a remarkably conservative group throughout the Mesozoic and Cenozoic eras.

Morphology

Structure of the posterior sector　Inarticulates lack structures homologous with the teeth and sockets of the Articulata, but the posterior sector of both valves is commonly flattened or otherwise marked off from the remainder of the valve to form a **pseudointerarea** (Figure 16.30).

The ventral pseudointerarea typically is a flattened subtriangular sector dorsal of the pedicle foramen in those forms that have the pedicle opening confined to the ventral valve (Figure 16.30, *A*). In contrast, the ventral pseudointerarea of the large group of inarticulates whose pedicle emerged from between the valves forms a triangular structure lying in the plane of commissure (Figure 16.30, *C*). A narrow depression, the **pedicle groove**, typically divides such pseudointerareas medially (Figure 16.30, *C*). Although the two types of pseudointerarea appear superficially dissimilar, they are homologous structures. Both were secreted primarily by the posterior sector of the ventral mantle.

Dorsal valves of most inarticulates possess marginal beaks. Generally, the dorsal pseudointerarea is also a triangular structure that lies close to the plane of commissure (Figure 16.30, *B* and *D*). It was deposited by the posterior sector of the dorsal mantle, which migrated anteriorly with growth of the animal.

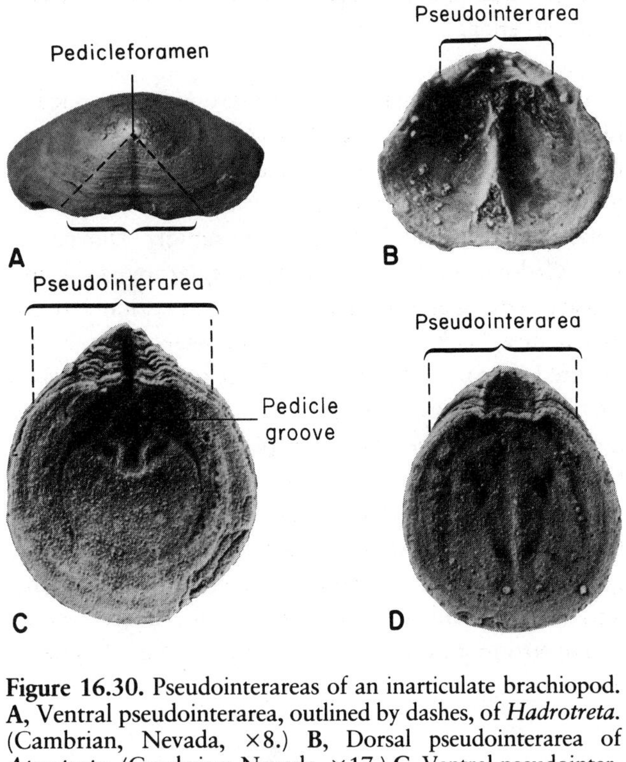

Internal features of the shell   Most inarticulate valves have rather featureless, internal surfaces. Muscle scars can be conspicuous but usually are faintly impressed and only rarely show more than weak indications of the adductor scars.

Characteristically the lophophore is unsupported by shelly projections from the dorsal valve. Hydrostatic pressure in the coelomic cavities of the lophophore gives

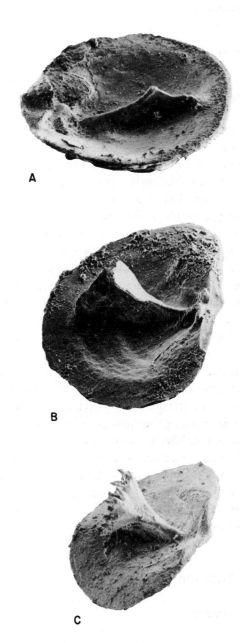

Figure 16.30. Pseudointerareas of an inarticulate brachiopod. **A**, Ventral pseudointerarea, outlined by dashes, of *Hadrotreta*. (Cambrian, Nevada, ×8.) **B**, Dorsal pseudointerarea of *Apsotreta*. (Cambrian, Nevada, ×17.) **C**, Ventral pseudointerarea of *Schmidtites*. (Ordovician, U.S.S.R., ×6.) **D**, Dorsal pseudointerarea of *Schmidtites*. (Ordovician, U.S.S.R., ×6.) (**C** and **D** from Rowell, A. J. In: Moore, R. C., editor. *Treatise on invertebrate paleontology, Part H.* New York and Lawrence, KS: Geological Society of America and University of Kansas Press; 1965, p. 265.)

Figure 16.31. Median structures in the dorsal valves of some acrotretide brachiopods. **A**, Simple median septum of *Conotreta*. (Ordovician, Nevada, ×45.) **B**, Septum bearing a triangular plate of *Torynelasma*. (Ordovician, Nevada, ×45.) **C**, Complex saddle-shaped plate of *Ephippelasma*. (Ordovician, Nevada, ×45.) (**A** from Krause F. F.; Rowell, A. J. *University of Kansas Paleontological Contributions*, 61: 11; 1975.)

it some measure of rigidity in living inarticulates and presumably did so in fossil inarticulates. Some lower Paleozoic Inarticulata have a **median septum**, a bladelike plate projecting from the inner surface of the dorsal valve (Figure 16.31, *A*). In its simplest form this structure could have given no support to the lophophore but would have separated the inhalant water currents. Variously shaped plates on the ventral edge of the median septum in some taxa, however, may have supported the posterior end of the organ (Figure 16.31, *B* and *C*).

The internal surfaces of both valves can show indications of the location of the mantle canals (Figure 16.13). The canals can have no direct influence on shell secretion, but the strip of secretory epithelium of the mantle opposite them commonly deposits the secondary shell layer more slowly than the tissue immediately lateral to the canals. When this happens, the location of the canals is marked by shallow grooves on the inner surfaces of the valves. The detailed form of the canals varies among different taxa of brachiopods (Figure 16.32), but their basic structure is the same and consists of one or two pairs of major trunks that branch toward the mantle margin.

The mantle canals serve several functions. They house the gonads in living calcareous-shelled inarticulates. In some other inarticulates, an ordered flow of coelomic fluid through the canals aids in respiration. The mantle canals of some infaunal forms are extended into the mantle cavity as numerous thin-walled, bulblike protrusions. These protrusions greatly increase the surface area of the canals and presumably improve respiratory efficiency.

Shell structure and composition

The shell material of most living inarticulate brachiopods, like those of the past, consists of layers of chitin and scleroprotein alternating with layers of apatite. The ultrastructure varies with different taxa, but in some of the lower Paleozoic forms, adjacent apatite layers were connected by rods of the same composition (Figure 16.33, *A*).

Although calcareous shells are thought to have developed polyphyletically, shell ultrastructure is remarkably similar among the three groups that have shells of calcite. Beneath a thin primary layer of acicular calcite crystals, the secondary shell is strongly laminated (Figure 16.33, *B*). The laminae consist of calcite tablets that were invested in a protein sheet in all except their growing edges (Figure 16.33, *B* and *C*). The evolution of these stocks involved not only the development of the ability to secrete calcite in forms whose ancestors had deposited apatite, but also included metabolic changes that controlled the composition of the organic components of the exoskeleton. Chitin, the principal organic material present in phosphatic-shelled inarticulates, is absent in calcitic shells, but a protein takes its place.

The fourth group of calcareous-shelled inarticulates is unusual because their shells were probably aragonitic. Although these forms are locally common in Silurian rocks and occur in association with other fossils whose calcite shells retain primary microfabric, they are typically coarsely recrystallized or are found only as molds.

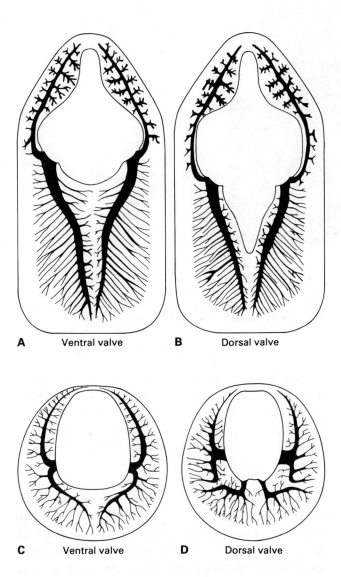

| **A** Ventral valve | **B** Dorsal valve |
| **C** Ventral valve | **D** Dorsal valve |

Figure 16.32. Mantle canal patterns of two modern inarticulates. **A** and **B**, *Lingula*. **C** and **D**, *Discinisca*. (From Williams, A.; Rowell, A. J. *Treatise on invertebrate paleontology, Part H.* New York and Lawrence, KS: Geological Society of America and University of Kansas Press; 1965, p. 24.)

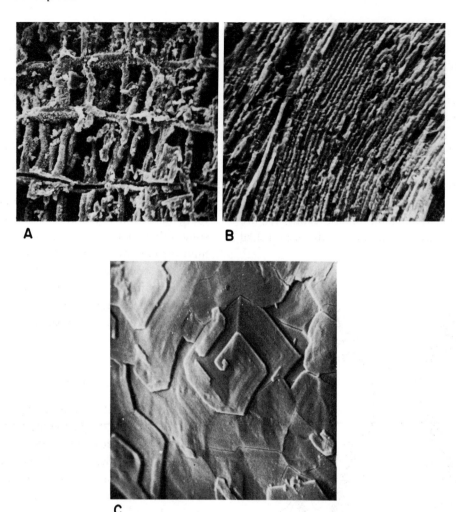

Figure 16.33. Scanning electron micrographs showing shell ultrastructure of some inarticulate brachiopods. **A,** Phosphatic-shelled acrotretide *Clistotrema* in cross section. (Ordovician, U.S.S.R., ×500.) **B,** Calcareous-shelled *Crania* in cross section showing laminar layer. (Holocene, Atlantic, ×5000.) **C,** Calcareous-shelled *Crania* with internal surface showing spiral growth of lamina. (Holocene, Atlantic, ×3500.) (**B** and **C** from Williams, A.; Wright, A. D. *Special Papers in Palaeontology* 7: pl. 6; 1970. Palaeontological Association.)

Classification

The class Inarticulata is divided into four orders named after representative genera: Paterinida, Obolellida, Acrotretida, and Lingulida. The International Code of Zoological Nomenclature, which guides in matters of the names of animals, does not address the question of ordinal names; consequently considerable variation in their treatment is found. During the past decade, however, the suffix "-ida" has been used consistently by brachiopod workers for the termination of ordinal names and the suffix "-idina" has been used for suborders. This standardization is useful because one can know whether an author is speaking of an order or a suborder merely by examining the end of the word. To avoid the repetitive use of the formal ordinal name, a vernacular equivalent is needed. We will follow the terminology of the *Treatise* brachiopod volume and use the suffixes "-ide" for orders and "-idine" for suborders: paterinide thus refers to brachiopods that are members of the order Paterinida, whereas acrotretidine refers to brachiopods that belong to the suborder Acrotretidina.

Each of the four orders of Inarticulata is relatively homogeneous. This makes ordinal identification of an inarticulate moderately easy but causes problems in trying to understand phylogenetic relations within the class. Stratigraphic range is of little help in phylogenetic speculation. Although the Paterinida seem to have appeared slightly earlier than the other three orders, the first appearance in the fossil record may mark the time that the group developed a mineralized shell rather than the time that the lineage originated.

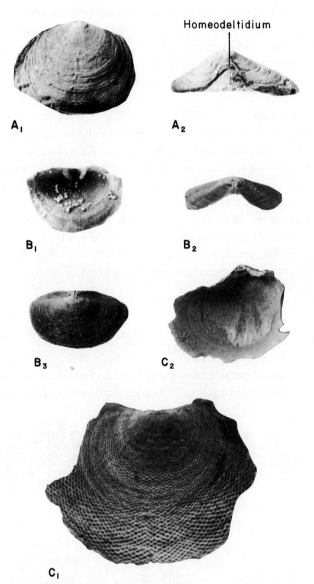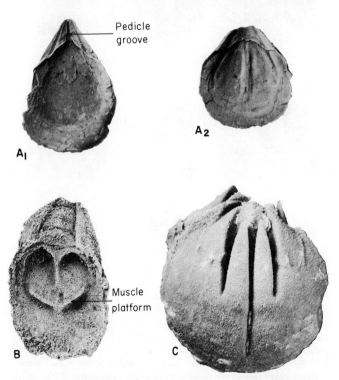

ORDER PATERINIDA (Figure 16.34)

Small inarticulates, typically less than 5 mm in length. Shell phosphatic with straight posterior margins. Both valves convex, ventral valve higher. Pseudointerareas each divided by open delthyrium (ventral) and notothyrium (dorsal), commonly partially closed by externally convex triangular plates (homeodeltidium and homeochilidium, respectively). Pedicle, if present, emerged from between these plates. Margins of delthyrium and notothyrium typically thickened internally. Probable muscle scars radiating outwards as faint triangular areas from beaks of both valves. (Early Cambrian to Middle Ordovician; about 14 genera.)

Although the order consists of only a few genera, their remains are a common component of Cambrian faunas. Its oldest members are the oldest brachiopods presently known. They occur in the lowest stage of the Lower Cambrian of Siberia.

ORDER LINGULIDA (Figures 16.1, *B*; 16.3; 16.10; 16.30, *C* and *D*; 16.35)

Small to large inarticulates, with adults in many living species 4 to 5 cm long. Shell of calcium phosphate and chitin in great majority; typically gently biconvex with circular, triangular, or elongate outline. Beaks marginal. Ventral and dorsal pseudointerareas well developed in lower Paleozoic forms, reduced to insignificance in living species. Ventral pseudointerarea divided by conspicuous pedicle

Figure 16.34. Representative Paterinida. **A**₁ and **A**₂, *Micromitra*. Ventral view with ventral valve posterior. (Cambrian, Canada, ×5.) **B**₁ to **B**₃, *Paterina*. Ventral valve interior, posterior and exterior. (Cambrian, Canada, ×10.) **C**₁ and **C**₂, *Dictyonina*. Dorsal valve exterior and oblique ventral valve interior showing musculature. (Cambrian, Nevada, ×30, ×10.) (**B**₁ to **B**₃ from Rowell, A. J. In: Moore, R. C., editor. *Treatise on invertebrate paleontology, Part H.* New York and Lawrence, KS: Geological Society of America and University of Kansas Press; 1965, p. 294. **C**₁ and **C**₂ from Rowell, A. J. *University of Kansas Paleontological Contributions* 98: 8; 1980.)

Figure 16.35. Representative Lingulida. **A**₁ and **A**₂, *Lingulella*. Ventral and dorsal valve interiors. (Cambrian, Greenland, ×9.) **B**, *Monomerella*. Latex mold of a cast of a ventral valve. (Silurian, Canada, ×1.) **C**, *Trimerella*. Dorsal view of natural cast of the interior of a complete shell. (Silurian, Ohio, ×1.)

groove. Dorsal pseudointerarea with broadly triangular median depression. Pedicle emerged from between posterior ends of valves in phosphatic-shelled species. (Early Cambrian to Holocene; about 85 genera.)

Many living lingulide species are locally abundant in tropical and subtropical areas of the world where they burrow into the sandy muds of the intertidal and shallow subtidal zones. Lingulides are the only living infaunal brachiopods, and some have utilized this mode of life since the Cambrian. Fossil specimens are sometimes found oriented normal to the bedding with anterior ends pointing upward in presumed life position similar to that of living representatives. Not all lingulides had an infaunal existence, however. Many are found in lower Paleozoic pelagic or hemipelagic black shales associated with graptolites. These rocks commonly lack any undoubted benthic fauna and are usually considered to have accumulated under anoxic bottom conditions. Presumably these lingulides were epiplanktic and were rafted into the area attached to seaweed.

One of the unusual evolutionary experiments is the

development of apparently aragonitic shells in a group of Ordovician and Silurian lingulides known as the Trimerellacea (Figure 16.35, *B* and *C*). This group is also unusual in two other respects. First, the two valves are crudely hinged. The hinge mechanism consists of a thickened plate in the dorsal valve that seats in a poorly defined groove immediately anterior to the ventral pseudointerarea. The structure would have restricted rotational movements between the valves and could have functioned as a fulcrum. The Trimerellacea are further unusual, although not unique, in the development of conspicuous muscle platforms (Figure 16.35, *B*) that are variously supported and raised above the floor of the valves.

Lingulides appeared in the middle of the Early Cambrian and range throughout the remainder of the Phanerozoic. Although minor differences are detectable in shell morphology, their gross aspect has shown little change. Some lingulides provide an example of a group that are well adapted to the niche that they occupy as shallow burrowers in intertidal and shallow subtidal environments. The limitations of their basic body plan has, however, precluded any successful adaptive radiation of the type that characterized the infaunal bivalved mollusks.

ORDER ACROTRETIDA (Figures 16.1, *A*; 16.30, *A* and *B*; 16.36; 16.37)

Small inarticulates, many less than 2 mm long. Shells either phosphatic (suborder Acrotretidina) or cal-

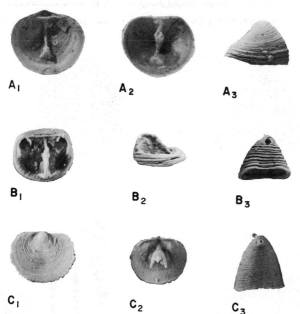

Figure 16.36. Representative Acrotretida. A_1 to A_3, *Ceratreta.* Dorsal valve interior, ventral valve interior, and ventral valve lateral. (Cambrian, Texas, ×12.) B_1 to B_3, *Rhysotreta.* Dorsal valve interior and lateral views and ventral valve oblique posterior view. (Ordovician, Alabama, ×10.) C_1 to C_3, *Ephippelasma.* Dorsal valve exterior and interior views and ventral valve posterior view. (Ordovician, Alabama, ×20.) (A_1 to A_3 from Rowell, A. J. In: Moore, R. C., editor; *Treatise on invertebrate paleontology, Part H.* New York and Lawrence, KS: Geological Society of America and University of Kansas Press; 1965. p. 279. **B** and **C** from Cooper, G. A. *Smithsonian Miscellaneous Collections* 127:pl. 17, 18; 1956.)

Figure 16.37. Representative Craniidina. A_1 to A_3, *Crania.* Ventral valve interior and dorsal valve exterior and interior. (Cretaceous, Sweden, ×2.) B_1 to B_3, *Isocrania.* Ventral valve exterior and interior and dorsal valve exterior. (Cretaceous, England, ×4.) From Rowell, A. J. In: Moore, R. C., editor. *Treatise on invertebrate paleontology, Part H.* New York and Lawrence, KS: Geological Society of America and University of Kansas Press; 1965, p. 289.)

careous (Craniidina). Phosphatic shelled forms typically circular in outline with convex dorsal valve and convex or subconical ventral valve; pedicle foramen characteristically near beak of ventral valve. Calcareous-shelled forms mostly attached by cementation of all or part of ventral valve; none known to have possessed pedicle. (Early Cambrian to Holocene; about 120 genera.)

The Acrotretida appear in the middle part of the Lower Cambrian and are represented in these rocks only by the phosphatic-shelled Acrotretidina. The calcareous-shelled craniidines are a later development that made their appearance in the Ordovician. Both suborders are still extant. If successis measured in terms of longevity, the Acrotretida are the second of two successful inarticulate stocks. One genus of this order (*Pelagodiscus*) is perhaps the most cosmopolitan of all living brachiopods and is found in deepwater parts of all the oceans.

ORDER OBOLELLIDA (Figure 16.38, *A* and *B*)

Small inarticulates with calcite shells otherwise superficially resembling some lingulides. Shell biconvex, subcircular in outline with marginal beaks. Pedicle opening variable in location. Pedicle either absent, or emerging from between valves, through

Figure 16.38. Representative Obolellida. **A₁** and **A₂**, *Obolella*. Ventral valve interior and dorsal valve interior. (Cambrian, Labrador, ×5.) **B**, *Trematobolus*. Ventral valve exterior showing anterior location of pedicle opening. (Cambrian, New Brunswick, ×3. (**A₁** and **A₂** from Rowell, A. J. *Journal of Paleontology* 36: Pl. 29; 1962. **B** from Rowell, A. J. In: Moore, R. C., editor. *Treatise on invertebrate paleontology, Part H.* New York and Lawrence, KS: Geological Society of America and University of Kansas Press; 1965, p. 293.)

tube at apex of ventral valve or through tear-shaped foramen in front of beak produced by anteriorly directed resorption. (Early to Middle Cambrian; 10 genera.)

Obolellida are geographically widespread and have a restricted stratigraphic range, although the order is not taxonomically diverse. It is known only in the upper two-thirds of the Lower Cambrian and the lowest Middle Cambrian. It gave rise to no known descendants and its origin is obscure; it may have arisen from some early lingulide.

Relations among the orders Among the four included orders of the class, morphologic features of the Lingulida and Acrotretida suggest that they may be cladistically closely related, a conclusion that is reinforced by consideration of the larval development of living representatives. Less is known about the Obolellida. Although they differ from contemporaneous Cambrian inarticulates in possession of a calcareous shell, their gross morphology is comparable to that of the lingulides and acrotretides.

The handful of genera that comprise the order Paterinida are more of a problem. They show similarity with many inarticulates in possessing a phosphatic shell, and the group is customarily referred to this class. In virtually every other feature of the shell, however, this assignment is anomalous. Their valves have straight posterior margins and possess open delthyria and notothyria, which have lateral borders thickened in a manner comparable to that of some early members of the class Articulata. Internal features are also unlike those of other inarticulates, lacking, for example, any indications of two pairs of adductor muscles. Its present taxonomic position is clearly unstable and the relationship of the order to the remainder of the phylum is not clear.

Biology

Reproduction and ontogeny As far as is known, the sexes are separate in all inarticulate brachiopods. Eggs and sperm are shed into the sea and reproduction is external. The breeding season apparently varies with the species and with latitude, but information is scanty. Evidence suggests that lingulides in warm climates breed throughout the year. No one female is capable of continuous reproductive activity, however, and any one individual spawns over a period of three or four months. Even during this interval, spawning occurs in bursts separated by resting periods. Different females seem to

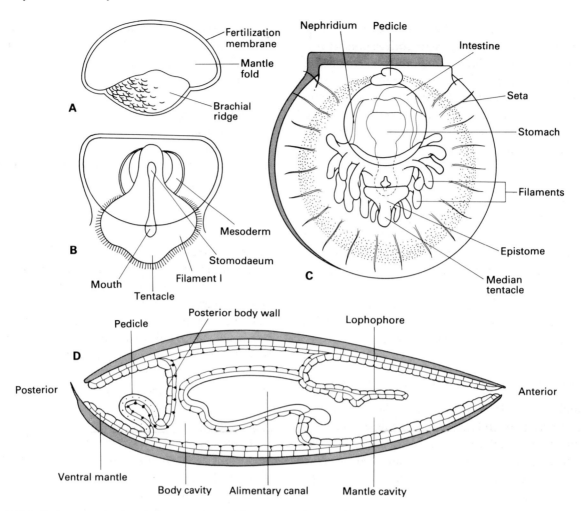

Figure 16.39. Early stages in the ontogenetic development of *Lingula.* **A,** Embryonic stage still enclosed in fertilization membrane. **B,** Early larva with one pair of lophophore filaments. **C,** Larva with eight pairs of lophophore filaments and small pedicle still enclosed between the valves. **D,** Median longitudinal section of larva with eight pairs of filaments showing origin of pedicle. (From Williams, A.; Rowell, A. J. In: Moore, R. C., editor. *Treatise on invertebrate paleontology, Part H.* New York and Lawrence, KS: Geological Society of America and University of Kansas Press; 1965, pp. 34, 47.

spawn at different times during the year. In more temperate climates, reproduction occurs during only part of the year.

The earliest stages of development are known only for the lingulides. While still enclosed in the fertilization membrane, the embryo becomes differentiated into two parts: a ringlike mantle fold, from which the two mantles subsequently develop, and an anterior ciliated brachial ridge, which later gives rise to the lophophore (Figure 16.39, *A*). The mantles grow forward to enclose the remainder of the animal. Shortly before the embryo ruptures its enclosing membrane to enter the larval phase of its development, it already shows many of the features of a brachiopod (Figure 16.39, *B*). The two mantles secrete a protegular shell that is initially continuous across the posterior margin of the animal but that sub-

sequently breaks to form two semicircular protegular valves. The earliest stage of the lophophore is also apparent with but a single pair of short lophophoral filaments separated by a median tentacle. A functional alimentary canal is present in the latest stages of the embryo.

The larval phase of lingulides is protracted. It is typically about three weeks but may be substantially lengthened if the larvae are unable to find suitable settling sites. The larvae swim during their planktic stage, using ciliary beat on their lophophores. The lophophore continues to grow and new filaments are added in pairs symmetrically about the median tentacle. The adult musculature and organs are elaborated during this period. Growth of the pedicle is initiated when the animals reach a stage of development with between six and 11 pairs of lophophore filaments. At first the pedicle is confined between the valves at the posterior end of the shell (Figure 16.39, *C*), developing as a blisterlike evagination of the posterior ventral mantle lobe near the posterior body wall. This evagination involves both the epithelium and the peritoneum lining the main coelomic space of the body cavity (Figure 16.39, *D*). The pedicle subsequently becomes elongated and sausagelike in external form but is protruded between the valves only immediately before settlement, when the animal has between 10 and 20 pairs of lophophore filaments.

The embryology of the two other living families of inarticulate brachiopods, both belonging to the order Acrotretida, is unknown. Their subsequent larval and postlarval development, however, is broadly comparable with that of the lingulides.

Class Articulata

Although articulate brachiopods are known in Lower Cambrian rocks, their distribution is patchy. Throughout the Cambrian they are typically rare organisms, but at some horizons they are relatively abundant, although never diverse. The class experienced a major radiation during the Ordovician, and by the end of that period all but one of the major groups of articulates had appeared. The overwhelming majority of post-Ordovician brachiopods are articulates, and representatives of the class are a major component of nearly all subsequent Paleozoic shelf faunas.

Given the limitations of being filter feeders enclosed in two valves, the articulates display a surprising variation in form and inferred mode of life. The majority were benthic, but a few were epiplanktic, attached to floating seaweed. Among the benthic forms, many were attached by their pedicle, but others lay free on the sea floor. Some were quasi-infaunal, low-level suspension feeders that had only the gape at the margin of their valves projecting above the soft sediment in which they burrowed. Other brachiopods were high-level suspension feeders attached by cementation of their ventral valve and with a shell form crudely resembling that of some solitary corals.

As previously mentioned, many brachiopods are homeomorphic with another member of the phylum. Although externally similar, homeomorphic pairs typically have markedly different internal features. Often these internal features must be understood before a brachiopod can be identified. The structures on the internal surface of a valve may be exposed by natural weathering at an outcrop. More commonly the specimen has to be prepared either by painstaking mechanical removal of adherent matrix or by grinding numerous parallel serial surfaces through the specimen. The latter technique results in destruction of the individual, but the shape of the internal features may be reconstructed from the sequence of cross sections. Occasionally the original calcite shell has been replaced by silica during diagenesis. If this happens in a carbonate rock, the matrix may be removed with dilute acid and the residual valves can reveal internal features in exquisite detail (such as in Figures 16.56, *B* and 16.67).

Morphology

Structure of the posterior sector Living articulate brachiopods differ most conspicuously from inarticulates in the posterior sectors of their shells. The differences are not limited to pedicle structure or the

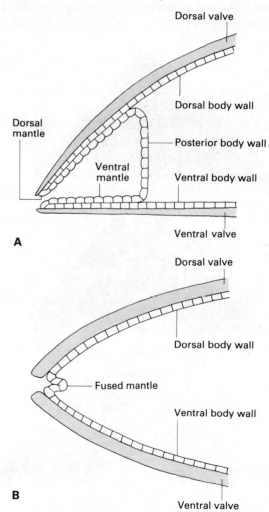

Figure 16.40. Generalized longitudinal sections of the posterior region, lateral of the pedicle, to show relationship of dorsal and ventral body walls and mantles. **A**, Inarticulate with posterior body wall separating mantles. **B**, Articulate with dorsal and ventral mantles fused along posterior margin.

development of articulation but are expressed, for example, in the relation between the posterior parts of the ventral and dorsal mantles (Figure 16.40). The mantles are fused lateral to the pedicle along the posterior margin of living articulates (Figure 16.40, *B*), whereas in inarticulates the corresponding mantle margins are separated by the posterior body wall (Figure 16.40, *A*).

The fused mantle margins are interrupted medially by the pedicle (Figure 16.21). The proximal end of the pedicle of living articulates forms a capsule that is depressed and sunk into the posterior part of the space between the valves (Figure 16.14). In fossils the forward limit of this pedicle capsule is indicated by the posterior margin of the ventral adductor muscle scars. Variation in the location of these scars reveals that the relation between shell and pedicle varies considerably in the evolution of articulates.

The distal end of the pedicle protrudes through a triangular delthyrium in the earlier stages of growth of

many articulate brachiopods. Continued growth of the animal may merely give rise to an increasingly larger delthyrium without any additional modification. This condition is especially common in lower Paleozoic brachiopods. Subsequent growth, however, may produce structures that modify the delthyrial opening. A wide variety of structures are known, the most common being a pair of **deltidial plates** (Figure 16.41, *A*). These subtriangular plates develop from the lateral margins of the delthyrium and extend medially, constricting the dorsolateral regions of the delthyrium. In some species, the tissue secreting the deltidial plates meets in the midline of the shell and fuses, thereafter laying down a single plate (Figure 16.41, *B*).

Some Paleozoic articulate brachiopods never possessed an open delthyrium. The pedicle never emerged from between the valves but was invariably confined to the ventral one. Even the smallest of such shells has a circular pedicle foramen at the apex of the ventral valve and a single plate that bridges the apical region of the delthyrium. Continued growth of the individual kept the apex of the delthyrium blocked by secretion of a single plate, the **pseudodeltidium** (Figure 16.41, *C*). Growth lines on the pseudodeltidium are uninterrupted medially and are continuous laterally with those on the interareas. Pseudodeltidium and interareas apparently were secreted by a continuous mantle margin, and the pedicle always lay ventral to it. Reconstruction of the relationship of tissue and shell in such brachiopods (Figure 16.42)

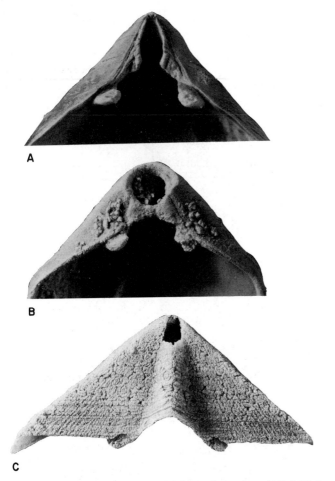

A

B

C

Figure 16.41. Modifications of delthyrial opening. **A,** Deltidial plates of *Hemithyris*. (Holocene, off coast of Washington, ×4.) **B,** Fused deltidial plates of *Sellithyris*. (Cretaceous, England, ×6.) **C,** Pseudodeltidium of *Tritoechia*. (Ordovician, Oklahoma, ×4.) (**C,** photo by G. A. Cooper).

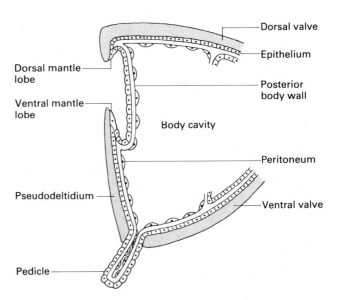

Figure 16.42. Diagrammatic reconstruction of possible shell and tissue relationships in the posterior region of a brachiopod with a pseudodeltidium. Pedicle interpreted as an extension of the ventral body wall.

suggests that they differed markedly from living articulates. Some of their features may have been more like those of inarticulates because they seem to have possessed a posterior body wall and mantle margins that were not fused posteriorly.

The interareas of many brachiopods, particularly Paleozoic forms, produce a straight posterior margin to the shell that approximately coincides with the hinge axis passing through the teeth. The posterior margin thus forms a **hinge line** (Figure 16.43, *B*) that assists in articulation by acting as a fulcrum. Brachiopods with a hinge line are said to be **strophic**. **Astrophic** shells lack a hinge line because their posterior margin is curved (Figure 16.43, *A*). Although both strophic and astrophic brachiopods are still living, the astrophic forms, which typically have stronger articulation, predominate.

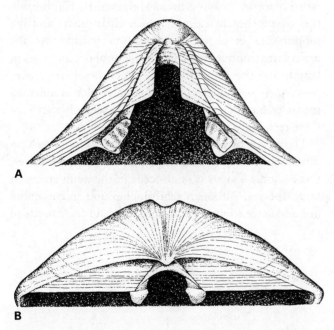

A

B

Figure 16.43. Astrophic and strophic posterior margins. **A,** Astrophic posterior margin of the rhynchonellide *Hemithyris*. **B,** Strophic posterior margin of the spiriferide *Eospirifer*. Note that posterior margin forms a hinge line. (From Jaanusson, V. *Smithsonian Contributions to Paleobiology* 3:35; 1971.)

Internal features of the shell Articulates may have complex internal projections and processes. These are usually associated either with articulation or with support of the fleshy lophophore. In the ventral valve the hinge teeth are often buttressed by **dental lamellae** (Figure 16.44), plates of secondary material that extend to the valve floor.

The greatest diversity of internal structure is typically developed in the dorsal valve. A series of plates associated with the sockets in the posterior region of this valve are

Dental lamella

Figure 16.44. Dental lamellae of the orthide *Enteletes* and interior view of ventral valve. (Permian, Texas, ×1.) (From Cooper, G. A.; Grant, R. E., *Smithsonian Contributions to Paleobiology* 24: 2973; 1976.)

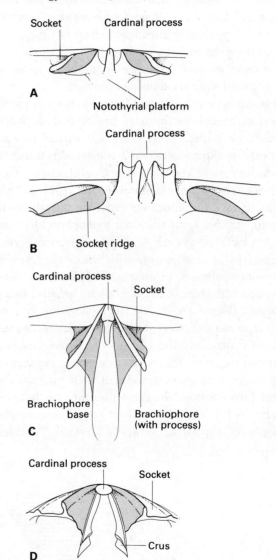

Socket   Cardinal process

A

Notothyrial platform

Cardinal process

B

Socket ridge

Cardinal process   Socket

Brachiophore base   Brachiophore (with process)

C

Cardinal process   Socket

D

Crus

Figure 16.45. Ventral views of the cardinalia. **A,** Relatively simple cardinalia with bladelike cardinal process of the orthide *Billingsella*. **B,** Simple cardinalia with bifid cardinal process to which diductors were attached of the strophomenide *Strophomena*. **C,** More advanced cardinalia with brachiophores of the orthide *Dalmanella*. **D,** Cardinalia with crura of the terebratulide *Dallithyris* (loop, which projects anterior of crura, is not shown). (From Williams, A.; Rowell, A. J. In: Moore, R. C., editor. *Treatise on invertebrate paleontology, Part H*. New York and Lawrence, KS: Geological Society of America and University of Kansas Press; 1965, p. 96, 97.)

collectively referred to as the **cardinalia**, from the Latin word for hinge (Figure 16.45). A complex and somewhat confusing terminology has been developed for the individual elements of the cardinalia, but only a few of the elements will be addressed here. In many dorsal valves a blade or more complex lobed structure occurs medially on the cardinalia; this is the **cardinal process** (Figure 16.45). When the cardinal process is only a thin blade (Figure 16.45, *A*) it merely separates the dorsal ends of the diductor muscles that are attached to the valve immediately lateral to it. In species having a more sturdily constructed process (Figure 16.45, *B* to *D*), the dorsal ends of the diductor muscles are attached directly to it, the cardinal process providing additional leverage to open the shell when the diductors contract.

Although a variety of plates and projections define the sockets of articulates, in many brachiopods a pair of bladelike or rodlike projections that extend forward anteriorly or anterolaterally is a conspicuous feature of the cardinalia. In the older brachiopod stocks, these processes, termed **brachiophores** (Figure 16.45, *C*), may have buttressed the alimentary canal but did not extend forward sufficiently to reach the lophophore. In many younger stocks, however, bladelike projections, **crura**, singular **crus**, are homologous with brachiophores, were long enough to reach the anterior body wall, and were embedded in the base of the lophophore on either side of the mouth (Figure 16.45, *D*). The lophophores of these younger stocks thus were supported to varying degrees. In some forms only the base of the lophophore was supported by crura; others had more extensive support from complex **loops** or **spiralia** of shell material that extend forward from the crura into the mantle cavity (Figures 16.1, *H* and 16.60). The shape of these complex lophophore supports, generally termed **brachidia**

(singular **brachidium**), provides information on the form of the lophophore of fossil articulates.

Growth and relation of the lophophore to its supports The two brachia that form the lophophore are attached to the anterior body wall. A food groove (Figure 16.46) runs the length of each brachium, with the two food grooves meeting medially at the mouth. The food groove is bounded by a lip on one side and by a row, typically a double row, of filaments on the other side (Figure 16.46). Growth of the lophophore involves addition of new filaments at the tip of each brachium and is associated with variably complex changes in the shape of the organ.

While the shell is still very small, the two brachia of all living articulates form a **trocholophe** (from the Greek word *trochos,* meaning circular) (Figure 16.47); the two tips of the brachia almost touch each other, and the lophophore is basically circular in outline. As the articulate continues to grow, the lophophore increases in length and the generative tip of each brachium moves posteriorly so that the original circular plan is indented medially. Such a lophophore is called a **schizolophe** (from the Greek word *schizo*, meaning split).

The adults of some small species of brachiopods have schizolophes or, more rarely, trocholophes. However, these simple shapes do not occur in the adult stages of large species. Although schizolophes and trocholophes are adequate to meet the respiratory and food needs of

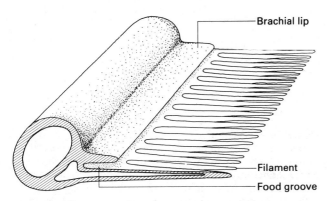

Figure 16.46. Diagrammatic section of a segment of a lophophore showing relationship of food groove, brachial lip, and filaments. Bases of filaments arranged in a double row. (Modified from Rudwick, M. J. S. *Journal of the Linnean Society of London* 44: 594; 1962.)

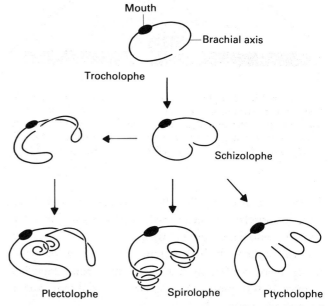

Figure 16.47. Ontogenetic pathways in the development of complex lobed or coiled lophophores. Diagrams show only the axes of the brachia. (Modified from Rudwick, M. J. S. *Journal of the Linnean Society of London* 44: 593; 1962.)

small brachiopods, more surface area and filaments are necessary to satisfy the demands of the larger masses of tissue associated with bigger individuals. This is a particular example of a well-known problem. Expressed in its simplest form, the metabolic needs of an organism are proportional to its volume and thus to some function of the third power of its length. The ability to meet these needs increases more slowly with an increase in size of the organism and may be proportional only to the area or length of some organ, which is itself some function of the second or first power of the length of the organism. Thus if there is no change in shape with continued growth, metabolic needs would tend to outstrip the ability to meet them.

Judging from what is known of living brachiopods, a schizolophe can supply the needs of a brachiopod that is a maximum of 3 or 4 mm in diameter. As it grows beyond this size, the size of the lophophore and the number of filaments it bears must increase more rapidly than is possible merely by increasing the diameter of the sub-circular lophophore. Living brachiopods, and by inference most fossil forms, increase the size of the lophophore by a folding or coiling of the organ. Three distinct pathways are known (Figure 16.47), and their end products are **ptycholophes** (from the Greek word *ptycho*, meaning fold), **spirolophes** (from the Greek word *speira*, meaning coil), and **plectolophes** (from the Greek word *plecto*, meaning twisted).

Ptycholophes are rarely supported by shelly projections from the cardinalia, but in some species the lophophore was buttressed by ridges projecting from the floor of the dorsal valve, and the disposition of these ridges gives some indication of the shape of the lophophore.

Spirolophes and plectolophes are commonly supported in articulate brachiopods by projections from the cardinalia. However, there is no close one-to-one correspondence between the shape of the lophophore and the shape of the lophophore support. In living brachiopods, plectolophes are supported by loops, and it is a reasonable assumption that fossil loop-bearing brachiopods also had plectolophes. In many living brachiopods, however, spirolophes are supported only by the tips of the crura, which are embedded in the posterior segment of the lophophore. In such forms the shape of the lophophore support is a poor guide to the shape of the lophophore. All brachiopods that have spiralia are extinct; even though specialists disagree as to the exact shape of the associated lophophores, it is accepted that they were spirally coiled.

Shell ultrastructure and composition All articulates have an exoskeleton consisting of a thin outer organic layer, the periostracum, underlain by primary and secondary shell layers composed of calcite. These are secreted in the manner previously discussed on p. 455. Differences between the various stocks of articulates are seen in the ultrastructure of the secondary layer and in the presence or absence of a tertiary layer.

Two principal types of secondary shell ultrastructure are recognized: fibrous and laminar. A **fibrous secondary layer** is the more common and apparently represents the primitive condition in the Articulata. It occurs in Early Cambrian articulates, is a persistent feature of many of the principal lineages, and is the characteristic ultrastructure of the majority of living articulates. During deposition of the fibrous layer, each cell secretes a long calcite rod with a distinctive shape (Figure 16.48, *A*). The epithelial cells are arranged in alternating rows, and only the posterior part of each cell deposits calcite; an anterior arcuate zone exudes an organic membrane. This membrane is fused with that of adjacent cells, and together they ensheath each calcite rod of the fibrous layer. The resultant fibers are inclined at an angle of about 10 degrees to the primary layer and in transverse section are seen to be regularly stacked (Figure 16.48, *B*). All the internal features of the shell, including the teeth, sockets, cardinalia, and loops or spiralia, are usually built up of the secondary layer. In these features the regularity of the fibers is commonly disrupted because of the effects of resorption and the development of new cells at secondary generative zones.

In a few fibrous-shelled articulates, a modification occurs in the secretory ability of cells that formerly were depositing the secondary fibrous layer. These cells cease exuding organic material, and the resulting shell consists of large prisms of calcite arranged approximately normal to the mantle (Figure 16.49). This prismatic calcite may occur as thin lenses within the secondary layer, indicating that the physiologic change was only temporary, or it may form a distinct, persistent tertiary layer covering much of the inner surface of the valve.

A **laminar secondary shell** is characteristic of one large order and occurs rarely in another. The basic units of this type of shell are thin laths of calcite (Figure 16.50), which are variably fused and amalgamated to build individual laminae oriented roughly parallel to the outer surface of the valve.

In many articulates the calcareous shell is essentially unmodified and is said to be **impunctate**. This term implies that modifications can occur of which two

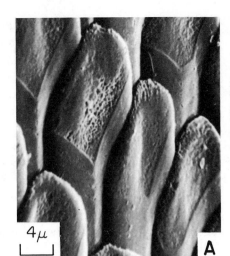

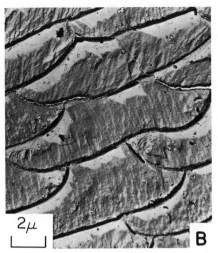

Figure 16.48. Ultrastructure of the fibrous secondary shell layer of the modern rhynchonellide *Notosaria*. A, Scanning electron micrograph of the inner surface of the valve showing exposed faces of fibers. B, Transmission electron micrograph of cross section of the secondary layer showing stacked fibers. (From Williams, A. *Lethaia* 1: 273; 1968.)

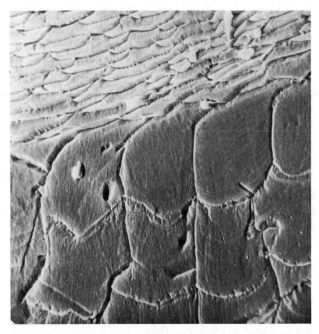

Figure 16.49. Ultrastructure of the tertiary prismatic layer of the modern terebratulide *Liothyrella*. Scanning electron micrograph shows boundary between secondary fibrous layer above and tertiary layer below. (×2100.) (From McKinnon D. I.; Williams, A. *Palaeontology* 17: pl. 24; 1974.)

common ones are known. Many fibrous shells have abundant, slender, cylindrical pores called **punctae** that extend from the inner surface of the valve to terminate immediately below the outer surface of the primary layer (Figure 16.51). Multicellular projections from the outer surface of the mantle occupied the punctae while the animal was alive. These projections, termed **caeca**, store a variety of organic compounds. Their presence apparently reduces the density of organisms that bore into the valves but has no effect on the frequency of attachment of epifauna. The distal surface of each caecum is connected by numerous fine organic strands to the periostracum (Figure 16.52). These fine strands pierce the thin canopy of calcite that caps the puncta. Shells possessing punctae are said to be **punctate**. Although not universally agreed, most brachiopod investigators believe that punctation arose independently on several occasions during evolution of the phylum.

The punctate condition does not occur in laminar shelled brachiopods, but many of them are **pseudopunctate**. The shell of pseudopunctate forms is penetrated by slender, coarsely crystalline rods of calcite (Figure 16.53) that are commonly raised above the inner surface of the valves as low tubercles. Pseudopunctate fibrous shells are relatively rare.

Classification

The class Articulata may be divided into six orders, into which the great majority of hinge-bearing brachiopods fit readily. These orders, named after relatively common genera, are Orthida, Strophomenida, Pentamerida, Rhynchonellida, Spiriferida and Terebratulida. Other possible ordinal groupings exist and have been proposed. Some of them differ only slightly from the ones used here, and the differences are relatively minimal. We would not want to defend the six major orders as representing the last word in brachiopod classification. We suspect that it is not completely consistent with the probable evo-

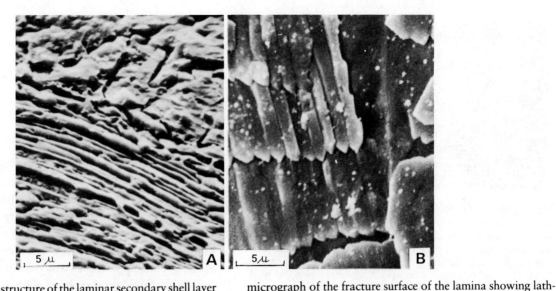

Figure 16.50. Ultrastructure of the laminar secondary shell layer of strophomenides. **A,** Scanning electron micrograph of cross section of the shell laminar layer beneath primary layer of *Strophomena*. (Ordovician, Oklahoma.) **B,** Scanning electron micrograph of the fracture surface of the lamina showing lath-like structure. (*Gacella*, Ordovician, Scotland.) (From Williams, A. *Lethaia* 3: 336, 337; 1970.)

lutionary history of the class, and at least one order may be polyphyletic. However, this classification is relatively simple, well known, and adequate as a utilitarian aid to communication.

ORDER ORTHIDA (Figures 16.1, *G*; 16.9, *B*; 16.12; 16.44; 16.45, C; 16.54)

Typically strophic articulates, having unequally biconvex valves with the ventral valve deeper; valves

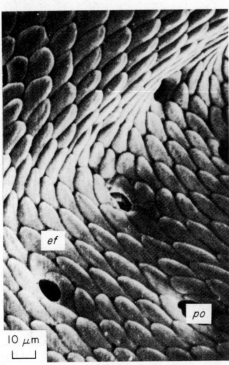

Figure 16.51. Scanning electron microphotograph of the internal surface of the punctate shell of the terebratulide *Waltonia* showing internal openings of punctae (*po*) and exposed faces (*ef*) of the fibers of the secondary layer. (From Owen, G.; Williams, A. *Proceedings of the Royal Society* 172: pl. 55; 1969.)

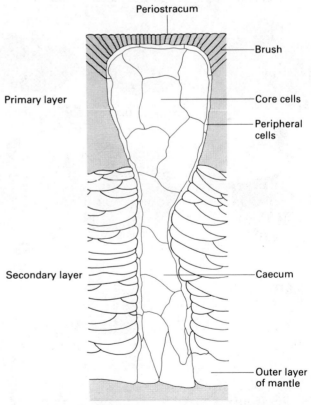

Figure 16.52. Diagrammatic longitudinal section of a caecum showing relationship to exoskeletal layers and mantle. (From Williams, A. *Special Papers in Palaeontology* No. 2, p. 27; 1968. © Palaeontological Association.)

Figure 16.53. Scanning electron microphotograph of an exfoliated dorsal valve interior of the strophomenide *Eoplectodonta* showing conical pseudopunctae projecting through a laminar secondary layer. The inner surface of the valve is toward the top. (Silurian, Sweden, ×250.) (From Brunton, C. H. C. *Bulletin British Museum (Natural History)* 21: pl. 3; 1972. ©Trustees of British Museum.)

usually ornamented by ribs; interareas well developed in both valves. Complexity of cardinalia variable: typical cardinalia consisting of (1) thickening of valve floor (notothyrial platform) carrying median blade or variously lobed cardinal process, and (2) bladelike brachiophores extending anteromedially from margins of notothyrium to form proximal borders of sockets. Brachiophores in some cardinalia supported by subvertical plates extending to valve floor. Sockets variously embedded in shell material or bounded by lateral plates. Shell impunctate or punctate, and in all but one small group fibrous. (Early Cambrian to Late Permian; 340 genera.)

The order Orthida may claim the distinction of including the oldest known articulate brachiopods, which are found in the middle part of the Lower Cambrian. The earliest forms are morphologically simple, but in their subsequent Paleozoic evolution orthides experimented with many of the features that were to become characteristic of other orders. The only major features that orthides failed to develop were cementation of the ventral valve, development of long, hollow spines, and the elaboration of complex lophophore supports.

The teeth of the earliest orthides are small but are located lateral to the delthyrial margins as is normal for the class. Cardinalia of early orthides are simple: the cardinal process is either absent or is an unmodified blade, and the brachiophores are curved parallel to the

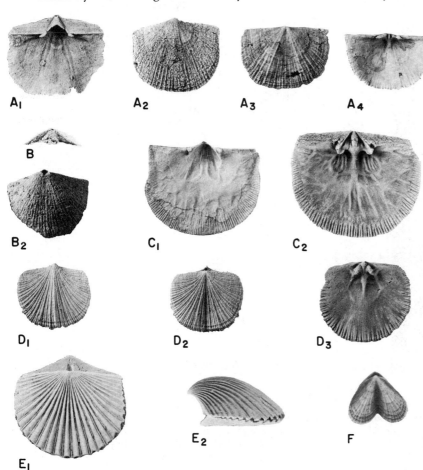

Figure 16.54. Representative Orthida. **A,** *Billingsella,* Ventral valve interior and exterior views and dorsal valve exterior and interior views. (Cambrian, Oklahoma, ×2.) **B,** *Nisusia.* Ventral valve posterior and exterior views. Note the supra-apical foramen and well-developed pseudodeltidium. (Cambrian, Queensland, ×1, ×2.) **C,** *Multicostella.* Ventral and dorsal valve interior views. (Ordovician, Tennessee, ×1.) **D,** *Onniella.* Ventral and dorsal valves exterior views and dorsal valve interior view. (Ordovician, Virginia, ×2.) **E,** *Hesperorthis.* Complete shell dorsal view and lateral view. (Ordovician, Oklahoma, ×2.) **F,** *Dicoelosia.* Complete shell, dorsal view. (Devonian, Oklahoma, ×1.5.) (**A** From Cooper, G. A. *Smithsonian Miscellaneous Collections* 117: pl. 1; 1952. **C** and **E** from Cooper, G. A. *Smithsonian Miscellaneous Collections* 127: pl. 12, 52; 1956. **D,** photos by G. A. Cooper. **F** from Amsden, T. W. *Bulletin Oklahoma Geological Survey* 78: pl. 1; 1958.)

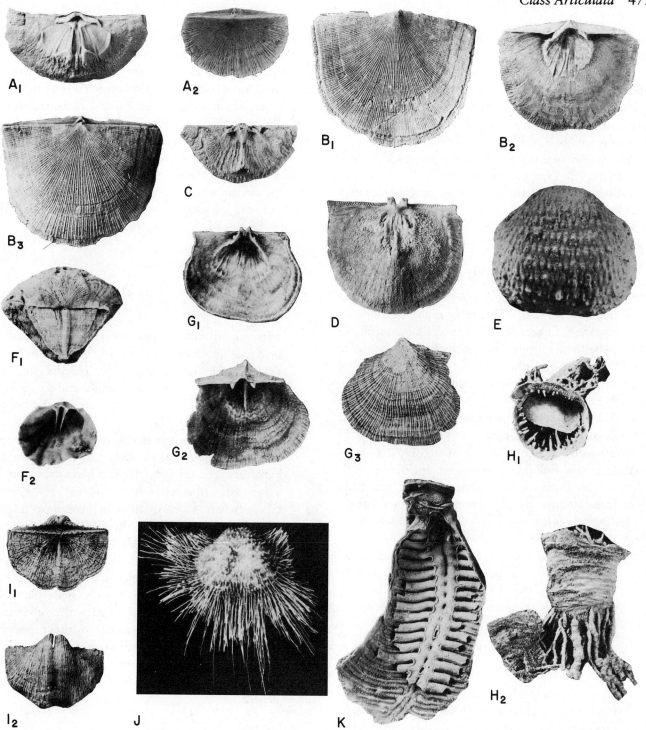

Figure 16.55. Representative Strophomenida. **A,** *Sowerbyella.* Dorsal valve interior view and dorsal view of complete shell. (Ordovician, Virginia, ×2.) **B,** *Strophomena.* Ventral valve exterior and interior views and complete shell dorsal view. (Ordovician, Oklahoma, ×2.) **C,** *Leptellina.* Dorsal valve interior. (Ordovician, Tennessee, ×3.) **D,** *Stropheodonta.* Dorsal valve interior. Note the row of denticles and sockets along the hinge line. (Devonian, New York, ×1.) **E,** *Juresania.* Ventral view of complete shell. (Pennsylvanian, Kansas, ×2.) **F,** *Meekella.* Posterior view of complete shell and ventral valve interior view. (Permian, Texas, ×1.5.) **G,** *Derbyia.* Dorsal and ventral valves interior views and ventral valve exterior view. (Permian, Texas, ×1.) **H,** *Cyclacantharia.* Dorsal view of complete shell, with dorsal valve partially open, and ventral valve from side. (Permian, Texas, ×1.) **I,** *Chonetinetes.* Complete shell dorsal and ventral views. Note spines confined to ventral interarea. (Permian, Texas, ×2.) **J,** *Waagenoconcha.* Ventral valve exterior. (Permian, Pakistan, ×1.) **K,** *Collemataria.* Complete shell viewed from dorsal side. (Permian, Texas, ×1.) (**A** to **C** from Cooper, G. A. *Smithsonian Miscellaneous Collections* 127: pl. 199, 255, 187; 1956. **D,** photo by G. A. Cooper. **F** and **G** from Cooper, G. A.; Grant, R. E. *Smithsonian Contributions to Paleobiology* 1974. **H, I,** from Cooper, G. A.; Grant, R. E. *Smithsonian Contributions to Paleobiology* 19: 1513, 1873; 1975. **J** from Grant, R. E. *Journal of Paleontology* 40: pl. 131; 1966. **K** from Cooper, G. A.; Grant, R. E. *Smithsonian Contributions to Paleobiology* 15: 689; 1974.)

hinge line, serving only to define the sockets (Figure 16.54, A_4).

Two groups that appear to be the ancestral stocks for the two major lineages of articulate brachiopods may be recognized among these early Orthida. Early representatives of the two groups are distinguished primarily by the presence or absence of a pseudodeltidium.

The majority of orthides possess open delthyria and notothyria (for example, Figure 16.54, C to F) and are the descendants of Early Cambrian forms with similar features. Evolution within this group involved elaboration of the cardinalia. The brachiophores of many orthide groups rotated to point anterolaterally and sockets became well defined. The cardinal process also underwent modification and, in some taxa, carried the dorsal ends of the diductor muscles on a bilobed projection (Figure 16.54, C_2). One large group developed a punctate shell, a characteristic that was subsequently to arise polyphyletically in many brachiopod orders (Figure 16.54, D). These punctate orthides were more successful than their impunctate cousins, which became extinct in the Devonian. Punctate forms survived to the end of the Paleozoic, and all Mississippian, Pennsylvanian, and Permian orthides are punctate.

The smaller group of orthides that have a pseudodeltidium (the Billingsellacea, Figure 16.54, A and B) show less evolutionary change. Their cardinalia remained primitive and like that of early members of the order (Figure 16.54, A_4). As a group, the Billingsellacea had only a short existence. They are known in rocks of Early Cambrian to Early Ordovician age. Their importance lies primarily in their phylogenetic position as ancestors to one of the most diverse groups of brachiopods, the members of the order Strophomenida.

ORDER STROPHOMENIDA (Figures 16.1, F; 16.11; 16.55)

Strophic articulates, typically with convex ventral valve and concave dorsal valve; characteristically pseudopunctate with laminar secondary shell. Wide variety of shell form, many for specialized modes of life either unknown or occurring only rarely in remainder of phylum. (Ordovician to Triassic; 865 genera.)

The Strophomenida include more genera than any other brachiopod order. Active debate continues over the possible inclusion of some Mesozoic and Cenozoic genera (suborder Thecideidina) within the order. If they are excluded, as in this chapter, the range of the Strophomenida is Ordovician to Triassic.

Four suborders are conventionally recognized within the order Strophomenida. Members of three of the suborders are easily distinguished because of characteristic morphologic features that are evolutionary novelties confined to their lineage. The fourth suborder, which includes the least modified strophomenides, is probably polyphyletic and originated from more than one billingsellacean species. If this suspicion subsequently is justified, the classification of the Strophomenida will need to be altered.

SUBORDER STROPHOMENIDINA (Figures 16.45, B; 16.55, A to D, F to G)

Typically strophic Strophomenida with semicircular or subquadrate outline. Interarea present on both valves, bigger on ventral valve; pseudodeltidium well developed. Cardinalia simple and in all but earliest Strophomenidina dominated by conspicuous cardinal process. Brachiophores, if present, serving only as socket ridges; one group with very thin space between valves and either concave dorsal valve and convex ventral valve or convex dorsal valve and concave ventral valve. This group typically with supra-apical pedicle foramen; ridges and low platforms on floor of dorsal valve of a few probably supported variably lobed ptycholophe or schizolophe. Conventional articulation replaced by row of denticles along hinge line in some. Second group characteristically with more space between valves and convex dorsal valve with convex or conical ventral valve; typically without pedicle foramen. Secondary shell layer laminar and pseudopunctate in most of suborder, laminar or fibrous impunctate in a few. (Ordovician to Triassic; 360 genera.)

SUBORDER CHONETIDINA (Figure 16.55, I)

Slightly modified Strophomenida, having concavo-convex shells resembling some of strophomenidines but with few, well-developed tubular spines limited to posterior margin of ventral interarea. Shell with pseudopunctate laminar secondary layer. (Ordovician to Permian; 120 genera.)

SUBORDER PRODUCTIDINA (Figures 16.1, *F*; 16.11; 16.55, *E, H,* and *J*)

More modified Strophomenida, having concavo-convex shells becoming more modified in some more bizarre forms into conical, often irregularly shaped, ventral valves with dorsal valve inserted within cone or perched on top as lid. Spines elaborately developed, many specialized in function, not restricted to hinge line. Pedicle present in juvenile stages of few; some adults attached by direct cementation of ventral beak and anchored by spines, others free on sea floor supported by spines. Teeth, sockets, and interareas lost in many. Shell with pseudopunctate laminar secondary layer. (Devonian to Permian; 350 genera.)

SUBORDER OLDHAMINIDINA (Figure 16.55, *K*)

Mostly highly modified Strophomenida, lacking fully developed dorsal valve so that mantle cavity not enclosed by shell. Ventral valve commonly cemented, varying from irregular cone to flattened plate, having internal, typical strongly lobed septa. Lophophore support a lobate plate with same outline as septa and resting upon them. Vestigial remainder of dorsal valve a small triangular plate behind lophophore support. Shell with pseudopunctate laminar secondary layer. (Pennsylvanian to Permian; 35 genera.)

ORDER PENTAMERIDA (Figures 16.1, *D*; 16.56 to 16.58)

Strophic or astrophic articulates, commonly strongly biconvex with incurved beaks. Majority with open delthyrium and concave, spoon-shaped structure, called a **spondylium**, in apical region of ventral valve formed by fusion of dental lamellae. Cardinalia typically with brachiophores supported by plates extending to floor of dorsal valve. Brachiophores in later genera having long processes extending toward mantle cavity, probably to support lophophore. Sockets defined by distinct plates. Impunctate, fibrous secondary shell layer. (Middle Cambrian to Devonian; 160 genera.)

Significant changes in the morphology of typical examples of the order took place during its evolutionary development. Such changes make the order relatively

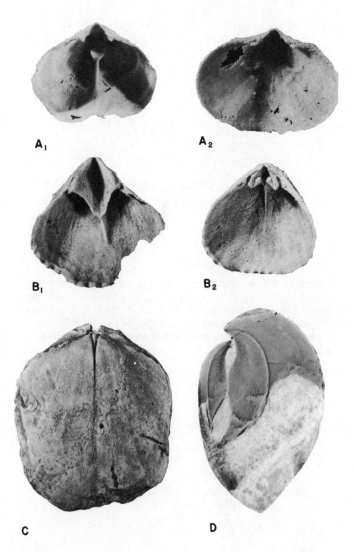

Figure 16.56. Representative Pentamerida. **A,** *Diaphelasma.* Ventral and dorsal interior views. (Ordovician, Oklahoma, ×3.5.) **B,** *Camarella.* Ventral and dorsal interior views. (Ordovician, Missouri, ×1.5.) **C,** *Pentameroides.* Dorsal view of internal mold. (Silurian, Iowa, ×1.5.) **D,** *Kirkidium.* Complete shell is split approximately down the median longitudinal plane of symmetry. The fracture is guided by plates buttressing brachiophores in the dorsal valve and by the spondylium and median septum in the ventral valve. (Silurian, Oklahoma, ×2.) (**C,** photo by G. A. Cooper. **D,** from Amsden, T. W. *Journal of Paleontology* 43: pl. 117; 1969.)

difficult to define unambiguously, devoid of numerous qualifying statements.

Early pentamerides, those of Cambrian and early Ordovician age, are not common. They differ little from some contemporary orthides, the undoubted ancestral group of the order. The most characteristic feature that distinguishes them is the persistent occurrence of a pronounced dorsal fold and ventral sulcus in the early pentamerides (Figure 16.56, *A*). Had the group not

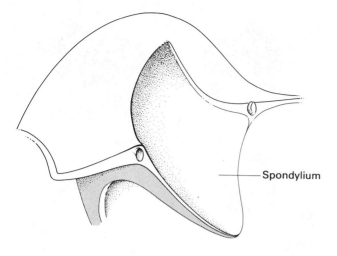

Figure 16.57. Posterior part of the ventral valve of the pentameride *Gypidula* showing spoonlike spondylium. (Modified from Williams, A.; Rowell, A. J. In: Moore, R. C., editor. *Treatise on invertebrate paleontology, Part H.* New York and Lawrence, KS: Geological Society of America and University of Kansas Press; 1965, p. 112.)

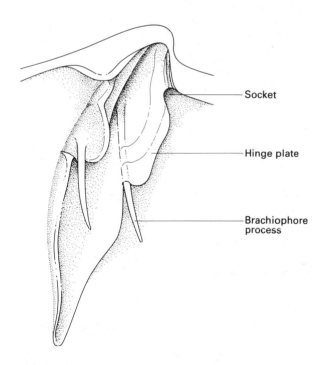

Figure 16.58. Oblique view of the posterior part of the dorsal valve of the pentameride *Gypidula* showing brachiophores supported by plates and extended anteriorly as brachiophore processes. (Modified from Williams, A.; Rowell, A. J. In: Moore, R. C., editor. *Treatise on invertebrate paleontology, Part H.* New York and Lawrence, KS: Geological Society of America and University of Kansas Press; 1965, p. 121.)

subsequently diversified, they would probably have been treated as unusual orthides.

Conspicuous changes had occurred in pentameride morphology by Silurian time, when the order first became abundant. Most Silurian and Devonian pentamerides have short, curved posterior margins and are thus astrophic. The beaks of many of these forms are incurved and the shells are commonly strongly biconvex. Internally, the spondylium typically is elaborately developed and may be supported by a median septum (Figure 16.56, *D*). The plates buttressing the brachiophores in the dorsal valve may be very long and brachiophore processes well developed. The later pentamerides thus resemble more modern brachiopods in having a lophophore supported by projections from the cardinalia, and in possessing a curved posterior margin. This resemblance, however, is an example of parallelism, since the remaining orders of brachiopods, two of which are still extant, were ultimately derived not from the later pentamerides but from the earlier Ordovician representatives of the order.

ORDER RHYNCHONELLIDA (Figures 16.1, C; 16.59)

Astrophic articulates having biconvex shells typically with dorsal fold and ventral sulcus; interareas limited to ventral valve, low; delthyrium partly closed by deltidial plates. Dental plates usually present in ventral valve. Outer plates separate welldefined sockets from crura that support base of lophophore. Extant species having pedicle with base sunk inside ventral valve; lophophore spirolophous. Secondary shell layer fibrous and impunctate, rarely punctate. (Ordovician to Holocene; 520 genera.)

Rhynchonellides have been remarkably conservative since their inception in the Ordovician. Only one group of Devonian to Permian rhynchonellides (the Stenoscismatacea) shows much departure from the basic pattern of the order in the development of an elongated spoonshaped structure in the dorsal valve on the crest of a median septum (Figure 16.59, *D*). The dorsal ends of the adductor muscles were seated on this plate, their ventral ends, together with those of the diductors, attached to a raised spondylium in the ventral valve. This combination of elevated platforms served to shorten the muscles so that they probably lacked the tendinous midsegment that is present when muscles are long.

Although the rhynchonellides never have been the dominant order of the phylum, they have been a sig-

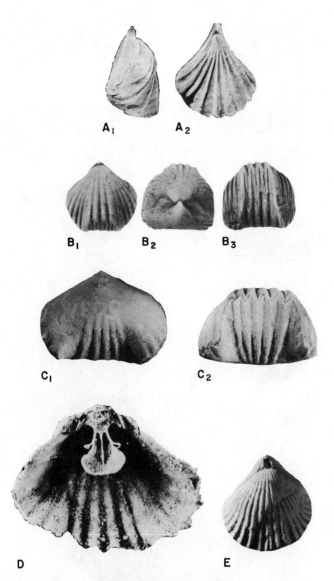

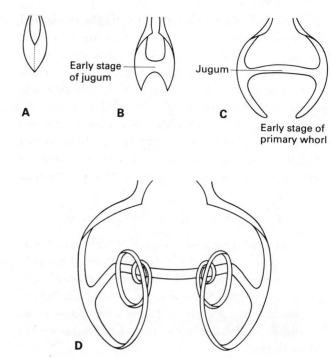

Figure 16.60. Diagrammatic representation of the growth of the spiralia in the dorsal valve of the atrypidine *Zygospira*. **A,** Earliest stage; fusion of two projections from cardinalia to form loop. **B,** Modification and initial appearance of jugum. **C,** Development of initial part of primary whorl and elaboration of jugum. **D,** Adult spiralia. (After Beecher, C. E.; Schuchert, C. *Proceedings of the Biological Society of Washington* 8: pl. 10; 1893.)

Figure 16.59. Representative Rhynchonellida. **A,** *Rhyncho-treta.* Lateral and dorsal views of complete shell. (Silurian, Indiana, ×1.5.) **B,** *Sphaerirhynchia.* Dorsal, posterior, and anterior views of complete shell. (Devonian, Oklahoma, ×1.3.) **C,** *Pugnax.* Ventral and anterior views of complete shell. (Mississippian, England, ×1.5.) **D,** *Stenoscisma.* Dorsal valve interior view. (Permian, Texas, ×2.) **E,** *Kallirhynchia.* Dorsal view of complete shell. (Jurassic, England, ×2.) (**B** from Amsden, T. W., *Bulletin Oklahoma Geological Survey* 78: pl. 6; 1958. **D** from Grant, R. E. *Smithsonian Miscellaneous Collections* 148: pl. 23; 1965.)

nificant component of marine faunas since Ordovician times. They are here regarded as the group ancestral to the Spiriferida.

ORDER SPIRIFERIDA (Figures 16.1, *E*; 16.9, *C*; 16.60 to 16.64)

Articulates having a variety of shell form, but all characterized by possessing spiralia to support lophophore, interarea commonly limited to ventral valve, and punctate or impunctate shell with fibrous secondary layer. (Ordovician to Jurassic; 720 genera.)

As far as is known, early growth stages of spiralia are similar among spiriferides whose adult lophophore supports differ markedly in both form and orientation. The earliest ontogenetic stage consists of two prongs that project forward from the cardinalia (Figure 16.60, *A*) and subsequently fuse anteromedially to form a loop. The spirally coiled parts of the lophophore support arise from the anterolateral margins of the early loop. The disposition of the first primary whorl of the spiralium and the number and configuration of subsequent whorls differ in different types of spiriferides. The **jugum**, a bridgelike structure connecting the primary whorls in many spiriferides, however, invariably is formed from the anteromedian segment of the initial loop (Figure 16.60, *B* and *C*). Growth of the jugum, like that of the remainder of the spiralium, involved a delicate balance between resorption and deposition of secondary shell.

Four major groups of spiriferides may be recognized, and representatives of all of them are relatively common. Here the four groups are regarded as suborders, but there is no universal agreement as to the relationship among them. Some authorities maintain that the oldest suborder, the Atrypidina, is best regarded as a group of spire-bearing rhynchonellides. Others argue that one suborder, the Spiriferidina, was derived from orthide ancestors and that evolution of spiralia is an example of convergence. These discrepancies have not been resolved.

SUBORDER ATRYPIDINA (Figure 16.61)

Astrophic spiriferides, typically biconvex. Jugum simple in most; spiralia variably directed medially, dorsomedially, or laterally; primary whorl commonly parallel to plane of commissure running subparallel to margin of valves. Shell impunctate, fibrous. (Middle Ordovician to Late Devonian; 140 genera.)

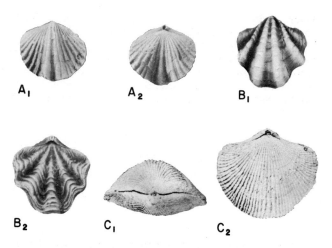

Figure 16.61. Representative Atrypidina. **A**, *Zygospira*. Ventral and dorsal views of complete shell. (Ordovician, Ohio, ×2.) **B**, *Atrypina*. Ventral and dorsal views of complete shell. (Devonian, Oklahoma, ×3.) **C**, *Atrypa*. Posterior and dorsal views of complete shell. (Devonian, Indiana, ×1.) (A from Copper, P. *Palaeontology* 20: pl. 37; 1977. B from Amsden, T. W. *Bulletin Oklahoma Geological Survey* 78: pl. 6; 1958.)

SUBORDER SPIRIFERIDINA (Figures 16.1, E; 16.9, C; 16.62)

Strophic spiriferides with biconvex shells. Ventral interarea typically well developed. Jugum simple or absent; spiralia well developed with many whorls directed laterally or posterolaterally; primary whorl near midline of shell, coiled in plane approximately parallel to plane of symmetry. Shell impunctate or

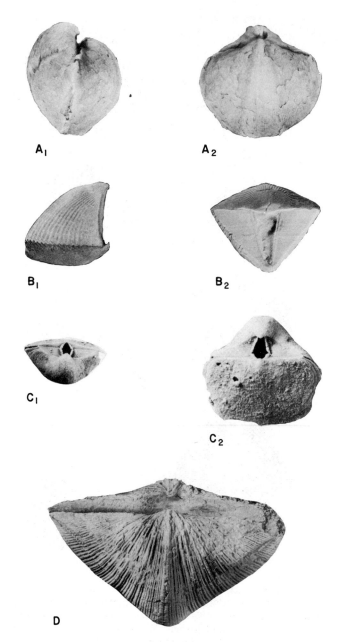

Figure 16.62. Representative Spiriferidina. **A**, *Eospirifer*. Lateral and dorsal views of complete shell. (Silurian, Oklahoma, ×2.) **B**, *Platyrachella*. Lateral and posterior views of complete shell. (Devonian, Iowa, ×2.) **C**, *Crurithyris*. Posterior and dorsal views of complete shell. (Permian, Texas, ×6.) **D**, *Neospirifer*. Dorsal view of complete shell. (Pennsylvanian, Kansas, ×1.) (A from Amsden, T. W. *Bulletin Oklahoma Geological Survey* 125: 51; 1978. C from Cooper, G. A.; Grant, R. E. *Smithsonian Contributions to Paleobiology* 21: 2459; 1976.)

punctate, secondary layer fibrous. (Silurian to Early Jurassic; 460 genera.)

SUBORDER ATHYRIDIDINA (Figure 16.63)

Biconvex, astrophic spiriferides. Valves ornamented

Figure 16.63. Representative Athyrididina. **A,** *Meristella.* Ventral, dorsal, and anterior views of complete shell. (Devonian, Oklahoma, ×1.) **B,** *Composita.* Ventral and dorsal vews of complete shell and view of spiralia through broken dorsal valve. (Permian, Texas, ×1.) (**A** from Amsden, T. W. *Bulletin Oklahoma Geological Survey* 78: pl. 10; 1958. **B,** from Cooper, G. A.; Grant, R. E. *Smithsonian Contributions to Paleobiology* 21: pl. 661; 1976.)

only by growth lines in most species, but some with large growth lamellae. Ventral interareas typically reduced. Jugum commonly elaborately developed, prongs or accessory lamellae from jugum may be expanded into secondary spires coiled coaxially with main spiralia. Spiralia directed laterally, well developed. Shell fibrous, impunctate. (Late Ordovician to Jurassic; 100 genera.)

SUBORDER RETZIIDINA (Figure 16.64)

Astrophic, ribbed, biconvex spiriferides externally resembling rhynchonellides. Primary whorl of spiralium near midline of shell, coiled approximately parallel with plane of symmetry; apices of spiralia laterally directed; jugum simple. Mostly punctate, some impunctate; secondary shell layer fibrous. (Late Silurian to Triassic; 20 genera.)

ORDER TEREBRATULIDA (Figures 16.1, *H* to *I*; 16.2; 16.9, *A*; 16.65 to 16.69)

Articulates having biconvex shells, typically astrophic, rarely strophic. Interareas confined to ventral valve, pedicle opening restricted by deltidial plates. Loop supporting part or all of lophophore. Shell punctate, secondary layer fibrous. (Early Devonian to Holocene; 540 genera.)

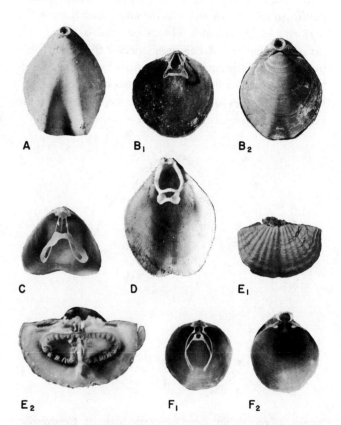

Figure 16.65. Representative Terebratulida. **A,** *Goniothyris.* Dorsal view of complete shell (Jurassic, England, ×1.) **B,** *Terebratula.* Dorsal interior view and dorsal view of complete shell. (Tertiary, Italy, ×0.6.) **C,** *Dallina.* Interior view of dorsal valve. (Holocene, Florida, ×0.6.) **D,** *Terebratulina.* Interior of dorsal valve. (Holocene, Japan, ×0.6.) **E,** *Argyrotheca.* Dorsal valve exterior and interior views showing dried schizolophe. (Holocene, Caribbean, ×2.5, ×3.) **F,** *Ecnomiosa.* Interiors of dorsal and ventral valves. (Holocene, Caribbean, ×0.6.) (**B** to **F,** photos by G. A. Cooper.)

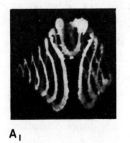  

Figure 16.64. Representative Retziidina; *Hustedia* with the spiralium free from the valve. Ventral, dorsal, and posterior views of complete shell. (Permian, Texas, ×4, ×1, ×1, ×1.) (From Cooper, G. A.; Grant, R. E. *Smithsonian Contributions to Paleobiology* 24: 3063; 1975.)

The terebratulides were the last order of articulate brachiopods to appear. The order is still extant and the majority of living brachiopods belong to it (for examples, Figure 16.65, *C* to *F*). The group is considered to have been derived neotenously from the Retiziidina. The early stages of loop development of Devonian terebratulides are similar to the early ontogeny of spiriferides, but the development of spirally coiled projections is suppressed.

Three major loop types are recognizable in the order. The ancestral type is spear shaped, termed **centronelliform** (Figure 16.66). It is the adult loop form in the majority of Devonian terebratulides and is found as a juvenile stage in all Paleozoic species that have been investigated. Subsequently both short-looped and long-looped species evolved from the ancestral stock. Evolutionary changes in the ontogenetic steps leading to the adult loop took place in both these groups.

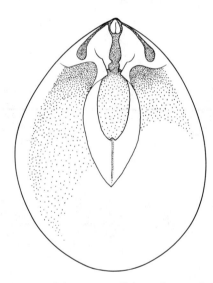

Figure 16.66. Adult centronelliform loop of *Centronella.* (Devonian, ×2.) (From Stehli, F. G. In: Moore, R. C., editor. *Treatise on invertebrate paleontology, Part H.* New York and Lawrence, KS: Geological Society of America and University of Kansas Press; 1965, p. 741.)

The early ontogenetic loop stage of Paleozoic short-looped terebratulides was centronelliform. Later growth of the loop involved extensive resorption to produce the adult short loop (Figure 16.67). The early spear-shaped loop stage was suppressed in ontogeny by Mesozoic time, and the juvenile loop was merely a small version of the adult loop. Judging from living short-looped species, this type of loop probably supported a plectolophe, but rigid support was provided only to the posterior segment of the organ.

Development of the adult loop in long-looped species typically involved a more complex sequence of events.

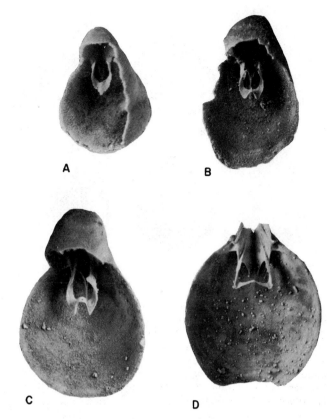

Figure 16.67. Ontogenetic development of a Paleozoic short-looped terebratulide, *Dielasma.* (Permian, Texas.) **A,** Juvenile stage with centronelliform loop. (×6.) **B** and **C,** Successive growth stages showing modification of loop by development of median fold and resorption. (×4.) **D,** Adult short loop. (×2.) (From Cooper, G. A.; Grant, R. E. *Smithsonian Contributions to Paleobiology* 24: 3107; 1976.)

The first stage in Paleozoic forms was the appearance of a centronelliform loop (Figure 16.68, *A*). A hood subsequently developed on the tip of this loop. Later growth and resorption grossly modified the shape of the structure. The dorsal descending branches of the adult loop were formed from the centronelliform loop, and the more ventrally located ascending branches and transverse bar were fashioned from the hood. (The terms descending and ascending apply to an internal view, as in a photograph or with the valve lying on a table.)

The hood of most Mesozoic and all Cenozoic long-looped terebratulides is carried on a separate median septum, and the posterior part of the descending branches are the only part of the loop originating from the cardinalia (Figure 16.69). These posterior parts of the descending branches grow forward and fuse with the anterior segments of the descending branches, which are developed from the flanks of the median septum. The ascending branches and transverse bar are fashioned from the hood as in the Paleozoic long-looped species.

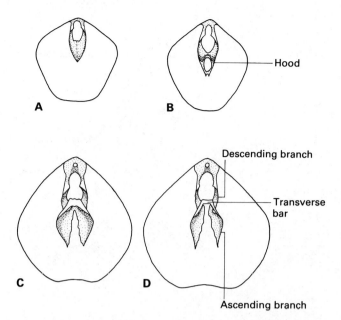

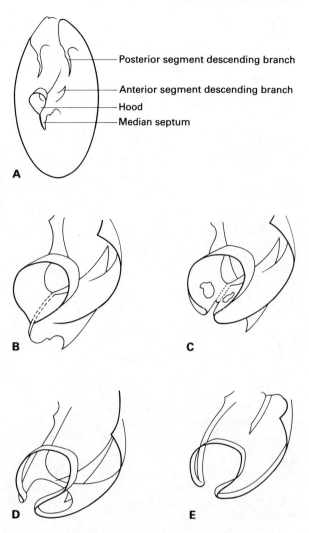

Figure 16.68. Ontogenetic development of a Paleozoic long-looped terebratulide, *Cryptacanthia.* (Pennsylvanian.) **A,** Development of centronelliform loop. (×10.) **B,** Hood present on tip of loop. (×8.) **C,** Enlargement of hood and formation of ascending branches and transverse bar. (×8.) **D,** Adult loop. (×5.) (Modified from Cooper, G. A. *Smithsonian Miscellaneous Collections* 134: 6; 1957.)

The resulting long loop differs little from that of Paleozoic forms and gives no hint that its origin was markedly different. The adaptive significance of this mode of loop development may be that it allows a complete lophophore support to be assembled more rapidly than one developing entirely from the cardinalia. In both cases, the long loops are thought to have offered rather complete support for a plectolophe.

Biology

Reproduction and ontogeny The sexes of most articulate brachiopods are thought to be separate. Only three living species are known to be hermaphroditic, but since the condition requires detailed examination to detect it, hermaphroditism may be more widespread in the class than is currently recognized.

The majority of articulate brachiopod species reproduce externally; eggs and sperm are shed directly into the sea. A few living species fertilize the eggs in the mantle cavity of the female, which then broods the young during early larval stages. The early development of the fertilized eggs is thus sheltered either in the mantle cavity or in specialized brood pouches.

The larval life of an articulate brachiopod is relatively brief, usually only a few hours and rarely as long as a

Figure 16.69. Ontogenetic development of a post-Paleozoic long-looped terebratulide. Compare with Figure 16.68. **A,** Hood supported by median septum. Descending branches grow from both cardinalia and septum. **B** to **D,** Progressive stages in the development of the adult loop by resorption of part of the hood and connections to median septum. **E,** Adult loop. (From Richardson, J. R. *Palaeontology* 18: 296; 1975.)

week. Unlike the inarticulates, the larvae are entirely without shells during all their planktic existence, but there are other more fundamental differences between representatives of the two classes. Early in its larval existence, an articulate develops a distinct polarity. The anterior end is densely ciliated, and the narrower posterior end of the pyriform larva loses its cilia (Figure 16.70, *A*). Subsequently, the mantle rudiment develops at the constriction between the anterior and posterior segments of the larva and grows posteriorly as a ring or skirt to cover the posterior rudiment (Figure 16.70, *B* and *C*). The posterior segment is the pedicle rudiment, a

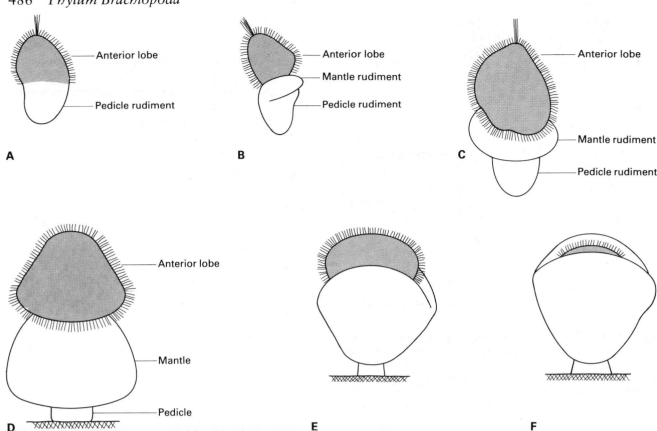

Figure 16.70. Larval development and early growth of the terebratulide *Waltonia*. **A,** Young larva after differentiation of the anterior lobe and pedicle rudiment. **B,** Subsequent development with early stage of the mantle rudiment. **C,** Growth of the mantle to cover the pedicle rudiment. **D,** Immediately after attachment, the mantle covers the pedicle. **E** to **F,** Successive stages in mantle reversal, with the mantle finally covering the anterior lobe. (Modified from Percival, E. *Transactions Royal Society New Zealand* 74: pl. 2, 3; 1944.)

primary feature of the larva. The anterior lobe subsequently gives rise to the lophophore and remaining organs of the animal, but only after undergoing a metamorphosis following settlement. No organs except the coelomic sacs are present in the larva.

Once settlement occurs, the organism becomes attached by the tip of the pedicle rudiment, and the mantles, which initially almost cover the pedicle, are turned inside-out during mantle reversal. The process is analogous to turning the finger of a glove inside-out. The mantles come to occupy a position such that they largely cover the anterior lobe (Figure 16.70, *D* to *F*). At this stage, the initial protegular shell is secreted. Meanwhile, extensive changes occur in the anterior lobe, which ultimately is modified to form the lophophore. The lophophore initially has short stubby filaments surrounding a subcentral mouth (Figure 16.71, *A*). Later growth involves further development of adult organs and the addition of lophophore filaments about the two antero-dorsally located generative zones of the organ.

Experiments and observations indicate that larvae can be site selective. Some species show marked preference for cryptic niches or will settle in a depression rather than a prominence on the sea floor. Other species are more eurytopic and will utilize the available habitat. Studies of living brachiopods off the coast of New Zealand have shown that some species are successful colonizers either attached to rock outcrops or lying free on a mud substrate. Furthermore, populations from these two habitats cannot always be distinguished morphologically.

Functional morphology and paleoecology of the phylum

A basic assumption in the functional and anatomic interpretation of brachiopods is that every feature has at least a potential meaning. All aspects of shell shape—rotundity or width, smoothness or ribs, spines and pits, internal plates and markings on the internal surfaces of valves, even accidental or pathologic conditions—can

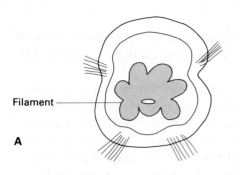

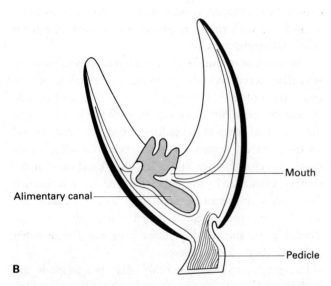

Figure 16.71. Early stages of the growth of the rhynchonellide *Notosaria.* **A,** View of lophophore shortly after mantle reversal with short stubby filaments. **B,** Diagrammatic sagittal section showing the early alimentary canal and incipient pedicle capsule shortly after mantle reversal. (From Williams, A.; Rowell, A. J. In: Moore, R. C., editor. *Treatise on invertebrate paleontology, Part H.* New York and Lawrence, KS: Geological Society of America and University of Kansas Press; 1965, p. 54.)

offer clues to the animal's habits and habitats. It may not be possible to deduce the meaning of each feature, since some may be nonadaptive, but nothing should be rejected a priori as meaningless.

Punctation

The significance of punctation was elusive for nearly a century. Various functions had been proposed, but only recently was it shown in living brachiopods that caeca within the punctae functioned for the storage of various proteins and lipids, and that their presence reduced predation by boring organisms. Paradoxically, punctation apparently afforded no protective advantage in Paleozoic shells where punctate forms seem to be as frequently bored as impunctate ones. Regardless of

function, punctation had or was correlated with a feature that had survival value because punctate stocks normally outlive homeomorphic impunctate stocks. The only major exception to this is three genera of punctate rhynchonellides that flourished for short intervals in the late Paleozoic and then went extinct. The main line of impunctate rhynchonellides survives to the present day.

Boring and encrustation of shells

Although predation by boring organisms and infestation of the shell by settling epibionts may be deleterious to the brachiopod, these misfortunes can be a boon to the paleoecologist. Some brachiopods are victimized while they are alive and others after death as their shells lie on the sea floor. This distinction becomes important because the position and orientation of borings or epibionts can be used to determine life positions and burial habits of fossil brachiopods. The normal reaction of the living brachiopod to relatively slow boring from the outside is the secretion of additional shell on the inside to prevent full penetration. Fossil shells are known in which barnacles have burrowed from without, and the inside of the shell is a mass of blisters where the brachiopod attempted to repair the damage. Gastropods, on the other hand, apparently drilled rapidly enough to prevent defensive shell secretion by the brachiopod, since normally the shell

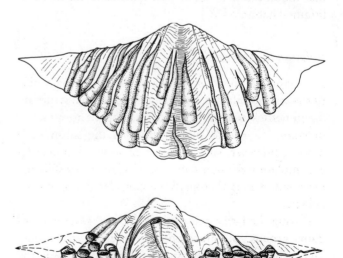

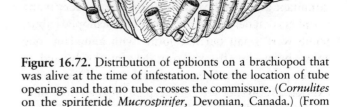

Figure 16.72. Distribution of epibionts on a brachiopod that was alive at the time of infestation. Note the location of tube openings and that no tube crosses the commissure. (*Cornulites* on the spiriferide *Mucrospirifer,* Devonian, Canada.) (From Schumann, D. *Palaeogeography, palaeoclimatology, palaeoecology.* Amsterdam: Elsevier Publishing Co.; 1967, p. 387.)

has just one hole and no internal blister. If both sides of the shell are riddled and no blisters are formed, the organism obviously was dead when its shell was infested.

Many fossil brachiopods are encrusted by bryozoans, serpulids, algae, and other small brachiopods that are either juveniles or adults of small species. If the encrustations are on the inside of the shell, the brachiopod was dead, but if the encrustations are on the outside, it might have been alive or dead but still articulated. A useful way to determine that the brachiopod was alive is to examine the commissure. If the encrusting organisms terminate abruptly at the commissure and none crosses it, the inference is that the brachiopod still was capable of gaping and hence was alive (Figure 16.72). Once it has been determined that the brachiopod was alive, close study of the epibionts can provide information about the living habits of the brachiopod. Corals, bryozoans, tube-secreting worms, and some brachiopods have preferred orientations with growth directions generally upward away from the substrate. If these attach to a living brachiopod in its life position, orientation of the epibionts is a clue to the orientation of the encrusted brachiopod. In addition, if the epibionts are confined to one valve it may be inferred that the other one was protected in some manner, normally by being buried. If a brachiopod is totally free of encrustation or boring, it may have lived buried or nearly buried, but lack of surface modification is not in itself sufficient evidence of an infaunal habitat.

Relation to substrate

As mentioned previously, brachiopods are normally part of the sessile epifauna, but continuing research is turning up numerous exceptions. Most brachiopods are attached in some manner to the substrate. If the substrate is mobile, however, such as a mat of seaweed, a floating accumulation of tar, a slab of pumice, or even a ship, the brachiopod might properly be considered vagrant, or pelagic.

During the latter half of the Paleozoic, several abundant groups of brachiopods contained free-living, unattached species. The spinose Productidina was the dominant group in the late Paleozoic and included many members that defy the general view of brachiopod habits. Some were small burrlike forms with abundant long spines protruding from all over the ventral valve and, in some species, even from the dorsal valve. These spines exerted a snowshoe effect in keeping the shell from sinking in the mud or sand. Judging from accumulations of silicified shells of this type in the Permian, some of these

productines were wafted along the seafloor like miniature tumbleweeds and thus cannot be considered fully sessile.

Some of the chonetidines may have had an unusual mode of life. They apparently preferred a near-shore environment, and their small flat shells, with long spines projecting from the ventral hinge region, could have been moved back and forth and repeatedly flipped over on their spines by the tides. In this respect they may have existed rather like some small living clams, perhaps going in and out with the tide, or subtidally, back and forth with oscillating currents.

The most surprising example of brachiopod mobility was discovered recently. A living terebratulide off the coast of southern Australia has a pedicle and muscle system designed not to attach the shell but to allow it to move up and down in the sediment and to push it around on the seafloor. Presence of a foramen in fossil brachiopods, therefore, is not an invariable indication that the shell was attached by the pedicle. Indeed, it is now known that many individuals of living species, which are attached by their pedicle when hard substrates are available, are able to adopt a free-lying mode of life when only mud bottoms are present.

Exceptions to the generality that brachiopods are largely epifaunal are common in the Paleozoic. One productidine genus (Figure 16.73) has been described as having spent the later part of its life almost completely buried. Even the dishlike dorsal valve was covered by sediment that acted as camouflage. When the valves gaped, only the narrow slit of the commissure opening would show, like the lair of a trapdoor spider. Undoubtedly many of the abundant genera of free-living productidines became partly buried during their lifetime and thus occupied a semi-infaunal habitat.

A completely infaunal mode of life is known only in the articulate lingulides. Fossil lingulides are occasionally found in the rock with an orientation like that of lingulides currently living: vertical with the anterior uppermost and the pedicle opening pointing downward (Figure 16.3). Scientists formerly thought that the animals burrowed into sediment using their pedicle and that the pedicle preceded the shell into the burrow. Observations in aquaria have shown that the pedicle is used only in the initial stages of burrowing as a strut to prop the animal at a high angle to the surface of the sediment. The burrow is then excavated by the shell, and rotary, sliding, and gaping movements of the two valves are employed to push the sediment aside, with active setae at the commissure pushing the grains up out of the burrow. Surprisingly, the lingulides burrow "head first," and the shell

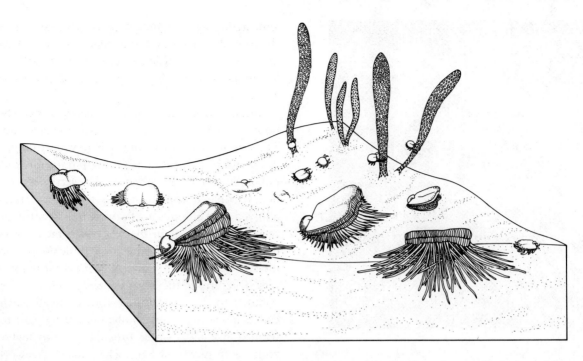

Figure 16.73. Reconstruction of habitat of the Permian productidine *Waagenoconcha*. After an initial period spent attached by clasping spines, individuals became free lying, and mature adults lived nearly completely buried by soft mud. (From Grant, R. E. *Journal of Paleontology* 40: 1068; 1966.)

precedes the pedicle into the burrow. To attain its final position, in which the shell is above the pedicle (Figure 16.3), the animal makes a U-shaped burrow and ends in a vertical position with the pedicle deepest and the anterior margin of the shell immediately beneath the sediment-water interface. Contraction of the pedicle subsequently may draw the shell deeper into the burrow as an escape mechanism. If the lingulide becomes fully exposed by current scour or some other action, it reestablishes itself by making a new U-shaped burrow. Many of these burrows have been found as trace fossils but only recently interpreted as the burrows of lingulides.

The habit of cementing directly to the substrate by some part of the ventral valve was widespread among articulate brachiopods in the late Paleozoic, but it is now confined to the inarticulate craniidines and one small group of articulates (the Thecideidina). Some productidines cemented themselves and then reinforced the attachment by great entanglements of rootlike spines (Figure 16.55, *H*).

The pedicle is capable of considerable modification as an instrument of attachment. A deepwater form was dredged from a depth of nearly 3000 m where the substrate was *Globigerina* ooze. The pedicle of this small shell is more than twice as long as the shell itself. The pedicle is abundantly branched at the end and along the main stalk, and the branches entangle the foraminifera both by wrapping around them and penetrating through them. One large group among the Strophomenida were long believed to be cemented by the ventral beak. However, strong evidence now exists that they were attached by a remarkable modification of the pedicle, whereby it was separated into its component linear fibers within the shell and protruded through numerous fine holes (Figure 16.74). The pedicle functioned much as the byssus of certain pelecypods.

Many brachiopod species occur in clusters. One is attached to the shell of another by means of a pedicle whose ends are splayed into numerous fibers, each capable of secreting a chemical that can dissolve calcium carbonate and etch its way into the surface of a neighboring shell. The patch of tiny holes thus produced is so distinctive that it has been recognized in terebratulides as old as the Jurassic, and it has been given the trace fossil name of *Podichnus* (Figure 16.75). The same ability to penetrate calcium carbonate allows the pedicle to attach itself to biogenic particles in the substrate, including foraminifera and larger particles.

Patchiness of distribution

The clusters produced by pedicle attachment and cementation point up another salient factor in brachiopod ecology, that is, patchiness of distribution. This tendency

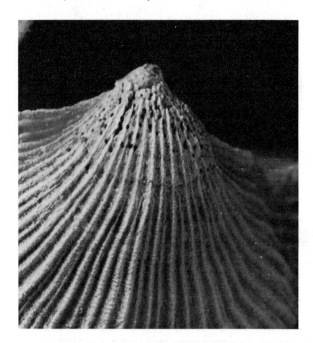

Figure 16.74. Openings of byssus-like pedicle of the strophomenide *Kiangsiella*. (Permian, Pakistan, ×10.)

Figure 16.75. Trace fossil *Podichnus* produced by pedicle attachment of another brachiopod to the shell of the terebratulide *Plectothyris*. (Jurassic, England, ×4.) (Photo by G. A. Cooper.)

can be observed in living populations as well as in the fossil record. Modern studies of deep-water brachiopods that require dredging report that many hauls will contain no brachiopods, but that some dredge hauls produce almost nothing but brachiopods. The same is true of shallow and tropical forms that can be collected by diving. Some promising recesses in reefs will have no brachiopods, but others contain hundreds per square meter.

Among fossils the tendency to patchiness can be both boon and bane for the collector and the paleontologist. The search can be frustrating or highly rewarding; patchy distribution can cause misleading pitfalls for studies of populations and of correlation. Nevertheless, the tendency to patchiness and the preference for attachment allowed brachiopods to be important contributors to reefy environments in the past when their diversity of form was much greater than now. Brachiopods occur in modern reefs but are not significant contributors to the reef structure. In the late Paleozoic, however, they contributed to the stable framework of small patch reefs called bioherms. These bioherms are thought to have formed in shallow water, probably in fairly high-energy zones with plenty of light. The major framework normally consists of organisms other than brachiopods, such as crinoids, algae, bryozoans or corals, but recent studies show that the pedicles of brachiopods are surprisingly strong, even when seemingly thin, so they can withstand normal ocean currents and waves.

Avoidance of light

Perhaps the safest generalization that can be applied to the habitat of living brachiopods is that they avoid light. Immediately prior to settlement, their larvae are photonegative. The majority of species avoid light by living at depths where the light intensity is low. In shallower tropical waters many species inhabit reefs but prefer shady places on the undersides of coral fronds or in caves. In cold boreal or austral waters without reefs, most shallow-water species prefer overhangs or recesses in ledges. Those attached to boulders and pebbles attach preferentially to the undersurfaces. A few species, however, do live in tidal pools and receive an abundance of light.

Sensitivity of mantle tissue

The functional significance of some shell features is fairly obvious, either because of attributes of the feature itself, like the rootlike spines of some productidines, or by analogy with comparable features in living forms whose function can be observed directly. For example, the mantle margin of living brachiopods is known to be light sensitive, although the animals do not have eye spots as

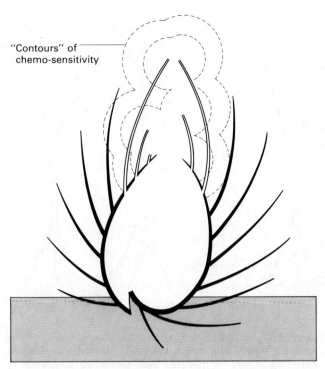

"Contours" of chemo-sensitivity

Figure 16.76. Diagrammatic interpretation of the long hollow spines of the rhynchonellide *Acanthothiris* as an early warning system for the animal. (From Rudwick, M. J. S. *Palaeontology* 8: 612; 1965.)

do some bivalve mollusks. A reasonable inference is that this sensivity was also present in fossil brachiopods.

Mantle sensitivity is not restricted to light. The animals also respond to noxious material in the water and to touch. The valves close abruptly when either reflex is activated. This property of the mantle margin affords an explanation of the functional significance of the long, open-ended spines found in some productidines and a few rhynchonellides. These spines extend beyond the commissure of the animal (Figure 16.76) and in life carried a ring of the mantle margin that was responsible for spine secretion. The mantle tissue at the tip of the spine was presumably sensitive like the remainder of the mantle margin, and its position enabled it to function as an early-warning system to the animal (Figure 16.76).

Feeding in conical productidines

It is diffficult to find a universally acceptable explanation of function for some structures, such as the dorsal valve of the conical productidines (the richthofeniaceans) (Figure 16.55, *H*). This valve is seated within a deeply conical ventral valve, and two contrasting functions have been proposed for it and the related feeding mechanisms of these brachiopods. All living brachiopods feed in

basically the same manner; the mechanism has been described previously (p. 454). For the system to function efficiently, it is essential that the incoming, food-laden water be kept separate from the outgoing water that has been filtered. This is achieved by the filaments that are disposed to produce ordered inhalant and exhalant currents. The only connection between the currents is through the filtering screen of filaments (Figure 16.77).

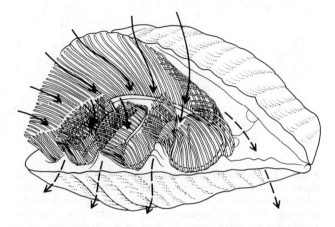

Figure 16.77. Filter feeding in the living terebratulide *Megathiris*. The inhalant water, represented by solid arrows, is separated from the exhalant water by the wall of filaments. (From Atkins, D. *Journal of the Marine Biological Association, United Kingdom* 39: 469; 1960.)

Overwhelming evidence exists that some of the Productidina possessed lobate ptycholophes, and at least a reasonable inference is that the lophophore of the conical forms was of the same type. If the group fed by ciliary pumping, like all living species, then the morphology may be interpreted functionally as in Figure 16.78, *A*. Inhalant water would have been drawn in laterally, and exhalant water would have escaped medianly. The dorsal valve would then have functioned merely to control communication between the mantle cavity and the sea by opening and closing in the manner normal for brachiopods.

An alternative and rather radical interpretation of the dorsal valve and feeding mechanism of the conical productidines is based on a method of functional analysis that has come to be known as the paradigmatic method. In the application of this method a function is postulated for a structure. Then, given the limitations of the material available to the organism, the form of the structure that would perform the postulated function with optimal efficiency is deduced. The specifications of this structure constitute a paradigm for the function. A close similarity between the paradigmatic structure and the structure in

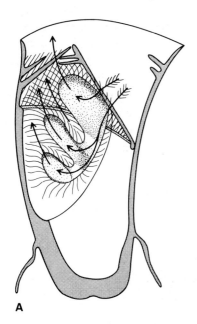
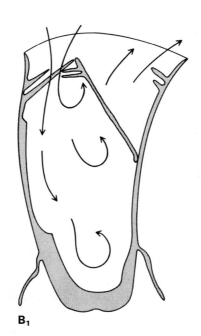
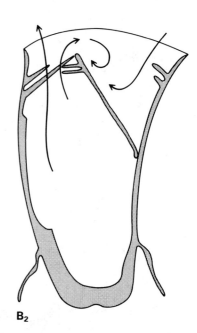

A   **B₁**   **B₂**

Figure 16.78. Alternative interpretations of feeding mechanisms in conical productidines. **A**, Conventional feeding mode using lophophore as a pump. Longitudinal section through a generalized form. Arrows depict current flow, inhalant feathered; exhalant plain. **B**, Feeding by flapping of dorsal valve. Longitudinal sections of shell. **B₁**, Dorsal valve opening drawing water into mantle cavity. **B₂** Dorsal valve closing expelling water. (**A** modified from Grant, R. E. *Journal of Paleontology* 46: 237; 1972. **B** from Rudwick, M. J. S.; Cowen, R. *Bolletino Società Paleontologica Italiana* 6: 150; 1968.)

the fossil organism suggests that the structure would have been capable of fulfilling the postulated function. The morphology of the conical productidines was considered to differ sufficiently from that of normal brachiopods, indicating that they may have used an entirely different method of feeding. The postulated function of the dorsal valve was that it flapped open and shut rhythmically. This movement of the valve would have induced tidelike currents of water in and out of the mantle cavity (Figure 16.78, *B*). Opening the valve would cause water to be sucked into the cavity; closing the valve would eject the water. In this model the role of the lophophore would be reduced to trapping particulate material brought in by the induced currents. The paradigm of this function bears close similarity to the dorsal valve of a conical productidine without a lophophore. In this particular study, a working model of an empty brachiopod was constructed, and rhythmic movement of the model dorsal valve did induce the postulated currents. Thus the study demonstrated that the dorsal valve was physically capable of having this function, and that the group may have had an unusual method of feeding if the lophophore was absent or greatly reduced, an unlikely possibility in a brachiopod.

In choosing between the two alternatives, one has to recognize that it is not possible to prove logically that either one was the feeding mechanism employed by these extinct forms. The paradigmatic method has been criticized for being too restrictive and essentially an engineering analysis of mechanical function and material strength. However, when other factors such as phylogenetic origin, metabolic efficiency, anatomy of missing soft parts that may be reconstructed such as a lophophore, and hydrodynamics of seawater are added to the purely mechanical and structural considerations, the paradigmatic method can be highly beneficial in providing insight into living habits of organisms that are too different from living ones to be analyzed by analogy alone. The application of the method is, of course, not restricted to brachiopods, although it has been used on several occasions in studies of the phylum.

Phylogeny and synoptic history of phylum Brachiopoda

Figure 16.79 summarizes the major features of the phylogenetic history of the brachiopods. As with any phylogeny, it is necessarily speculative and subject to modification as additional information becomes available. The data on which Figure 16.79 is based have been discussed throughout the chapter, but the figure illustrates several features that merit emphasis.

It has previously been argued that in spite of the conspicuous differences between the two classes, the

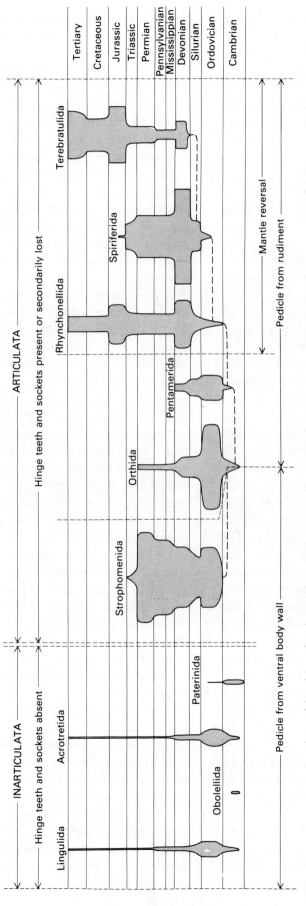

Figure 16.79. Principal orders of brachiopods: their stratigraphic ranges and major anatomic or developmental features. Column width is roughly proportional to generic diversity.

phylum is a phylogenetic entity and is holophyletic. Some of the differences between living inarticulates and articulates result from divergent evolutionary histories. Cambrian and Early Ordovician members of the two classes probably resembled each other more closely than do their living descendants. Evidence suggests that many of the distinguishing features between them resulted from rapid evolutionary change in the Articulata during their Ordovician radiation.

The principal features differentiating Cambrian inarticulates and articulates initially were the characteristic articulation and musculature of the latter group together with a calcite shell having a fibrous secondary layer. Other features that permit discrimination of living members of the two classes were not developed. For example, the lophophore was unsupported by projections from the cardinalia in both groups. Moreover, if interpretation of the relationship between tissue and shell is correct (p. 455), those articulates that developed pseudodeltidia also had a pedicle like that of the inarticulates, being an evagination of the ventral body wall. Such morphology was characteristic of the billingsellacean orthides and strophomenides (Figure 16.79).

Many early articulates lack pseudodeltidia, however, and have unmodified delthyria and notothyria. This condition is characteristic of the young stages of living articulates and is invariably associated with a pedicle that arose as a primary larval segment. Thus such early articulates may have a pedicle with a comparable origin. Their pedicle, however, could not have been identical with that of living rhynchonellides or terebratulides. The pedicle must have been more superficial, its base not depressed to form a capsule between the valves, because, in such forms, the ventral musculature extended to the beak and preoccupied the needed space (Figures 16.8 and 16.54, C). Such space first became available with the evolution of the Rhynchonellida, whose shell musculature did not extend to the ventral beak. In living rhynchonellides the pedicle capsule is seemingly an expression of incomplete recovery from mantle reversal and is detectable immediately after settling (Figure 16.71). Thus the presence of a pedicle capsule may reflect the development of mantle reversal at the close of larval development. If this supposition is correct, mantle reversal did not occur among any of the orthides or pentamerides. The larvae of these two orders must have had mantle rudiments that grew forward to cover the anterior lobe as they do today among living inarticulates. We can only guess at the selective advantages of mantle reversal. The advantages would seem to benefit primarily the larvae. Modern articulate larvae swim by ciliary beat

on the anterior lobe, and presumably this would be more effective if the lobe were not partly ensheathed by the mantle rudiment.

Many of the features useful in discriminating between living articulates and inarticulates apparently arose in several stages (Figure 16.79). It is certainly anatomically impossible for the orthides to have had pedicle morphology identical with that of modern articulates; there was simply no room for a pedicle capsule.

The Orthida occupy the central position in the phylogeny of the Articulata (Figure 16.79). Two major lineages are recognizable, and both of them originated from the orthides. The relative successes and failures of these two lineages, as reflected in the relative dominance of different groups of brachiopods, are biostratigraphically useful.

Given a rudimentary knowledge of the brachiopods, an alert observer can identify relative ages of strata, at least to the level of period, especially in the Paleozoic. Mesozoic rocks present more difficulty, both because brachiopods are relatively rare and because they are externally less variable, requiring a more specialized knowledge of their morphology and distribution.

The Cambrian Period was dominated by fairly simple orthides (Figure 16.54, A and B) and tiny acrotretide inarticulates (Figures 16.30, A and B and 16.36, A). The latter are difficult to see in the rock, but if a moderate size sample of Cambrian limestone is placed in dilute acetic acid, the limestone will be etched away, leaving a residue of phosphatic brachiopods, mostly acrotretidines and lingulides.

The Ordovician brachiopod fauna is much more diverse, with more kinds of both inarticulates (Figures 16.10; 16.31; and 16.36, B and C) and articulates. The inarticulates are small as in the Cambrian and can be obtained by acid treatment of limestone, but some larger lingulides can be seen easily on the outcrop. Among the articulates, the orthides (Figures 16.12 and 16.54, C to E) still predominate, but they are joined by numerous Strophomenida (Figure 16.55, A to C), mostly rather flat forms lacking spines. Rhynchonellides are abundant for the first time, but other groups that appear first in the Ordovician and come into prominence later are rare.

Silurian brachiopods are more difficult to characterize; there are fewer genera than in the Ordovician, partly because the Silurian was a shorter period. Large pentamerides (Figures 16.1, D and 16.56, C and D) are the most distinctive, but they occur in restricted facies. The impunctate orthides declined greatly in the Silurian, becoming extinct in the Early Devonian, but punctate orthides are abundant. In addition, primitive spiriferidines (Figure 16.62, *A*) and athyrididines come into some abundance, and the inarticulates are rare constituents of the fauna.

The Devonian was the period of greatest expansion of the brachiopods. The inarticulates are generally too rare to be helpful in quick recognition of rocks of this system; when found, they are not conspicuously different from those of the remainder of the Phanerozoic. The Spiriferida come into prominence for the first time (Figures 16.1, *E* and 16.9, *C*), especially as represented by the Atrypidina (Figure 16.61, *B* and *C*), which are so distinctive and abundant in the Lower and Middle Devonian that they may be said to typify the Devonian brachiopod fauna. Strophomenide brachiopods (Figure 16.55, *D*) remain abundant but fairly simple in form, with only a hint of the variety and spinosity that they developed later. Rhynchonellides (Figure 16.59, *B*) are also abundant and terebratulides (Figure 16.66) are numerically significant for the first time in their geologic history.

The Mississippian and Pennsylvanian witnessed another decline in the diversity of brachiopods, but the costate Spiriferidina with their laterally directed spires became abundant and varied. The spinose Strophomenida, belonging to the Productidina (Figure 16.1, *F*), began to occupy the dominant position they would hold for the remainder of the Paleozoic. Some of the largest brachiopods known are found in the Mississippian. The largest Chonetidina occur in the Mississippian and terebratulides attained a size unequalled until the present time. In North America, the Pennsylvanian has a rather restricted suite of brachiopods dominated by productidines (Figure 16.55, *E*), chonetidines, and spiriferides.

The Permian was the period of greatest morphologic and ecologic experimentation by brachiopods since their original diffusion in the early Paleozoic. The Strophomenida increased their predominance in the fauna through great diversity in the productidines (Figures 16.11 and 16.55, *H* and *J*), chonetidines (Figure 16.55, *I*) and oldhaminidines (Figure 16.55, *K*). A suite of brachiopods containing Productidina with a large range in size and arrangement of their characteristic spines, together with cone-shaped richthofenaceans and deeply lobate oldhaminidines, can safely be identified as Permian in age.

For all practical purposes the Strophomenida can be considered to be absent from Triassic strata. The only numerically significant groups of brachiopods are the Spiriferida, Rhynchonellida, and Terebratulida. In general, brachiopods are rare in Triassic rocks, although they seem to be fairly abundant in certain parts of southern Europe and the Soviet Union. This partly reflects a lack of comprehensive studies of Triassic brachiopods but also is a real consequence of the scarcity of Triassic marine sediments.

By Jurassic times, even the Spiriferida had declined to insignificance, leaving only the rhynchonellides and terebratulides. From then on the terebratulides became dominant, although Jurassic faunas are locally rich in a great variety of both rhynchonellides (Figure 16.59, *E*) and terebratulides (Figures 16.1, *H* and 16.9, *A*), especially the short-looped kinds. It is impossible to evoke recognition of Jurassic brachiopods with words alone; one must study either pictures or actual specimens to be able to identify a Jurassic suite. The most that can be said is that the Jurassic is typified by numerous biplicate terebratulides (Figure 16.65, *A*) and large coarse-ribbed rhynchonellides (Figure 16.59, *E*).

Brachiopods are again comparatively rare in Cretaceous rocks, but the terebratulides have nearly replaced the rhynchonellides and are the predominant element. Again, one must see the specimens or photographs of them (Figure 16.1, *I*) to differentiate between a Cretaceous collection and one from the Jurassic or the Cenozoic.

By the beginning of the Cenozoic Era, the brachiopod fauna was essentially modern in aspect with mostly terebratulides (Figure 16.65, *B* to *F*), the controversial thecidedines, and a few rhynchonellides. From the low point at the end of the Cretaceous, the phylum has been recovering gradually and now numbers more than 300 species in about 100 genera, roughly comparable to an average "instant" in time in the Paleozoic. The phylum occupies a variety of habitats from the tropics to the cold waters of the Arctic and especially the Antarctic. Oceanographic research in recent years has revitalized study of living brachiopods by discovering many hitherto unknown forms in previously unexplored geographical areas, leading to the surprising conclusion that Brachiopoda, long neglected by zoologists as unimportant and in decline, are well and thriving in modern seas.

Supplementary reading

Cooper, G. A.; Grant R. E. Permian brachiopods from west Texas (Parts 1–6). *Smithsonian Contributions to Paleobiology* Nos. 14, 15, 19, 21, 24, 32; 1972–1977. Probably not a book that you will read from cover to cover, but an epic monograph describing the best-known and possibly the most spectacularly preserved brachiopod fauna yet recovered.

Hyman, L. H. *The invertebrates*, vol. 5. New York: McGraw-Hill Book Co.; 1959, pp. 516–609. General account of the biologic aspects of the phylum.

Rudwick, M. J. S. *Living and fossil brachiopods*. London: Hutchinson and Co. Ltd; 1970. General overview of the phylum written to emphasize functional morphology.

Williams, A. Scanning electron microscopy of the calcareous skeleton of fossil and living Brachiopoda. pp. 37–66. In: Heywood, V. H., editor. *Scanning electron microscopy. Systematic and evolutionary applications*. New York: Academic Press, Inc.; 1971.

Williams, A.; Hurst, J. M. Brachiopod evolution, pp. 79–121. In: Hallam, A., editor. *Patterns of evolution as illustrated by the fossil record*. Amsterdam: Elsevier Scientific Publishing Company; 1977. Comprehensive account of the evolution of the phylum, which is regarded as monophyletic.

Williams, A.; Rowell, A. J. Brachiopod anatomy, pp. 6–57. In: Moore, R. C., editor. *Treatise on invertebrate paleontology. Part H*. New York and Lawrence, KS: University of Kansas Press and Geological Society of America; 1965. The most recent summary of brachiopod softpart morphology in English.

Williams, A.; Rowell, A. J. Morphology p. 57–138. In: Moore, R. C., editor. *Treatise on invertebrate paleontology, Part H*. New York and Lawrence, KS: University of Kansas Press and Geological Society of America; 1965. Relatively comprehensive account of brachiopod morphology. The section on shell ultrastructure, which was written before scanning electron microscopes were widely available, is outdated.

Wright, A. D. Brachiopod radiation, pp. 235–252. In: House, M. R., editor. *The origin of major invertebrate groups*. New York: Academic Press, Inc.; 1979. Stimulating discussion of the lower Paleozoic radiation of the brachiopods, in which it is argued that the brachiopods are polyphyletic.

17

Phylum Bryozoa

Richard S. Boardman
Alan H. Cheetham

Part I Phylum overview

Bryozoa, sometimes called Ectoprocta, are the only phylum in which all the known species form **colonies** (Figure 17.1). A colony consists of minute members called **zooids**, which are physically connected, asexually produced, and therefore assumed to be genetically identical. Some or all of the zooids in a colony are organized as functional and morphologic units comparable to solitary animals in that they are capable of feeding and performing other vital functions such as digestion and sexual reproduction. Feeding zooids in a colony have tentacles and a complete U-shaped digestive tract (Figure 17.2). **Body walls** of colonies and their zooids enclose fluid-filled **body cavities**. The cellular layers of the body walls are derived from extensively reorganized triploblastic cellular layers of bryozoan larvae. Body cavities are regarded either as coeloms throughout or as combinations of coeloms and pseudocoels. The phylum is considered to be at the triploblastic coelomate grade (Chapter 2).

Bryozoa have been consistently distinguished from other invertebrates and plants only since the late 19th century. Bryozoan colonies can be similar in appearance to colonies of some hydrozoans or corals or even some plants such as mosses or calcareous algae. The distinguishing characteristics of the phylum are in the minute morphologic features of the zooids. The zooids of bryozoans are just large enough to be seen by the unaided eye; thus their morphology can be observed closely only with a microscope.

Colonies range in size from microscopic to more than 1 m. The smallest colonies consist of just a few zooids and the largest contain tens of millions of zooids. The shapes of colonies vary widely (Figure 17.1). Most colonies are attached to surfaces of hard material such as rock, shell, or sand grains. Others adhere to soft objects such as seaweed or freshwater plants.

Many colonies are low **encrusting** growths composed wholly or initially of a single layer of zooids in contact with a substrate (Figure 17.1, *H*). Colonies can become multilayered nodular masses by new encrusting layers budding from and growing over the original or by repeated vertical budding of single zooids. **Massive** growths are formed by larger nodular colonies.

Erect colonies arise from a small encrusting base or are attached by rootlets, and most are branching (Figure 17.1, *A*, *B*, and *D* to *G*). They can be heavily calcified and more or less rigid (Figure 17.1, *B*, *D*, and *F*) or flexible, either through a reduction or absence of calcification throughout whole colonies or through alternation of calcified segments and uncalcified joints (Figure 17.1, *A*). Branches of some erect colonies are connected regularly to form netlike expansions (Figure 17.1, *G*).

Free-living colonies (Figure 17.1, *C*) are unattached or only initially or intermittently attached and commonly have discoid or conical shapes with orifices opening on convex upper surfaces.

Measured by abundance and diversity, Bryozoa are a major phylum of invertebrates. They are restricted to aquatic environments and are widely distributed; the great majority of modern species are marine, occurring in shallow shelf seas to abyssal depths of the oceans and from polar regions to the tropics. Some live in estuarine

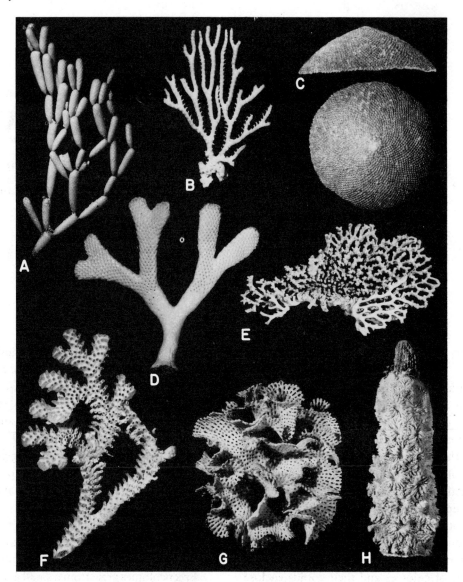

Figure 17.1. Colonies of Holocene marine Bryozoa. **A,** Erect with branches of alternating calcified segments and uncalcified joints. (*Microporina*, Bering Sea, ×1.) **B,** Erect and fanlike with narrow branches. (*Idmidronea*, Long Island Sound, ×2.5.) **C,** Free-living conical. (*Cupuladria*, Gulf of Mexico, ×2.5; side view above; top view below.) **D,** Erect with flat branches. (*Cystisella*, North Atlantic, ×2.5.) **E,** Erect with irregularly anastomosing branches. (*Frondipora*, Mediterranean, ×1.) **F,** Erect with wide branches. (*Diaperoecia*, Australia, ×2.5.) **G,** Erect and netlike with regularly connected branches. (*Hornera*, Australia, ×1.) **H,** Low, encrusting a twig. (*Lichenopora*, Gulf of California, ×2.5.)

environments of variable salinity, and a few live in freshwater lakes and streams from near sea level to high altitudes.

The fossil record of the phylum extends over the last 500 million years (Ordovician to Holocene). Throughout most of that time, marine Bryozoa have been numerous and widely distributed geographically, comprising large numbers of species with up to several thousand living at any one time. Bryozoa are the most abundant fossils in many marine limestones and calcareous shales and mudstones. Freshwater Bryozoa have virtually no known fossil record.

Introductory morphology

Morphology of zooids

Zooids within a colony can differ distinctly in morphology and functions even though a colony is genetically uniform. Some or all zooids in a colony are **feeding zooids** (Figure 17.2). **Polymorphs** are zooids that differ in morphology from the common feeding zooids. Many polymorphs have been observed to perform specialized functions such as reproduction, support, connection, cleaning, or defense instead of or in addition to feeding.

Some polymorphs consist of little more than body cavities enclosed by body walls.

In addition to body cavities and body walls, feeding zooids have a protrusible tentacle-bearing feeding organ called the **lophophore**, an alimentary canal, muscles, a nervous system, and **funicular strands** (tissues connecting digestive organs to body walls). Egg- and sperm-producing structures of simple form are present in some feeding or nonfeeding zooids in all colonies. Special organs for excretion, circulation, and respiration are apparently absent throughout the phylum, and these functions therefore are not understood.

The lophophore (Figure 17.2, *E*) consists of a ring of hollow **tentacles** and a supporting **tentacle sheath**. The tentacle sheath encloses the tentacles when they are retracted and turns inside out to support the tentacles when they are protruded for feeding (Figure 17.2, *C* and *D*). Tentacles are ciliated, and the cilia produce feeding currents directed toward the mouth, which is centered at the base of the ring of tentacles. The mouth opens into the U-shaped alimentary canal, which ends at an anus on the side of the protruded tentacle sheath below the ring of tentacles.

Protrusion of tentacles involves hydrostatic pressure on the fluid filling the body cavity. Pressure is produced in various ways in different groups of Bryozoa by muscles modifying the shapes of parts of the body cavity. Retraction is by direct and swift contraction of **retractor muscles** throughout the phylum.

Body walls of a feeding zooid (Figure 17.2, *A*) include an orificial and supporting walls. An **orificial wall** bears or defines an opening, termed the **orifice**, through which tentacles are protruded. Supporting walls include basal, vertical, and frontal walls. A **basal wall** floors a zooidal body cavity and generally parallels the orificial wall. **Vertical walls** are oriented entirely or in part at high angles to basal and orificial walls providing depth, length, or both to zooidal body cavities. **Frontal walls**, present in many but not all Bryozoa, provide front sides to zooids more extensive than orificial walls alone.

Extrazooidal parts

Colonies in some Bryozoa grow **extrazooidal parts** that are outside zooidal boundaries throughout life (Figure 17.3). These parts of colonies have body cavities and enclosing body walls, are usually larger than zooids, and can be colonywide in extent. Extrazooidal parts lack feeding organs, and their skeletons generally provide support to the colony.

Body walls of colonies

Body walls of colonies include walls of zooids and of any extrazooidal parts. They consist of one or two cellular layers adjacent to the body cavity and one or two noncellular layers on the opposite side of the body cavity (Figure 17.4, *A*). An outer layer of cells called the **epidermis** secretes the noncellular layers. An inner layer of cells called the **peritoneum** lines those body cavities regarded as coeloms and is absent on one side of those body cavities regarded as pseudocoels. Noncellular layers can consist of a membranous, organic outermost layer, called the **cuticle**, or a calcified skeletal layer, or both. Most colonies have all three possible noncellular arrangements in different body walls (Figure 17.4, *B* and *C*). All connected calcareous parts of a zooid form the zooidal skeleton (zooecium). The skeleton of the colony (zoarium) is formed by a combination of zooid skeletons and any extrazooidal skeleton. All skeletal layers of colonies are secreted on the side of the epidermis opposite the body cavity and therefore are considered exoskeletal.

In manner of growth and position in a colony, body walls are of two different kinds (Figure 17.4). **Exterior walls** are the outermost body walls of colonies that expand in growing regions to increase the volume of the body cavity. Exterior walls include cuticle as their outermost protective layer. **Interior walls** grow by infolding epidermal and peritoneal layers into the body cavity to partition the preexisting space established by the exterior walls. Interior walls are isolated from the environment and commonly lack protective cuticles.

Reproductive morphology

In marine Bryozoa, most colonies originate by sexual reproduction, starting with the production of gametes by parent colonies. In most species, fertilized eggs are retained through embryonic development to the larval stage by the parent colony in modified zooids or special brood chambers (Figure 17.5). **Brood chambers** are parts of body cavities or water-filled spaces partly enclosed by body walls of modified zooids or extrazooidal parts of colonies. Fertilized eggs are released directly into the water in a few species and undergo development away from the parent colony.

Larvae of marine Bryozoa are ciliated and generally capable of swimming. Most lack digestive tracts and have a comparatively short motile stage (lecithotrophic), often only a few hours. After settlement, a larva undergoes extensive reorganization of tissues to produce the first zooid, the **ancestrula**, of a colony (Figure 17.6).

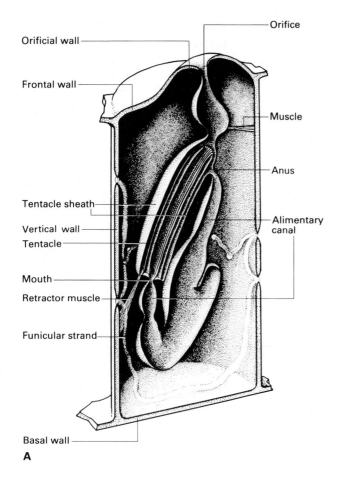

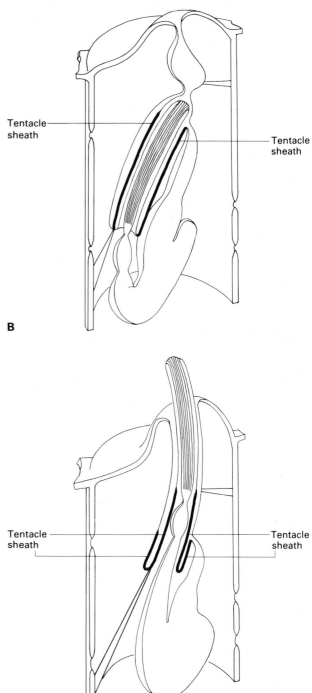

A

Orificial wall

Orifice

Frontal wall

Muscle

Anus

Tentacle sheath

Alimentary canal

Vertical wall

Tentacle

Mouth

Retractor muscle

Funicular strand

Basal wall

B

Tentacle sheath

Tentacle sheath

C

Tentacle sheath

Tentacle sheath

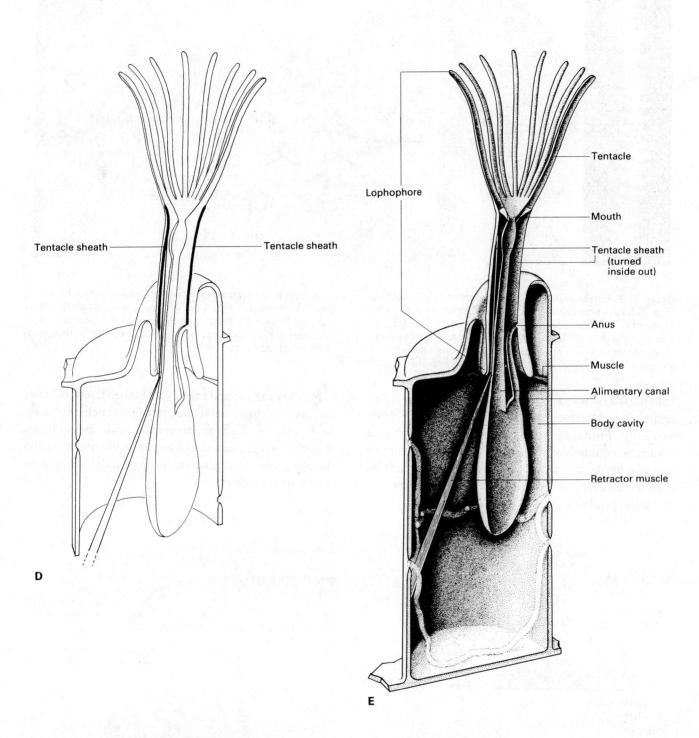

Figure 17.2. Sections through the body walls and organs of a generalized bryozoan feeding zooid. (Approximately ×50.) **A** and **B,** The tentacles are retracted within the protection of the body walls. **C** and **D,** Stages in protrusion of tentacles with the tentacle sheath turning inside out. **E,** The tentacles are protruded for feeding.

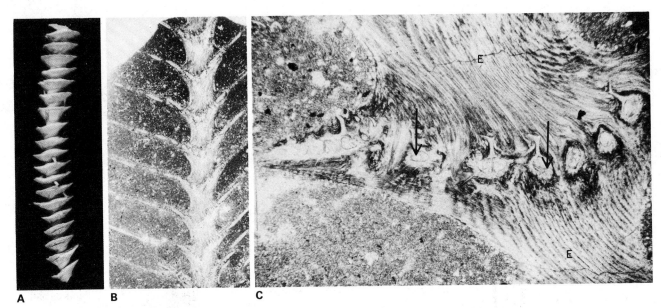

A   B   C

Figure 17.3. Extrazooidal skeletal structure of an erect colony of a tubular bryozoan (*Archimedes,* Upper Mississippian, Alabama). **A,** The spiral extrazooidal axial structure as commonly seen in the field with fragile outer netlike expansions of colony broken off. (×0.5.) **B,** A section through the center of the complete colony showing cut edges of outer netlike expan- sions consisting of feeding zooids embedded within the extra- zooidal structure. (×4.) **C,** The same section showing zooids (arrows) within the extrazooidal skeleton (*E*) where they were necessarily nonfunctional after growth of the extrazooidal skeleton of the colony. (×30.)

In most freshwater Bryozoa, few colonies originate by sexual reproduction. Most colonies originate from asexually produced, encapsulated, resistant resting bodies called **statoblasts** (Figure 17.7), which can survive unfavorable conditions such as drying and freezing from season to season. When favorable conditions return, a statoblast produces the first zooid of a new colony.

In many marine and freshwater Bryozoa, new colonies can also develop from fragments of preexisting colonies. Such colonies lack an ancestrula and, like colonies originating from statoblasts, are asexually produced and therefore can be assumed to be genetically identical to their parent colonies.

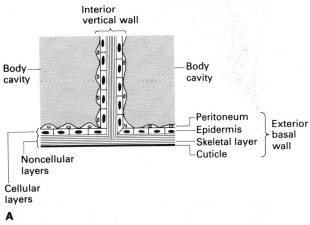

A

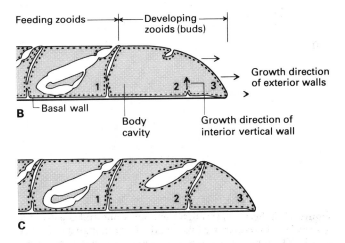

Figure 17.4. Sections through approximately box-shaped body walls that enclose the body cavities of feeding zooids of bryozoan colony. **A,** Arrangement of cellular (peritoneum, epidermis) and noncellular (skeleton of interior wall and skeleton and cuticle of exterior wall) layers of body walls. **B** and **C,** Successive growth stages in the development of the exterior and interior body walls (cellular layers shown by single dotted line; skeletal layer by unshaded areas; cuticle by outermost continuous line of exterior walls). The body cavities of zooids 2 and 3 are confluent in **B** and not in **C** after the interior vertical wall is completed. The exterior frontal walls lack skeletal layer so are membranous.

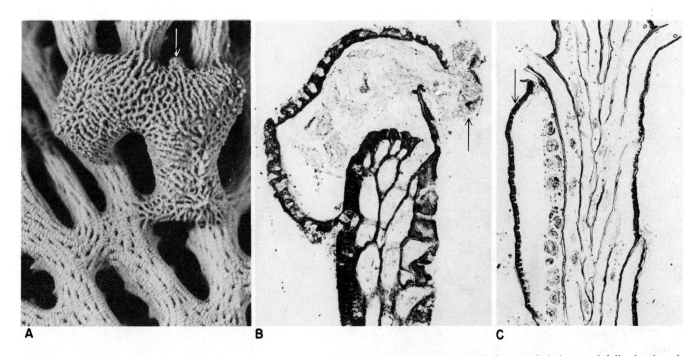

A   **B**   **C**

Figure 17.5. Brood chambers of Holocene tubular Bryozoa. **A,** A brood chamber on the back side of an erect colony of anastomosing branches; the aperture (arrow) through which larvae are discharged faces toward the front of the colony. (*Hornera*, Australia, ×30.) **B,** A section through the center of the branch and brood chamber of another hornerid from Australia, showing tubular zooids below and fully developed larvae with one larva (arrow) leaving the chamber through the aperture. (×45.) **C,** A section through the center of the branch of an erect colony of tubular zooids showing the brood chamber of a modified zooid (arrow) containing developing embryos. (*Mecynoecia*, France, ×45.)

Features of the phylum

Colony growth

All growth within sexually produced colonies is from the ancestrula, is physically connected, and involves asexual reproduction of zooids and any extrazooidal parts. Growth in asexually produced colonies is similar but can be traced back to the ancestrula of a parent colony.

Growth in marine Bryozoa begins with expansion of the exterior membranous body wall to provide additional colony volume. The expanding colony space is differentiated into developing zooids, called **buds,** by the formation of interior body walls partitioning the body cavity (Figures 17.4 and 17.8). Zooidal organs develop from body walls of buds. Buds become zooids when organs first become complete and functional.

Budding in marine Bryozoa results in the addition of zooids at the edges or ends or over the surfaces of colonies, farther and farther from the ancestrula; this is termed the **distal** direction (Figure 17.9). Thus the newest zooids showing the youngest stages of development are distalmost in the colony. Progressively older stages of zooids are marked by increasing morphologic complexity. The course of development of a zooid is its **ontogeny.** From the distalmost zooids toward the ancestrula, termed the **proximal** direction, zooids are progressively older. Until reaching a stage of full development, these zooids show progressively more complex morphology (Figure 17.9).

Sources of morphologic variation within a colony

Zooids in a colony might be expected to be morphologically identical because of their assumed genetic uniformity. In any bryozoan colony, however, zooids can be observed to differ in some morphologic features. This morphologic variation follows different distributional patterns within a colony and is attributed to four sources: ontogeny of zooids, astogeny of the colony, polymorphism of zooids, and microenvironment.

Ontogeny of zooids within colonies produces a pattern of increasing morphologic complexity proximally from growing extremities, as just described. **Astogeny** occurs only in colonies and is the sequential development of the asexual generations of zooids, beginning with the ancestrula. Astogeny is expressed in a bryozoan colony by two recognizable morphologic zones (Figure 17.9), the zone of change and the zone of repetition. A proximal **zone of change** includes the ancestrula, and generations

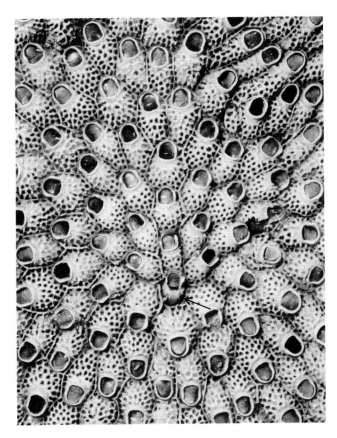

Figure 17.6. Ancestrula (arrow) and successive asexually produced zooids in an encrusting colony of a bryozoan from the Holocene, with approximately box-shaped zooidal body walls. (*Cryptosula*, California, ×25.)

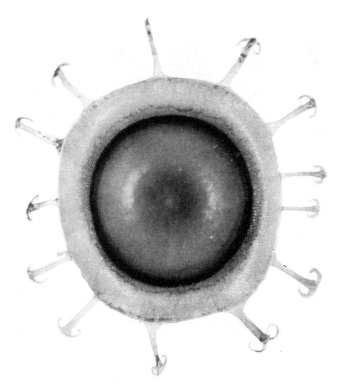

Figure 17.7. A statoblast of a freshwater bryozoan, (×60.) (Courtesy T. S. Wood.)

of zooids immediately following that differ from each other in size and complexity. These morphologic differences are transitional between the morphology of the ancestrula and that of the zooids of the zone of repetition. The **zone of repetition** follows the zone of change distally and is composed of zooids that repeat the morphology of the last generation of the zone of change. The zone of repetition is characterized by many generations of similar zooid morphologies and therefore becomes much the larger of the two zones in most colonies.

Polymorphism is expressed as the repeated occurrence in a colony of one or more kinds of zooids (polymorphs) in addition to the ordinary feeding zooids (Figure 17.10). Polymorphism may be recognized in the same generation of zooids in the zone of change or in any zooids at the same ontogenetic stage in the zone of repetition.

The morphology of zooids within a colony is the result of a series of reactions between the uniform genotype of the colony and the environmental conditions surrounding the colony as it grows. Environmental conditions may affect the whole colony or parts of it. Morphologic differences produced by localized environmental conditions

affecting parts of a colony are termed **microenvironmental** (Figure 17.11). These differences include any morphologic variation that does not follow the patterns of ontogenetic, astogenetic, or polymorphic differences. A few of the microenvironmental conditions that can produce morphologic variation in bryozoan colonies are crowding by growth of the colony itself or competitive growth of other organisms; irregularities in the substrate; encrustation by organisms or sediment on the colony; differential turbulence; and various forms of breakage and boring.

The assumed genetic uniformity of a bryozoan colony provides an advantage in understanding sources of morphologic variation not normally available in the study of solitary animals. Solitary animals generally reproduce only sexually, and therefore morphologic differences between individuals are caused by genetic as well as environmental factors. Morphologic variation among zooids within a bryozoan colony can be recognized and used without the "background noise" of genetic differences.

Colony control of function and morphology

Neither a bryozoan colony nor its member zooids can be compared exactly either functionally or morphologically to a solitary animal.

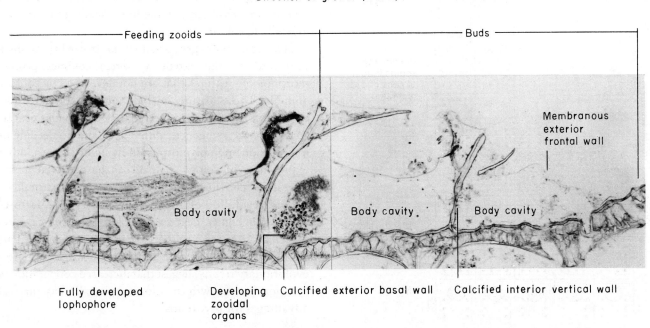

Direction of growth (distal)

Feeding zooids — Buds

Membranous exterior frontal wall

Body cavity — Body cavity — Body cavity

Fully developed lophophore — Developing zooidal organs — Calcified exterior basal wall — Calcified interior vertical wall

Figure 17.8. A median longitudinal section through the body walls and organs of zooids and buds in a growing region of a living encrusting colony. Zooids have approximately box-shaped body walls. (*Metrarabdotos,* Ghana, ×80.) Expansion of the exterior walls of the bud provided more body cavity into which interior walls grew as partitions. Organs of feeding zooids developed within the separated zooidal spaces.

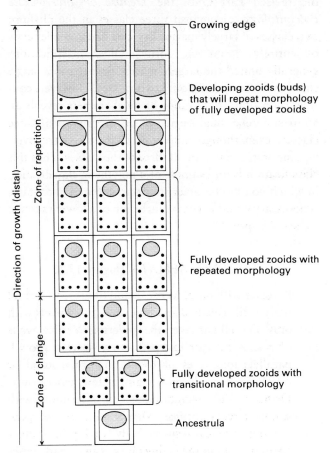

Direction of growth (distal)

Zone of repetition

Zone of change

Growing edge

Developing zooids (buds) that will repeat morphology of fully developed zooids

Fully developed zooids with repeated morphology

Fully developed zooids with transitional morphology

Ancestrula

Figure 17.9. A hypothetical bryozoan colony showing morphologic differences among zooids in a distal growing region (ontogeny) and in the region of the ancestrula (astogeny).

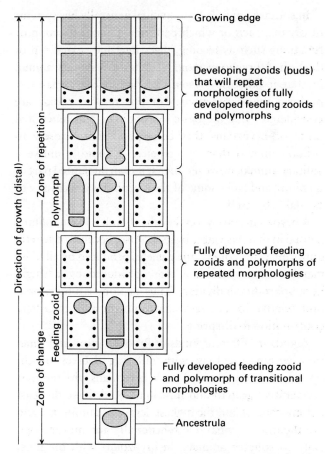

Direction of growth (distal)

Zone of repetition

Polymorph

Zone of change

Feeding zooid

Growing edge

Developing zooids (buds) that will repeat morphologies of fully developed feeding zooids and polymorphs

Fully developed feeding zooids and polymorphs of repeated morphologies

Fully developed feeding zooid and polymorph of transitional morphologies

Ancestrula

Figure 17.10. A hypothetical bryozoan colony showing morphologic differences between feeding zooids and polymorphs recognizable in zones of change and repetition. (Compare with Figure 17.9 in which polymorphs are absent.)

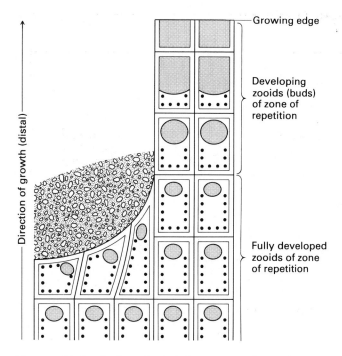

Figure 17.11. A hypothetical bryozoan colony showing morphologic differences in feeding zooids produced by crowding in growth around an obstruction.

In solitary animals an individual generally has systems of organs, each of which performs a major function or functions such as feeding, digestion, and sexual reproduction. A bryozoan zooid is similar to a solitary animal in that it possesses systems of organs that separately perform major functions. For this reason, zooids are considered to be the basic morphologic and functional units of bryozoans that correspond more closely to solitary animal than do colonies. A zooid is unlike a solitary animal, however, in that it is an integrated part of a colony and lacks some of the morphology necessary to be viable by itself.

A bryozoan colony, on the other hand, is like a solitary animal in that it is viable by itself. Also, because of genetic uniformity of a colony, the gametes produced by all of its member zooids are like those produced by a single hermaphroditic individual. Thus a colony is a functional unit relative to its overall environment and its genetic contribution to offspring.

Zooids in different kinds of Bryozoa grow and function within a colony with differing degrees of cooperation among themselves. In bryozoans with the least cooperative growth and function, zooids have the least colony control and the highest degree of autonomy, and are the most comparable functionally and morphologically to solitary animals. In bryozoans with the most cooperative growth and function, zooids show the highest degree of colony control and the least autonomy, and are the least comparable functionally and morphologically to solitary animals.

In practice, the degree of colony control in bryozoans is estimated by the extent to which colonies possess morphologic features that are not developed by solitary animals. These features include interior zooidal walls, communication between zooidal body cavities, extrazooidal parts, astogenetic differences, and nonfeeding or nonreproducing polymorphs. In living bryozoans, other expressions of colony control can be observed in cooperative behavior in feeding or in response to stimuli.

In the evolution of major groups of Bryozoa, features expressing higher degrees of colony control have replaced those expressing more zooidal autonomy (see Part II, "Additional Concepts," for further discussion). These evolutionary trends suggest that increased cooperation in function and growth in colonies generally has survival advantages for bryozoans.

Classification

Modern classifications of the Bryozoa, including those in the revised Part G of the *Treatise on Invertebrate Paleontology,* distinguish three classes in the phylum: two classes of largely marine Bryozoa and a smaller class of entirely nonmarine forms. Older classifications generally united the largely marine Bryozoa in a single class, thus recognizing only two classes. However, consideration of a broad range of 49 characters reveals an approximately equal degree of distinctiveness of three classes, even though numerous characters have overlapping states in two or all three of the classes. Thus this classification is an example of a polythetic classification in which no one character state or combination of states is necessary or sufficient for inclusion of a lower taxon in a class (Chapter 6).

CLASS STENOLAEMATA (Figure 17.12)

Bryozoa with basal, vertical, and (if present) frontal body walls calcified and inflexible; body walls with only one cellular layer, the epidermis. Vertical walls of zooids interior except for few species; growth parallels long axes of zooids to form conical or tubular zooids. Frontal walls in few genera only. Orificial walls membranous, most terminal with simple circular orifice. Membranous sac of peritoneum divides zooidal body cavity into inner coelom containing alimentary canal and outer pseudocoel. Annular muscles of sac protrude lopho-

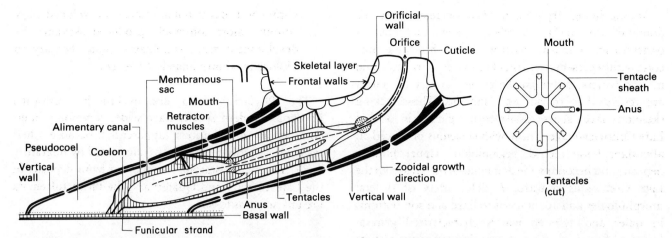

Figure 17.12. A tubular feeding zooid of a stenolaemate bryozoan with an exterior calcified frontal wall. (Approximately ×70.) **A,** Longitudinal section through the body walls and organs. **B,** View looking into the circular tentacle crown protruded in feeding position. The tentacles have been removed just above their bases.

phore. Tentacles arranged in circle around mouth. In most post-Triassic genera, embryos brooded within body cavities where they undergo asexual multiplication. (Early Ordovician to Holocene; approximately 750 genera.)

Among fossil Bryozoa, the Stenolaemata include the overwhelming majority of genera in rocks ranging from Early Ordovician to the mid-Cretaceous, an interval of nearly 400 million years. Stenolaemates are the most abundant macrofossils in some Paleozoic rock units, and colonies are large and numerous enough to be significant rockbuilders in some calcareous shales, mudstones, and limestones. During the Late Cretaceous, the stenolaemates began to lose their predominance within the phylum to the class Gymnolaemata, although they can still be found in large numbers in many marine communities.

CLASS GYMNOLAEMATA (Figure 17.13)

Bryozoa with basal and vertical body walls calcified in most; frontal walls calcified in many; all body walls uncalcified in few. Body walls include two cellular layers, peritoneum and epidermis. Vertical walls of zooids generally combination of exterior walls parallel to zooidal growth direction and interior walls perpendicular to growth direction; together with basal and frontal walls form box- or sac-shaped enclosure for body cavity. Frontal walls present in all; all, part, or derivative of frontal wall remains uncalcified and flexible in lophophore protrusion by means of attached muscles. Orificial walls subterminal in most, formed from movable folds of body wall commonly reinforced to form

lidlike structure closing orifice. Orifices slitlike or puckered. Tentacles arranged in circle around mouth. Embryos brooded singly in water-filled chambers outside body cavity in most; brooded in body cavity in few; not brooded in some. (Late Ordovician to Holocene; approximately 1050 genera.)

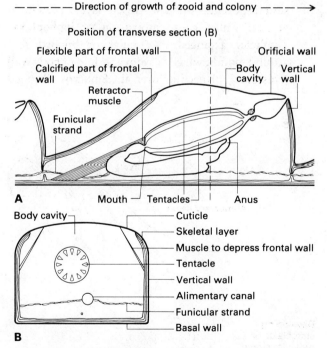

Figure 17.13. A feeding zooid of a simple gymnolaemate bryozoan with calcified roughly sac-shaped walls. (Approximately ×80.) **A,** Median longitudinal section (parallel to the length of the zooid) through the body walls and organs. The calcified part of the frontal wall protects retracted organs. **B,** Transverse section (perpendicular to the length of the zooid). Muscles depress the flexible part of the frontal wall to produce hydrostatic pressure for protruding the tentacles. The shape of the tentacle row is similar to that of stenolaemates (Figure 17.12).

Among living Bryozoa, the Gymnolaemata are the dominant class both in number of genera and in abundance in many marine environments. This is the only class with representatives that live in fresh, brackish, and marine waters, although the great majority of genera are restricted to marine environments. Those lacking skeletons have a sparse fossil record beginning in rocks of Late Ordovician age. Those with skeletons produced an abundant fossil record, beginning in Upper Jurassic deposits and becoming the dominant bryozoans from the Late Cretaceous onward. A rich variety of skeletal morphologies and adaptations to hard and soft bottoms in quiet and agitated waters characterized gymnolaemates found in calcareous marine sediments of Late Cretaceous, Tertiary, and Quaternary ages.

CLASS PHYLACTOLAEMATA (Figure 17.14)

Bryozoa with no skeleton. Body walls include two cellular layers, peritoneum and epidermis. Vertical walls tubelike and limited to outer ends of zooids so that zooidal organs suspended in confluent body cavity of colony. Vertical walls exterior or combination of exterior and interior; parallel to zooidal growth direction. Orificial walls terminal, bearing simple porelike orifices. Muscle layers present in vertical walls; cause protrusion of lophophore. Tentacles arranged in circular or bilobed row around mouth, which has movable fold of body wall (epistome) projecting over it. Embryos brooded within body cavity but infrequently developed. New colonies most commonly produced asexually by development from statoblasts. (Late Tertiary to Holocene; approximately 10 genera.)

The Phylactolaemata are exclusively freshwater Bryozoa. Modern species have wide, commonly intercontinental distributions apparently because their resistant statoblasts are easily transported by migratory birds. Except for statoblasts reported from deposits of the Late Tertiary and Quaternary, the Phylactolaemata are unknown as fossils.

Origin and relationships to other phyla

The earliest known bryozoans are well-calcified stenolaemates of the Early Ordovician. They are characterized by few species of highly variable morphologies. By the Late Ordovician, numerous morphologically distinct groups of species had diverged from earlier Ordovician stocks. This divergence was expressed by segregation of the original character states and by addition of new states resulting in many new taxa, considered to be genera and higher taxa. The presumed origin of the gymnolaemates, represented only by rare molds of soft-bodied colonies, also occurred during the early period of divergence before the end of the Ordovician Period. Phylactolaemates are not known from the Paleozoic.

Because of the sporadic fossil record of the apparently entirely uncalcified gymnolaemates in rocks of the

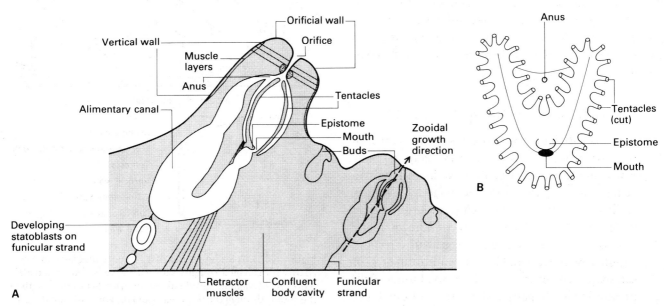

Figure 17.14. A feeding zooid and buds of a phylactolaemate bryozoan. (Approximately ×60.) **A,** Median longitudinal section through the body walls (no skeletal layer) and organs. **B,** View looking into a bilobed tentacle crown protruded in feeding position. The tentacles have been removed just above their bases.

Ordovician to Middle Jurassic, and because the first appearance of both gymnolaemates and stenolaemates was in the Ordovician, no convincing evidence exists of their possible phylogenetic relationships. Both might have evolved independently from an unknown ancestor, or either class might have evolved from the other. Uncalcified gymnolaemates most likely gave rise to calcified gymnolaemates later in the Jurassic (see Part II, "Additional Concepts," for further discussion). The absence of a Paleozoic fossil record for the phylactolaemates suggests that they appeared later than the other two classes. However, since the living species lack skeletons, there is no actual proof for this supposition.

The fossil record is even less adequate to establish phylogenetic relationships among phyla than among classes. As a result, relationships among phyla are most commonly inferred from morphologic similarities of living representatives. With this approach, bryozoans usually have been compared to the phyla Phoronida and Brachiopoda, or to the phylum Entoprocta.

The Phoronida, Brachiopoda, and Bryozoa are now grouped together informally as the lophophorates because they are all coelomate and possess lophophores. Although the lophophorates are an informal group, they are generally believed to have a common ancestor. Of the three lophophorate phyla, living phoronids are considered by some investigators to show the most primitive characteristics. The assumed common ancestor of the three groups is therefore considered to have been comparable to phoronids. The phoronids, however, are soft bodied and have a possible fossil record extending back only to the Devonian (Chapter 2). Also, they compare most closely to the phylactolaemates among the Bryozoa but possess additional kinds of organs. The lack of significant fossil records for both the phoronids and the phylactolaemates makes any early phylogenetic relationships between phoronids and bryozoans speculative and unconvincing. Bryozoa are much less similar to brachiopods than they are to phoronids, and no intermediate forms are known. There seems to be no direct fossil evidence for the origins of the three lophophorate phyla.

The Entoprocta were once included in the phylum Bryozoa. They have a ring of nonretractable tentacles that is not regarded as a lophophore and a body cavity that is regarded as a pseudocoel (Chapter 2). The entoprocts also are soft bodied and have no fossil record to help in establishing phylogenetic relationships. Their relationship to the Bryozoa is therefore no better understood than that of the Bryozoa to the phoronids.

Part II Additional concepts

Class Stenolaemata

Richard S. Boardman

The wealth of Bryozoa in rocks of Ordovician through mid-Cretaceous age belongs almost entirely to the class characterized by elongate tubular or conical zooids, the Stenolaemata. The only exceptions known in Paleozoic rocks are minute uncalcified gymnolaemates that formed netlike colonies inset in a substrate and preserved as external molds. By the Late Cretaceous the stenolaemates were overtaken in diversity and numbers by calcified gymnolaemates but currently continue to thrive in many marine communities.

Details of skeletal structures are commonly well preserved in fossil stenolaemates and provide much biologic information and many taxonomic characters. Skeletal microstructures and configurations of colony interiors permit broad inferences concerning the general nature of soft parts, modes of growth, and functional morphology. However, little morphologic detail is displayed on outer surfaces of colonies. Many colonies with similar external appearances have morphologically different interiors, a condition called **external homeomorphy**.

It is necessary to make oriented thin sections or peels of colonies to observe their internal morphology. Planes of **longitudinal sections** (Figure 17.15) are parallel to zooidal lengths; **tangential sections** parallel planes tangential to colony surfaces and cut outer ends of zooids at high angles; and **transverse sections** are perpendicular to the direction of colony growth. Three-dimensional relationships and microstructural details of colony interiors can be determined with certainty from this combination of sections.

Morphology

Morphology of zooids Stenolaemates are characterized as tubular Bryozoa because vertical walls of zooids typically form elongate tubes enclosing body cavities (Figure 17.15). Vertical walls are interior walls; that is, they partition existing body cavities of colonies. The calcified layers of vertical walls are uninterrupted, except in stenolaemates with **communication pores** (Figures 17.17, *B*; 17.19, *B*; and 17.20). Many pores are open and presumably permit transfer of nutrients among neighboring zooids. The portion of body cavity into which

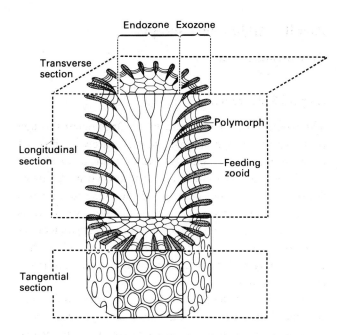

Figure 17.15. A segment of a branch of a dendroid colony cut to show the planes of oriented sections used to study stenolaemates. All zooid walls here are vertical walls.

zooidal organs retract is the **living chamber** (Figures 17.16; 17.17, *A* and *B*; and 17.18, *C*). Most **skeletal apertures** of zooids are simple terminal openings just under membranous orificial walls (Figures 17.19, *A* and 17.20).

Intrazooidal skeletal structures are formed immediately adjacent to zooidal soft parts and are assumed to have significant functions, although not all of these functions are known. **Diaphragms** are either membranous or skeletal partitions that extend transversely completely across body cavities. **Basal diaphragms** (Figures 17.16 and 17.17, *A*) floor living chambers. **Terminal diaphragms** at outer ends of body cavities (Figure 17.17, *A*) seal living chambers from the environment during dormant periods. **Hemisepta** are shelf-like skeletal projections from chamber walls that occur either singly or in one or two pairs (Figure 17.18, *A*). Feeding organs retract behind hemisepta in a modern species (Figure 17.18, *C*). In one Paleozoic order, apertures and living chambers of feeding zooids are complicated by a **lunarium** (Figure 17.21). This trough-like skeletal structure is generally microstructurally distinct from the remainder of the zooidal wall. Lunaria extend throughout the outermost lengths of vertical walls of feeding zooids and project above colony surfaces.

A **membranous sac** (Figures 17.18, *C*; 17.19; 17.20; and 17.22) surrounds the digestive and reproductive systems in feeding zooids of modern stenolaemates and divides the living chamber into two parts (Figures 17.19, *A* and 17.20). A recent study indicates that body walls of probably all stenolaemates have only one cellular layer, the epidermis, and that the membranous sac is the peritoneum. Body cavities within sacs are surrounded by peritoneum (possibly a mesoderm) and are considered to be coeloms. Body cavities outside of sacs are pseudocoels lined by peritoneum on one side and epidermis on the other.

Membranous sacs are fastened by attachment ligaments (Figure 17.19, *A*) to chamber walls and contain annular muscles. Contraction of the annular muscles

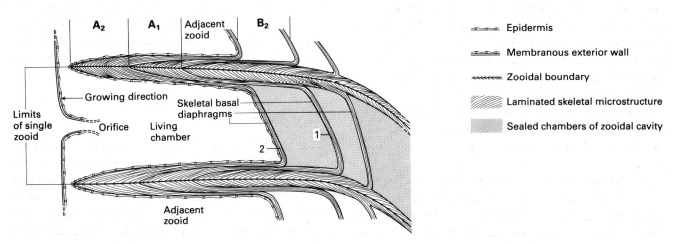

Figure 17.16. Longitudinal section of the outermost part (exozone) of a hypothetical Paleozoic zooid with its organs left out of the living chamber. The living chamber is sealed off from the inner portions of the zooidal body cavity by the last basal diaphragm (2) to form. The epidermis deposits laminae of skeletal parts approximately parallel to itself and skeletal surfaces in most stenolaemates. (From Boardman, R. S. In: Robison, R. A., editor. *Treatise on invertebrate paleontology, Part G,* (revised), *Bryozoa,* vol. 1. Boulder, CO, and Lawrence, KS: Geological Society of America and University of Kansas Press; 1983.)

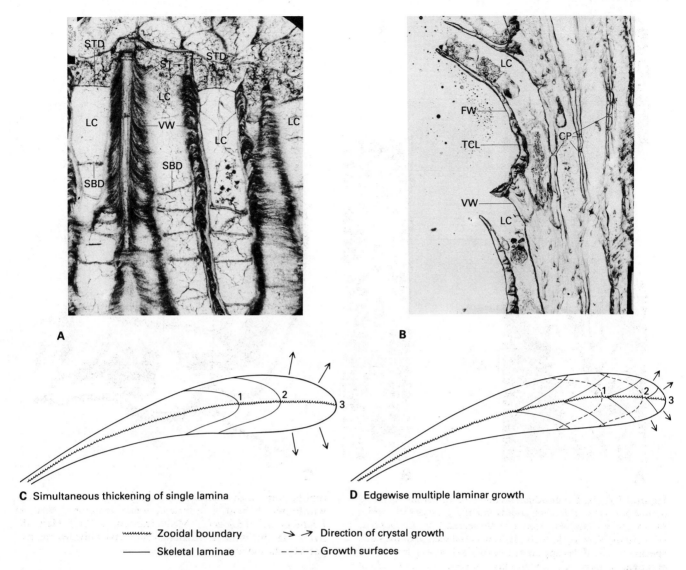

Figure 17.17. Longitudinal sections showing the living chambers (*LC*) and laminated microstructure of the vertical walls (*VW*). **A,** Skeletal basal diaphragms (*SBD*) and terminal diaphragms (*STD*) of feeding zooids and aligned styles (*ST*) of *Heterotrypa* (a free-walled trepostome). (Ordovician, Kentucky, ×66.) The living chambers are outward from last basal diaphragms to form. The vertical walls have laminae that are parallel to the skeletal surfaces at the growing edges. **B,** Frontal walls (*FW*), retracted tentacles (*TLC*) in the living chamber, and communication pores (*CP*) in the vertical walls of fixed-walled tubuliporate. (Holocene, Antarctica, ×30.) Laminae of the vertical walls are at high angles to the skeletal surfaces. **C,** Adjacent vertical zooidal walls showing laminae (*1* and *2* of earlier growth stages) parallel to the growth and skeletal surfaces (*3*). **D,** Vertical wall laminae at high angles to the growth and skeletal surfaces. (C and D from Boardman, R. S. In: Robison, R.A., editor. *Treatise on invertebrate paleontology, Part G* (revised), *Bryozoa*, vol. 1. Boulder, CO, and Lawrence, KS: Geological Society of America and University of Kansas Press; 1983.)

reduces the volume of a sac, slowly forcing the digestive organs and lophophore outward just enough to free the tentacles for feeding. The tentacles can be withdrawn quickly by relaxation of the annular muscles and contraction of powerful retractor muscles.

Composition and microstructure of skeleton   All stenolaemates have calcareous skeletons. All skeletons are calcitic except for one reportedly aragonitic species in the Triassic.

The microstructure of most stenolaemates is **laminated** (Figure 17.17, *A* and *B*), that is, made up of layers of flattened crystals surrounded by a thin organic matrix. In most stenolaemates, the laminae (Figure 17.17, *A*) are approximately parallel to skeletal surfaces and to the layers of epidermal cells that secreted them (Figure 17.16). These laminae thus record skeletal surfaces of past growth stages (Figure 17.17, *A* and *C*). Laminae in vertical walls of some stenolaemates are at high angles to epidermal layers and skeletal surfaces (Figure 17.17, *B*

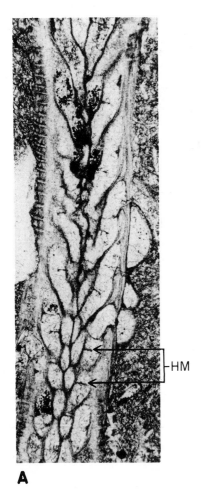

A

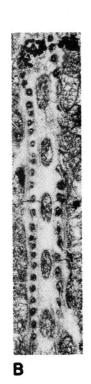

B

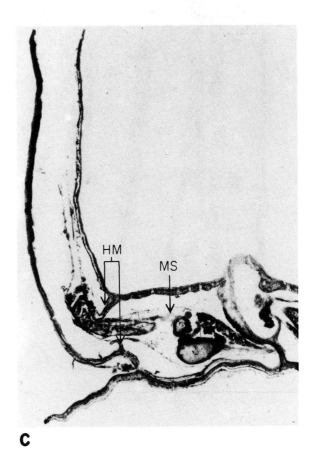

C

Figure 17.18. A, Longitudinal section of cryptostome showing hemisepta (*HM*) of feeding zooids and aligned styles (the series of spines on the left side of the section) in *Orthopora.* (Devonian, New York, ×50.) **B,** Tangential section of the same specimen. The styles appear as rows of dark spots lateral to the elliptical zooids. **C,** Longitudinal section of a zooid of tubuliporate with tentacles and the alimentary canal partly withdrawn behind a hemiseptum-like structure (*HM*) in *Diaperoecia.* (Holocene, Mediterranean, ×100.) Here the living chamber is the entire zooidal body cavity divided into two parts by the membranous sac (*MS*).

and *D*) and as a result do not record past growth surfaces. This orientation results from edgewise growth of a number of laminae simultaneously at the growing edges of walls.

Laminae of vertical walls of adjacent zooids display bilateral symmetry in section because they were grown simultaneously on both sides from adjacent chambers (Figure 17.16). **Zooidal boundaries** are therefore necessarily along centers of bilateral symmetry of the combined vertical walls between adjacent zooidal body cavities (Figure 17.17, *C* and *D*).

Granular microstructure occurs in part or all of the skeletons of some stenolaemates. **Hyaline microstructure** is nearly transparent without indications of laminae and occurs in body walls of one order and in minor parts of skeletons in the other orders. Neither microstructure reveals much about skeletal growth because of its uniform texture.

Colony organization   Stenolaemate colonies are divided into two zones, the endozone and exozone, based on ontogenetic changes in zooids (Figures 17.15 and 17.23). The **endozone** is the inner region of a colony occupied by inner ends of zooids where they generally have a combination of thin vertical walls, zooidal growth directions at low angles to colony surfaces or colony growth directions, and relative scarcity of intrazooidal skeletal structures. The **exozone** is the remaining outer region of the colony where zooidal growth directions are at high angles to colony growth directions or colony surfaces, and zooids generally have a combination of thicker vertical walls and concentrations of intrazooidal skeletal structures.

Most stenolaemate colonies are **free-walled,** that is, covered by membranous exterior walls not attached to outer ends of vertical walls of zooids (Figures 17.19 and 17.22). The membranous covering wall of a colony is

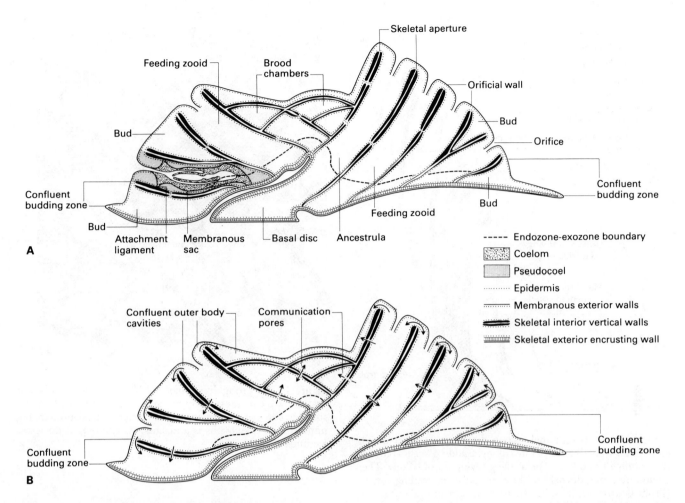

Figure 17.19. Longitudinal section through the center of a small button-shaped colony of a free-walled tubuliporate. Budding occurs in two sites: the endozone, which is on the encrusting wall of the colony in the confluent budding zone that extends around the colony margin; and in the exozone above the encrusting colony wall between established zooids. **A,** The coelom and pseudocoel shown in only one feeding zooid. **B,** The straight arrows show the presumed transfer of nutrients through communication pores. The curved arrows indicate nutrient transfer through the confluent outer body cavities.

held in place by attachment ligaments within zooids. Lack of attachment of covering wall to outer ends of vertical walls produces a **confluent outer body cavity** connecting all outermost skeletal surfaces of a free-walled colony. Apparent advantages of outer body cavities are colony-wide distribution of nutrients and the possibility of growth of all outermost colony surfaces throughout life (Figure 17.19, *B*). Parts of colonies suffering accidents are commonly regenerated under the sealed membranous covering by overgrowth of new zooids originating from adjacent undamaged zooids (Figure 17.24).

All extrazooidal skeletal structures occur under confluent outer body cavities, nourished by neighboring feeding zooids. In addition to the support structures illustrated in Figure 17.3, important extrazooidal structures include **skeletal vesicles** between zooids that characterize a major group of Paleozoic Bryozoa (Figure

17.29, *A* to *D*). More common are solid, elongate, and spinelike structures of several kinds called **styles. Aligned styles** extend between and are generally parallel to feeding zooids, ending as spines on colony surfaces (Figures 17.17, *A* and 17.18, *A* and *B*). They raise exterior membranous colony walls above skeletal apertures (Figure 17.22) apparently to increase volumes of confluent outer body cavities and to facilitate colony-wide communication.

Many of the more robust free-walled colonies develop **maculae** (Figure 17.21), which are generally evenly spaced clusters of polymorphs, or extrazooidal structures, or a combination of both. Polymorphs in maculae can be larger or smaller than feeding zooids between maculae and commonly have thicker skeletal walls. Most maculae form prominences on colony surfaces, but some are flush or depressed.

Stenolaemate colonies are described as **fixed-walled** if

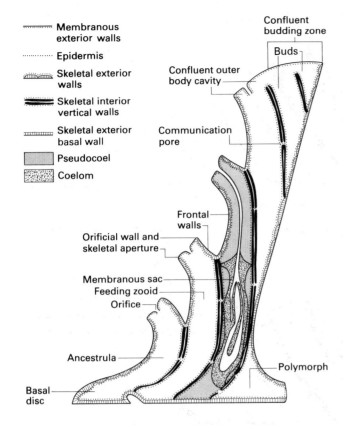

Legend:
- ——— Membranous exterior walls
- .... Epidermis
- Skeletal exterior walls
- Skeletal interior vertical walls
- Skeletal exterior basal wall
- Pseudocoel
- Coelom

Labels: Confluent budding zone; Buds; Confluent outer body cavity; Communication pore; Frontal walls; Orificial wall and skeletal aperture; Membranous sac; Feeding zooid; Orifice; Ancestrula; Polymorph; Basal disc

Figure 17.20. Longitudinal section through the center of an erect fixed-walled tubuliporate colony. Budding occurs only in the confluent budding zone at the growing tip of colony. The coelom and pseudocoel are shown in only one feeding zooid. (From Boardman, R. S. In: Robison, R. A., editor. *Treatise on invertebrate paleontology, Part G* (revised), *Bryozoa*, vol. 1. Boulder, CO, and Lawrence, KS: Geological Society of America and University of Kansas Press; 1983.)

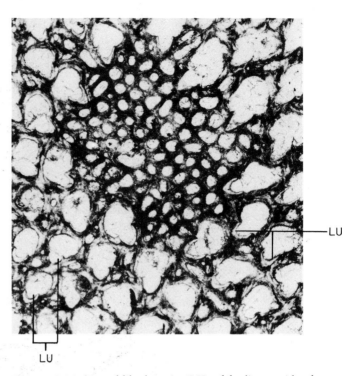

Figure 17.21. Troughlike lunaria (*LU*) of feeding zooids of a cystoporate cut transversely by tangential section to show the smaller radius of the curvature. The cluster of small, presumably nonfeeding polymorphs is a macula. (*Crepipora,* Ordovician, Kentucky, ×30.)

their orificial walls are attached at skeletal apertures of feeding zooids so that confluent outer body cavities between zooids are eliminated (Figures 17.17, *B* and 17.20). The great majority of fixed-walled stenolaemates have orificial walls attached to frontal walls (Figure 17.20). Skeletal layers of frontal walls are attached to outer ends of vertical zooidal walls and are terminated at apertures. The outermost cuticles of fixed-walled colonies are attached to the entire outer surfaces of calcareous layers of frontal walls so that outer confluent body cavities are eliminated. Frontal walls necessarily contain skeletal apertures because they are the outermost and last skeletal walls of zooids to form.

Stenolaemates have developed many kinds of erect growth habits. **Dendroid colonies** (Figure 17.23, *F*) are branching, and the branches have a circular cross section (Figures 17.15 and 17.25). **Bifoliate colonies** have flattened, ribbon-shaped branches in which zooids grow back to back from central planar colony walls (Figure 17.30). **Frondose colonies** have flattened, leaf-shaped branches, some of which are also bifoliate. **Reticulate**

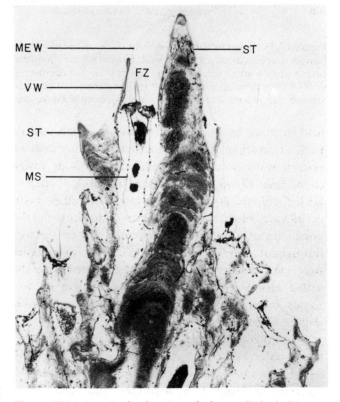

Figure 17.22. Longitudinal section of a free-walled tubuliporate showing aligned styles (*ST*) supporting a membranous exterior wall (*MEW*) above feeding zooids (*FZ*) with vertical walls (*VW*) and membranous sacs (*MS*). (*Densipora,* Holocene, Australia, ×100.)

colonies form broad netlike fronds or cones (Figure 17.32, *D*) made of closely spaced parallel branches that either anastomose (Figure 17.32, *E*) or are connected by crossbars (Figure 17.32, *C* and *F*) to form the network. **Pinnate colonies** (Figure 17.32, *A*) have two sets of smaller branches extending in opposite directions from a central branch.

Classification

The class Stenolaemata includes four to six orders, depending on the classification. Five orders are recognized here; the orders are generally distinct morphologically so that most genera can be placed in them without difficulty. The phylogenetic significance of the

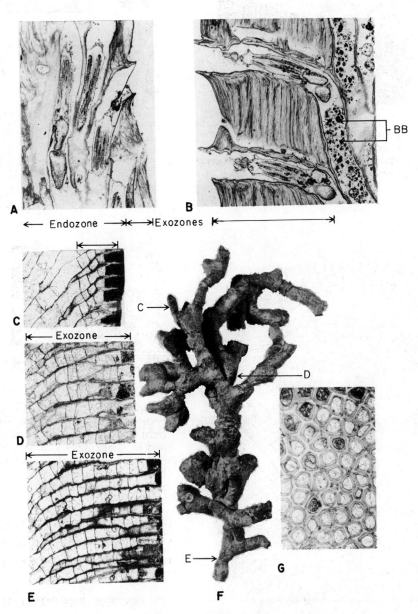

Figure 17.23. Ontogeny. **A,** Young stage from a distal part of a free-walled tubuliporate colony with a narrow exozone and most of the retracted organs in endozone. (*Hornera*, Holocene, Arctic Ocean, ×66.) **B,** Old stage from proximal part of the same colony with retracted organs in the wide exozone. The inner ends of the zooids are filled with brown bodies (*BB*) from past degenerations. **C,** Young stage from position *C* in the trepostome colony *F* has a narrow exozone and three to four basal diaphragms in each feeding zooid. A short living chamber is to the right of the last (farthest to right) basal diaphragm. (×13.) **D,** Intermediate stage from position *D* in colony *F*. **E,** Oldest stage with the widest exozone and most basal diaphragms from position *E* at the base of the colony *F*. **F,** Dendroid colony of *Leptotrypella*. (Devonian, New York, ×0.2.) **G,** Tangential view from the same colony showing a scarcity of small polymorphs. (×20.)

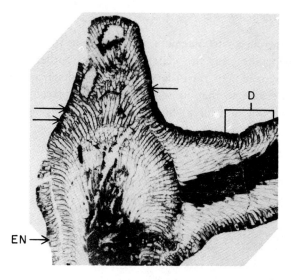

Figure 17.24. Longitudinal section showing overgrowth on the original branch of a trepostome. The overgrowth is both encrusting (*EN*) and dendroid (*D*) and originated from adjacent zooids (arrows). (*Leptotrypella*, Devonian, New York, ×1.7.)

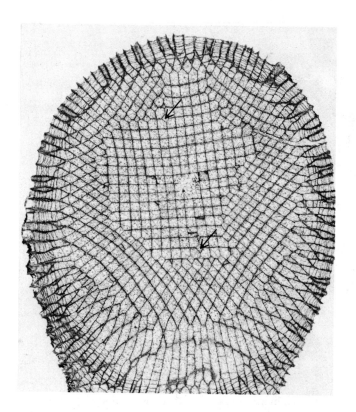

Figure 17.25. Transverse section of a dendroid colony showing zooids with square cross sections in a regular pattern, which is rare in trepostomes. The buds, which are the smallest squares (see arrows) at the inner ends of feeding zooids, are centered at the corners of four older (larger) zooids of *Rhombotrypa*. (Ordovician, Indiana, ×9.)

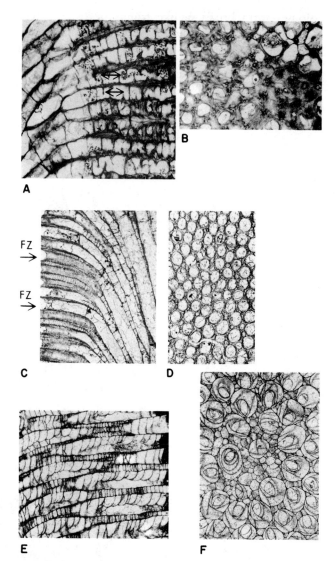

Figure 17.26. Ordovician Trepostomata, ×15. **A** and **B**, Sections showing alternating skeletal plates (arrows), which form sinuous living chambers in *Hemiphragma* (Canada). **C** and **D**, Diaphragms in inner ends of feeding zooids (*FZ*) and throughout the length of small polymorphs (arrows) in *Parvohallopora* (Ohio). **E** and **F**, Curved skeletal plates in feeding zooids of *Prasopora* (New York).

orders can be questioned, however, and a polythetic study of all the orders based on all the genera could well make major improvements in the classification.

ORDER TREPOSTOMATA (Figures 17.23, C to G; 17.24 to 17.27)

Free-walled stenolaemates of typically robust colonies with maculae. Colonies mostly dendroid, others encrusting or massive, few frondose. Feeding zooids budded throughout endozones in almost all taxa. Most feeding zooids long with basal dia-

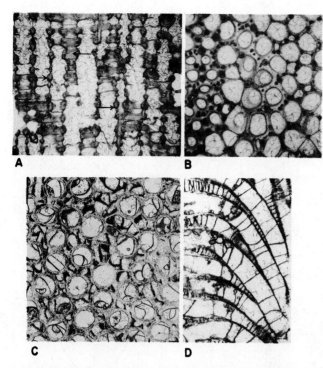

Figure 17.27. Trepostomata, ×15. **A** and **B**, Longitudinal and tangential sections showing annular thickenings (arrows) in vertical walls of *Tabulipora* (Permian, Timor). **C** and **D**, Sections showing large feeding zooids with cavities lined by spines in *Hallopora* (Silurian, Tennessee). (Compare with *Parvohallopora* in the Ordovician shown in Figure 17.26, **C** and **D**.)

phragms; positioned adjacent to each other without apparent pattern and with only one type of aligned style intervening; apertures closely spaced. Body cavities of feeding zooids in exozones generally equidimensional in cross section, either circular or polygonal. Most zooidal skeletons laminated in exozones, laminae parallel to growth surfaces. Many with small polymorphs between maculae, with or without diaphragms. (Early Ordovician to Triassic; about 200 genera.)

The first known appearance of the Trepostomata is in rocks considered to be from the latest Early Ordovician. The earliest Middle Ordovician genera are of two types: those with generalized skeletal morphology of a few taxonomic characters poorly defined and those with many well-defined characters. One genus with many characters has skeletal shelves that alternate from opposite sides of chamber walls to give living chambers a complex sinuous shape (Figure 17.26, *A*), suggesting more specialized forms having considerably earlier evolution.

During the Middle and Late Ordovician, trepostomes evolved rapidly to their maximum diversity and were the dominant bryozoans of that time (Figure 17.28). They

are the most abundant megafossil group in some Ordovician beds, and their robust colonies, in addition to their abundance, made them significant rock formers in many calcareous shales and limestones. Trepostomes were greatly reduced in the Silurian, made a partial recovery in the Devonian, and then suffered a gradual decline to two or three genera in the Triassic Period.

ORDER CYSTOPORATA (Figures 17.21; 17.29)

Free-walled stenolaemates with robust colonies generally encrusting or massive; more delicate colonies dendroid or bifoliate; most with maculae. Budding positions of feeding zooids highly variable; single zooids bud from extrazooidal vesicles in few. Feeding zooids short to long, diaphragms typically sparse to absent. Most feeding zooids or groups of zooids separated at least partly by extrazooidal skeletal vesicles so that apertures widely spaced and arranged in patterns at colony surfaces. Styles on extrazooidal vesicles in few genera, but aligned styles absent. Nearly all living chambers and skeletal apertures modified by lunaria unique to order. Most zooidal skeletons granular, some with transversely laminated walls in exozones. Polymorphs restricted to maculae in most genera. Communication pores in vertical walls in few genera, hemisepta in few others. (Early Ordovician to Triassic; about 100 genera.)

The earliest known cystoporates occur in the Early Ordovician. Three Early Ordovician genera now placed in the cystoporates have generalized morphology that seems to be transitional between the Cystoporata and the Trepostomata. The few characters of these Early Ordovician forms occur in some cystoporates or trepostomes of younger Ordovician beds, suggesting that the two orders had a common origin or that one gave rise to the other.

The taxonomic diversity of the cystoporates increased slowly through the early Paleozoic and reached a maximum of about 27 genera in the Middle Devonian. Diversity dropped progressively through the late Paleozoic until only one undescribed species is known from the Triassic.

ORDER CRYPTOSTOMATA (Figures 17.18, *A* and *B*; 17.30; 17.31)

Free-walled stenolaemates typically with delicate colonies, either dendroid or bifoliate. Maculae in

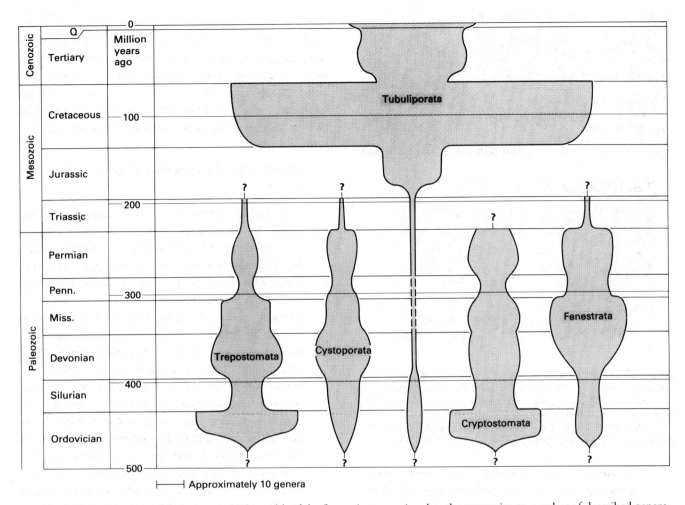

Figure 17.28. Range chart of Stenolaemata. The width of the figures is proportional to the approximate number of described genera for each system.

many bifoliates and in few dendroids. Budding of most feeding zooids either from central planar walls, linear axes, or less commonly axial zooids or axial bundles of zooids. Feeding zooids typically short, elliptical in cross section in exozones. Diaphragms commonly absent and hemisepta in few. Aligned styles in dendroids commonly more than one type, absent in most bifoliates. Feeding zooids in many separated by solid extrazooidal skeleton in exozones so that zooidal apertures widely spaced and arranged in rhombic or longitudinal patterns at colony surfaces. Zooidal skeletons in exozones with laminae paralleling growth surfaces. Many with small polymorphs lacking diaphragms in exozones. (Early Ordovician to Late Permian; about 90 genera.) genera.)

Bifoliate and delicate jointed dendroid cryptostomes have been reported in rocks of the latest Early Ordovician. Both growth habits seem relatively specialized, suggesting considerably earlier evolution.

The bifoliate cryptostomes (Figure 17.30) evolved rapidly in Middle Ordovician time and reached their maximum diversity during the Late Ordovician. They diminished rapidly in diversity during the Silurian and apparently became extinct during the Pennsylvanian.

The dendroid cryptostomes are represented in Ordovician and Silurian rocks largely by a few delicate jointed genera (Figure 17.31, *E* to *J*). These declined during the Devonian, and unjointed dendroids (Figure 17.31, *A* to *D*) became abundant, reaching their maximum diversity in the Mississippian. The dendroids are considered to have become extinct near the end of the Permian.

ORDER FENESTRATA (Figure 17.32)

Free-walled stenolaemates typically with delicate colonies, either reticulate or pinnate. Budding of feeding zooids from back walls, zooidal apertures open on front sides of branches only, commonly in longitudinal rows. Most feeding zooids short; skeletons hyaline or both hyaline and laminated.

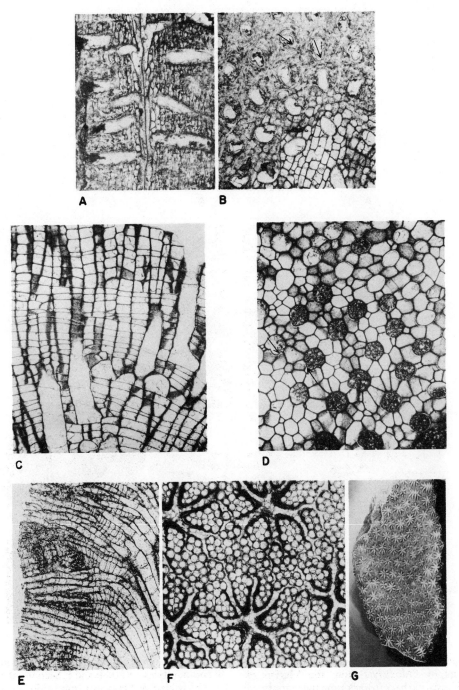

Figure 17.29. Cystoporata. **A** and **B**, Longitudinal and tangential sections showing lunaria (arrows) and an extrazooidal cystoidal structure between zooids in a typical upper Paleozoic bifoliate genus. (*Dichotrypa*, Mississippian, Illinois, ×20.) **C** and **D**, Sections of a massive colony of *Pinacotrypa* showing coarse cystoidal structures between long, widely spaced zooids with lunaria (arrows). (Silurian, Oklahoma, ×20.) **E** to **G**, *Constellaria* from the Ordovician of Ohio. **E** and **F**, Sections showing star-shaped maculae composed of cystoidal structure and clusters of zooids between rays (×13.) **G**, External view showing evenly spaced star-shaped maculae. (×1.)

Diaphragms absent in feeding zooids; hemisepta in some. Extrazooidal skeleton laminated on back sides of colonies around to front sides between zooidal apertures of most. Styles abundant, hyaline, projecting through laminated extrazooidal skeleton commonly from all sides of zooids. Polymorphs un-

common and varied. (Early Ordovician to Triassic; about 100 genera.)

Poorly preserved reticulate Fenestrata are reported in latest Early Ordovician rocks. Some of the early reticulate species placed in the fenestrates have zooidal

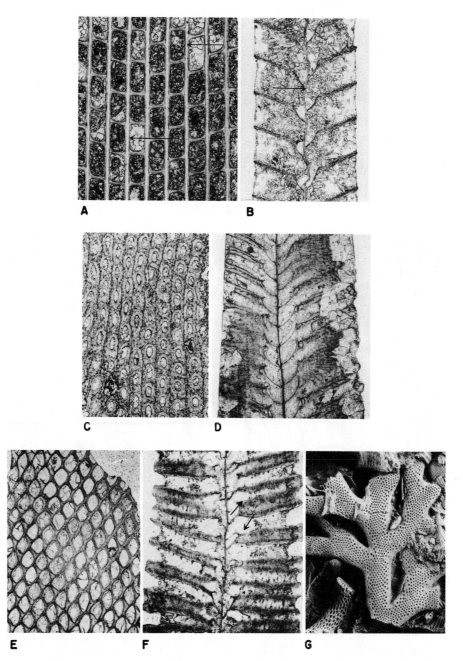

Figure 17.30. Bifoliate Cryptostomata. **A** and **B**, Tangential and longitudinal sections of *Ptilodictyia* from the Silurian of Sweden, with zooids growing back to back from the central wall (**B**, arrow) and arranged in longitudinal rows separated by strong vertical ridges (**A**, arrows). (×20.) **C** and **D**, Sections of *Stictopora* from the Ordovician of Minnesota, with zooids arranged in longitudinal rows and surrounded by small inter-zooidal styles. (×20.) **E** and **F**, Sections of *Escharopora* from the Ordovician of Ohio showing zooids in rhombic arrangement and hemisepta (arrows) and small spines in vertical walls. (×20.) **G**. External view with zooids in rhombic arrangement in *Graptodictya*. (Ordovician, Ohio, ×3.5.)

skeletons comparable to trepostomes in shape and microstructure (Figure 17.32, *E*). Other early species have characters transitional to those of younger, more typical fenestrates, suggesting that the Trepostomata gave rise to the Fenestrata. Another equally reasonable origin of the Fenestrata is through the earliest jointed cryptostomes. Some of these cryptostomes have zooidal apertures on only one side of slender branches (Figure

17.31, *I* and *J*) and zooidal skeletal characters transitional to those of some early reticulate Fenestrata. Detailed evolutionary studies are necessary to produce a convincing hypothesis.

Reticulate fenestrates increased in abundance through the early Paleozoic and reached their maximum development during the Mississippian when their delicate colonies became so abundant in some limestones and

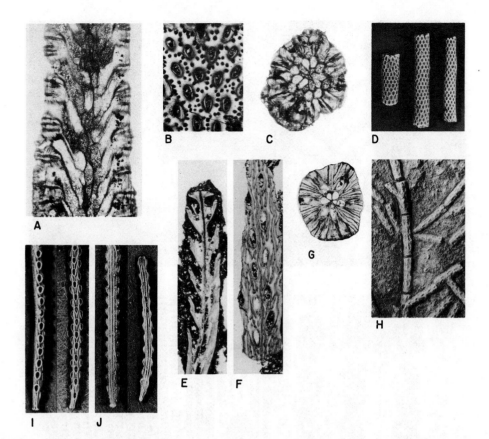

Figure 17.31. Dendroid Cryptostomata. **A** to **D**, Longitudinal, tangential, and transverse sections (×15) and specimens (×2.5) of *Saffordotaxis* (Mississippian, Kentucky). **E** to **H**, Jointed colony of *Arthroclema* (Ordovician, Canada). Sections (×25) and external view (×5.) **I** and **J**, External views of segments of jointed colonies. The joints are intact at the lower ends. (*Arthrostyloecia*, Ordovician, Virginia, ×10.) **I**, Front sides with apertures. **J**, Back sides lacking apertures.

shales that they became rock formers. Pinnate Fenestrata (Figure 17.32, *A*) had their major development in the late Paleozoic and are abundant in some Permian rocks. The Fenestrata apparently became nearly extinct by the end of the Permian, and only a few reticulate specimens are reported from the Triassic.

ORDER TUBULIPORATA (Figures 17.17, *B*; 17.18, *C*; 17.22; 17.33; 17.34)

Stenolaemates with colonies either free-walled, fixed-walled, or a combination; fixed-walled colonies with frontal walls unique within class. Growth habits of colonies widely variable; few with maculae; aligned styles absent in nearly all. Budding positions and arrangements of feeding zooids typically in regular patterns. Feeding zooids short to long; skeletal basal diaphragms absent in nearly all. Exterior terminal diaphragms in many feeding zooids unique within class; apertures or cluster of apertures widely spaced in most, separated by frontal walls or solid extrazooidal skeleton in exozones. Zooidal skeletons laminated, granular, or laminated and either granular or hyaline; laminae in vertical walls either parallel or uniquely at high angles to growth surfaces; communication pores in vertical walls of most. Polymorphs varied and common, typically clustered in different patterns on erect colonies. Larval brooding in body cavities of large polymorphs or larger extrazooidal brood chambers common. (Early Ordovician to Holocene; about 250 genera.)

Tubuliporates (until recently called Cyclostomata) first appear in Lower Ordovician rocks. Most formed small, insignificant fixed-walled colonies scattered through the Paleozoic, and few genera are recognized. Communication pores were absent in the Paleozoic except for one Ordovician genus so that each fully formed zooid was skeletally closed off from others within a colony and lived essentially as a sessile solitary animal.

Communication pores appeared in vertical zooidal

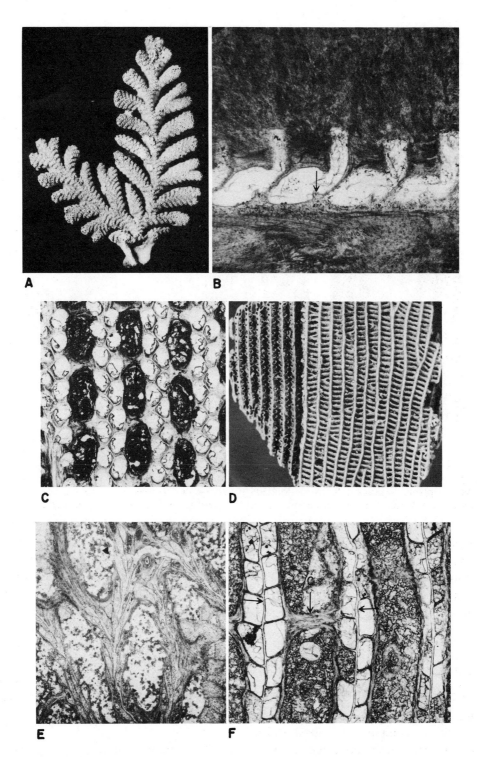

Figure 17.32. Fenestrata. **A,** The front side of a pinnate colony of *Acanthocladia.* (Permian, Texas ×2.) The front side has zooidal apertures; the back side has none. **B,** Longitudinal section of feeding zooids immersed in supporting skeleton showing hemisepta (arrow) in a reticulate colony of *Lyroporella.* (Mississippian, Illinois, ×66.) **C,** Deep tangential section through living chambers parallel to a frond of a reticulate colony of *Fenestella* showing zigzag walls between zooidal chambers. Cross bars (arrow) connect erect branches. (Devonian, Michigan, ×30.) **D,** Exterior view of the front side of a colony of *Unitrypa* showing zooidal apertures (to left) and the extrazooidal skeletal lattice (to right) covering apertures. (Devonian, Ohio, ×6.) **E,** Deep tangential section of a reticulate phylloporinid colony from the Ordovician of Oklahoma, showing anastomosing branches and long zooids. (×20.) **F,** Deep tangential section of the reticulate genus *Fenestella* from the Silurian of Tennessee, showing straight planar walls (arrows) between zooids and extrazooidal crossbar (*c*) with no zooids. (×30.)

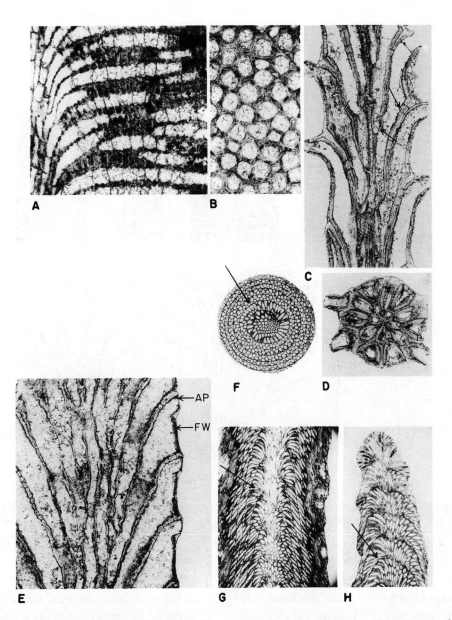

Figure 17.33. Jurassic Tubuliporata. **A** and **B**, Longitudinal and tangential sections showing the annular thickenings and communication pores of the vertical walls and thin basal diaphragms in *Ceriocava*. (France, ×20.) **C** and **D**, Longitudinal and transverse sections of a delicate fixed-walled tubuliporate with the communication pores in vertical walls (arrows). (*Entalophora*, England, ×50.) **E**, Longitudinal section showing the communication pores, frontal walls (*FW*), and closed apertures (*AP*) of *Bisidmonea*. (France, ×30.) **F** to **H**, Transverse and two longitudinal views showing spiralling growth surfaces (arrows) in *Terebellaria*. **H** is the growing tip of the colony. (France, ×4.5.)

walls in the Jurassic, presumably establishing physiologic communication among zooids in fixed-walled colonies and improving communication in free-walled forms. Both fixed- and free-walled tubuliporates began a proliferation of new morphology and taxa with the appearance of communication pores.

During the Cretaceous Period, the tubuliporates underwent an evolutionary explosion, quadrupled the number of genera to about 175, and reached their maximum diversity. At the end of the Cretaceous, the tubuliporates apparently went into decline and averaged approximately 50 genera at any one time until the Holocene when a few more are known.

As presently classified, the Tubuliporata are the only order of stenolaemates in Jurassic and younger rocks. Paleozoic species have been studied using sections and many internal characters of colonies. Only the few exterior characters have been routinely studied in post-Paleozoic species. As a result, morphologic comparisons have not been made. Comparable morphology in newly made sections suggests that some free-walled Jurassic stenolaemates evolved from other Paleozoic orders (See

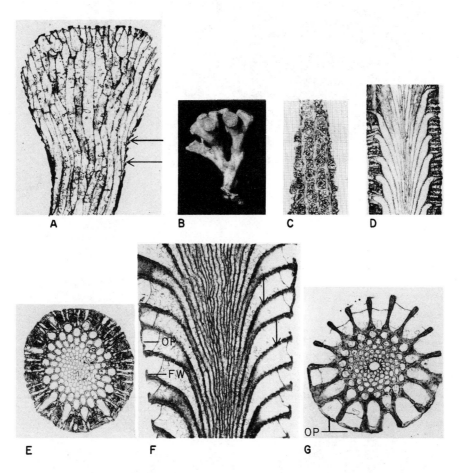

Figure 17.34. Cretaceous Tubuliporata. **A** and **B**, A section (×15) and specimen (×2) of a stalked colony of *Corymbopora* (France). Small polymorphs (arrows) line the sides of colonies; feeding zooids open at the top. **C** to **E**, Sections of *Petalopora* showing the regular zooidal pattern that is suggestive of den-droid cryptostomes (Figure 17.31, **A** to **C**). (France, ×15.) **F** and **G**, Longitudinal and transverse sections showing slender zooidal tubes in a spiral pattern in the endozone, communication pores (arrows), short frontal walls (*FW*), and opercula (*OP*) in *Meliceritites*. (France, ×25.)

the section, "Evolution and Stratigraphic Distribution"). If true, the present concept of the Tubuliporata is polyphyletic, and a major reclassification is required.

Biology

Founding of colonies Most stenolaemate colonies are founded by sexually produced larvae. Larvae of the living members of the class are completely ciliated (Figure 17.35, *A*) and lecithotrophic (Chapter 2) so that their swim periods are short, measured in minutes to a few hours. When metamorphosis occurs, parts of a larva turn inside out to produce an adhesive organ in contact with the substrate (Figure 17.35, *B*) and an exterior cuticle (Figure 17.35, *C*). The exterior cuticle covers the body and is calcified on its inner side to produce the basal disc (Figure 17.35, *D*) of the ancestrula, the first adult member of the colony (Figure 17.35, *F*).

Budding of zooids Asexual reproduction of zooids is initiated by the growth of interior vertical walls of buds into expanding confluent body cavities. Where budding is localized at distal ends and edges of colonies, these cavities are termed **confluent budding zones** (Figures 17.19 and 17.20). They are established by the swelling of membranous exterior colony walls over those actively growing regions.

Individual buds of most stenolaemates are centered on growing edges or corners of interior vertical walls of two to four adjacent zooids (Figure 17.25) or on colony-wide structures. Buds therefore cannot be related to single-parent zooids. Growing edges of vertical walls of buds and adjacent older zooids develop simultaneously and advance evenly into confluent body cavities.

In free-walled colonies, budding occurs in endozones at distal ends or margins and in some growth habits in exozones on all outermost colony surfaces (Figure 17.19,

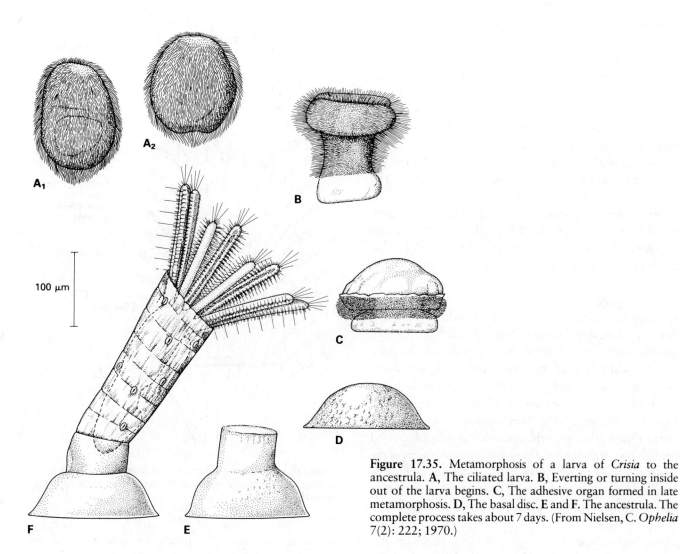

Figure 17.35. Metamorphosis of a larva of *Crisia* to the ancestrula. **A,** The ciliated larva. **B,** Everting or turning inside out of the larva begins. **C,** The adhesive organ formed in late metamorphosis. **D,** The basal disc. **E** and **F.** The ancestrula. The complete process takes about 7 days. (From Nielsen, C. *Ophelia* 7(2): 222; 1970.)

A). The confluent outer body cavities that provide buds with expanding spaces in which to grow occur everywhere above basal encrusting walls.

In the fixed-walled colonies of tubuliporates (Figure 17.20), frontal walls develop early in zooid ontogeny, closing off the confluent outer body cavities. As a result, confluent budding zones and most budding are limited to growing ends and margins of colonies where buds have not yet developed frontal walls.

Growth of zooids Ontogeny in free-walled stenolaemates is expressed chiefly by the progressive lengthening of exozonal portions of zooids. Exozones widen progressively from the youngest zooids in distal parts of colonies to the oldest zooids located proximally (Figure 17.23).

Zooidal chambers in some genera are apparently just long enough (Figure 17.18, *C*) to hold retracted organs in the same position throughout life. Growth continues only by minor additions to outermost membranous walls and outer ends of skeletal walls.

Many stenolaemates have longer zooidal chambers in which retracted positions of organs advanced with substantial skeletal growth. Outward growth of zooids is enough for retracted positions of organs to shift outwardly from endozones of younger zooids (Figure 17.23, *A*) to exozones in older zooids, causing inner ends of chambers to be vacated (Figure 17.23, *B*). In many extinct genera the vacated ends are separated from living chambers by basal diaphragms (Figures 17.16 and 17.23, *C* to *E*).

In living bryozoans most organs of feeding zooids periodically undergo **degeneration-regeneration cycles.** The degenerated organs of a dormant zooid form a brownish, shapeless mass of dead cells, membranes, and other organic material called a **brown body.** During the regeneration phase, new organs are produced and brown bodies are either ejected from zooids or are retained in proximal ends of longer zooidal body cavities (Figure 17.23, *B*).

The outward shift of zooidal organs and the spacing of basal diaphragms in fossils apparently resulted from the degeneration-regeneration cycles. While organs were degenerated, the vertical zooidal walls apparently continued to grow (A_2 in Figure 17.16). The laminated skeletal microstructure indicates that the newest growth of vertical wall (A_2) and the outermost diaphragm (2) grew simultaneously as a single skeletal unit. The distance (B_2) between the last two diaphragms (*1* and *2*) approximates the distance (A_1) that the vertical wall grew in the previous cycle. When the newest organs regenerated, they were displaced outwardly by the distance (A_2) that the vertical wall grew during the newest cycle.

Growth of dendroid colonies   Dendroid colonies develop elongated branches basically because the thin zooidal walls of endozones grow faster than the thick zooidal walls of exozones. Most dendroid trepostome colonies added to that growth differential a growth and resorption cycle of exozones at the growing tips of branches (Figure 17.36). After a period of budding and rapid thin-walled growth, exozones developed around branch tips. Varying lengths of the outermost parts of the exozones were then resorbed. The cycle was completed by a new budding and growth of thin-walled endozone on the remnants of the partially resorbed exozones.

Sexual reproduction   Colonies of living stenolaemates are bisexual. Feeding zooids are either bisexual or unisexual with both sexes of zooids occurring in the same colony. Sperm cells develop within membranous sacs of zooids and are released through the ends of tentacles in at least two species of stenolaemates, similar to the sperm release more generally observed in gymnolaemates. Eggs also develop within membranous sacs and must be fertilized internally. The means by which released sperm enter body cavities of maternal zooids is not known, but cross breeding is assumed.

The embryology of modern stenolaemates is characterized by **embryonic fission**. One or rarely two primary embryos develop in a fertile zooid. Primary embryos divide to form secondary embryos, and in some species secondary embryos divide to form tertiary embryos. One primary embryo reportedly can produce as many as 100 embryos. Adult colonies resulting from fission of a primary embryo presumably are clones, all with the same genetic makeup.

Brood chambers are all enclosed body cavities in steno-

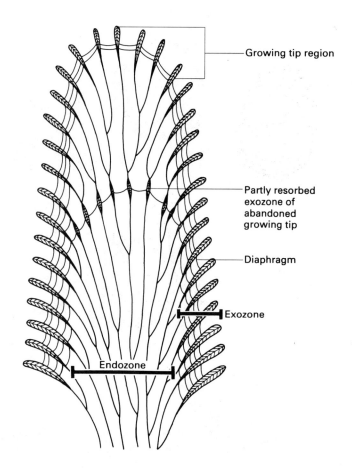

Figure 7.36. Idealized diagram of a longitudinal section of a dendroid trepostome showing the resorption cycle of branch growth.

laemates and must be relatively large to hold the secondary and tertiary embryos and subsequent larvae produced from a single primary embryo. In many taxa the brood chambers are single inflated polymorphs (Figure 17.5, *C*). In others the chambers are extrazooidal on or just below colony surfaces (Figure 17.5, *A* and *B*). It would seem that stenolaemates lacking large chambers of some sort could not brood and undergo any significant amount of fission as described here. Most Paleozoic species lack skeletal indications of large chambers and probably did not have embryonic fission in their reproductive cycles.

Ecology and functional morphology

Stenolaemates are potentially useful paleoenvironmental indicators partly because of the varied growth habits that have been observed to characterize particular habitats in living species. Also, many erect fossil colonies are found more or less intact, indicating little or no transportation after death.

Stable and unstable species Bryozoans can be divided into **stable species**, capable of producing only one growth habit, and **unstable species**, capable of producing two or more growth habits. Unstable species can adjust to different habitats. For example, in quiet water some unstable species develop erect colonies capable of extracting nutrients from water above the substrate. Where currents or waves are so strong that erect colonies are unable to grow, the same unstable species might develop low encrusting colonies. An erect stable species, however, is unable to live in higher energy environments without special adaptations to resist breakage, such as flexible skeletons. Occurrences of fossil species having one growth habit and others having more than one habit suggest that the two kinds of species have been present throughout the history of the phylum.

Feeding process In the feeding process, zooids set up incoming currents with the cilia on their extended tentacles. Most of that current passes between tentacles down to the colony surface. Colonies, therefore, must have some method for water to escape from colony surfaces without interrupting the inflowing currents.

Regions on living stenolaemates where zooids are not feeding or are absent will necessarily function as an outlet because outflow is unopposed. For example, a centrally located brood chamber or cluster of nonfeeding polymorphs surrounded by feeding zooids in small button-shaped colonies form a central outlet (Figure 17.37). Cooperative action of feeding zooids produces centripetal currents strong enough to sweep clay-sized particles from colony surfaces.

Maculae lacking feeding zooids in colonies of the

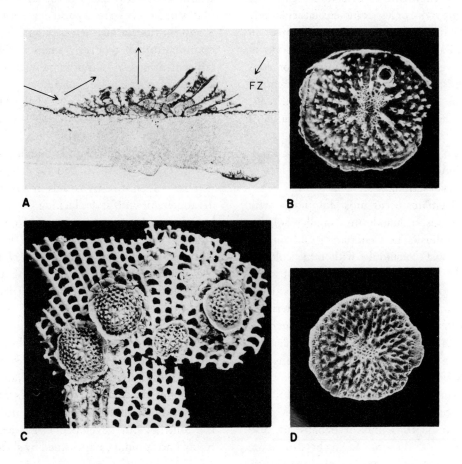

Figure 17.37. Feeding currents. **A,** Longitudinal section through the center of a button-shaped tubuliporate colony that cuts a row of low polymorphs on the left and in the center and longer feeding zooids (*FZ*) on right. (*Disporella*, Holocene, Puerto Rico, ×22.) Currents (arrows) produced by feeding zooids bring water and nutrients in around the margins of the colony and out through the center of the colony. **B,** Another *Disporella* from the same locality showing short nonfeeding polymorphs clustered in the center of the colony and between radial rows of the longer feeding zooids. **C,** Permian cystoporate colonies on the back side of a fenestrate with radially arranged feeding zooids that must have had feeding currents similar to *Disporella*. (Texas, ×7.) **D,** Silurian cystoporatid with comparable growth habit. (Tennessee, ×3.5.)

Paleozoic (Figures 17.21 and 17.29, *E* to *G*) must have been outlets when surrounding zooids were feeding. Regular spacing, which is characteristic of maculae, probably contributed to equalizing currents for intervening feeding zooids.

Colonies with slender branches of one to several feeding zooids at any one level generally lack maculae. Zooids apparently are not concentrated enough to cause outflow problems from these delicate colonies. Reticulate and pinnate colonies have feeding zooids only on one side of branches so that currents flow from front to back sides without opposition (Figure 17.32, *A* and *D*).

Physical and biologic environments   All stenolaemates apparently have been marine. Paleozoic stenolaemates occur in most impure calcareous rocks. They are most commonly found in calcareous shales or in thin-bedded limestones alternating with shales or mudstones. Where abundant in these lithologies, erect colonies generally predominate and are more or less complete, either largely unbroken or in broken segments concentrated in lenticular masses. Paleozoic stenolaemates also occur in fine-grained massive limestones and are a major constituent of the transported debris of many calcarenites. Bryozoans acted as frame builders in some fossil reefs, such as many of the Permian reefs and reef knolls of Texas.

Stenolaemates do not occur in most Paleozoic sandstones, siltstones, and fissile shales due in part to lack of firm surfaces for larval settling.

Paleozoic stenolaemates occur most commonly with articulate brachiopods, echinoderms, smaller solitary corals, and some mollusks. Low encrusting and massive bryozoans occur most commonly with large colonial corals. Most rocks composed predominantly of algal, sponge, stromatoporoid, inarticulate brachiopod, or foraminiferal remains contain little bryozoan material.

Post-Paleozoic stenolaemates seem to have much the same environmental requirements as their Paleozoic ancestors, occurring most abundantly in calcareous lithologies and rarely or not at all in sands and some clays, especially those that accumulated rapidly. Mollusks generally replace brachiopods in abundance in post-Paleozoic bryozoan deposits, but bryozoans are also found where brachiopods do occur. Living bryozoans are common constituents of reefs and occur most abundantly in forereef zones and on spreading fringes where they contribute to the organic debris.

Evolution and stratigraphic distribution

Ordovician stenolaemates   Bryozoa are unknown in the Cambrian and extremely rare in Lower Ordovician rocks. They appear in large numbers representing few species in some of the oldest Middle Ordovician rocks and become biostratigraphically significant with an explosion of new taxa through the Middle and Upper Ordovician (Figure 17.28). Families typically appear fully differentiated in the Middle Ordovician as if they had evolved for long periods in other regions and had migrated into the areas collected.

Trepostomes are the most abundant and diversified of the orders of Bryozoa in Ordovician rocks. Their colonies are characteristically massive or thickly branched, and many have abundant small polymorphs with closely spaced diaphragms (Figure 17.26, *C* and *D*). Some of the more commonly occurring genera in the Ordovician are characterized by feeding zooids with curved skeletal plates in closely spaced parallel repetition (Figure 17.26, *E* and *F*), alternating shelves (Figure 17.26, *A* and *B*), or a combination of relatively small zooidal diameters and concentrations of diaphragms in inner ends of zooidal cavities (Figure 17.26, *C* and *D*).

Bifoliate cryptostomes also evolved to their maximum diversity during the Ordovician. One of the major groups is marked by a combination of zooids in a pronounced longitudinal alignment and abundant small interzooidal styles (Figure 17.30, *C* and *D*). The other large group usually is identified by zooids in a pronounced rhombic arrangement with styles lacking (Figures 17.30, *E* to *G*).

Other orders are generally less noticeable on Ordovician outcrops. One exception is the cystoporate with star-shaped maculae (Figure 17.29, *E* to *G*). Fenestrates are characterized by colonies of relatively delicate anastomosing branches containing feeding zooids with elongate living chambers (Figure 17.32, *E*).

Silurian stenolaemates   In the Silurian, stenolaemates characteristically developed smaller and more delicate colonies than those of the Ordovician and were reduced in numbers of genera. Trepostomes with diaphragms in inner ends of zooidal cavities continued to be abundant in the Silurian but developed larger feeding zooids, and many had zooidal cavities lined by skeletal spines (Figure 17.27, *C* and *D*).

Silurian bifoliate cryptostomes are characterized by a combination of zooids arranged in longitudinal rows and separated by distinct skeletal ridges (Figure 17.30, *A* and *B*).

The fenestrates are characterized from the Silurian through the remainder of the Paleozoic by relatively short zooids and cross bars connecting vertical branches. In the Silurian, the branches are relatively straight, and zooidal chambers are separated longitudinally by planar axial walls (Figure 17.32, *F*).

Devonian stenolaemates In the Devonian, cystoporates and trepostomes again developed more robust colonies (Figure 17.23, *F*) comparable in size but differing taxonomically from those of the Ordovician. Many trepostome genera differed from Ordovician forms most noticeably by the absence of small polymorphs between maculae. Where present, polymorphs are usually widely scattered (Figure 17.23, *G*) and have few or no diaphragms. Cystoporates reached their maximum diversity with well-developed lunaria and encrusting, massive, or bifoliate colonies.

The fenestrates increased in diversity and abundance in the Devonian. Hemisepta in feeding zooids of fenestrates became common for the first time and were prevalent throughout the remainder of the Paleozoic (Figure 17.32, *B*). Zooidal chambers were separated longitudinally by zigzag walls in many species (Figure 17.32, *C*). Fenestrates with extrazooidal skeletal lattices supported on spines in front of zooids (Figure 17.32, *D*) were common, and few of these lattices survived into the Mississippian.

Upper Paleozoic and Triassic stenolaemates Upper Paleozoic stenolaemates are dominated by a great profusion of delicate, fragile-appearing colonies. For example, the cystoporates were primarily slender bifoliate colonies of different shapes (Figure 17.29, *A* and *B*). Slender dendroid cryptostomes became diversified and abundant enough to be useful stratigraphically in the Devonian (Figure 17.18, *A* and *B*) and reached their maximum abundance and number of genera in the Mississippian (Figure 17.31, *A* to *D*).

Fenestrates were the most abundant of the fragile bryozoans in upper Paleozoic rocks. They are characterized by a wide range of coarseness of the reticulate spacing between branches from species to species. The reticulate spacing is largest in many of the Pennsylvanian and Permian species. Fenestrates also developed massive extrazooidal supports that occur abundantly from the Upper Mississippian through most of the Permian in different regions. The supports usually are corkscrew shaped (Figure 17.3, *A* to *C*). In the Permian, reticulate species with ornate extrazooidal skeletal lattices reappeared, and colonies of pinnate fenestrates became locally abundant (Figure 17.32, *A*).

Trepostomes were progressively less common in the upper Paleozoic, but some developed distinctive annular thickenings of vertical walls to give a beaded appearance in sections (Figure 17.27, *A* and *B*).

Triassic Bryozoa are rare and not generally useful stratigraphically. Most are holdovers from the Paleozoic, including several genera of trepostomes and cystoporates and a fenestrate reported from the Soviet Union.

Jurassic stenolaemates The Jurassic Period marked the first appearance of several major new morphologic features that have been considered characteristic of the order Tubuliporata since that time. Perhaps the most important of these are communication pores that appeared in vertical zooidal walls in the Lower Jurassic. Less long-lasting were the distinctive new zooidal arrangements within colonies, such as a continuously spiralling, overgrowing surface to form elongate colonies (Figure 17.33, *F* to *H*).

New Jurassic structures were combined with others apparently inherited from Paleozoic stocks. A few dendroid, free-walled genera (Figure 17.33, *A* and *B*) have long zooids arranged without apparent pattern, annular thickenings of vertical walls, and thin basal diaphragms, all presumably inherited from Paleozoic trepostomes (Figure 17.27, *A* and *B*). In Jurassic species, new communication pores were added to vertical walls located in the thinner-walled intervals between annular thickenings (Figure 17.33, *A* and *B*). These generally robust forms have continued to the present. One Jurassic and Cretaceous group apparently added to those structures narrowed zooidal cavities in endozones and frontal walls to become fixed-walled species (Figure 17.33, *E*). Zooidal arrangements (Figure 17.34, *C* to *E*) comparable to those of Paleozoic cryptostomes also occur in Cretaceous stenolaemates now classified as tubuliporates. The presence of small button-shaped cystoporates in the Paleozoic (Figure 17.37, *C* and *D*) and comparable tubuliporates with lunaria-like structures (Figure 17.37, *A* and *B*) suggest that cystoporates also have survived to the present.

Most fixed-walled Jurassic tubuliporates (Figure 17.33, *C* and *D*) presumably evolved from the few encrusting fixed-walled Paleozoic species. The addition of communication pores to vertical walls in the Jurassic apparently provided physiologic connections among zooids of fixed-walled colonies, allowing them to flourish for the first time. Their diversity increased, and they developed new erect growth habits to become major parts of bryozoan faunas from the Jurassic and onward.

Cretaceous stenolaemates   An explosion of new genera produced the maximum post-Paleozoic diversity of stenolaemates during the Cretaceous. Cretaceous genera are characterized in part by dendroid colonies of free- or fixed-walled slender branches. Most had geometrically regular arrangements of zooids within branches (Figure 17.34, *C* to *G*). Some of the dendroid forms that apparently evolved from Jurassic stocks and had relatively slender zooidal tubes in endozones (Figure 17.33, *E*) added opercula to the apertures of fixed walls (Figure 17.34, *F* and *G*).

Small button-shaped colonies with single radial arrangements of feeding zooids (Figure 17.37) are common from the Cretaceous onward. Circular or radial arrangements of feeding zooids raised on stalks (Figure 17.34, *A* and *B*) are also characteristic of the Cretaceous.

Cenozoic stenolaemates   Stenolaemates were greatly reduced in diversity during the Late Cretaceous, and by the Paleocene only about a fifth of the 175 or so Cretaceous genera remained. The decline of the stenolaemates occurred while the gymnolaemates were greatly increasing in diversity and abundance (see the next section), suggesting that the stenolaemates generally competed unsuccessfully with these newly expanding bryozoans.

In present-day seas, stenolaemates have apparently made a moderate comeback in numbers of genera. They are more common and diversified in the Mediterranean and Pacific Ocean basins than in the Atlantic Ocean.

Class Gymnolaemata

Alan H. Cheetham

The Gymnolaemata are the most morphologically varied of the three classes of Bryozoa. The simplest gymnolaemates are uncalcified and have no polymorphs and no astogenetic change. They occur as flat, encrusting colonies in which simple zooids grow and function almost as autonomously as solitary animals, each closely dependent on the substrate. The most complex gymnolaemates have elaborately calcified zooids, many with calcite and aragonite layers in the same skeleton. Their polymorphs are commonly of two or more kinds, functions, and positions within a colony, and colonies show astogenetic changes in one or more zones. These properties give the colony itself some of the integrated morphology and cooperative functions characteristic of a miniature tree or a vagile solitary animal, lessening its association with the substrate. Between these extremes,

an array of morphologies developed through time in a general pattern of increasing complexity, enabling the Gymnolaemata to become the most successful class of bryozoans at present, adapted to aquatic habitats from freshwater to abyssal marine.

The greater part of gymnolaemate diversity evolved just since the mid-Cretaceous, about 100 million years ago. The nearly 1000 post-Early Cretaceous genera proliferated at a rate similar to that of other rapidly evolving stocks such as mammals. This part of the gymnolaemate record is represented by abundant, well-preserved fossils in many parts of the world. The earlier part of the record spans much more time, starting by the Late Ordovician, but includes only 10 genera represented by scattered external molds.

Morphology

Zooidal morphology   Body walls of zooids in the Gymnolaemata typically form box-shaped units with pronounced bilateral symmetry (Figure 17.38). The plane of symmetry bisects frontal, orificial, basal, and two of the four vertical walls; it contains the principal growth direction of the zooid.

Vertical walls in most gymnolaemates are calcified and include a pair of longer **lateral walls** on either side of the plane of symmetry and a pair of shorter **transverse walls** bisected by the plane. Usually lateral walls are exterior walls with outermost cuticular layers (Figure 17.40, *A*), and transverse walls are interior walls without cuticles (Figure 17.40, *B*).

Vertical walls are interrupted by **communication organs** (Figure 17.38, *B*) consisting of several types of cells, some of which apparently transmit substances from zooid to zooid. These organs connect adjacent zooids through minute pores in thinned portions of walls called **pore plates**. Communication organs are attached to zooidal organs by funicular strands that may also function in transmitting substances.

An orificial wall in most Gymnolaemata consists of a single distally directed flap with reinforced margins (Figure 17.38, *A*). This movable lidlike structure, the **operculum**, is closed by a pair of muscles when tentacles are retracted.

Frontal walls are involved directly or indirectly in tentacle protrusion. Either a part of the frontal wall (Figure 17.39, *A* to *F*) or a saclike **ascus** derived from it (Figure 17.39, *G* and *H*) forms a flexible membrane to which **parietal muscles** are attached. These muscles pull inward, producing pressure within the zooid that causes the lophophore to protrude through the orifice.

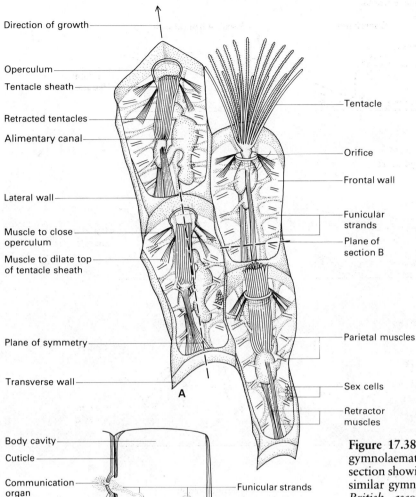

Direction of growth

Operculum

Tentacle sheath

Retracted tentacles

Alimentary canal

Lateral wall

Muscle to close operculum

Muscle to dilate top of tentacle sheath

Plane of symmetry

Transverse wall

A

Tentacle

Orifice

Frontal wall

Funicular strands

Plane of section B

Parietal muscles

Sex cells

Retractor muscles

Body cavity

Cuticle

Communication organ

Pore plate

Basal wall

B

Funicular strands

Lateral wall

Figure 17.38. A, Feeding zooids of a lightly calcified living gymnolaemate, *Membranipora*, from Britain. **B,** Transverse section showing a communication organ in the lateral wall of a similar gymnolaemate. (**A** from Hayward, P. J.; Ryland, J. S. *British ascophoran bryozoans.* London and Cambridge, England: Linnean Society of London and Estuarine and Brackish-Water Sciences Association; 1979. **B** modified from Banta, W.C. *Journal of Morphology* 135 [2]; 1971.)

All zooids having an ascus and many lacking one possess a protective and supportive structure, the **frontal shield,** that can be a calcified part of the frontal wall, termed a **gymnocyst** (Figure 17.39, *A*), or a calcified wall underlying or overlying the frontal wall. An underlying frontal shield, termed a **cryptocyst** (Figure 17.39, *C* and *D*), is an interior wall partitioning the body cavity of a zooid. An overlying frontal shield can consist of discrete hollow tubular or flattened extensions of body wall with contained body cavities called **spines** (Figure 17.39, *E* and *F*), or a series of spinelike extensions fused at their tips and intermittently along their lengths (Figure 17.42, *A*), known as **costae** (singular, **costa**).

Composition and microstructure of skeleton The great majority of living and fossil gymnolaemates have skeletons composed of calcareous material in an organic matrix. The calcareous material may be wholly calcite or

aragonite or calcite and aragonite in discrete layers (Figure 17.40, *C* and *D*). Within a species, skeletal composition is consistently all calcite, all aragonite, or mixed, but within a genus or higher taxon, different species can have different compositions.

Skeletal microstructure can be categorized, with much variation, into two major types: lamellar and spherulitic. Both microstructures commonly occur in the same zooidal skeleton.

Lamellar microstructure (Figure 17.40, *A*, *B*, and *D*) is best developed in calcite skeletons and consists of tablets arranged in more or less continuous layers subparallel to surfaces of walls next to the secreting epidermis. Individual layers can continue from basal walls into lateral walls (Figure 17.40, *A*) or transverse walls (Figure 17.40, *B*). Outer basal wall layers beneath interior vertical walls can continue from zooid to zooid as common skeleton (Figure 17.40, *B*).

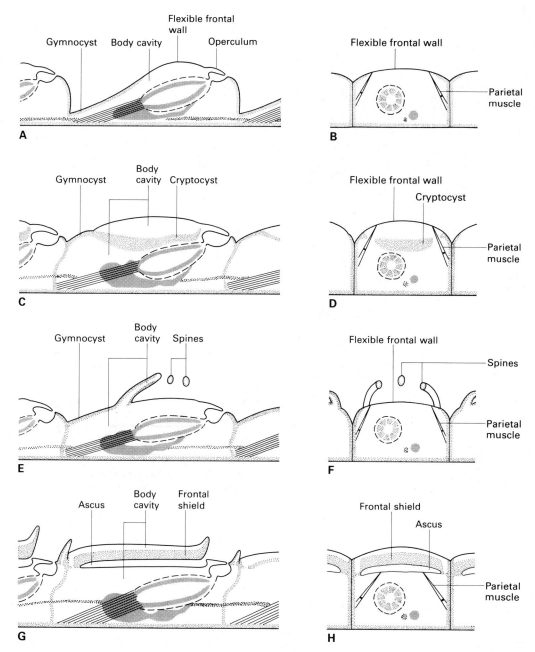

Figure 17.39. Longitudinal (left) and transverse (right) sections showing frontal walls, parietal muscles, and frontal shields in living calcified gymnolaemates. **A** and **B**, A simple gymnolaemate with part of the frontal wall flexible by means of attached parietal muscles and part calcified to form a gymnocyst. **C** and **D**, A complex gymnolaemate with the frontal wall as in **A** and **B**, but the flexible part is underlain by a cryptocyst. **E** and **F**, A complex gymnolaemate with the frontal wall as in **A** and **B**, but the flexible part is overarched by spines. **G** and **H**, A complex gymnolaemate with the ascus flexible by means of attached parietal muscles and overlain by a frontal shield and extension of zooidal body cavity. (From Cheetham, A. H.; Cook, P. L. In: Robison, R. A., editor. *Treatise on invertebrate paleontology, Part G* (revised), *Bryozoa*, vol. 1. Boulder, CO, and Lawrence, KS: Geological Society of America and University of Kansas Press; 1983.)

Spherulitic microstructure consists of needles arranged in conical or fan-shaped arrays. Conical arrays (Figure 17.40, *D*) are more or less at right angles to surfaces of walls next to the secreting epidermis and are well developed in either aragonite or calcite skeletons. Fan-shaped arrays (Figure 17.40, *E*) are parallel to wall surfaces and are limited to outermost skeletal layers of exterior walls deposited next to outermost cuticle. They are best developed in calcite skeletons.

Polymorphs   In addition to ordinary feeding zooids, colonies of most gymnolaemates include one or more

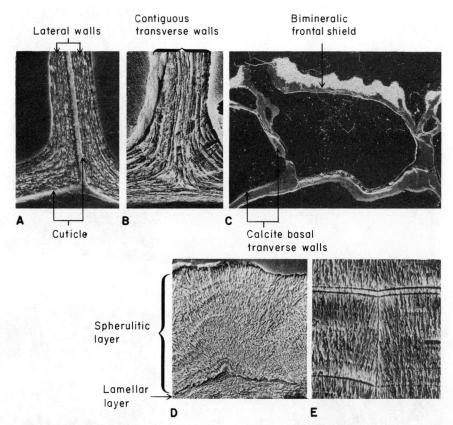

Figure 17.40. Microstructure and composition of skeletons of living gymnolaemates. **A,** Transverse section through the basal-lateral parts of two zooids. Calcite lamellae and the outermost cuticles of the basal walls continue into the lateral walls. (*Schizoporella,* Gulf of Mexico, ×275.) **B,** Longitudinal section through the basal parts of two zooids. Outer calcite lamellae and the outermost cuticle of the basal wall continue from one zooid to the next. Inner calcite lamellae continue from the basal wall into the transverse walls. (*Schizobrachiella,* Britain, ×525.) **C,** Longitudinal section through a zooid with the skeleton stained with a solution specific for aragonite (light areas). The frontal shield has underlying calcite and overlying aragonite layers. The other walls are made entirely of calcite. (*Metrarab-* *dotos,* Gulf of Mexico, ×45.) **D,** Transverse section through the frontal shield with underlying lamellar calcite and overlying spherulitic aragonite. (*Schizoporella,* Gulf of Mexico, ×800.) **E,** Outer surface of the skeleton of the basal wall with spherulitic calcite. (*Petraliella,* Gulf of Mexico, ×1200.) (**A** from Sandberg, P. A. *Micropaleontology* 17 [2]; 1971. **B** from Sandberg, P. A. In: Woollacott, R. M.; Zimmer, R. L., editors. *Biology of bryozoans.* New York: Academic Press, Inc.; 1977. **C** from Sandberg, P. A. In: Boardman, R. S.; Cheetham, A. H.; Oliver, W. A., Jr., editors. *Animal colonies.* Stroudsburg, PA.: Dowden, Hutchinson, and Ross; 1973. **D** from Sandberg, P. A. *Journal of Paleontology* 49 [4]; 1975.)

kinds of polymorphs. Polymorphs commonly have nonprotrusible or no lophophores and thus are incapable of feeding. Some have protrusible lophophores different in size, shape, number of tentacles, or other features from those of ordinary feeding zooids, and these may be capable of feeding.

Kenozooids (Figure 17.41, *A*) are polymorphs that occur in both calcified and uncalcified gymnolaemates and consist of body walls enclosing body cavities, which generally contain only funicular strands attached to communication organs. There is no equivalent of lophophore, orifice, or orificial wall. Kenozooids commonly function as small space-filling structures of variable size and shape or as long, tubular anchoring rootlets.

In calcified gymnolaemates, the most characteristic polymorphs are **avicularia** (Figure 17.41, *B* to *D* and 17.42, *A*), in which the equivalent of an operculum, the **mandible** (Figure 17.41, *B*), is relatively larger and more intricately reinforced than opercula of ordinary feeding zooids. Along one edge, the mandible hinges on a skeletal bar (Figures 17.41, *D* and 17.42, *A*) or on toothlike skeletal protuberances (Figure 17.41, *C*). On its hinge, the mandible opens and closes by action of two sets of muscles apparently equivalent to those in the opercular region of a feeding zooid (Figure 17.38). When closed, the free end of the mandible rests on a beaklike skeletal rim, which may or may not closely approximate the mandible in shape. In all but a few genera the lophophore underlying the mandible is rudimentary and non-

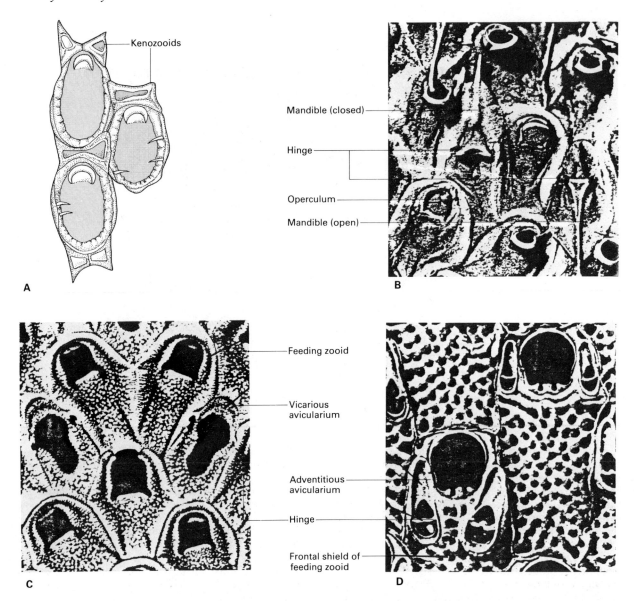

Figure 17.41. Polymorphs in living calcified gymnolaemates. **A,** Space-filling kenozooids of irregular shapes intercalated among feeding zooids in *Conopeum*. (Britain, ×50.) **B,** Vicarious avicularia and feeding zooids in *Smittipora*. (North Carolina, ×45.) When closed (upper left), the mandible extends well beyond the skeletal parts of the avicularium. **C,** Same colony with noncalcified parts removed showing the boundaries and skeletal parts of the avicularia and feeding zooids. (×45.) **D,** Adventitious avicularia on the frontal shields of feeding zooids in *Petraliella*. (Gulf of Mexico, ×65.) (**A** from Ryland, J. S.; Hayward, P. J. *British anascan bryozoans.* London and Cambridge, England: Linnean Society of London and Estuarine and Brackish-Water Sciences Association; 1977.)

protrusible. The action of the mandible serves in cleaning, defense, locomotion, and possibly other functions.

Avicularia can occupy two kinds of positions in a colony. **Vicarious avicularia** are intercalated among ordinary feeding zooids as though taking their place in the budding pattern (Figures 17.41, *B* and *C*, and 17.42, *A*). **Adventitious avicularia** are placed on feeding zooids (Figure 17.41, *D*), becoming virtual appendages of those zooids in extreme cases. Both vicarious and adventitious avicularia occur in the same colony in some genera.

Brood chambers are found in the great majority of living and fossil gymnolaemates. They generally consist of a water-filled space enclosed by body walls except for an opening through which a fertilized egg is deposited and the larva escapes (Figure 17.42, *B*). Zooids whose body walls are modified to enclose a brood chamber and the zooid that deposits eggs in it, called a **maternal zooid** (Figure 17.42, *A* to *C*), are all polymorphs. The body walls enclosing a brood chamber can be contained within the maternal zooid and uncalcified, or can be extended from the distal margin of the maternal zooid or from

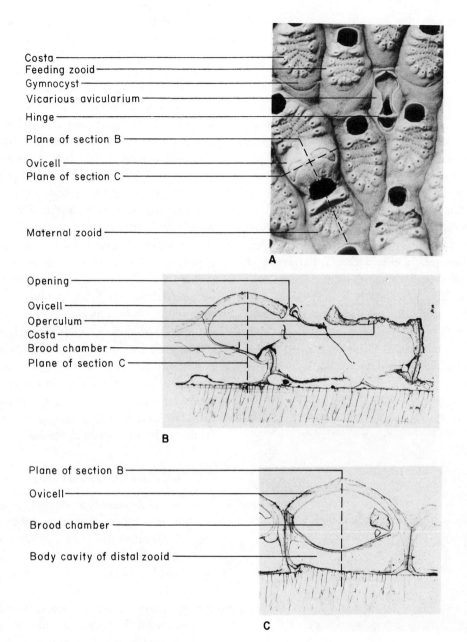

Costa
Feeding zooid
Gymnocyst
Vicarious avicularium
Hinge

Plane of section B

Ovicell
Plane of section C

Maternal zooid

A

Opening
Ovicell
Operculum
Costa
Brood chamber
Plane of section C

B

Plane of section B
Ovicell
Brood chamber
Body cavity of distal zooid

C

Figure 17.42. Feeding zooids and polymorphs in a living calcified gymnolaemate. (*Figularia,* Mediterranean.) **A,** The vicarious avicularium, maternal zooid, and ovicell occur among ordinary feeding zooids. (×30.) The frontal shields of feeding zooids consist of costae within marginal gymnocysts. **B,** Longitudinal section through the maternal zooid, ovicell, and part of a distal zooid (to the left). (×60.) **C,** Transverse section through the ovicell and underlying part of the distal zooid. (×60.)

distal zooids and generally calcified. The structure formed by extended body walls is called an **ovicell** (Figure 17.42, *A* to *C*).

Classification

The class Gymnolaemata is considered to include two orders of highly unequal diversity. The more diversified order is here considered to comprise three suborders, each of which is as distinct morphologically and as diversified taxonomically as an order in the class Stenolaemata.

Classification of the Gymnolaemata at both the order and suborder levels emphasizes characters of feeding zooids that are the basic elements of all colonies and the only ones in simplest members of the class. Characters of polymorphism, colony growth habit, and (if applicable) skeletal composition and microstructure are more variable, generally with overlapping states among taxa. Many overlaps apparently arose through convergent evolution and thus can only be used at lower levels of classification.

ORDER CTENOSTOMATA (Figure 17.43, A to C)

Uncalcified gymnolaemates having feeding zooids with parietal muscles attached to flexible frontal walls. Orificial walls of one or more flaps or ringlike, rarely opercula. Polymorphs, where present, kenozooids of various form usually connecting or supporting feeding zooids individually or in groups, or brooding zooids. Brooding, where present, most commonly within body cavity, possibly in water-filled brood chambers in two genera. Extrazooidal parts unknown. Growth habits include encrusting, erect and flexible, completely immersed in calcareous substrates such as shells or in organic material, and in one genus, free-living. (Late Ordovician to Holocene; about 50 genera, three-fourths of which are known only living.)

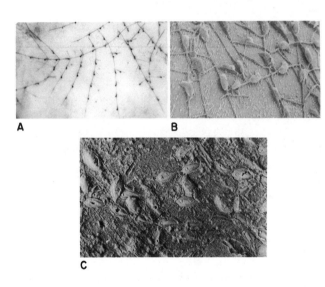

Figure 17.43. Fossil boring and nonboring ctenostomes. **A** and **B**, *Cookobryozoon*. (Pliocene, Australia, ×5.) **A**, The boring lies in a gastropod shell that was treated to make it translucent. **B**, Polyester cast made from a boring to show the form of feeding zooids and connecting stolons. (Miocene, California, ×20.) **C**, *Arachnidium*. (Middle Jurassic, Germany, ×10.) A mold of a soft-bodied colony on the attachment area of a shell of an overgrowing oyster. The orificial walls of the zooids are ringlike with porelike orifices. The specimen is photographed so that the zooids appear to stand up in relief. (**A** and **B** from Pohowsky, R. A. *Bulletins of American Paleontology* 73 (301); 1978. **C**, from Voigt, E. *Neues Jahrbuch für Geologie und Paläontologie* 153 [2]; 1977.)

Fossils assigned to the Ctenostomata are all external molds found in small numbers at relatively few levels in marine deposits of Paleozoic, Mesozoic, and Cenozoic ages. Most were produced as borings by soft-bodied colonies growing **immersed** within calcareous substrates (Figure 17.43, *A*). The method of boring was apparently chemical but is not understood even in living species. Some paleontologists consider fossil ctenostome borings to be trace fossils (Chapter 21), and a dual nomenclature exists for some species. Borings conform closely to the forms of zooids. This is best observed in artificial casts (Figure 17.43, *B*) made by filling borings with polyester resin and dissolving the calcareous substrate.

Immersed colonies consisted of widely spaced feeding zooids with orifices opening separately to the surface of the substrate. Within the substrate, feeding zooids were connected by networks of **stolons**. In most living species, stolons are tubular kenozooids with body cavities connected to those of feeding zooids only by communication organs. In living species of one genus, stolons are extensions of feeding zooids themselves.

The only fossils of nonboring colonies assigned with confidence to the order are a few molds produced by overgrowth of soft-bodied colonies by shelled organisms such as oysters (Figure 17.43, C). These fossils have been found in Jurassic and Cretaceous deposits and are closely related to living nonboring species.

ORDER CHEILOSTOMATA

Calcified gymnolaemates having feeding zooids with parietal muscles variously attached to flexible parts of frontal walls or to ascus. Orificial walls opercula, except in few genera. Polymorphs, where present, include avicularia; supporting, connecting, or anchoring kenozooids; and feeding or nonfeeding sexual or brooding zooids. Brooding commonly in water-filled spaces enclosed by ovicells; embryos released without brooding in few genera. Extrazooidal parts present or absent. Growth habits include single and multiple layered encrusting; erect and rigid, jointed, or flexible; and free-living. (Late Jurassic to Holocene; about 1000 genera, the majority known from both fossil and living species.)

The calcareous skeletons of cheilostomes are readily preserved and commonly only slightly altered in fossilization. As a result, fossils of this order are abundant in deposits of the Late Cretaceous to Pleistocene in many parts of the world. In marine sediments of these ages deposited near outer margins of continental shelves or at equivalent depths off oceanic islands or on sea mounts, cheilostomes can comprise a major biotic component of limestones or calcareous clays. In rocks of the Late Jurassic and Early Cretaceous, cheilostomes are both less common and less diversified, evidently as a reflection of their gradual establishment at the beginning of their history.

SUBORDER ANASCA (Figures 17.44, *A* to *E*; 17.45, *A, B,* and *D* to *G*)

Cheilostomes with feeding zooids having parietal muscles attached to flexible parts of frontal walls. Frontal shields, where present, spines, gymnocysts, cryptocysts, or combinations. Polymorphs most commonly vicarious avicularia, ovicell-bearing zooids, and kenozooids; less commonly adventitious avicularia. Extrazooidal skeleton commonly developed in free-living colonies. Growth habits include all known in order. (Late Jurassic to Holocene; about 400 genera.)

All pre–Late Cretaceous cheilostomes are assigned to the Anasca, and this suborder remained dominant through Late Cretaceous and Paleocene time. Later Tertiary and Quaternary anascans have maintained high diversity and abundance, although they have lost dominance to the Ascophora.

SUBORDER CRIBRIMORPHA (Figure 17.45, *C*)

Cheilostomes with feeding zooids having parietal muscles attached to flexible parts of frontal walls covered by overarching, discontinuous frontal shields composed of costae with gaps between. Marginal gymnocysts generally present. Polymorphs most commonly vicarious avicularia and ovicell-bearing zooids; less commonly adventitious avicularia and kenozooids. Extrazooidal skeleton apparently developed over frontal shields of zooids in some extinct genera. Growth habits include encrusting, less commonly rigidly erect, and rarely free-living. (Late Cretaceous to Holocene; about 100 genera.)

Although locally abundant in Upper Cretaceous and Cenozoic deposits, the Cribrimorpha have never been as abundant and diversified as the other suborders. Generic diversity peaked within the Late Cretaceous and has been maintained at a relatively low level throughout the Cenozoic.

SUBORDER ASCOPHORA (Figure 17.46, *A* to *E*)

Cheilostomes with feeding zooids having parietal muscles attached to ascus. Frontal shields continuous, not costate, and present in all; in most with overlying body cavity and outer membranous wall; in few with only adjacent outermost cuticle. Polymorphs most commonly adventitious avicularia and ovicell-bearing zooids; less commonly vicarious avicularia and kenozooids. Extrazooidal skeleton common in many genera with rigidly erect colonies. Growth habits include all known in order. (Late Cretaceous to Holocene; about 500 genera.)

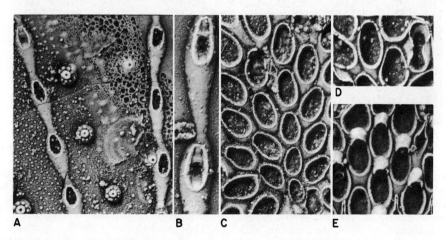

Figure 17.44. Lower Cretaceous anascan cheilostomes from Texas. **A,** *Pyriporopsis* with simple uniserially arranged zooids encrusting an echinoid. (×17.) **B,** Two zooids in which membranous parts of frontal walls had calcified to preserve form of opercula. (×30.) **C,** *Wilbertopora* with simple multiserially arranged zooids in the zone of change beginning with the ancestrula (lower center). (×30.) **D,** Another colony with a vicarious avicularium similar in form to ordinary feeding zooids. (×30.) **E,** A colony from a higher stratigraphic level with vicarious avicularia less similar in form to the feeding zooids and numerous zooids with ovicells. (×30.)

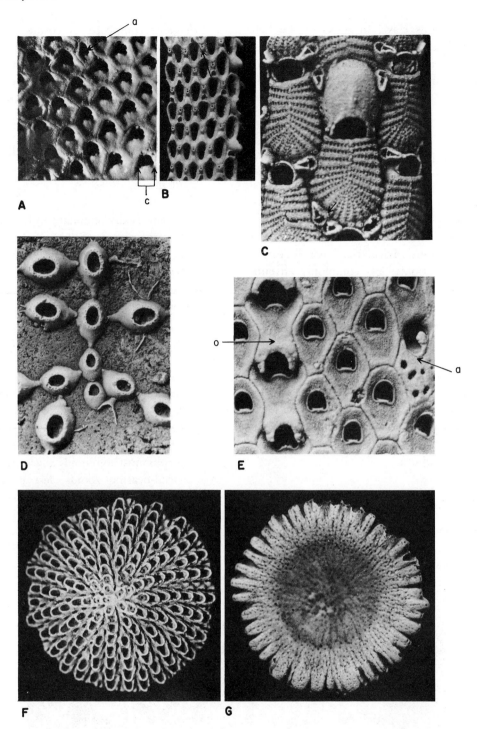

Figure 17.45. Upper Cretaceous and lower Tertiary anascan and cribrimorph cheilostomes. **A**, An anascan with extensive cryptocysts (*c*), notched (arrows) for passage of the parietal muscles, and vicarious avicularia (*a*). (*Hagenowinella*, Upper Cretaceous, The Netherlands, ×13.) **B**, An anascan with narrow cryptocysts and small adventitious avicularia (arrow) on gymnocysts. (*Stamenocella*, Upper Cretaceous, The Netherlands, ×13.) **C**, A cribrimorph with numerous thin costae and small vicarious avicularia (arrow). (*Castanopora*, Paleocene, New Jersey, ×30.) **D**, A uniserial anascan colony encrusting an oyster. The distinct zone of change begins with a small oval ancestrula. (*Allantopora*, Paleocene, New Jersey, ×20.) **E**, An anascan with extensive cryptocysts notched for passage of the parietal muscles and a large vicarious avicularium (*a*) and ovicells (*o*). (*Coscinopleura*, Paleocene, New Jersey, ×30.) **F**, The upper side of a free-living anascan colony with feeding zooids having narrow cryptocysts and small vicarious avicularia in rows between feeding zooids. (*Lunulites*, Eocene, Alabama, ×13.) **G**, The basal side of the same colony as in **F** with a thick extrazooidal skeleton in radiating sectors. (×13.) (**A** and **B** from Voigt, E. *Grondboor en Hamer*. Losser, The Netherlands: Nederlandse Geologische Vereniging; 1979.)

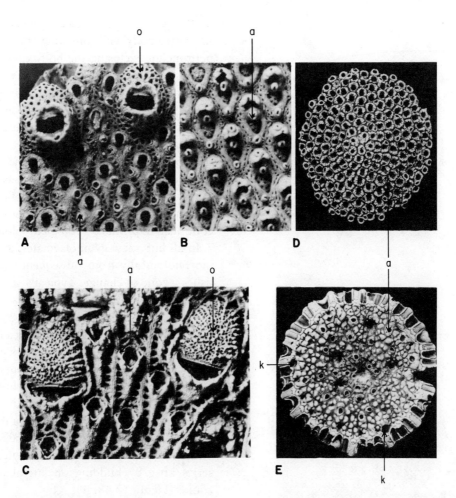

Figure 17.46. Tertiary ascophoran cheilostomes. **A,** *Schizostomella.* Feeding zooids with the orifice proximally notched for the opening to the ascus. The frontal shield has marginal perforations. The avicularia (*a*) are small and adventitious, and the ovicells (*o*) are large. (Eocene, France, ×30.) **B,** *Adeonellopsis.* Feeding zooids with a large perforated central disc that forms the opening to the ascus, a frontal shield with marginal perforations, and small adventitious avicularia (*a*). (Oligocene, Alabama, ×30.) **C,** *Metrarabdotos.* Feeding zooids with marginally perforate frontal shields, orifices hidden by collarlike extensions, small adventitious avicularia (*a*), and large ovicells (*o*). (Miocene, Dominican Republic, ×30.) **D,** *Mamillopora.* Upper side of a loosely anchored, rooted colony with feeding zooids that have reduced frontal shields and small adventitious avicularia (*a*). (Pliocene, Jamaica, ×10.) **E,** Basal side of same colony as in **D** with openings of rootletlike kenozooids (*k*) and basal walls of feeding zooids supporting adventitious avicularia (*a*).

The Ascophora appeared last in the fossil record but by Eocene time had achieved dominance which has been maintained to the present. Major differences in ontogeny of ascus and frontal shield (see discussion of ontogeny that follows) suggest the Ascophora may be polyphyletic, but evidence from the fossil record is still insufficient to test this hypothesis.

Biology

Founding of colony In most Gymnolaemata, colonies are founded most frequently or exclusively by sexually produced larvae. Larvae in this class are of two types: planktotrophic and lecithotrophic (Chapter 2). Planktotrophic larvae (Figure 17.47, *A* and *B*) are similar to those of many other metazoans in that they have functional alimentary canals and are able to live in the plankton for extended periods of time. These larvae are covered by bivalved shells composed of cuticle. They are found in a few genera of Ctenostomata and Anasca. Lecithotrophic larvae (Figure 17.47, *C* and *D*) are more specialized but characterize most Ctenostomata and Cheilostomata. These larvae have incomplete or no alimentary canal and generally lack a shell. They have short motile stages, as little as a few hours in many.

Once attached to a substrate, larvae of both types undergo similar patterns of metamorphosis, including complex reorganization of tissues. Unlike most other metazoans, they lose all traces of a larval gut and other structures derived from embryonic endoderm (Chapter

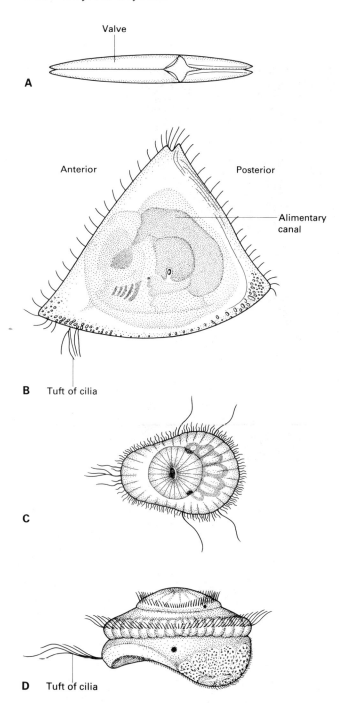

Valve

A

Anterior

Posterior

Alimentary
canal

B Tuft of cilia

C

D Tuft of cilia

Figure 17.47. Top and side views of gymnolaemate larvae. **A** and **B**, A bivalved planktotrophic larva of anascan *Membranipora*. (Britain, ×80.) **C** and **D**, A lecithotrophic larva of ascophoran *Microporella*. (France, ×110.) (**A** and **B** from Ryland, J. S.; Hayward P. J. *British anascan bryozoans*. London and Cambridge, England: Linnean Society of London and Estuarine and Brackish-Water Sciences Association; 1977. **C** and **D** from Hayward, P. J.; Ryland, J. S., *British ascophoran bryozoans*. London and Cambridge, England: Linnean Society of London and Estuarine and Brackish-Water Sciences Association; 1979.)

2). Reorganization produces a hemispherical or flattened sac with a body wall consisting of cuticle, epidermis, and peritoneum enclosing the body cavity. Body walls and body cavities of all zooids and any extrazooidal parts of a colony form by expansion and partitioning of the sac.

The expanding sac first forms one or more founding zooids, usually a single zooid, the ancestrula (Figures 17.6; and 17.48, *A*). Parts of the sac forming the distal transverse and lateral walls of the ancestrula now swell to produce buds. Body cavities of buds become cut off from that of the ancestrula by ingrowth of interior walls. Buds become zooids that produce new buds, and the process is continued to produce a complete colony.

In all but a few gymnolaemates, an ancestrula differs morphologically from subsequent zooids and so establishes a zone of change. Typically several generations of intermediate morphology lead to a zone of repetition.

Growth of colony   Budding typically produces **lineal series** consisting of distal sequences of zooids bounded laterally by exterior walls and in contact with each other along interior walls. This arrangement is best seen in uniserial colonies in which a single lineal series is isolated (Figure 17.48, *A*). Each of the one or more lineal series of a uniserial colony is produced by the repeated process of budding from distal margins of parent zooids. New lineal series arise by budding on lateral margins of parent zooids (Figure 17.48, *A*).

In multiserial colonies more typical of gymnolaemates, lineal series are less easily observed because lateral walls of zooids in adjacent series are in contact and contain communication organs (Figure 17.48, *B*). Except for their lateral communications, these series grow in the same way as those in uniserial colonies. Lateral contact is along outermost cuticles of exterior walls, in which openings are dissolved so that cells can connect to form communication organs.

In multiserial growth, cooperation is evident in the opening of lateral communications and in the production of smooth growing edges consisting of coordinated distal ends of numerous lineal series (Figure 17.48, *B*). In encrusting colonies, subcircular growing edges commonly develop by lineal series wrapping around the proximal margin of the ancestrula and by the outward growth in all directions from these inner zooids (Figure 17.48, *B*).

Ontogeny of zooids   The bounding walls of zooids are established early in ontogeny during bud expansion. The orificial wall and zooidal organs then develop from parts

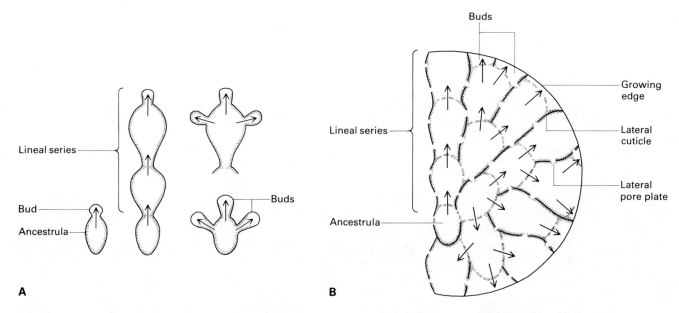

Figure 17.48. Growth of colonies in gymnolaemates. Arrows indicate directions of budding. **A**, Stages in growth of a single lineal series (left) and a branching lineal series (right) of a uniserial colony. **B**, A multiserial colony with the lineal series in contact along the cuticles of the lateral walls of zooids. The growing edge becomes coordinated around the whole periphery by wrapping of the lineal series around the ancestrula. (From Cheetham, A. H.; Cook, P. L. In: Robison, R. A., editor. *Treatise on invertebrate paleontology, Part G* (revised), *Bryozoa*, vol. 1. Boulder, CO, and Lawrence, KS: Geological Society of America and University of Kansas Press; 1983.)

of these body walls. Once a zooid becomes functional in the Ctenostomata and the simplest Cheilostomata, there may be little further change except for cyclic regeneration of organs. In many Cheilostomata, however, growth and continued modification of frontal shields can produce profound changes in the appearance and function of zooids farther from the growing edge.

In the Anasca, the most profound ontogenetic changes take place in cryptocysts (Figure 17.49, *A*). Gymnocysts calcify early in bud expansion and thus are little developed beyond proximal ends of zooids. Further extension is limited by the necessity for part of the frontal wall to remain flexible in tentacle protrusion. Cryptocysts can extend distally to the operculum, leaving openings for passage of parietal muscles already attached to the overlying flexible frontal wall. Unlike a gymnocyst, a cryptocyst is overlain by cellular body wall layers and body cavity below the outermost cuticle of the frontal wall. Therefore it commonly receives calcareous layers on its frontal side throughout zooid life.

In the Ascophora, frontal shields commonly grow into the zooidal body cavity below the membranous frontal wall in the same way as an anascan cryptocyst (Figure 17.49, *B*). Unlike an anascan cryptocyst, however, this kind of frontal shield in an ascophoran grows all the way to the operculum before the parietal muscles appear. At or near the proximal margin of the operculum, part of the membranous frontal wall then extends by infolding under the frontal shield to form an ascus. Finally, parietal muscles attach to the ascus floor. The difference in the ontogenetic sequence in which frontal shield and parietal muscles develop in these anascans and ascophorans is an example of **heterochrony**. The process of heterochrony, or change in sequence of development, is thought to be a source of some large-scale morphologic changes by which higher taxa might suddenly evolve.

In many other ascophorans, a frontal shield and ascus of similar appearance to those in the first group are produced in different ways, one of which is illustrated in Figure 17.49, *C*. Most of these frontal shields can continue to receive calcareous layers on outer sides throughout zooid life.

Sexual reproduction   Colonies become sexually mature when some zooids, usually in zones of repetition but proximal to growing edges, begin to produce gametes. These are formed in simple structures within body cavities of zooids (Figure 17.38), both sperm and eggs in the same colony. Sexual zooids in the Gymnolaemata commonly retain protrusible lophophores from which gametes are released.

Sexual reproduction is simplest in nonbrooding gymnolaemates. Eggs and sperm are produced in the same zooids, which are ordinary feeding zooids without

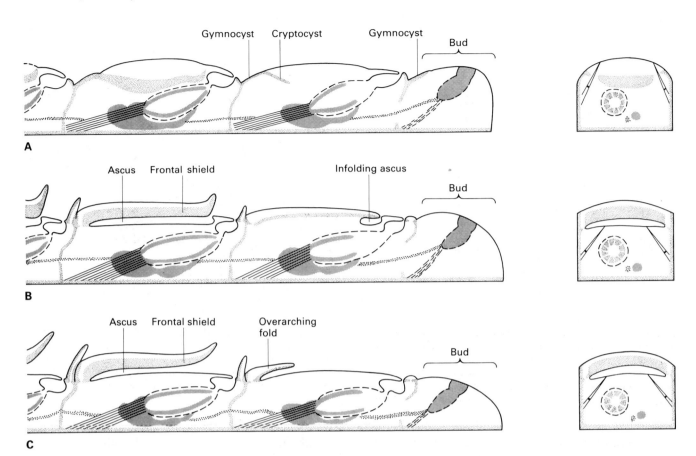

Figure 17.49. Ontogeny of feeding zooids in Cheilostomata. Longitudinal (left) and transverse (right) sections through zooids and buds at growing edges. **A,** An anascan with a gymnocyst formed at the proximal end of the bud and a cryptocyst extending into the body cavity. The cryptocyst continues to thicken after other zooidal structures are complete. **B,** An ascophoran with the frontal shield developing as a cryptocyst beneath which the ascus infolds. **C,** An ascophoran with the frontal shield developing as part of an outfold overarching the frontal wall, which becomes transformed into the ascus floor. (From Cheetham, A. H.; Cook, P. L. In: Robison, R. A., editor. *Treatise on invertebrate paleontology, Part G* (revised), *Bryozoa,* vol. 1. Boulder, CO, and Lawrence, KS: Geological Society of America and University of Kansas Press; 1983.)

skeletal expression of sexual function. Sperm are released through tentacle tips and travel in feeding currents to other lophophores, presumably in different colonies. Eggs are extruded through an opening on the distal part of the lophophore. Fertilization takes place just before or after the egg has been extruded. Fertilized eggs are shed into the water in larger numbers, where embryos develop into planktotrophic larvae.

In brooding gymnolaemates, eggs are produced by maternal zooids usually associated with ovicells. Sperm can be produced by maternal zooids, nonmaternal zooids, or both. In most species, release of gametes is apparently similar to that in nonbrooding gymnolaemates. Once fertilized, eggs are extruded into brood chambers to undergo embryonic development into lecithotrophic larvae. Generally, only one embryo occupies a brood chamber at a time. The lophophore of the maternal zooid may degenerate after depositing an egg and subsequently regenerate to deposit another after the brood chamber is evacuated by the first larva.

Ecology and functional morphology

Distribution Gymnolaemata occur throughout the broad range of aquatic habitats available to the phylum but are characteristic of marine environments of normal salinity and generally low sedimentation rate. A few species have been found at salinities almost twice that of normal sea water, and a few, all ctenostomes, inhabit fresh water. A larger number of species of both orders live in brackish water.

Ctenostomata and all three suborders of Cheilostomata are found in all major regions of the world's oceans. The number of genera increases toward the tropics, especially in the Ascophora, but more gradually than in the Mollusca. Many gymnolaemate species and genera are restricted to one or a few contiguous provinces, but even at these taxonomic levels, geographic distributions tend to be wider, more eurygeographic (Chapter 7), than in other benthic invertebrates.

Many warm-water marine species have east-west distributions virtually circling the globe (*Nellia* in Figure 17.50, *A*). Such **circumtropical** distributions were achieved in the Tertiary (Figure 17.50, *B*), when tropical climates extended farther north and south and marine connections existed through Central America and the Middle East. Some cheilostomes with narrower present-day distributions (*Poricellaria* in Figure 17.50, *A*) also were circumtropical in the Tertiary, at least at the generic level.

Most circumtropical cheilostomes live in shallow water, some only at depths less than 100 m. Most produce lecithotrophic larvae with motile stages much too brief to permit their drifting as plankton across broad expanses of ocean. Rafting of colonies attached to floating objects such as seaweed has been suggested as a means of wide dispersal. A few circumtropical species produce planktotrophic larvae capable of swimming for up to 2 months, but these species also grow on drifting seaweed.

Wide north-south distributions are common in cool-water marine gymnolaemates. Species found in shallow water in both polar and tropical regions must tolerate high temperatures as well. Others have wide distributions, although apparently limited to cold water, by occurring at greater depths toward the equator. For example, an ascophoran (Figure 17.52) found in the Arctic as shallow as 70 m occurs only below 450 m at the equator.

Most marine gymnolaemates live at depths of less than 200 m, but some have been found living well below 5000 m. Deep-sea gymnolaemates include some species and genera restricted to bathyal or abyssal depths as well as those found in shallow water in cold-water regions. Both kinds of deep-sea cheilostomes have been found in some Tertiary deposits in oceanic regions. Most fossil cheilostomes, however, apparently were shallow-water species, including many in Tertiary sediments cored by the Deep Sea Drilling Project at oceanic sites now more than 1000 m deep. These fossils together with other benthic invertebrates have figured in interpretations of changing water depths during Cenozoic crustal evolution of the world's oceans.

Encrusting colonies   Encrusting colonies are found in the broadest range of freshwater to marine habitats. Species with this growth habit are the most abundant gymnolaemates in brackish water, the intertidal zone, and in many shallow shelf environments.

Gymnolaemates can encrust both upper and lower surfaces of substrates. Where physical conditions are severe, as in intertidal habitats, or where competition with light-loving organisms is intense, as in reefs, protected and shaded undersurfaces are more heavily encrusted by gymnolaemates. Preference for undersurfaces can be related to larval behavior. When first released, lecithotrophic larvae are attracted to light (photopositive) but, in many species, become attracted to shade (photonegative) before settlement. Colonies growing on confined undersurfaces such as rock cavities are ordinarily flat and single layered. Those growing on exposed upper surfaces can either remain flat or become nodular, multilayered, and in some species massive. Both cavity-filling and massive cheilostomes can be minor rock-forming components of reefs.

Larvae of many encrusting species have other behavioral properties, resulting in preferential settlement on particular kinds of substrates such as a species of seaweed. Preferences can differ with habitat, and in a given habitat more than one species may settle preferentially on the same kind of substrate. Ten or more cheilostome species commonly can be found on a single shell or coral undersurface.

Settlement by different species on limited substrates may lead to competition for space. Colonies of the same species may avoid competition by ceasing to grow at points of contact or in some species fusing and growing as one. At contacts between different species, one commonly overgrows the other. Species have different abilities to overgrow each other, generally expressed as **competitive networks** (Figure 17.51) in which no species is capable of excluding all others from a substrate.

Erect colonies   An erect growth habit provides a means of escaping space limitations of encrusting growth. Many living gymnolaemates grow erect, and fragmentary remains of erect cheilostomes dominate most bryozoan-rich Upper Cretaceous and Tertiary marine deposits.

Erect colonies are in contact with more water relative to substrate occupied by the colony base. Some ctenostomes reach a meter or more from the point of attachment. Cheilostomes generally do not exceed heights of 10 to 15 cm. Basal attachments of even large colonies commonly are less than 1 cm².

Erect colonies can gain access to food and other resources away from the substrate only if they resist breakage and deformation under stress induced by water movement or contact with vagile animals such as

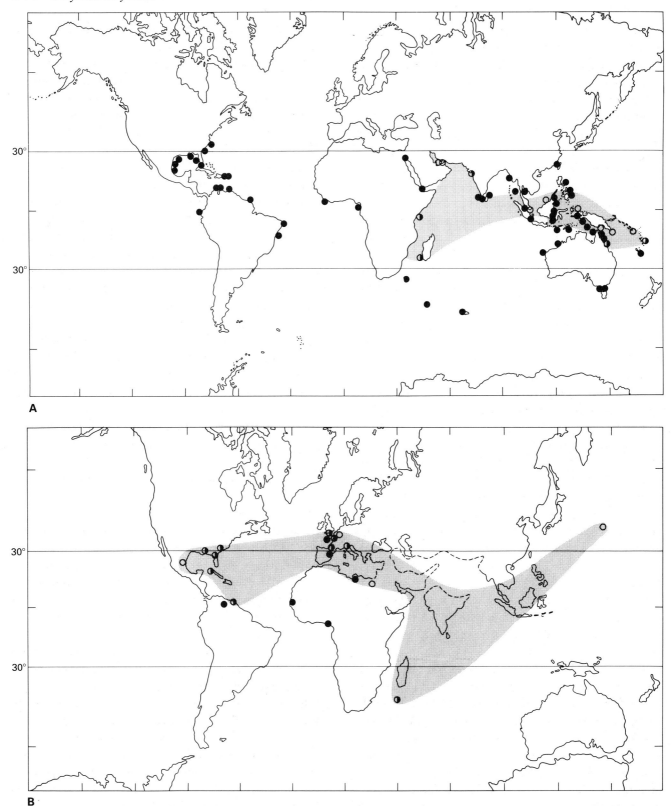

A

B

Figure 17.50. Modern and fossil distributions of two warm-water anascans, *Nellia* (●) and *Poricellaria* (○). (◑ denotes joint occurrence.) **A,** Holocene distributions of a single species of each genus. *Poricellaria* has only one Holocene species; its distribution is indicated by shading. **B,** Eocene distributions of the same species of *Nellia* as in **A** and of all known species of *Poricellaria* (distribution shaded). (Modified from Lagaaij, R. *Geologie en Mijnbouw* **48** [2]; 1969, and Lagaaij, R; Cook, P. L. *Atlas of palaeobiogeography.* New York: Elsevier Scientific Publishing Co.; 1973. Base map for **B** from Smith, A. G.; Briden, J. C.; Drewry, G. E. In: Hughes, N. F. *Special Papers in Palaeontology,* No. 12; 1973.)

predators. The rigid skeletons of Cheilostomata resist deformation through mechanical strength but are brittle. Many cheilostomes, however, can reinforce their skeletons toward the base of the colony where risk of breaking is highest. Ontogenetic thickening of frontal shields on outer surfaces by overlying cellular layers produces tapering branches (Figure 17.52) similar to those of trees. This calcification can seal zooidal orifices, changing the function of proximal parts of a growing colony from nutritive to supportive as in a growing tree. However, the surface area of the growing colony increases more rapidly than its girth. The consequently increasing risk of breakage may explain why cheilostomes with rigidly erect growth habits are more abundant in protected, often deeper habitats where water is less turbulent than in most intertidal and shallow shelf environments.

Erect colonies with jointing or general reduction or absence of calcification resist breakage through their flexibility but may lose some advantages of erect growth by lying prostrate. Species with flexible growth habits are

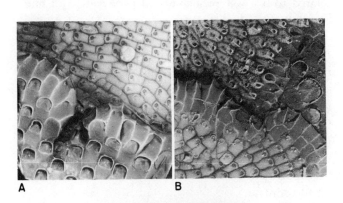

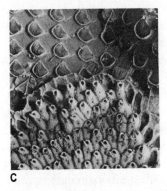

Figure 17.51. Overgrowth among living cheilostomes encrusting a single coral undersurface at 15 m depth. (Jamaica, ×5.) The three species form a competitive network such that the anascan *Steginoporella* overgrows (toward top) the ascophoran *Stylopoma* (**A**), *Stylopoma* overgrows the ascophoran *Reptadeonella* (**B**), and *Reptadeonella* overgrows *Steginoporella* (**C**). Note that if any one of the overgrowth abilities were reversed, one species would be able to exclude the other two. (From Buss, L. W. Jackson, J. B. C. *American Naturalist* 113: 228. ©1979. The University of Chicago.)

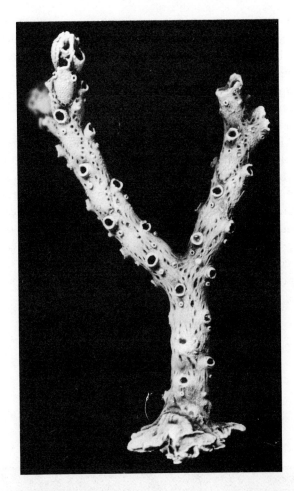

Figure 17.52. Rigidly erect branching colony of the ascophoran *Tessaradoma*. (Holocene, North Atlantic, ×20.) The zooidal orifices open all around cylindrical, distally tapering branches but are nearly sealed with calcareous deposits proximally.

common in many shallow-water habitats. Where substrates suitable for encrusting are rare, jointed species can become dominant. Disarticulated segments of jointed colonies of anascans predominate in some Holocene submarine deltaic deposits such as those off the Rhone and Orinoco Rivers.

Free-living colonies Dependence on a substrate is further lessened in free-living colonies. This is the dominant growth habit on sandy and muddy bottoms at depths from a few meters to about 200 m in many warm-water regions.

Free-living species commonly show three major adaptations to fine-grained, often soupy sea bottoms. First, larvae are capable of initiating growth on a minute substrate, such as a sand grain or test of smaller foraminiferid, commonly incorporated into the colony and surrounded by skeletal layers. Second, these species frequently reproduce by fragmentation, thus partly avoiding the necessity for metamorphosis on a substrate. Third, some colonies are capable of movement, which can be important in avoiding burial where deposition is rapid or sediments are shifting. The first two of these adaptations are also shown by morphologically similar species that are anchored in loose sediment by rootletlike kenozooids (Figure 17.46, *D* and *E*).

Many free-living cheilostomes can support themselves on long mandibles of vicarious avicularia projecting beyond the colony periphery (Figure 17.53, *A*). Coordinated movement of mandibles enables a colony to right itself when overturned and in some species can produce sustained locomotion (Figure 17.53, *B*). Mandibles of avicularia away from the periphery can sweep off sediment. These colonies thus approach the functional organization of a vagile solitary animal.

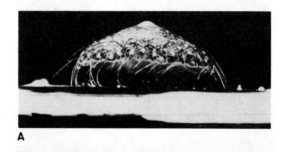

A

B

Figure 17.53. Locomotion in the living anascan *Selenaria*. (Australia, ×3.) **A,** Lateral view of a colony moving on peripheral avicularian mandibles over glass. **B,** One colony climbing over another on a gravel bottom. (From Cook, P. L.; Chimonides, P. J. *Cahier de Biologie Marine* **19:** 154; 1978.)

Evolution and stratigraphic distribution

Ctenostomata   The fossil record of the Ctenostomata has become better understood through modern studies, but sporadic distribution still makes even the most basic evolutionary trends problematic. The Paleozoic, Mesozoic, and pre-Holocene Cenozoic have each yielded only five to eight genera (Figure 17.54). All five Paleozoic genera were borers immersed in calcareous

substrates similar to that shown in Figure 17.43, *A* and *B*. Their specialized morphology and mode of life suggest that these genera were not close to the ancestral stock from which the numerous nonboring Gymnolaemata descended.

The oldest known nonboring ctenostome, found in the Middle Jurassic of Germany (Figure 17.43, *C*), had a simpler morphology. Its encrusting uniserial colonies consisted of zooids apparently of the same form throughout. Lack of polymorphism and astogenetic change, together with predominantly exterior walls, suggest that zooids in this simple ctenostome could have functioned almost as autonomously as solitary animals.

Unless a morphology with this low level of integration is postulated as the starting point, evolution in the Ctenostomata must be regarded as a sequence of decreases and increases in complexity, all represented by the handful of fossil genera. Such a pattern contrasts with that in the Cheilostomata. Apparent absence of simple ctenostomes until the mid-Mesozoic seems more likely related to a lower probability of preservation of nonboring, soft-bodied colonies.

Early Cheilostomata   The rich and well-preserved fossil record of the Cheilostomata (Figure 17.54) begins with a simple, predominantly uniserial, encrusting anascan found in the Upper Jurassic of England and the Lower Cretaceous of England, United States, and Antarctica (Figure 17.44, *A* and *B*). Its morphology is so similar to that of the Middle Jurassic nonboring ctenostome that this appears to be the evolutionary starting point for the Cheilostomata.

The early anascan differed little more from its presumed ctenostome ancestor than by having calcified walls and distinct opercula. The form of the operculum is preserved where flexible parts of frontal walls of some zooids had calcified and lost their function (Figure 17.44, *B*). The calcified walls, apparently calcitic as in living anascans with similar zooidal morphology, were virtually limited to exterior walls, including vertical walls and gymnocysts. Lack of brooding structures suggests that these anascans produced planktotrophic larvae as do living species with similar zooidal morphology.

At stratigraphically higher levels within the Lower Cretaceous, the uniserial anascan is found with others of more complex morphology. These species were also encrusting, but with predominantly multiserially arranged zooids (Figure 17.44, *C* to *E*). Colonies had distinct zones of change (Figure 17.44, *C*) and in some, vicarious avicularia were present. Avicularia initially were little different in form from feeding zooids (Figure

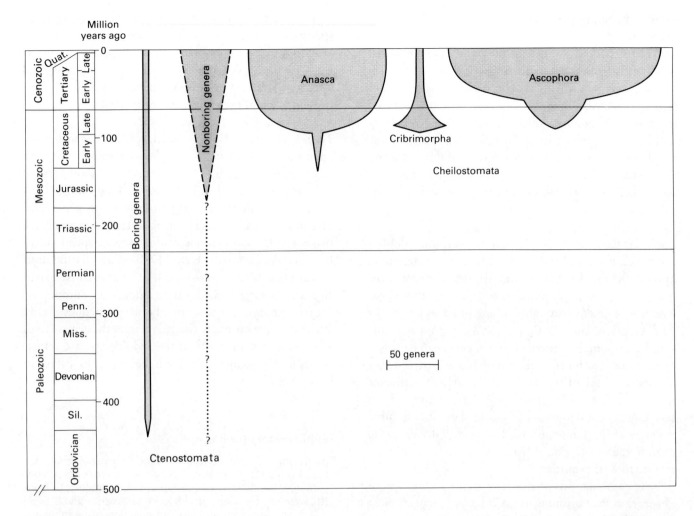

Figure 17.54. Range chart of the class Gymnolaemata. For nonboring Ctenostomata, the pre-Holocene diversity and pre-Middle Jurassic range are hypothetical.

17.44, *D*), but before the mid-Cretaceous had become more specialized (Figure 17.44, *E*). The presence of ovicells in many colonies (Figure 17.44, *E*) furnishes evidence that some Anasca were producing lecithotrophic larvae before the mid-Cretaceous.

Early Cretaceous evolution in the Cheilostomata thus was marked by gradual increase in integration of colonies into more cooperatively functioning units with more specialized components. This took place with relatively little change in the morphology of ordinary feeding zooids.

Late Cretaceous proliferation From the few anascans of the Early Cretaceous, the Cheilostomata expanded to more than 100 genera in the Late Cretaceous. This proliferation included further increases in levels of integration and even more striking modifications in zooidal morphology. Early in the Late Cretaceous, all three suborders were present (Figure 17.54).

Proliferation was most pronounced in the Anasca. Many genera developed extensive cryptocysts with only notches or perforations left for passage of parietal muscles (Figure 17.45, *A*). Avicularia became less like feeding zooids in many lineages, while either remaining vicarious (Figure 17.45, *A*) or in a few genera becoming adventitious (Figure 17.45, *B*). A large number of anascans developed erect growth habits, and by the end of the Cretaceous a few jointed and free-living species had appeared.

The Cribrimorpha and Ascophora appeared near the beginning of the Late Cretaceous and both suborders underwent some diversification in morphology and hence in numbers of genera. By the close of the Late Cretaceous, however, neither of the two suborders had diversified nearly as much as the Anasca. Most cribrimorphs retained encrusting growth habits and relatively low levels of integration as in their vicarious avicularia. Many ascophorans grew erect and had generally high levels of integration, especially in possessing adventitious avicularia.

Skeletons wholly of calcite apparently predominated in all three suborders in the Late Cretaceous. Cheilostomes commonly are well preserved in Upper Cretaceous chalks in which other invertebrates, inferred to have had aragonitic skeletons, are represented only by molds. In the upper part of the Upper Cretaceous some free-living anascans, inferred to have had mixed calcite-aragonite skeletons, occur. Their apparently aragonitic basal skeleton was dissolved leaving the calcitic upper side intact.

Early Tertiary turnover The Late Cretaceous proliferation had produced most generic-level character states known in the Cheilostomata. In the Cenozoic, the number of genera expanded to several times that in the Cretaceous chiefly through shuffling of states into many new combinations. Within a genus, morphologically simple, presumably primitive states of some characters came to be combined with more complex, presumably advanced states of others (Figure 17.45, *D*). In some genera, such combinations could have resulted from convergent redevelopment of simple states, but in other genera were probably produced by marked differences in evolutionary rates of different characters, a process termed **mosaic evolution**.

New combinations of character states did not appear abruptly at the beginning of the Tertiary to replace those in the Cretaceous. Turnover in genera was greatest between the Paleocene and Eocene, well into the Tertiary. Paleocene cheilostomes include many genera of Anasca (Figure 17.45, *D* and *E*) and some Cribrimorpha (Figure 17.45, *C*) and Ascophora that range upward from the Cretaceous but not into the Eocene. In contrast, Eocene cheilostomes include many genera of Anasca and Ascophora (Figure 17.46, *A*) and a few Cribrimorpha that range upward into the Pliocene, Pleistocene, or Holocene but are unknown in the Cretaceous. There are relatively few Eocene and Oligocene genera that have not survived to the present. Long stratigraphic ranges may well be related to the low probability of extinction of genera with wide geographic distributions, such as those shown in Figure 17.50.

Beginning in the Eocene, the Ascophora underwent proliferation of the same intensity as that of Anasca in the Late Cretaceous. Jointed and loosely anchored rooted forms were added to the range of ascophoran growth habits by adaptive convergence with Anasca. Many new generic combinations of states were assembled from characters of frontal shields, orifices, avicularia, and ovicells.

In both the Anasca and Ascophora, well-preserved specimens of either mixed calcite-aragonite or just aragonite are common in calcareous clays as old as Eocene. Many other anascans and ascophorans, and apparently all cribrimorphs, have continued to the present to build skeletons entirely of calcite.

Later changes Upper Tertiary and Quaternary cheilostomes (Figure 17.46, *C* to *E*) show fewer and more gradual changes at the generic level than those of the Cretaceous and Lower Tertiary. Many changes were regional in extent, resulting from shifts in geographic distributions rather than from episodes of extinction and origination. Contractions from circumtropical distributions (*Poricellaria* in Figure 17.50, *A* and *B*) and shifts toward the equator as the result of cooling climates particularly changed cheilostome faunas in Europe and North America. Those of the tropical Indo-Pacific Province have probably changed least at the generic level. For many genera, parts of the Indo-Pacific are refugia (Chapter 7), which may well be the sites of future extinctions.

Supplementary reading

Boardman, R. S. General features of the class Stenolaemata. In: Robison, R. A., editor. *Treatise on invertebrate paleontology, Part G* (revised), *Bryozoa*, Vol. 1. Boulder, CO and Lawrence, KS: Geological Society of America and University of Kansas Press; 1983. Introduction to morphology, modes of growth, and classification of the Stenolaemata.

Boardman, R. S.; Cheetham, A. H. Degrees of colony dominance in stenolaemate and gymnolaemate Bryozoa. In: Boardman, R. S.; Cheetham, A. H.; Oliver, W. A., Jr., editors. *Animal colonies, development and function through time.* Stroudsburg, PA: Dowden, Hutchinson, & Ross; 1973. Interpretation of evolutionary trends in the Stenolaemata and the Gymnolaemata based on the concept of morphologic integration.

Boardman, R. S.; Cheetham, A. H.; Cook, P. L. Introduction to the Bryozoa. In: Robison, R. A., editor. *Treatise on invertebrate paleontology, Part G* (revised), *Bryozoa*, Vol. 1. Boulder, CO and Lawrence, KS: Geological Society of America and University of Kansas Press; 1983. Details of the concepts presented here.

Borg, F. Studies on recent cyclostomatous Bryozoa. *Zoologiska Bidrag fran Uppsala* 10: 181–507; 1926. Classic paper that began modern zoologic studies of stenolaemate Bryozoa, unfortunately difficult to understand.

Cheetham, A. H.; Cook, P. L. General features of the class Gymnolaemata. In: Robison, R. A., editor. *Treatise on invertebrate paleontology, Part G* (revised), *Bryozoa*, Vol. 1. Boulder, CO and Lawrence, KS: Geological Society of America and University of Kansas Press; 1983. Introduction to morphology, modes of growth, and classification of the Gymnolaemata.

Cumings, E. R.; Galloway, J. J. Studies of the morphology and histology of the Trepostomata or monticuliporoids. *Geological Society of America Bulletin* 26: 349–374; 1915.

Classic paper that began modern paleobiologic studies of Paleozoic Bryozoa.

Duncan, H. Bryozoans. Pp. 783–799. *Treatise on marine ecology and paleoecology*, Vol. 2. New York: Geological Society of America, Memoir 67; 1957. Excellent summary of the ecology of Bryozoa for that time.

Ryland, J. S. *Bryozoans*. London: Hutchinson University Library; 1970. Introduction to morphology, biology, and ecology of bryozoans, particularly the Gymnolaemata.

Ryland, J. S. Physiology and ecology of marine bryozoans. *Advances in marine biology* 14: 285–443; 1976. Comprehensive review of biologic processes relevant to ecology of marine bryozoans, particularly the Gymnolaemata.

Schopf, T. J. M. Patterns and themes of evolution among the Bryozoa. In: Hallam, A., editor. *Patterns of evolution*. Amsterdam: Elsevier Pub. Co.; 1977. Interpretation of evolutionary trends in Bryozoa, especially the Gymnolaemata, with a model based on ecologic theory.

Wood, T. S. Morphological review of the Phylactolaemata. In: Robison, R. A., editor. *Treatise on invertebrate paleontology, Part G* (revised), Bryozoa, Vol. 1. Boulder, CO and Lawrence, KS: Geological Society of America and University of Kansas Press; 1982. General review of the class.

Woollacott, R. M.; Zimmer, R. L., editors. *Biology of bryozoans*. New York: Academic Press, Inc.; 1977. Review papers on most aspects of bryozoan biology, emphasizing the Gymnolaemata.

18

Phylum Echinodermata

James Sprinkle
Porter M. Kier

Part I Phylum overview

James Sprinkle

Almost anyone who has walked along a seashore picking up shells knows what a starfish or a sea urchin looks like. These interesting animals belong to the phylum Echinodermata, which includes complex invertebrates that are common in many different marine environments. Echinoderms are a medium-sized phylum having five living classes (Figure 18.1) with about 6000 living species, but they are much less diverse today than other well-known invertebrate phyla such as mollusks or arthropods. Echinoderms have a long and detailed fossil record with more fossil species than are living today and a large number of extinct classes. Complete fossil echinoderms are sometimes as beautiful and complex as living representatives; they are eagerly sought by amateur fossil collectors and professional paleontologists. Because of this widespread interest, the fossil record of echinoderms, although incomplete, is nearly as well studied as that of any other invertebrate group.

A preserved or living starfish (Figure 18.2) shows many of the diagnostic features that can be used to identify an echinoderm. The most obvious feature is the five-sided, or **pentameral, symmetry** that the animal shows. Echinoderms are one of the few animal groups that have a radial symmetry based on five or multiples of five. Also obvious are the separate calcareous **plates** or **ossicles** that are sutured together to make up the skeleton of the starfish. The underside of each arm contains a long food-collecting **ambulacral groove**, which has soft, movable, suckered projections called **tube feet**. These are the

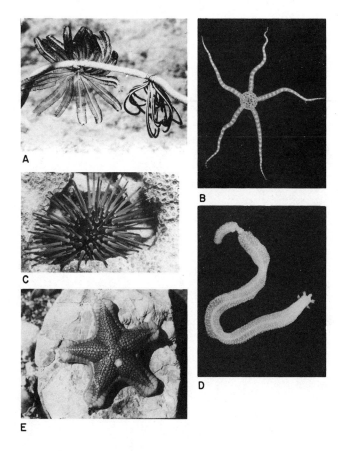

Figure 18.1. Examples of the five classes of echinoderms living today. **A**, The sea lily (crinoid) *Cenometra* from Australia attached to a coral. **B**, The brittle star (ophiuroid) *Ophiomyxa* from Puerto Rico. **C**, The sea urchin (echinoid) *Echinometra* from Belize in its burrow. **D**, The sea cucumber (holothurian) *Chiridota*. **E**, The starfish (asteroid) *Asterodon* on a rock. (**A**; courtesy D. L. Meyer. **B**, courtesy K. Sandved and David Pawson. **C**, courtesy Porter M. Kier. **D**, courtesy N. Coleman and David Pawson. **E**, courtesy David Pawson.)

550

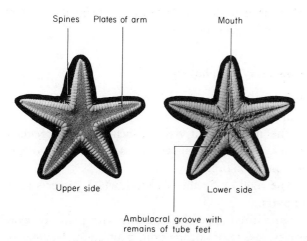

Spines   Plates of arm   Mouth

Upper side

Lower side

Ambulacral groove with
remains of tube feet

Figure 18.2. Preserved specimen of the living starfish *Astropecten* showing the five large arms with their plates and spines and dried-up tube feet of the water vascular system in the ambulacral grooves. (×0.6.)

external parts of the tubular **water vascular system** that hydraulically controls the movement of tube feet and is diagnostic of most echinoderms. Another distinctive feature of the echinoderm is the rough spiny surface of the animal, which gives the phylum its name (from the Greek *echinos*, meaning spiny, and *derma*, meaning skin).

Present-day echinoderms almost always live in normal marine environments because they have little tolerance for salinity variations. They range from the intertidal zone to the bottom of oceanic trenches. Several classes such as starfish (asteroids), sea urchins (echinoids), and sea lilies (crinoids) (Figure 18.1) are common in shallow, near-shore shelf environments. Other classes, such as brittle stars (ophiuroids) and sea cucumbers (holothurians) (Figure 18.1), not only occur in shallow water environments but are the dominant macroinvertebrates

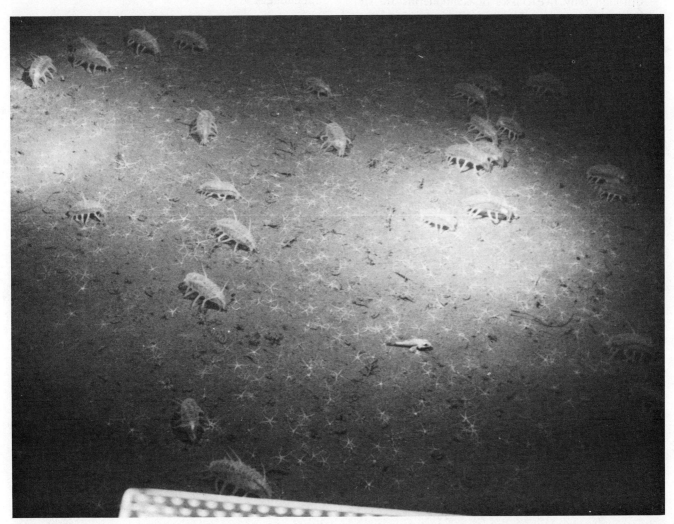

Figure 18.3. Deep-sea photograph taken from the U.S. Navy bathyscaphe *Trieste* at about 1200 m in the San Diego Trough shows a large group of sea cucumbers (*Scotoplanes*) walking on their tube feet across a soft mud bottom covered by hundreds of small brittle stars (probably *Amphilepis*). (From Barham, E. G.; Ayer, N. J., Jr.; Boyce, R. E. *Deep-Sea Research* 14: 798; 1967. Official photograph, U.S. Navy.)

in many deep-sea environments such as abyssal plains and oceanic trenches. In some places these deep-sea echinoderms almost completely cover the ocean floor (Figure 18.3).

The numerous fossil genera and species that have been described probably represent only a small percentage of the total diversity of fossil echinoderms that have existed. At many times in the past, echinoderms may have been as common and diverse in favorable environments as they are today. In fact, some rock units, such as the Middle Mississippian (Lower Carboniferous) Burlington Limestone in the Central United States measuring about 40 m thick, are composed mostly of echinoderm fragments. Millions of echinoderms lived and died in this area for thousands of years to produce this deposit. Many other limestone units composed mostly of echinoderm fragments are known from other parts of the geologic record.

Although only five classes of echinoderms are still living, at least 15 other classes are known exclusively from the fossil record, making a total of 20 echinoderm classes. At times in the past, as many as 11 or 12 classes were living together in a single region. The fossil record for the phylum goes back to the Early Cambrian (and perhaps even to the late Proterozoic), giving this phylum a long and complex history.

Introductory morphology

The body of most echinoderms can be divided into three major parts. They are:

1 The internal soft tissues, including the coelomic cavities, digestive system, water vascular system, gonads, and other vital organs.

2 The rigid or flexible skeleton, which protects these soft parts.

3 Specialized appendages, either with plates or axial ossicles, used for locomotion, protection, feeding, or attachment.

The internal soft-part anatomy will be discussed first, followed by a description of the body skeleton and its appendages.

Internal anatomy

Echinoderms are an advanced phylum with a fairly complex internal anatomy (Figure 18.4). The interior of

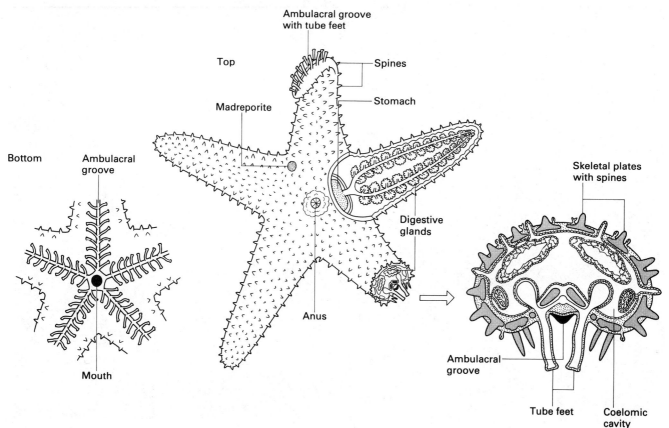

Figure 18.4. Generalized morphology of a living starfish (asteroid). Note the coelomic cavities and extensive digestive glands that extend into the large arms, the loosely connected skeletal plates with spines, and the ambulacral groove on the oral side of each arm with tube feet from the water vascular system. (Arm cross section redrawn from Buchsbaum, R. *Animals without backbones.* Chicago: University of Chicago Press; 1948, p. 302.)

the body is made up of fluid-filled coelomic cavities occupied by several organ systems. These systems first appear in the larvae as several tiny internal fluid-filled pouches. One of these becomes the water vascular system (or hydrocoel) and two others become coelomic cavities in the adult. The fluid in coelomic cavities is similar in composition to sea water and contains floating amoeboid cells that aid in transporting materials and removing wastes. The pressurized, fluid-filled coelomic cavities help to maintain the shape of the echinoderm.

The mouth, which is usually centrally located on one side of the body (Figure 18.4), is the entrance to the digestive system. Most echinoderms have a one-way digestive system with a mouth, esophagus, either a stomach with digestive glands or a long gut, and an anus. Brittle stars and a few starfish are an exception; they have no anus, and the small amount of waste material is expelled through the mouth. Digestion and absorption of nutrients take place in the long gut or in long, pouched digestive glands that branch off a central enlarged stomach (Figure 18.4). Waste material is usually eliminated from the anus as fecal pellets.

Most echinoderms have a water vascular system (Figure 18.5). This coelomic system of water-filled tubes consists of several parts. First, a **circumoral ring** surrounds the esophagus and usually is connected by a tube called the **stone canal** to an opening to the exterior. If that opening is single, it is called a **hydropore**, and if multiple, it is called a **madreporite**. Second, branching off the circumoral ring are five main tubular **radial canals**, each one extending radially outward to the end of an **ambulacrum** (Figures 18.2 and 18.4), a specially plated food-

gathering area on the body or on an arm, bearing an ambulacral groove and tube feet. Each ambulacrum is ciliated and leads to the mouth. If the radial canals themselves pass out through the skeleton at some point and extend down ambulacral grooves external to the plates, the water vascular system is termed **open**. If the radial canals remain internal and have numerous branches that extend to the exterior through single or paired pores through the skeleton, the system is termed **closed**. The third part of the system includes numerous short branches from the radial canal that lead to the external tube feet (Figure 18.5). Bulbous **ampullae** attached to the base of the tube feet help control water pressure in this part of the water vascular system. Tube feet may be clustered together for better coordination, and some are finely branched. The hydraulically driven tube feet are used for feeding, respiration, movement, and sensory perception. They may or may not have terminal suckers, depending on their function. They usually have extensive muscles and nerves to control movement.

The distinctive water vascular system with its tube feet is found only in echinoderms and is well developed in all living classes except for some sea cucumbers, which lack most or all of the tube feet. A water vascular system was probably present in many fossil echinoderms because they have the characteristic single or paired pores through the ambulacral plates. Tube feet and radial water vessels are actually preserved in a few Cambrian and Ordovician specimens. Some authors have argued, however, that several Paleozoic groups lacked tube feet because no evidence of pores has been found.

The nervous system is relatively simple in echino-

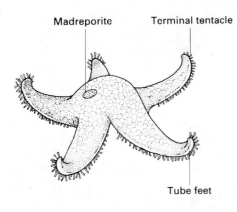

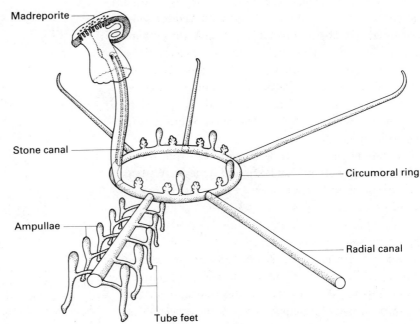

Figure 18.5. Parts of the water vascular system in a living starfish. Tube feet and ampullae are drawn for only one arm. (From Nichols, D. In: Boolootian, R. A., editor. *Physiology of Echinodermata.* New York: Interscience Publishers Inc.; 1966, p. 221.)

derms. Several enlargements, or ganglia, are present in the coelomic cavity with one surrounding the esophagus. Trunk nerves extend outward from the ganglia to the extremities, and an extensive nerve network is located in the skin, tube feet, and appendages. This network responds to stimuli and controls movement of appendages or the whole animal. Echinoderms usually do not move very fast, and this type of nervous system is adequate for their needs.

In most echinoderms, the sexes are separate, and most kinds have one to five gonads for production of eggs or sperm, suspended in a coelomic cavity. The gonads usually open to the exterior through one or more small pores called **gonopores** (see Figure 18.8, *A*), although in brittle stars the gonads open into specialized brooding pouches. Sea lilies are unusual because they have external gonads attached to the food-gathering appendages near the main body of the animal. Some species of each living class have developed the habit of brooding their young.

Skeleton

The skeleton is one of the most distinctive and useful features for identifying living as well as fossil echinoderms. It is of mesodermal (middle layer) origin and is therefore an endoskeleton. The skeleton is covered by a somewhat ciliated epidermis, or skin, that forms the outermost surface of the animal. Two basic skeletal designs are present in echinoderms. The first is a boxlike or saclike body called a **theca** (also called **calyx**, **test**, or **disk**) made up of relatively large sutured or imbricate plates surrounding the main body that protects the internal soft parts. Plates can either be tightly sutured together to form a rigid theca, loosely sutured for some movement, or imbricate to form a flexible (or perhaps expandable) theca. The total number of skeletal plates ranges from about 100 up to several hundred thousand. In the second skeletal design, microscopic spicules, or **sclerites**, (Figure 18.6) are developed in the body wall to strengthen it. Several million sclerites may be present in a single echinoderm.

Echinoderm plates or ossicles range in size from microscopic to around 30 mm, and echinoid spines may attain a length of around 150 mm. They are microporous calcite structures with a high (4% to 16%) Mg^{2+} content and optically appear to be single crystals under the petrographic microscope. The plates consist of a meshwork of microscopic girders and struts collectively termed **stereom**, in life imbedded in mesodermal tissue termed **stroma** (Figure 18.7). In some plates, the stereom and stroma occupy almost equal volumes; in other plates,

either the stereom or the stroma is greater. Each plate is secreted by stromal cells. When an echinoderm becomes fossilized, the **stromal galleries** are filled by the addition of secondary calcite, which is precipitated with the same crystallographic orientation as the original stereom. The result is that each fossilized plate is a solid single crystal with well-developed calcite crystal cleavage, which makes fossil echinoderm fragments easy to identify in hand specimen (cleavage planes on a broken edge) or in thin section (the microporous internal structure).

Echinoderm skeletal plates are connected by muscles, ligaments (nonelastic connective tissue), or interlocking processes from the stereom. Muscles occur between loosely sutured or imbricate plates and in movable appendages. Ligaments suture skeletal plates tightly and also attach immobile appendages. Stereom galleries on the edges of plates where ligaments are attached have a distinctive aligned appearance (Figure 18.7, *B*), which can be distinguished from galleries containing muscles. In those forms with very rigid tests, calcite rods or processes project across the plane of the suture and interlock, producing the nonflexible sutures.

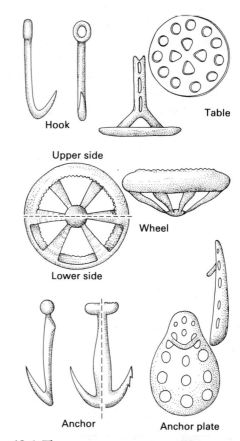

Figure 18.6. The most common types of microscopic sclerites found in living holothurians (greatly enlarged). (From Frizzell, D. L.; Exline, H. *Missouri School of Mines and Metallurgy Bulletin 89*; 1956.)

Many starfish and brittle stars have loosely connected or imbricate thecal plates that fall apart quickly after death. Other echinoderms such as sea urchins have thecal plates that are tightly sutured and almost fused, yielding a rigid skeleton that is not easily disarticulated after death. A rigid skeleton provides good protection but must have openings for the animal to function. Those normally found in a rigid skeleton include a mouth, anus, hydropore (or sievelike madreporite), one to five gonopores, often single or paired pores for ambulacral tube feet or for **podia** (small thin outpouchings of the skin), and openings through **articular facets** (the surfaces where large appendages carrying coelomic systems are attached (Figure 18.8, *B*).

Sclerites are microscopic ossicles (0.05 to 2.0 mm in size) commonly having a sieve, wheel, hook, or anchor shape (Figure 18.6). Sclerites are usually not found together but stiffen the body wall and the walls of internal organs. They are abundant in most sea cucumbers (Figure 18.1, *D*) and are also known to occur in some other echinoderms. Several authors have used different types of sclerites to identify and classify groups such as sea cucumbers that are poorly represented in the fossil record.

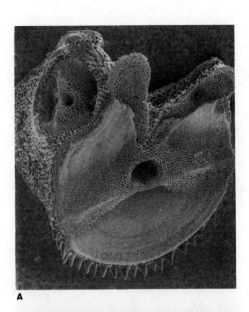

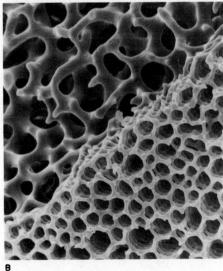

Figure 18.7. Scanning electron micrographs of stereom and stromal canals in an arm plate of the living crinoid *Nemaster*. **A,** Whole arm plate showing microporous nature and other morphologic features. (×53). **B,** Close-up of the same arm plate showing stereom with large labyrinthine stromal canals (for muscles) at the upper left versus smaller aligned stromal canals (for ligaments) at lower right (×1000). (From Macurda, D. B., Jr., Meyer, D. L. *University of Kansas Paleontological Contributions,* Paper 74; 1975.)

Pentameral symmetry

The skeletal parts of most echinoderms are arranged in a fivefold radial pattern of pentameral symmetry. The radial segment of a test having an ambulacrum is termed a **ray**, whether the ambulacra are on arms as in starfish, crinoids, or brittle stars, or recumbent on tests as in sea urchins. Few other animals have pentameral symmetry, and this feature serves to identify most echinoderms. Evidence of pentameral symmetry was already present in the arrangement of the ambulacra of some of the earliest fossil echinoderms, and it later extended to the arrangement of other thecal plates and skeletal structures such as respiratory organs. This symmetry may reflect the optimal number of small calcite plates in the late embryo as it starts to secrete the adult skeleton or the number of radial water vessels growing outward at that time. The larvae of most living echinoderms show bilateral symmetry so the pentameral symmetry that develops later during metamorphosis is probably secondary. Some echinoderms, such as mobile or burrowing sea cucumbers and sea urchins plus some mobile or current-dwelling extinct echinoderms, have a tertiary bilateral symmetry superimposed on pentameral symmetry.

An individual echinoderm will sometimes deviate from a normal pentameral arrangement of skeletal parts (and internal structures) and instead have three, four, or six sides (Figure 18.9). These abnormal individuals represent only about 0.1% to 2% of most large echinoderm collections. Abnormal individuals make prized specimens because of their rarity and because they may yield information on how skeletal growth is controlled.

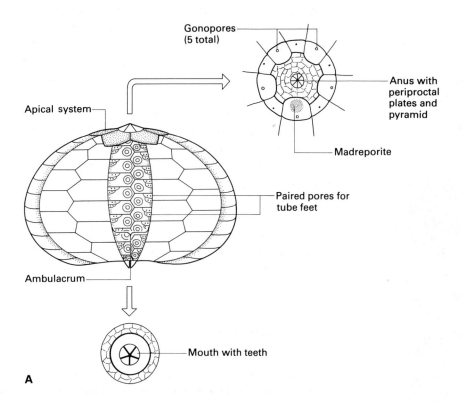

A

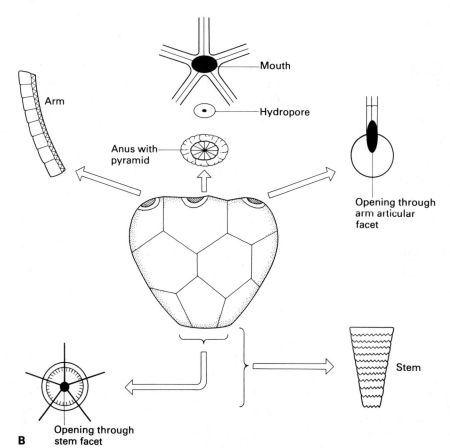

B

Figure 18.8. Typical openings in echinoderm skeletons. **A,** An echinoid with a mouth (at bottom), anus, madreporite, five gonopores (at top), and numerous paired pores for tube feet through the ambulacral plates. **B,** A generalized fossil crinoid with a mouth and anus on the summit plus openings through articular facets where arms and stem are attached.

Figure 18.9. Four-sided specimen of the echinoid *Echinometra* from Florida (top view with the anus in the center); note that the four ambulacral areas with pores are almost symmetrically arranged (×1.5). (From Kier, P. M.; Grant, R. E. *Smithsonian Miscellaneous Collections*, Publication 4649; 1965.)

Appendages

Several distinctive types of large- or small-plated appendages are commonly attached to the exterior of the theca of living and fossil echinoderms. These include appendages that serve mainly for protection (spines and pedicellariae), appendages that collect food (arms and brachioles), and appendages that are used for attachment (stem or holdfast).

Spines occur on the plates of many echinoderms. They are either external projections of the plate itself or separate, elongate, movable ossicles (Figure 18.10) mounted on a ball-shaped base. Muscles and nerves rim the base to control the movement of the spine. These spines serve both for protection from predators and for locomotion; sometimes they also are adapted for other functions. Two echinoderm classes (sea urchins and starfish) also have external, specialized, tiny-plated pincerlike, stalked structures called **pedicellariae** attached to the test (Figure 18.10). These serve to remove small foreign objects that might settle on the test.

Arms include a wide variety of food-gathering appendages that are major radial extensions of the body and coelomic systems, as well as the skeleton. Five arms (or some multiple of five) are usually present. They carry ambulacral food grooves and tube feet and are primarily used for feeding and sometimes movement. The arms of starfish are large, hollow, and make up most of the body (Figures 18.1, *E* and 18.4). The arms of brittle stars are thin, snakelike, and strongly differentiated from the central disk (Figure 18.1, *B*). In sea lilies, arms merge into the calyx, are often extensively branched, and bear smaller appendages called **pinnules** (see Figure 18.13). Some extinct Paleozoic forms have as few as two arms.

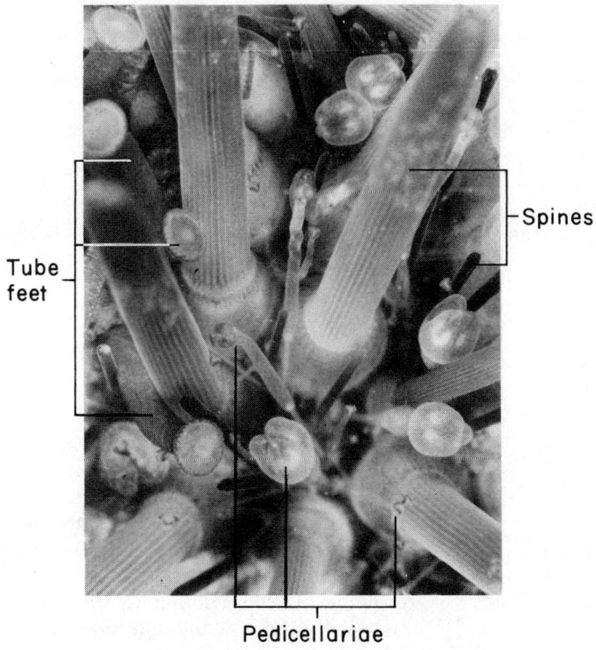

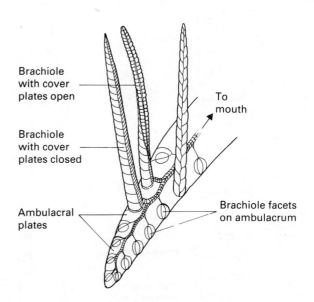

Figure 18.10. Greatly enlarged view of the surface of a living echinoid showing the suckered tube feet (at left), several sizes of spines, and at least three types of pedicellariae. (Courtesy Porter M. Kier.)

Figure 18.11. Morphology of brachioles attached to the ambulacral plates of an extinct echinoderm; note the biserial plating, absence of branching or smaller appendages, and the food groove on one side protected by a biserial set of tiny cover plates that open during feeding.

Brachioles are smaller, erect food-gathering appendages mounted on the recumbent ambulacral areas of five or six classes of extinct Paleozoic echinoderms. They are thin, biserially plated, unbranched appendages bearing a protected ambulacral groove on one side (Figure 18.11). Brachioles do not have major extensions of the coelomic cavity and may have lacked radial water vessels and tube feet, instead using mucus secretion and cilia to collect food.

A **stem** is a long or short attachment appendage made up of numerous disk-shaped **columnals** (stem segments) stacked on top of each other (Figures 18.12, *A* and 18.13). A stem extends from a facet on the lower part of the theca or calyx down to the seafloor, where it is usually cemented or rooted in place. Columnals have a central hole or **lumen** containing branches of the coelomic systems and often have smaller columnal-bearing appendages called **cirri**, composed of smaller columnals, attached to them. The disk-shaped columnals are held together by muscles or ligaments between faces. The faces also have interlocking zigzag **crenulae** to limit relative movement between adjacent columnals. Stems are found in most fossil and some living sea lilies and in several other classes of extinct suspension-feeding echinoderms. A **holdfast** is a more primitive attachment appendage than a stem, composed of numerous small polygonal plates surrounding a large central lumen (Figure 18.12, *B*). It is found in several classes of Cambrian echinoderms. A stem or holdfast allows echinoderms to raise their thecae and food-gathering appendages above the substrate for suspension feeding and prevents them from being carried away by currents.

Classification

The phylum Echinodermata is here subdivided into five subphyla and 20 classes (Table 18.1). The classification is based primarily on the skeletal morphology of the adult stages. There is some controversy concerning the position of several groups in this classification and the taxonomic level they should be given. Most of these controversies involve fossil echinoderms that are poorly known or difficult to interpret. Some dispute also exists as to whether starfish and brittle stars (both living today) should be assigned to separate classes or to the same class. Classification is never a static subject; it keeps changing as new specimens are found and described, as previously known forms are reinterpreted, and especially as new ideas about the relationships of various groups are proposed. The following classification is the one prefer-

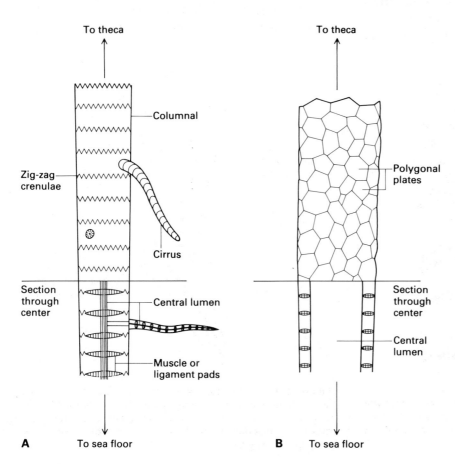

Figure 18.12. Morphology of a stem and holdfast in attached echinoderms. **A,** A columnal-bearing stem found in many Paleozoic echinoderms such as crinoids showing disklike columnals, smaller cirri attached to the stem, and small central lumen carrying coelomic systems from the calyx. **B,** A multiplated holdfast found in two classes of Cambrian echinoderms showing numerous small irregular plates and large central lumen.

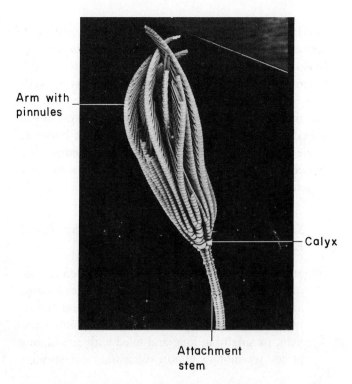

Arm with pinnules

Calyx

Attachment stem

Figure 18.13. Living stemmed crinoid (*Isocrinus*) from deep water south of Cuba showing the attachment stem with a few broken cirri, the reduced calyx, and the long uniserial arms with numerous pinnules (×0.9). (Courtesy Porter M. Kier.)

red by this author, but inasmuch as it includes his biases and present knowledge, it should not be considered the final word on the subject.

SUBPHYLUM CRINOZOA (Figures 18.1, A; 18.13)

Echinoderms with globular, tightly sutured calyces; most have long erect arms with open water vascular system and numerous tube feet; many arms pinnulate. Most have theca of several circlets of plates showing well-developed pentameral symmetry. Mouth nearly central on upper surface, anus lateral; most lack hydropores, gonopores, and accessory respiratory structures. Most attached to substrate either by long columnal-bearing stems with distal attachment structures or by cirri branching directly from base of theca. (Middle Cambrian to Holocene; approximately 1025 living and fossil genera in two classes.)

Crinoids are the only living representatives of the Crinozoa and are the only group of attached suspension-feeding echinoderms now available to help paleontologists interpret the many other extinct classes that

Table 18.1. Preferred classification of phylum Echinodermata

Classification	Geologic range and number of genera
Subphylum Crinozoa	
Class Crinoidea	Middle Cambrian, Early Ordovician–Holocene; about 1005 genera
Class Paracrinoidea	Early Ordovician–Early Silurian; 13 to 15 genera
Subphylum Blastozoa	
Class Blastoidea	Middle Ordovician?, Middle Silurian–Late Permian, about 95 genera
Class Rhombifera	Early Ordovician–Late Devonian; about 60 genera
Class Diploporita	Early Ordovician–Early Devonian; about 42 genera
Class Eocrinoidea	Early Cambrian–Late Silurian; about 30 to 32 genera
Class Parablastoidea	Early to Middle Ordovician; 3 genera
Subphylum Asterozoa	
Class Asteroidea	Early Ordovician–Holocene; about 430 genera
Class Ophiuroidea	Early Ordovician–Holocene; about 325 genera
Subphylum Homalozoa	
Class Stylophora	Middle Cambrian–Middle Devonian; about 32 genera
Class Homoiostelea	Middle Cambrian–Early Devonian; 12 to 13 genera
Class Homostelea	Middle Cambrian; about 3 genera
Class Ctenocystoidea	Middle Cambrian; 2 genera
Subphylum Echinozoa	
Class Echinoidea	Late Ordovician–Holocene; about 765 genera
Class Holothuroidea	Middle Cambrian?, Middle Ordovician–Holocene; about 200 genera
Class Edrioasteroidea	Early Cambrian–Middle Pennsylvanian; about 35 genera
Class Ophiocistioidea	Early Ordovician–Early Mississippian; 6 genera
Class Helicoplacoidea	Early Cambrian; 3 genera
Class Cyclocystoidea	Middle Ordovician–Middle Devonian; 8 genera
Class Edrioblastoidea	Middle Ordovician; 1 genus

apparently had a similar way of life. Crinoids were a large and successful class in the Paleozoic and dominated the fossil record of echinoderms during this interval with about 750 genera. They almost disappeared during the Permo-Triassic extinction and are now a somewhat smaller group today with about 165 living genera and 650 species.

SUBPHYLUM BLASTOZOA (Figure 18.14)

Echinoderms with globular, tightly sutured thecae, bearing erect brachioles usually mounted on specialized ambulacral areas. Thecae with numerous irregularly arranged plates in most earlier forms, later types having fewer plates in circlets with well-developed pentameral symmetry. Mouth central on upper surface, anus lateral; most with hydropore, gonopore, and accessory thecal respiratory structures. Most attached to substrate by stem composed of columnals or by holdfast composed of many plates in early forms. (Early Cambrian to Permian; approximately 233 genera in five classes.)

Blastozoans were suspension feeders occupying much the same ecologic niche as crinozoans. Several classes of blastozoans competed with the dominant crinoids throughout much of the Paleozoic. Primitive Middle Cambrian blastozoans were the first echinoderms to develop a true stem. Other stem-bearing blastozoan classes (including the rhombiferans, diploporans, and blastoids) became fairly common during the middle Paleozoic, but the last blastozoans became extinct at the end of the Permian. Several authors have argued that advanced blastozoans had a water vascular system with tube feet similar to those of living crinoids. However, some major differences in ambulacral morphology between these two groups may imply that tube feet were absent in all five blastozoan classes.

SUBPHYLUM ASTEROZOA (Figures 18.1, B, and E, and 18.15)

Star-shaped, loosely built to tightly sutured echinoderms with five or more large radial arms bearing ambulacral grooves. Arms either large and hollow with major extensions of coelomic systems (starfish), or small but long and filled with articulating plates resembling minute vertebrae and strongly

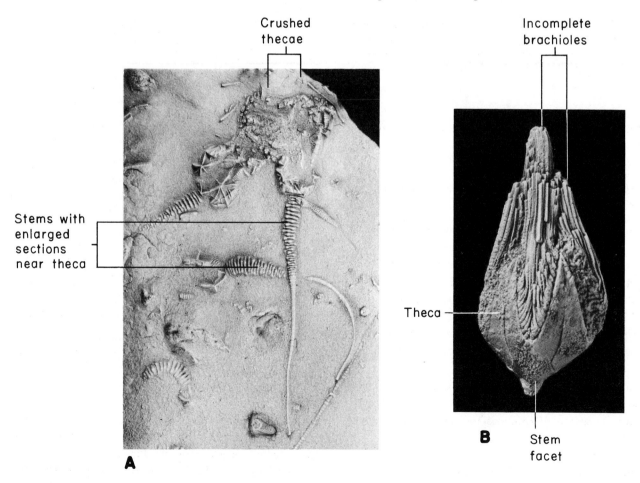

Figure 18.14. Blastozoan echinoderms. **A,** Several complete specimens of the rhombiferan *Cheirocystis* from the Middle Ordovician of New York; note crushed thecae and enlarged stems near thecal attachment. (×1.7.) **B,** A specimen of the blastoid *Pentremites* from the Upper Missisippian (Lower Carboniferous) of northern Alabama showing many incomplete brachioles attached to the ambulacra. (×4.1.)

Figure 18.15. Asterozoan echinoderms. A small slab from the Middle Devonian of central New York containing a large specimen of the asteroid *Devonaster* and a small specimen (white arrow) of the ophiuroid *Encrinaster,* both preserved as natural molds. (×0.8.)

differentiated from central disk containing most of coelomic systems (brittle stars). Ambulacral grooves on arms contain open water vascular system with numerous tube feet. Plates small, numerous; pentameral symmetry usually well-developed. Mouth central on lower surface; anus (if present) on upper surface, and madreporite in various positions. Most have accessory respiratory structures. (Early Ordovician to Holocene; approximately 755 genera in two classes.)

Asterozoans include two large classes of living echinoderms, asteroids (starfish) (Figure 18.2) and ophiuroids (brittle stars) (Figure 18.15). The fossil record of these classes is not very well-known because the skeleton tends to disintegrate after death, and the individual plates are usually small. Only about 25% of all asterozoan genera are known to occur as fossils. Living asteroids and ophiuroids are distinct from each other in their morphology, but as they are traced back into the Paleozoic, they converge with each other. Some Ordovician forms are difficult to assign to one class or the other. Zoologists working with modern echinoderms tend to classify living asteroids and ophiuroids as separate classes because of their obvious morphologic differences; this classification is followed here. However, some students of Asterozoa, impressed with the convergent morphology early in the fossil record, classify these large groups as subclasses of the single class Stelleroidea, along with a third small subclass of very primitive early Paleozoic starfish.

Most asterozoans are mobile benthic crawlers; a large number are carnivores or scavengers, but some are suspension feeders and others are detritus feeders.

SUBPHYLUM HOMALOZOA (Figure 18.16)

Echinoderms with flattened, asymmetrical to bilaterally symmetrical, tightly sutured to expandable thecae; one or more armlike or taillike appendages usually present; no attachment appendage. Armlike appendage with protective cover plates, often with pores for tube feet. Mouth at base of feeding appendage along midline of theca in symmetrical forms; anus lateral or at other end of theca. Small, lateral accessory opening (hydropore or gonopore) in some; porelike or slitlike respiratory structures in others. (Middle Cambrian to Middle Devonian; approximately 49 genera in four classes.)

Homalozoans (originally called carpoids) are a diverse group of flattened echinoderms showing no trace of pentameral symmetry. Some forms have large plates covering the entire theca; others have a frame of large, elongate **marginal plates** and thin, flexible, plated membranes called **central areas** covering the upper and lower sides (Figure 18.16). These latter forms were probably anal pumpers, using this mechanism to respire and perhaps swim by jet propulsion. Homalozoans probably wriggled or swam slowly across the bottom or burrowed through the top layer of organic-rich sediment.

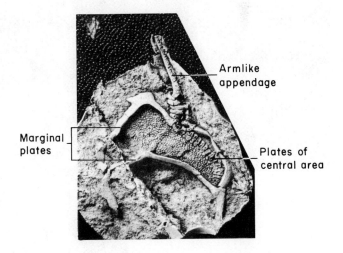

Figure 18.16. Homalozoan echinoderm. This latex cast of the stylophoran *Cothurnocystis* from the Upper Ordovician of Scotland shows the flattened theca with elongate marginal plates surrounding the tiny-plated central area on this side and the long armlike appendage (top) used for feeding. (Latex cast courtesy G. Ubaghs.)

Most were detritus or suspension feeders. The subphylum contains four small classes, all of which are extinct.

SUBPHYLUM ECHINOZOA (Figures 18.1, C, and D; 18.17)

Globular, flattened, cylindrical or fusiform echinoderms having tests ranging from tightly sutured or reduced to sclerites; also having closed water vascular system usually with tube feet protruding through pores in ambulacral areas. Most lack plated feeding appendages (arms or brachioles) or attachment appendages (stems). Skeleton with numerous plates, in most, organized into well-developed pentameral symmetry. Most having mouth and anus at opposite ends of theca and hydropore or madreporite and one to five gonopores near mouth. (Early Cambrian to Holocene; approximately 1018 genera in seven classes.)

Echinozoans include a variety of body plans but use the same type of water vascular system. Four of the seven classes are small and extinct; these were present only in the early and middle Paleozoic. A somewhat larger class, called seated-stars or edrioasteroids, occurs sporadically throughout much of the Paleozoic. The two major living classes, sea urchins (echinoids, Figure 18.17, A) and sea cucumbers (holothurians, Figure 18.17, B), have very different fossil records because of their contrasting skeletal development. Echinoids, which mostly have a tightly sutured skeleton, were not very common or diverse during the Paleozoic but became much more abundant and diversified during the Mesozoic and early Cenozoic when they were the dominant echinoderms. They are still a major group today with about 280 living genera. Most holothurians have a much-reduced skeleton made up of sclerites and are sparsely represented in the fossil record, although they are presently a widespread and diverse group with about 180 genera.

Most echinozoans sit on or move slowly over the substrate, but some form burrows and a few are permanently attached. Many forms are detritus feeders or herbivores, but a few are suspension feeders or carnivores.

Reproduction and life cycle

Many echinoderms have a mass spawning, usually at yearly intervals. During the few weeks just before spawning, much metabolic activity is diverted into producing ripe eggs or sperm. When one individual in a group begins spawning, this triggers (probably by chemical stimulus) nearby individuals to spawn, thus assuring external fertilization.

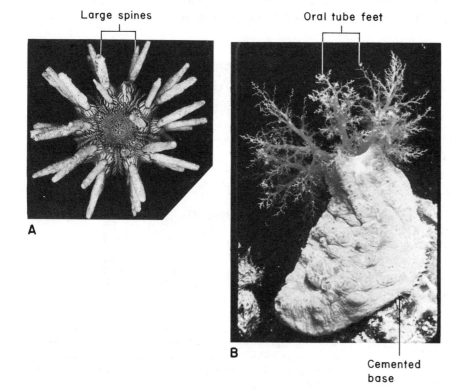

Large spines

Oral tube feet

A

B

Cemented base

Figure 18.17. Living echinozoan classes. **A,** Oral view (with mouth in center) of the echinoid *Eucidaris* from the Caribbean; note the two sizes of spines and the near-perfect pentameral symmetry. (×0.7.) **B,** The attached holothurian *Pentacta* from the coast of western Australia with its oral tube feet extended for feeding. (**A,** Courtesy Porter M. Kier. **B,** Courtesy M. Coleman and David Pawson.)

In many echinoderms, the fertilized eggs develop into free-floating planktic larvae and remain in this free-floating stage for several days or weeks. Echinoderm larvae are bilaterally symmetrical in marked contrast to the radial symmetry of most adults. The larvae either subsist on yolk from the egg or feed on nutrients or minute plankton in the sea water. The free-floating larvae are widely distributed during this stage and may even cross oceanic barriers if carried by favorable currents. The larvae of some deep-sea echinoderms remain near the seafloor and undergo development there. Other shallow- and deep-water echinoderms protect their eggs in specialized brooding chambers on or inside the theca where the eggs are fertilized and develop into larvae or

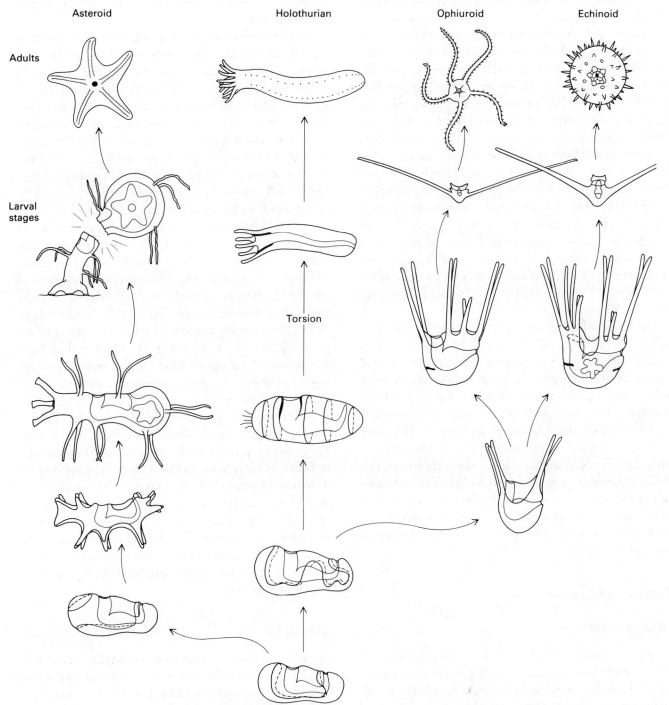

Figure 18.18. Larval stages of four of the five living echinoderm classes showing ophiuroids and echinoids (right) and asteroids and holothurians (left) with similar larval stages. However, these groups are not classified together based on their adult morphology and fossil record. (From Nichols, D. *Echinoderms*, 4th ed. London: Hutchison University Library; 1972.)

immature adults ready to settle. Brooding structures have been found in a few Tertiary echinoids and in one Pennsylvanian (Late Carboniferous) blastoid.

Echinoderm larvae have several characteristic features that mark them as deuterostomes. In their early stages, they resemble the larvae of other deuterostomes such as hemichordates and chordates. This larval resemblance is in contrast to the adult morphology that is quite different in these three phyla.

Within echinoderms, the larval types of the five living classes do not match the grouping into subphyla, which is based on adult morphology and known fossil record. The later larval stages of echinoids and ophiuroids (assigned to different subphyla) are similar but differ considerably from the corresponding larvae of starfish and holothurians, which are also assigned to different subphyla (Figure 18.18). Both of these larval types are somewhat different from the larvae of living crinoids. The later larval stages of echinoderms apparently are themselves adapted to certain environmental conditions and ways of life and have developed specific morphologic features independent of the adult morphology and the known evolutionary history. Unfortunately, little is known about the larval stages of fossil echinoderms because the larvae either have no ossicles or such small ossicles that they are almost never found.

At the end of the larval stage, the free-swimming larva settles or attaches to the sea floor and undergoes metamorphosis into a juvenile echinoderm. Several major changes take place at this time, including secretion of the first skeletal plates, adoption of the adult body form, and outward growth of the five radial water vessels from the newly formed circumoral ring. During later growth, juveniles become more and more like adults in thecal proportions and numbers of plates (Figure 18.19). Most echinoderms grow rapidly as juveniles and more slowly as adults. They usually become sexually mature after one to two years; individuals of many species live eight to ten years, but some forms have been known to live even longer.

Features of the phylum

Skeletal growth

The porous mesodermal nature of echinoderm skeletal plates determines how growth of the skeleton can occur. As an individual increases in size, its skeleton can grow in at least two ways. New plates are added to specific areas, and established plates become larger by the addition of new layers of calcite to sutured margins (and sometimes to the exterior or interior). Resorption of plates is known in a few forms but is uncommon, and major modifications of plate shape (such as those indicated by vertebrate bones during growth) occur mostly on small plates that have just been inserted. Echinoderms do not molt as do arthropods because their skeleton is an endoskeleton with living tissue (the epidermis) on the outside; in addition, plates grow continuously as the animal increases in size.

An echinoderm with a rigid, tightly sutured skeleton must maintain a close fit between all the plates as it grows. This is probably done by a fitting process, in which the tightness of fit determines the growth rate of adjacent plates. In some ways, the plates act like soap bubbles in a confined area, spreading out (growing) to close any available gaps and maintaining tight, straight, or slightly curved sutures where there is little room. Sutures of unequal length around a plate indicate differential rates of growth from the plate origin.

Growth lines are often preserved either on the outside or inside surfaces of echinoderm plates and extend through the plates (observable when the plates are made translucent) (Figure 18.20). These lines may represent either tiny increments of growth or periodic cessation of growth. Sometimes growth lines are grouped together into larger **growth bands** separated by major breaks shown by pigmentation, surface relief, or composition of the plate (Figures 18.20 and 18.21). Many authors have considered these larger bands to be annual markers caused by seasonal climatic changes or some once-a-year activity such as spawning.

When an echinoderm skeleton is damaged, an irregular set of new plates rapidly fills in the damaged area. A spine grows back rapidly from the broken stump to produce a repaired spine (Figure 18.22) almost indistinguishable from the original. Echinoderms are also proficient at regrowing broken appendages, such as arms, or regenerating damaged internal organs. The most extreme form of regeneration is shown by holothurians that can discard some of their viscera as part of an escape mechanism to confuse a predator. These discarded organs are subsequently regrown.

Ways of life

Living echinoderms occupy many ecologic niches (Figure 18.23) available to benthic marine invertebrates. Fossil echinoderms apparently favored ways of life not common in living echinoderms. For example, epifaunal suspension and detritus feeders apparently were dominant among Paleozoic echinoderms. Possible reasons for this include the high primary productivity of algae in the early Paleozoic, the relatively few infaunal organisms plowing up the sediment on which the echinoderms were living, and the lack of efficient predators. Infaunal detritus feeders and heavily armored herbivores and carnivores became more common in the Mesozoic and Cenozoic. This change may have occurred because of decreasing primary productivity, increasing competition for space and resources, increased disruption of the bottom sediment by infaunal burrowers, and the

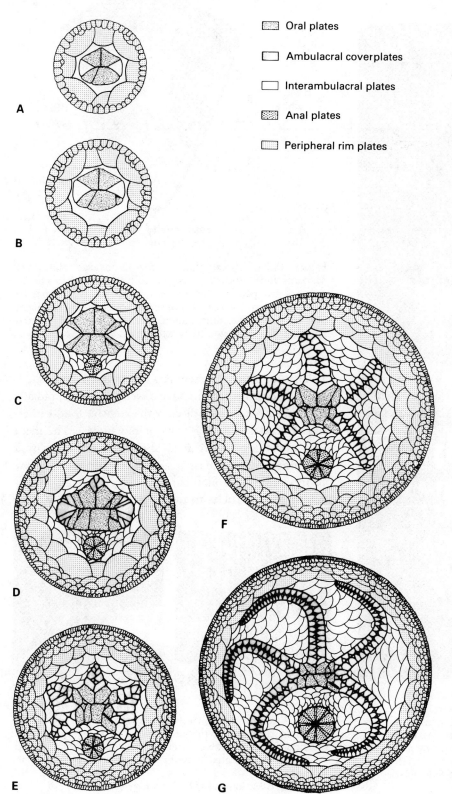

Oral plates

Ambulacral coverplates

Interambulacral plates

Anal plates

Peripheral rim plates

Figure 18.19. Growth stages after metamorphosis in the Ordovician edrioasteroid *Isorophus.* **A** to **E** are juvenile stages ranging from 0.5 mm to 3.5 mm in diameter; **F** and **G** are adolescent and adult stages, 5 mm to 20 mm in diameter. Note the great increase in number of plates, outward growth and curvature of the ambulacra, and changes in proportions of various thecal regions. (From Bell, B. M. *Journal of Paleontology* 50: 1010–1011; 1976. Courtesy New York State Museum.)

presence of more efficient predators, especially fish, on unprotected echinoderms.

Many Paleozoic and some living echinoderms were attached, high-level, epifaunal suspension feeders; groups such as the crinoids, blastoids, and many other

Figure 18.20. Two columns of interambulacral plates in the living echinoid *Strongylocentrotus*, which were immersed in a heavy liquid to bring out the internal growth lines in the translucent plates. New plates are added to the top of each column as triangular wedges and continue to grow into nearly rectangular plates as the relative position of the plates on the surface of test is changed. (From Raup, D. M. *Journal of Paleontology* 42; 1968.)

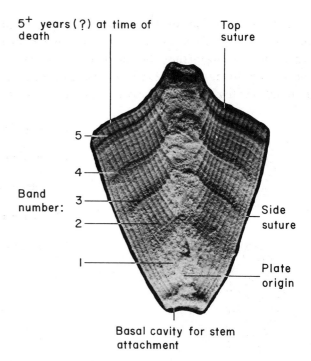

Figure 18.21. Thecal plate of the Ordovician blastozoan *Meristoschisma* showing tiny growth lines parallel to sutured margins grouped into five or more larger bands that may be yearly markers. Surface relief of these growth bands is brought out by low-angle illumination from below. (×7.) (From Sprinkle, J. T. *Museum of Comparative Zoology Special Publication;* 1973.)

Paleozoic groups represent this way of life (Figure 18.23, *B*). These forms tend to have a well-developed, radially symmetrical, compact theca with extensive food-gathering appendages (either arms or brachioles). The theca and appendages were held above the sea floor by a medium to long holdfast or stem. These forms fed in slow to moderate horizontal currents, which brought microscopic food particles to them and carried away wastes.

Figure 18.22. Scanning electron micrographs of regenerated spines in the living echinoid *Arbacia*. **A**, Close-up of fully regenerated spine showing stereom rods and stromal canals. (×40.) **B**, Regenerated spine tip growing from broken spine stump. (×20.) (From Davies, T. T.; Crenshaw, M. A.; Heatfield, B.M. *Journal of Paleontology* 46; 1972.)

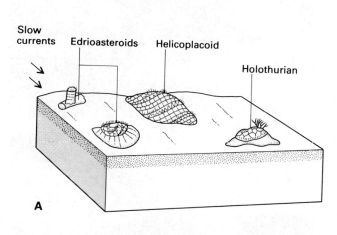

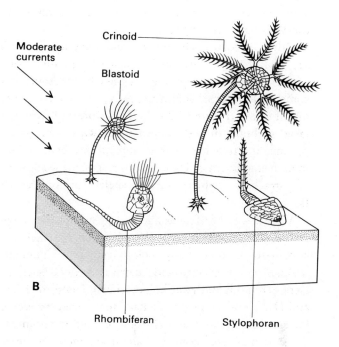

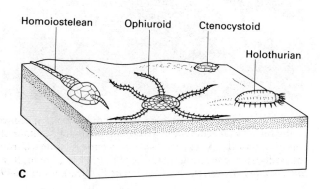

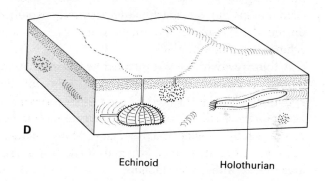

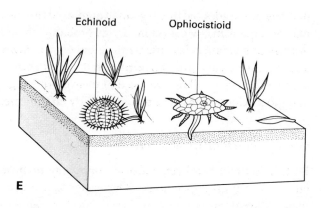

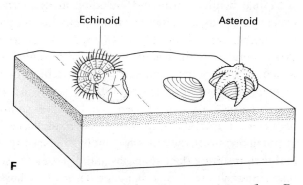

Figure 18.23. Ways of life of fossil and living echinoderms. **A,** Low-level, epifaunal, suspension feeders sat directly on the sea floor and used short appendages or tube feet to capture microscopic food carried by slow bottom currents. **B,** High-level, epifaunal suspension feeders used specialized appendages to capture microscopic food carried by currents above the sea floor. **C,** Mobile, epifaunal detritus feeders collected organic material from the top layer of sediment on the sea floor. **D,** Mobile, infaunal detritus feeders burrow through and ingest the bottom sediment to extract organic material. **E,** Mobile, epifaunal herbivores eat algae or seaweed growing on the sea floor. **F,** Mobile, epifaunal carnivores prey on other invertebrates living on the sea floor. (From Sprinkle, J. T. *University of Tennessee Studies in Geology* No. 3; 1980.)

There was an evolutionary trend toward lengthening and improving the holdfast or stem to get further off the substrate and also toward improving and elaborating the food-gathering appendages.

Other Paleozoic echinoderms and some living forms are stationary or attached, low-level, epifaunal suspension feeders; groups such as the edrioasteroids and some diploporans lived in this way, as do some living holothurians (Figure 18.23, *A*). These forms often have a large globular or disk-shaped theca with symmetrical plating, and tube feet or short appendages serve as their main feeding organs. These groups rest on the sea floor or are attached to objects such as shells lying on the sea floor.

A third way of life was adopted by groups that were either mobile, epifaunal detritus feeders (such as homalozoans and many holothurians) or infaunal, shallow- to deep-burrowing detritus feeders (such as some echinoids and some holothurians) (Figure 18.23, *C* and *D*). These forms usually have specialized structures for sorting organic material from the bottom sediment. Because they are mobile, bilateral symmetry is often superimposed on the basic pentameral symmetry. The burrowing mode of life was not used by many Paleozoic echinoderms but was much more extensively exploited during the Mesozoic and subsequently by advanced echinoids and holothurians.

A fourth way of life was adopted by groups that are mobile benthic herbivores, including many of the epifaunal echinoids (Figure 18.23, *E*). These forms move about on the sea floor looking for plant material that they can detach by chewing and then ingest. Many epifaunal echinoids currently feed on kelp and can destroy a large area of kelp within a few months.

Mobile benthic carnivores including many starfish, some ophiuroids, and some epifaunal echinoids, adopted yet another way of life (Figure 18.23, *F*). These forms slowly crawl over the seafloor seeking other animals on which to prey. Some echinoids use their teeth to break through shells of other invertebrates to get at the soft parts. Starfish have developed the highly organized behavior of eating pelecypods or other mollusks by gripping them with their tube feet, pulling the shell apart, and then inserting their stomachs and digesting the soft parts. Starfish are now a major predator of mollusks, corals, barnacles, and other echinoderms (Figure 18.24).

Certain ways of life are not commonly used by adult echinoderms. These include the planktic, or floating, mode adopted by a few holothurian genera; the nektic, or swimming, mode used by a few deep-sea holothurians and some stemless crinoids part of the time; and the

Figure 18.24. The living starfish *Oreaster* preying on a specimen of the burrowing echinoid *Meoma* exposed on the sea floor off southern Florida. **B,** The starfish is retracting its stomach from the badly etched echinoid test. (From Kier, P. M.; Grant, R. E. *Smithsonian Miscellaneous Collections* Publication No. 4649; 1965.)

parasite mode adopted by a few small ophiuroids in other marine organisms. Most echinoderms probably are not floaters or swimmers because they can only move slowly and would be easy prey for active predators. Parasites tend to be very small and reduced in their morphology, and most echinoderms have remained fairly large in size.

Gregariousness of echinoderms

Both living and fossil echinoderms commonly occur in large gregarious associations. Whenever a single complete echinoderm is found at a fossil locality, the area should be carefully examined for other specimens; often dozens or even hundreds of additional specimens can be found in a single bed over a small area. Many types of echinoderms probably occur in groups for the following reasons:

1   The local microenvironment may be especially favor-

able because of abundant food, desirable attachment sites, or protection from wave or current activity.

2 Fertilization during spawning is more likely if many individuals are living together.

3 Clustering of echinoderms may provide some protection against predators.

Preservational factors can also play a part in producing isolated occurrences of abundant, complete fossil echinoderms. Dead echinoderms rapidly decay and fall apart when left exposed on the sea floor for even a few days, producing numerous plates but no complete specimens. In contrast, numerous well-preserved individuals on a single bedding plane usually represent instantaneous kills by underwater mud slides or storm-deposited sediment that overwhelmed living individuals and buried them intact before decay could set in.

Occurrence of fossil echinoderms

Fossil echinoderms are most abundant in rock units representing offshore shelf environments or epicontinental seas with shallow to moderate water depths and normal marine salinity. Echinoderms are also abundant in and around fossil reefs and banks, especially in the Paleozoic, and their disarticulated plates and columnals may be a major constituent of the detrital fill and adjacent flank beds. Deeper-water slope and basinal deposits usually do not have abundant echinoderms. Although some echinoderms are abundant in the deep sea at present (Figure 18.3), deposits of these oceanic environments are rarely preserved in the fossil record. Echinoderms are usually rare in most shore face, tidal flat, and deltaic deposits, although exceptions are known. Echinoderms usually did not live in these near-shore environments because of the unfavorable and rapidly deposited sediment (coarse sand and soft mud), the great environmental variability (water depth, current activity, and temperature) and especially the periodic influx of fresh water from the nearby land, producing rapid salinity variations.

Fossil echinoderms seem to occur most commonly in shale and thin-bedded shaly limestone and are often complete, showing excellent preservation in these rock types. Massive limestone beds sometimes contain complete echinoderms, but disarticulated plates are usually more common in this matrix. Complete specimens may be difficult to recover from massive limestone beds unless they weather from the limestone surface or are silicified and can be extracted by acid etching. Complete echinoderms, excepting later Cenozoic sand

dollars, are uncommon in most sandstone deposits, although disarticulated plates may be present.

Origin of phylum and classes

Echinoderms are a well-marked phylum, having no close living or fossil relatives in other phyla. The distinctive skeleton, development of a water vascular system, and pentameral symmetry separate most echinoderms from all other invertebrates. Larval stages link echinoderms to the other major deuterostomes, the hemichordates and chordates. However, this relationship cannot be very close because the adults are so different. Echinoderms probably branched off from lineages leading to the other deuterostomes well before the beginning of the Phanerozoic.

Most echinoderm classes are also well-marked. They tend to appear in the fossil record already fully formed and different from any possible ancestral group. Only a few cases of intermediates or transitional forms between echinoderm classes are known. However, parallelism and convergence are common in Paleozoic echinoderms. A small group in one class often developed a morphology that was similar to that of another class. For example, a few echinozoans apparently developed a morphology similar to high-level suspension-feeding crinozoans or blastozoans. This seems to imply that there may have been only one or a few most-favorable designs for echinoderms in a particular way of life, but several ways of developing the morphologic features necessary to achieve each of these designs.

The fossil record for many echinoderm classes is long. All of the living classes extend back to the Ordovician, and several extinct classes range throughout much of the Paleozoic (Figure 18.25). However, most classes in the early Paleozoic were short-lived and confined to only a small part of the record. They probably never became very diverse nor geographically widespread, and they may not even have been abundant when alive. In part, this pattern may result from our poor knowledge of the fossil record in the earliest Paleozoic because of limited collecting in this interval. However, most of this pattern probably reflects the environmental opportunities of early echinoderms.

Several authors have complained about the large number of echinoderm classes that have been described (Figure 18.25). This number includes many Cambrian and Ordovician classes with distinct morphologies but low diversity and short fossil records. Should these small groups really be ranked as separate classes? This problem

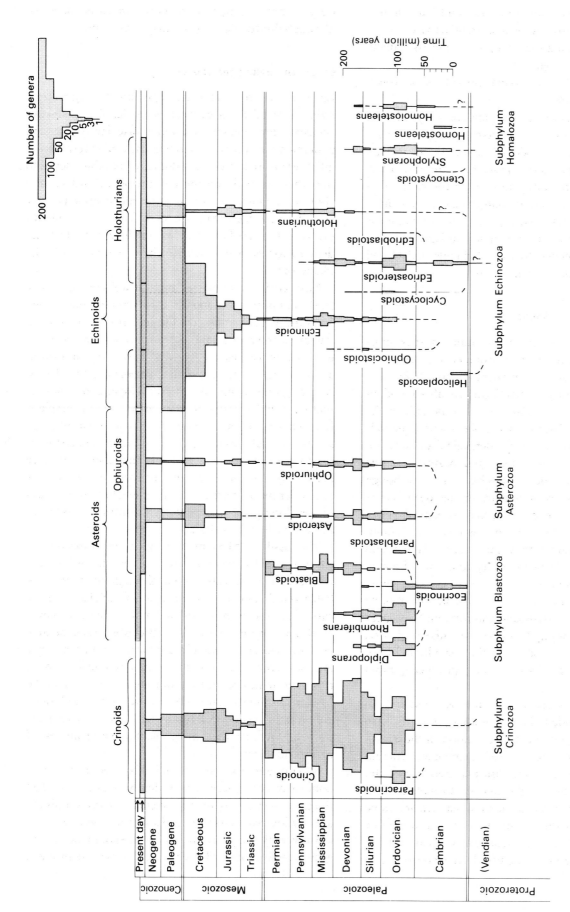

Figure 18.25. The fossil record of echinoderm classes showing their stratigraphic ranges and generic diversities (see scale at top). Note the large number of classes in the early Paleozoic, the decrease in number of classes throughout the late Paleozoic, the dominance of the Paleozoic record by blastozoans and crinozoans (especially crinoids), the dominance of the Mesozoic and Cenozoic record by echinoids, and the diversity of the five echinoderm classes living today. (From Sprinkle, J. T. *University of Tennessee Studies in Geology* No. 3; 1980.)

relates to criteria used for definition of higher taxa; many authors require a taxon at the class level to have very different morphology from other classes and considerable diversity and success through time. Some early echinoderm groups certainly have different morphology but apparently never became very diverse and did not last long. However, most echinoderm investigators (including this author) believe that morphology is the most important factor in classifying these early echinoderms and that the short time range and lack of success of these groups is not critical to their taxonomic rank.

The numerous small and short-lived groups of echinoderms known only from the early Paleozoic diversified in an environment that apparently contributed to high productivity and had few metazoan competitors or predators. This resulted in extensive experimentation toward development of new and different ways of life. Many of these apparently experimental forms were relatively inefficient in exploiting their ways of life and were short-lived as competition and predation increased. This pattern suggests that the late Proterozoic, Cambrian, and Ordovician represent unique periods of low competition in the fossil record—times when the shallow-water marine environment was first being colonized by metazoans with skeletons. Echinoderms were one of the early phyla diversifying under these conditions, and they experimented with many different ways of life, only some of which were successful over time.

Evolutionary history of echinoderms

Echinoderms appear to have first evolved in the late Proterozoic but have left little record during that interval, implying that they were perhaps soft-bodied without a plated skeleton. The earliest possible echinoderm may be the unusual form called *Tribrachidium* from the Ediacara fauna of Australia (Figure 18.26). A few authors have postulated that this genus may be an edrioasteroid-like echinoderm belonging to the subphylum Echinozoa (Figure 18.25).

In the Early and Middle Cambrian, several echinoderm classes appeared suddenly with no obvious ancestors or close relatives (Figure 18.25). Four of the subphyla apparently had already differentiated and crossed the Proterozoic-Cambrian boundary as separate lineages. Most early classes show primitive morphology, low diversity, and a limited geographic and stratigraphic range. At least three classes of echinoderms occur in the Early Cambrian, increasing to seven to eight classes in the Middle Cambrian. The Late Cambrian record of echinoderms is not as well-known but appears to have lower diversity with only four classes. The lower diversity may represent a consolidation phase in the Cambrian radiation or an environmental or preservational bias. Suspension- and detritus-feeding echinoderms (Figure 18.23) were most common in the Cambrian.

Echinozoans and homalozoans include seven classes in the Early to Middle Cambrian, but none of these became abundant and three soon became extinct (Figure 18.25). Crinozoans and blastozoans each produced only one class with low to moderate diversity in the Cambrian, probably because they were not efficient as high-level suspension feeders, using a short multiplated holdfast (Figure 18.23).

Beginning in the Early Ordovician and continuing into the Middle Ordovician, a second radiation of echinoderms took place, resulting in the appearance of many new classes and new ways of life at this time (Figure 18.25). This Ordovician echinoderm radiation produced much higher diversities at the generic and specific levels; many Middle Ordovician faunas have 15 to 20 echinoderm species occurring together in contrast to Cambrian echinoderm faunas, which include only two to three species. The independent development of a long columnal-bearing stem plus other morphologic improvements in both crinozoans and blastozoans probably led to their major diversification during the

Figure 18.26. Cast of the puzzling fossil *Tribrachidium*, which may be an echinoderm resembling an edrioasteroid from the late Proterozoic Ediacara fauna of South Australia (×1.7). (From Glaessner, M. F. *Pre-cambrian animals.* Copyright © 1961 by Scientific American, Inc. All rights reserved.)

Early to Middle Ordovician and resulted in the appearance of four to five new classes at that time. Stemmed echinoderms, especially the crinoids, were the most common and diverse members of the phylum from the Ordovician to the late Paleozoic (Figure 18.25). The asterozoans (both asteroids and ophiuroids) first appeared in the Early Ordovician with no obvious ancestors and were present (but not very diverse) throughout the rest of the Paleozoic. Echinoids and holothurians appeared in the Middle Ordovician, but remained fairly small classes in the Paleozoic along with continuing edrioasteroids and homalozoans. No new echinoderm classes, with the possible exception of blastoids appeared in the fossil record after the Middle Ordovician; the same is true in several other phyla. This implies that all the major ways of life for echinoderms were tried early in their evolutionary history. Later changes have been within the successful major groups (classes) established during the early Paleozoic radiation.

The Middle Ordovician marks the high point of echinoderm diversity at the class level with 17 classes present. Diversity started to decrease by the Late Ordovician (14 classes), and this decrease continued throughout the rest of the Paleozoic (Figure 18.25). Several classes disappeared during the Devonian, and others dropped out one by one during the upper Paleozoic. These unsuccessful classes tended to be small groups apparently having inefficient designs and too much competition in their way of life. However, echinoderm diversity at the generic and specific level remained high in the remaining successful classes through the Mississippian (Early Carboniferous), when crinoids, blastoids, and echinoids reached their maximum Paleozoic diversity in the widespread epicontinental seas of that period (Figure 18.25). Diversity at the generic and specific level decreased considerably in the Pennsylvanian (Late Carboniferous) and Permian, but some groups such as crinoids remained common.

The extinction at the Permo-Triassic boundary severely affected suspension-feeding stemmed echinoderms. Blastoids, representing the last of the blastozoans, became extinct, and crinoids almost disappeared with only a single genus known from the Early Triassic. Asteroids, ophiuroids, echinoids, and holothurians appear to have survived the extinction at the Permo-Triassic boundary without much change, but they were rare and not very diverse during this interval (Figure 18.25). All classes that did survive are still living.

Echinoids became the dominant class of echinoderms in the Mesozoic and early Cenozoic (Figure 18.25). They were a small class during the Paleozoic, probably living as slow-moving epifaunal herbivores and scavengers. However, once past the Permo-Triassic boundary, echinoids started to diversify and increase their morphologic diversity. Other groups that had competed with echinoids apparently disappeared during the Permo-Triassic extinction, giving the surviving echinoids a distinct advantage. By the Jurassic, several new groups of epifaunal echinoids had appeared, and the first heart-shaped echinoids, which were shallow-burrowing detritus feeders, had evolved. These new morphologies and ways of life led to a major radiation during the Jurassic, Cretaceous, and early Cenozoic (Figure 18.25). Echinoids then decreased somewhat in diversity, but with 280 living genera are still a common and diverse group.

Asteroids, ophiuroids, holothurians, and the surviving crinoids diversified moderately during the Mesozoic and Cenozoic, and except for crinoids are probably more common today than they have been at any previous time. Asteroids and ophiuroids are the most diverse echinoderms now living (Figure 18.25). Modern holothurians also are diverse and adapted to many habitats. Mesozoic (and later) crinoids differ from Paleozoic crinoids, and a separate subclass has been established for them. Today crinoids are fairly diverse in some favorable environments, such as coral reefs, but based on the number of fossils now known, were probably much more common and varied during the Paleozoic.

Part II Additional concepts

James Sprinkle

This section reviews morphology, ecology, and evolutionary history of the 13 most important living and fossil echinoderm classes with figures of some of the more common and distinctive genera. In each subphylum, classes are listed in the order of diversity, except within the Echinozoa where the Echinoidea are mentioned last. All 20 echinoderm classes recognized here are listed with their stratigraphic ranges and generic diversities in Table 18.1, and their fossil records are indicated in Figure 18.25.

Subphylum Crinozoa

Subphylum Crinozoa contains two classes of arm-bearing, stemmed echinoderms of very different size and importance. Crinoids are the largest class of echinoderms known from the fossil record and are still living today; paracrinoids are a relatively small extinct class known only from the early Paleozoic (Figure 18.25). Members of both of these classes were stemmed, medium- to high-level suspension feeders characterized by an open water vascular system.

CLASS CRINOIDEA

Crinozoa with calyx composed of conical, globular, or bowl-shaped cup below arm attachments and flat to highly domed tegmen above; cups with several circlets of small to large plates, most with well-developed pentameral symmetry and no respiratory slits or pores through plates; tegmen with few to many small plates. Arms short to long, uniserial or biserial; some repeatedly branched, many with smaller uniserial pinnules, and all with open water vascular system bearing tube feet in food grooves. Gonads external on lower pinnules or arms in living forms. Calyx either permanently attached to substrate by long columnal-bearing stem or free-living with numerous cirri branching from base of calyx for crawling and temporary attachment. (Middle Cambrian, Early Ordovician to Holocene; about 1005 genera.)

Morphology

A typical Paleozoic crinoid is divided into three morphologic regions (Figure 18.27): the plated calyx or theca that encloses most of the soft parts (vital organs) of the animal and has the most diagnostic plating for classification, the food-gathering arms that branch off the side or top of the calyx, and a stem or group of cirri that attach the base of the calyx to the substrate. The soft parts differ in a few respects from those of most other echinoderms (Figure 18.27, *B*). There are several coelomic cavities and a well-developed open water vascular system with a branch out to each arm. Crinoids have a complete digestive system with a central mouth, a long U-shaped and sometimes coiled gut, and a lateral anus, usually on the calyx summit. Oral and aboral nerve systems send branches out to the arms and stem, and low in the calyx is an internal **chambered organ** near the stem (or cirri) attachment area that may control movement and balance (Figure 18.27, *B*).

The calyx is divided into a lower part below the arm attachments called the **cup** (also called the dorsal or aboral cup) and a capping portion containing the mouth and anus called the **tegmen** (Figure 18.27, *B*). The cup is

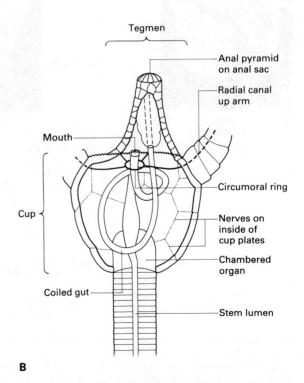

Figure 18.27. A, Well-preserved specimen of the crinoid *Cyathocrinites* (a dicyclic inadunate) from the middle Paleozoic showing the attachment stem, the large plated calyx that con- tains most of the viscera, and five long branched arms (×0.9). **B,** Inferred soft-part anatomy in this fossil is based on a comp- arison with living crinoids. (**A,** Courtesy Porter M. Kier.)

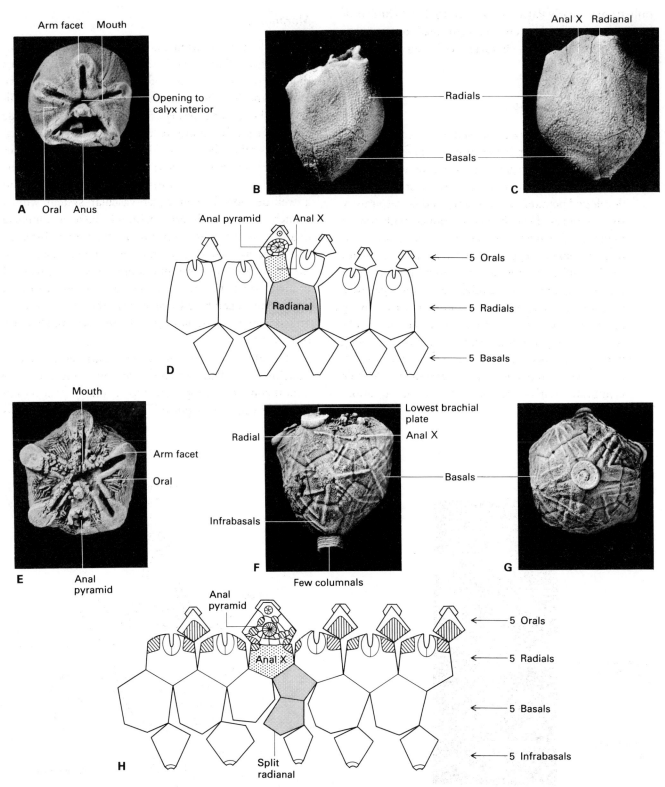

Figure 18.28. Monocyclic and dicyclic cups in simple Paleozoic crinoids. **A** to **C**, Top and two side views of the monocyclic inadunate *Hybocrinus* from the Middle Ordovician of southern Oklahoma. Note only two circlets of cup plates (basals and radials), extra plates on the anal side, and the oral plates plus the mouth, anus, and arm facets on the simple tegmen (×2). **D**, Side layout of thecal plating in *Hybocrinus*. **E** to **G**, Top, side, and bottom views of the dicyclic inadunate *Carabocrinus* from the Middle Ordovician of southern Oklahoma showing three circles of cup plates (infrabasals, basals, and radials), few stem segments attached to cup, and few arm brachials and part of anal pyramid on summit (×2.1). **H**, Side layout of thecal plating in *Carabocrinus*.

usually made up of relatively few tightly sutured plates arranged in several well-defined circlets. Two major types of cups are known in crinoids: **monocyclic cups**, in which only one circlet of plates is present below the plates bearing the arms (Figure 18.28, *A* to *D*), and **dicyclic cups**, in which two circlets are present (Figure 18.28, *E* to *H*).

The topmost circlet of plates that bear facets for the attachment of arms are called **radials** (Figure 18.28). As their name suggests, they are located in a radial position. Some early crinoids have horizontally divided radials in some or all rays; the lower part is called an **inferradial**, and the upper part is called a **superradial**. The next circlet of plates below the radials and offset from them (in an interradial position) are called **basals**. If a third circlet of plates is present (found in dicyclic cups), they are situated below and offset from the basals (in a radial position) and are termed **infrabasals**. Many crinoids have a full complement of either five basals or five infrabasals in their lowest circlet of cup plates, but others have reduced the number of plates in this lowest circlet to four, three (common), two, or even one, primarily by fusion. On the posterior or anal side, a few extra plates are usually present to form an **anal series**, a group of plates supporting the anal opening. If the lowest of the additional plates also lies below or supports a radial, it is called a **radianal** (Figure 18.28). The next plate (sometimes the only other cup plate) in the anal series above the radianal is called **anal X**. If the first plate in the anal series lies above and between two radials (instead of below a radial) on the posterior side, it is usually called a **primanal**. Some Paleozoic crinoids also have lower arm plates incorpor-

ated into the top of the rigid cup so that the cup extends well above the radials; other Paleozoic crinoids have groups of small fused plates called **interbrachials** inserted between the major circlets in the rigid cup.

In primitive types, the tegmen is made up of few to many plates above the radials. If only five small plates are present on the tegmen around the mouth, they are called **orals** and are located in an interradial position between the food grooves (Figure 18.28, *A* and *D*). Many forms have additional tegmen plates around the anal opening and below the orals. The simplest anal structure is a low-domed cover with small triangular plates over the anal opening called an **anal pyramid** (Figures 18.27, *B*

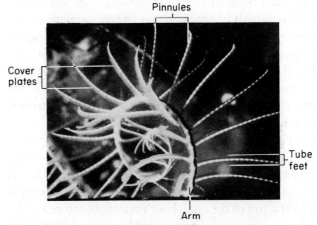

Figure 18.30. Pinnules, tube feet, and cover plates on a coiled arm of the living crinoid *Nemaster* from 18 to 24 m of water at Grand Bahama Island. Only the largest set of tube feet on the pinnules can be seen, along with the tiny cover plates, which show up as white spots along the edge of the pinnules. (From Macurda, D. B., Jr. *Hydro-Lab Journal* 3, Plate 3, Figure 1; 1975.)

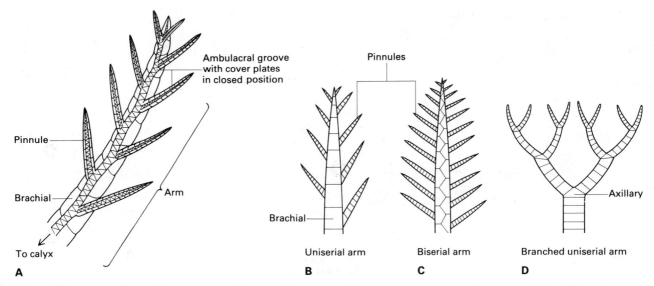

Figure 18.29. Arm morphology in crinoids. **A**, Part of arm showing brachial plates, smaller pinnules attached to the brachials, and the food grooves with protective cover plates. **B** to **D**, Examples of uniserial, biserial, and extensively branched arms.

and 18.28, *D*). The anal structure is commonly modified into an elongate cylindrical **anal tube** or an expanded multiplated **anal sac** (Figures 18.27, *B* and 18.33, *D*), which bears the anal pyramid on its top or side.

An arm is an erect feeding appendage made up of numerous **brachials** that are cylindrical or wedge-shaped plates bearing the ambulacral groove on one side (Figure 18.29). Smaller uniserial pinnules commonly are attached to the brachials. Arms can be **uniserial**, composed of a single row of brachials (Figure 18.29, *B*), or **biserial**, composed of a double row of alternating brachials (Figure 18.29, *C*). Arms can be unbranched, can branch a few or many times (Figure 18.29, *D*), and can either be short with only 10 to 12 brachials or long with hundreds of brachials and attached pinnules. Each branch point in an arm has a specially shaped triangular or pentagonal brachial called an **axillary** (Figure 18.29, *D*). If an arm branches into two equal segments at each axillary, this is termed **isotomous** branching; if the branches are unequal in size or length, this is termed **heterotomous** branching. Some crinoids restrict movement in the lower arms by having a small curved projection, or **patelloid process**, extending from each brachial into a notch on the brachial directly below (Figure 18.33, *E*). In crinoids that have the lower arms incorporated into the top of the cup above the radials and below the tegmen, these modified arm plates are called **fixed brachials**.

The arms and pinnules are provided with numerous tiny tube feet branching off the open radial water vessels (Figure 18.30). The ambulacral grooves on the pinnules, arms, and tegmen have numerous tiny **cover plates** to protect them and the enclosed tube feet. The cover plates are open while feeding and shut at other times (Figure 18.29). Ciliated ambulacral grooves carry the collected food down the pinnules and arms to the radial attachment facet and then between the orals or beneath the tegmen plates to the mouth, near the center of the tegmen. The lower pinnules near the base of the arms bear the gonads in living crinoids, and this was probably true in most or all fossil crinoids because no gonopore has been found on the tegmen.

Crinoid stems are usually composed of short, disk-shaped columnals that can be round, pentagonal, star shaped, elliptical, or other shapes in cross section (Figure 18.31). Smaller cirri often branch off a stem and are made up of columnals that are very small in diameter. Columnals in a stem or cirrus have a central opening, or lumen, through which canals and nerves from the coelemic cavity extend. Most columnals in a crinoid stem are sutured together with a zigzag or crenulate contact. The columnals in many stems may be differentiated by larger and smaller sizes. The larger columnals may have cirri branching from them and are called **nodals**. In contrast, the columnals with smaller diameters, which are often thinner, lack cirri and are called **internodals** (Figure 18.31).

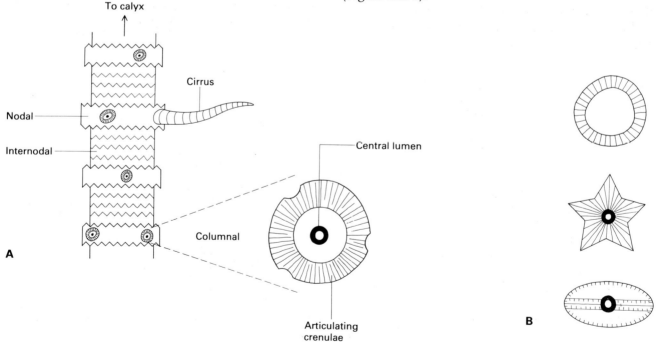

Figure 18.31. Stem morphology in crinoids. **A,** Segment of a stem showing different sized columnals, cirri branching off nodals, and small central lumen and articulating crenulae on columnal face. **B,** Examples of cross-sectional shapes of some fossil crinoid columnals.

Crinoid stems usually end in some type of structure to attach the stem and calyx to the seafloor or other objects. This can be a branching root system, a mat of cirri arising from the lower columnals, a multiplated or solid disk for cementing to some object, or a solid structure resembling a grapnel or anchor. A few Paleozoic crinoids have a large bulblike structure at the end of their stems, which either sat on the seafloor to stabilize the crinoid or was filled with gases for flotation. Stems in some primitive crinoids have multipart columnals made up of four to five segments; the earliest reported crinoid from the Middle Cambrian has a primitive holdfast of numerous plates, which is an elongate extension of the lower calyx used to attach the crinoid to an object on the substrate (Figure 18.32). Crinoids apparently developed a columnal-bearing stem from this primitive holdfast either in the Late Cambrian or the Early Ordovician.

Classification

Crinoids are usually subdivided into four major subclasses (Table 18.2) based on several features, par-

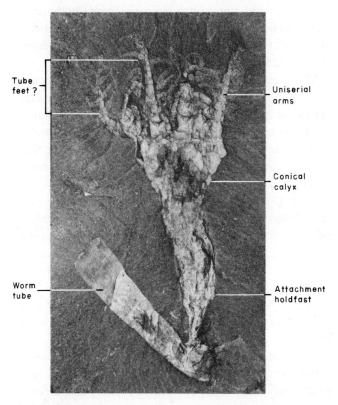

Figure 18.32. *Echmatocrinus,* the earliest known crinoid from the Middle Cambrian Burgess Shale of southeast British Columbia, Canada, has a multiplated holdfast attached to a worm tube. Note the poorly organized calyx and the short arms with smaller appendages that have been interpreted as tube feet (×0.7). (From Sprinkle, J. T. *Museum of Comparative Zoology Special Publication;* 1973.)

ticularly the arrangement of plates in the calyx, structure of the arms, and nature of the tegmen (Figure 18.33). Three of these subclasses—the inadunates, flexibles, and camerates—are mostly Paleozoic, appearing in the Early or Middle Ordovician. Two of these groups became extinct during the Permo-Triassic extinction; the third (inadunates) dwindled on into the Middle Triassic. A fourth subclass, the articulates, arose from inadunates in the Triassic and survives today.

Inadunates The inadunates are the largest and most successful group of Paleozoic crinoids (about 480 genera), and appear to have been the ancestral group for both flexibles, which appeared just after the earliest inadunates in the Ordovician, and articulates, which appeared much later in the Triassic. Inadunate crinoids (Figures 18.27; 18.28; and 18.33, *A* and *D*) are characterized by a cylindrical to bowl-shaped monocyclic or dicyclic cup with the relatively few plates firmly joined; a simple lightly plated tegmen with orals and anal plates sometimes extended into a large anal sac (Figure 18.33, *D*); free arms (no fixed brachials) that are primitively uniserial and later become biserial and pinnulate; and a stem usually composed of one-piece columnals often bearing cirri. Inadunates range from the Early Ordovician to the Middle Triassic, although only a single genus survived the Permo-Triassic boundary; they were common throughout the entire Paleozoic.

Camerates Camerates, a large subclass of distinctive Paleozoic crinoids (about 210 genera), are of unknown ancestry and apparently did not evolve into any other crinoid groups. Camerate crinoids (Figure 18.33, *B* and *C*) have a large conical or globular monocyclic or dicyclic cup with the plates firmly joined; fixed brachials and interbrachials commonly incorporated into the top of the cup, thus increasing the number of cup plates in most forms; a heavily-plated tegmen with many small plates usually covering the mouth and food grooves and often developed posteriorly into a high anal tube; uniserial or biserial arms almost always bearing pinnules; and a stem with one-piece columnals often bearing cirri. Camerates range from the Early Ordovician to the Late Permian, were most common in the Silurian and Mississippian, and are rare after that time.

Flexibles Flexibles are a fairly small subclass of Paleozoic crinoids (about 60 genera) that appear to have evolved from inadunate ancestors in the Middle Ordovician and survived to the Late Permian. Flexible crinoids (Figure 18.33, *E*) are characterized by a conical to bowl-

Table 18.2. Crinoid subclasses

Feature	Inadunates	Camerates	Flexibles	Articulates
Range	Ordovician–Triassic	Ordovician–Permian	Ordovician–Permian	Triassic–Holocene
Diversity	About 480 genera	About 210 genera	About 60 genera	About 255 genera (about 165 living)
Cup plating	Monocyclic or dicyclic; cup plates firmly joined	Monocyclic or dicyclic; cup plates firmly joined	Dicyclic (one or three infrabasals); cup plates loosely joined	Plating often reduced (lower cup plates and top of stem may be fused)
Arms	Uniserial or biserial (Miss. on); pinnules rare until Devonian; lower arms free	Uniserial or biserial; pinnules; lower arms in cup	Uniserial; no pinnules; lower arms in cup; petalloid process on many brachials	Uniserial; pinnules; lower arms free
Tegmen	Usually simple; orals plus other plates; some have anal sac	Large; many tightly, sutured plates; anal tube common	Flexible; many plates	Usually unplated
Stem	Many shapes; cirri usually present	Usually circular to elliptical; cirri usually present	Circular stem; no cirri present	Circular to pentagonal stem; cirri usually present; many forms stemless
Examples	Figure 18.33, *A* and *D*	Figure 18.33, *B* and *C*	Figure 18.33, *E*	Figure 18.33, *F* and *G*

shaped dicyclic cup made up of loosely sutured plates; lower arms and interbrachials incorporated into the top of the cup; a flexible tegmen with many small plates; uniserial arms lacking pinnules but having a patelloid process on the lower brachials; and a stem made up of circular one-piece columnals lacking cirri. Flexibles reached their maximum diversity in the Silurian but remained a fairly minor group compared to the more successful inadunates and camerates.

Articulates A major crinoid subclass, the articulates (about 255 genera, 165 living) first appeared in the Early Triassic apparently derived from advanced inadunate ancestors. Articulate crinoids (Figure 18.33, *F* and *G*) usually have a reduced dicyclic cup with the infrabasals lost or fused during growth and an unplated or lightly plated tegmen. Arms are usually long and uniserial, almost always branch on the second brachial, and bear many pinnules. Stems either are circular to pentagonal with cirri or absent with only cirri attached to a fused circular plate, called the **centrodorsal**, at the base of the cup. Articulates range from the Triassic to the Holocene, and nearly two-thirds of the known genera are still living.

Evolution The earliest known crinoid is a single Middle Cambrian genus from western Canada (Figure 18.32). Morphologically it is primitive, with an irregularly plated cup, short uniserial arms, and a multiplated holdfast instead of a stem. It cannot be assigned with certainty to any of the Paleozoic subclasses. More advanced crinoids with well-organized cups and columnal-bearing stems

(Figure 18.33, *A* and *B*) appeared in the Ordovician. Inadunates and camerates occur in the Early Ordovician, and flexibles occur slightly later in the Middle Ordovician, when crinoids became the dominant echinoderm class (Figure 18.25).

Many early inadunates are characterized by small conical cups, often with divided radials and several anal plates on the posterior side, long but rather simple branched uniserial arms usually without pinnules, and a stem made up of columnals divided into four or five cryptic segments (Figure 18.33, *A*). Some other early inadunates have large cup-shaped to globular cups (Figure 18.28) and fairly short arms. Early camerates have large conical cups with numerous fixed brachials and interbrachials, long uniserial or biserial pinnulate arms, and long stems with round to pentagonal columnals (Figure 18.33, *B*). Early flexibles are much like early dicyclic inadunates except that the cup plates are loosely joined, and the lower arms are incorporated into the top of the cup separated by one to a few large interbrachials.

All these subclasses, except the Articulata, were fairly common in the early and middle Paleozoic and continued to diversify and occupy new ways of life. For example, some Silurian camerates developed specialized slotlike structures on the tegmen, into which the shortened arms could fold for protection (Figure 18.33, *C*), probably for living in high-energy environments. Other early and middle Paleozoic inadunates developed bilaterally symmetrical cups and arms and lived recumbent on the sea floor.

Crinoids continued to diversify in the middle Paleozoic, reaching a peak in diversity and abundance in the Mississippian (Early Carboniferous) when favorable shallow marine environments were widespread on several continents. Dicyclic inadunates with a cup- or bowl-shaped cup and simplified plating plus long multibranched arms (some with pinnules) (Figure 18.27) became common along with long-armed flexibles with reduced cup (Figure 18.33, *E*) and monocyclic camerates with a simplified cup (fewer interbrachials and fixed brachials) and long pinnulate arms. Crinoids remained fairly common during the Pennsylvanian (Late Carboniferous) and Permian, but camerates became much less common than they had been earlier. Dicyclic inadunates with a bowl-shaped cup, greatly reduced or hidden infrabasals, reduced number of anal plates, a large anal sac, and long pinnulate arms (Figure 18.33, *D*) dominated these late Paleozoic faunas.

Crinoids almost became extinct at the Permo-Triassic boundary (Figure 18.25), but a few dicyclic inadunates survived (one genus has been reported from the Middle Triassic) and gave rise to the articulates that first appeared in the Early Triassic and became more abundant in the Jurassic and Cretaceous. Stemmed articulates with a small reduced cup and long branched arms (Figure 18.33, *F*) were most common during the Mesozoic, but some unstemmed floating forms with a large bulbous cup and no stem (Figure 18.33, *G*) were also present during

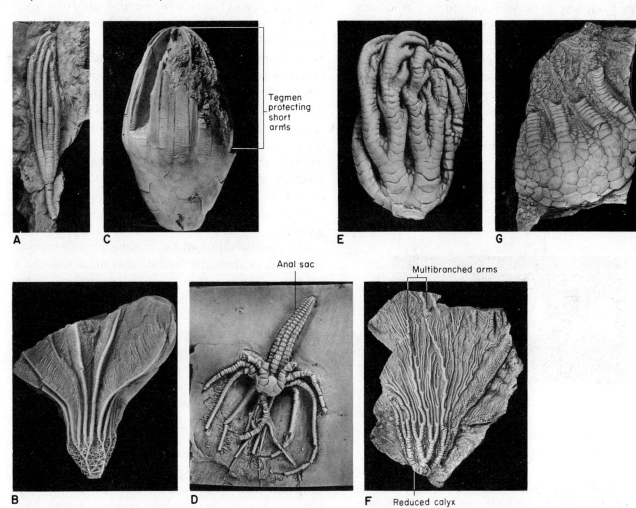

Figure 18.33. Examples of fossil crinoids (see also Figures 18.27, 18.28, 18.32, and 18.34). **A**, *Ectenocrinus*, a monocyclic inadunate crinoid with divided radials and long uniserial arms. (Upper Ordovician, southwestern Ohio, ×2.) **B**, *Glyptocrinus*, a dicyclic camerate crinoid with long pinnulate arms. (Upper Ordovician, southwestern Ohio, ×0.8.) **C**, *Eucalyptocrinites*, a specialized dicyclic camerate crinoid having a vaultlike tegmen with slots into which the short arms fold for protection. (Middle Silurian, southern Indiana, ×0.7.) **D**, *Aesiocrinus*, a dicyclic inadunate crinoid with a large anal sac. (Middle Pennsylvanian [Upper Carboniferous], western Missouri, ×1.) **E**, *Onychocrinus*, a dicyclic flexible crinoid with hidden infrabasals and five large arms with patelloid processes. (Middle Mississippian [Lower Carboniferous], western Indiana, ×0.7.) **F**, *Pentacrinus*, an articulate crinoid with a small calyx (mostly missing) and repeatedly branching arms. (Lower Jurassic, England, ×0.9.) **G**, *Uintacrinus*, an unusual free-floating articulate crinoid with a large bulbous calyx. (Upper Cretaceous, western Kansas, ×0.8.)

the Cretaceous. Fossilized remains of articulate crinoids became less common during the Cenozoic, but nearly 650 species are living today. Most shallow-water living articulates are stemless swimmers or crawlers (see Figure 18.1) that use their cirri to cling to other objects while feeding. These crinoids are most common on present-day reefs. The abundance of predators today in most shallow-water environments has probably favored the stemless mobile crinoids that can hide when conditions are unfavorable or predators are active. Living articulates with stems (Figure 18.13) are more common in deep-water environments but are less diverse and widespread than stemless forms.

One of the more unusual types of crinoids represented by a few inadunates in the Paleozoic and a few articulates in the late Mesozoic are tiny forms called **microcrinoids** (Figure 18.34), which range from 0.2 to 2 mm in size. These crinoids have very simple plating and may be paedomorphic forms, or crinoids that have become sexually mature in a size range and morphology resembling the early juvenile stages of their ancestors. Microcrinoids are abundant in some middle and late Paleozoic rock units along with more normal-sized crinoids.

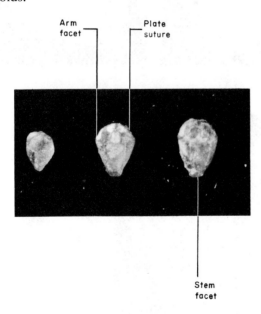

Figure 18.34. Unidentified microcrinoids from the Lower Mississippian (Lower Carboniferous) of northwestern Montana showing ovoid shape, some plate sutures, and facets where tiny arms and a stem were attached. (×15.)

CLASS PARACRINOIDEA

Crinozoa with lens-shaped to globular theca with numerous irregularly arranged plates. Food-gathering system developed either as asymmetrical recumbent ambulacra with erect uniserial pinnules or asymmetrical erect, uniserial arms with pinnules; two to five ambulacral branches or arms present. Mouth offset on summit, anus nearby usually opposite stem facet, hydropore and gonopore just below mouth. In many, slitlike pores extend part way into or between thecal plates; others with very thin plates without pores. Stem usually curved near theca, small to fairly large, attached to three basals. (Early Ordovician to Early Silurian; 13 to 15 genera.)

Paracrinoids are a relatively small class of ambulacrum or arm-bearing echinoderms. They have a lenticular to globular theca composed of numerous irregularly arranged plates. The two to five food-gathering structures are asymmetrical and developed as either recumbent ambulacra (Figure 18.35) or erect arms, each type bearing smaller, uniserial erect appendages that are probably pinnules. All have a curved stem of unknown length. Most genera have either slitlike pores extending part way through the plates or thin thecal plates to aid in respiration. The anus (and not the mouth) is usually opposite the stem facet and is covered by a domed anal pyramid; the mouth is nearby on the thecal margin at the junction of the ambulacra or arms. A small slitlike hydropore is also present near the mouth, but no evidence of tube feet has been found in the ambulacral grooves.

The offset position of the mouth, the asymmetrical ambulacra or arms, and the distinctive thecal shape have led different authors to postulate that paracrinoids either were "droopers" that had an erect stem (Figure 18.35, *B*) or rested directly on the seafloor with a buried trailing stem (Figure 18.35, *A*) probably in a bidirectional current field. The asymmetrical ambulacra or arms (Figure 18.35) are a key identifying feature. In nearly all paracrinoids, the food groove is located on the right side of the ambulacrum or arm (looking away from the mouth), while pinnules branch off the left side, one per ambulacral or arm plate.

The morphology of paracrinoids does not compare well with that of other crinozoans. It has recently been proposed that paracrinoids be put in a separate subphylum, but this has not been widely accepted. Most genera are known only from North America, and they are very abundant in some Middle Ordovician deposits.

Subphylum Blastozoa

This extinct subphylum contains five classes of brachiole-bearing echinoderms, most of which were stemmed, high-level suspension feeders. They are usually charac-

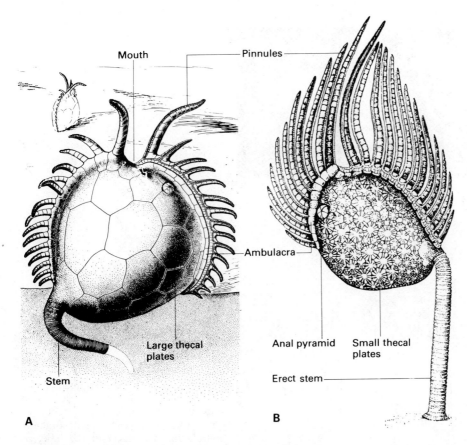

Figure 18.35. Different life positions proposed for two paracrinoid genera. **A,** *Platycystites* from the Middle Ordovician of southern Oklahoma, reconstructed as sitting on the sea floor with short trailing stem. Note two asymmetric ambulacra with erect pinnules and large, thin, thecal plates lacking pores. (×1.6.) **B,** *Amygdalocystites* from the Middle Ordovician of southeastern Ontario, Canada, reconstructed as erect on a short attached stem. Note longer pinnules on recumbent ambulacra and numerous thecal plates with pores. (×1.6.) (**A** drawn by M. Blos for J. W. Durham. **B** from Parsley, R. L.; Mintz, L. W. *Bulletins of American Paleontology* 68 [288]: 17; 1975.)

terized by a compact theca with well-developed pentameral symmetry, respiratory structures through the thecal plates, and a reduced water vascular system perhaps lacking ambulacral tube feet. Blastozoans were fairly common throughout most of the Paleozoic and worldwide in their distribution.

CLASS BLASTOIDEA

Blastozoa with conical, bud-shaped, or globular theca with four circlets of plates in well-developed pentameral symmetry. Five short to long ambulacra, developed in radial sinuses, underlain by lancet plates that support two types of side plates in sets, each bearing long erect brachiole. Mouth central on summit; anus lateral on summit with hidden gonopore. Foldlike hydrospires suspended in coelomic cavity alongside ambulacra; hydrospires cross suture between radials and deltoids; eight to ten groups of hydrospires (some reduced or absent on anal side); hydrospires either open along length as slits or closed off, opening as series of pores between side plates plus a larger porelike spiracle on summit. Stem usually long but small in diameter. (Middle Ordovician?, Middle Silurian to Late Permian; about 95 genera.)

Blastoids are the largest and most advanced class of blastozoan echinoderms. They have a conical to globular theca with several circlets of well-organized plates in well-marked pentameral symmetry (Figure 18.36). The lowest plates around the stem facet are **basals**; three basals (two larger, one smaller) are usually present. Above the basals are five medium to large **radials**, each of which supports an ambulacrum in a large central sinus. Above the radials and between the ambulacra are five or more **deltoids** surrounding the central mouth on the summit (Figure 18.37). Five well-organized ambulacra extend from the mouth into the radial sinuses. Each ambulacrum has a central **lancet plate** supporting two types of **side plates**; each set of side plates bears a single brachiole (Figures 18.11; 18.36; and 18.37). Most blastoids have a long columnal-bearing stem used for attachment, but a few forms were apparently stemless and lay recumbent on the seafloor.

Blastoids have well-developed respiratory structures called **hydrospires** (Figure 18.37), which are thin pleat-like folds of calcite stereom and stromal tissue extending from the overlying plates (radials and deltoids) into the coelomic cavity alongside and underneath the ambulacra. Open hydrospires are exposed along their entire length as one or more vertical slits on the plate surface; closed hydrospires are more hidden with a series of pores

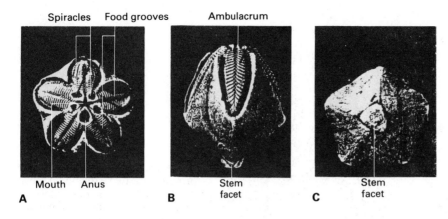

A — Spiracles, Food grooves, Mouth, Anus
B — Ambulacrum, Stem facet
C — Stem facet

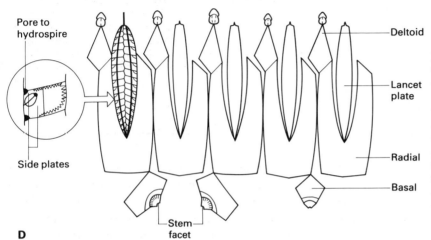

Pore to hydrospire / Side plates / Deltoid / Lancet plate / Radial / Basal / Stem facet

D

Figure 18.36. Morphology of blastoids. **A** to **C**, Top, side, and bottom views of a well-preserved specimen of *Pentremites* from the Upper Mississippian (Lower Carboniferous) of southern Illinois showing well-developed pentameral symmetry, wide and highly organized ambulacra, stem facet at the base, and summit openings (leading to the hydrospires) around the central mouth. **D**, Side-layout plating diagram for *Pentremites* showing the four main circlets of plates and the side plates restored for one ambulacrum.

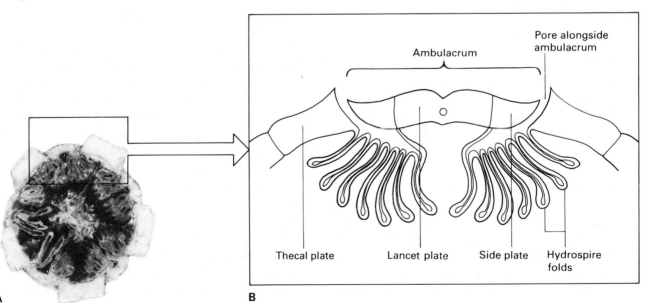

A

B — Ambulacrum, Pore alongside ambulacrum, Thecal plate, Lancet plate, Side plate, Hydrospire folds

Figure 18.37. Hydrospires in a blastoid. **A,** Sectioned specimen of *Pentremites* from the Lower Pennsylvanian (Upper Carboniferous) of northeastern Oklahoma showing hydrospire folds extending into coelemic interior; note that two groups of folds at lower left (anal side) are differently developed, an unusual feature. (×2.8.) **B,** Enlargement of one hydrospire group showing them centered below edge of ambulacrum.

along the margins of the ambulacra (apparently inlet areas) leading through the underlying folds to a large **spiracle** on the summit (an outlet) (Figure 18.36). Spiracles from adjacent ambulacra are sometimes combined and may also include the anus, which is on one side of the summit (Figure 18.36). The contrast between open and closed hydrospires is used at present to classify blastoids into two major groups.

A few poorly preserved specimens and plates that are referred to the blastoids occur in the Middle Ordovician. Otherwise, blastoids range from the Middle Silurian to the Late Permian. Their distribution was nearly worldwide during most of this time. Blastoids apparently

evolved in the Ordovician from a brachiole-bearing ancestor with reduced numbers of plates. They first appeared with all the characteristics normally considered diagnostic of this class. Early blastoids apparently competed with ecologically similar rhombiferans and replaced this group during the Silurian and Devonian (Figure 18.25). They had fewer thecal plates, more highly developed pentameral symmetry, more advanced respiratory folds, and better developed ambulacra than rhombiferans.

During their history, blastoids evolved a wide range of thecal shapes with either short or long ambulacra (Figures 18.38). All Silurian blastoids and many Devonian blastoids had a conical theca with short ambulacra (Figure 18.38, *A*), but bud-shaped to globular forms with long ambulacra (Figure 18.38, *D*) had evolved by the Middle Devonian. Globular blastoids underwent a major expansion in the Mississippian (Figure 18.38, *C, G,* and *H*) when blastoids were most abundant, but forms with both long and short ambulacra continued on until the Late Permian (Figure 18.38, *B* and

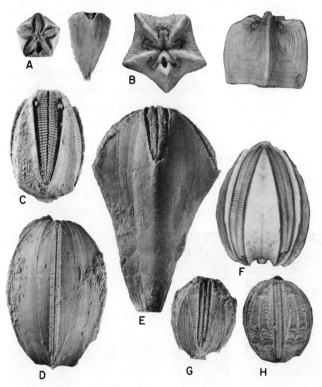

Figure 18.38. Blastoids. **A,** Top and side views of *Heteroschisma.* (Middle Devonian, southern Ontario, Canada, ×1.8.) Note the short ambulacra and open hydrospire slits. **B,** Top and side views of *Timoroblastus* showing unusual thecal shape and very short ambulacra. (Upper Permian, Timor, Indonesia, ×1.2.) **C,** *Pentremites,* side view. (Middle Mississippian [Lower Carboniferous], southeastern Iowa, ×1.8.) **D,** *Placoblastus,* side view. (Middle Devonian, northeastern Michigan, ×1.2.) **E,** *Drophocrinus,* side view. (Lower Carboniferous [Mississippian], Ireland, ×1.2.) **F,** *Deltablastus,* side view. (Upper Permian, Timor, Indonesia, ×1.2.) Note strong growth lines. **G,** *Lophoblastus,* side view. (Middle Mississippian [Lower Carboniferous], Missouri, ×1.8.) **H,** *Schizoblastus,* side view. (Middle Mississippian [Lower Carboniferous], southeastern Iowa, ×1.8.) Note long deltoids. (From Macurda, D. B., Jr. In Moore, R. C., editor. *Treatise on invertebrate paleontology, Part S.* New York and Lawrence, KS: Geological Society of America and University of Kansas Press; 1967.)

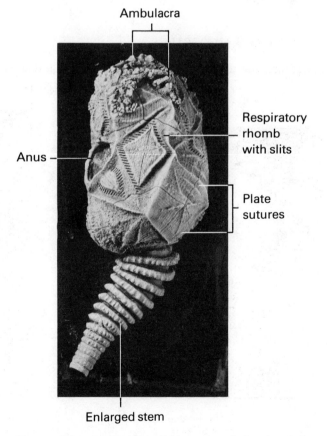

Figure 18.39. Well-preserved specimen of the rhombiferan genus *Glyptocystella* from the Middle Ordovician of southern Oklahoma showing the thecal plates, enlarged stem attached to base of theca, anal opening at left, short ambulacra with trace of brachioles on summit, and respiratory rhombs through thecal plates. (×2.2.)

Figure 18.40. Plating diagram for the typical rhombiferan *Glyptocystites* from the Middle Ordovican of northern Michigan. Side layout of thecal plates showing ridges, long ambulacra nearly reaching base of theca, and rhombs with slits. *M*, mouth; *A*, location of anus; *h*, hydropore slit; *g*, gonopore. (From Kesling, R. V. *University of Michigan Museum of Paleontology* 17; 1961.)

F). Blastoids and crinoids competed for dominance as high-level suspension feeders throughout the middle and late Paleozoic until the end of the Permian when blastoids became extinct.

CLASS RHOMBIFERA

Blastozoa with globular to flattened theca, four to five circlets of plates usually with moderately developed pentameral symmetry. Two to five ambulacra usually extending to edge of summit or down over thecal plates. Ambulacra of two types of flooring plates in sets, each set supporting an erect brachiole. Ambulacral system sometimes reduced to one or two large brachioles near mouth or developed as erect armlike appendages. Mouth central on sum-

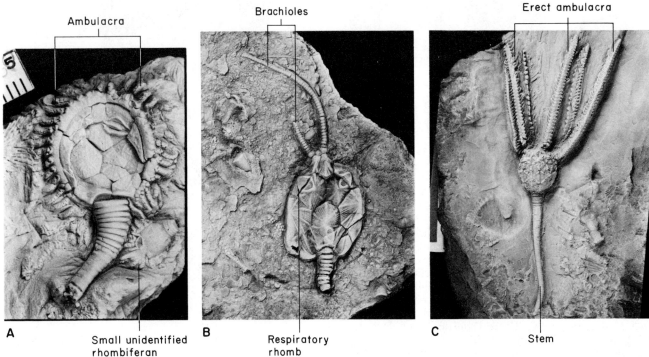

Figure 18.41. Rhombiferans (also see Figures 18.14, *A* and 18.39). **A**, *Pseudocrinites*, a lens-shaped rhombiferan with two long ambulacra bearing numerous short brachioles (note tiny unidentified rhombiferan at lower right). (Middle Silurian, England, ×2.) **B**, *Pleurocystites*, a flattened rhombiferan with only two long brachioles and three small rhombs. (Middle Ordovician, southeastern Ontario, Canada, ×1.3.) **C**, *Caryocrinites*, a nearly complete specimen with a long stem, compact theca, and erect ambulacra with short brachioles developed as pinnulate arms. (Middle Silurian, western New York, ×0.9.)

mit; anus on side of theca, sometimes surrounded by numerous small plates; hydropore and gonopore just below mouth. Foldlike respiratory rhombs usually suspended in coelomic cavity between adjacent thecal plates; surface openings at rhombs long slits, short slits, or pores in rhombic arrangement; usually three or more rhombs present. Medium to long stem attached to base of theca; stem often enlarged near theca with large lumen, other end often unattached. (Early Ordovician to Late Devonian; about 60 genera.)

Rhombiferans (Figure 18.39) are a medium-sized group of early-to-middle Paleozoic brachiole-bearing echinoderms. Their most distinctive feature is the respiratory **rhombs** developed in adjoining plates. They are rhombic in shape (Figure 18.39) and usually consist of rows of slitlike or porelike openings, half on each plate and leading into internally projecting pleated folds of thin calcite stereom that connects slits on adjacent plates. A few rhombiferans have different respiratory structures in which canals from the coelomic interior extend across sutures through horizontal ducts or slits within the overlying plates.

Rhombiferans have a globular, elongate, lens-shaped, or flattened theca with numerous plates usually arranged into four to five well-defined circlets (Figure 18.40). Food-gathering brachioles arise from the summit or from

recumbent ambulacra extending to the edge of the summit (Figure 18.39) or even down the outside of the theca. The ambulacra are well-organized, usually with two types of **flooring plates** supporting each brachiole. A stem of columnals was used for attachment or possibly swimming. Many rhombiferans have the stem enlarged just below the theca and composed of alternating larger and smaller columnals, which are overlapped for flexibility and enclose a large lumen (Figure 18.39 and 18.41, *A* and *B*). These may have been free-living swimmers that were able to flex the distal part of the stem using muscles in the lumen.

Rhombiferans range from the Early Ordovician to the Late Devonian and apparently were distributed worldwide. They seem to have evolved from an earlier brachiole-bearing form that had standardized the thecal plating into four or five circlets. This transition probably took place in the latest Cambrian or earliest Ordovician. Rhombiferans apparently competed with other Ordovician suspension-feeding echinoderms and may have been partly responsible for their decline. Rhombiferans were themselves replaced in the Silurian and Devonian by more efficient blastoids (Figure 18.25). Most Early and Middle Ordovician rhombiferans were cyclindrical to globular forms with short ambulacra and many rhombs (Figure 18.39). Later forms developed other thecal shapes, elongated the ambulacra (Figure 18.41, *A*), and reduced the number of rhombs to three (figure 18.42, *A* and *B*) in standardized positions.

A few rhombiferans became morphologically similar to camerate crinoids by developing a globular theca bearing long erect ambulacra with biserial brachioles that look like pinnulate arms and a plated tegmen covering the food grooves leading to the mouth (Figure 18.41, *C*). However, these forms retained the typical rhombiferan thecal plate arrangement and rhombs. Some of these rhombiferans were fairly common in the Late Ordovician and Silurian and are an example of convergent evolution between members of distantly related classes in separate subphyla.

CLASS DIPLOPORITA

Blastozoa with globular to elongate theca, some with numerous irregularly shaped plates, others with regularly arranged plates showing pentameral symmetry. Small, erect, brachiole-like appendages attached around central mouth on summit or along three to five simple food grooves extending down theca. Anus at edge of summit or on side of theca; hydropore and gonopore below mouth. Some or

Elliptical diplopores

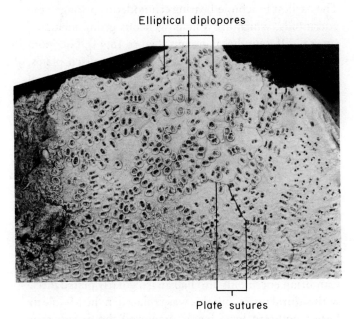

Plate sutures

Figure 18.42. Partial specimen of the diploporan *Mesocystis* from the Lower Ordovician of the western U.S.S.R. showing numerous well-preserved diplopores. Note elliptical rim (for papula) around each pair of pores and orientation of diplopores perpendicular to plate margins. (×2.6.)

most thecal plates have diplopores extending to interior. Some have medium to long stem attached to four basals; others stemless, either attached to substrate on lower surface of theca or recumbent on substrate. (Early Ordovician to Early Devonian; about 42 genera.)

Diploporans are a medium-sized group of apparent blastozoan echinoderms. Their most diagnostic feature is the paired pores, called **diplopores**, that penetrate some or most of the thecal plates, one or more pairs to a plate (Figure 18.42). Each diplopore had a raised rim that appears to have housed an external bulblike structure, or papula, used for respiration. This respiratory structure may have evolved from single pores on plate sutures that perhaps became trapped within a plate then separated into paired pores for better control of in-and-out circulation of coelomic fluid within each papula.

Diploporans also have a globular to elongate theca usually composed of numerous irregularly to regularly arranged plates (Figure 18.43). Some diploporans have a normal columnal-bearing stem (Figure 18.43, *A*), but others were stemless, either using the flattened lower surface to attach to the substrate or lying recumbent on the sea floor (Figure 18.43, *B*).

Diploporans range from the Early Ordovician to the Early Devonian and have a nearly worldwide distri-

bution. For many years, diploporans were classified together with rhombiferans in the class Cystoidea, but in the late 1960s this group was separated into two separate classes because the thecal respiratory structures are so different. The feeding appendages of diploporans are unfortunately poorly known and may not have been typical brachioles because they appear to be uniserial in several genera. However, these appendages are much smaller than crinozoan arms and appear to lack extensions of the coelomic systems. If these appendages were in fact derived from brachioles, diploporans probably evolved from a blastozoan ancestor in the latest Cambrian or earliest Ordovician.

CLASS EOCRINOIDEA

Blastozoa with globular to flattened theca, with numerous irregularly to regularly arranged plates; poor to well-developed pentameral or bilateral symmetry. Food-gathering system composed of long erect brachioles mounted on two to five simple ambulacra. Mouth usually central on summit, anus lateral, hydropore and gonopore usually just below mouth. Sutural pores present in early forms; later forms have very thin thecal plates. Attachment appendage short to long holdfast in early representatives; stem having columnals in later forms. (Early Cambrian to Late Silurian; 30 to 32 genera.)

The earliest brachiole-bearing echinoderms belong to the eocrinoids, which are a heterogeneous group including several different thecal designs (Figure 18.44). Eocrinoid thecae have irregularly to fairly well-organized plates, two to five simple ambulacral areas bearing long biserial brachioles, a primitive holdfast of many plates or a columnal-bearing stem, and in many early forms **sutural pores** for respiration between the thecal plates.

Eocrinoids are among the earliest echinoderms to appear in the fossil record and are the most diverse class of echinoderms at generic and specific levels in the Cambrian. They were apparently the ancestral group for all other blastozoan classes. Eocrinoids were the first group to develop a true columnal-bearing stem late in the Middle Cambrian, about 30 million years before crinoids made the same transition. All Early and most Middle Cambrian eocrinoids also had an irregularly plated theca with sutural pores, but this was replaced in the Middle to Late Cambrian by a better organized theca with thin plates lacking pores. Most eocrinoids were attached high-level suspension feeders, but a few genera were apparently unattached or recumbent on the sea floor.

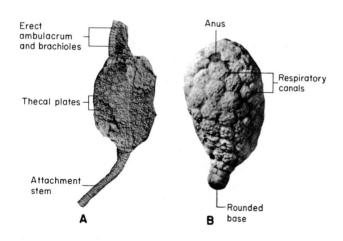

Figure 18.43. Diploporans. **A,** *Eumorphocystis* from the Middle Ordovician of southern Oklahoma showing the numerous small thecal plates, columnal-bearing stem, and brachiole-like appendages attached to erect ambulacra developed as arms. (×0.6.) **B,** *Holocystites* from the Middle Silurian of southeastern Indiana. Note the thecal shape and rounded base (no stem present) and trace of respiratory canals on many of the plates. (×0.8.)

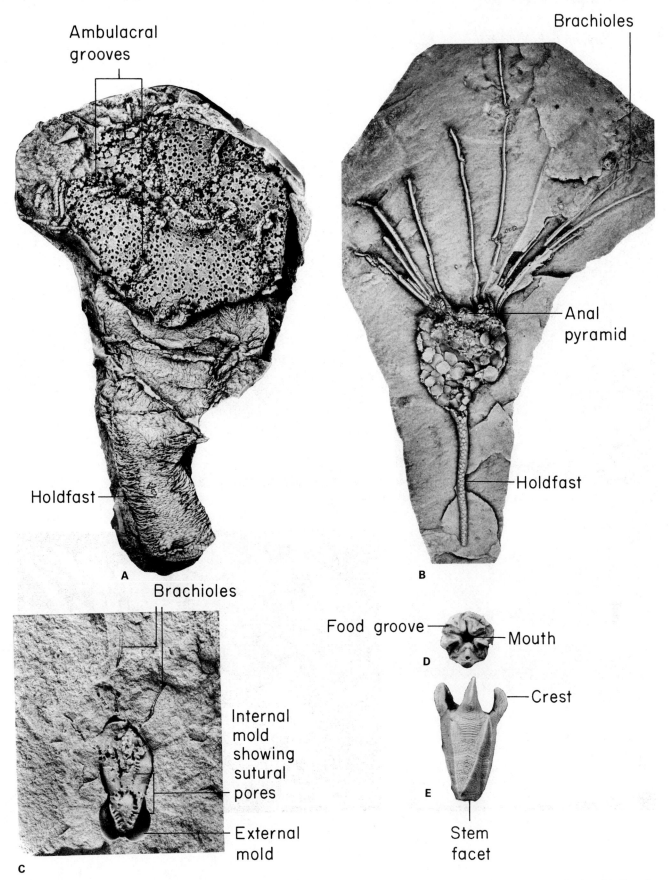

Figure 18.44. Eocrinoids. **A**, *Kinzercystis* (latex cast) from the Lower Cambrian of southeastern Pennsylvania showing numerous imbricate plates on conical theca and a cylindrical holdfast and sutural pores between plates on the summit, which also shows five ambulacral grooves and scattered brachioles. (×3.2.) **B**, *Gogia* (latex cast) from the Middle Cambrian of western Alberta, Canada. Note the long, thin, multiplated holdfast, numerous irregular thecal plates having a few sutural pores, anal pyramid at the edge of the summit, and long brachioles. (×1.7.) **C**, *Lichenoides* (internal and external molds) from the Middle Cambrian of Czechoslovakia showing the bulbous lower plates, traces of sutural pores, and a few brachioles. (×2.4.) **D** and **E**, *Stephanocrinus* (top and side views) from the Middle Silurian, western New York. Note the few thecal plates with growth lines, "crests" extending above the summit, and the central mouth, lateral anus and five food grooves on the summit. (×1.8). This problematic genus (and its relatives) have been assigned to the blastoids and most recently the crinoids, but probably represent either the youngest eocrinoids or a separate class. (**A** and **B** from Sprinkle, J. T. *Museum of Comparative Zoology Special Publication*; 1973.)

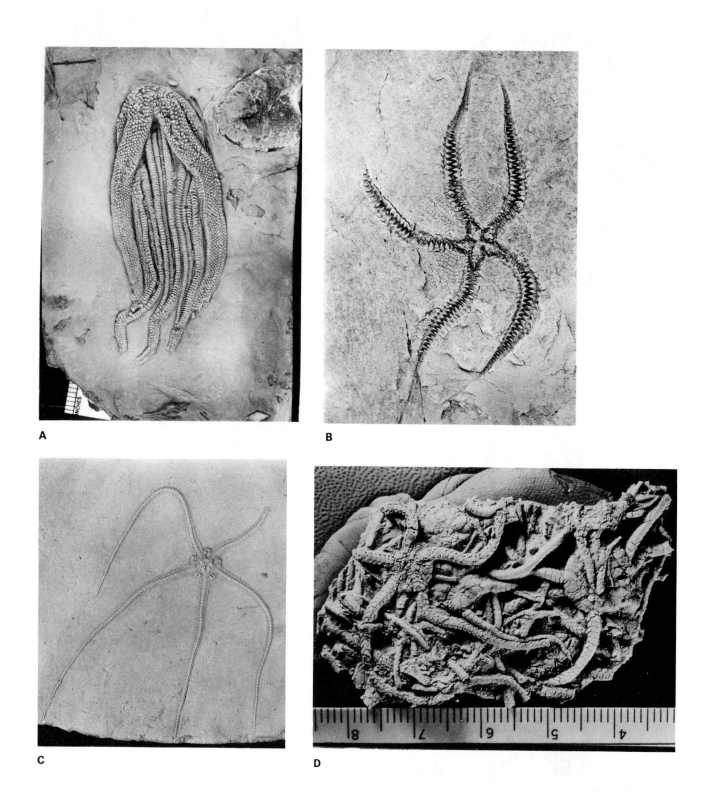

Figure 18.45. Asteroid and ophiuroids (see also Figure 18.15). **A**, *Salteraster* (asteroid), with arms folded together. (Upper Ordovician, southwestern Ohio, ×2.) **B**, *Taeniaster* (ophiuroid). (Middle or Upper Ordovician, southern Pennsylvania, ×1.1.) Note barely visible disk. **C**, *Geocome* (ophiuroid). (Lower Jurassic, Germany, ×1.6.) Note long delicate arms. **D**, *Ophiura* (ophiuroid); a mass of specimens with intertwined arms. (Lower Cretaceous, central Texas, ×1.8.) (**A** to **C**, courtesy of Porter M. Kier.)

Subphylum Asterozoa

The large and successful subphylum Asterozoa contains two classes of star-shaped echinoderms, the asteroids (or starfish) and the ophiuroids (or brittle stars). Both classes are characterized by the distinctive star-shaped body, an open water vascular system, and usually a mobile benthic mode of life.

CLASS ASTEROIDEA

Asterozoa with star-shaped to pentagonal body with five to 25 large arms containing extensions of coelomic cavity; tests with well-developed pentameral symmetry and numerous small plates. Mouth central on lower surface, gut short and straight, anus (if present) usually on upper surface with madreporite adjacent. Lower side of each arm contains ambulacral groove with open radial water vessel and numerous, large tube feet. (Early Ordovician to Holocene; about 430 genera, 310 living.)

Asteroids are a large class of living echinoderms that are not well represented in the fossil record. They range from the Early Ordovician to the Holocene and are distributed worldwide. They are currently most abundant in shallow-water environments. Asteroids evolved from an unknown ancestor in the Late Cambrian or Early Ordovician, and the class includes early Paleozoic forms known as somasteroids, which are somewhat intermediate between true asteroids and ophiuroids. Asteroids were relatively uncommon during the Paleozoic (see Figures 18.15 and 18.25) but survived the Permo-Triassic extinction without much change. They were more common in the late Mesozoic and Cenozoic and are one of the major living echinoderm classes. The plates of most asteroid skeletons are held together by connective tissue, allowing considerably flexibility in life. As a result, the skeletons disarticulate readily and complete asteroids are uncommon in the fossil record (Figure 18.45, *A*). A few authors have tried to use disassociated ossicles to identify fossil asteroids with partial success.

Some asteroids are detritus or suspension feeders, but many forms are carnivores or scavengers on prey that includes bivalves, gastropods, other echinoderms (Figure 18.24), barnacles, and corals. Asteroids may have been the first echinoderm carnivores, and many asteroid groups have probably used this method of feeding throughout their entire history.

CLASS OPHIUROIDEA

Asterozoa with body consisting of large central disk with five long, thin, flexible, and sometimes branched arms. Disk of numerous small plates with

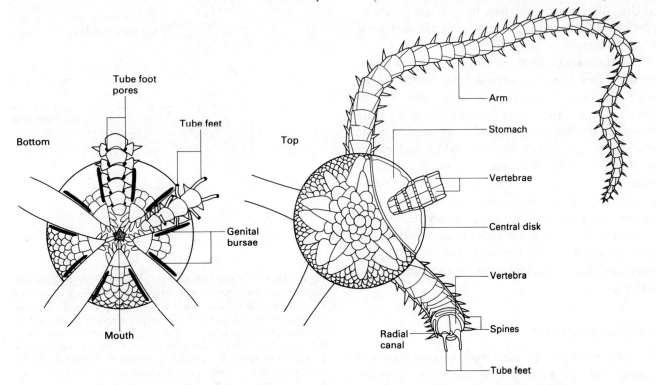

Figure 18.46. Morphology of a living ophiuroid showing the central disk and long thin arms with vertebrae and enclosing plates, as well as tube feet in ambulacral groove.

pentameral symmetry well-developed; mouth central on lower side, leading into large blind gut with no anus. Madreporite variously developed on oral side or absent. Ambulacral groove on lower surface on each arm contains open radial water vessel and numerous small tube feet; arms articulated by axial row of large uniserial plates called vertebrae surrounded by other small plates and spines. (Early Ordovician to Holocene; 325 genera, 280 living.)

Ophiuroids are a large class of star-shaped echinoderms. The central mouth leads into a large stomach filling much of the disk. There is no anus, and waste material is regurgitated through the mouth. Unlike asteroids, the coelomic cavities and stomach are confined to the central disk. The arms are made of specialized ossicles called **vertebrae** running down the axis of each arm (Figure 18.46). Four longitudinal series of plates enclose the vertebrae. Unlike most other arm-bearing echinoderms, ophiuroids can move their arms rapidly with somewhat snakelike movements. Tube feet are used for feeding, slow movement, and respiration. Specialized, pouchlike structures called **genital bursae** are located alongside the bases of the arms on the lower side of the disk and appear to serve both respiratory and reproductive functions (Figure 18.46). Viviparity and brooding of young are common. In some species, dwarf males live permanently on the oral surface of the central disk of the females.

Ophiuroids range from the Early Ordovician to the Holocene and have a worldwide distribution. Many forms are especially common in deep-water environments (see Figure 18.3). They first appeared in the Early Ordovician descended from an unknown ancestor at the same time as the earliest asteroids, which had similar morphology. Ophiuroids were uncommon during the Paleozoic (Figure 18.25) but survived the Permo-Triassic extinction with little change. They are only slightly more common fossils in the Mesozoic (Figure 18.45) and Cenozoic but are a major echinoderm group today. Most forms are either mobile, benthic suspension or detritus feeders, although some ophiuroids are also carnivores on small prey.

Subphylum Homalozoa

This extinct subphylum contains four small classes of asymmetrical to bilaterally symmetrical, flattened, bottom-dwelling echinoderms that have several different types of swimming or feeding appendages. They have

scattered occurrences throughout the early and middle Paleozoic and were probably distributed worldwide, although most specimens are known from Europe, North America, and North Africa.

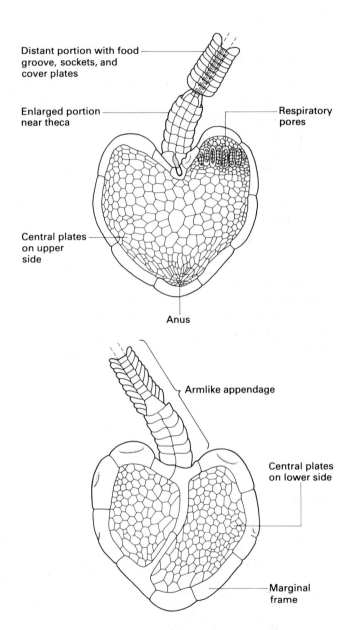

Figure 18.47. Thecal and arm morphology in the stylophoran *Phyllocystis* from the Lower Ordovician of France. Note the elongate marginal plates making up the thecal frame, top and bottom central areas plated with small polygonal plates, and an armlike appendage with an enlarged portion near the theca and the distant portion bearing a food groove and tube foot sockets protected by small overlapping cover plates. (From Ubaghs, G. in: Moore, R.C., editor. *Treatise on invertebrate paleontology, Part S, Echinodermata,* New York and Lawrence, KS: Geological Society of America and University of Kansas Press; 1967.)

CLASS STYLOPHORA

Homalozoa with flattened, boot-, heart-, or lozenge-shaped theca, grading from asymmetrical to nearly bilaterally symmetrical. Theca plated either with elongate thick plates around margin with small thin plates covering upper and lower sides or with only large thin plates; some forms having large spines, paddles, or knobs on margin. Large, armlike appendage attached to one end of theca; short proximal section greatly enlarged, hollow, built of rings composed of four plates; elongate uniserial section slowly tapering and bearing a food groove that is protected by two rows of overlapping cover plates. Mouth apparently at base of armlike appendage, anus near opposite margin of theca, one or more other small openings near mouth; early forms have rows of elongate pores or folds laterally on one side of theca. (Middle Cambrian to Middle Devonian; about 32 genera.)

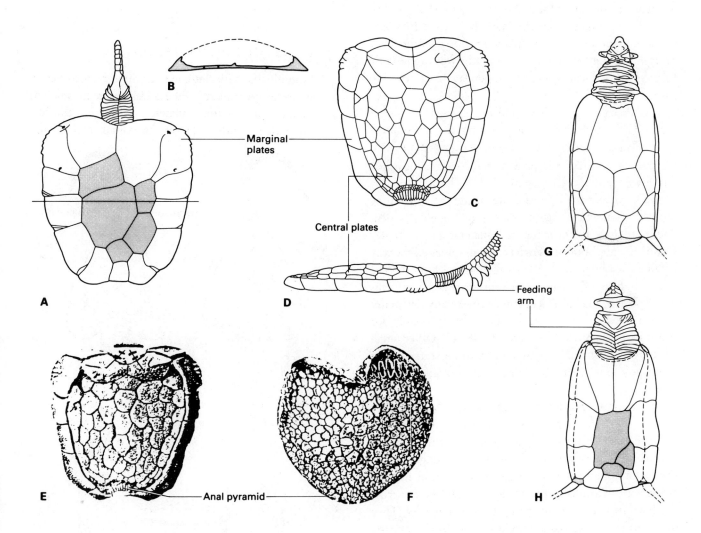

Figure 18.48. Stylophorans (see also Figure 18.16). **A** to **E**, *Mitrocystites.* Bottom, cross section, top, and side reconstructed views plus a latex cast of the lower surface of the actual specimen. (Middle Ordovician, Czechoslovakia). Note the large spine-bearing plates in the arm. (**A** to **D**, ×2; **E**, ×2.2.) **F**, *Phyllocystis.* Upper face showing elliptical respiratory structures (top) and the anal pyramid (bottom). (Lower Ordovician, France, ×2.2.) **G** and **H**, *Enoploura.* Top and bottom views showing large plates covering most of the theca. (Upper Ordovician, southwestern Ohio.) (From Ubaghs, G. in: Moore, R. C., editor. *Treatise on invertebrate paleontology, Part S. Echinodermata.* New York and Lawrence, KS: Geological Society of America and University of Kansas Press; 1967.)

Stylophorans are a medium-sized group of homalozoan echinoderms ranging from the Middle Cambrian to the Middle Devonian. They are distinguished by a single erect appendage attached to a flattened, irregularly shaped, almost bilaterally symmetrical theca (Figure 18.47). The appendage usually has an enlarged, multi-plated, cylindrical region near the theca with a large axial cavity, a conical plate (stylocone) in a medial position, and a long, slowly tapering terminal region that may bear spines or ridges (Figure 18.48, *D* and *H*). It carries an apparent central food groove with lateral sockets (Figure 18.47) protected by tiny cover plates on each side. This appendage appears to represent a feeding arm with the food groove bearing a radial water vessel that gave off branches to tube feet located in the lateral sockets. The armlike appendage may also have been used for locomotion, assuming that the enlarged region near the theca housed strong muscles. The mouth was apparently located at the facet where this appendage is attached to the theca.

The theca in most stylophorans consists of either medium- to large-sized plates, covering the entire upper and lower sides, or large elongate marginal plates that strengthen the theca and small plates, covering the upper and lower central areas (Figure 18.48). The anal opening is apparently at the opposite end of the theca from the mouth. Sometimes large paddlelike or spinelike appendages are attached near the anal opening (Figure 18.48, *G* and *H*), and some genera have small spines all the way around the margin. A few early forms have sutural pores between the plates in one of the central areas. Later forms have specialized elliptical openings covered by tiny plates in one of the central areas. They were probably for respiration (Figures 18.47 and 18.48, *F*). Other small openings, which may represent the hydropore and one or more gonopores, open on the thecal margin near the mouth.

Stylophorans are unusual echinoderms because they show no evidence of pentameral symmetry. They appear to have been either suspension-feeding benthic echinoderms, holding the elongate armlike appendage up into the water column (Figure 18.23, *B*), or detritus feeders, swinging the armlike appendage back and forth across the bottom sediments to stir up organic particles. They probably moved across the seafloor either by short bursts of wiggling motion, using the appendage for locomotion, or by jetting water from the anal pyramid. Stylophorans apparently were specialized echinoderms that were not very successful, becoming extinct by the middle Paleozoic.

Recently, however, it has been argued that stylophorans were an intermediate group between echino-

derms and chordates and should be classified as primitive vertebrates. The same argument has proposed that the armlike appendage was in fact a "tail" and that internal structures observed in stylophoran thecae resemble features found internally in primitive vertebrates. However, other investigators dispute these morphologic comparisons. Stylophorans clearly show the typical echinoderm plate structure with its single-crystal calcite and microporous nature, and some features of the armlike appendage are almost identical to the arms or erect ambulacra of other echinoderms known to have a water vascular system and tube feet. Most authors therefore have rejected the concept of stylophorans as early chordates and prefer to regard them as unusual but true echinoderms.

CLASS HOMOIOSTELEA

Homalozoa with flattened, asymmetrical to nearly bilaterally symmetrical theca built of many small to medium-sized thin plates. A large, tail-like appendage with enlarged proximal region and

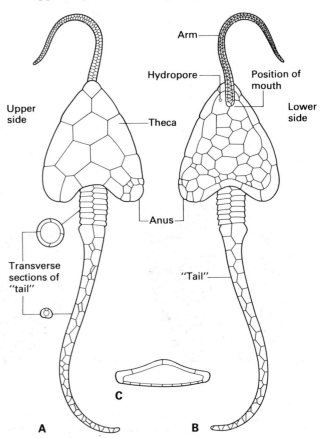

Figure 18.49. **A** to **C**, *Iowacystis*, a homoiostelean from the Upper Ordovician of northeastern Iowa showing the upper and lower sides of the theca with its arm and tail appendages and a cross section (**C**) through the theca, (From Kolata, D. R.; Strimple, H. L.; Levorson, C. O. *Palaeontology* 20; 1977.)

sometimes spines attached to one end of theca; smaller, biserial, armlike appendage with cover plates attached near other end. Mouth apparently at base of armlike appendage, anus lateral near other end of theca, small opening (hydropore?) on projecting ossicle near mouth. (Middle Cambrian to Early Devonian; 12 to 13 genera.)

Homoiostelea is a relatively small class of homalozoan echinoderms ranging from the Middle Cambrian to Early Devonian. They have a flattened asymmetrical to almost bilaterally symmetrical theca (Figure 18.49) bearing a large taillike appendage at one end and a smaller armlike appendage slightly to one side near the other end. The mouth is inferred to be at the base of the arm, and an anus is present laterally on the thecal margin near the tail.

Homoiosteleans are the only class of homalozoans that have both arm and taillike appendages. The tail has an enlarged proximal section near the theca that probably housed muscles, allowing the animal to wriggle or swim for a short distance (Figure 18.49). Homoiosteleans sometimes occur in large numbers and have an almost worldwide distribution.

Subphylum Echinozoa

This subphylum Echinozoa contains seven classes of echinoderms characterized by a "closed" water vascular system with numerous pores for tube feet passing through the skeleton. Echinoids are the dominant class in the fossil record and along with sclerite-bearing holothurians are the only echinozoans now living.

CLASS HOLOTHUROIDEA

Usually free-living Echinozoa with an elongate, flexible body. Mouth at one end surrounded by oral tube feet, anus at other end. Five rays, marked by large tube feet, usually extend along body, but ambulacral grooves lacking. Skeleton in most consists of microscopic sclerites in body wall; others have small imbricate plates covering body. (Middle Cambrian?, Middle Ordovician to Holocene; about 200 genera.)

Holothurians are a large class of living echinoderms that are present in both shallow- and deep-water environments; about 200 fossil and living genera are known. They are distinctive echinoderms, since most are elongate and flexible with a poorly developed skeleton that is usually not visible externally. Instead of plates, most holothurians have microscopic sclerites in the body wall. Some living holothurians may contain between 10 to 20 million sclerites in each individual. These sclerites can be wheel, table, hook, or anchor-shaped (Figure 18.6). Sclerites in holothurians may be immature plates that do not develop beyond the earliest juvenile stages, thus representing the reduced vestiges of a more elaborate ancestral skeleton. A few holothurians have the body covered by imbricate plates.

Holothurians usually have an elongate body with five or fewer rays defined by rows of tube feet extending from the oral region to the other end of the body (Figure 18.50). They have a few calcareous plates around the mouth to support the large specialized oral tube feet, which are used for feeding (Figure 18.51). Holothurians usually pick up detrital particles from the sea bottom using mucus and wipe the mucous strands into the mouth. The mouth opens into an elongate curved gut that loops once before reaching the anus, which is usually at the other end of the body. Several internal structures open into an enlargement called the **cloaca**, which is situated just inside the anal opening. Branched structures called **respiratory trees** open here (Figure 18.50). Most holothurians are anal pumpers, pumping water into and out of the respiratory trees through the cloaca. A small madreporite usually opens internally at the oral end of the body, and the single gonad has an external gonopore.

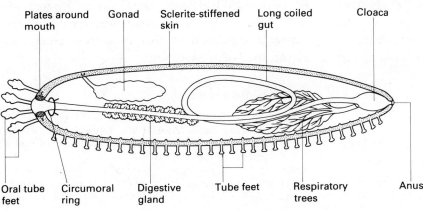

Plates around mouth / Gonad / Sclerite-stiffened skin / Long coiled gut / Cloaca

Oral tube feet / Circumoral ring / Digestive gland / Tube feet / Respiratory trees / Anus

Figure 18.50. Internal morphology of a generalized living holothurian. Note the external rows of tube feet plus modified oral tube feet, coiled gut, respiratory trees opening into cloaca, and other internal structures.

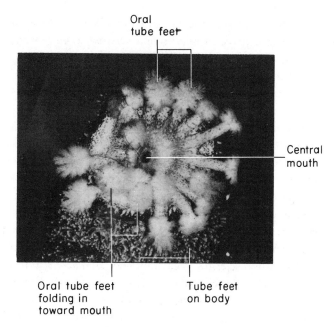

Oral
tube feet

Central
mouth

Oral tube feet
folding in
toward mouth

Tube feet
on body

Figure 18.51. Tufted oral tube feet located around the central mouth in the living holothurian *Holothuria*. (×1.2.) Note several tube feet (below and to the left of the mouth) bending in to carry food particles to mouth. (Courtesy David Pawson.)

The rows of tube feet on the body are used primarily for locomotion and perhaps respiration. Because the radial canals remain internal with the branches terminating in tube feet that individually extend to the exterior, holothurians are classified as members of the Echinozoa.

Holothurian-type sclerites are known as far back as the Middle Cambrian, although some investigators question whether the earliest of these might have belonged to other echinoderms. Sclerites became more abundant during the middle and late Paleozoic and extend through the rest of the Mesozoic and Cenozoic. A few complete holothurians have been found in the Devonian, Pennsylvanian, and Jurassic.

Holothurians today are a diverse group of echinoderms that have apparently developed several unusual designs not found in other echinoderms (Figures 18.1, *D*, and 18.17, *B*). For example, several pelagic holothurians exist that swim slowly, an unusual adaptation. A few holothurians with a plated test attach to hard substrates, using the unplated lower surface in a manner similar to Paleozoic edrioasteroids.

CLASS EDRIOASTEROIDEA

Echinozoa with discoidal, globular, or cylindrical theca. Rounded upper surface contains five straight to curved radiating ambulacra with small cover plates. Mouth central, with nearby hydropore in same interambulacrum as the lateral anus. Imbricate plates often present between ambulacra, usually with a peripheral rim of tiny plates around margin of upper surface. Concave lower surface often attached to objects on substrate and may not have been plated. (Early Cambrian to Middle Pennsylvanian; about 35 genera.)

Edrioasteroids are a medium-sized group of Paleozoic echinozoans that lived in shallow-water near-shore environments. They have a discoidal, club-shaped, globose, or cylindrical theca with a rounded upper surface and a flat to concave lower attachment surface (Figure 18.52). The mouth is central on the upper surface and protected by several oral plates. Five straight or curved ambulacra with numerous cover plates radiate out from the mouth. Between the ambulacra, the theca is built of **interambulacral plates**, with many discoidal forms having a **peripheral rim** of tiny plates (Figure 18.52). The anus is lateral in position on the upper surface and protected by a pyramid of small plates; a slitlike hydropore lies near the mouth (Figure 18.52, *A*) in the same interambulacrum. Edrioasteroids usually lived attached to some type of firm substrate such as a hardground or to a living or dead organism such as a brachiopod, bryozoan, or crinoid stem lying on the sea floor (Figure 18.52, *C*, *F*, and *G*).

Edrioasteroids range from the Early Cambrian to Middle Pennsylvanian and occur primarily in North America and Europe. The Australian late Proterozoic *Tribrachidium* may belong to this class (Figure 18.26). Edrioasteroids are not common in the fossil record, but sometimes hundreds or even thousands of specimens are found in a single favorable bedding plane. They apparently were attached, low-level suspension feeders using tube feet or open ambulacral grooves in feeding. Most forms seem to have had a water vascular system because there are pores for tube feet in the ambulacra, but some show no evidence of these structures. A few edrioasteroids had their lower surface covered with imbricating plates. The surface was capable of considerable expansion so that the edrioasteroid could elevate itself off the substrate.

Supplementary reading

Boolootian, R. A., editor. *Physiology of Echinodermata*. New York: Wiley Interscience; 1966. General and summary articles by many authors, dealing mostly with the ecology and physiology of living echinoderms.

Broadhead, T. W.; Waters, J. *Echinoderms, notes for a short course*. Knoxville, TN: University of Tennessee Studies in Geology 3; 1980. Summary articles on fossil and living

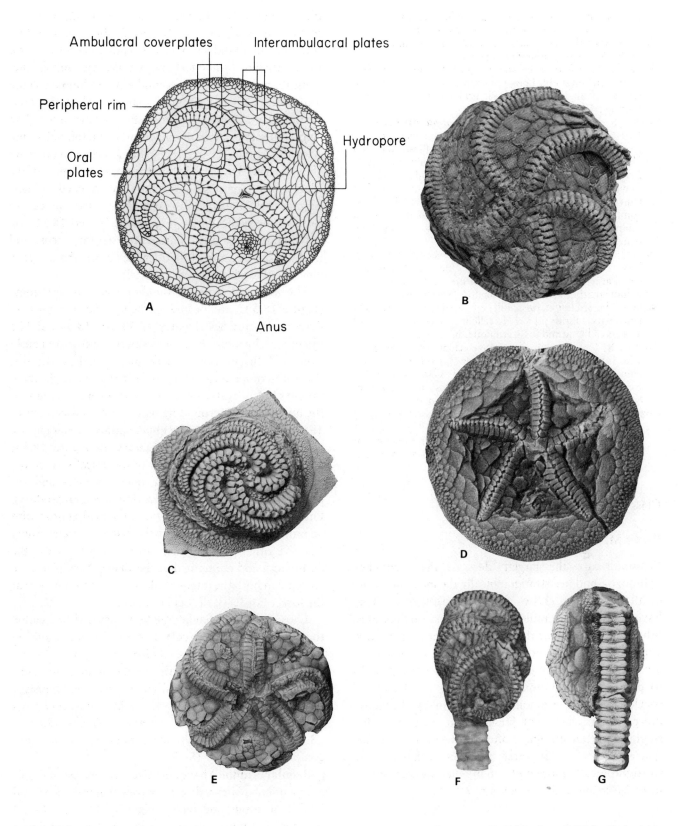

Figure 18.52. Edrioasteroid morphology and examples. **A,** Morphologic features of the Ordovician edrioasteroid *Carneyella.* (×3.8.) **B,** *Foerstediscus.* (Middle Ordovician, southern Minnesota, ×2.3.) **C,** *Streptaster* attached to a bryozoan fragment. (Upper Ordovician, southeastern Indiana, ×2.3.) **D,** *Cryptogoleus.* (Middle Ordovician, southeastern Ontario, Canada, ×3.8.) **E,** *Edriophus.* (Middle Ordovician, southeastern Ontario, Canada, ×1.1.) **F** and **G,** *Carneyella* attached to a crinoid stem. (Upper Ordovician southwestern Ohio, ×2.3.) (From Bell, B. M. *New York State Museum and Science Service, Memoir 21;* 1976.)

echinoderm groups and their ways of life, usefulness in the fossil record, classification, and geologic history.

Clark, A. *Starfishes and related echinoderms*, 3rd ed. London: British Museum (Natural History); 1977.

Hyman, L. H. *The invertebrates*, vol. 4, Echinodermata. New York: McGraw-Hill Book Company, Inc.; 1955. Detailed review of all living and fossil echinoderm groups by a leading zoologist.

Moore, R. C., editor. *Treatise on invertebrate paleontology, Part U, Echinodermata 3* (Asterozoa, Echinozoa); *Part S, Echinodermata 1* (General Sections, Blastozoa and Homalozoa); Moore, R. C.; Teichert, C., editors, *Part T, Echinodermata 2* (Crinozoa). New York, Boulder, CO, and Lawrence, KS: Geological Society of America and University of Kansas Press; 1966, 1967, 1978. The most complete work on the morphology and systematics of fossil echinoderms; contributions by many echinoderm paleontologists.

Moore, R. C.; Lalicker, C. G.; Fisher, A. G. *Invertebrate fossils.* New York: McGraw-Hill Book Co., Inc.; 1952. Contains six chapters on echinoderms; still one of the best single-volume textbooks for identifying fossil specimens.

Nichols, D. *Echinoderms.* 4th ed. London: Hutchinson University Library; 1972. Excellent general review of all aspects of living and fossil echinoderms.

Paul, C. R. C. Evolution of primitive echinoderms. Pp. 123–158. In: Hallam, A. editor. *Patterns of evolution.* Amsterdam: Elsevier Scientific Publishing Company; 1977. Excellent summary of general patterns in echinoderm evolution, especially among earliest echinoderms.

Ubaghs, G. Early Paleozoic echinoderms. *Annual Review of Earth and Planetary Sciences* 3: 79–98; 1975. Excellent review of earliest echinoderms; best short summary of arguments why stylophorans should be classified as echinoderms and not chordates.

Class Echinoidea

Porter M. Kier

Echinoids are the largest class of the subphylum Echinozoa and are stratigraphically the most important of all echinoderm classes in post-Paleozoic rocks. They have evolved more rapidly since the Triassic than most other invertebrates with the development of many new and distinctive morphologic types. Many of these types are still living, and enough is known of their living habits to permit interpretations of the living habits of fossil species. Finally, echinoids have so many distinctive morphologic characters preserved in their shell that single or partial specimens often can be identified to the species. Because of these qualities, echinoids are stratigraphically more useful than other invertebrates in some Mesozoic and Cenozoic strata.

Morphology

External anatomy Echinoids have a globular to flattened calcareous skeleton or test composed of many plates (Figure 18.53). These plates are arranged radially in two kinds of columns that are double in most echinoids. The first kind has one or two pores in each plate for the extrusion of tube feet and is called the **ambulacrum** (plural: ambulacra). The second kind of double column alternates with the ambulacra and is called the **interambulacrum**. Some Paleozoic species have more than two columns in each ambulacrum or interambulacrum. Movable spines (Figure 18.54) are attached to **tubercles** (Figures 18.53 and 18.55) on the exterior of most of the plates. The largest spines usually are attached to interambulacral plates. Most echinoids carry pincerlike organs, the pedicellariae (Figures 18.10 and 18.54), on small stalks. The pedicellariae are used for defense and feeding, but because of their fragile nature, these organs generally have been lost from fossils.

The mouth and its surrounding tissue, or **peristome** (Figure 18.53), are located on the lower or ventral side. **Peristomial notches** (Figures 18.53 and 18.56), slitlike openings through the test formerly and erroneously called gill slits, may occur on the margin of the peristome. Pressure-compensating organs extrude through these notches for the water vascular system. In many echinoids, the anus and its surrounding tissue, or **periproct** (Figures 18.53 and 18.57), is on the upper surface surrounded by a circlet of plates called the **apical system** (Figures 18.53; 18.57; and 18.58). Five plates of the apical system, the **genital** plates, are above the interambulacra and are perforated by ducts for the extrusion of genital products. One of the genital plates serves as the madreporite with its numerous pores opening to the water vascular system. The five other plates of the apical system are above the ambulacra and are termed **ocular plates**. New plates are secreted in both the interambulacra and the ambulacra at the lower border of the ocular plates.

Those echinoids having the periproct enclosed within the apical system are circular in marginal outline and are termed **regular** (Figures 18.53; 18.55; and 18.56). Echinoids in which the periproct has shifted out of the apical system have a bilateral symmetry and are generally elongated antero-posteriorly with the periproct posterior. They are called **irregular** (Figures 18.61 to 18.64). Their apical systems contain fewer than five genital plates (Figure 18.58).

Regular echinoids have ambulacra composed of simple independent plates called **primaries** (Figures 18.55 and 18.59) or **compound** plates (Figures 18.56 and 18.60) composed of more than one plate partially covered by a large tubercle.

The ambulacra of both regular and irregular echinoids are continuous from the apical system to the peristome.

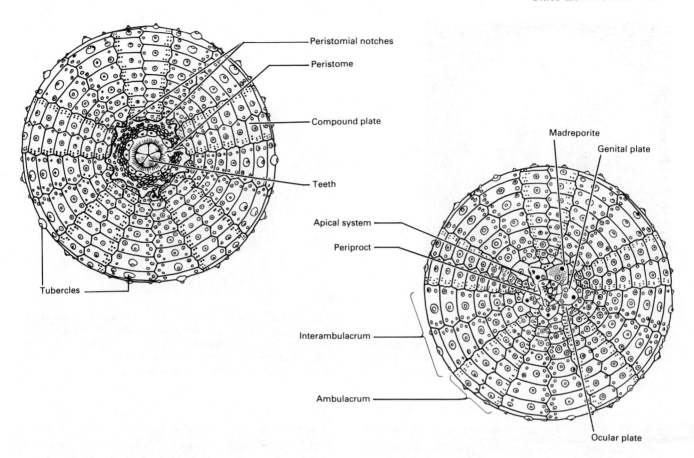

Figure 18.53. External anatomy of an echinoid, *Echinus*. (Top surface by Durham, J. W., and bottom after MacBride, E. W. From: Durham, J. W., *et al.,* In: Moore, R. C., editor. *Treatise on invertebrate paleontology, Part U, Echinodermata 3.* New York and Lawrence, KS: Geological Society of America and University of Kansas Press; 1966.)

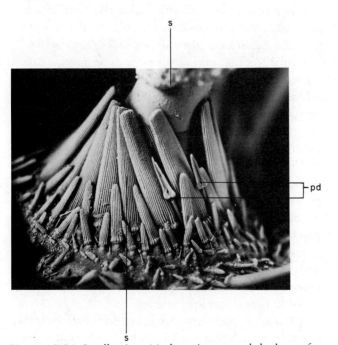

Figure 18.54. Small spines (*s*) clustering around the base of a large spine shielding its muscles. Two large and many small pedicellariae (*pd*) are visible. (*Cidaris,* Holocene, ×8.5.)

Figure 18.55. Side view of a regular cidaroid echinoid, *Cidaris*. Note the large tubercles (*tc*) and primary ambulacral plates (*pa*) (Eocene, North Carolina, ×4.)

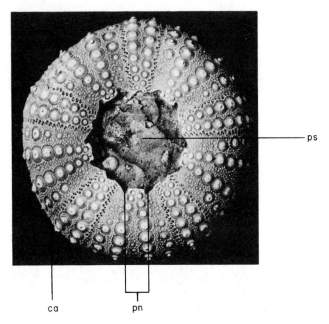

ca   pn

Figure 18.56. Bottom view of a regular noncidaroid, *Echinotiara,* showing peristome (*ps*), peristomial notches (*pn*) and compound ambulacral plates (ca). (Lower Jurassic, Saudi Arabia, ×3.)

In most irregular echinoids the pores are much larger on the upper surface. The adapical areas characterized by these large pores, often elongated into slits, are termed **petals** (Figures 18.61, *B*, and 18.62 to 18.64). Flattened tube feet extend through these pores. In some species of

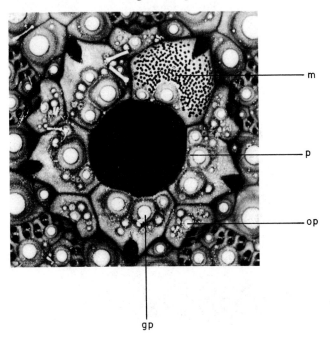

gp

Figure 18.57. Apical system of a regular noncidaroid, *Psammechinus,* showing large madreporite (*m*) with many perforations, the four other genital plates (*gp*) and five ocular plates (*op*). The periproct (*p*) is surrounded by the plates of the apical system. (Miocene, Virginia, ×10.)

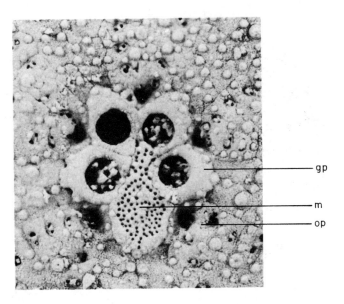

Figure 18.58. Apical system of a spatangoid, *Eupatagus,* showing presence of only four genital plates (*gp*) and the ocular plates (*op*). One of the genital plates is enlarged, extending across the apical system, and is perforated by pores for the madreporite (*m*). (Eocene, North Carolina, ×11.)

heart urchins, these petals are depressed into the test (Figure 18.63). The anterior ambulacrum on the upper surface may have pores of different shape, size, and number from those in the paired petals (Figures 18.61 and 18.63). Tube feet of the paired petals are respiratory and those of the anterior ambulacra are sensory or are used to maintain a burrow to the surface of the substrate.

Many regular and irregular echinoids have enlarged and crowded ambulacral pores on their lower surface near the peristome, forming **phyllodes** (Figure 18.62). The tube feet that extrude through these pores are used for feeding and adhesion to the substrate. Some irregular echinoids have inflated interambulacra at the edge of the peristome, forming **bourrelets** (Figure 18.62). The many spines attached to the bourrelets push food into the mouth.

All echinoids except the order that includes the sand dollars (the clypeasteroids) have only one pore or pair of pores called a **porepair** in each ambulacral plate and none in the interambulacra. In addition to the normal porepairs, the clypeasteroids have many small **accessory pores** (Figure 18.64) in the ambulacral and some interambulacral plates. The small accessory tube feet that extend through these pores are used to grasp sand grains or for transporting food particles.

All regular and some irregular echinoids have five calcareous teeth (Figures 18.53 and 18.65) held in place by beaklike jaws called **pyramids** (Figure 18.65). The combined pyramids and teeth are called **Aristotle's**

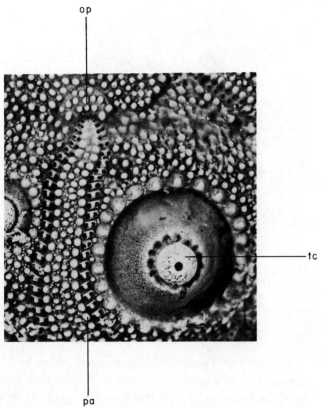

Figure 18.59. View of part of upper side of a regular, *Pseudocidaris*, showing ocular plate (*op*), primary ambulacral plates (*pa*) and large tubercle (*tc*). (Jurassic, Saudi Arabia, ×6.5.)

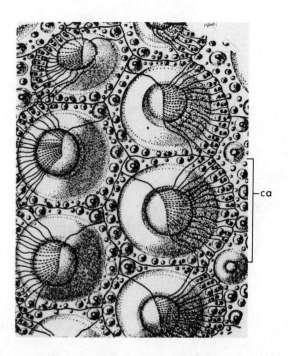

Figure 18.60. Part of ambulacrum of a living regular, *Hetero-centrotus*, showing compound plates (*ca*) (×6.) (From Hawkins, H. L. *Philosophical Transactions of the Royal Society of London*, series B, 209: 337–480, pls. 61–69; 1959.)

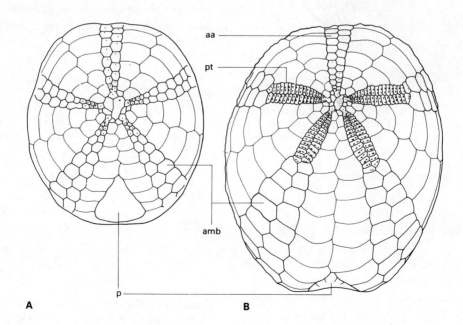

Figure 18.61. Top views of a living spatangoid, *Meoma*, showing an immature test (**A**) lacking petals in the ambulacrum (*amb*) and having its periproct (*p*) on the upper surface. In a more mature specimen (**B**), petals (*pt*) are well developed except in the anterior ambulacrum (*aa*) and the periproct has shifted to the posterior margin. (×9.)

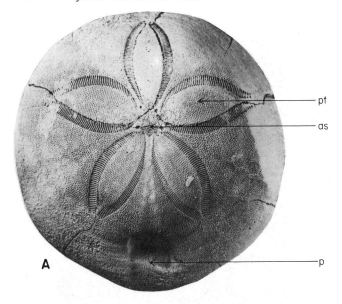

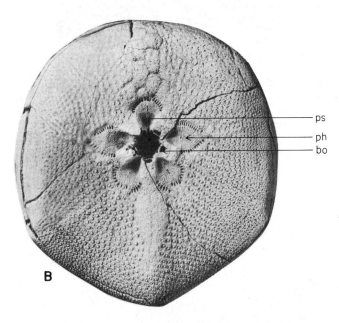

Figure 18.62. Top (**A**) and bottom (**B**) views of an irregular cassiduloid, *Hardouinia.* Note on the top view the location of the periproct (*p*) outside the apical system (*as*) and the broad petals (*pt*) formed by slitlike pores. On the lower side are the peristome (*ps*) and phyllodes (*ph*) with enlarged pores and the bourrelets (*bo*). (Upper Cretaceous, Mississippi, ×1.5.)

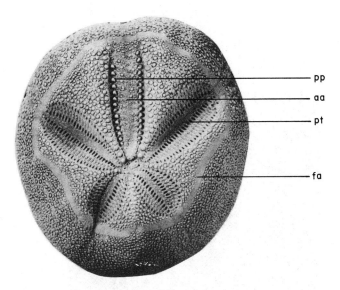

Figure 18.63. Spatangoid, *Hemiaster,* showing some of the features that enabled the echinoid to live buried, including depressed petals (*pt*), fasciole (*fa*), and specialized porepairs (*pp*) in the anterior ambulacrum (*aa*). (Upper Cretaceous, Texas, ×2.5.)

on the plastron (Figure 18.67). Some sand dollars have a hole, or **lunule** (Figure 18.68) near the edge of their tests. Sand and nutrients pass through the lunule to food grooves on the oral surface (Figure 18.68) that transport the material to the mouth.

Internal anatomy The gut is the most conspicuous internal organ, extending from the mouth to the anus on the upper or posterior side (Figure 18.69). It is twisted in a loop. The ring canal (circumoral ring) of the water vascular system encircles the gut just above the Aristotle's

lantern. This structure is attached by muscles to raised calcareous processes lying inside the test at the edge of the peristome.

Many heart-shaped, irregular echinoids called heart urchins or spatangoids have **fascioles**, narrow bands of small ciliated spines encircling part of the test. These fascioles (Figure 18.63) create currents for respiration, feeding, or excretion. Heart urchins have greatly enlarged posterior interambulacral plates on the ventral (lower) surface, forming a broad platform or **plastron** (Figure 18.66). Spines used for locomotion are attached

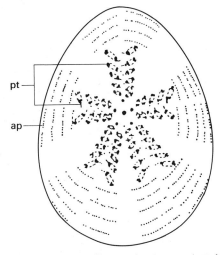

Figure 18.64. Top view of living clypeasteroid, *Echinocyamus,* showing larger pores in petals (*pt*) for respiratory tube feet and smaller pores for accessory tube feet (*ap*). (×12.) (From Durham, J. W. In: Moore, R. C., editor. *Treatise on invertebrate paleontology, Part U, Echinodermata 3,* New York and Lawrence, KS: Geological Society of America and University of Kansas Press; 1966.)

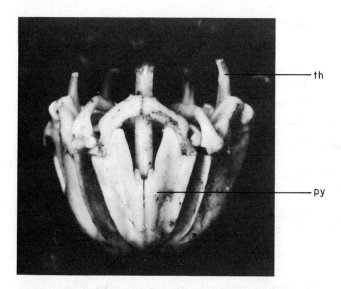

Figure 18.65. Pyramids (*py*) and teeth (*th*) of Aristotle's lantern of *Psammechinus*. (Miocene, Virginia, ×6.5.)

lantern or the peristome. The stone canal connects it with the madreporite. The radial water canals branch from the ring canals and pass up on the interior surface of the test in the ambulacral areas to terminate at the ocular pore. Branches from the radial canal form ampullae and tube feet. The gonads are suspended by mesenteries from the inner surface of the interambulacra, and each is connected by a duct to a pore in the genital plate. Features of the internal anatomy are rarely used in the classification of echinoids.

Classification

Most investigators accept the classification of the echinoids up to the ordinal level employed in Part U of the

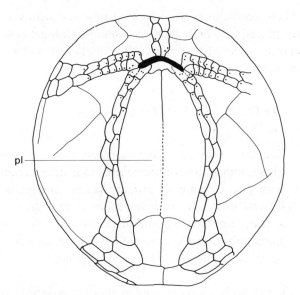

Figure 18.66. Plate arrangement of lower side of a spatangoid, the heart urchin *Linthia*, showing large plates of plastron (*pl*). (Eocene, North Carolina, ×0.6.)

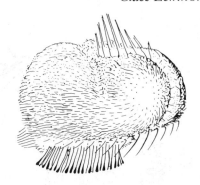

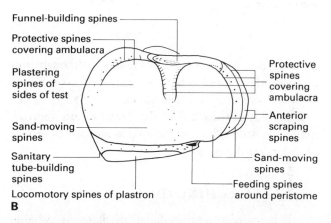

Figure 18.67. Side views of spatangoid, *Echinocardium*, showing types of spines with spines attached (**A**) and denuded test showing areas of origin and different types of spines (**B**). (From Nichols, D. *Philosophical Transactions of the Royal Society of London,* series B; Biological Sciences 242 (693): 347–437; 1959.

Treatise on Invertebrate Paleontology, except that more recent discoveries have resulted in the removal of the botriocidaroids from the class Echinoidea. Considerable disagreement exists on the composition of the superorders and subclasses. Uncertainty as to the origins of the orders makes it difficult at this time to formally delineate these higher taxa. For this reason, many investigators formally or informally divide the echinoids into two

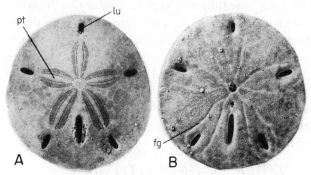

Figure 18.68. Sand dollar, the clypeasteroid *Mellita*, showing the lunule (*lu*) and petals (*pt*) on the upper side (**A**) and food grooves (*fg*) on the bottom (**B**). (Pliocene-Pleistocene, North Carolina, ×0.7.)

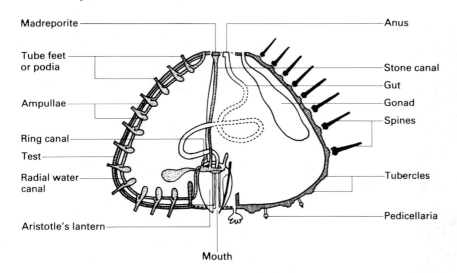

Madreporite

Tube feet or podia

Ampullae

Ring canal

Test

Radial water canal

Aristotle's lantern

Mouth

Anus

Stone canal

Gut

Gonad

Spines

Tubercles

Pedicellaria

Figure 18.69. Cutaway view showing the internal anatomy of an echinoid.

groups, the regulars and irregulars. Although this division does not reflect the phylogeny in all taxa, it is the most convenient for this discussion.

Echinoids are divided in the *Treatise* into approximately 20 orders, but only seven major orders or groups of orders are discussed here.

CLASS ECHINOIDEA

Echinozoans with globular to flattened skeleton composed of series of interlocking plates bearing spines; five series of ambulacral plates alternating with five series of interambulacral plates; mouth on lower surface, jaws and teeth present or absent, but all echinoids descended from forms having them; anus on upper surface within apical system or in posterior interambulacrum on upper or lower surface; radial canals internal. (Late Ordovician to Holocene; approximately 765 genera.)

Regular echinoids Regular echinoids are easily distinguished from irregular echinoids by their circular test (when viewed from above), nearly perfect pentameral symmetry, and the location of the anus within the apical system in the center of the upper surface. Plates are thin to thick and may be imbricated. Ambulacra have two or more columns of plates. Interambulacra have one or more columns of plates and are all similar. Spines are generally long, and an Aristotle's lantern occurs in all taxa. The group of approximately 355 genera ranges from the Late Ordovician to the Holocene. All Paleozoic echinoids were regular.

ORDER ECHINOCYSTITOIDA
(Figure 18.71)

Regular echinoids with thin imbricating plates;

ambulacrum and interambulacrum with two or more columns of plates; one or five genital plates. (Late Ordovician to Permian; 24 genera.)

Echinocystitoida include many echinoids with very flexible tests that were often flattened. The lower side in later species differs from the upper in having widened ambulacra with larger pores. The tube feet that extended through these pores were probably used for feeding.

ORDER PALAECHINOIDA (Figure 18.72)

Regular echinoids with thick plates, not strongly imbricating; ambulacrum with two or more columns of plates, interambulacrum with one or more columns; five genital plates. (Late Silurian to Permian; 10 genera.)

These echinoids typically have a rigid test with very thick plates, frequently in many columns. They are most common in the Mississippian where they occur in large numbers at a few localities.

ORDER CIDAROIDA (Figures 18.55; 18.73; 18.74)

Regular echinoids with moderately thick to thin plates, imbricating or not imbricating; ambulacrum with two columns of primary plates, interambulacrum with one or more columns in Paleozoic species, only two in post-Paleozoic commonly with single large tubercle on each plate; five genital plates. (Early Mississipian to Holocene; 60 genera.)

The cidaroids are considered by most investigators to be the only echinoids to survive the Paleozoic and are

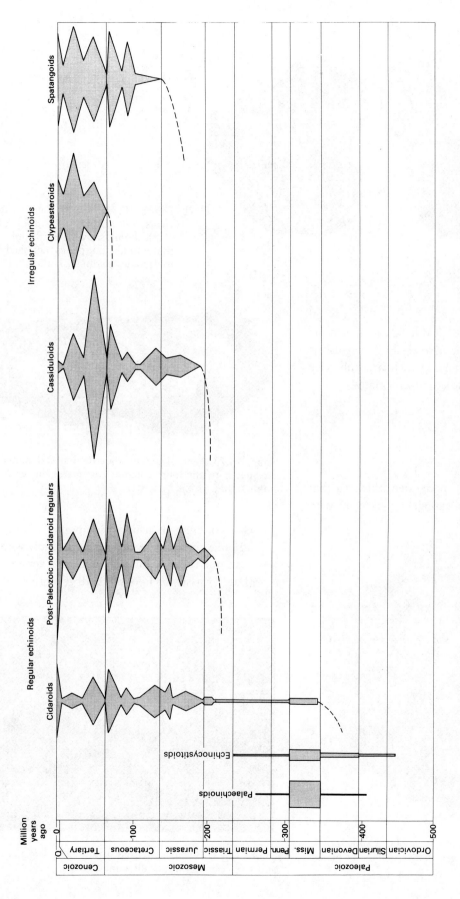

Figure 18.70. Stratigraphic range and relative abundance of species of the most common groups of echinoids.

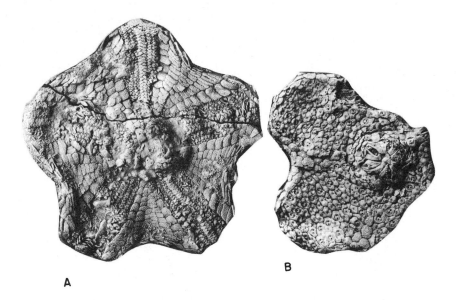

Figure 18.71. Top (**A**) and bottom (**B**) views of an echinocystitoid echinoid, *Pronechinus*, showing the flexible test with the ambulacral plates enlarged on the lower side. (Permian, Turkey, ×1.5.)

considered to be ancestral to all other groups of post-Paleozoic echinoids. Cidaroids are easily distinguished from most other regular echinoids by their large tubercles and spines and by their simple ambulacral plates.

GROUP: POST-PALEOZOIC NONCIDAROID REGULAR ECHINOIDS (Figure 18.56)

Regular echinoids with moderately thick to thin plates, usually not imbricating; ambulacrum with

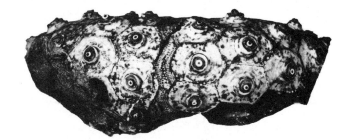

Figure 18.73. Pennsylvanian cidaroid from Oklahoma, *Archaeocidaris*, with four columns of plates in each interambulacrum. (×0.6.)

two columns of usually compound plates, interambulacrum with two columns generally with more than one tubercle on each plate; five genital plates. (Late Triassic to Holocene; 258 genera.)

Figure 18.72. Typical palaechinoid echinoid, *Melonechinus*, showing its rigid test composed of many columns of plates. (Mississippian, Missouri, ×0.6.)

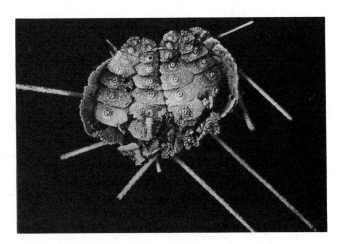

Figure 18.74. Permian cidaroid from Texas, *Miocidaris*, with only two columns in each interambulacrum. (×2.)

These echinoids are the most common of living regular echinoids with 330 species as compared to 144 cidaroid species. Most are easily distinguished from cidaroids by their smaller and more numerous tubercles and compound ambulacral plates.

Irregular echinoids Adult irregular echinoids are distinctive in their usually elongate test and in the position of the anus outside and generally posterior to the apical system. They usually have petals on the upper surface. Each ambulacrum and interambulacrum has two columns of plates, and the posterior interambulacrum differs from the others. Test plates are thin and not imbricating, spines are generally short, and an Aristotle's lantern is absent in adults of most species except for the sand dollars. The group of approximately 410 genera ranges from the Early Jurassic to the Holocene. They underwent a spectacular radiation in the Mesozoic and are much more common as fossils than regular echinoids.

ORDER CASSIDULOIDA (Figure 18.62)

Irregular echinoids with five petals, no accessory pores, no food grooves; phyllodes and bourrelets present; one to four genital plates; no plastron; no fascioles; Aristotle's lantern absent in adults. (Early Jurassic to Holocene; 69 genera.)

Cassiduloids are easily distinguished from the spatangoids, having bourrelets and five respiratory petals and lacking fascioles and a plastron. They were common in the Jurassic through the Eocene but are rare at present.

ORDER SPATANGOIDA (Figures 18.61; 18.63; 18.66)

Irregular echinoids usually with only four respiratory petals and with the tube feet of the anterior petal modified for other functions; anterior ambulacrum often sunken with smaller pores; no pores for accessory tube feet, no food grooves; phyllodes present, no bourrelets; one to four genital plates; plastron; commonly fascioles; no Aristotle's lantern. (Early Cretaceous to Holocene; 147 genera.)

The spatangoids include the heart urchins and are distinguished by commonly having the anterior ambulacrum differently modified, having a well-developed plastron, and usually by having fascioles. The more deeply burrowing species often have tests composed of very thin plates.

ORDER CLYPEASTEROIDA (Figure 18.68)

Irregular echinoids usually with flattened test often reinforced internally; five petals; accessory pores; no phyllodes or bourrelets; one genital plate; no plastron or fascioles; Aristotle's lantern with broad teeth. (Paleocene to Holocene; 73 genera.)

The clypeasteroids are a distinct group that include the sand dollars. They have accessory tube feet, a feature that distinguishes them from other echinoids. They occur in great numbers in Cenozoic deposits and reached their zenith in the Miocene.

Biology

Echinoids are unisexual, having the male reproductive organs in one individual and the female reproductive organs in another. Generally the sex cannot be determined from the test except in brooding echinoids (Figure 18.75) where the females may have brood pouches or larger genital pores. The female releases a large number of eggs while the male simultaneously releases the sperm. Fertilization ensues immediately, and development begins. After 4 to 6 weeks, the planktic pluteus larvae undergo metamorphosis and settle to the seafloor, and

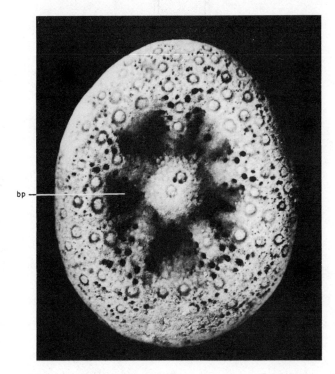

Figure 18.75. Brood pouches (*bp*) in *Pentedium* from the Eocene of Georgia. (×18.)

within a few months the echinoid has many adult features.

An immature echinoid exhibits many features of its ancestors. For example, an immature spatangoid such as *Meoma* has its periproct within its apical system. As the spatangoid grows, the periproct shifts posteriorly (Figure 18.61, *A*) to its adult position on the posterior margin of the test (Figure 18.61, *B*). Likewise, petals are absent on young specimens (Figure 18.61, *A*), the peristome is much larger, the test higher and rounder, and the spines longer and less numerous. In all these characters, the most immature spatangoid resembles its regular ancestor.

The calcareous test of an echinoid, including its spines, is of mesodermal origin and covered by thin epidermis (Figure 18.76). Therefore an echinoid can increase its size not only by adding new plates but also by increasing the size of existing plates and spines. A broken spine can either be dropped and replaced by a new one, or growth can resume at the broken end and a new tip produced. Likewise, a hole in the test can be repaired by new growth along the edges of the broken plates.

Echinoids are preyed upon by fish, crabs, starfish, mammals, and birds. They are eaten in large numbers by people who consider their eggs a delicacy. Snails drill holes in the test and extract the soft parts. Tests of many fossil spatangoids have small holes 1 to 4 mm in diameter, the result of such predation.

Functional morphology

Tube feet　Tube feet of echinoids are versatile organs. They serve in locomotion, anchoring, feeding, sensory reception, respiration, and funnel building.

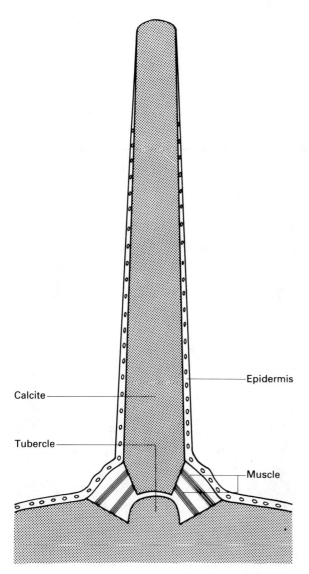

Figure 18.76. Diagrammatic section through a spine and part of the test of a regular echinoid showing the presence of an epidermis over the spine and test.

Calcite

Tubercle

Epidermis

Muscle

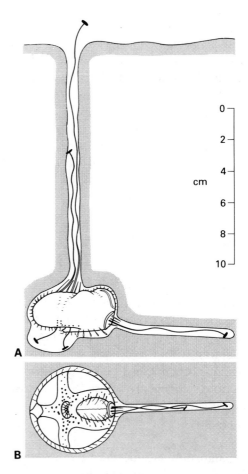

Figure 18.77. Spatangoid echinoid, *Echinocardium*, lives deeply buried, using its long tube feet to maintain its burrow. **A**, Side view; **B**, bottom view. (From Nichols, D. *Philosophical Transactions of the Royal Society of London*, Series B; Biological Sciences 242 (693): 347–437; 1959.)

The most common tube feet in regular echinoids have small suckers at their tips (Figure 18.10) used for both grasping and respiration. They are thin walled and divided longitudinally by a membrane separating oxygenated from deoxygenated fluid. Their suckers are often used to hold detritus over the top of the animal. Tube feet around the mouth in phyllodes have large suckers for feeding and for holding the echinoid firmly to the substrate.

Irregular echinoids have many different kinds of tube feet. The respiratory tube feet in petals (Figures 18.62, 18.63 and 18.68) are much broader and flattened, producing more area for gaseous exchange. The phyllodal tube feet in spatangoids have a large disk at the tip that secretes adhesive mucus. They collect large amounts of fine, organic-rich sediment that is then passed to the mouth. Spatangoids also have special tube feet to dig funnels to the surface of the substrate (Figure 18.77) and to maintain a sanitary drain behind the anus.

Although tube feet are not normally preserved in fossils, their number, nature, and function can usually be determined by study of the pores through which the tube feet extended. For example, a tube foot used for respiration always has a pair of pores often joined by a furrow. These pores are elongated into slits especially in cassiduloids (Figure 18.62) and sand dollars (Figure 18.68). Suckered tube feet used for grasping have smaller pores, paired in the regulars and less advanced irregulars. These pores are usually surrounded by a differentiated area. This is the site for attachment of the retractor muscles.

Spines Spines of a regular echinoid are primarily for protection from predators and for walking. In the cidaroids, the spines are of two sizes: the large spines are for protection and locomotion; the small spines cluster around the base of the large spines to shield the retractor muscles (Figure 18.54). They also arch over the ambulacral pores and protect contracted tube feet. Most of the advanced regular echinoids have much shorter and more numerous spines. These echinoids use their tube feet, which extend beyond the spines, to hold plant debris and shells over the test for camouflage or to reduce incident light. Some regular echinoids living in shallow water and exposed to large waves have spines expanded at the tips to form a protective canopy over the test. Some tropical echinoids have poison in the tips of their spines. Finally, many regular echinoids use their spines to excavate holes in hard substrate. They then brace themselves with their spines to prevent removal by predators and wave action.

Irregular echinoids have smaller and more numerous spines. Besides being used for locomotion, they protect the animal from abrasion by sediment particles and also help to maintain a space for water circulation between the test and the burrow. Spines attached to the lower side at the plastron (Figure 18.67) may be curved and paddle shaped and are used for locomotion.

Although spines are not often preserved in fossil echinoids, their corresponding tubercles provide considerable information as to the nature, number, and function of the attached spines.

Pedicellariae (Figure 18.54) are modified spines. Their pincers prevent small organisms from attacking or settling on the test and also catch food. Larger animals such as starfish are repelled by the poison produced by glands on some pedicellariae.

Fascioles Fascioles are composed of crowded tiny spines covered with cilia. They are found only in heart urchins (spatangoids). These fascioles produce currents that increase the amount of oxygenated water crossing the petals and aid in feeding and in sweeping away excreta. Although these spines are not preserved on fossils, the presence of a fasciole is indicated by a narrow tract of very small tubercles (Figure 18.63).

Mouth All regular echinoids have their mouth or peristome central on the lower surface of the test. The opening is large; within it are the jaws and teeth that tear off pieces of organic material, which then are passed into the gut.

The mouth of an irregular echinoid is much smaller and is frequently anterior. It usually lacks teeth; tube feet pass large amounts of sediment to the mouth, and organic material is extracted in the intestine. The sand dollar, however, has teeth that are much larger than those in a regular echinoid. They scrape organic material off sand grains and crush small organisms.

Anus The anus in a regular echinoid is on the top of the test within the apical system. It is a small opening reflecting the small amount of excreta produced by a regular echinoid.

The anus in an irregular echinoid is generally much larger because of the large amount of ingested sediment that is ejected after digestion of organic material. The posterior position of the anus (often on the lower side) prevents this debris from fouling the respiratory tube feet. Debris is even more channeled in some by the placement of the anus in a deep trough.

Summary The life habits of most fossil echinoids

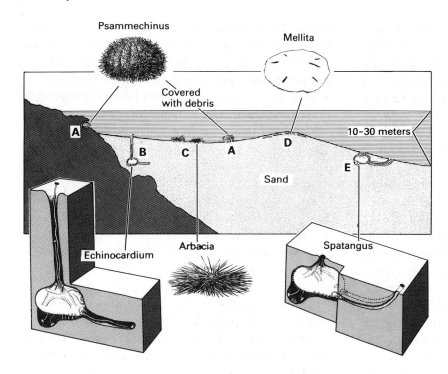

Figure 18.78. Probable life positions of five echinoid species from the Pliocene Yorktown Formation, Virginia. (From Kier, P. M. *Smithsonian Contributions to Paleobiology.* No. 13: fig. 1; 1972. Figures of *Spatangus* and *Echinocardium* are based on the figures and descriptions of Nichols, D. *Philosophical Transactions of the Royal Society of London,* Series B; Biological Sciences 242 (693): 347–437; 1959.)

(Figure 18.78) can be predicted by the often obvious relation between morphology and its known function in living species. One can assume that a regular urchin (like *Arbacia* in Figure 18.78) with long spines lived on the surface of the substrate or in cavities in the rock using its jaws to graze. An irregular echinoid having petals but lacking fascioles and a deep anterior groove (like *Mellita* in Figure 18.78) presumably lived partially buried in well-aerated sediment. An irregular echinoid with fascioles around the petals and anus, a deep anterior groove, and sunken petals (like *Echinocardium* in Figures 18.77 and 18.78) probably lived deeply buried in sand or clay. The long tube feet extending upward from the ambulacrum in the anterior groove would maintain a funnel to the surface of the substrate. The cilia in the fasciole around the petals would generate currents drawing water down the funnel and over the respiratory petals. The fasciole around the anus would dispel water and excreta posteriorly away from the echinoid.

Evolution and biostratigraphy

Regular echinoids Paleozoic echinoids are rare as fossils and of little value stratigraphically, occurring in relatively large numbers only at a few Mississippian localities in the United States and western Europe. The first echinoids, the echinocystitoids, appeared in the Ordovician (Figure 18.70) and are characterized by flexible tests (Figure 18.71). The Ordovician and Silurian species were small with few plates, but in the later Paleozoic they became very large with many columns of interambulacral and ambulacral plates. Some echinoids in the Mississippian and later periods (Figure 18.71) had flattened tests with tube feet on the lower side much larger than those on the upper side. Another group, the palaechinoids, arose in the Silurian. The test was rigid and the plates very thick. They became common in the Mississippian and are typified by *Melonechinus* (Figure 18.72). They had short spines and high tests and probably lived on the surface of the sea floor.

The first cidaroid appeared in the Mississippian. These echinoids first had a slightly flexible test that gradually became more rigid. They had only two columns of plates in each ambulacrum, but some Mississippian and Pennsylvanian species had four columns in each interambulacrum (Figure 18.73). By the Permian, all cidaroids had only two interambulacral columns in each area (Figure 18.74). These echinoids also could not burrow. All echinoids became extinct by the end of the Paleozoic except for the cidaroids, which survived into the Mesozoic and still exist. During the Triassic, the cidaroids dominated, but in the Jurassic a great change occurred. Among regular echinoids (Figure 18.56), the ambulacral plates compounded in many species permitting echinoids to have large spines attached to the ambulacra and to have more tube feet, thus increasing their ability to respire and gather food.

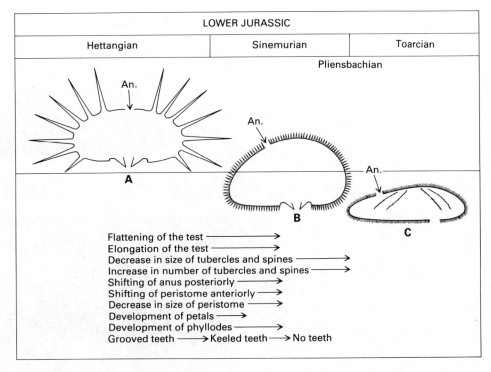

Figure 18.79. Evolution of the irregular echinoid showing the changes that enabled the echinoid to live buried in the substrate. **A,** A Hettangian (earliest Jurassic) regular echinoid, *Diademopsis.* **B,** The earliest known irregular, *Plesiechinus,* from the Sinemurian (middle Early Jurassic). C, A cassiduloid, *Galeropygus,* from the Toarcian (upper Early Jurassic). These changes occurred in less than 15 million years. (From Kier, P.M. *Palaeontology* 25 (1): 1–9, pls. 1–2; 1982.)

Irregular echinoids The most significant post-Paleozoic development was the evolution (Figure 18.79) of the irregular echinoids during the Early Jurassic. In only 10 to 15 million years, the periproct shifted posteriorly out of the apical system; the test became elongated; the spines and their tubercles became much smaller; the ambulacral pores on the top of the test enlarged forming petals; the teeth vanished; and the peristome moved anteriorly. All these changes enabled echinoids to live buried in the sediment, a major new niche. The resulting rapid evolution produced many new morphologic types.

The cassiduloids (Figure 18.62) first appeared in the Middle Jurassic and were characterized by phyllodes and bourrelets. They are particularly useful as index fossils because of parallel evolution that occurred throughout the order. For example, Early Cretaceous and older cassiduloids had two pores in each ambulacral plate beyond the petals, but later species descended from different stocks had only one. Similarly, the apical system in cassiduloids of different lineages changed at the end of the Cretaceous from four genital plates to only one. Finally, the number of pores in the phyllodes decreased as their tube feet became more specialized and fewer tube feet were needed. This change not only occurred in different stocks of the cassiduloids but also in species of the same genus. A Middle Jurassic *Pygurus* (Figure 18.80, *A*) has far more phyllodal pores than a Late Cretaceous species (Figure 18.80, *D*). As a result, a single specimen of *Pygurus* can be very useful in dating strata.

The spatangoids (Figures 18.61, 18.63, and 18.81) appeared in the Early Cretaceous and were even better adapted to burrowing (Figure 18.81), especially in finer sediments. Their fascioles enabled them to create the currents necessary for respiration in a deep burrow (Figure 18.77). They generally lacked a petal in the anterior ambulacrum, instead having long tube feet (Figure 18.77) that maintained a funnel to the surface. Their phyllodal tube feet terminated in a large mucus-secreting disk that permitted spatangoids to feed on finer sediment than cassiduloids. Finally, many of them bore curved, paddle-shaped spines (Figure 18.67) particularly efficient in burrowing.

Evolutionary trends within the spatangoids include sinking of the petals and the anterior ambulacrum into deep grooves on the test, a decrease in the thickness of the plates, an increase in the number of fascioles, enlargement of the plastron, and greater differentiation of the spines and tube feet.

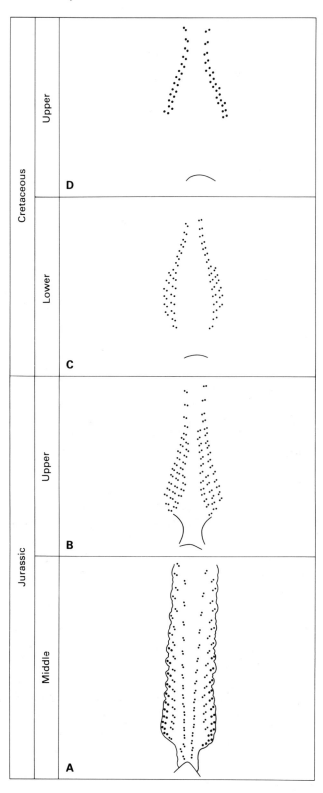

Figure 18.80. Evolution of the phyllode in the cassiduloid *Pygurus* showing the decrease in the number of pores in four species. From Durham, J. W. In: Moore, R. C., editor. *Treatise on invertebrate paleontology, Part U, Echinodermata 3.* New York and Lawrence, KS: Geological Society of America and University of Kansas Press; 1966.)

Figure 18.81. Burrowing of a living spatangoid, *Paraster*, at 15 m depth of sea in the Caribbean. Note fasciole (*fa*).

Fossil echinoids are much more common in the Mesozoic than in the Paleozoic. Although rare in the Triassic, they are abundant and increasingly useful stratigraphically in the Jurassic and even more so in the Cretaceous. Jurassic species are common in Europe, northern Africa, and the Middle East. Cretaceous echinoids are especially numerous in Texas, Europe, and northern Africa.

The increase in the number of fossil echinoids in the Mesozoic may not be because more echinoids were living then than in the Paleozoic. Rather the increase may result from the development of the irregular echinoids, which were much more likely to be preserved because of their burrowing lifestyle. The regular echinoids of the Paleozoic lived on the surface of the substrate where predators and turbulence could break up their tests. Furthermore, unless rapidly covered by sediment, the plates of echinoid tests dissassociated soon after death because of decomposition of the tissue that bound them together.

The last significant development in echinoids was the evolution of the clypeasteroids including the sand dollars (Figure 18.68) in the early Cenozoic. The clypeasteroids with their accessory tube feet, flattened tests, and well-developed petals were especially well-adapted to living in great numbers in sand. They first appeared in the Paleocene of Africa, and during the Eocene spread all over the world.

Cenozoic echinoids are prolific in the southeastern United States, the Caribbean, western Europe, northeastern Africa, the Middle East, India, and Australia.

Supplementary reading

Durham, J. W., et al. In: Moore, R. C., editor. *Treatise on invertebrate paleontology, Part U, Echinodermata 3.* New York and Lawrence, KS: Geological Society of America and University of Kansas Press; 1966. Excellent compilation with illustrations of all fossil and living genera; also an extensive glossary.

Jackson, R. T. Phylogeny of the echini, with a revision of Paleozoic species. *Memoirs of the Boston Society of Natural History* 7, 1912. All Paleozoic echinoid species are well illustrated.

Kier, P. M. Evolutionary trends in Paleozoic echinoids. *Journal of Paleontology* 39 (3): 436–435; 1965.

Kier, P. M. Evolutionary trends and their functional significance in the post-Paleozoic echinoids. *Journal of Paleontology* Memoir 5, 48 (3); 1974.

Mortensen, T. *A Monograph of the Echinoidea*, 5 volumes. Copenhagen: C. A. Reitzel; 1928–1951. All described living species and fossil genera are included.

Nichols, D. Changes in the Chalk heart-urchin *Micraster* interpreted in relation to living forms. *Philosophical Transactions of the Royal Society of London*, series B, Biological Sciences. 242: 347–437; 1959. Functional morphology of living spatangoids is described and related to the evolution of *Micraster*.

Smith, A. A functional classification of the coronal pores of regular echinoids. *Palaeontology* 21: 159–789; 1978.

Smith, A. The structure, function, and evolution of tube feet and ambulacral pores in irregular echinoids. *Palaeontology* 23: 39–83; 1980. Description of tube feet and pores of many living echinoids. With the information in the two articles by Smith it is possible to suggest the function of the tube feet of fossils by a study of their pores.

Smith, A. *Echinoid palaeobiology*, Special Topics in Palaeontology, London: Allen & Unwin, Inc.; 1984. A review of the paleobiology of fossil echinoids including a revised classification.

19

Phylum Hemichordata (Including Graptolithina)

William B. N. Berry

Part I   Phylum overview

The Hemichordata include two minor groups of living marine invertebrate animals that are not, despite the name, closely related to true chordates. The wormlike acorn worms, or enteropneusts (Figure 19.1), and the colonial or aggregated, bryozoanlike pterobranchs (Figures 19.2, and 19.3) are commonly grouped as classes of the phylum Hemichordata. A major extinct group of fossils called graptolites (Figures 19.4 to 19.6) bears certain similarities to the hard parts of some pterobranchs. Graptolites therefore are considered possible relatives of the pterobranchs and are included in the phylum Hemichordata as the class Graptolithina. The relationship of the extinct graptolites with the pterobranchs is debatable, however, which will be discussed in some detail. The question of graptolite relationships is one of the most intriguing puzzles in invertebrate paleontology; the answer has involved much detailed morphologic investigation.

Hemichordates were once thought to be closely allied to the chordates because living members possess what was thought to be a notochord or notochord analogue, a rodlike structure considered similar to the vertebrate backbone. This structure in the hemichordates actually has been shown to be a form of pouch connected with the digestive tract. It is an elongate, tubular structure that projects forward from a position near the mouth. Absence of a notochord in the Enteropneusta and in pterobranchs has resulted in treatment of the hemichordates as a phylum separate from phylum Chordata.

Living hemichordates are coelomate with the body and coelom divided into three regions. Their nervous system is both dorsal and ventral in the epidermis, and most have paired gill slits. The embryo and early larva of hemichordates and echinoderms share some similarities (they are both deuterostomes), suggesting a common origin for both phyla.

Graptolites are the fossil remains of colonial marine organisms that lived from the Cambrian into the Pennsylvanian, a span of about 200 million years. Some graptolites formed complexly branched colonies (Figure 19.4, *A*); other colonies formed from a simple linear series of interconnecting tubes (Figures 19.4, *B* and 19.5). The fossil remains of linear colonies range from about 3 to 5 mm to more than 1 m in length and from little more than 0.1 to 1 to 2 mm in width. The many-branched colonies may range from 5 to 10 cm in breadth and 10 to 15 cm in height.

No clearly identifiable soft parts of the graptolite animal have been recovered. The anatomy of the animals and the mode of colony growth and function therefore must be deduced from the skeletal or hard parts preserved in the fossil record and from their comparison with living animals that are thought to be related.

The skeletal or hard parts in modern hemichordates are proteinaceous in composition. Apparently, graptolite skeletal materials also were proteinaceous when the animals were living. Relatively few graptolites have been preserved uncrushed (Figure 19.6), and most of these have been recovered from limestones and cherts. Acids may be used to dissolve the surrounding rock matrix to free the uncrushed graptolites for study. These graptolites yield much detailed anatomic and some chemical information. Some graptolites have been preserved uncrushed because pyrite crystals formed in the hollow spaces of the colony interiors during diagenesis. Even though the original skeletal material of these specimens

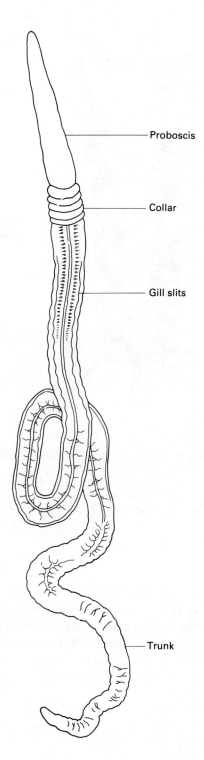

Figure 19.1. Acorn worm *Dolichoglossus* showing major external features of the body.

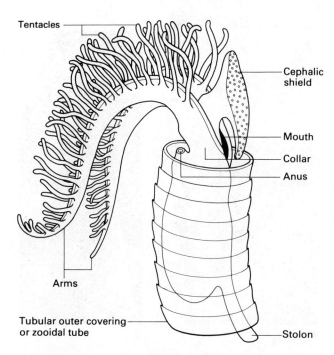

Figure 19.2. Zooid of the pterobranch *Rhabdopleura* showing the principal external features. (From Bulman, O. M. B. In: Teichert, C., editor. *Treatise on invertebrate paleontology, part V (revised).* Boulder, CO and Lawrence, KS: Geological Society of America and University of Kansas Press; 1970.)

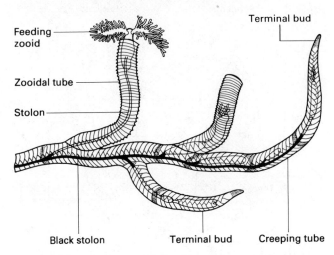

Figure 19.3. Part of the pterobranch *Rhabdopleura* colony showing the creeping tube, black stolon inside the creeping tube, terminal buds, and feeding zooids. (From Bulman, O. M. B. In: Teichert, C., editor. *Treatise on invertebrate paleontology, part V (revised).* Boulder, CO and Lawrence, KS: Geological Society of America and University of Kansas Press; 1970.)

commonly has been highly altered to carbonaceous material, many fine anatomic details may be preserved.

The original skeletons of a few graptolites have been preserved crushed but not carbonized significantly. Such specimens commonly are found on the surfaces of light-colored, fine-grained rocks and usually are tan or amber in color. Skeletal growth increments and other morphologic features may be seen on these specimens.

Most graptolites are preserved on rock surfaces as diagenetically flattened, carbonaceous films that resemble tiny hacksaw blades (Figure 19.4, *B*). Even the carbonaceous film has been destroyed in many specimens,

A   **B**

Figure 19.4. Appearance of two types of graptolites as seen in hand specimens obtained in field collecting. **A,** Dendroid (benthic, seaweedlike) graptolite. (×2.5.) **B,** Graptoloid (planktic) graptolite. (×5.)

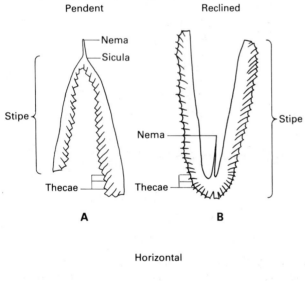

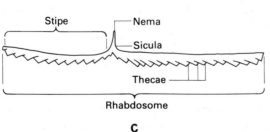

Figure 19.5. Three planktic graptolite rhabdosomes showing major morphologic features. **A,** *Didymograptus* (pendent or tuning fork form). **B,** *Isograptus* (a reclined rhabdosome). **C,** *Didymograptus* (horizontal form).

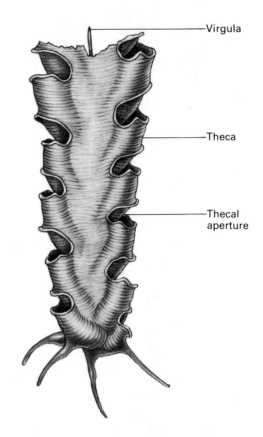

Figure 19.6. Planktic graptolite rhabdosome (a biserial scandent form) freed from its rock matrix. Comparison with Figure 19.4 indicates the morphologic detail that may be seen in specimens freed from the rock matrix compared with those that may be observed in flattened specimens.

leaving only an impression of the former colony. The crushed and carbonized specimens have yielded most of the information leading to an understanding of the general features of graptolite evolution because they are so abundant. As a result they have proven of most value in the relative dating of certain sedimentary and metamorphic rocks.

Introductory morphology

Similarities of some morphologic details of skeletal parts between a living class of hemichordates (the pterobranchs) and the extinct class of graptolites suggest that at least some of their soft parts also could have been comparable. Softpart morphology of modern pterobranchs is emphasized in the following section primarily as a means of suggesting what the graptolite animal could have looked like and how it might have lived.

Morphology and growth of pterobranchs

The body of a hemichordate may be divided into three regions: a **proboscis** or **cephalic shield**, a **collar**, and a **trunk** (Figures 19.1 and 19.2). Pterobranchs form colonies or aggregates of soft-bodied **zooids** (Figures 19.2 and 19.3) housed in proteinaceous tubular outer coverings. The cephalic shields of pterobranchs are essentially discoidal, muscular, and highly ciliated on outer surfaces. Two to nine arms arise from the body just behind the cephalic shield, and each arm has tentacles, which in turn bear cilia. Both arms and tentacles are hollow, each containing an extension of the coelom. In this respect, pterobranch arms and tentacles resemble those of brachiopod and bryozoan lophophores.

The mouth (Figure 19.2) is located at the base of the arms in a position beneath the cephalic shield so that the shield can partly cover the mouth. Beating of the cilia moves food particles between the tentacles and down the arms to the mouth, while unsorted particles are flicked off the tentacle tips. Diatoms, radiolarians, and crustacean larvae have been found in pterobranch stomachs.

The trunk of pterobranch zooids is saclike and contains a U-shaped gut with the anus located on the dorsal side of the collar between the cephalic shield and trunk (Figure 19.2). Gametes (sex cells) develop within the coelom of the trunk. Sexes are separate, and union of egg and sperm takes place in the female zooid. A developing larva is released from the female, and a floating larval stage forms. The floating larval form settles to the substrate, and a zooid develops.

In the genus that forms true colonies (*Rhabdopleura*, Figures 19.2 and 19.3), the zooids are connected by soft, fleshy, tubular **stolons**, also called stalks. The initial, sexually produced zooid generates a stolon from which other zooids in a colony bud asexually.

Colony development begins with growth of the soft, tubular stolon that bears a unique individual called the **terminal bud** (Figure 19.3) at its growing tip. The terminal bud secretes proteinaceous material to form a **creeping tube** as it advances over the substrate. **Growth increments** of the creeping tube are deposited in regular annular bands, and each band is composed of two equal parts or half segments (Figure 19.3). (See Figures 19.8 and 19.9 for comparable growth increments in graptolites.) These half segments are formed alternatively to the left and right, resulting in a characteristic zigzag pattern developed at the junction where the half segments overlap and join.

Feeding zooids (Figure 19.3) bud from the stolon housed in the creeping tube behind the terminal bud to form a linear series. As each zooid develops, it becomes sealed off from the soft growing end of the stolon by a partition in the creeping tube. As this occurs, the stolon in the creeping tube becomes black in appearance (Figure 19.3) as a result of the development of a hard, black rind or hull that forms to encase the stolon. The black aspect of the stolon, from which feeding zooids developed, has given rise to the appellation **black stolon** for the stolonal system in the creeping tube. The black stolon is housed within and commonly is located at the bottom of the creeping tube.

After each zooid has budded from the stolon in the creeping tube, it resorbs a pore in the proteinaceous material of the creeping tube and grows through it. As the new zooid grows away from the creeping tube, it develops its tubular outer covering of proteinaceous material. Glands on the cephalic shield and cilia on the shield and tentacles appear to be involved in secretion of the tubular outer covering of the zooid. The outer covering, which may be called the **zooidal tube** (Figures 19.2 and 19.3), has annular growth increments that are complete rings. A single oblique suture line marks the beginning and end of formation of each ring. Growth of the zooidal tube away from the creeping tube results in a relatively elongate tubular structure that is connected to and is growing or developing at an angle to the creeping tube (Figure 19.3). The creeping tube always is closely pressed to the substrate. Each zooid inside its individual zooidal tube has its own stalk or stolon by which it is attached to the black stolon encased inside the creeping tube. The animal may pull itself up to the lip of its zooidal tube using its cephalic shield.

Zooidal tube walls have been examined with electron microscopes to determine their actual structure. They are formed from a meshwork of randomly oriented fibers in an electron lucent matrix. Certain fibers are themselves composed of two fine fibrils woven together in the form of a double helix. Other fibers do not seem to have any finer structure.

Rhabdopleuran zooids may degenerate following sexual reproduction or when unfavorable conditions occur. Dormant buds have been found in some colonies. These buds may permit the colony to regenerate after sustaining severe environmental stress, during which the colony degenerates.

Morphology of graptolites

When a graptolite is first observed, little more than the general shape of the skeletal material secreted by the whole colony may be noted. Some graptolites are so flattened or such vague imprints that general shape is about all that may be recognized. In most graptolites, however, tiny, sawtoothlike denticles may be seen as well (Figure 19.4, *B*).

Members of the graptolite colony secreted a hard part or skeleton, the **rhabdosome** (Figures 19.5 and 19.6). Rhabdosomes originate in a minute, conical **sicula** (Figures 19.7 and 19.8), which presumably housed an individual, a zooid, produced by sexual reproduction. One or more linear series of interconnecting or overlapping tubular structures, the **thecae**, developed from the sicula (Figures 19.5 to 19.7). Each linear series of connected thecae forms a **branch**, or **stipe** (Figure 19.5). Some colonies have many stipes; others have only one. Each theca is presumed to have housed an individual zooid, a member of the colony. The sicula apparently gave rise asexually to one or more zooids. These zooids in turn gave rise to one or more zooids.

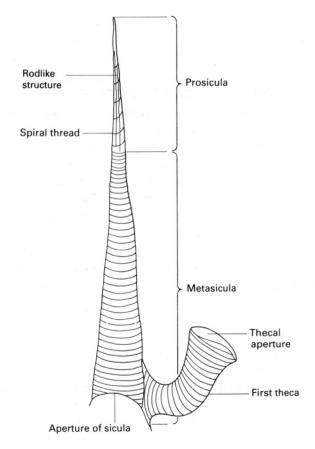

Figure 19.7. Sicula and first theca of a planktic graptolite rhabdosome showing major characters of sicula and first theca.

Siculae have two distinct parts (Figures 19.7 and 19.8). The apical part of the conical sicula, the **prosicula**, is formed from thin membranous material that commonly includes a threadlike structure, the **spiral thread**, that

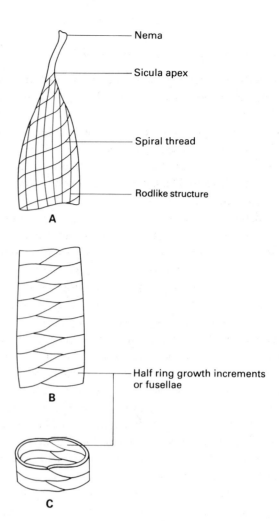

Figure 19.8. A, Prosicula showing spiral thread and rodlike fibers in periderm and the nema. **B,** Section of the metasicula showing overlap of the half ring growth increments of the periderm. **C,** An oblique view into a part of the metasicula showing half ring growth increments or fusellae in the fusellar tissue of the periderm.

spirals down the length of the cone from its apex. Some **rodlike structures** oriented normal to the cone aperture may be present. The apertural part of the sicula, the **metasicula**, is formed of discrete growth increments called **fusellae** that have flattened or beveled edges and appear to have a half ring shape (Figure 19.8, *C*). The edges of the half ringshaped elements overlap to some extent. The pattern formed where they overlap has a zigzag shape or aspect (Figure 19.8, *C*). Fusellae also occur in the walls of thecae throughout rhabdosomes.

A slender tubular structure, the **nema**, extends from the sicula apex (Figures 19.5 and 19.8, *A*). The nema may be termed the **virgula** (Figures 19.6 and 19.9) if it is located within the colony (such colonies are formed of two stipes arrayed back to back or side by side) or within the wall of the rhabdosome. The nema or virgula seems to be hollow in most graptolites.

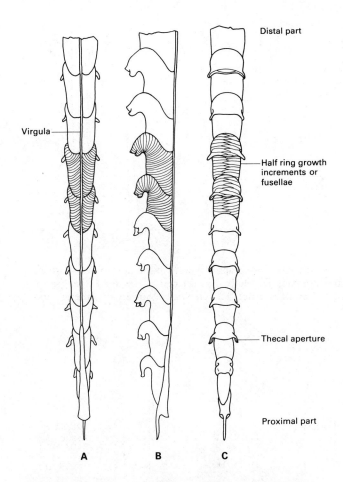

Thecae seen in any one colony may have different sizes and may exhibit different shapes. Many graptolite colonies are characterized by having thecae of two distinct sizes. The larger thecae are the **autothecae**, and the smaller thecae are called **bithecae** (Figures 19.12, *B* and 19.14). In some colonies, shape differences may be expressed by those thecae nearer to the sicula (this is the **proximal** part (Figure 19.9) of the colony), which have different width-to-length ratios than those thecae more distant from the sicula (those in the **distal** part of the rhabdosome (Figure 19.9)). Thecae may range in relative width-to-length proportions throughout the length of the rhabdosome, and thecal walls may exhibit markedly different degrees of curvature in different parts of the rhabdosome. Thecae also may range markedly in shape among graptolite genera, and these differences may be used in the identification of different genera and species (Figure 19.10).

Thecal **apertures** also may range widely in shape (Figures 19.10 and 19.11); they may be circular to constricted. Some have adjacent spines, hoods, or flanges (Figure 19.9 to 19.11). Apertures in some rhabdosomes open outward close to the outside margin, whereas apertures in other rhabdosomes open into small depressions or excavations situated away from the margin (Figure 19.10).

Graptolite rhabdosomes composed of four or fewer stipes commonly are oriented such that the sicula apex points upward. Using that orientation, the rhabdosome is termed **pendent** (Figures 19.4, *B* and 19.5, *A*) if the stipes are directed downward and the thecal apertures essentially face each other. Those rhabdosomes in which the stipes are developed normal to the sicula are termed **horizontal** (Figure 19.5, *C*). The rhabdosome is said to be

Figure 19.9. Uniserial scandent planktic graptolite *Monograptus* showing position of the virgula and shape of the thecae. **A,** View looking at the dorsal or back side of the rhabdosome illustrating the position of the virgula inside the periderm and nature of the half ring growth segments. **B,** A side view showing thecal shape and aspect of half ring growth segments. **C,** View looking at the thecal apertures or the ventral side of the rhabdosome showing aspects of overlap of the half ring growth segments.

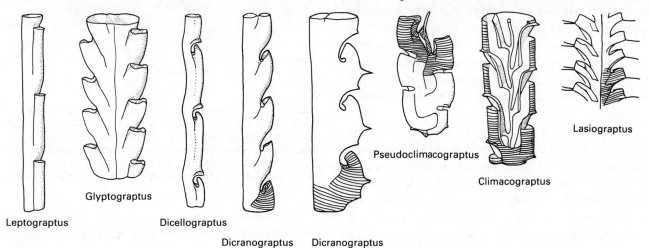

Figure 19.10. Some variations in thecal shape among some of the planktic graptolites, the graptoloids. (From Bulman, O. M. B. In: Teichert, C., editor. *Treatise on invertebrate paleontology, part V (revised), Graptolithina.* Boulder, CO and Lawrence, KS: Geological Society of America and University of Kansas Press; 1970.)

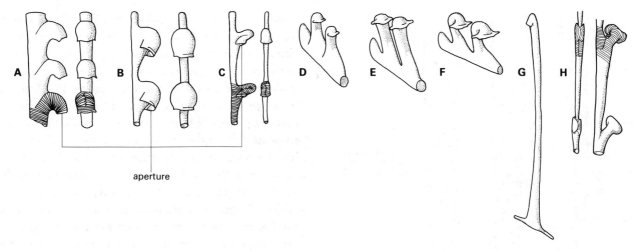

aperture

Figure 19.11. Variations in thecal and apertural form and shape in different monograptid type planktic graptolites. **A** to **E**, *Monograptus*. **F**, *Oktavites*. **G**, *Rastrites*. **H**, *Cucullograptus*. (From Bulman, O. M. B. In: Teichert, C. editor.) *Treatise on invertebrate paleontology, part V (revised), Graptolithina*. Boulder, CO and Lawrence, KS: Geological Society of America and University of Kansas Press; 1970.)

reclined if the stipes are directed upward but do not touch (Figure 19.5, *B*). The rhabdosome is termed **scandent** (Figures 19.6 and 19.9) if the stipes are oriented upward and the virgula is enclosed. Some scandent rhabdosomes are formed from two stipes that touch either side by side or back to back. These rhabdosomes are described as being **biserial scandent** (referring to the two rows of thecae) (Figure 19.6). A single series of thecae oriented upward from the sicula may be described as **uniserial scandent** (Figure 19.9). Triserial and even quadriserial (three and four rows of thecae in contact) arrangements of thecal rows rarely occur among the graptolites.

The preserved filmy material that comprises the fossilized graptolite rhabdosome is the **periderm**. Graptolite periderm has been examined closely, using both light and electron microscopes, in the hope that some aspects of graptolite peridermal structure might indicate potential relationships between graptolites and living organisms. Electron microscope studies have suggested that graptolite periderm is composed of fibrils, some of which are similar to collagen fibrils known in a number of animal groups.

The periderm from all thecae as well as that from the metasicula part of the sicula is formed from two basic tissue layers (Figure 19.12). The inner of the two layers is the **fusellar tissue**. The half ring growth increments, the fusellae, seen in the metasicula and in all thecae are formed from fusellar tissue. The outer layer of tissue is the **cortical tissue**. Cortical tissue has the superficial appearance of being formed from numbers of bandage like structures, actually called **bandages**, piled on top of each other (Figure 19.13).

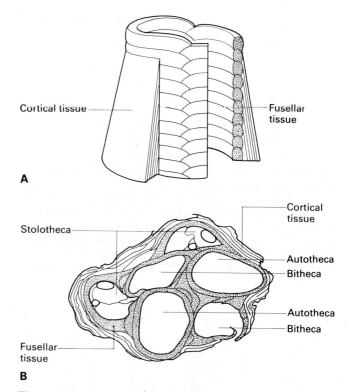

Cortical tissue — — Fusellar tissue

A

Stolotheca —

Cortical tissue

Fusellar tissue —

Autotheca
Bitheca

Autotheca
Bitheca

B

Figure 19.12. Portions of thecal periderm showing positions of cortical and fusellar tissue and the fusellae or half ring growth increments of fusellar tissue. (From Bulman, O. M. B. In: Teichert, C., editor. *Treatise on invertebrate paleontology, part V (revised), Graptolithina*. Boulder, CO and Lawrence, KS: Geological Society of America and University of Kansas Press; 1970.)

Electron microscope examination of fusellar and cortical tissues has shown both to be formed from two or more major components called fabrics. A common fabric in cortical tissue is composed of relatively long fibrous elements or fibrils clustered in parallel rows to form the

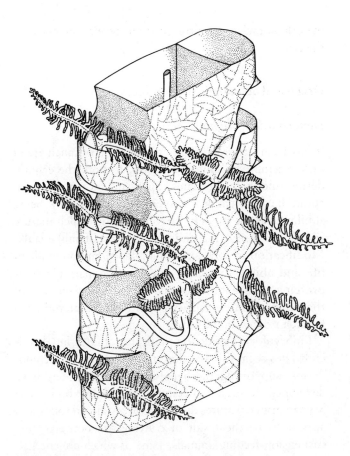

Figure 19.13. Part of a biserial scandent rhabdosome of *Climacograptus* showing cortical tissue bandagelike structures and a hypothetical reconstruction of graptoloid zooids. (From Crowther, P.; Rickards, B. *Geologica et Paleontologica* 11: 16; 1977.)

bandages. When layered one on top of another, bandages give the cortical tissue a laminated aspect. Cortical tissue bandages may extend across several half-ring growth increments, and they may be distributed randomly over the outer part of the periderm (Figure 19.13).

Those graptolites that were planktic in life have relatively fewer sets of bandages in the cortical tissue than do those graptolites that were benthic. Cortical tissue in graptolites that were members of the benthos may be relatively thicker than the fusellar tissue, whereas cortical tissue may be somewhat thinner than fusellar tissue in most planktic graptolites. The bandages formed of long fibers appear to have given the rhabdosomes a relatively high degree of flexibility and strength.

Classification

The phylum Hemichordata includes two classes with living representatives and, tentatively, one extinct class.

CLASS ENTEROPNEUSTA (Figure 19.1)

Solitary, soft-bodied hemichordates popularly called acorn worms with elongate wormlike bodies without proteinaceous outer covering. Commonly 1 to 50 cm in length, longest exceeds 2 m. Elongate proboscis connected to collar by narrow stalk; collar short and cylindrical, overlapping proboscis stalk. Funnel-shaped mouth near proboscis stalk; no arms. Long trunk behind collar, contains 10 to 80 pierced gill slits. Alimentary canal straight, extends entire length of trunk to anus. (Holocene; about 12 genera and 70 species known.)

Acorn worms live in shallow marine environments burrowed in mud, although a few live under stones or shells. Burrowers commonly form U-shaped tubular structures, similar to those of some marine annelids, with openings at each end. Acorn worms feed by taking in water, sediment, and included food particles (organic detritus or minute organisms). The food particles are extracted and wastes are excreted in the form of ropelike castings. One can see these castings on tidal flats in many parts of the world.

Many acorn worms are capable of a certain degree of regeneration. In these species, cutting the animal in two parts across the body results in regeneration of the front part by the rear part. Front parts, however, will not regenerate missing rear parts.

Although commonly seen in modern marine environments, enteropneusts have no clearly demonstrable fossil record. Certain piles of castings or tubular burrows found in rocks, particularly mudstones, formed in nearshore marine environments could be fossilized remains of enteropneust activity.

CLASS PTEROBRANCHIA (Figures 19.2; 19.3)

Hemichordates in which zooids occur attached to each other in colonies or unattached in aggregates; secrete proteinaceous, tubular outer coverings; are 1 to 3 mm long (colonies or aggregates may be several centimeters in diameter); develop shieldlike discoidal proboscis, two to nine tentacle-bearing, hollow, curved arms, and rotund saclike bodies. Alimentary canal U-shaped, gill slits two or none. (Cambrian to Holocene; 3 genera and 20 species known.)

Colonies or aggregates commonly encrust shells, rocks, or other firm substrates. Most are found in relatively deep cold water in the Southern Hemisphere. In the Northern Hemisphere, they have been dredged from Norwegian fjords and the North Sea as well as from Bermuda.

Fossil pterobranchs are rare. They have been recorded from rocks of the Cambrian, Ordovician, Silurian, Jurassic, Cretaceous, and the Eocene. Fossil species clearly resemble modern species in growth habit and gross aspects of shape and structure of tubular outer coverings.

CLASS GRAPTOLITHINA (Figure 19.4)

Tentatively considered hemichordates; colonies of one to many branches or matlike encrustations; zooids commonly in linear series or in series of clusters, some in irregular aggregates. Branches may be about 5 to 10 cm long and 0.5 to 2 mm wide. Skeletal parts apparently proteinaceous in composition in life. Zooids connected to each other by stolons. (Cambrian to Pennsylvanian; approximately 240 genera and 1800 species.)

Graptolites appear to have exploited many different benthic habitats. In addition, the major group was planktic. The members of the planktic group were by far more numerous and diverse than any of the benthic forms. The planktic forms commonly are the only fossils found in the **graptolitic biofacies**, which includes primarily dark, organic-rich shales and mudstones that developed under anoxic conditions in which benthic organisms could not live. These graptolite-bearing black shales and mudstones are especially well developed in rock successions that formed from sediment deposited on the outer parts and slopes of platforms or shelves of the Ordovician, Silurian, and Early Devonian.

A few graptolites occur in rocks developed from sediment deposited in shallow- to moderate-depth shelf sea or platform environments. Bottom waters in these environments were oxygenated enough to permit benthic life to flourish. Such sediments commonly bear the remains of brachiopods, bryozoans, crinoids, nautiloids, trilobites, and other benthic marine invertebrates, and they are therefore designated the **shelly fossil biofacies**. Such rocks generally are light-colored carbonates or fine- to medium-grained terrigenous rocks. The lithologic as well as biologic aspect of these rocks contrasts sharply with that of the graptolite-bearing black shales and mudstones. Markedly fewer graptolite taxa occur in the shelly fossil biofacies than have been found in the graptolitic biofacies. A few of the same graptolites occur in both biofacies, permitting certain correlations between the two.

Origin and taxonomic relationships

Hemichordate origins

It has been suggested that a modern pterobranch zooid may resemble an animal ancestral to both the echinoderms and hemichordates. Such an ancestral animal could have resembled a pterobranch, possibly being similar to those modern pterobranchs in which mature zooids are free living and have a short stemlike stalk. Modification of such an animal, depending on mode of life and ability to secrete calcium carbonate plates or proteinaceous material, could have led to development of the echinoderms on the one hand and to pterobranch hemichordates on the other.

Embryologic evidence indicates that the hemichordates seem to have closest affinity with the echinoderms, supporting the hypothesis that they may have developed from an ancestor common to both. The bryozoan-like features displayed by certain pterobranchs appear to be the result of convergence between tiny, suspension-feeding animals, some of which are encrusting in mode of life.

Chordate affinities of the hemichordates are suggested by the presence of gill slits and the dorsal nerve cord located in the acorn worm collar (Figure 19.1), which is somewhat similar to the hollow dorsal nerve cord in chordates. Absence of a notochord and many differences in body structures indicate that the hemichordates should not be included in the phylum Chordata.

Graptolite origins   To what living organisms could graptolites possibly be related? The answer to that question has perplexed paleontologists for more than two centuries. Indeed, graptolites have held a certain fascination for students of fossils because their potential relationships to modern organisms are so obscure. The clues to graptolite relationships and their origins are faint because graptolites are the remains of long extinct organisms, and all that is available from them is certain hard skeletal materials. The elements of the search for graptolite relationships exemplify the painstaking, exhaustive paths of inquiry that the paleontologist must follow in an attempt to answer a potentially unanswerable question.

Since graptolite morphology and peridermal structure are the only evidence available for close examination and comparison with features of modern organisms, details of graptolite morphology and peridermal structure have been described and will be discussed more thoroughly than might seem necessary. However, these details are at the heart of the quest for understanding graptolite origins and relationships. The ensuing discussion will suggest a method by which fossil objects, which only faintly resemble objects from the domain of living organisms, may be analyzed for their relationships.

The graptolites with encrusting modes of life have yielded fossil remains that are more similar to the hard parts secreted by colonial pterobranchs, the rhabdopleurans, than they are to hard parts of any other known organism. The creeping tube portion of the rhabdopleuran colony (Figure 19.3) is formed from half ring growth segments of a relatively thin proteinaceous material. Housed within the creeping tube and commonly embedded in the lower wall of the creeping tube is the black stolon from which the zooids bud. The creeping tube portions of some encrusting graptolites are virtually identical in the characters of both the creeping tube and the enclosed black stolon. Many of the encrusting graptolites have a black stolon housed within a tubular outer cover as do certain of the highly branched graptolites

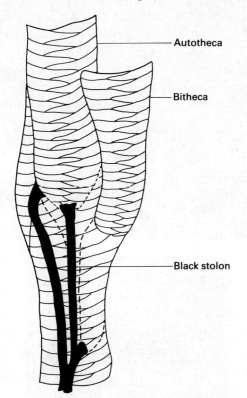

Figure 19.14. Portion of the dendroid *Dendrograptus* showing an autotheca and bitheca and the black stolon inside the thecal periderm. (From Bulman, O. M. B. In: Teichert, C., editor. *Treatise on invertebrate paleontology, part V (revised), Graptolithina.* Boulder, CO and Lawrence, KS: Geological Society of America and University of Kansas Press; 1970.)

(Figure 19.14). The black stolons among the graptolites not only appear similar; they also seem to have had comparable functions as the stolons in the colonial pterobranchs, the rhabdopleurans. The black stolon and its position within an outer tube are the most marked similarities between the graptolites and the rhabdopleurans. They constitute the most compelling evidence for suggesting that they were related.

The general appearance of the bandages (Figure 19.13) formed from clusters of long fibers in the cortical tissue of graptolite periderm suggests that they could have been 'painted' on by some organ or organs. Bandages have been found lining the inside surfaces of thecal walls of some graptolites, suggesting that zooids had the ability to secrete fibrous, bandagelike structures both inside and outside the thecae. The addition of bandages could have been used for added wall strength as well as for repair of wall damage.

How were the bandages formed? Study of the activities of modern *Rhabdopleura* zooids (Figure 19.3) suggests a potential answer to this question. *Rhabdopleura* zooids have certain cilia on both their tentacles and cephalic shields that have paddle-shaped tips. It seems likely that these cilia are involved in secretion of the zooidal tube. Zooids have been observed extending partly outside of the tubular covering to secrete peridermal material on the outside of the tube. The cilia on the cephalic shield and tentacles with the paddle-shaped tips have an exceptionally sticky adhesive material on them that apparently contributes to the periderm. Although the structure of the peridermal materials in graptolites and rhabdopleurans differs, the graptolite zooids may have functioned in a manner similar to rhabdopleuran zooids in that they both climbed part way out of their zooidal tubes to 'paint' fibrous fabric on the outside of the tube (Figure 19.13). The bandages secreted by the graptolite zooids could have been deposited in a sweep of tentacles and cephalic shield. Successive sweeps could have resulted in the several sets of bandages seen in cortical tissue of some graptolite periderm.

The morphologic similarities between certain encrusting graptolites and the colonial pterobranch *Rhabdopleura* suggest that the graptolites are more closely related to colonial pterobranchs than to any other known extant animal. Graptolite zooids could have resembled rhabdopleuran zooids and could have carried out a part of the secretion of the zooidal tube in a similar manner. The morphologic similarities between encrusting graptolites and the hard parts of rhabdopleurans, however, are the primary reasons for tentatively grouping graptolites with the pterobranchs in the phylum Hemichordata.

Part II Additional concepts

Classification of graptolites

Graptolites may be divided into three generalized types based on apparent mode of life. The majority of the graptolites that have been found seem to have been planktic because they have widespread distribution and commonly occur in dark, organically rich rocks deposited in anoxic environments. Those graptolites with a bushlike aspect (Figures 19.4, *A*; 19.15; and 19.16) appear to have been benthic and relatively similar to certain seaweed in growth form. Other graptolites appear to have been encrusting, based on their skeletal structures.

Using characters such as different sizes of thecae and the presence or absence of a hard, black stolonal system, graptolites may be classified into at least the following six orders: Dendroidea, Camaroidea, Crustoidea, Stolonoidea, Tuboidea, and Graptoloidea. Most dendroids apparently have been sessile benthic colonies with conical or bushlike habits. They seem to have lived in relatively shallow waters with other shallow-water marine organisms. The camaroids, crustoids, stolonoids, and tuboids are minor orders of mostly encrusting graptolites known primarily from relatively fragmentary specimens recovered from the Tremadoc (latest Cambrian to earliest Ordovician) cherts in Poland. A few specimens, mostly impressions, recovered from other localities have been assigned with some reservation to one or another of these orders. The graptoloids were planktic and are the major order of graptolites.

ORDER DENDROIDEA (Figures 19.4, *A*; 19.14 to 19.16)

Attached, branching graptolites characterized by two sizes of thecae, the larger autothecae and smaller bithecae. Rhabdosomes commonly with many branches; branches in some connected by rodlike structures formed from clusters of thecae. Autothecae and bithecae linked by stolonal system; in

Figure 19.15. Fragment of the dendroid *Dictyonema* rhabdosome showing thickened basal part of rhabdosome and branching pattern.

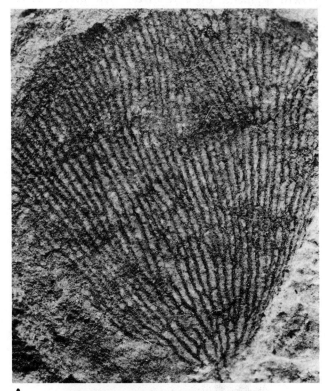

Figure 19.16. Dendroid *Dictyonema*. **A,** Specimen shows branching pattern of stipes. **B,** Specimen illustrates thickened stemlike structure at the initial part of the colony.

most species, stolons included hard, black, organic substance, probably proteinaceous; stolons housed in periderm (Figure 19.14). (Middle Cambrian to Pennsylvanian; about 30 genera.)

Several hundred thecae may be present on each stipe. Thecae tend to be essentially tubular, and thecal apertures, particularly those of autothecae, may bear spines or flanges.

That part of an autotheca and bitheca housing the stolon has been called the stolotheca. The designation is confusing because the stolotheca did not house a third type of zooid. It may have housed an immature zooid of the type that secreted the autotheca once it became mature. The term stolotheca commonly is limited to the initial part of the autotheca, that part enclosing the stolon.

The area about the sicula apex is thickened in most dendroid colonies, and in some, the thickened area resembles a stem. Most dendroid colonies are thought to have been attached to the sea floor by the stemlike material that is at or projects from the sicula apex (Figure 19.16, *B*).

Rhabdosomes that have been preserved only partly crushed or filled with sediment indicate the possibility that some colonies were conical in life, others were fanlike, and still others were shrublike. Local environmental conditions apparently have exerted some influence on colony shape and size.

Dendroid graptolites appear in rocks of about the mid-Cambrian and range sporadically into strata of the Pennsylvanian. Some genera range through much of the mid-Cambrian to Mississippian interval with little structural change.

ORDER CAMAROIDEA (Figure 19.17)

Encrusting graptolites that include both autothecae and bithecae and hard, black stolonal systems. Characterized by distinctive autothecae differentiated into two parts: inflated, bulbous structure, apparently encrusting part of the colony; and small, erect tubular part arising from bulbous part. Bithecae small, tubular, distributed irregularly over mat of bulbous parts of autothecae and encrusting tissue of colonies. Thecal apertures unadorned, circular openings. (Latest Cambrian to Early Ordovician; about 5 genera.)

Most genera of the Camaroidea are known only from rocks of the Tremadoc. Most of the recorded specimens are from one locality in Poland.

ORDER CRUSTOIDEA (Figure 19.18)

Encrusting graptolites similar to camaroids except autothecae not so closely spaced, unique autothecal apertures complexly infolded and borne on short but distinct neck. (Middle Ordovician to Late Silurian; about 7 genera.)

All crustoids are known only from fragmentary specimens, most of which are from glacial erratic boulders in Poland. Nearly all specimens are thought to be from the mid-Ordovician. A few come from rocks thought to be of the Late Ordovician and Late Silurian.

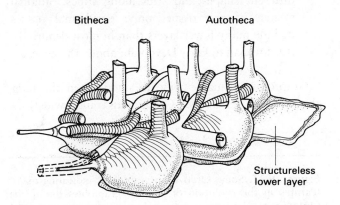

Figure 19.17. Diagrammatic reconstruction of a camaroid based on *Bithecocamara*. (From Bulman, O. M. B. In: Teichert, C., editor. *Treatise on invertebrate paleontology, part V (revised), Graptolithina*. Boulder, CO and Lawrence, KS: Geological Society of America and University of Kansas Press; 1970.)

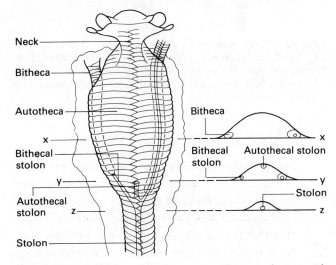

Figure 19.18. Diagrammatic restoration of a part of a crustoid rhabdosome, based on *Bulmanicrusta*. Transverse sections at *x, y,* and *z.* (From Bulman, O. M. B. In: Teichert, C., editor. *Treatise on invertebrate paleontology, part V (revised), Graptolithina*. Boulder, CO and Lawrence, KS: Geological Society of America and University of Kansas Press; 1970.)

ORDER STOLONOIDEA (Figure 19.19)

Sessile or encrusting graptolites characterized by exceptional development of stolons, divided irregularly into groups or clusters. (Latest Cambrian, Ordovician(?); possibly 2 genera.)

One or often several stolons may be encased by normal periderm with its half ring growth increments. Parts of the peridermal material lack the half ring growth increments, leading to the supposition that stolonoids were likely encrusting in life. Stolonoids are known only from fragmentary specimens, most of which are from Tremadoc rocks in Poland. Possible stolonoids have been found in Early and Late Ordovician rocks.

Figure 19.19. Stolonoid *Stolonodendrum*. (From Bulman, O. M. B. In: Teichert, C., editor. *Treatise on invertebrate paleontology, part V (revised), Graptolithina.* Boulder, CO and Lawrence, KS: Geological Society of America and University of Kansas Press; 1970.)

ORDER TUBOIDEA (Figure 19.20)

Sessile, benthic, mostly encrusting graptolites; many characterized by autothecae and bithecae clustered into bundles, budding relatively irregular. Many with discoidal bases formed either from part of autotheca or from both autothecae and bithecae; erect tubular parts of thecae from discoidal bases of autothecae or both autothecae and bithecae; small, conical structures termed conothecae with an opening at the apex in several tuboid genera. No stolons found. (Latest Cambrian to Silurian; about 13 genera.)

Since most tuboid colonies seem to have had a basal discoidal structure that could have adhered to the substrate, most are considered to have been encrusting during life. Although much of the colony appears to have formed an essentially matlike structure, either parts of single autothecae or, relatively rarely, parts of autothecae and bithecae developed from the basal matlike structure. Many sets of closely clustered thecae arise from the basal structure in some colonies. Five of the genera are known only from Tremadoc rocks in Poland. Most tuboids are either fragmentary specimens or carbonaceous imprints.

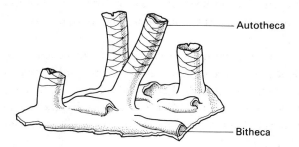

Figure 19.20. Diagrammatic reconstruction of a tuboid, based on *Idiotubus*, showing erect portions of autothecae arising from basal, matlike, encrusting portion of colony that is formed from basal parts of authothecae and the bithecae.

ORDER GRAPTOLOIDEA (Figures 19.4, *B*; 19.5 to 19.11; 19.13; 19.21; 19.22)

Planktic graptolites characterized by one type theca equivalent to autotheca of other graptolites, on few stipes. In some colonies, autothecae of progressively different shapes and sizes along stipes. Stolonal system of soft tissues only; peridermal tissues include many fewer layers than in most dendroids. (Ordovician to Early Devonian; about 185 genera.)

Graptoloids usually come to mind when the term graptolite is designated. Graptoloids range widely in

Figure 19.21. Some Ordovician graptolites indicating major features in the evolution of planktic graptolites during the Ordovician. The names and numbers on the left side of the diagram refer to a succession of zones used in western North America developed from analysis of the overlapping stratigraphic ranges of many species obtained from several stratigraphic sections in a number of areas in western North America.

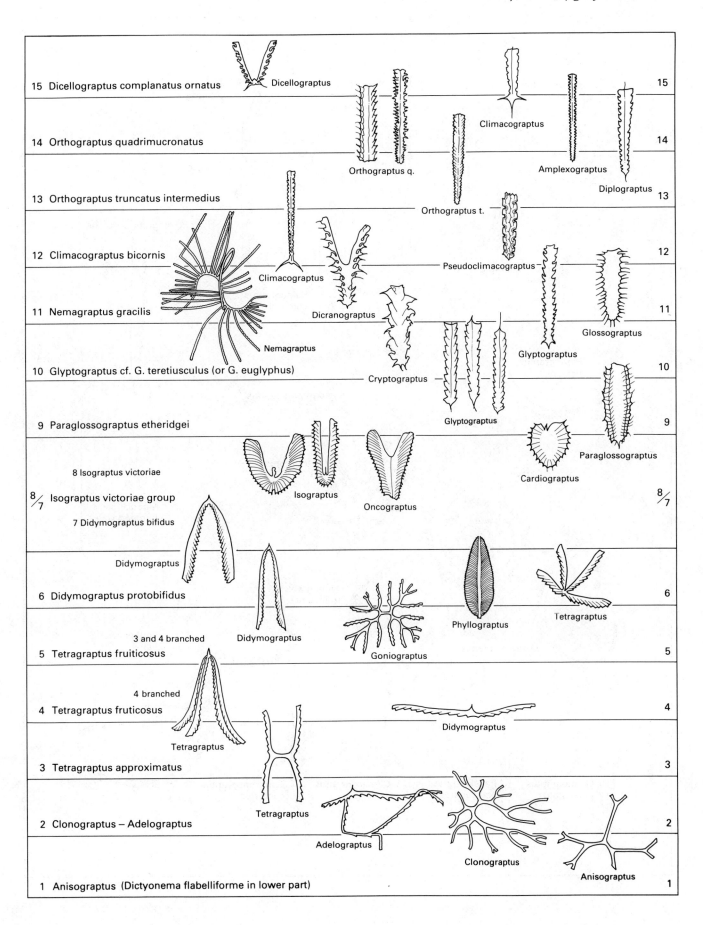

15 Dicellograptus complanatus ornatus — Dicellograptus — 15

14 Orthograptus quadrimucronatus — Orthograptus q. — Climacograptus — Amplexograptus — Diplograptus — 14

13 Orthograptus truncatus intermedius — Orthograptus t. — 13

12 Climacograptus bicornis — Climacograptus — Pseudoclimacograptus — 12

11 Nemagraptus gracilis — Dicranograptus — Nemagraptus — Glyptograptus — Glossograptus — 11

10 Glyptograptus cf. G. teretiusculus (or G. euglyphus) — Cryptograptus — Glyptograptus — 10

9 Paraglossograptus etheridgei — Cardiograptus — Paraglossograptus — 9

8 Isograptus victoriae

8/7 Isograptus victoriae group — Isograptus — Oncograptus — 8/7

7 Didymograptus bifidus

Didymograptus

6 Didymograptus protobifidus — Didymograptus — Goniograptus — Phyllograptus — Tetragraptus — 6

3 and 4 branched

5 Tetragraptus fruiticosus — Didymograptus — 5

4 branched

4 Tetragraptus fruticosus — Tetragraptus — Didymograptus — 4

3 Tetragraptus approximatus — Tetragraptus — 3

2 Clonograptus – Adelograptus — Tetragraptus — Adelograptus — Clonograptus — Anisograptus — 2

1 Anisograptus (Dictyonema flabelliforme in lower part) — 1

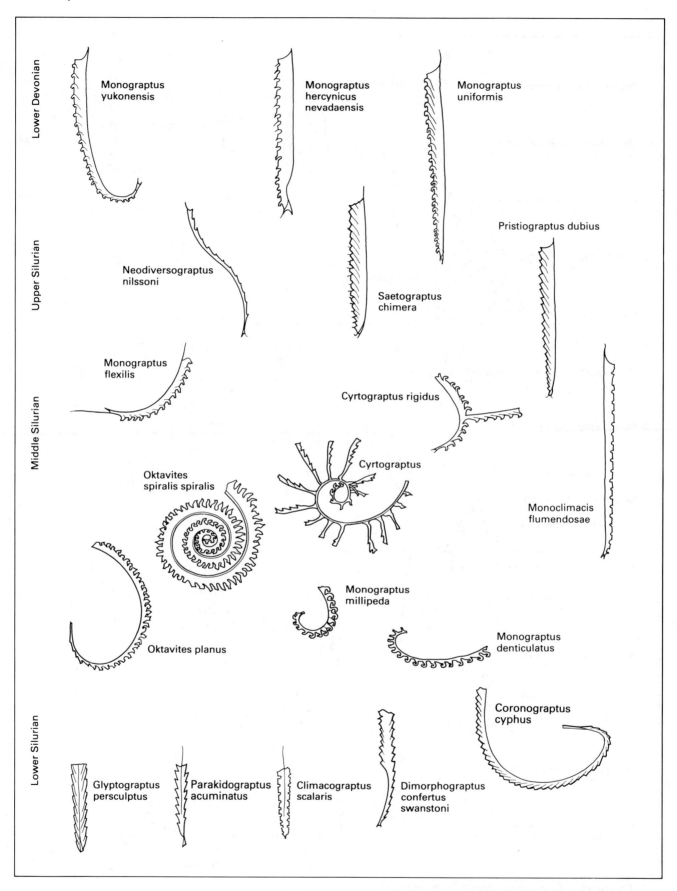

Lower Devonian

Monograptus
yukonensis

Monograptus
hercynicus
nevadaensis

Monograptus
uniformis

Upper Silurian

Neodiversograptus
nilssoni

Saetograptus
chimera

Pristiograptus dubius

Middle Silurian

Monograptus
flexilis

Cyrtograptus rigidus

Cyrtograptus

Oktavites
spiralis spiralis

Monoclimacis
flumendosae

Monograptus
millipeda

Oktavites planus

Monograptus
denticulatus

Coronograptus
cyphus

Lower Silurian

Glyptograptus
persculptus

Parakidograptus
acuminatus

Climacograptus
scalaris

Dimorphograptus
confertus
swanstoni

shape and size. Some colonies are formed from a simple linear series of thecae (Figure 19.9); others have several stipes. Some are tightly coiled (Figure 19.22); others have tuning fork (Figures 19.4, *B* and 19.5, *A*) or petallike shapes (Figure 19.21). Most colonies are a few centimeters in length, but some achieve lengths of more than a meter.

Graptoloids range from Early Ordovician through Early Devonian (Figures 19.23, and 19.24). Although classification of the graptoloids is viewed differently by different students of the subject, about 180 to 230 genera may be distinguished. Many more generic names have been proposed, some of which may become stabilized with time.

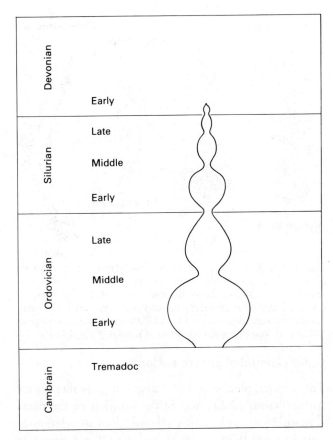

Figure 19.24. Diagrammatic indication of the general features of planktic graptolite history. Width of the figure suggests the relative number of graptoloid taxa at any time in relation to the number at other times.

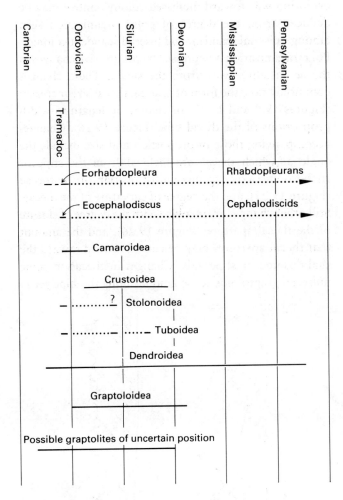

Figure 19.23. Stratigraphic range chart indicating the geologic age range of rhabdopleurans, cephalodiscids, and graptolites.

Figure 19.22. Some Silurian and Early Devonian graptoloids indicating certain major features in planktic graptolite development in that time span. It may be noted that, by agreement of the International Stratigraphic Commission, the first occurrences of *Parakidograptus acuminatus* and *Monograptus uniformis* are taken as the bases of the Silurian and Devonian Systems, respectively. Members of the *Monograptus yukonensis* group are among the youngest graptoloids known.

Problematic graptolites

A number of fossil objects have been recorded from early Paleozoic rocks that have been assigned to the graptolites with reservation. Nearly all these objects are highly carbonized films, revealing little more than a stemlike structure bearing tubular protuberances that might be considered potential thecae. Certain of these structures may be fossilized hydroids or even pterobranchs. Others might have been algal. Some of these objects found in Siberia (Figure 19.25) have attracted attention because they were found in Cambrian rocks. These particular fossils appear to have what may be autothecae. Potential half ring growth elements of the periderm have been recognized in some of them.

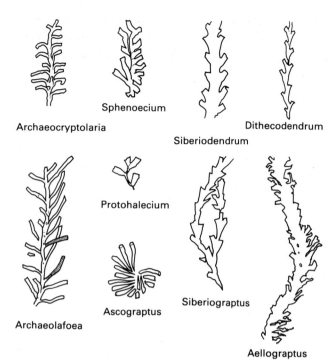

Figure 19.25. Certain Cambrian fossils from Asiatic Soviet Union that potentially may be graptolites of uncertain taxonomic position. (From Bulman, O. M. B. In: Teichert, C., editor. *Treatise on invertebrate paleontology, part V (revised), Graptolithina.* Boulder, CO and Lawrence, KS: Geological Society of America and University of Kansas Press; 1970.)

Colony control of growth and form

The stratigraphic record of changes in graptolite colony form (Figure 19.21) and of the position of the nema (Figure 19.5) and virgula (Figure 19.6) in relation to colony form (Figures 19.21 and 19.22) indicates that these structures, or at least tissues from which they formed, had some influence on colony growth and form. Tissues related to nema formation would have had little relationship to those related to zooidal budding. The virgula, a structure that is an extension of the nema in graptolites with biserial and uniserial scandent rhabdosomes, may have been formed by tissues that were somewhat more closely related to colony form or to colony function. The tissues from which the virgula in the biserial scandent rhabdosomes (Figure 19.6) formed would have been relatively close to and possibly even confluent with tissues involved in soft tissue stolon formation as zooids budded serially one after another. In the uniserial scandent rhabdosomes (Figure 19.9), the virgula lies within the peridermal wall of the rhabdosome, indicating that it may well have been even more closely related to zooidal development and secretion of thecal tubes.

The virgula may have had some relationship to colony function and control, perhaps by enhancing the physiologic connection among the zooids, inasmuch as the tissues connected with the virgula probably extended from the sicula directly. The trend in evolutionary history of the graptoloids may have been toward increased cooperation in colony growth and control of colony function. The uniserial and biserial scandent colonies had relatively long geologic age ranges, suggesting that if increased cooperation in colony function and cooperation among zooids did develop certain survival advantages would have existed for these tiny colonial animals.

Additional evidence that the trend in graptoloid evolution was toward increased colony control may be deduced from the degree of polar organization seen among uniserial scandent and biserial scandent colonies. Polar organization is expressed by progressive changes in thecae distally, away from the sicula. These changes commonly take the form of gradual thecal size increases (Figures 19.9 and 19.27) or changes in length-to-width proportions of the thecal tube (Figure 19.26). Changes accompanying those of proportion and size include the angle at which thecae are inclined from the stipe or rhabdosome axis, the amount of overlap of the thecae (Figure 19.27), and the degree of curvature of the thecae. Still other changes are involved with the degree of closure of the thecal apertures (Figure 19.26), and the amount that thecal apertures may be curved inward toward the rhabdosome or stipe axis. Changes in thecae in some uniserial graptoloids are so gradual along the stipe yet so

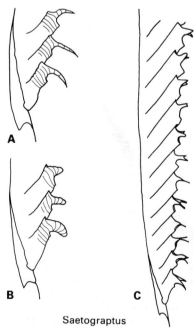

Figure 19.26. Uniserial scandent graptoloid *Saetograptus* showing spinosity of thecal apertures. (From Bulman, O. M. B. In: Teichert, C., editor. *Treatise on invertebrate paleontology, part V (revised), Graptolithina.* Boulder, CO and Lawrence, KS: Geological Society of America and University of Kansas Press; 1970.)

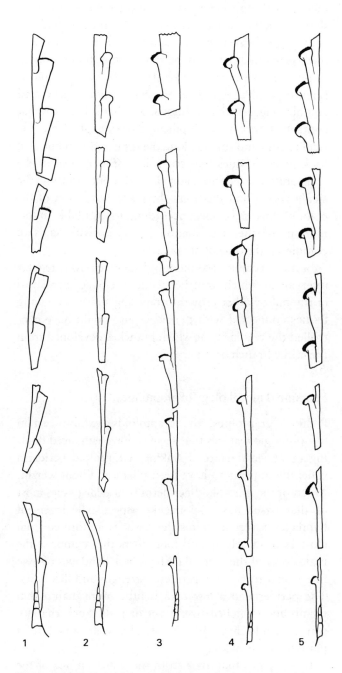

Figure 19.27. Uniserial scandent graptoloids showing changes in the amount of thecal overlap and thecal shape from proximal to distal ends of the rhabdosomes. All specimens are species of *Cucullograptus*. (From Bulman, O. M. B. In: Teichert, C., editor. *Treatise on invertebrate paleontology, part V (revised), Graptolithina*. Boulder, CO and Lawrence, KS: Geological Society of America and University of Kansas Press; 1970.)

Figure 19.28. Two uniserial scandent monograptid graptoloids showing changes in thecal form from the proximal to distal part of the rhabdosome. (From Bulman, O. M. B. In: Teichert, C., editor. *Treatise on invertebrate paleontology, part V (revised), Graptolithina*. Boulder, CO and Lawrence, KS: Geological Society of America and University of Kansas Press; 1970.)

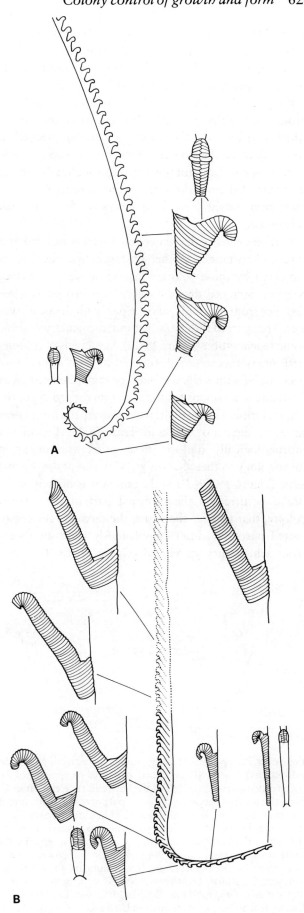

marked at opposite ends of the stipe that both a size and a structural gradient can be inferred (Figure 19.28).

The presence of these zooidal gradients suggests that zooidal development was controlled by the colony. Colony control could have been directed through the production of hormonal secretions from the sicula, which spread along the colony as budding proceeded. Not only could the stolonal connections between zooids have been pathways for hormonal spread, but the tissues that extended from the sicula, which related to virgula formation, might have been pathways for hormonal substances.

Further evidence that the colonies were controlled by a spread of hormones or other substances from the sicula is indicated by those colonies that show signs of having been so damaged that the colony was broken at some distance from the sicula, and the part with the sicula was lost. That part of the colony without the sicula may show some regeneration (Figure 19.29). Specimens of colonies with many thecae serially arrayed to form a stipe have been found with a regenerated stipe that developed away from and in an opposite direction from the growth of the original stipe. Thecae of the regenerated parts form linear series (to the left of breaks in Figure 19.29, *A* and *C*) of morphologically uniform thecae that are similar in morphology to thecae at the growing end of the original stipe (Figure 19.29, *B*) at the time it was broken. If the break occurred in the proximal part of the original colony, that is, near the sicula, the thecae in the regenerated part may differ conspicuously from the thecae from which they developed (Figure 19.29, *C*), since

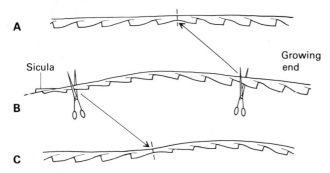

Figure 19.29. Features of rhabdosome regeneration. When the normal rhabdosome (**B**) is cut in its proximal part, thecae of the regenerated part (to the left of the line indicating the cut in **C**) have the form of thecae in the distal part of the rhabdosome at the time that breakage took place. When the normal rhabdosome is broken in the distal part, the regenerated thecae (to the left of the line indicating the cut in **A**) have essentially the same form as distal thecae of the original rhabdosome at the time that the break occurred. (From Bulman, O. M. B. In: Teichert, C., editor. *Treatise on invertebrate paleontology, part V (revised), Graptolithina.* Boulder, CO and Lawrence, KS: Geological Society of America and University of Kansas Press; 1970.)

thecae in the proximal part of most colonies differ in shape from thecae in the growing end of the colony. If the colony was broken in the distal part, then the thecae of the regenerated part may be similar morphologically to those from which they developed in the original colony (Figure 19.29, *A*). The evidence from regenerated colonies suggests that hormonal or other substances that controlled thecal development did emanate from the sicula zooid and spread along the stipe as thecal budding took place. Tissues that controlled development of the regenerated part of the colony lacked connection with the sicula zooid, and thecae in the regenerated part of the colony thus developed according to available information for thecal formation in the growing ends of colonies at the time of breakage.

Branched colonies commonly display relatively regular intervals at which branches were developed. Spread of hormonal or other growth-promoting substances along tissues connected with the sicula also could have controlled the regularity at which branches developed from the initial branch or stipe.

Functional morphology of graptoloids

Planktic graptolites, the graptoloids, exhibit certain morphologic features that seem to have enhanced floating or at least retarded sinking. Graptoloid periderm generally appears to have been pliant and light weight. Much of the graptoloid periderm is fusellar tissue; most of that tissue has a spongelike aspect. The layers of bandages covering the fusellar tissue were formed from clusters of closely spaced long fibers that comprise the major part of the cortical tissue. These bandages of long fibers tend to give the periderm strength and flexibility. The periderm in a few taxa is little more than a thin membrane stretched over a set of rods. Such rhabdosomes would have been especially well suited to a planktic mode of life.

Long spines that arise from the outer surface of the periderm in rhabdosomes of certain taxa and a loosely woven, meshlike structure (Figure 19.30) found outside the main periderm in a few taxa could have aided floating. Of potentially greater significance for floating were vanelike structures found at the tips of the nema in some immature colonies (Figure 19.31) and at the tips of the virgula of many different taxa, primarily those with biserial scandent rhabdosome form (Figure 19.32). Dissection of these bulblike and vanelike structures indicates that they could have been hollow. If these structures had been filled with gas, they would have been significant aids in floating. It is possible that a gas-filled sac or bulb of soft tissue was present but not preservable

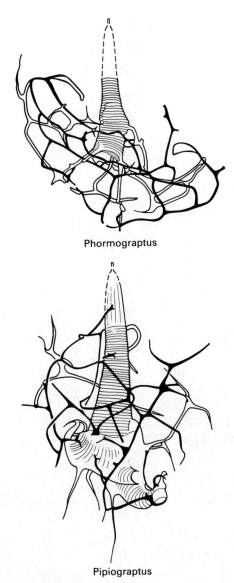

Phormograptus

Pipiograptus

Figure 19.30. Mesh that forms outside the main portion of the rhabdosome in certain graptoloids. (From Bulman, O. M. B. In: Teichert, C., editor. *Treatise on invertebrate paleontology, part V (revised), Graptolithina.* Boulder, CO and Lawrence KS: Geological Society of America and University of Kansas Press; 1970.)

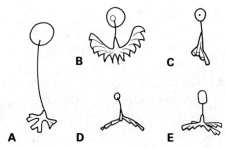

Figure 19.31. Discs at the tips of the nema in immature rhabdosomes. **A**, *Dictyonema.* **B** and **C**, *Tetragraptus.* **D**, *Adelograptus.* **E**, *Staurograptus.* (From Bulman, O. M. B. In: Teichert, C., editor. *Treatise on invertebrate paleontology, part V (revised), Graptolithina.* Boulder, CO and Lawrence, KS: Geological Society of America and University of Kansas Press; 1970.)

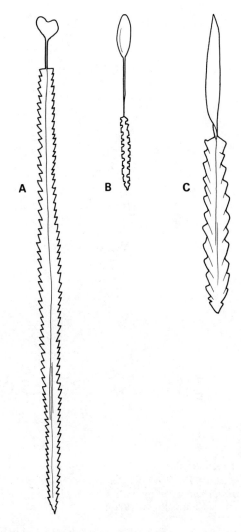

Figure 19.32. Discoidal or vanelike structures at the tips of the virgula in certain rhabdosomes. **A**, *Diplograptus.* **B**, *Climacograptus.* **C**, *Petalograptus.*

at the tips of the nema or virgula in nearly all graptoloids. The bulbs seen at the tips of the virgula in mature colonies potentially could have been used as vanes to control orientation of the colony as it was moved about by wind and water motion.

A few graptoloids have helicoidally coiled or spiral-shaped rhabdosomes (Figure 19.33). If the zooids in these colonies acted in concert, they might have generated enough activity to cause the whole colony to spin. Alternatively, such concerted activity might have pumped water past the zooids of a relatively stationary colony. Either spinning and resultant self-propulsion or a pumping action would have permitted the sampling of a somewhat greater parcel of water for food than would have been possible for colonies lacking these capabilities. Branches arise from the main stipe in certain spirally coiled graptoloids. Both the self-propulsion and the

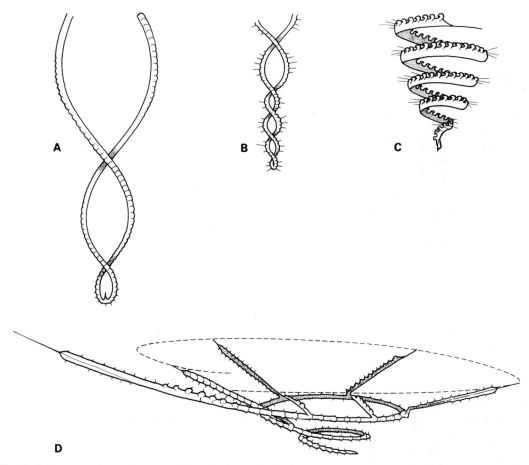

Figure 19.33. Rhabdosomes of certain graptoloids, the form of which suggests that they could have rotated spirally in life. **A**, *Dicellograptus*. **B**, *Dicranograptus*. **C**, *Monograptus*. **D**, *Cyrtograptus*. (From Bulman, O. M. B. In: Teichert, C., editor. *Treatise on invertebrate paleontology, part V (revised), Graptolithina*. Boulder, CO and Lawrence KS: Geological Society of America and University of Kansas Press; 1970.)

pumping activity of the colony could have been enhanced by the presence of more zooids in the colony and the resultant extra activity they provided.

Graptoloid zooids in many colonies possibly could have acted in concert to position the colony in the water column. At least some graptoloids may have experienced diurnal migrations similar to those of many modern zooplankton. Indeed, zooids in some graptoloid colonies may have functioned collectively similar to groups of zooids in certain siphonophore colonies to provide self-propulsion action for the colony. If so, certain graptoloids could have been nektic in life.

Thecal apertures in many graptoloids are spinose, hooded, or bear flanges, some of which may be highly curved. Thecae in many other graptoloid taxa are curved or bent such that apertures face inward toward the stipe (Figure 19.11). These variations in form presumably provided some protection for the zooids. Potentially, zooids remained within thecae having complexly infolded apertures. Presumably, they pumped nutrient-bearing water into the thecae and waste-laden water out.

Graptoloid zooids housed in thecae in which the apertures were unadorned with spines or flanges and the thecae were not markedly curved probably could have passed in and out of the thecae in an action similar to that of rhabdopleuran zooids.

Paleoecology of graptoloids

Study of the distribution of planktic graptolites, the graptoloids, has shown that these fossil organisms were distributed in ancient seas in patterns similar to those exhibited by certain modern planktic and a few nektic organisms. Graptoloid occurrences in relation to other organismal remains and rock types and their depositional environments reveal a number of different patterns.

Analysis of the occurrence of graptoloids with benthic faunas and of the distribution of graptoloids in relation to shore lines, shelf areas, and possible slopes of Ordovician and Silurian seas suggests that graptoloids are richest in diversity in rocks formed on the outer parts of shelves and on shelf slopes. Many modern plankton, notably planktic foraminifera, are richest in diversity in these waters in today's oceans. This pattern suggests that water over shelf edges is enriched with nutrients from a distance below the surface by upwelling. Local upwelling of marine waters along ancient shelf margins similarly may have generated the nutrient supply needed by phytoplankton. These organisms, in turn, could have provided a food resource for the planktic graptolites.

A second distributional pattern exhibited by certain graptoloids seems to reflect the depth at which they lived. Some graptoloids apparently lived at or close to the sea surface because they are found in many different rock types deposited in shallow shelf sea environments as well as on shelf margins and slopes. These surface floaters occur in the deposits of shelf margins and slopes with other graptoloids that occur only in these deposits. This graptoloid occurrence pattern suggests that those graptoloids found only in shelf margin-shelf slope deposits lived at some depth within the oceanic water column. Their remains were mixed in with those of the near-surface dwellers when the colonies died. Graptoloids apparently had a depth zonation similar to that among certain modern zooplankton.

Graptoloids found in rocks formed in areas near prograding shore line or deltaic environments differ from those of the same age from more offshore areas. These near-shore graptoloids may have been limited to certain water masses that had physical and chemical properties that differed from those of more open oceans. The near-shore or inner-shelf water masses could have differed from those of more open ocean conditions in range of temperature variation and salinity because near-shore waters are more strongly influenced by fresh waters from lands and warmer surface waters in some seasons of the year. Similar distribution patterns of planktic faunas restricted to specific water masses are known in modern oceans. Indeed, certain plankton associations are used as indicators of particular shelf sea water masses. More broadly, the distribution patterns exhibited by planktic graptolites may be used in concert with evidence of depositional environments to document environments in the ancient oceans.

Evolution and utility of graptolites

Only graptolites with a planktic mode of life are common and widely found. Many of them are restricted to early Paleozoic black shales that rarely bear remains of other organisms. The greatest utility of graptolites for the geologist seems to be that their shapes changed relatively rapidly through time and that they therefore may be used in precise rock correlations, particularly those of the lower Paleozoic black shales.

Those graptolites with attached or encrusting modes of life are rare and commonly must be freed from their rock matrix to be identified properly. Nearly all dendroid graptolites appear to have lived attached to the sea floor in shallow shelf seas and so competed with other benthic suspension feeding organisms for food and space. Aside from a few localities from which a number of different dendroids have been recovered, dendroid species are known primarily from fragmentary specimens that are relatively few in number and from rocks with many different lithologic aspects. Black, carbonized films, commonly only fragments of colonies, comprise most of the dendroid fossil record. Most dendroids are, therefore, difficult to identify precisely and are of little significance in correlations or in paleoecologic reconstructions other than in comprising part of the total biota that lived in a particular place.

Well-preserved specimens of all kinds of graptolites freed from the rock matrix will continue to provide information concerning potential phyletic relationships and colony function. However, the graptoloids will be of continued use in correlations and studies of patterns in the geographic distribution of organisms because they are so widely found and so numerous in many early Paleozoic rock suites in most parts of the world.

Evolutionary trends of graptoloids

Planktic graptolites lived in a number of different environmental settings and in different faunal associations as did other organisms. Graptoloid evolutionary development was influenced by aspects of the physical and biological environment in which the graptoloids lived. The interplay between ecologic relationships and evolutionary development noted here may intrigue the inquisitive student into a further look at the role that ecologic relationships had on the general evolutionary history of organisms.

Graptoloid taxa that appear to have floated at or near the ocean surface tend to be found widely in many geographic areas. In addition, they seem to have had large populations because large numbers of individuals of such taxa may be collected at most localities. Their numbers are significantly greater in any given area than are the numbers of individuals of taxa that floated below the surface.

Planktic taxa that floated near the surface seem to have lived longer geologically than did those taxa that floated at some depth beneath the surface. The near-surface floating taxa also appear to have been ancestral to more taxa, including those that floated at some depth beneath the surface, than were the taxa that floated at depth. Opportunities for separation into relatively small, local populations presumably abounded among taxa with widespread, large populations. Such local populations could have led to development of new taxa. The interplay of local environmental conditions and the genetic makeup of these local populations could explain the relatively large number of founding or initial members of completely new lineages that apparently arose from relatively long-lived, near-surface dwelling taxa.

The first major step in graptoloid evolution occurred in the Early Ordovician and included the appearance of colonies with many branches (shown in the first few zones of Figure 19.21). These colonies appear to have developed from certain dendroids by loss of bithecae and of the hard, black rind from the stolonal system connecting the zooids. The loss of bithecae could be taken to indicate that the graptoloid zooids were hermaphroditic, whereas sexes were separate in dendroid and other benthic graptolites, assuming that the autothecae and bithecae housed zooids of different sexes. The loss of a hard black rind from the stolonal system could have been a modification to planktic life, since it would have resulted in less weight for a floating colony to support. The overlapping aspect (Figure 19.10) of graptoloid thecae is similar to that seen in autothecae in many dendroids, suggesting that the graptoloids had a stolonal system that was similar to that in dendroids except that it lacked the hard, black rind.

Graptoloids with few branches (shown in zones 3 to 7 of Figure 19.21) replaced many of those with many branches in the next step in graptoloid evolution. If it is assumed that the sicula aperture was oriented downward and the sicula apex pointed upward toward the sea surface in life, the growth direction of thecae and stipes in some colonies was downward in the first (geologically oldest) of the colonies with few branches (Figure 19.21). That the sicula apex was oriented upward is supported by the presence in some colonies of small vesicular bodies that could have been floats at the tip of the sicula apex or nema that extends from the sicula apex (Figure 19.31). If they were floats, the sicula apex would have been oriented upward in life.

Colonies with thecae opening more or less downward were replaced by colonies exhibiting a wide diversity in colony form in which thecae appear to have opened more or less upward (scandent rhabdosomes, Figure 19.21) or at least to one side (such taxa occur in strata of the age of zones 8 to 15 of Figure 19.21). Graptoloids with biserial scandent rhabdosome form were most abundant for much of the latter part of the Ordovician (the time span of zones 9 to 15 in Figure 19.21). Graptoloids with uniserial scandent rhabdosome form (popularly called monograptids) (Figures 19.9 and 19.22) generally replaced the biserial form during the early part of the Silurian. The uniserial scandent rhabdosome was the most prominent colony form for most of the Silurian and the Early Devonian.

Biogeography of graptoloids

Although many graptoloid species have worldwide distribution, they were distributed into faunal provinces and regions at certain times in graptoloid history. Graptoloid faunal provincialism was marked during the early part of the Ordovician when two distinct faunas occupied different parts of the world's oceans. The Pacific Faunal Region is characterized by the presence of certain scandent genera, particularly *Cardiograptus* and *Isograptus* (see zones 8/7 in Figure 19.21). These taxa and their associates are found in North America, Argentina, Asiatic Soviet Union, China, and Australia. Remanent magnetism data and the distribution of extensive carbonate rock deposits of the Early Ordovician indicate that the Pacific Faunal Region lay within a belt of tropical environments and probably straddled the Ordovician equator.

The fauna of the Atlantic Faunal Region is characterized by certain tuning fork graptoloids that are pendent forms within the genus *Didymograptus* (see Figure 19.4, *B* and zones 5 and 6 in Figure 19.21) and certain of the oldest biserial scandent graptoloids. The fauna has been found in Britain, western Continental Europe, Bolivia, Peru, and North Africa. Graptoloids became rare in South America and North Africa during the latter part of the Ordovician; no Late Ordovician graptoloids have been recorded from these areas.

At the time of the last occurrence of graptoloids in the Ordovician, in what may have been the southern part of the Atlantic Faunal Region (based on remanent magnetism data), sedimentologic evidence indicates that these areas became sites of cold climates and, in the latest Ordovician, glaciation. Remanent magnetism evidence and the presence of glacial deposits of the Late Ordovician in North Africa demonstrate that what is now North Africa lay in a polar (probably South Polar) region during the Late Ordovician. The paleoclimatic evidence indicates that graptoloids disappeared from oceans situated under cold climatic regimes during the Late Ordovician.

Graptoloid diversity in numbers of different taxa is significantly less in the Atlantic Faunal Region than in the Pacific Faunal Region. The patterns of markedly greater diversity of taxa in warm tropical seas and virtual exclusion of graptoloids from polar seas is remarkably analogous to the distribution of planktic foraminifers in modern oceans. Temperature appears to have been as major an influence on graptoloid faunal regionalism in the Ordovician as it is today on foraminiferan distribution.

Correlation of events in geologic or evolutionary history in one region with those in the other are obviously rather difficult because the faunas found in rocks of potentially the same age may be so markedly dissimilar. Faunal provincialism and regionalism form another hurdle for geologists to overcome in their attempt to reconstruct earth history on a global scale.

Graptoloid zones

Planktic graptolites not only floated or drifted widely in Ordovician, Silurian, and Early Devonian seas, and passed through many changes in colony form, but thecae secreted by each individual animal also changed in shape. These shape changes have led to recognition of several hundred species of planktic graptolites. The life spans of many of these species appear to have been relatively short according to their known stratigraphic ranges in many different stratigraphic sections. The patterns of appearances of new species and extinctions of others that may be recognized from collecting through many stratigraphic sections in Ordovician, Silurian, and Early Devonian strata in many places in the world has led to recognition and use of graptolite zones in these systems (Figure 19.21). The zones are thought to be relatively short chronologic divisions of about 1 to 5 million years duration. They are recognized by and based on unique associations of graptolite species found to occur in only a limited stratigraphic interval. It is the patterns generated by appearances and extinctions that led to recognition of those unique associations on which the zones are founded.

Commonly, a zonal sequence is built up through stratigraphic collecting of many different graptolite species in many stratigraphic sections over a very broad geographic area (such as, for example, the western part of North America). A zonal sequence may be used as a sort of standard of reference with which to compare graptolite associations obtained from isolated exposures or newly explored stratigraphic sections. The zones may be designated by a number or by the name of a graptolite species that occurs in the zone (see Figure 19.21).

Because planktic graptolites occur widely, they are useful to geologists for making correlations between rock sequences in widely separated areas. Graptolites are found in many metamorphic rock terrains, and they have proven to be valuable tools to structural geologists not only in dating rocks in such terrains but also in making correlations between metamorphosed and nonmetamorphosed rock sequences. Graptolites have been used in complexly deformed areas to document ages of different parts of potential stratigraphic successions and thus have aided in establishing a section and documenting structural patterns as well as geologic histories. Thus, although graptolites may have little allure for paleobiologists because they are found most commonly as films or imprints, they have proven to be important tools to geologists in working out the geologic structures and histories of highly deformed terrains of older Paleozoic rocks.

Supplementary reading

Barrington, E. T. W. *The biology of Hemichordata and Protochordata*. San Francisco: W. H. Freeman & Co.; 1965. General review of the biology of Hemichordates.

Berry, W. B. N. Graptolite biostratigraphy: A wedding of classical principles and current concepts. In: Kauffman, E.; Hazel, J. E., editors. *Concepts and methods of biostratigraphy*. Stroudsburg, PA: Dowden, Hutchinson & Ross, Inc.; 1977. Discussion of both graptolite zonation and biogeography and ecology.

Bulman, O. M. B. In: Teichert, C., editor. *Treatise on invertebrate paleontology, part V (revised), Graptolithina*. Boulder, CO and Lawrence, KS: The Geological Society of America and the University of Kansas Press; 1970. Summary work on graptolite morphology and taxonomy.

Elles, G. L.; Wood, E. M. R. *Monograph of British graptolites, parts I–XI*. London: Palaeontographical Society Monograph; 1901–1918. Basic reference to graptoloid (planktic) graptolite morphology and taxonomy as well as a discussion of graptolite zones.

Urbanek, A. Organization and evolution of graptolite colonies, pp. 441–514. In: Boardman, R. S.; Cheetham, A. H.; Oliver, W. A., Jr. editors. *Animal colonies; their development and function through time*. Stroudsburg, PA: Dowden, Hutchinson, & Ross, Inc.; 1973. Thorough analysis of colonial organization and function of graptolite colonies.

20

Phylum Conodonta

David L. Clark

Part I Phylum overview

Conodonts are an extinct group of marine animals whose skeletal parts consisted of microscopic mineralized **elements**, many of which were arranged in patterns or **apparatuses** (Figure 20.1). More or less complete apparatuses are called **natural assemblages**. Some conodonts contained a single type of skeletal element, but most possessed a number of differently shaped elements. Many of the elements occurred in symmetrical pairs (that is, right and left members) (Figure 20.1). These elements, disarticulated from internal soft body parts, are the only parts of conodonts commonly preserved. The geologic importance of conodonts is based on an understanding of these elements.

Conodont elements are known worldwide and have been recovered from a variety of marine sedimentary rocks ranging in age from Late Proterozoic through the Triassic. Conodonts evolved rapidly, and elements found in lower Paleozoic rocks are considerably different from those recovered from rocks of later Paleozoic and the Triassic. The demonstration of rapid evolution in many conodonts has made them superb geologic tools. They are used with great success in the worldwide biostratigraphy of Upper Cambrian through Triassic rocks.

Conodonts are assumed to have been metazoans at a coelomate level of organization. The exact biology of conodonts has not been determined, and almost all that is known about the animals is based on study of their mineralized elements. Clusters of elements, fused diagenetically, as well as elements occurring on bedding planes (Figure 20.2) in symmetrical associations are interpreted as natural assemblages from single animals. Assemblages provide evidence that conodonts were bilaterally symmetrical animals that were probably small, possibly only a few centimeters in greatest dimension. Because the important soft parts of conodonts can only be inferred from a few very unusual occurrences, details of size and shape, mode of life, and relationship to other organisms are few or lacking altogether. Recent work related to conodont paleoecology has produced a surprising amount of data on possible habits and environmental relationships.

Introductory morphology

Soft parts

Whole conodonts with body outlines, soft-part impressions, and mineralized elements have only recently been discovered. Earlier, specimens with body outlines and elements were reported, but whether the soft-part impressions were those of conodonts or only of conodont predators is debated. For example, several specimens approximately 5 cm in length have been recovered from the Upper Mississippian of Montana (Figure 20.3). These specimens include impressions of small sac-shaped bodies with possible muscle and organ impressions and groupings of conodont elements. The elements have been interpreted as either part of a filter-feeding digestive system of a conodont or the stomach contents of a conodont predator. The poor impressions suggest that these were anteroposteriorly elongate organisms, but

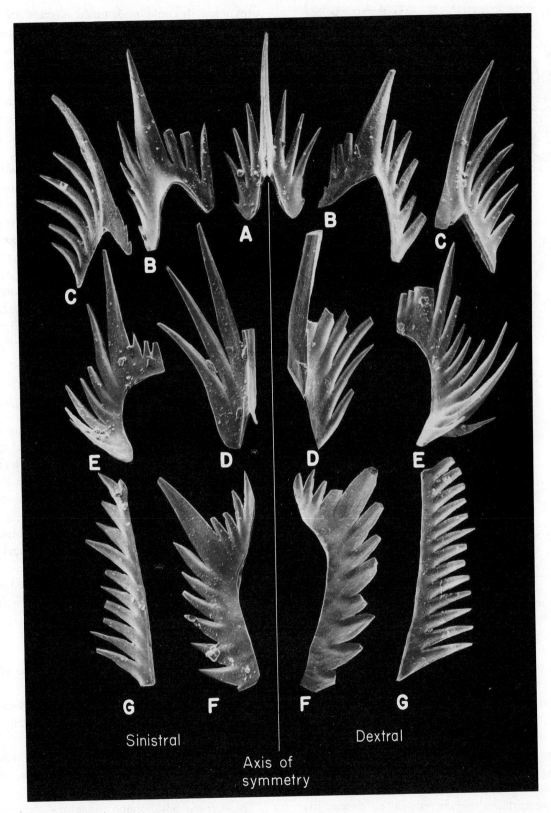

Figure 20.1. Hypothetical conodont apparatus based on disassociated specimens reconstructed using association analysis. This multielement grouping has a bilaterally symmetrical element (*A*) and sinistral and dextral elements around an inferred line of symmetry. Element arrangement is hypothetical but, in part, is based on known naturally occurring assemblages and may be typical of that found in the symmetry series of many conodonts. This apparatus represents a species of *Ellisonia* from the Upper Permian of west Texas. (Scanning electron micrographs, ×150.)

A

B

Figure 20.2. Naturally occurring assemblages of conodont elements. **A,** A resorted assemblage. *a* and *d,* S elements; *b,* M elements; *c,* P elements. (Upper Paleozoic, Montana, ×42.) **B,** A slightly disturbed assemblage of *Idiognathodus. p,* P elements; *o,* M elements; *h* and *s,* S elements. (Pennsylvanian, Illinois, ×32.) (**A** from Scott, H. W. *Journal of Paleontology* 16: Pl. 37; 1942; **B** from Rhodes, F. H. T. *Journal of Paleontology* 26: Pl. 126; 1952.)

details of a head or possible appendages are lacking.

A single specimen from the Middle Cambrian Burgess Shale of western Canada is similarly intriguing but quite distinct morphologically from the Montana specimens (Figure 20.4). This animal was approximately the same size as the Montana specimens but unlike them was compressed dorsoventrally and had a distinct head. The head included element-like structures that may have been part of a ventral feeding apparatus. The element-like structures are not definitely identifiable as true conodont elements. The comparative morphology of the feeding apparatus of this specimen has been used as evidence that it is related to lophophorates. If the elementlike struc-

tures definitely could be identified as conodont elements, the specimen from Canada would become important for biologic studies. Preservation is not good, however, and the specimen can be interpreted only as a possible conodont.

Valid conodont body impressions with assemblages of elements were found in 1982. Only a few specimens are known. The specimen in Figure 20.5 is elongate, wormlike, 4 cm in length, has impressions of a midline, terminal tail-like structure and, most important, an almost complete assemblage of ramiform and pectiniform elements. The elements are arranged in the head region of the impression. The occurrence of elements and body impression constitutes a natural association, and the elements may have been part of some grasping or tentacle structure associated with a mouth. Specimens are from the Lower Carboniferous (Mississippian) of Scotland and occur with body impressions of shrimplike crustaceans. Discovery of additional well-preserved specimens will be important for interpretation of conodont paleobiology.

Morphology and microstructure of skeletal elements

Even with the very limited data concerning the conodont's size, shape, and soft parts, the skeletal apparatus is partially understood. The microscopic elements are composed of a form of carbonate apatite known as francolite. This is a sturdy mineral, and the elements can be easily freed from sedimentary rocks with organic-acid treatment. Elements recovered from acid residues almost always are disassociated from original apparatus structure. The natural assemblages found as associations on bedding planes or as fused clusters have served as models for interpretation of the more common disassociated elements.

Three basic types of conodont elements have been recognized as follows (Figure 20.6):
1 **Coniform** elements—cone-shaped structures consisting of a base and a cusp
2 **Ramiform** elements—structures that include a main cusp flanked by posteroanterior and/or laterally directed processes that commonly possess denticles
3 **Pectiniform** elements—structures that commonly bear denticles on platforms of laterally expanded processes

These three basic shape categories comprise many varieties, including two types of coniform, seven types of ramiform, and 15 types of pectiniform elements (illustrated in figures throughout the chapter).

Conodont elements are constructed principally of

A

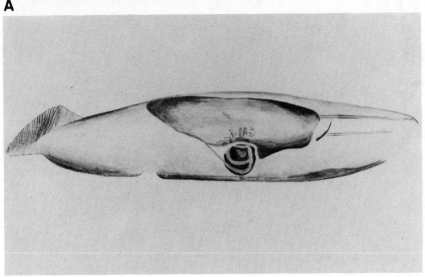

B

Figure 20.3. Possible conodont predator (**A**) and interpretation of morphology (**B**). Several similar specimens have been found associated with excellent fossil fish. Resorted assemblages of conodont elements (**B**) in dark internal organ. This specimen has been suggested to be a conodont by some specialists, but others argue that it may represent a conodont predator. Notice tail fin, uncertain head, and possible internal organs. Uncertainty concerning the origin of the elements associated with the fish and their apparatus structure makes positive interpretation difficult. (*Lochriea wellsi*, Upper Mississippian, Montana.) (From Melton, W. G.; Scott, H. W. In: Rhodes, F. H. T., editor. *Geological Society of America Special Paper* 141: figs. 10, 15; 1973. Courtesy H. W. Scott.)

laminations of carbonate apatite. The **lamellae** are formed by outer accretion around an initial point of growth (Figure 20.7). In addition, a **basal plate** or **cone**, of the same mineral composition but with more organic matter and different in appearance, has been found attached to the lower surface of many conodont elements and may represent the attachment material that connected element to flesh in the conodont body (Figures 20.7 and 20.8). This structure evidently was not present on the geologically oldest conodont elements.

Between and cutting across lamellae, and occasionally occurring as bubblelike structures in the elements, is nonlaminated lighter-colored material called **white matter**. This is chemically about the same as the lamellae and often is uniformly developed in similar taxa. White matter is nearly opaque and therefore appears darker than lamellae in transmitted light passing through thin

sections (Figures 20.7 and 20.9, *A* and *B*); its function is unknown. Lamellae and white matter are the most important internal features of elements.

In the major class of conodonts, the conodontophorids, two different kinds of basal structures of elements were grown during accretion of lamellae. The kinds of basal structures originated by different growth patterns from the original points of calcification of elements. One pattern resulted in a **basal cavity** (Figure 20.10, *A*) and the other in a **pit** (Figure 20.10, *B*). In construction of elements with a basal cavity, the cavity continued to increase in size during the entire ontogeny of the element by addition of lamellae around cavity margins.

In contrast, during formation of a pit, the maximum pit size was reached early in growth. As subsequent lamellae were added externally, each younger lamella

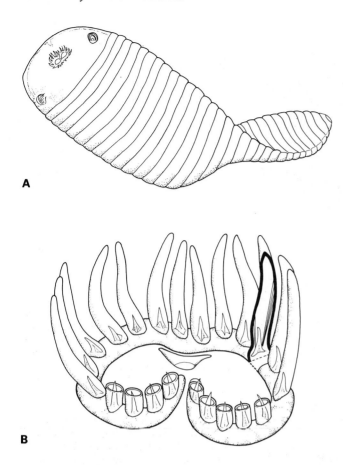

Figure 20.4. Fossil, *Odontogriphus,* bearing conodont-like elements. **A,** Reconstruction shows ventral and part of dorsal surface plus feeding apparatus with conodont-like elements. **B,** Reconstruction of feeding apparatus. The different positioning and presumed function of elements between this reconstruction and that of Figure 20.3 suggest that one or both may not be conodonts. (Middle Cambrian Burgess Shale fauna of British Columbia.) (From Conway Morris, S. *Palaeontology* 19: 206; 1976.)

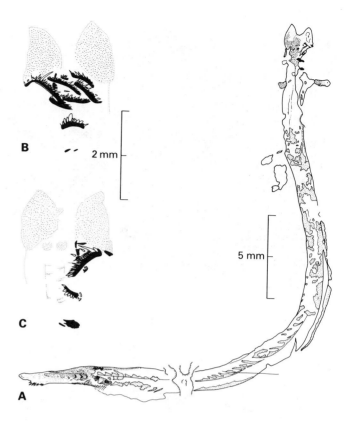

Figure 20.5. A, Drawing of conodont body showing elements in head region. Note midline and finlike structure on posterior. **B** and **C** show details of head region in original (**C**) and counterpart (**B**) of original. Specimen from Lower Carboniferous of Scotland. (From Briggs, D. E. G.; Clarkson, E. N. K.; Aldridge, R. J. *Lethaia* 16: 5; 1983. Courtesy Richard Aldridge, University of Nottingham.)

stopped short of a complete overlap of previously formed lamellae in the basal portion and left the pit size unmodified.

Some evolutionary significance can be attached to the fact that elements with basal cavities are more common in lower Paleozoic rocks, while elements with pits are more common in upper Paleozoic and Triassic rocks.

Externally, elements have a variety of ornamentation, including nodes, ridges, furrows, and fine striations (Figure 20.11). Certain elements also have reticulate patterns. This micro-ornamentation has been used in determining taxonomic relationships among some conodont taxa.

Organizational plan of element apparatuses

During the early investigations of conodont elements, each discrete element was treated as a separate taxonomic unit. The name that was applied to a particular element identified the supposed whole animal. In later studies, this name represented the distinctive morphologic shape of the element as well. It is now recognized that many conodonts contained more than a single kind of element. Furthermore, different assemblages of elements commonly contain some similarly shaped elements.

These ideas were confirmed by the discovery of natural assemblages a number of years ago (Figure 20.2). These associations of various kinds of elements in definite patterns on bedding planes also confirmed that most conodonts were bilaterally symmetrical and that many elements occurred in "right" and "left" symmetry pairs.

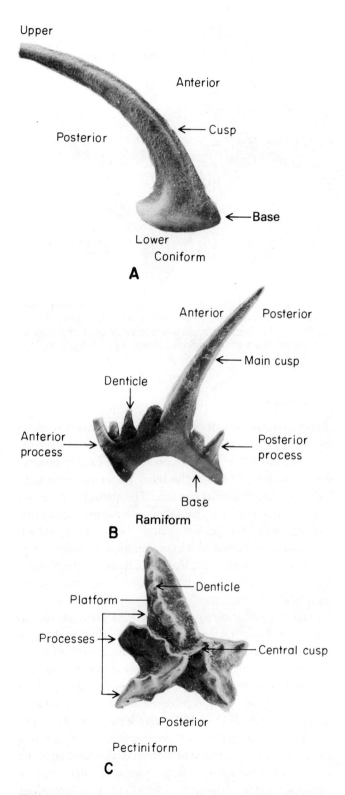

Upper

Anterior

Posterior

Cusp

Base

Lower

Coniform

A

Anterior   Posterior

Main cusp

Denticle

Anterior
process

Posterior
process

Base

Ramiform

B

Denticle

Platform

Processes

Central cusp

Posterior

Pectiniform

C

Figure 20.6. Three basic shapes of conodont elements. **A,** Coniform element with anterior, posterior, upper, and lower orientation; base and cusp are indicated. (×120.) **B,** Ramiform element with two branches or processes, anterior and posterior, of main cusp. The processes possess discrete denticles. (×130.) **C,** Pectiniform element showing lateral expansion in platform growth and various lateral processes with low denticle rows. (×125.)

More recently, diagenetically fused clusters have been found that confirm this pattern. In addition, imaginative research by a number of conodont investigators has resulted in reconstruction of the apparatus composition of many conodonts from collections of disassembled, discrete elements. Symmetry series in elements (Figure 20.12) and similarity in internal structure (for example, white matter distribution) as well as numerical associations have formed the basis for reconstructions. The numerical associations are based on constant co-occurrence of elements in definite ratios in different samples.

Evidence from all these assemblages, natural and reconstructed, supports the idea that the number of major skeletal patterns for conodonts was relatively small. Some skeletal patterns, termed **unimembrate,** consisted of apparatuses with a single morphologic type of element, and other skeletal patterns, termed **multimembrate,** consisted of apparatuses with two or more kinds of elements. A maximum of seven morphologically distinct element types are known in a single apparatus (Figure 20.1). It is not known how many complete skeletal apparatuses may have existed in a single conodont.

Because the majority of conodont work is done with disarticulated elements, a system has been designed that puts the elements into hypothetical groupings that approximate the original apparatuses. Positions within different skeletal assemblages presumed to be analogous have been established arbitrarily as **P** (principal), **M** (medial), and **S** (symmetry) series (Figure 20.13). Figure 20.2 shows a natural assemblage, the P, M, and S elements of which can be determined. The positions and kinds of P, M, and S elements found loose are not known but are established hypothetically (Figure 20.13) by comparison with known assemblages.

P positions commonly are occupied by pectiniform or specialized ramiform elements, M positions by ramiform or coniform elements, and S positions by coniform or ramiform elements. P positions have been further differentiated into Pa and Pb, and S positions into Sa, Sb, Sc, and so on, depending on the number of elements and symmetry arrangement (Figure 20.13; see also Figure 20.1 for hypothetical symmetry series arranged from scattered elements). Ramiform elements generally are most numerous in given multimembrate apparatuses (Figure 20.13). The relatively few unimembrate apparatuses commonly consist of pectiniform or coniform elements that are arbitrarily said to occupy the P position. Although this definition may only approximate the real arrangement, it provides a uniform model for specialists.

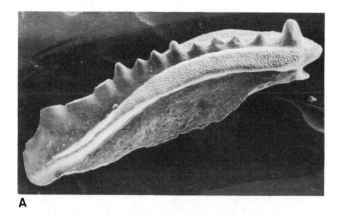

Figure 20.7. Idealized cross section of a pectiniform element and its attached basal plate (finely dotted). Growth lamellae are indicated by parallel lines, contact zone between element and basal plate by lines crossing lamellae, and white matter by coarse dots. Growth lamellae are continuous from element to basal plate, although basal plate commonly is broken from element. (*Palmatolepis*, Upper Devonian, Germany.) (From Müller, K. J.; Nogami, Y. *Memoirs Faculty of Science Kyoto University, Series of Geology and Mineralogy* 38: fig. 2; 1971.)

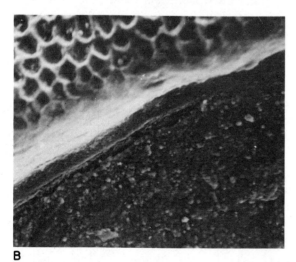

Figure 20.8. A, Pectiniform element showing main denticulate part above and basal plate below. (×100.) **B,** Enlargement showing reticulate pattern of platform and structureless pattern of basal plate. (×1000.) (*Neogondolella*, Upper Permian, Texas.)

Classification

Most investigators of conodonts concur that this group of animals, known from disarticulated skeletal elements and natural assemblages, represents an extinct invertebrate phylum. What is the basis for considering conodonts to be a distinct phylum? The answer is based on the uniqueness of conodont elements, apparatus arrangement, and geologic range. There are no known organisms that have such a combination of morphology, chemistry, arrangement, and occurrence. In the absence of comparative data, the phylum designation seems practical.

The classification of disarticulated elements of an extinct phylum has proven difficult. Early classification attempts were based strictly on the morphology of separate elements. For example, each different coniform element was given a distinct binomial name, and all coniform types were considered to be closely related. This approach also was followed for most ramiform and pectiniform elements. This form-species approach was practical but contributed little to an understanding of the apparatus structure or biologic interpretations, or to the biologic species concept. More recently, conodont elements have been interpreted taxonomically in terms of actual assemblages and numerical associations of disarticulated elements within a hypothetical apparatus based on P, M, and S assignments. The result has been the definition of multi-element assemblages. It is also supposed that one conodont bore at least one and perhaps

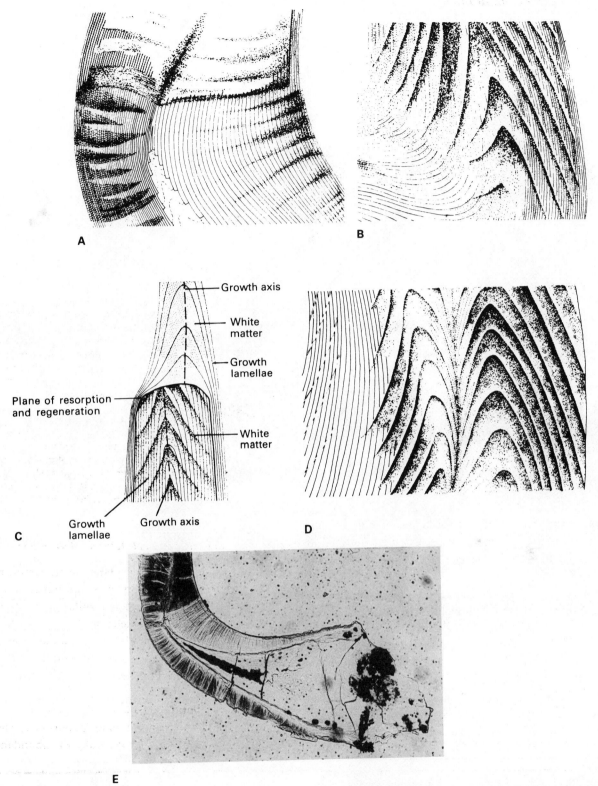

A

B

Growth axis

White matter

Growth lamellae

Plane of resorption and regeneration

White matter

C

Growth lamellae   Growth axis

D

E

Figure 20.9. Relative orientation of white matter and growth lamellae of conodont elements in various evolutionary stages of development, as seen in diagrammatic thin sections. Growth lamellae are indicated by subparallel lines, growth axes by central dashed lines, and white matter by dotted areas signifying opaqueness in thin section. **A,** Regions of white matter in element of *Cordylodus* cross vertical growth lamellae at irregular intervals. White matter is broadest at element margin and thins toward the middle. (Lower Ordovician, Wyoming, ×410.) **B,** Chevron stripes of white matter form stacked cones in three dimensions cutting growth lamellae in element of *Ligonodina*. (Upper Devonian, Iowa, ×230.) **C,** Example of internal structure before and after resorption and regeneration. In lower part of element (*Neoprioniodus*), growth lamellae are crossed by cone-shaped regions of white matter similar to those in **B.** In upper regenerated part of element, growth lamellae and white matter are less dense and have different orientation. (Upper Devonian, Iowa, ×275.) **D,** White matter crossing growth lamellae in M-shaped pattern in element of *Ligonodina*. (Lower Mississippian, Iowa, ×320.) **E,** Photograph of white matter (black stripes) in Lower Ordovician *Cordylodus*. (×150.)(From Müller K. J.; Nogami, Y. *Memoirs Faculty of Science Kyoto University, Series of Geology and Mineralogy* 38: fig. 7 and pl. 7, fig. 3; 1971.)

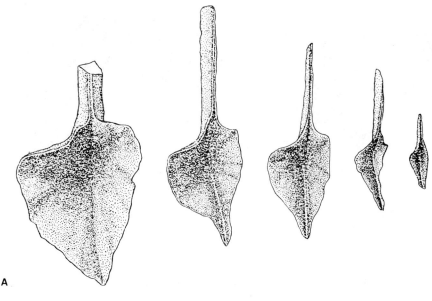

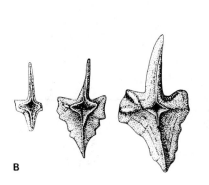
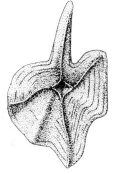
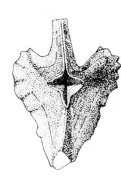

Figure 20.10. Basal opening of conodonts. **A**, Basal cavity type. Five specimens of various sizes of a P element of *Gnathodus* showing change in size of basal cavity during growth. (All specimens, Lower Carboniferous, Italy, ×40.) **B**, Pit type. Five specimens of various sizes of P elements of *Ancryodella* showing that size of pit remains essentially constant during growth. (All specimens, Upper Devonian, Michigan, ×30.) (**A** from Clark, D. L.; Müller, K. J. *Journal of Paleontology* 42: 563; 1968. **B** from Müller, K. J.; Clark, D. L. *Journal of Paleontology* 41: 908; 1967.)

several apparatuses. The knowledge of multi-element species (in contrast to form-species based on separate elements) has led to recognition of several major and minor taxonomic categories.

The two major classes of conodonts can be differentiated on the basis of slightly different element structure and chemistry. One of these groups, the Paraconodontida, includes the Proterozoic, most of the Cambrian, and a few of the Ordovician conodonts. The other group, the Conodontophorida, includes the Upper Cambrian and most younger conodonts. The Conodontophorida include at least two important orders whose differences are based on different assemblage structures and elements.

CLASS PARACONODONTIDA (Figure 20.14)

Conodonts having elements mostly coniform with large, deep basal cavity, no white matter, few widely spaced lamellae, and a significant percentage of organic matter in skeleton. Mostly unimembrate apparatuses but simple multimembrate apparatuses known. (Late Proterozoic to Middle Ordovician; approximately 15 genera, mostly pre-Ordovician.)

Paraconodontida are the smaller class of conodonts, relatively unknown and of minor importance. This class is known worldwide, but generally the abundance and

Figure 20.11. Surface ornamentation in various elements. **A** and **B**, Smooth surface of P element (pectiniform) of *Furnishius*, a hibbardellid. (Lower Triassic, Utah; **A**, ×150.) **B**, Enlargement of **A** showing unsculptured surface of one denticle. (×1000.) **C** and **D**, Finely striated surface on ramiform element of *Ellisonia*. (Upper Permian, Texas.) **C** shows lateral surfaces. (×225.) **D**, Enlargement of **C** showing main denticle and fine striations. (×900.) **E**, Coarse striations on coniform *Scolopodus* element. (Middle Ordovician, Wisconsin, ×120.) **F**, Ridges and nodes on P element (pectiniform) of *Siphonodella*. (Lower Mississippian, Nevada, ×100.) **G** to **I**, Coarse reticulations on surface of *Epigondolella* (pectiniform). **H**, Enlargement of posterior tip showing reticulations on edge and finer structure on inner margin. **I**, Enlargement of **G** showing fine reticulations. (Middle Triassic, Africa; **G**, ×200; **H**, ×900; **I**, ×2000.) **J** to **K**, Coarse regular reticulations on *Neogondolella*, a pectiniform element from the Lower Triassic of Nevada. **J** shows regular coarse pattern. (×100.) **K**, Enlargement of **J**. (×2000.)

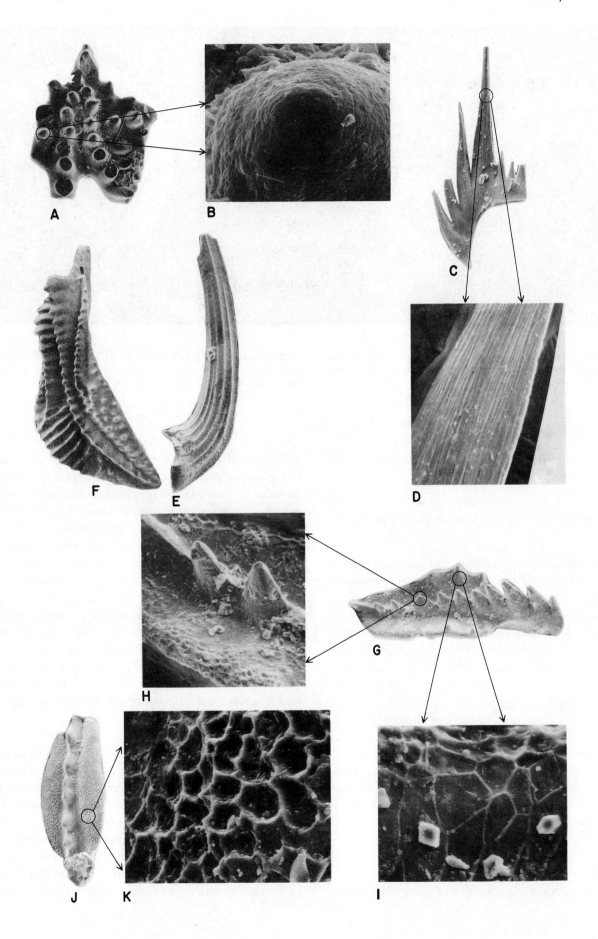

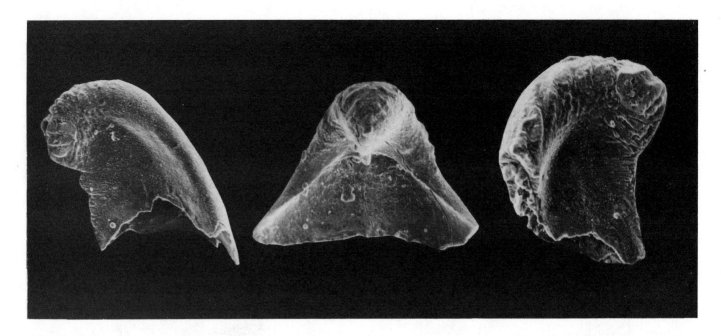

Figure 20.12. Symmetry transition in *Fryxellodontus*. Center photograph shows bilaterally symmetrical element and sinistral and dextral specimens on either side. This is a simple apparatus and elements have a large basal cavity and a hook-shaped tip that has different positions on different elements. (Lower Ordovician, Utah, ×300.)

stratigraphic importance of the group is less than for the advanced conodonts.

CLASS CONODONTOPHORIDA (Figures 20.15 to 20.19)

Conodonts having coniform, ramiform, and pectiniform elements distinguishable by numerous closely spaced lamellae. White matter and basal plate present. Basal cavity of elements generally shallow, or specialized-type pit present (see Part II). Exterior of elements characterized by weak to strong ornamentation. Unimembrate to seven-member apparatuses present. (Cambrian–Upper Triassic; approximately 215 genera.)

Conodontophorida are the major class of conodonts and include most of the geologically important conodonts. This class includes conodonts with great stratigraphic value. More than 140 Paleozoic and Triassic faunal zones are recognized on members of this class. Many genera are worldwide and abundant.

Origin

The origin of conodonts, like that of many of the higher categories of organisms, is not known. No organisms, living or fossil, have skeletal structures the same as that of conodonts. Comparison of conodont elements to supposed analogous structures in other organisms has resulted only in identification of groups that definitely are not related. Thus during the past 100 years, conodont elements have been considered to be related to conulariids (Cnidaria); to the copulatory structures of Nematoda; to the radular teeth or other parts of cephalopods or gastropods (Mollusca); to worm jaws or parts (Annelida), gnathostomulids, crustaceans or parts of other arthropods; to the teeth of cyclostome fish, sharks, ostracoderms, crossopterygii, placoderms, or higher fishes (Osteichthyes); and even to calcareous algae and vascular plants. None of the arguments put forth to support these relationships is convincing.

Because the conodont elements and their organization into skeletal apparatuses constitute the principal evidence for biologic relationships (or lack of biologic relationships), a search for conodont ancestors or relatives could lead to other organisms that have similar carbonate apatite elements. Carbonate apatite is not a common substance in the mineralogic skeletal parts of other organisms, but some brachiopods as well as worm groups, some arthropods, tunicates, and vertebrates have carbonate apatite skeletal parts. Although all of these

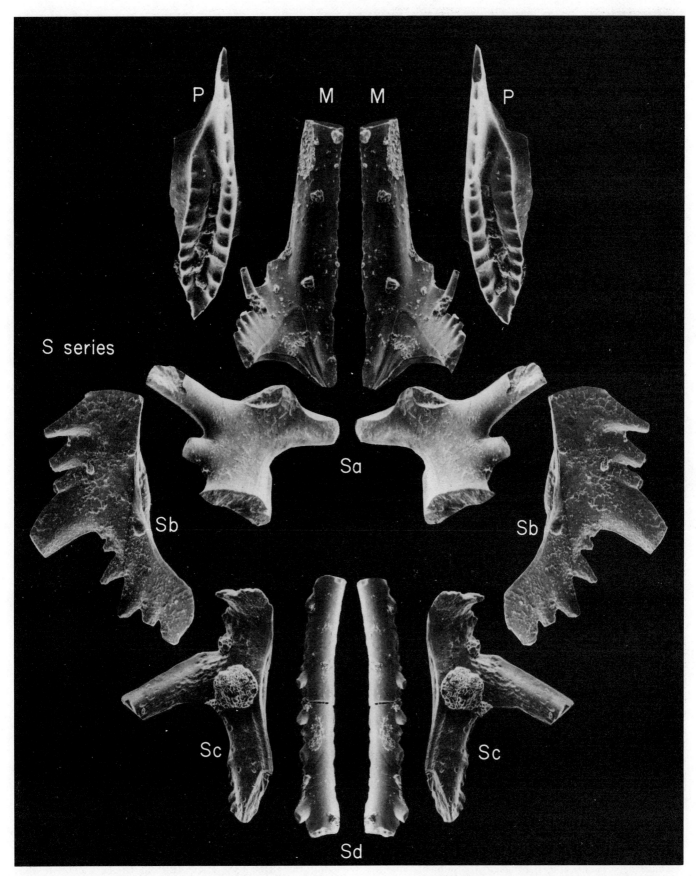

Figure 20.13. Reconstructed apparatus illustrating positional scheme of elements. P elements are *principal* elements, commonly pectiniform as here, or specialized ramiform. M elements are *medial*, ramiform as here, or coniform. S elements are *symmetry series*, all ramiform as here, but commonly coniform. This is a basic apparatus assemblage of P, M, and S elements found in the conodontophorid conodonts. The reconstruction is hypothetical. (*Adetognathus*, Pennsylvanian, Utah, ×160.) In this reconstruction, sinistral and dextral positions are attained by reverse printing of negatives.

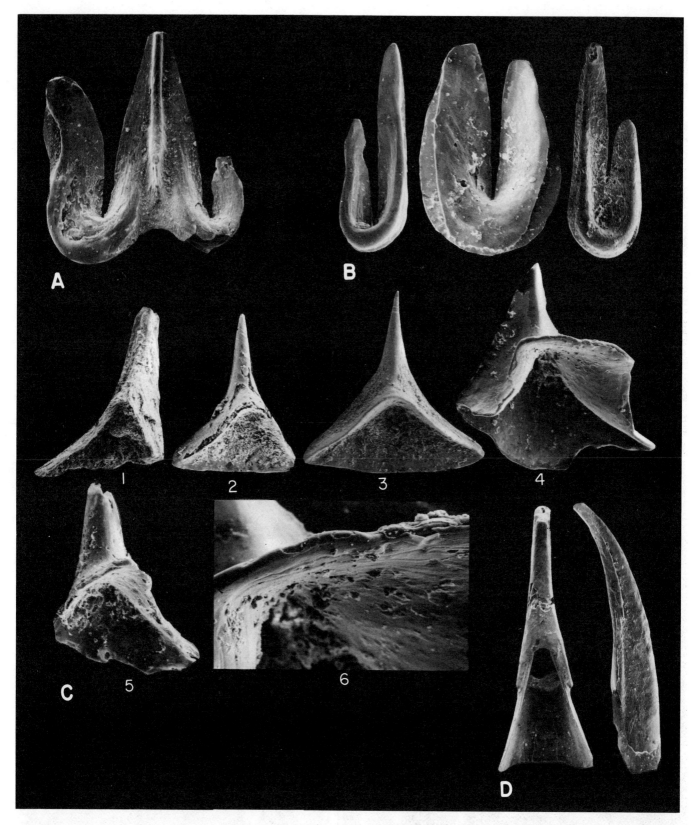

Figure 20.14. Representative elements of the Paraconodontida (primitive conodonts) from the Middle and Upper Cambrian. **A,** Specialized coniform element of *Westergaardodina*. (Middle Cambrian, Nevada, ×100.) **B,** Symmetry transition of coniform elements of *Westergaardodina*. (Upper Cambrian, Sweden, ×120.) **C,** Symmetry transition of coniform elements of *Furnishina* from the Upper Cambrian of Sweden. Symmetrical element in center (*3*) with sinistral and dextral elements shown on either side, ×150. 6 is an enlargement of the base of the first dextral element of *4*, showing growth lamellae. (×420.) **D,** Coniform element of *Hertzina* from Upper Cambrian of Sweden, posterior and lateral views. Notice deep basal cavity. (×80.)

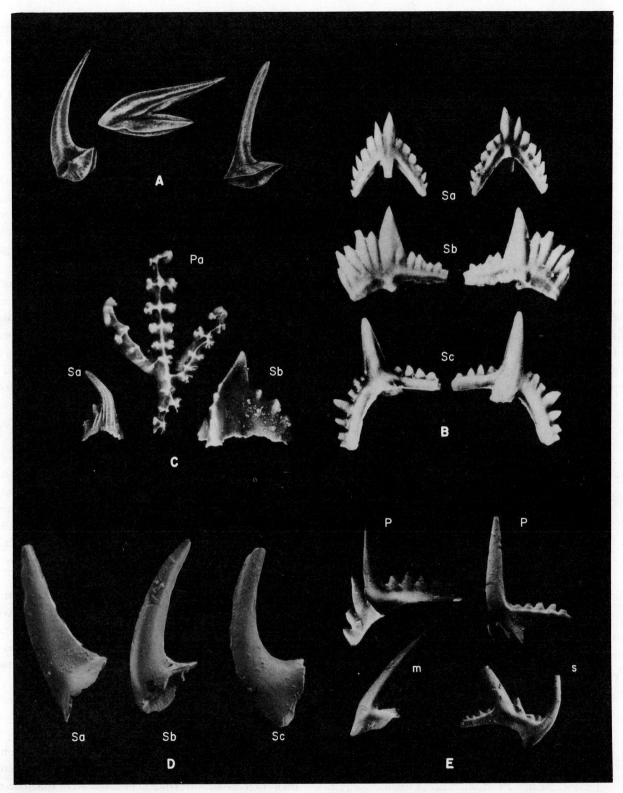

Figure 20.15. Representative elements of the Distacodontina members of the Conodontophorida (advanced conodonts). **A,** The three kinds of coniform elements (multimembrate) of *Drepanoistodus* from Lower Ordovician of Sweden. Series shows no conspicuous symmetry series. (×45.) **B,** Three of the kinds of elements (multimembrate) of *Rhipidognathus* from the Middle Ordovician of Tennessee. These S elements, shown here in symmetry pairs, are all ramiform elements representative of complete S series. (×34.) **C,** Three kinds of elements (multi- membrate) of *Pedavis* from the Lower Devonian of Nevada. Pectiniform (Pa) and ramiform elements (Sa, Sb). (×40.) **D,** Three kinds of elements (multimembrate) of *Scandodus*, from the Lower Ordovician of Sweden. S positions of coniform elements as labeled. (×71.) **E,** Four kinds of elements (multi- membrate) of *Phragmodus* with M and S and two kinds of P elements shown, all ramiform. (Middle Ordovician, Kentucky, ×73.) (**A, B, D,** and **E,** courtesy of S. M. Bergström; **C,** courtesy of G. Klapper.)

groups are "potential" relatives, there is little to compare other than chemically similar hard parts, and a relationship certainly is not confirmed. Conodonts currently are considered to be a distinct phylum with no known relatives.

Part II   Additional concepts

Classification of Conodontophorida orders

Two orders of the class Conodontophorida are recognized, Distacodontina and Hibbardellina. Evolution as reflected by assemblages is fairly well known for both orders. No single feature is diagnostic of each group, however, and the classification at the order level is tentative. As more information concerning evolutionary relationships among assemblages is developed, greater confidence of order assignment will be possible. Genera and species are based on assemblage composition and element modification within assemblages. However, very similar elements may occur in different genera or orders.

ORDER DISTACODONTINA (Figures 20.15 and 20.16)

Conodontophorids characterized by unimembrate apparatuses of coniform or ramiform elements or by multimembrate apparatuses of coniform or ramiform elements or by a combination. Also included are multimembrate forms with coniform, ramiform, and primitive pectiniform elements. (Cambrian to Devonian; approximately 85 genera.)

Distacodontina include at least six superfamilies that are important for the lower Paleozoic, particularly the Lower Ordovician. The widespread Ordovician seas contained abundant distacodontids and included a diversity of taxa never exceeded.

ORDER HIBBARDELLINA (Figures 20.17 to 20.19)

Conodontophorids characterized by multimembrate apparatuses with ramiform elements only, with ramiform and pectiniform elements, or with unimembrate apparatuses consisting of pectiniform elements. (Middle Ordovician to Late Triassic; approximately 70 genera.)

Hibbardellina include four superfamilies and many of the most important post-Ordovician genera. Much of the biostratigraphy of the upper Paleozoic and Triassic is based on members of this order.

The skeletal patterns for some 60 coniform, ramiform, and pectiniform taxa have not been identified so they have not been assigned to orders. They range from the Ordovician through the Triassic.

Mode of growth and function of skeletal apparatus

Although skeletal element apparatuses are partially understood from study of elements and assemblages, their function is not. Evidently, certain primitive Cambrian elements were borne externally by the conodont as indicated by the position of growth surfaces (Figure 20.20, A and B). Many Cambrian and all post-Cambrian elements were embedded in a secretory tissue during all or most of their life (Figures 20.20, C and 20.21), and many broken or partly resorbed elements were repaired (Figure 20.24). It has been suggested that the advanced conodont element could be extruded from a fleshy pocket (Figure 20.21). If whole skeletal apparatuses could be similarly extruded, then a grasping or even defense function would be tenable.

If the skeletal apparatus was internal, as proposed by other investigators, the skeleton may have functioned as a support for feeding tentacles or of some other food-gathering mechanism. One elaborate reconstruction of the skeleton has been interpreted to have been such a support (Figure 20.22). Certainly, most ideas on conodont function visualize the element apparatuses as some type of physiologic support. The rapid evolution of conodont apparatuses and the elements perhaps can be best understood using this general assumption.

Three basic patterns of growth and development of elements have been described (Figure 20.20) as follows:
1 Primitive—Early, Middle, and Late Cambrian elements have deep internal cavities, and growth was by lamellae additions on the inner surface of the element only (Figure 20.20, A).
2 Intermediate—Cambrian and some Ordovician elements added lamellae both internally and partially on the outer surface as well (Figure 20.20, B).
3 Advanced—elements at the end of the Cambrian, had internal cavities generally much smaller, basal attachment plates had formed, and lamellae were added on the entire external surface of the elements so that each new layer covered all or most of the preceding layers (Figure 20.20, C). The advanced pattern of development continued throughout the Paleozoic and Triassic.

The Paraconodontida are characterized by the primitive and intermediate growth patterns. All Conodon-

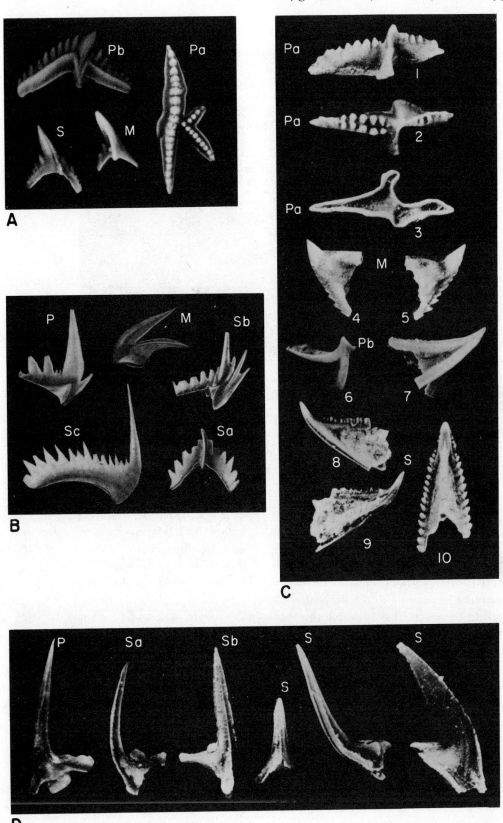

Figure 20.16. Representative elements of the Distacodontina members of the Conodontophorida (advanced conodonts). Positional notations indicated. **A**, A multimembrate apparatus of pectiniform (*Pa*) and ramiform elements. (*Pterospathodus*, Lower Silurian, Oklahoma, ×40.) **B**, A multimembrate apparatus of ramiform elements of the conodont *Microzarkodina*. (Lower Ordovician, Sweden, ×75.) **C**, A multimembrate (five member) apparatus of pectiniform and ramiform elements of *Icriodella* (Middle Ordovician, Kentucky, ×50.) *1*, Lateral; *2*, upper; *3* lower view of Pa element; *4* and *5*, lateral view of M element; *6*, upper; *7*, lateral view of Pb element; *8*, lateral view of S element with two denticulate processes; *9*, lateral and *10*, upper views of S element with three denticulate processes. **D**, A multimembrate apparatus of coniform elements of *Eoneoprioniodus*. (Ordovician, Oklahoma, ×40.) (**A**, courtesy G. Klapper; **B** to **D**, courtesy S. M. Bergström.)

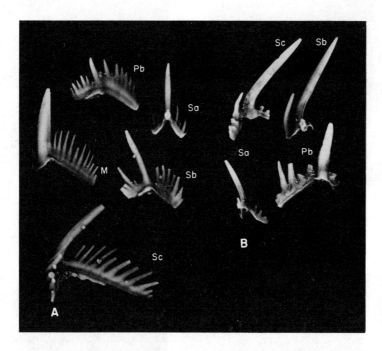

Figure 20.17. Representative elements of the Hibbardellina Conodontophorida (advanced conodonts). **A,** A multimembrate apparatus of ramiform elements of *Hibbardella.* (Upper Devonian, Australia, ×40.) Positional scheme noted. **B,** A multimembrate apparatus of ramiform elements of *Idioprioniodus.* (Upper Pennsylvanian, Kansas, ×40.) This is an apparatus consisting of S position elements only. (Courtesy G. Klapper.)

tophorida are characterized by the advanced growth pattern.

Another example of the importance of growth differences occurs in Triassic rocks. During the Middle and Late Triassic, in particular, several groups of conodonts bore similar pectiniform elements in the P position of their apparatuses. In late stages of growth, the elements are difficult to distinguish from one another. Earlier growth stages show significant differences, however (Figure 20.23). One group, the paragondolellids (Figure 20.23), have no lateral platform during earliest growth but have a prominent platform later. In contrast, neogondolellids with adult elements closely similar to the paragondolellids have a laterally placed platform during early growth stages as well. Recognition of these differences in the P elements suggests different biologic relationships. Both neogondolellid and paragondolellid groups range from the Middle Triassic until extinction during the Late Triassic.

Another interesting growth feature of many elements was regeneration following resorption or some other kind of destruction of parts of elements. Well-preserved elements commonly show interrupted lamellae (Figures 20.24 and 20.25), resulting from parts of elements having been broken or resorbed and then new lamellae added to "regenerate" the structure. New lamellae were added on outer surfaces of many elements, supporting the interpretation that elements were flesh-covered internal structures. Pectiniform P elements of the Late Devonian *Ancryodella* (Figure 20.24) developed a complete row of denticles during the earliest growth stages. During an intermediate growth stage, two to five denticles commonly were resorbed. Later growth stages showed remarkable ability to repair denticles, although the new denticles often were somewhat different from the originals. Thus one or two denticles commonly were regenerated to replace four or five denticles that had been removed. The consistency of this process in the same apparatus suggests that this was not a chance pathologic event but was a normal function of the apparatus. Perhaps stress during a particular developmental stage resulted in resorption of carbonate apatite to be used elsewhere in the apparatus, only to be replaced by regeneration at a later stage. Solving the problem of the function of the apparatus could answer the question of why regeneration was necessary.

Figure 20.18. Representative elements of the Hibbardellina Conodontophorida (advanced conodonts). **A,** A unimembrate or bimembrate apparatus of coniform elements of *Parachirognathus.* (Lower Triassic, Utah, ×90.) This apparatus consisted of S position elements. **B,** A pectiniform element of a unimembrate *Epigondolella* (Middle Triassic, Africa, ×150.) Only the P element is known in this specialized species. **C,** Pectiniform element of *Furnishius,* upper surface. (Lower Triassic, Nevada, ×100.) This probably was a unimembrate species. **D,** Lower surface of pectiniform element of *Furnishius.* (Lower Triassic, Nevada, ×140.) **E,** Sinistral and dextral pectiniform elements of *Furnishius.* (Lower Triassic, Utah, ×200.) **F,** Lower surface of P pectiniform element of *Platyvillosus,* showing basal cavity. (Lower Triassic, Utah, ×180.) **G,** Upper surface of *Platyvillosus* showing nodose ornamentation. (Unimembrate apparatus, ×260.)

Figure 20.19. Representative elements of the Hibbardellina Conodontophorida (advanced conodonts). **A,** Multimembrate apparatus of pectiniform and ramiform elements of *Mesotaxis*. (Upper Devonian, Australia, ×40.) Two P elements (Pa and Pb), M, and S series represented in this six-member apparatus. **B,** Multimembrate apparatus of *Polygnathus*. (Upper Devonian, New York, ×40.) **C,** Two elements of a six element (multimembrate) apparatus of *Palmatolepis*. (Upper Devonian, Australia, ×45.) Pa and M elements only shown. Complete apparatus like that of *Mesotaxis* (**A**) with Pb, and S series. Pa element is common fossil of Upper Devonian and basis for evolution chart of Figure 20.31. **D,** Multimembrate apparatus of pectiniform and ramiform elements of *Kockelella*. (Silurian, Oklahoma, ×40.) *1* is lower and *2* is upper view of Pa element; other coniform elements positional scheme shown. (Courtesy G. Klapper.)

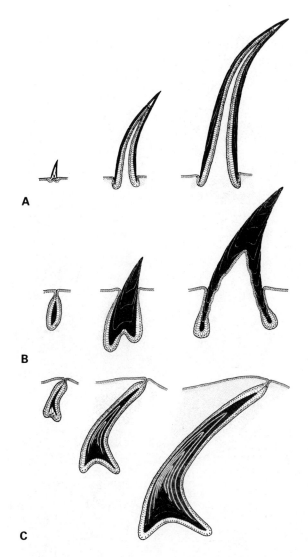

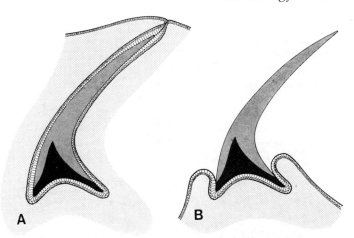

Figure 20.21. Diagrammatic reconstruction of a coniform distacodontid element in possible rest position enclosed in secreting tissue (**A**), and extruded position for possible grasping function (**B**). (From Bengtson, S. *Lethaia* 9: 203; 1976.)

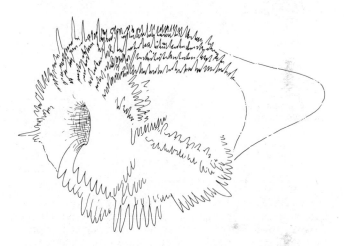

Figure 20.22. Hypothetical reconstruction of conodont with elements enclosed in secreting tissue of a lophophore-like structure. In this reconstruction, elements could serve as supports for food grooves directing nutrients to central mouth. (From Lindström, M. *Palaeontology* 17: 741; 1974.)

Figure 20.20. Inferred relationship of conodont elements to secreting tissues in primitive (**A**), intermediate (**B**), and advanced (**C**) conodonts. **A,** The most primitive elements (Proterozoic to Upper Cambrian) had lamellae added internally only, and secreting tissue probably lined a deep basal cavity. **B,** In intermediate conodonts (Upper Cambrian to Middle Ordovician), the secreting tissue extended from the basal cavity on to part of the outer surface of the element. **C,** Advanced elements (Upper Cambrian to Triassic) had lamellae added only on all outer surfaces and must have been covered by tissue during most or all of the growth of the element. (From Bengtson, S. *Lethaia* 9: 201; 1976.)

Paleoecology

Ecologic adaptation

The problem of determining the ecologic requirements of an extinct group of organisms with unknown zoologic affinity and enigmatic parts is a challenge to conodont investigators.

Because identical sequences of conodont elements have been identified in different sediment types in widely separated areas, most paleontologists early concluded that conodonts were pelagic and ubiquitous in the marine environment. Modern paleoecologic studies suggest that conodonts probably had a range of environmental adaptations that extended from the benthic to the pelagic realm.

Conodonts commonly are rare in stromatolitic facies, interpreted to be extremely shallow water. The presence of conodonts in some stromatolitic facies and not in others probably is related to water energy level, salinity, nutrients, and water temperature.

Conodonts generally are rare to absent in sediments interpreted to represent far-offshore and deep basinal environments. Scarcity of nutrients in surface waters

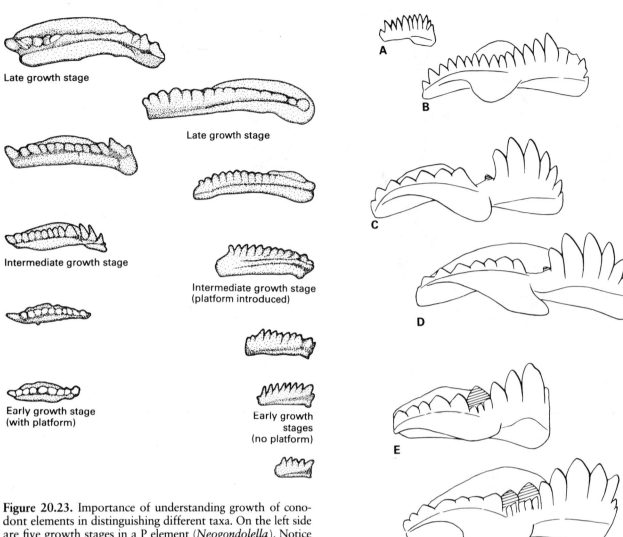

Late growth stage

Late growth stage

Intermediate growth stage

Intermediate growth stage
(platform introduced)

Early growth stage
(with platform)

Early growth
stages
(no platform)

A

B

C

D

E

F

Figure 20.23. Importance of understanding growth of conodont elements in distinguishing different taxa. On the left side are five growth stages in a P element (*Neogondolella*). Notice that at all stages of growth the P element is pectiniform with lateral growth of a platform on either side of the carina. On the right, note that earliest stage of growth of *Paragondolella* is ramiform with no lateral platforms. The lateral growth developing into a true pectiniform type occurs later so that in late growth stage, pectiniform elements of both *Neogondolella* (left) and *Paragondolella* are similar. (From Mosher, L. C. *Journal of Paleontology* 42: 950; 1968.)

Figure 20.24. Regeneration shown diagrammatically in P element (pectiniform). Lateral view shows pectiniform element in all stages of growth from early ontogeny when denticles are complete (**A** and **B**) to the stage when the central denticles are resorbed or broken (**C** and **D**), to the stage when regeneration of one or two denticles overgrows missing series of denticles (lined areas in **E** and **F**). (*Ancyrodella,* Upper Devonian, Michigan, ×36.) (From Müller, K. J.; Clark, D. L. *Journal of Paleontology* 41: 905; 1967.

(except in upwelling areas) as well as a factor related to greater depth may explain this.

Conodonts commonly are most abundant in sediments bearing stenohaline organisms. Alternatively, tolerance for slightly higher salinity (but not hypersalinity) has been suggested for certain Ordovician, Mississippian, and Permian taxa. No brackish water occurrences have been reported.

Some information is available concerning temperature preference for conodonts. Plots of conodont species' occurrence for the Paleozoic periods show abundance zones parallel to paleomagnetically determined equators.

It has been suggested that Permian species preferred near-equatorial belts (warmer) rather than higher latitudes (colder). In support of this are data from Australia that indicate that conodonts are absent from Permian sediments that accumulated proximally to cold glacial waters.

Ordovician conodonts additionally can be shown to have a distribution that can be explained on the basis of water temperature. It has been demonstrated that

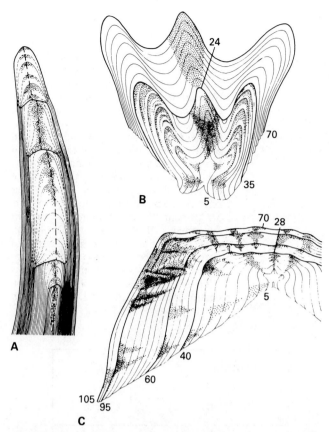

Figure 20.25. Diagrammatic thin section drawings showing internal structure of elements demonstrating resorption and subsequent regeneration. Heavier lines mark zones of discontinuity in elements at point of resorption and later regeneration. **A,** Coniform element illustrating four resorptions and subsequent regenerations. Growth lamellae and dotted zones of white matter interrupted. (*Scolopodus,* Lower Ordovician, Sweden, ×220.) **B,** Pa element (pectiniform). Prominent discontinuity in center of specimen shows area where resorption and subsequent regeneration occurred. Note that in the center of the specimen only 24 of the 35 deposited lamellae are still present (only a few of which are shown). (*Polygnathus,* Upper Devonian, Iowa, ×275.) **C,** Pa element (pectiniform). Three zones of resorption and resulting discontinuities illustrated but more conformable growth during regeneration than in other elements shown. Numbers indicate 35 missing lamellae. (*Idiognathodus,* Upper Pennsylvanian, Iowa, ×220.) (From Müller, K. J.; Nogami, Y. *Memoirs Faculty of Science Kyoto University, Series of Geology and Mineralogy* 38: fig. 17; 1971.)

throughout the Ordovician Period there were at least two main faunal provinces in the Northern Hemisphere, a North American-Midcontinent fauna and an Anglo-Scandinavian fauna. Species of the two provinces are quite different. The Anglo-Scandinavian fauna may have developed in temperate or even cold waters and occupied a wide bathymetric range, whereas most of the North American fauna were restricted to relatively shallow water. Although there is always danger in using a single ecologic factor to explain absence or presence of

organisms, such data support the idea that most conodonts may have had stenothermal tolerances narrow enough to locate the approximate distribution of broad isothermal regions.

Conodont feeding patterns have not been established. Some studies have concluded that conodonts used their tissue-covered skeletal apparatus as an aid in food gathering and fed on microplankton strained from the water. In addition, there is evidence from an observed correlation of trace fossils and conodonts that conodonts flourished on or above sediment with abundant nutrients. Conodonts are rare in association with sediment with a low nutrient supply (Figure 20.26). Because most conodonts probably were pelagic, the implication is that a relationship exists to quantity of microscopic nutrients passing through the water column on its way to incorporation in the sediment. Where this supply was limited (that is, far from shelf, over abyssal plains, away from upwellings), conodonts were rare or absent. The absence of conodonts in Permian deep-water anaerobic basins may be related to the nutrient factor.

Conodonts are known to occur with almost all invertebrates, fishes, and even algae. This is not surprising for a pelagic group whose remains would settle to the seafloor from whatever part of the water column was its normal habitat. Conodonts retain a rather close association with fish and cephalopods, both of which included predominantly pelagic members. The successful application to conodont research of a depth-stratification model common for pelagic animals is consistent with the idea that most conodonts were pelagic (Figure 20.27).

Evolutionary trends and biostratigraphy

The principal use of conodonts in solving paleontologic and geologic problems is in biostratigraphy. This is because the progressive modification of conodont elements and their skeletal apparatuses from the Proterozoic through the Triassic produced a superb array of easily identifiable morphotypes. Although some provincialism is recognizable, especially in the Ordovician and Carboniferous, and biofacies dependency is understood, a sequence of fossil zones has been defined that generally has worldwide utility. The 140 zones of Paleozoic and Triassic taxa constitute one of the most widely applicable biostratigraphic zonations in invertebrate paleontology (Figure 20.28).

The oldest conodonts reported are Paraconodontida from the Proterozoic Erathem of the Soviet Union (Siberian Platform and Kazakhstan). The conodonts are coniform elements that occur in a section of Proterozoic-

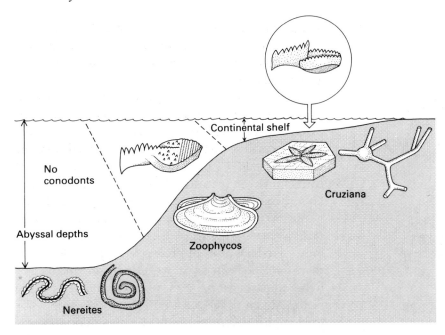

Figure 20.26. Pennsylvanian-Permian conodont biofacies interpreted from the Oquirrh Formation in Utah. In rocks characterized by a shallow water *Cruziana* ichnofauna (see Chapter 21), *Adetognathus* is the dominant conodont (P element shown). In rocks found in slightly deeper water and characterized by *Zoophycos* ichnofauna, *Idiognathodus* is the dominant conodont (P element shown), but in deep water *Nereites* rocks, no indigenous conodonts are present. If the ichnofauna is nutrient dependent, the conodont distribution may also be related to abundance of nutrients. Absence of conodonts in deep-water, nutrient-poor sediments suggests definable ecologic controls. (Based on Chamberlain, C. K; Clark, D. L. *Journal of Paleontology* 47: 663–682; 1973.)

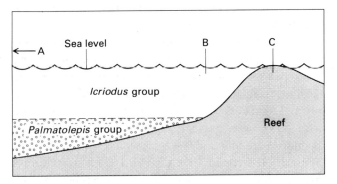

Figure 20.27. Depth stratification model for conodonts. In this diagram, a vertical stratification of conodonts is shown. This suggests why only the *Icriodus* group would be found in sediments deposited between points *B* and *C*, but how both the *Icriodus* and *Palmatolepis* groups would be found in those deposited between points *A* and *B*. The distribution of many conodonts has been explained on the basis of this model. (From Seddon, G.; Sweet, W. C. *Journal of Paleontology* 45: 873; 1971.)

Triassic	22
Permian	15
Pennsylvanian	~8
Mississippian	16
Devonian	43
Silurian	11
Ordovician	22
Cambrian	~3

Figure 20.28. Number of conodont zones or assemblages of species with biostratigraphic use in each of the Paleozoic and Triassic systems.

Cambrian boundary strata. This occurrence confirms that conodonts were one of the early animal groups to have mineralized skeletal parts.

Early and Middle Cambrian conodonts were also paraconodontids that bore only coniform or slightly modified coniform elements that are interpreted to have been in unimembrate skeletal patterns (Figure 20.14, *A*). The elements had large and widely spaced lamellae and deep basal cavities. The paraconodontids have not yet been organized into a workable Cambrian zonal scheme.

Paraconodontid elements have a larger percentage of carbon (10% to 15%) than do the younger elements. A

mineralogic evolution has been suggested for conodont elements from a Cambrian and probable Proterozoic condition of relatively high organic content to considerably less carbon in conodontophorids of Upper Cambrian and younger strata. A few representatives of the paraconodontids survived until the Middle Ordovician (Figure 20.29).

The more advanced Conodontophorida first appeared in the Late Cambrian (distacodontids) and bore coniform

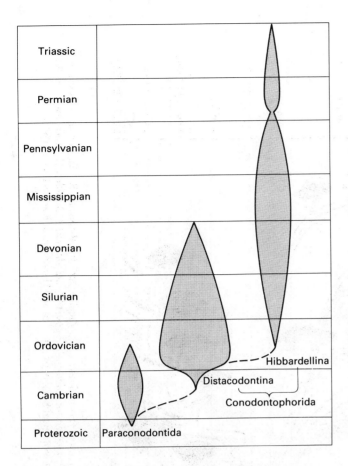

Figure. 20.29. Geologic range of major groups of conodonts but with unclassified genera not indicated. Width of range lines is proportional to abundance.

types (Figure 20.15, *A* and *D*). Most elements were coniform and ramiform. Pectiniform elements were rare.

Middle and Upper Ordovician conodonts (Figures 20.15, *B* and 20.16, *C* and *D*) are characterized by a variety of multimembrate skeletal patterns, and approximately 22 zones of faunal sequences based on species of advanced conodontophorids have been recognized in Ordovician rocks. Midcontinent and Atlantic-European provinces reportedly have somewhat different zonations for the Ordovician.

Following a Middle Ordovician peak of evolution (Figure 20.30), numerous conodonts became extinct and Early Silurian taxa are much fewer. Silurian conodonts are characterized by multimembrate patterns of ramiform elements closely related to those of the Ordovician, but primitive pectiniform elements became common. Some 11 zones of conodonts have been recognized in the Silurian.

The hibbardellids range from the Ordovician to the Triassic (Figures 20.17 to 20.19). Many of the skeletal patterns of the later Paleozoic and Triassic were established during the Devonian. Following a period of low diversity in the Silurian, conodonts showed a second evolutionary burst in the Devonian (Figure 20.30). Devonian conodonts are characterized by specialized ramiform and pectiniform elements in multimembrate

elements that were arranged in unimembrate apparatuses. At least one Late Cambrian multimembrate apparatus with two element types is known. All elements had minor amounts of organic material, and lamellae were all formed by outer accretion and were closely spaced and numerous. These characteristics distinguish Late Cambrian conodontophorids from Early and Middle Cambrian paraconodontids.

From the Middle Ordovician to the Devonian, all variations of multimembrate apparatuses appeared. The greatest diversity of elements and apparatuses was reached during the Ordovician when more than half the known conodont genera were in existence (Figure 20.30). This is comparable to the pattern of rapid diversification early in evolutionary history that is illustrated in other groups of organisms such as echinoderms.

Many Early Ordovician conodonts were unimembrate or had simple coniform symmetry series, but multimembrate skeletal systems had evolved and a few Early Ordovician genera had up to seven different element

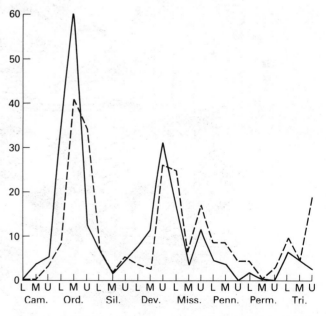

Figure 20.30. Approximate numbers of genera of conodonts that appeared in each period (solid line) compared with total number that became extinct during the same period (dashed line). Genera are based on discrete elements rather than complete apparatuses but are representative of evolution of group. (From Clark, D. L. In: Lindström, M.; Ziegler, W., editors. *Geologica et Palaeontologica* SB1: 151; 1972.)

Figure 20.31. Evolution of P element of *Palmatolepis* during the Late Devonian. Each distinctive P element has been interpreted to represent a distinct species. Various zonal names are indicated as vertical axis. The rapid evolution of the P element has enabled the recognition of a very fine zonation in Upper Devonian rocks. (From Helms, J.; Ziegler, W. In: Robison, R. A., editor. *Treatise on invertebrate paleontology, part W, supplement* 2. Boulder, CO and Lawrence, KS: Geological Society of America and University of Kansas Press; 1981.)

patterns. Pectiniform elements are numerous, and most of the important Devonian apparatuses contained at least one distinctive form (Figure 20.31).

The evolutionary burst in the Devonian has resulted in the establishment of 43 zones and faunal sequences. The Upper Devonian zonation is generally regarded as a model of biostratigraphy. More than 1000 names have been proposed for Upper Devonian taxa, and more than 100 species have been organized into 31 Upper Devonian zones. To a considerable extent, this zonation is based on fairly well understood ancestor-descendant relationships among species of one genus (Figure 20.31).

While Silurian conodonts bore ramiform, pectiniform, and coniform elements, coniform elements had disappeared and pectiniform elements in the P apparatus position, complemented by M and S ramiform types, formed conodont skeletal patterns by the end of the Devonian. This basic pattern characterized both Mississippian and Pennsylvanian conodonts. Even though conodonts were a declining group following the Late Devonian "burst," the newly evolved taxa have been organized into 16 Mississippian and at least eight Pennsylvanian zones.

The lowest point of conodont evolution (except for the Late Triassic extinction) was reached during Early Permian, and this produced a dramatic change in all younger faunas (Figure 20.30). Most conodonts became extinct, and all post-Early Permian taxa evolved from a very few ancestors. Representatives of this final surge of conodont evolution have been used to define 15 Permian and 22 Triassic zones that have worldwide application. Many of the Permian and Triassic species were unimembrate, and the final members of the lineage had small-size, single-element apparatuses. This final evolutionary burst was relatively short lived. Apparently all conodonts were extinct by the close of the Triassic, and their extinction was nearly synchronous worldwide in beds just below the Triassic-Jurassic boundary. All post-Triassic reports of conodonts have proven to be based on reworked or stratigraphically disturbed specimens.

Color alteration of elements

An important application of conodonts as a tool in geologic interpretations is the use of color alteration of elements and its relationship to depth and duration of sediment burial and geothermal gradient of the rocks in which the elements occur. Normal unaltered conodont elements are pale yellow (color alteration index of 1). Intensely altered elements are colorless (alteration index

of 8). The gradation from 1 to 8, easily identified under a binocular microscope, can be related to thermal activity that the element experienced due to depth of burial or other geothermal events that have modified the sediment. Controlled laboratory tests suggest that the color change is due primarily to fixed carbon alterations in the element and that the color change occurs within a fixed range of temperature. The color change is irreversible.

Field applications of the color alteration index (CAI) result in maps that delineate the thermal history of rocks. Because petroleum does not occur in rocks where geothermal history is extreme, maps showing CAI contours provide definition of areas where oil and gas production could or could not be expected.

Supplementary reading

Babcock, L. C. Conodont paleoecology of the Lamar Limestone (Permian), Delaware Basin, West Texas. In: Barnes, C. R., editor. *Conodont paleoecology. Geological Association of Canada Special Paper* 15: 279–294; 1976. Detailed description of statistical analysis of conodont data as an aid to understanding Permian conodont paleoecology.

Barnes, C. R.; Fahrhaeus, L. E. Provinces, communities, and the proposed nektobenthic habit of Ordovician conodontophorids. *Lethaia* 8: 133–149; 1975. A suggestion that some Ordovician conodonts were adapted to living above the water-sediment interface.

Behnken, F. H. Leonardian and Guadalupian (Permian) conodont biostratigraphy in western and southwestern United States. *Journal of Paleontology* 49: 284–315; 1975. Description of the establishment of Permian biostratigraphy with conodonts.

Bengtson, S. The structure of some Middle Cambrian conodonts, and the early evolution of conodont structure and function. *Lethaia* 9: 185–206; 1976. Interesting interpretation of conodont element microstructure and position and function.

Bergström, S. M. Ordovician conodonts. pp. 47–58. In: Hallam, A., editor. *Atlas of paleobiogeography.* Amsterdam: Elsevier Pub. Co., 1973. Summary of Ordovician "provincialism" defines different but contemporaneous faunas.

Chamberlain, C. K.; Clark, D. L. Trace fossils and conodonts as evidence for deep-water deposits in the Oquirrh Basin of central Utah. *Journal of Paleontology* 47: 663–682; 1973. Study showing relationship between trace-fossil biofacies and conodont faunas; suggests nutrient distribution and water depth as controlling factors for conodonts.

Clark, D. L. Early Permian crisis and its bearing on Permo-Triassic conodont taxonomy. In: Lindström, M.; Ziegler, W., editors. *Geologica et Palaeontologica* SB1: 147–158; 1972. Documentation of evolutionary crises for conodonts throughout the Paleozoic—an Early Permian event almost caused extinction of conodonts.

Clark, D. L. Conodont biofacies and provincialism. *Geological Society of America Special Paper* 196; 1984. A summary of ecologic factors used in biofacies analysis and quantitative treatment of conodont elements.

Clark, D. L.; Miller, J. F. Early evolution of conodonts. *Geological Society of America Bulletin* 80: 125–134; 1969. Traces morphologic and chemical changes in Cambrian to

Ordovician conodonts and documents changes from para-conodontids to conodontophorids.

Clark, D. L.; *et al.* In: Robison, R. A., editor. *Treatise on invertebrate paleontology, part W, Conodonta.* Boulder, CO and Lawrence, KS: Geological Society of America and University of Kansas Press; 1981. Detailed description of morphologic and positional schemes in use for conodonts and classification of conodont genera.

Conway Morris, S. A new Cambrian lophophorate from the Burgess Shale of British Columbia. *Palaeontology* 19: 199–222; 1976. Interesting description of a single specimen that bore conodont-like elements.

Epstein, A. G.; Epstein, J. B.; Harris, L. D. Conodont color alteration—an index to organic metamorphism. *United States Geological Survey Professional Paper* 995; 1977. Explanation of theory and application of conodont element color change and thermal factors responsible.

Lindström, M. The conodont apparatus as a food-gathering mechanism. *Palaeontology* 17: 729–744; 1974. Proposal that conodont elements supported fleshy lophophore-like structures.

Mosher, L. C. Are there Post-Triassic conodonts? *Journal of Paleontology* 41: 1554–1555; 1967. Discusses reports of conodonts in Jurassic and Cretaceous rocks, and demonstrates that the specimen must be reworked.

Müller, K. J.; Nogami, Y. Über den Feinbau der Conodonten. *Memoirs of the Faculty of Science, Kyoto University, Geology and Mineralogy Series* 38; 1971. Detailed description of microstructure of conodont elements.

Scott, H. W. New Conodontochordata from the Bear Gulch Limestone (Namurian, Montana). *Publications of the Museum of Michigan State University, Paleontologic Series* 1: 85–99; 1973. Report on specimens of body impressions with conodont elements as well. Concludes that specimens were whole conodont but does not consider the idea that they were conodont predators.

Sweet, W. C.; Bergström, S. M., editors. Symposium on conodont biostratigraphy. *Geological Society of America Memoir*, 127; 1971. Best worldwide summary of Paleozoic and Triassic conodont biostratigraphy.

21

Trace Fossils

Richard G. Osgood, Jr.

Definition of ichnology

Ichnology is the study of tangible evidence of the activity of an organism, other than the production of body parts, that reflects a behavioral function (Figure 21.1, *A* and *B*). Biologically produced sedimentary structures are known as **trace fossils** (also called **ichnofossils**, or fossil Lebensspuren). Trace fossils include, for example, the tracks of trilobites and the trails of gastropods, as well as burrows made below the surface of soft sediments by a host of other organisms. Also included in trace fossils are borings in hard rock, coprolites (fossilized fecal pellets), and other fecal material. Specifically excluded are markings that might result from the movement of a dead animal being swept across the substrate because such markings do not represent organism activity.

Ichnology is a very broad field, which by no means lies solely within the realm of paleontology. The prospective researcher must have some knowledge of paleontology, sedimentology, and zoology. This chapter stresses only the paleontologic and paleoecologic value of trace fossils of invertebrate origin.

Comparison of trace fossils and body fossils

Because many tracks and trails are accurate representations of particular body parts such as legs or podia, trace fossils often intergrade imperceptibly into certain types of **body fossils**, which in this case are molds and casts of invertebrates. The key to distinguishing molds and casts from trace fossils involves the ability of the researcher to ascertain movement of body parts. For example, if the negative impression of a trilobite with

intact appendages is found, it is the ventral mold of a body fossil. Conversely, if there is evidence that the appendages were being employed for digging as the mold was made, the structure is a trace fossil.

Generic and specific classification of trace fossils

Distinctive trace fossils are given Latin or Greek generic and specific names ("ichnogenus" and "ichnospecies"), although they are not assigned to families, orders, and so on for reasons to be mentioned later. Criteria for classification are morphologic and ideally should not include subjective interpretations about the behavioral function represented by the trace fossil or about the originator of the structure.

Advantages of trace fossils over body fossils

Trace fossils have the following advantages over body fossils:
1   They are often preserved in clastic rocks (sandstones), where body fossils may be rare.
2   Diagenesis can destroy or distort body fossils but has little effect on trace fossils and may even enhance their preservation. For example, some U-shaped burrows in the Upper Ordovician rocks of Ohio are coated with iron oxide; these stand out clearly in outcrops.
3   Because trace fossils are made on or within the sediment, they are not transportable. For paleoenvironmental interpretations, it is critical to determine whether body fossils actually lived where they are preserved.

A

B

Figure 21.1. *Flexicalymene* (Upper Ordovician, near Cincinnati, Ohio, ×1.2.) **A,** This remarkable specimen is one of the few examples in the geologic record where the trace maker is preserved in situ. **B,** Ventral view of specimen in **A** showing the bilobed trace fossil *Rusophycus* with the bilobed furrow made by the motion of the trilobite's appendages. (From Osgood, R. G. *Paleontographica Americana* 6 [41]; 1970.)

Identity of the trace maker

When dealing with molds or casts of body fossils, it is normally possible to identify the fossil taxonomically. Regrettably, in most cases the exact taxonomic identity of the tracemaker remains a mystery. In only a few rare examples has the body fossil been found in situ with the trace fossil. Figure 21.1, *A* shows the dorsal side of the Ordovician trilobite *Flexicalymene*. Figure 21.1, *B* is the ventral side of the same specimen, which is the cast of the bilobed burrow made by the motion of the trilobite's appendages. The cast is the trace fossil *Rusophycus*. The

trilobite died and was preserved in its burrow. *Rusophycus* is one of the most common trace fossils found in Paleozoic rocks, but only three specimens are known where the trilobite occurs with the burrow. All are from the Upper Ordovician strata of the Cincinnati area. In all other specimens, the burrow was a temporary structure that the trilobite abandoned.

The previous example stands out because of its extreme rarity. In most other cases, the trace maker is absent; its identification is either tentative or impossible. This is especially true when one considers the large number of infaunal worms, such as polychaete annelids, balanoglossid hemichordates, and sipunculids. The three are assigned to different phyla, yet they can leave very similar traces. It is obvious that trace fossils can make the greatest contribution where the trace maker can be identified. Unfortunately, however, this is the rare exception.

The reasons for naming trace fossils, even if the animal that made the trace can be identified, are pragmatic. Trace fossils have intrinsic values; they are paleo-environmental indices, and they reflect the behavior of extinct organisms. Experience has demonstrated that trace fossils that are not formally named tend to become 'lost' in the literature. For example, burrows of lingulid brachiopods have been mentioned by at least nine authors as 'lingulid burrows,' but these informal citations are not recorded in the latest compendium on trace fossils. It was for this reason that the name *Lingulichnus* was proposed for lingulid burrows in 1976. It is essential to always make sure that the name of the animal is not confused with the name of its burrow or trail.

Preservation of trace fossils

There is some disagreement among specialists regarding both the mechanics of preservation and the terminology applicable to trace fossil preservation. To avoid confusion, only one set of terms is presented here.

Some trace fossils are preserved in **full relief** totally within the sediment (Figure 21.2, upper left) or as **semi-reliefs** at the juncture of two different rock types such as sandstone and shale (Figure 21.2, upper right). In both cases, the clarity of preservation depends on the grain size and the cohesive properties of the sediment.

Full reliefs may be formed by sediment-eating (detritus-feeding) organisms mining the sediment at some depth below the surface. These organisms pass a large volume of sediment through their digestive systems, extracting the food substances and excreting the rest. The

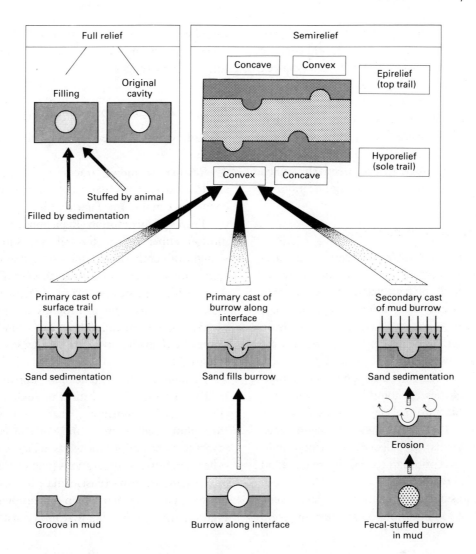

Figure 21.2. Various types of preservation of trace fossils. The dark stippling indicates mud; the light stippling indicates silt or fine sand; terminology of semireliefs is relative to middle sand or silt bed. (From Seilacher, A. In: Imbrie, J. and Newell, N. D., editors. *Approaches to paleoecology.* New York: John Wiley & Sons; 1964, p. 297.)

burrow may become packed with fecal pellets. However, if the organism's anus is positioned at the depositional interface, the sediment will pass out of the burrow and leave an empty tunnel. It is possible that the burrow will collapse after the organism has mined out the sediment, leaving an ill-defined structure that probably cannot be identified. Conversely, if the sediment is sufficiently cohesive, the open burrow might later be filled by sediment passively sifting into it. It is advantageous if the material that fills the burrow has a different color or grain size than the original sediment.

Not all trace fossils preserved as full reliefs are produced by feeding activities. Various members of the infaunal community may move through the sediment without feeding or depositing fecal material. The likelihood of well-defined preservation of such burrows is low,

however, because most of them collapse and only disturbed bedding is recognizable. **Bioturbation** is the term generally applied to the mixing of sediment by organisms, and a sediment that is thoroughly bioturbated but without distinctive trace fossils is said to have a **bioturbate texture**. Only trace fossils that show distinct structures are given names.

Generally, trace fossils preserved as semireliefs are more valuable to ichnologists because their quality of preservation is higher. They also are easier to study because they are found on the upper or lower surfaces of beds. In contrast, the three-dimensional geometry of full relief burrows often must be determined by time-consuming serial sections.

Semireliefs most frequently occur as concave depressions on the upper surface (**concave epireliefs**) or as convex bodies on the underside of beds (**convex hypo-reliefs**) (see Figure 21.2). Convex epireliefs and concave hyporeliefs are rare. Concave epireliefs and convex hyporeliefs that show fine detail usually are found in fine-grained sandstones or siltstones (see Figure 21.3). The fissility of most shales causes them to crumble, thereby obliterating any trace fossils. However, if the rock bearing the trace fossil is too coarse-grained, detail will be lost. Many trace fossils are not effectively studied with anything more powerful than 5× magnification.

Suprageneric classification of trace fossils

Because it is usually impossible to identify the trace-making organism with any degree of certainty (arthropod traces being one of the few exceptions), the normal nomenclatorial hierarchy (class, order, family, and so on) is inapplicable. Most specialists classify trace fossils on the basis of their geometry, which in turn reflects the behavior of the organism.

The most commonly employed classification contains five behavioral categories. It should be emphasized that this is a nonhierarchic, informal classification. As will be demonstrated later, these groupings are useful when employing trace fossils as paleoenvironmental indicators. With a few exceptions, trace fossils are not good time indicators. The reason is that they are the result of reactions of organisms to environmental parameters, such as energy level, food distribution and turbidity, and

therefore possess characters that may be shared by diverse organisms. As a result, many trace fossils have long geologic ranges.

Behavioral classification of trace fossils

The following lists the behavioral classification of trace fossils:

1 **Resting or hiding traces** (Cubichnia): Fillings of shallow excavations that usually mirror the morphology of the trace maker (see Figure 21.4).

2 **Traces of locomotion** (Repichnia): Evidence of locomotion either on or through the substrate; usually straight or gently curved, indicating that the animal was simply moving from one site to another (Figure 21.5).

3 **Dwelling traces** (Domichnia): Domiciles of infaunal animals, such as some arthropods and many wormlike phyla; may be vertical, unbranched cylindrical burrows, or may be U-shaped structures of filter-feeding organisms (Figure 21.6).

4 **Feeding traces** (Fodinichnia): Burrows of sediment feeders, usually with a distinct three-dimensional aspect (Figure 21.7), indicating that food was dispersed throughout the sediment and not concentrated in thin layers; may be multibranched as in Figure 21.7 or take on other geometric configurations (Figures 21.8 and 21.9).

5 **Grazing burrows** (Pascichnia): Burrows of sediment feeders that are primarily two-dimensional in aspect (see Figure 21.10), indicating that food must have been dis-

Figure 21.3. *Asaphoidichnus* (Upper Ordovician, Cincinnati, Ohio, ×1.8). Enlargement of a portion of the trackway produced by the trilobite *Isotelus*. Note the trifid imprints preserved as concave epirelief. (From Osgood, R. *Paleontographica Americana* 6 [41]; 1970.)

Figure 21.4. *Asteriacites* (Upper Ordovician, northern Kentucky, ×1.4), preserved as convex hyporelief. The hiding trace of a starfish; the scratch marks were made by the tube feet. (From Osgood, R. G. *Paleontographica Americana* 6 [41]; 1970.)

tributed in thin layers; usually appear on bedding planes as a spiral or as a series of tight **S**-shaped curves or some other very regular geometric pattern, demonstrating feeding activity rather than locomotion; only rarely does the trail cross over a previously formed segment. Apparently the organism was able to recognize through chemical or tactile receptors that it was in close proximity to another trail, and it reacted accordingly. Obviously it is pointless for a sediment-feeding organism to graze over the same area twice.

Figure 21.5. *Diplichnites* (Upper Ordovician, Cincinnati, Ohio, ×1). Specimen is the convex hyporelief of a trilobite trackway. The configuration of the imprints indicates that the midaxial line of the body was angled slightly to the right of the direction of movement. The direction of movement is from bottom to top. (From Osgood, R. G. *Paleontographica Americana* 6 [41]; 1970.)

Figure 21.6. *Diplocraterion* (Upper Ordovician, near Cincinnati, Ohio, ×1.6). Cross-section view of a U-shaped burrow. The parallel arcs between the arms of the burrow represent previously formed bases of the **U**. The burrow was lengthened with time, but the reason for the expansion near the base is unclear. The **U** tube probably served as the domicile for a filter (suspension) feeding organism. (From Osgood, R. G. *Paleontographica Americana* 6 [41]; 1970.)

Figure 21.7. *Chondrites.* Schematic drawing of three-dimensional feeding burrow. The organism, a sediment feeder, probably lay with its anus at the depositional interface and mined down into the substrate until it could 'stretch' no further. Apparently, it then contracted the body and repeated the mining process several times into previously unexploited areas. *Chondrites* is one of the most common ichnogenera. (From Simpson, S. *Geological Society of London Quarterly Journal* 112: 475–499; 21–24; 1957, Teichert, C., editor. *Treatise on invertebrate paleontology, part W, supplement 1.* Boulder, CO; and Lawrence, KS: Geological Society of America and University of Kansas Press; 1975.)

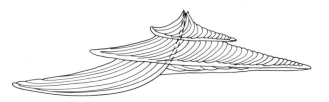

Figure 21.9. Schematic cross section showing three-dimensional aspect of *Zoophycos*. Because of the large size of the trace, it is extremely difficult to collect complete specimens. (From Teichert, C., editor. *Treatise on invertebrate paleontology, part W, supplement 1.* Boulder, CO, and Lawrence, KS: Geological Society of America and University of Kansas Press; 1975.)

Figure 21.8. *Zoophycos* (Mississippian, northeastern Ohio, ×0.8). Bedding plane view with the characteristic rooster tail configuration. However, this represents only a portion of the entire structure. In all likelihood, it is the burrow of a sediment feeder but the exact mode of formation is not understood. Compare with Figure 21.9.

Trace fossils as paleoenvironmental indicators

Probably the most valuable information that trace fossils provide to geologists is in paleoecology. The concept of using trace fossils for paleoenvironmental studies was established by Seilacher some 20 years ago and refined in 1980. Seven **ichnofacies** are recognized, each characterized by a distinct group of trace fossils exhibiting uniform behavioral patterns (Figure 21.11) and each representing different ecologic parameters. Below the intertidal zone, the main controlling factors for three ichnofacies appear to be food distribution and feeding habits. Within the intertidal zone, wave and current action and the nature of the substrate are important in distinguishing between three different ichnofacies. The seventh ichnofacies is characterized by trace fossils of nonmarine origin. Generally, near-shore shallow-water faunas were dominated by filter (suspension) feeders, whereas deeper water organisms are mostly sediment (deposit) feeders. It should be recognized that other

Figure 21.10. *Taphrelminthopsis* (Eocene, Guipuzcoa region, northern Spain, ×0.2). Preserved as a convex hyporelief. Spiral forms are found in flysch deposits and interpreted as a grazing trace. Compare with Figures 21.12. (Courtesy T. Peter Crimes, University of Liverpool.)

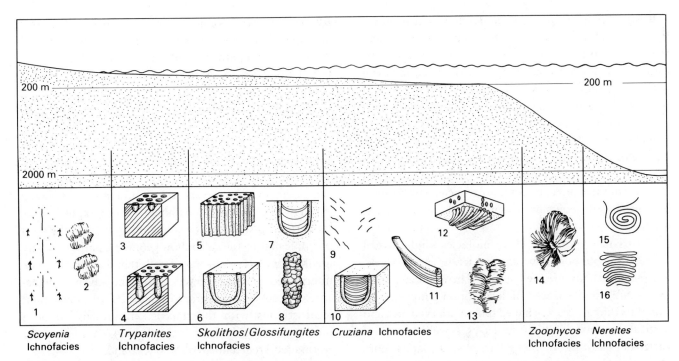

| *Scoyenia* Ichnofacies | *Trypanites* Ichnofacies | *Skolithos/Glossifungites* Ichnofacies | *Cruziana* Ichnofacies | *Zoophycos* Ichnofacies | *Nereites* Ichnofacies |

Figure 21.11. Bathymetric distribution of common ichnogenera and their respective ichnofacies. *1, Kouphichnium* (xiphosurid tracks); *2, Isopodichnus; 3,* echinoid borings; *4,* pholad (pelecypod) borings; *5, Skolithos; 6, Arenicolites (Glossifungites* ichnofacies); *7, Diplocraterion* (with uneven U-shaped arcs); *8, Ophiomorpha (Skolithos* ichnofacies); *9, Diplichnites* (trilobite tracks); *10, Diplocraterion* (with evenly spaced U-shaped arcs); *11, Teichichnus; 12, Phycodes; 13, Rusophycus; 14, Zoophycos; 15, Spirorhaphe; 16, Helminthoida.* (Modified from Crimes, T.P. In: Frey, R. W., editor. *The study of trace fossils.* New York: Springer-Verlag; 1975; and Frey, R. W.; Seilacher, A. *Lethaia* 13: 183–208; 1980.)

environmental variants such as high or low energy levels, which cause subaqueous erosion or deposition and turbidity of the water, can also be determining factors at any depth in certain cases. For example, it has been pointed out that a sheltered lagoon with a low food production may provide the same environmental conditions for the trace fossil producing animals that occur in a far deeper offshore area. If true, the resultant trace fossils may be identical. Nevertheless, with some exceptions and modifications, the ichnofacies model has stood the test of time. It should be emphasized that it is conceptual; the ichnofacies are not separated by sharp boundaries, and mixed assemblages are known. Each of the ichnofacies is named after a common trace fossil in that assemblage as indicated in the following list:

1 **Scoyenia ichnofacies:** The rocks are clastics and the trace fossils are nonmarine; they include isopod burrows and xiphosurid trackways. This ichnofacies is not common in the geologic record. Examples are the Juniata Formation (Ordovician) and the Catskill complex (Devonian) of the Appalachians.

2 **Trypanites ichnofacies:** The substrate is lithified and pockmarked by vertical borings that may be cylindrical, vase-, tear-, or U-shaped. Organisms responsible for the borings are mainly marine filter feeders such as pelecy-

pods, echinoids, and polychaetes. A specific example is a Miocene siltstone containing Pleistocene borings found near Newport Bay, California. The siltstone was beveled off by wave action to form a marine terrace, extensively bored, and later covered by sand. The *Trypanites* ichnofacies intergrades with the firm, but not lithified, substrate of the *Glossifungites* ichnofacies.

3 **Glossifungites ichnofacies:** This ichnofacies is characterized by a firm but unlithified substrate. It usually lies within the intertidal zone. Burrows are vertical and moderately deep; U-shaped tubes may be present. Certain strata in the Eocene of Georgia are representative of this ichnofacies.

4 **Skolithos ichnofacies:** In one way, the *Skolithos* ichnofacies is the soft-sediment counterpart of the *Glossifungites* facies. Soft-sediment substrates in the intertidal environment shift due to current or wave activity. Many times the trace fossils show evidence that the trace maker was forced to deepen or elevate the burrow to compensate for scouring or deposition, producing U-shaped tubes with uneven arcs caused by abandoned sections of tubes. Vertical, cylindrical burrows are also characteristic. One of the best examples of the *Skolithos* ichnofacies in the United States is the Silurian Tuscarora Formation of New Jersey.

It is useful at this point to explain how soft-sediment burrows can be distinguished from borings into a consolidated substrate. Although there is always some uncertainty, burrows made in unconsolidated sediment will usually show some evidence of deflection of the beds, whereas the surrounding material in borings is less likely to be disturbed. An animal moving normal to the bedding planes is somewhat analogous to a bullet passing through a piece of wood. The grain of the wood will be deflected toward the direction of movement. Also carbonate grains (seen in thin section) commonly are truncated by borings; burrows in soft sediment of course would not cut across grains.

5 **Cruziana ichnofacies:** This ichnofacies is indicative of the shallow subtidal zone and moderate energy levels. Sediments usually are well sorted. The dominant trace fossils are simple traces of locomotion, three-dimensional feeding traces, U-shaped tubes with even arcs and vertical burrows of filter feeders, as well as several varieties of resting or hiding burrows. U-shaped tubes with evenly spaced parallel arcs (Figure 21.11) are probably a response to normal growth of the animal, thus necessitating a lengthening (deepening) of the burrow. Uneven arcs (Figure 21.11) are more likely to represent a response to submarine scouring or deposition where the organism was compelled to deepen or elevate the burrow to maintain equilibrium. Grazing traces are conspicuously absent in this ichnofacies. Obviously, protection from predators is a problem in a subtidal zone with sufficient light penetration. Examples from the geologic record are numerous and include the Upper Ordovician of Ohio and the Silurian (Clinton) of central New York state.

6 **Zoophycos ichnofacies:** The *Zoophycos* ichnofacies is characterized as being below wave base and below the depth where oscillation ripple marks could form. Sediments are poorly sorted clastics; turbidites are usually absent. Many times the only trace fossil present is the very complex feeding trace, *Zoophycos*, which, when viewed from above resembles a rooster tail on the bedding plane surface (Figure 21.8). However, cross sections reveal the complicated three-dimensional nature of the structure (Figure 21.9). There is almost general agreement that *Zoophycos* is a feeding structure, but no consensus exists as to its exact mode of formation. There is also some disagreement as to the validity of *Zoophycos* as a bathymetric indicator. It occurs in rocks believed to be deposited below wave base in parts of the Jurassic of Germany, but it has been reported from presumed shallower environments of the Mississippian of Ohio and the Carboniferous of the Soviet Union. Moreover *Zoo-*

phycos has been found in extremely deep water (for example, Holocene cores taken from water depths of 3800 m off Valparaiso, Chile). A monographic study of the genus is necessary to realize its potential in paleoecology. *Zoophycos* is simply too broad as defined today.

7 **Nereites ichnofacies:** Sequences deposited in bathyal or abyssal environments are noted for an abundance of grazing traces and an absence of traces of concealment. In deep-water environments characterized by intermittent sedimentation, it is more likely that food would be concentrated in thin layers than randomly distributed through the sediment. Although photographs of the modern deep-sea floor reveal current activity in the form of ripple marks, they also show geometric patterns nearly identical with those of trace fossils that had already been interpreted as grazing traces before the advent of widespread deep-sea photography. Figure 21.12 shows spiral fecal strands with the tracemaker, thought to be a balanoglossid *in situ* at a depth of 4735 m near the Kermadec Trench north of New Zealand.

Rock units containing numerous grazing traces are usually marked by turbidites. Excellent examples are found in certain Tertiary flysch deposits of Europe and the Pennsylvanian of the Ouachita Mountains in the United States.

Most trace fossil studies include rocks that embrace only one or two ichnofacies. No examples are known where all six ichnofacies are present in one area, but suites containing three ichnofacies have been reported. In a portion of the Pennsylvanian of the Ouachita Mountains in Oklahoma, there is a vertical transition from the *Cruziana* to the *Zoophycos* to *Nereites* ichnofacies. Lateral changes of ichnofacies can also be delineated in the Ouachitas. In addition, there is, in the Ordovician of northern Iraq, a vertical sequence toward deeper water where ichnofossils seem to provide the only index to bathymetry.

Trace fossils and the history of invertebrate life

The rise of the Metazoa

One of the most dynamic areas of study in paleobiology is the apparent rapid evolution and radiation of the Metazoa in the late Proterozoic. Trace fossils have played a role in this unfolding drama. Tubelike structures of questionable biologic origin have been reported from the 2 to 2.5 billion-year-old Medicine Peak Quartzite of Wyoming. These are the oldest potential trace fossils found to date if it can be demonstrated that they are indeed organic.

Figure 21.12. Holocene, somewhat disorganized spiral grazing(?) trace with the trace maker, probably a balanoglossid, in situ. Water depth was 4735 m near Kermadec Trench in the southwest Pacific Ocean. (×0.6.) (Courtesy C. D. Hollister, Woods Hole Oceanographic Institution.)

However, uncontested trace fossils that predate the famous soft-bodied Ediacara fauna of Australia (650±50 million years B.P.; late Proterozoic) are rare and of simple morphology. One of the most complex pre-Ediacarian ichnogenera is a single specimen of a trail characterized by a central furrow and transverse grooves and ridges. This form (Figure 21.13) was found 2000 m below the base of the Cambrian in Australia and may be the trail of a primitive mollusk. By contrast, the Ediacara fauna is associated with at least six distinct trace fossils. One form exhibits complex meanders.

Thick sections that straddle the Proterozoic-Paleozoic boundary have been studied in Australia, Canada, Greenland, Norway, the Soviet Union, and the United States. In most of these examples, beds low in the section yield only a few simple traces such as vertical burrows or uncomplicated surficial indications of locomotion. Slightly below the base of the Cambrian, the trace fossils diversify in number and complexity. The Cambrian beds exhibit even greater diversification and complexity. Moreover, these transitions occur in rocks of varying lithology, thereby demonstrating that rock type is not the only controlling factor. The research conducted to date on trace fossils seems to confirm the studies on body fossils. The Metazoa may not have evolved until shortly before the beginning of the Paleozoic. Further research on trace fossils could prove to be quite valuable especially in clastic sequences where preservation of body fossils might be poor.

Trace fossils during Phanerozoic time

Only rarely are soft-bodied, wormlike animals preserved as body fossils. Yet several phyla of these organisms are common today and were probably present throughout much of the Phanerozoic. It is likely that many trace fossils with a cylindrical cross section were made by such animals. Therefore, these trace fossils increase our knowledge of the spectrum of invertebrate life by their presence alone. Admittedly, a trace fossil may not yield much morphologic information, but any record is better than no record.

Trace fossils occasionally do provide us with morphologic data. One Late Ordovician specimen (Figure 21.14) is preserved as a convex hyporelief. Its most striking feature is the presence of 21 paired lobes that resemble the parapodia of some Holocene polychaetes. In fact, an argument could be made for calling the specimen a body fossil (cast) rather than a trace fossil. However, the ill-defined ovoid mass visible near the left margin of Figure 21.14 appears to represent the head of the organism. Apparently the animal arched off the substrate and burrowed straight down, leaving most of the length of the imprint undisturbed but obliterating any remnant of the head. Later, silt filled the depression. This is one of the few trace fossils that can be identified as being of annelid origin.

Trace fossils also can provide morphologic data for parts of body fossils that are not generally preserved.

Trilobite appendages are rarely preserved in the fossil record because they were weakly sclerotized. Even when the appendages are found intact, the terminal claws are usually missing. If one can identify the trilobite responsible for a given trackway, something can be learned about the trilobite's claws. For example, it is possible to identify the trackways of adult *Isotelus* trilobites in the Ordovician because of the large size of the animal, which

Figure 21.14. *Walcottia* (Upper Ordovician, Cincinnati, Ohio, ×2), preserved as convex hyporelief. This probably was made by a polychaete annelid. Casts of parapodia are evident near the middle of the specimen. The ovoid area at the left is thought to be the anterior end distorted by movement. (From Osgood, R. G. *Paleontographica Americana*, 6 [41]; 1970.)

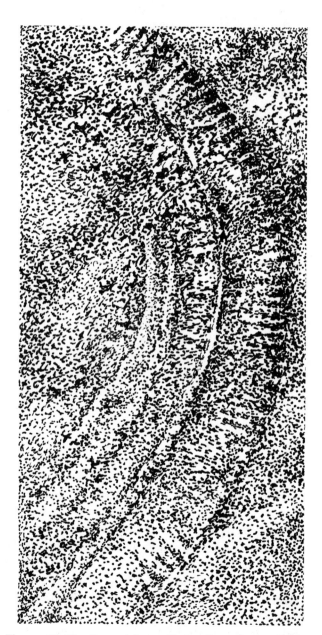

Figure 21.13. *Bunyerichnus* (Upper Proterozoic, Flinders Range, South Australia, ×1). Concave epirelief of a complex trail possibly made by a mollusk. (From Glaessner, M. F. *Lethaia* 2; 1969; and Teichert C., editor. *Treatise on invertebrate paleontology, part W, supplement 1*; Boulder, CO and Lawrence, KS: Geological Society of America and University of Kansas Press; 1975.)

Figure 21.15. *Cruziana* (Upper Cambrian, northern Wales, ×0.8), preserved as convex hyporelief. This is the filling of a furrow produced by a half-buried trilobite moving parallel to the depositional interface. Direction of movement is bottom to top as shown by the coarse scratch marks which forms a V directed posteriorly. For mode of formation, compare with Figure 21.16, C. (Courtesy T. Peter Crimes, University of Liverpool.)

is reflected by the width of the trackway (Figure 21.3). *Isotelus* possessed three long hairlike claws at the distal end of each walking leg. Trilobite burrows have yielded similar data for other genera.

The elongate, bilobed *Cruziana* (Figure 21.15) is preserved as a convex hyporelief, and it was created by the filling of a furrow in which a presumed trilobite moved horizontally through the mud substrate. As interpreted (Figure 21.16), *Cruziana* was formed by the appendages sweeping sediment toward the midventral line and to the

Figure 21.17. *Climactichnites* (Upper Cambrian or Lower Ordovician, near Perth, Ontario, ×0.2). Bedding plane view of a large sandstone slab showing several of the tractor tread markings. (From Logan, W. E. *Geology of Canada*; 1863; courtesy Geological Survey of Canada.)

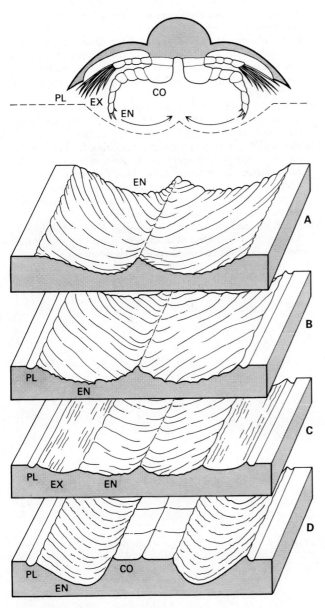

Figure 21.16. Mode of formation of various types of *Cruziana*. The endopodite (*EN*) portion of the walking leg is responsible for most of the excavation, but evidence of the coxae (*CO*), expodites (*EX*) and pleurae (*PL*) may be present. (From Seilacher, A. In: Crimes, T. P.; Harper, J. C., editors. *The Geological Journal, Special Issue 3*: 451; 1970.)

posterior of the trilobite. This is probably contrary to the direction of food movement (see Chapter 13). The morphology of *Cruziana* can vary depending on the orientation of the body. If the animal was digging primarily with the anterior appendages and the body axis was inclined toward the anterior, the only markings found would be those made by the endopodites (Figure 21.16, *A*). However, if the posterior appendages were employed and the body was tilted posteriorly, evidence of pleurae, exopodites, endopodites, and coxae might be present (Figure 21.16, *B* to *D*). It is possible that both anterior and posterior appendages were used in excavation. Note the similarity between Figure 21.15 and 21.16, *C*.

The affinities of most trace fossils are more enigmatic than those just mentioned. One puzzling species of extraordinarily large size (Figure 21.17) occurs only in the Upper Cambrian rocks of New York, Wisconsin, and adjacent Canada. Its maximum width of 15 cm equals or exceeds that of any known Cambrian body fossil. The trail, preserved as a concave epirelief, begins with a shallow, ovoid dishlike mass (not shown in Figure 21.17) and then continues as a series of chevron-shaped ridges separated by depressions exhibiting very delicate rills. It is a locomotion trail, but speculation about the organism responsible is widespread. Gastropods have been known to produce somewhat similar traces, and it is possible that this species was made by soft-bodied marine nudibranch gastropods. Nudibranch body fossils, however, have not been found in Paleozoic rocks.

Behavioral patterns of organisms can be inferred from the morphology of extinct body fossils and from observing Holocene analogues when present. Trace fossils, however, can provide some direct evidence of behavior. For example, one might infer from hard part morphology that some trilobites could burrow, but trace fossils provide the proof (see Figure 21.1, *A* and *B*). The convex hyporelief *Rusophycus* proves that *Flexicalymene* burrowed. The burrowing traditionally has been interpreted as a defense reaction to protect the vulnerable ventral surface of the trilobite.

Diversity of Phanerozoic trace fossils

Studies of trace fossils from shallow marine environments show that their diversity has not changed markedly since the Cambrian. Genera appear and disappear, but the overall diversity remains constant. It is somewhat easier to identify the trace maker of Mesozoic and Cenozoic ichnofossils because of Holocene analogues. Conversely, grazing traces from the *Nereites* ichnofacies portray a marked increase in diversity beginning in the Cretaceous and extending through Tertiary time. Two possible explanations for this increase are the rapid radiation of the angiosperms in the late Mesozoic with a resultant influx of detrital plant material, or the breakup of Pangea in the early Mesozoic. With the Mesozoic and Cenozoic sea floor spreading and enlarging the deep ocean basins, it is possible that new niches were created.

Supplementary reading

Crimes, T. P.; Harper, J. C., editors. Trace fossils 2. *Geological Journal Special Issue* 9. Liverpool: Seel House Press; 1977. Collection of papers that provides a sampling of trace fossil research; three papers discuss trace fossils and the rise of the Metazoa.

Frey, R. W., editor. *The study of trace fossils.* New York: Springer-Verlag; 1975. Covers nearly all aspects of trace fossil research; excellent resource volume for readers with some knowledge of paleontology, paleoecology, and stratigraphy.

Frey, R. W.; Seilacher, A. Uniformity in marine invertebrate ichnology. *Lethaia* 13: 183–208; 1980. General article that contains a revision of the various ichnofacies.

Häntzschel, W. Trace fossils and problematica. In: Teichert, C., editor. *Treatise on invertebrate paleontology, part W, Miscellanea, Supplement 1*. Boulder, CO, and Lawrence, KS: Geological Society of America and University of Kansas Press; 1975. All trace fossil genera described and illustrated; the bibliography is especially valuable.

Seilacher, A. Bathymetry of trace fossils. *Marine Geology* 5: 413–428; 1967. One of a series of papers by this author that has established trace fossils in paleoecology.

Index to Genera

Page numbers in *italics* refer to figures and tables.

Subject Index

Page numbers in **boldface** refer to definitions of terms boldfaced in the text; page numbers in *italics* refer to figures and tables.

681

Taxonomic hierarchy used in this book

Kingdom
    Subkingdom
        Phylum   Division (Coccolithophorida)
           Subphylum
              Superclass
                Class
                    Subclass
                      Order
                          Suborder
                            Superfamily
                          Family

                                Genus   Form Genus (Coccolithophorida)
                                          'Ichnogenus' (Trace fossils)
                                Species   'Ichnospecies' (Trace fossils)